Springer

Berlin
Heidelberg
New York
Barcelona
Hong Kong
London
Milan
Paris
Singapore
Tokyo

C. Bartsch H. Bartsch D. E. Blask
D. P. Cardinali W. J. M. Hrushesky
D. Mecke (Eds.)

The Pineal Gland and Cancer

Neuroimmunoendocrine Mechanisms
in Malignancy

With 82 Figures and 27 Tables

Springer

ISBN 3-540-64051-7 Springer-Verlag Berlin Heidelberg New York

Library of Congress Cataloging-in-Publication Data
The pineal gland and cancer: neuroimmunoendocrine mechanisms in malignancy/Christian Bartsch ... [et al.]. p.cm. Includes bibliographical references and index.
ISBN 3540640517 (hardcover : acid-free paper)
1. Carcinogenesis-Congresses. 2. Pineal gland-Congresses. 3. Melatonin-Physiological effect-Congresses. I. Bartsch, Christian.
RC268.5 .P56 2001   616.99'4071--dc21   00-035760

Springer-Verlag Berlin Heidelberg New York
a member of BertelsmannSpringer Science + Business Media GmbH

© Springer-Verlag Berlin Heidelberg 2001
Printed in Germany

Cover design: E. Kirchner, Heidelberg
Typesetting: Fotosatz-Service Köhler GmbH, 97084 Würzburg

Printed on acid-free paper   SPIN 10661997   21/3130/op 5 4 3 2 1 0

Dedicated to all those suffering from cancer

In memoriam Dr. Dinshah K. Mehta (1903 – 1993)

With deep thanks to our parents,
and to our academic teachers
Prof. Dr. G. P. Talwar, Prof. Dr. D. Gupta,
Prof. Dr. T. H. Lippert, and Prof. Dr. D. Mecke

Christian and Hella Bartsch

Preface

Three years ago, most authors contributing to this book gathered at the Heinrich Fabri Institute of the University of Tübingen at Blaubeuren near Ulm in Germany for the third conference on "Pineal Gland and Cancer". In 1987, the late Derek Gupta organized the second meeting and published the first book on the topic, 10 years after Vera Lapin, as part of the 25th anniversary celebrations of the Vienna Cancer Research Institute, had held the first meeting. It was in Vienna during the 1930s and 1940s that W. Bergmann and P. Engel demonstrated that pineal extracts possess growth inhibitory properties on experimental rodent tumors and R. Hofstätter reported favorable results when these extracts were given to cancer patients. In the 1970s, Vera Lapin and others reported that surgical removal of the pineal gland (pinealectomy) stimulates experimental tumor growth rendering fundamental support for an involvement of the pineal gland in malignancy. A focal question of past and present research in this field is whether the pineal gland exerts its tumor inhibitory activity primarily or exclusively via melatonin. Currently, it appears that the action of melatonin on experimental tumor growth critically depends on the circadian timing of its administration as well as on the type and stage of cancer, and that primarily highly differentiated tumor cells are controlled. Initial clinical applications of the pineal hormone for incurable cancers raise hopes for a promising future use, particularly when combined with other therapies (e. g., interleukin-2) increasing their efficacy and reducing their toxicity. Such studies, however, require proper chronobiologic design and replication as well as complementary laboratory studies to better understand the mechanisms involved and to integrate these findings into our general knowledge about the functions of melatonin.

It is generally accepted that melatonin acts as a chemical signal of the main circadian oscillator of the body located in the suprachiasmatic nuclei, which are tuned to environmental photoperiods. Retinal impulses thus convey temporal information, about time of day as well as season, to the cells and functional systems of the body including the neuroimmunoendocrine network. At the cellular level, melatonin may provide free radical scavenging properties that are histologically relevant to prevent oxidative damage to the cell and its nucleic acids. It is a matter of current debate whether pineal-derived melatonin, due to its low concentrations in the circulation, meaningfully neutralizes free radicals. Because of its high lipophilicity, melatonin freely passes into cells and can thus be concentrated in cellular as well as subcellular compartments. Thus, melatonin exists at considerably higher concentrations in these sites of action than in plasma. It is also conceivable that melatonin produced abundantly in extra-pineal tissues, e. g., within the enterochromaffin cells of the gastrointestinal tract, may function locally to scavenge free radicals being converted to a kynurenine derivative. Pineal melatonin, on the other hand, functioning as a temporal signal and acting via specific receptors is metabolized in the liver mainly to 6-sulphatoxymelatonin. Thus the prime function and metabolic fate of melatonin derived from the pineal gland are fundamentally different from the extra-pineal hormone. The overlap of biological actions as well as of metabolic fates of melatonin, within different compartments of the body, along with inadequate circadian control of experiments may help explain why experimental studies with melatonin sometimes lead to puzzling results, which are difficult to reconcile with the original hypothesis. Because pineal melatonin is primarily a temporal signal, it is understandable that its effects on tumor growth are critically dependent on the biological time of its administration (with respect to the cell cycle, to the time of day and season), and thus melatonin can either inhibit, stimulate, or have no effect upon the host–cancer balance. These cellular actions appear to be receptor-mediated and coupled to other hormone-response pathways, particularly those of the gonadal steroids crucially involved in the control of tumors of the reproductive tract. This relationship between melatonin and sex steroids may explain why melatonin primarily affects well-differentiated malignant tumor cells still possessing specific receptors. It is, however, currently unclear how the growth of some undifferentiated tumors of endocrine origin and leukemias may be stimulated by melatonin. This possibility raises concern about any uncontrolled human use of the pineal hormone as is currently advocated in some countries due to unjustified claims of it being a "wonder drug" preventing aging and even cancer. The observations of both an inhibitory and stimulatory action of melatonin on tumor growth, however, indicate that melatonin can exert profound control over cellular growth and differentiation, raising hopes for a future use of melatonin to prevent or control human cancer. This idea absolutely requires chronobiologically adequate research providing better insight into the mechanisms of action including the biological time dependence. In the meantime, patience will be required not only by those actively involved in research and cancer treatment but also by those evaluating the validity of such studies for further financial support. Patience without properly designed clinical chronotherapeutic trials using melatonin, will, however, not be rewarded but will rather cause additional confusion. Finally, it has to be stressed that apart from melatonin effects, alternative routes of the pineal gland controlling tumor growth seem to exist. These routes may include as yet unidentified low molecular weight pineal compounds and autonomic nervous system effects.

We hope that this book will help those working in this field to link the pineal gland and cancer and encourage others from diverse areas to join and thus enrich this field of research. The interplay between the pineal gland and malignancy has to be viewed in the chronobiologic framework of the largely epigenetic, highly complex, and non-linearly organized psychoneuroimmunoendocrine network as well as progressing genomic aberrations within the tumor. Therefore, scientific support is required from biomathematicians, informaticians, and even theoretical physicists being familiar with system theory, (bio)cybernetics, and other areas suitable to adequately describe complex multidimensional problems. The use and development of advanced models will help to channel the current flood of experimental findings and to gradually develop more effective (onco)therapeutic strategies with minimal side effects. For such purposes, not only the pineal gland but also other areas of the diencephalon should be considered, such as the hypothalamus, as well as subcortical structures, particularly the limbic system, because tumor growth is also likely to be perceived and responded to by higher nervous centers. Among the cited structures of the brain, the pineal gland is most accessible at present because of the clear melatonin signal. Therefore, it is logical to focus current investigations in brain–tumor interactions on the link between the pineal gland and cancer.

We would like to express our deepest gratitude to all those who kindly supported the Blaubeuren conference and this book. First of all, it has to be mentioned (and we apologize) that due to very limited funding almost all participants had to pay their own travel expenses. Without their conviction concerning the importance of the topic this conference had never been possible. Mrs. Dipl. Biochem. Anita Buchberger, Mrs. Angela Karenovics, Mrs. Heike Pfrommer, and Mrs. Ellen Lehmann worked hard organizing the meeting and thus paved its success. We are also grateful to the administration of the University of Tübingen who allowed us to utilize the excellent facilities of the Heinrich Fabri Institute at Blaubeuren for the conference sessions as well as for accommodating the participants. Financial support for this book was obtained foremost from the Deutsche Telekom AG and the Margarete Markus Memorial Grant for Cancer Research of the Margarete Markus Charity. Additional support was received from the Daimler-Chrysler Company, Eli Lilly Ltd. in England, IBL GmbH, Jenapharm GmbH, Merckle-Ratiopharm AG in Germany, Servier International, France, and some donors who wish to be anonymous. At this point we would also like to highlight the pleasant and professional co-operation with the team of the Springer-Verlag in Heidelberg. Finally, we would like to thank the Deutsche Forschungsgemeinschaft for funding a project on "The Role of Melatonin in the Neuroendocrine Control of Gynecological Neoplasias" at the Tübingen University Women's Hospital (1990–1997) and the National Cancer Institue of the National Institutes of Health for its support of melatonin/cancer studies in the United States.

The Editors
September 2000

Contents

11   Melatonin and Colon Carcinogenesis

**25   Could the Antiproliferative Effect of Melatonin Be Exerted
       Via the Interaction of Melatonin
       with Calmodulin and Protein Kinase C?**

Section V: Oncotherapeutic Potential of Melatonin

**26   Efficacy of Melatonin in the Immunotherapy
       of Cancer Using Interleukin-2**

27   Melatonin Cancer Therapy

Section VI: Electromagnetic Fields and Cancer: The Possible Role of Melatonin 509

28   Circadian Disruption and Breast Cancer

Richard G. Stevens 511

29   Breast Cancer and Use of Electric Power: Experimental Studies on the Melatonin Hypothesis

Wolfgang Löscher 518

30   Magnetic Field Exposure and Pineal Melatonin Production (Mini-Review)

Brahim Selmaoui, *Yvan Touitou* 534

Editors and Corresponding Authors

Editors

Priv.-Doz. Dr. Christian Bartsch
Centre for Research in Medical
and Natural Sciences
University of Tübingen
Ob dem Himmelreich 7
72074 Tübingen, Germany

Dr. Hella Bartsch
Centre for Research in Medical
and Natural Sciences
University of Tübingen
Ob dem Himmelreich 7
72074 Tübingen, Germany

David E. Blask, Ph. D., M. D.
Senior Research Scientist
Laboratory of Experimental
Neuroendocrinology/Oncology
Bassett Research Institute
One Atwell Road
Cooperstown, NY 13326, USA

Prof. Daniel P. Cardinali, M. D., Ph. D.
Department of Physiology
Faculty of Medicine
University of Buenos Aires
Paraguay 2155
1121 Buenos Aires, Argentina

Prof. William J. M. Hrushesky, M. D.
Director of Research (151)
University of South Carolina
WJB Dorn VA Medical Center
6439 Garners Ferry Road
Columbia, SC 29209, USA

Prof. Dr. Dieter Mecke
Director
Institute of Physiological Chemistry
University of Tübingen
Hoppe-Seyler-Str. 4
72076 Tübingen, Germany

Corresponding Authors *

Anisimov, Vladimir N.
Department of Carcinogenesis
and Oncogerontology
N. N. Petrov Research Institute
of Oncology
Pesochny-2
189646 St. Petersburg, Russia

Antón-Tay, Fernando
Autonomous Metropolitan University
Iztapalapa
Department. of Reproductive Biology
Av. Michoacán y Purísima s/n
Col. Vicentina Iztapalapa
México 09340, D. F. México

Bartsch, Christian
Centre for Research in Medical
and Natural Sciences
University of Tübingen
Ob dem Himmelreich 7
72074 Tübingen, Germany

Bartsch, Hella
Centre for Research in Medical
and Natural Sciences
University of Tübingen
Ob dem Himmelreich 7
72074 Tübingen, Germany

Benson, Bryant
Department of Cell Biology and Anatomy
The University of Arizona
Tucson, AZ 85724-5044 ,USA

Blask, David E.
Senior Research Scientist
Laboratory of Experimental
Neuroendocrinology/Oncology
Bassett Research Institute
One Atwell Road
Cooperstown, NY 13326, USA

Callaghan, Brian D.
Department of Anatomical Sciences
University of Adelaide
Adelaide, South Australia, 5005,
Australia

Cardinali, Daniel P.
Department of Physiology
Faculty of Medicine
University of Buenos Aires
Paraguay 2155
121 Buenos Aires, Argentina

Conti, Ario
Centre for Experimental Pathology
Institute of Pathology
in Selva 24
6601 Locarno, Switzerland

Cornélissen, Germaine
Halberg Chronobiology Center
University of Minnesota
Mayo Hospital, Room 735 (7th floor)
Minneapolis Campus, Del Code 8609
420 Delaware Street S. E.
Minneapolis, MN 55455, USA

Cos, Samuel
Department of Physiology
and Pharmacology
School of Medicine
University of Cantabria
Cardenal Herrera Oria s/n
39011 Santander, Spain

Guerrero, Juan M.
Department of Medical Biochemistry
and Molecular Biology
The University of Sevilla
School of Medicine
Sanchez Pizjuan 4
41009 Sevilla, Spain

---

* In *italic type* in the table of contents and on the title page of the chapter.

Halberg, Franz
Halberg Chronobiology Center
University of Minnesota
Mayo Hospital, Room 715 (7th floor)
Department of Laboratory Medicine
Minneapolis Campus, Del Code 8609
420 Delaware Street S. E.
Minneapolis, MN 55455, USA

Hill, Steven M.
Department of Anatomy SL49
School of Medicine
Tulane University Medical Center
1430 Tulane Avenue
New Orleans, LA 70112-2699, USA

Hrushesky, William J. M.
Director of Research (151)
University of South Carolina
WJB Dorn VA Medical Center
6439 Garners Ferry Road
Columbia, SC 29209, USA

Jung, Detlev
Institute of Occupational,
Social and Environmental Medicine
Johannes Gutenberg-University Mainz
Obere Zahlbacher Str. 67
55131 Mainz, Germany

Karasek, Michal
Chair of Pathomorphology
Laboratory of Electron Microscopy
Medical University of Lodz
Czechoslowacka 8/10
92-216 Lodz, Poland

Khavinson, Vladimir Kh.
Institute of Bioregulation
and Gerontology
3 Dinamo Pr.
197110 St. Petersburg, Russia

Kvetnoy, Igor M.
Laboratory of Experimental Pathology
Medical Radiological Research
Center
4 Korolev Street
249020 Obninsk , Russia

Lapin, Vera
Seitenberggasse 7
1170 Wien, Austria

Liebmann, Peter M.
Institute of Pathophysiology
Medical Faculty, University of Graz
Heinrichstr. 31 a
8010 Graz, Austria

Lissoni, Paolo
Division of Radiotherapy
S. Gerardo Hospital
Via Donizetti 106
20052 Monza (Milano), Italy

Löscher, Wolfgang
Department of Pharmacology,
Toxicology and Pharmacy
School of Veterinary Medicine
Bünteweg 17
30559 Hannover, Germany

Maestroni, Georges J. M.
Centre for Experimental Pathology
Institute of Pathology
in Selva 24
6601 Locarno, Switzerland

Reiter, Russel J.
Department of Cellular
and Structural Biology
The University
of Texas Health Science Center
Mail Code 7762
7703 Floyd Curl Drive
San Antonio, TX 78229-3900, USA

Röschke, Joachim
Department of Psychiatry
University of Mainz
Untere Zahlbacher Str. 8
55131 Mainz, Germany

Schauenstein, Konrad
Institute of Pathophysiology
Medical Faculty, University of Graz
Heinrichstr. 31a
8010 Graz, Austria

Stevens, Richard G.
Department of Community Medicine
UConn Health Center
263 Farmington Ave.
Farmington, CT 06030-6325, USA

Touitou, Yvan
Department of Biochemistry
Faculty of Medicine
47–83 boulevard de l'Hôpital
75651 Paris Cedex 13, France

Vollrath, Lutz
University of Mainz
Department of Anatomy
Becherweg 13
55128 Mainz, Germany

Zisapel, Nava
Department of Neurobiochemistry
Tel Aviv University
Ramat Aviv
Tel Aviv 69978, Israel

Section I
Significance of the Pineal Gland and Its Hormone Melatonin

1 Some Historical Remarks Concerning Research on the Pineal Gland and Cancer

Vera Lapin

Mr. Chairman, ladies and gentlemen, dear fellow pinealogists,

First of all, I would like to thank Drs. Christian and Hella Bartsch for their kind invitation to this conference. I am delighted that I can still attend and listen to your presentations on research work concerning the pineal gland and cancer. I am also glad that I, too, could contribute to this research. It was only 20 years ago that I organized a small workshop on this subject.

For a long time, the function of the pineal gland was a mystery. All of us know the significance of this organ as "the third eye" and its role in different mythologies and religions. You might even know that Descartes called this gland "the seat of the soul".

We can relate our modern research on the pineal gland to systematic investigations of this organ at the turn of the century. Research on the link between the pineal gland and cancer has had a long tradition in Austria. Engel presented, in 1936 at the meeting of the Viennese Society of Endocrinologists, an extensive lecture, reviewing all scientific activities concerning the pineal gland since 1896. In 1934, Engel and Bergmann reported their results in experimental cancer treatment with pineal extracts or whole gland implants. In the following years, Toygarli and Hutschenreiter investigated the effect of pineal substances in veterinary medicine. Hofstätter, Janovetz, Sander, and Schmid performed similar research in human medicine. In the meantime, and perhaps until today, the real significance of the role of the pineal gland in oncogenesis remained obscure. Its role in different processes such as sexual and reproductive function, circadian rhythms, immunity and aging, kept posing new questions rather than offering answers (for review see Lapin 1976).

My own work began in the early 1970s, at the University Institute for Cancer Research in Vienna, upon a suggestion by Prof. Wrba to investigate the influence of ablation of the pineal gland in experimental animals on the growth of neoplasma. In those days we were slowly beginning to understand the function of thymus and bursa Fabricii and the differentiated functions of T and B lymphocytes. Our knowledge of receptors, neuronal and tissue mediators was non-existent. It was yet to be acquired.

I wanted to determine the role the thymus and the pineal gland played in carcinogenesis. I started my experimental work with newborn rats, as the function of their thymus and pineal glands is immature. In order to obtain a suitable experimental model, I developed a technique for simultaneous "thymectomy" and "pinealectomy" in experimental animals. For further studies I used three experimental cancer models:

1. Transplantable tumors "Yoshida sarcoma"
2. Tumor induction by chemical carcinogenic substances
3. Viral induction of tumors by Polyomavirus

By using the chemical carcinogen DMBA and/or transplantable tumors, I could provoke specific leukemia-like dissemination of malignant cells in rats, whose both pineal gland and thymus had previously been removed. I also established that the "viral" induction of kidney sarcoma by Polyoma virus was positive only in thymectomized rats, but not in the pinealectomized ones. Hence my interpretation of this finding was that this kind of carcinogenesis was not affected by the endocrinium. In other words, the absence of the pineal gland has a specific effect on carcinogenesis, which could be interpreted as proof that the pineal gland would be involved in the regulation of the central neuroendocrinological mechanism which, in turn, would influence the immuno-logical control of carcinogenesis. Certainly, such a hypothesis was in need of further research (Wrba et al. 1975; Lapin 1976, 1978; Lapin and Ebels 1976).

Of course, my work was not without difficulties, which might sound so familiar to many of you. There was lack of support, not a material one, but, much more important to a researcher, there was no enough encouragement for my work. Nobody seemed to be interested in my research. From time to time, I did report about my progress, but in the end this was a one-person department, one-person job, and one-person involvement. Slowly, first through the literature, and later also personally in some cases, I made acquaintance with many fine pineal researchers: Drs. Reiter, Cardinali, Wurtman, Fraschini, Miline, Quay, Tapp, Dilman, Benson, Aubert, just to name a few. Soon I met Dr. Ebels from Utrecht and in cooperation with her I investigated the effect of non-melatonin sheep pineal fractions on experimental tumors. Stimulated by such contacts, I decided to pursue my own experimental work and, in 1976, I published a review article about the pineal gland and malignancy.

At that time I received a kind letter from two German students who were under-taking studies in India and had a very keen interest in several aspects of pineal research. They were Christian Bartsch and Hella Trompelt.

In 1977, I organized the first conference on the pineal gland and oncogenesis, entitled "Pineal Gland as a New Approach to Neuroendocrine Control Mechanisms in Cancer". And the rest, I think, you all remember better than I do.

Now I would like to wish you all the very best for your research. Let this meeting be a successful one! Thank you.

References

Lapin V (1976) Pineal gland and malignancy. Österr Z Onkol 3:51–60
Lapin V (1978) Effects of reserpine on the incidence of 9,10-dimethyl-1,2-benzanthracene induced tumors in pinealectomized and thymectomized rats. Oncology 35:132–135
Lapin V, Ebels I (1976) Effects of some low molecular weight sheep pineal fractions and melatonin on different tumors in rats and mice. Oncology 33:110–113
Wrba H, Lapin V, Dostal V (1975) The influence of pinealectomy and of pinealectomy combined with thymectomy on the oncogenesis caused by polyoma virus in rats. Österr Z Onkol 2:37–39

2 Biology of the Pineal Gland and Melatonin in Humans

Lutz Vollrath

2.1 Introduction

The preliminaries of this conference have revealed that there is a fair amount of scepticism among scientists regarding possible interrelationships between the pineal gland hormone melatonin and cancer. I am sure that by the end of this conference we will be in a better position to judge for ourselves whether or not the scepticism is justified. In my view, despite the scepticism, scientists have a moral obligation towards the public, and especially the hundreds of thousands of cancer patients, to explore even the remotest possibilities to prevent, cure, or at least alleviate the suffering of this disease. On the other hand, it is mandatory not to waste public money, but to discontinue a specific branch of research once we are convinced that it is no longer promising.

This is not the place to consider in detail the reasons for the scepticism. My impression is that three factors play a role. The first is that a menacing and multi-factorial disease such as cancer is being related to a single organ and in particular to one whose function remained mysterious for a long time. The second is the successful attempt of the media and some scientists to promote melatonin as a wonder drug (Reppert and Weaver 1995; Turek 1996; Wirz Justice 1996; Kendler 1997; Weaver 1997; Bergstrom and Hakanson 1998). This has recently discredited melatonin research to a certain degree. Thirdly, the results dealing with interrelationships between the pineal gland and cancer have not been so spectacular as to incite research in this area on a large scale. Compared to other areas of the biosciences, the number of scientists working in this field is relatively small.

The aim of the present contribution is to deal with the basics of the hormone melatonin and of the organ in which most of the hormone is synthesized, the pineal gland. As, for obvious reasons, only certain aspects of the gland can be studied in humans, a short overview dealing with the gland in vertebrates in general is given first. More details may be obtained from specialized books and review articles (Wurtman et al. 1968; Quay 1974; Reiter 1981; Vollrath 1981; Vaughan et al. 1986a; Arendt 1995; Korf et al. 1998).

2.2 The Pineal Gland in Animal Species

The mammalian pineal gland is a neuroendocrine transducer (Wurtman and Axelrod 1965; Wurtman and Anton Tay 1969) in which neuronal signals originating in the retina-SCN (suprachiasmatic nuclei)-system are transformed into a hormonal message, the message being the circadian synthesis and release of the indolalkylamine melatonin (MEL). Although MEL is without doubt the major active principle of the pineal gland, a variety of peptides have been extracted from pineal tissue with antigonadotropic (Benson and Ebels 1978) and antitumor activities (Bartsch et al. 1987). Moreover, peptides derived from proenkephalin, prodynorphin, and pro-opiomelanocortin are present (Schröder et al. 1988).

2.2.1 Structural Aspects, Innervation, and Pinealocyte Receptors

The mammalian pineal gland is a tiny and simple-structured organ. In humans it has the shape of a pine cone, measuring ca. 10 mm in length and weighing approximately 100 mg. It is part of the diencephalon and projects from the epithalamus backwards, overlying the tectal plate of the mesencephalon (Fig. 2.1). Anteriorly, the pineal gland is attached to the epithalamus by a stalk containing commissural fibers running from the epithalamus to the gland. In common laboratory rodents, the pineal gland has a more complex form than in humans, usually extending to the calvarial roof. Here, the pineal gland may be dumb-bell-shaped, e.g., in guinea pigs, or may consist of two distinct parts, the large superficial pineal and the small deep pineal part lying intercalated between the habenular and posterior commissures (e.g., in rats).

The main cellular component of the pineal gland is the pinealocyte. In rats, pinealocytes amount to 82% of the cells present, the remainder being mainly astrocyte-like glial cells (12%) (Wallace et al. 1969). Nerve cell bodies are not regularly present. Pinealocytes are neuron-like cells in that they consist of a cell body with cell processes of variable lengths, often forming bundles. The ultrastructure of the cytoplasm is not very revealing. There are few, if any, electron-dense secretory granules. This is not

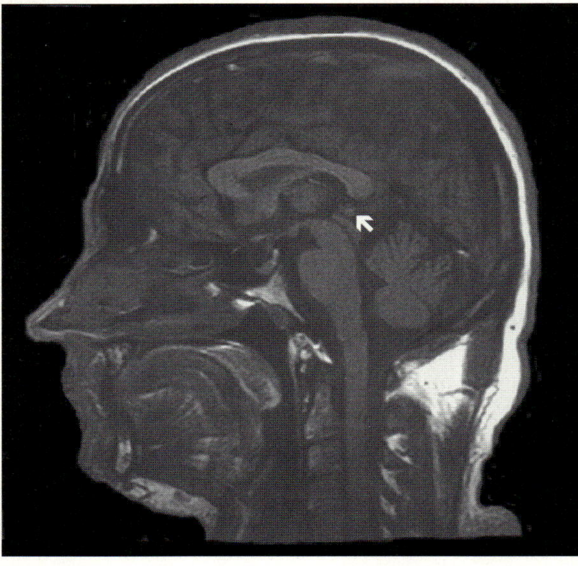

Fig. 2.1. Median-sagittal section through the head of an adult human (magnetic resonance image) to show the location of the small diencephalic pineal gland, indicated by an *arrow*, overlying the mesencephalon. (Courtesy of Dr. K. Blasinger of Radiologische Gemeinschaftspraxis im Fort Malakoff Park, Dr. B. Buddenbrock, Dr. K. Blasinger and Dr. P. Benz, Mainz, Germany)

surprising as MEL is not stored in granules and not linked to a carrier protein, but is released into the bloodstream soon after its synthesis. There is usually little rough and smooth endoplasmic reticulum, indicating that peptide and protein synthesis for export is not very pronounced and that functions residing in the smooth ER such as steroid metabolism and detoxification are likewise inconspicuous. In the terminals of pinealocyte processes, electron-lucent vesicles may be abundant. They contain the proteins that are needed for exocytosis as well as transmitters for intercellular signalling, e. g., glutamate (Redecker and Veh 1994) and L-aspartate (Yatsushiro et al. 1997), the latter of which inhibits MEL synthesis (Yamada et al. 1997). The vesicles take up glutamate, but not MEL, serotonin, noradrenaline, γ-aminobutyric acid (GABA), or acetylcholine (Moriyama and Yamamoto 1995).

That pinealocytes are neuron-like cells is further illustrated by the fact that they can be stained with antibodies against certain neuronal markers such as neurofilament protein, synaptophysin (Vollrath and Schröder 1987; Schröder et al. 1990), synaptotagmin I, synaptobrevin II, syntaxin I (Redecker 1996; Redecker et al. 1997), and rab 3 (Redecker 1995). Moreover, pinealocytes contain synaptic organelles, the synaptic ribbons (Vollrath 1981; McNulty and Fox 1992; Bhatnagar 1994; Wagner 1997), which are restricted to certain sensory and nerve cells. These organelles are basically plate-like structures (Jastrow et al. 1997) with prominent day < night (Vollrath 1973) and seasonal changes in number and size (Karasek et al. 1988; McNulty et al. 1990). Synaptic ribbons are thought to transport synaptic vesicles to the presynaptic membrane for exocytosis and may play a role in intercellular communication between pinealocytes together with gap junctions (Moller 1976; Taugner et al. 1981). Some of the pinealocytes exhibit action potentials, even in vitro (Schenda and Vollrath 1998 a, b). These electrically active cells appear to interact to form an intrapineal network which may monitor the intrapineal neurotransmitter levels (Schenda and Vollrath 1999).

In submammalian vertebrates (Collin 1971; Oksche 1971; Korf et al. 1998), the gross anatomy of the pineal gland is more complicated than in mammals. Here the pineal gland is part of a complex that consists of the *pineal organ proper* plus the *parapineal organ* in lampreys and fishes, the *frontal organ* in certain amphibian species, and the *parietal eye* in certain reptiles, the lizards. The intrinsic cells of the pineal complex are mainly photoreceptors and modified photoreceptors with rudimentary outer segments, which explains the direct light sensitivity of these organs. Cells resembling mammalian pinealocytes are present to varying degrees. In the avian pineal gland, modified photoreceptors predominate. Avian and teleost pineal glands have a built-in circadian oscillator so that MEL continues to be formed in a circadian manner in vitro.

The kinship of mammalian pinealocytes to photoreceptors is highlighted by the observation that in some of them, photoreceptor proteins are demonstrable, e. g., rhodopsin (Korf et al. 1985 a; Huang et al. 1992), S-antigen (Korf et al. 1985 b, 1986 a, b, 1990), recoverin (Korf et al. 1992; Schomerus et al. 1994), and peripherin (Lerchl et al. 1998 a). The calcium-binding protein calretinin has also been demonstrated immunocytochemically in pinealocytes of different mammalian species including human (Novier et al. 1996).

The question of a possible functional heterogeneity of mammalian pinealocytes has not been definitively answered. In particular, it is not known whether all of the pinealocytes are engaged in MEL synthesis. Immunocytochemistry suggests that this may not be the case (Freund et al. 1977).

2.2.1.1 Innervation (Korf and Moller 1984; Korf 1996)

The pineal gland receives nervous input from several sources, but, except for sympathetic innervation which is responsible for the circadian synthesis of melatonin, the functional significance of the nervous connections is obscure.

Sympathetic Innervation. The mammalian pineal gland is densely innervated by postganglionic noradrenergic sympathetic nerve fibers coming from the superior cervical ganglia of the sympathetic trunk. The sympathetic fibers destined for the pineal gland follow the internal carotid nerve, giving off the pineal nerve. The nerve fibers terminate mainly in the perivascular spaces of the gland and usually do not form direct synapses on pinealocytes. These fibers are of utmost importance because they convey circadian and photic information to the pineal gland and regulate MEL synthesis. The circadian information comes from the circadian oscillator of the body, the SCN which is linked to the retina via the retino-hypothalamic tract (Reuss 1996). From the SCN, a multisynaptic neuronal pathway extends via the hypothalamic paraventricular nuclei, lower brain stem, spinal cord, intermediolateral columns of the spinal gray matter, and preganglionic cholinergic sympathetic neurones to the superior cervical ganglia mentioned above. Some photic information reaches the gland via central, nonsympathetic fibers (see below). The reason for the circadian synthesis of MEL is that during the day, the sympathetic nerve fibers release much less noradrenaline (3 fmol) than during the night (12.5 fmol; rat) (Drijfhout et al. 1996c). Recently, the interesting observation has been made that in genetically mutant anophthalmic rats lacking a complete visual system, the MEL rhythm was entrained to light/darkness and could be suppressed by exposure to continuous light, indicating that extravisual light perception exists in these animals (Jagota et al. 1999). Also in mice lacking rod and cones, light has been shown to inhibit MEL synthesis (Lucas et al. 1999).

Parasympathetic Innervation. Hard data on parasympathetic cholinergic innervation of the mammalian pineal gland are scarce (Wessler et al. 1997; Korf et al. 1998). Recently, it has been shown that cholinergic mechanisms regulate Ca influx into pinealocytes (Schomerus et al. 1995; Korf et al. 1996), cause depolarization, activate L-type Ca^{2+} channels (Letz et al. 1997), and increase the firing rates of spontaneously active pineal cells (Letz et al. 1997; Schenda and Vollrath 1998b). Moreover, acetylcholine (ACh) prevents MEL synthesis by inhibiting noradrenaline release from intrapineal sympathetic nerve fibers (Drijfhout et al. 1996a) and by triggering glutamate release from pineal stores (Yamada et al. 1998a, b). The source of ACh may not necessarily be the parasympathetic innervation of the gland. In the rat, pinealocytes synthesize relatively large amounts of ACh, the synthesis being distinctly higher at night than day (Wessler et al. 1997).

Central Innervation. There is abundant morphological evidence for neuronal processes entering the pineal gland directly from the epithalamus via the habenular and posterior commissures. Recently, a paired connection from the pretectal area to the pineal gland has been described in humans (Sparks 1998). Electrophysiological studies involving electrical stimulation have clearly shown that interactions are demonstrable between the epithalamus and the pineal gland, and vice versa (Reuss 1987). Particularly striking are electrophysiological experiments (Thiele and Meissl 1987), showing that after sympathectomy of the pineal gland, pineal cells lying near the stalk still respond to light, even of different wavelengths, and to darkness. These results illustrate that photic information does not only reach the pineal gland via the peripheral sympathetic nervous system mentioned above, but also via the epithalamus. Apparently, the latter photic input does not suffice to induce circadian rhythmicity of MEL synthesis since sympathectomy is well known to abolish the circadian rhythm in the pineal gland (Moore et al. 1967).

Peptidergic Innervation (Korf et al. 1998). There is abundant evidence for the presence of a variety of neuropeptides in intrapineal nerve fibers, e. g., NPY, CPON, substance P, somatostatin, VIP, LHRH, CGRP, vasopressin, and oxytocin. Whereas most of the intrapineal NPY is present in postganglionic sympathetic fibers, the origin of the remainder of the NPY fibers and of the other peptide-containing fibers is obscure. They may come from various diencephalic and mesencephalic nuclei or from the pterygopalatine and trigeminal ganglia (Reuss 1999).

2.2.1.2 Pinealocyte Receptors

With respect to MEL synthesis, β_1- and α_1-*adrenoreceptors* of pinealocytes are the most important ones. While β_1-adrenoreceptor density shows a somewhat variable, low-amplitude day/night rhythm with peaks either late in the day or at night, α_1-adrenoreceptors do not appear to exhibit such a rhythm (Pangerl et al. 1990). β_1-Adrenoreceptor mRNA expression is greatly enhanced at night (Pfeffer et al. 1998).

The existence of *nicotinic ACh receptors* in the pineal gland has been postulated from results obtained by immunocytochemistry (Reuss et al. 1992), radiolabelled ligand binding (Stankov et al. 1993), calcium imaging (Schomerus et al. 1995), and patch-clamp studies (Letz et al. 1997). Evidence for *muscarinic ACh receptors* is also available (Taylor et al. 1980; Govitrapong et al. 1989). There are also receptors for VIP, PHI, PACAP, NPY, vasopressin, and oxytocin (Korf et al. 1998).

2.2.2 Melatonin (MEL) Synthesis and the Regulation of Its Day/Night Rhythm

The hormone MEL was discovered and given its name by the dermatologist Aaron Lerner and co-workers (Lerner et al. 1958), who studied pigment disorders of the skin and were interested in characterizing the pineal compound that leads to melanin aggregation in dermal melanophores, described in 1917 (McCord and Allen 1917) in frogs. Lerner and co-workers (Lerner et al. 1959) extracted MEL from 250,000 bovine pineal glands and characterized it to be 5-methoxy-*N*-acetyltryptamine (Fig. 2.2). We now have a fairly complete picture of how MEL synthesis proceeds and how it is regulated.

Pinealocytes take up the amino acid tryptophan from the circulation and convert it to 5-hydroxytryptophan by tryptophan hydroxylase. 5-Hydroxytryptophan is decarboxylated by aromatic L-amino acid decarboxylase to 5-hydroxytryptamine (serotonin). Serotonin is *N*-acetylated by arylalkylamine-*N*-acetyltransferase (AA-NAT) yielding *N*-acetyl-5-hydroxytryptamine which is then *O*-methylated by hydroxyindole-*O*-methyltransferase (HIOMT) to the final product 5-methoxy-*N*-acetyltryptamine (melatonin).

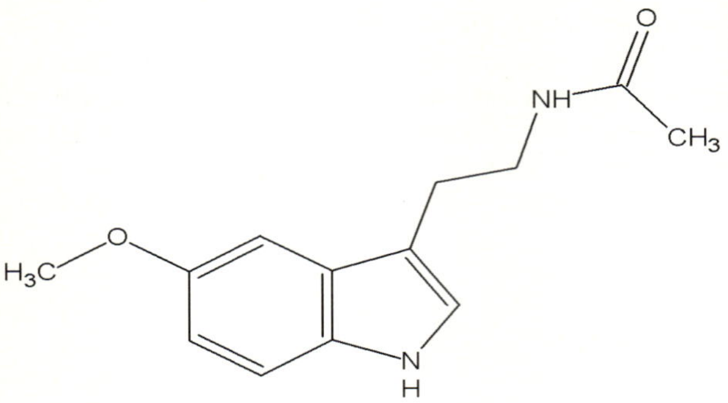

Fig. 2.2. Structural formula of melatonin (5-methoxy-*N*-acetyltryptamine), a derivative of the amino acid tryptophan

For experimental studies involving animals, it must be noted that certain strains of mice are not capable of synthesizing MEL because of a gene-related defect of AA-NAT and HIOMT or of HIOMT only (Ebihara et al. 1986, 1987; Goto et al. 1989). Although the greatest amount of MEL produced in the body stems from the pineal gland, MEL is also synthesized in the retina, expressing a circadian rhythm even in vitro (Tosini and Menaker 1996; Tosini and Menaker 1998), ciliary body (Martin et al. 1992), inner ear (Biesalski et al. 1988), and intestine (Lee and Pang 1993). Y79 human retinoblastoma cells also synthesize MEL (Janavs et al. 1991). MEL formation is not restricted to verte-brates. It has been demonstrated in insects (Vivien Roels et al. 1984; Wetterberg et al. 1987; Finocchiaro et al. 1988) and in plants (Dubbels et al. 1995). In the unicellular alga *Gonyaulax polyedra,* MEL formation undergoes a circadian rhythm with enhanced formation in darkness (Hardeland et al. 1996b).

Soon after the discovery of MEL, the day < night rhythm of its synthesis was detect-ed, the regulation of which has been the subject of intense research for more than two decades. As mentioned above, during nighttime there is an increased release of nor-adrenaline (NA) from intrapineal postganglionic sympathetic nerve fibers (Drijfhout et al. 1996c), regulated by the above-mentioned circadian oscillator, the hypothalamic SCN. NA acts upon cell membrane-bound β_1- and α_1-adrenoreceptors of pinealocytes, that interact with G_s proteins to stimulate adenylate cyclase activity and cAMP forma-tion. The increased cAMP levels stimulate AA-NAT activity through transcriptional, translational, and posttranslational events, involving cAMP-inducible protein kinase A (PKA), immediate early genes (IEGs), and transcription factors. Currently, intense research is underway to elucidate the function of the IEGs *c-fos, jun-B,* and *fra-2* as well as the transcription factors FRA-2, CREB (cyclic AMP response element binding protein), and ICER (inducible cyclic AMP early repressor) (Stehle et al. 1993; Stehle 1995). AA-NAT mRNA increases more than 150-fold at night in rats (Roseboom et al. 1996), and the AA-NAT protein is posttranslationally regulated by proteasomal proteolysis (Gastel et al. 1998). Moreover, the increased cAMP levels stimulate trypto-phan hydroxylase activity (Ehret et al. 1991) and the formation of β_1-adrenoreceptor mRNA (Pfeffer et al. 1998). Following an 8-h phase advance of the light/dark cycle, it takes 10 days for the MEL rhythm to adjust completely (Drijfhout et al. 1997).

MEL synthesis exhibits a day < night rhythm independently of whether the animals are diurnally or nocturnally active (Reiter 1982). Three types of rhythms are distin-guished (Reiter 1986). In the type I rhythm, the nocturnal increase occurs rather late and exhibits a short-term peak only (Syrian hamster, Mongolian gerbil); in type II, MEL begins to increase slowly at the day/night transition reaching a peak near the middle of the dark, followed by a drop, reaching daytime levels at about the time of lights on (albino rat, ground squirrel, Turkish hamster); in the third type, there is a rapid increase of MEL synthesis at the beginning of the dark period followed by a long high plateau and a drop roughly at light onset (e.g., white-footed mouse, cotton rat, Djungarian hamster, cat, sheep).

2.2.3 The Influence of Light and Darkness on MEL Formation

The day < night rhythm of MEL secretion is often interpreted in the sense that darkness stimulates MEL synthesis and light inhibits it. This conclusion is only partly correct because the day < night rhythm is still present in animals that are blind or that are kept under constant darkness. Under these conditions the rhythm becomes free-running with a period slightly longer than 24 h since the synchronizing effect of the ambient lighting regime is lacking and since the pineal gland is driven by the endogenous circadian rhythm of the SCN. In common laboratory animals it is not possible to stimulate MEL synthesis during daytime by exposure to darkness. In contrast, there is ample evidence that short pulses of light given at night acutely suppress MEL formation (Klein and Weller 1972a; Kanematsu et al. 1994; Drijfhout et al. 1996b). Light pulses as short as 1 ms are effective (Vollrath et al. 1989). The irradiance required to suppress MEL synthesis varies considerably between species. In albino rats, $0.0005\ \mu W/cm^2$ is sufficient (Webb et al. 1985), but in Richardson's ground squirrel a light irradiance of $1850\ mW/cm^2$ is required (Reiter et al. 1983). The suppressive effect of light depends also on the wavelength. In hamsters, blue light is most effective (Brainard et al. 1984; Reiter 1985). In rats, white light is most effective followed by green, blue, and red light (Zawilska et al. 1995). Red light cannot be regarded as "safe" light when studying nocturnal MEL production in albino rats because high intensities have been shown to have a suppressive effect (Sun et al. 1993). Red light given at 0400 hours is more effective in suppressing MEL synthesis than when given at 2400 hours (Honma et al. 1992b). In some species also UV light depresses MEL synthesis (Brainard et al. 1986; Benshoff et al. 1987; Zawilska et al. 1998) and entrains the SCN (Amir and Robinson 1995). Whether or not MEL synthesis recovers after a light pulse depends on the time point light is given: the earlier the animals cease to be exposed to light, the greater the chance for MEL synthesis to recover that night (Illnerova and Vanecek 1979).

The depressive effect of light given at night is brought about by an abrupt decrease of NA release from the intrapineal nerve fibers (Drijfhout et al. 1996b) followed by a decrease of pineal cAMP levels and AA-NAT activity (Klein and Weller 1972b; Klein et al. 1978), the AA-NAT protein undergoing reversible proteasomal proteolysis (Gastel et al. 1998). In the morning decline of MEL synthesis, ICER appears to be involved (Stehle et al. 1993; Takahashi 1994).

2.2.4 The Function of MEL in Non-Human Mammals

The role of MEL is threefold. First, it provides feedback to the circadian clock, the SCN, and helps to adjust the clock (Cassone et al. 1986, 1987; Mason and Brooks 1988; Stehle et al. 1989; McArthur et al. 1991; Reuss 1996). Second, during pregnancy and in the newborn, it sets the circadian phase of the descendants (Deguchi 1975, 1978; Takahashi and Deguchi 1983; Weaver and Reppert 1986; Yellon and Longo 1987, 1988; Zemdegs et al. 1988; Horton et al. 1989; McMillen et al. 1995; Shaw and Goldman 1995). Third, in photoperiodic mammals, MEL adjusts the physiology of the body to the seasonal requirements. According to Reiter (1982), the pineal gland ensures that the young are born at a season when they can best survive. This goal is achieved because the pineal gland measures day/night length and adjusts MEL secretion accordingly. When the nights become longer in the autumn/winter, the enhanced nocturnal MEL secretion is prolonged. This MEL message is interpreted differently by different photoperiodic species. In Syrian hamsters, the gonads and the accessory organs become atrophic. In this species, MEL is clearly antigonadotropic and at least 12.5 h of light per day are necessary to keep the gonads in a functioning state. In contrast, the nocturnal prolongation of MEL synthesis in sheep is the signal to begin with reproduction, MEL here being progonadotropic (Arendt et al. 1983). Timed infusion of MEL to intact or pinealectomized animals has unequivocally shown that it is the duration of the elevated nocturnal MEL secretion that mediates the seasonal changes and not the amplitude (Bartness et al. 1993).

2.2.5 Influence of Magnetic and Electromagnetic Fields on the Pineal Gland

As a lack of MEL has been accused of influencing the development of cancer and as magnetic (MFs) and electromagnetic fields (EMFs) have been reported to depress MEL levels, this aspect is briefly mentioned.

That the pineal gland is susceptible to changes in the ambient *static MF* was first shown in our laboratory by electrophysiological in vivo studies. In guinea-pigs exposed to artificially generated earth-strength magnetic fields (0.5 Oe) for 17 min, single-unit recordings revealed that after a lag phase of 2 min a proportion of pinealocytes responded with more than 30% decrease in the firing rate, which persisted as long as the recordings of individual cells could be maintained (Semm et al. 1980). Later, we showed that magnetically sensitive cells are also present in the rat pineal gland (Reuss et al. 1983), and that in rats and other laboratory rodents, excluding Syrian hamsters, nocturnal MEL formation is abruptly inhibited by the MF (Welker et al. 1983; Olcese et al. 1985; Olcese and Reuss 1986). In the rat, MEL formation is depressed even when the horizontal component of the artificial MF is changed as little as 5° (Welker et al. 1983). The depressive effect of MF stimuli on the pineal gland has been confirmed by several other groups (Cremer Bartels et al. 1983, 1984; Rudolph et al. 1988; Lerchl et al. 1990, 1991; Richardson et al. 1993; Schneider et al. 1994). Zero-compensation of the earth's MF did not affect pineal MEL secretion (Khoory 1987).

In mammals, there is compelling evidence that MFs affect the pineal gland via the retina. In intact rats, the MF stimulus decreased retinal dopamine and NA levels, but did not affect retinal MEL (Olcese et al. 1988). In cone-dominated pigmented ground squirrels, retinal dopamine levels decreased following the MF stimulus, whereas an increase was observed in the rod-dominated golden hamster (Olcese and Hurlbut 1989) in which the MF stimulus was without effect on pineal MEL secretion (Olcese and Reuss 1986). In acutely blinded rats, MFs had no effect on pineal MEL secretion (Olcese et al. 1985), and in rats with degenerate retinas due to constant light exposure, MFs did not affect retinal catecholamine levels in contrast to intact animals (Olcese et al. 1988). Moreover, MF effects were more readily demonstrable when the experiments were carried out under dim red light in contrast to complete darkness (Reuss and Olcese 1986). In Mongolian gerbils, retinal pigmentation may play a role as MF effects were demonstrable in albinotic but not pigmented strains (Stehle et al. 1988). An involvement of the retina is also suggested by the fact that in quail, MF exposure was followed by a decrease in retinal HIOMT and that in humans visual acuity decreased at nighttime (Cremer Bartels et al. 1983). There are also hints that MF may affect the mammalian pineal gland and pinealocytes in vitro (Lerchl et al. 1991b; Richardson et al. 1992; Rosen et al. 1998).

In contrast to the earth-strength MFs just mentioned, *strong* static MFs as used in nuclear magnetic resonance imaging (NMRI) are without effect on pineal MEL secretion (Reuss et al. 1985), but they seem to reduce the stimulatory β-adrenergic effect on the pineal gland (LaPorte et al. 1990).

Considerable effort has been devoted to studying the effects of exposure to 50 Hz/60 Hz *pulsed MFs, EFMs,* and to *pulsed direct current MFs.* Most often a *decrease* of MEL secretion or serum MEL levels has been reported (Wilson et al. 1981, 1986; Kato et al. 1993, 1994; Grota et al. 1994; Yellon 1994; Selmaoui and Touitou 1995; Rogers et al. 1995b; Mevissen et al. 1996). In a considerable number of studies, *no effects* could be demonstrated (Lee et al. 1993; Bakos et al. 1995; Rogers et al. 1995a; Truong and Yellon 1997; John et al. 1998). *Inconsistent results* were also reported (Löscher et al. 1998; Reiter et al. 1998), but when an effect was seen it was invariably a decrease of MEL levels. An *increase* of MEL levels was found in rats during daytime (Jentsch et al. 1993) and in the brook trout at night (Lerchl et al. 1998b). In one study, an increase of intra-pineal NA levels has been described (Zecca et al. 1998). To what extent the above inconsistencies are due to the methodology used (e.g., short-term versus long-term exposure; species differences; type of magnetic field: intensity, continuous versus intermittent, square-wave versus sinusoidal; time of day, season, etc.) is not known.

High-frequency EMFs (e.g., 900 MHz, as used in mobile phones) are apparently without effect on MEL synthesis (Vollrath et al. 1997). Relevant studies in humans will be mentioned below.

2.3 The Human Pineal Gland and MEL

2.3.1 General Aspects

For obvious reasons, the *human* pineal gland has not been as extensively studied as that in experimental animals. Yet, sufficient evidence has accumulated to conclude that in humans, MEL synthesis and its regulation are basically similar to that in other mammals. That the human pineal gland is an important source of MEL is indicated by the fact that the hormone is clearly demonstrable in postmortem pineal glands (Arendt et al. 1977; Kopp et al. 1980; Beck et al. 1982), there is a close parallelism between pineal and serum MEL levels (Arendt et al. 1977), and that after pinealectomy serum MEL levels are extremely low (Neuwelt and Lewy 1983; Murata et al. 1998).

Also, the two most important enzymes of MEL synthesis, AA-NAT and HIOMT, have been demonstrated in the pineal gland (Smith et al. 1977, 1981; Bernard et al. 1995) together with HIOMT mRNA (Donohue et al. 1993; Rodriguez et al. 1994). The AA-NAT gene has been mapped to chromosome 17q25 (Coon et al. 1996).

Innervation and Receptors. Also in humans, pineal MEL synthesis is regulated by preganglionic and postganglionic sympathetic neurones coming from the intermediolateral column of the spinal cord and the superior cervical ganglia of the sympathetic trunk, respectively. As the ganglia receive their input from spinal cord segments C8–T3, patients with cervical spinal lesions down to that level or with sympathectomy have low MEL levels without a diurnal rhythm, whereas in patients with spinal injuries in the thoracic or lumbar regions (T2–L2), MEL levels are usually normal with a clear day/night rhythm (Kneisley et al. 1978; Li et al. 1989; Bruce et al. 1991).

As MEL synthesis is regulated by NA released from intrapineal postganglionic sympathetic nerve fibers, it is relevant that β_1- and β_2-*adrenergic receptors* have been demonstrated in human postmortem pineal glands (Little et al. 1996). Hence, MEL synthesis can be inhibited by administration of β-adrenergic receptor antagonists such as atenolol (Cowen et al. 1983; Arendt et al. 1985b; Rosenthal et al. 1988; Cagnacci et al. 1992, 1994; Deacon et al. 1998) and propranolol (Wetterberg 1978; Lewy et al. 1985; Rommel and Demisch 1994; Mayeda et al. 1998).

It is not clear which role α-*adrenoceptors* play with respect to MEL synthesis/release. Clonidine, an α_2-adrenoceptor agonist, reduced plasma MEL levels, perhaps by acting on inhibitory presynaptic α_2-adrenoceptors (Lewy et al. 1985, 1986). In another study, clonidine and the α_2-antagonist yohimbine were without effect on nocturnal MEL levels, although an increased excretion of 6-sulfatoxymelatonin was noted between 1800 and 2200 hours under yohimbine (Kennedy et al. 1995).

As the MEL-regulating NA is catabolized by monoamine oxidase (MAO)(Goridis and Neff 1972), MEL levels can be increased by administering nonselective MAO inhibitors, e.g., tranylcypromine (Oxenkrug et al. 1986; Oxenkrug et al. 1988) or MAO-A inhibitors such as clorgyline (Murphy et al. 1986) or brofaromine (Bieck et al. 1988). The MAO-B inhibitors deprenyl (Murphy et al. 1986) and pargyline (Bieck et al. 1988) are without effect.

MEL secretion in humans is likely to be affected by many different substances given the fact that binding sites/receptors in the pineal gland have been described for benzodiazepines (Suranyi Cadotte et al. 1987) as well as for luteinizing hormone (LH), follicle-stimulating hormone (FSH), estrogen, and androgens (Luboshitzky et al. 1997). Thus, the benzodiazepine alprazolam (McIntyre et al. 1988, 1993) and the prostaglandin antagonists ibuprofen and indomethacin (Surrall et al. 1987) decreased nocturnal plasma MEL levels, whereas chlorpromazine led to an increase (Smith et al. 1979). In women, sex steroids do not affect MEL secretion (Delfs et al. 1994).

2.3.2 Release and Fate of MEL

There is general consensus that MEL synthesized in the pineal gland is not stored in secretory granules, but is readily released into the bloodstream. In view of the close topographical relationship of the pineal gland to cerebrospinal fluid (CSF)-containing compartments, it has been studied whether MEL is released into the CSF. Evidence against this route include the following facts: (a) MEL levels are lower in CSF than in serum (Arendt et al. 1977; Vaughan et al. 1978; Tan and Khoo 1981), (b) there is no gradient of MEL levels between CSF from cranial and lumbar sources (Brown et al. 1979), and (c) MEL levels in CSF and plasma change correspondingly after oral administration of MEL (Young et al. 1984). A recent study has challenged the view that MEL is mainly released into the bloodstream, because it was found that MEL levels were lower in serum than CSF (Rousseau et al. 1999).

The question whether MEL release is *pulsatile* has not been definitively answered. Episodic MEL secretion (Weitzman et al. 1978; Weinberg et al. 1979; Fevre-Montange et al. 1981; Follenius et al. 1995) with 8–12-min peaks (Vaughan et al. 1979a,b) or four peaks and troughs during nighttime (Claustrat et al. 1986) could not be verified (Bojkowski et al. 1987b; Trinchard Lugan and Waldhauser 1989) using sampling frequencies of 0.5 or 3, 8, and 10 min, respectively. Most recently, more evidence for pulsatile MEL release has been obtained (Geoffriau et al. 1999).

Between 50 and 75% of the MEL released into the bloodstream binds reversibly to plasma proteins (Cardinali et al. 1972; Morin et al. 1997; Di et al. 1998; Kennaway and Voultsios 1998), especially to α_1-acid glycoprotein and less so to albumin (Morin et al. 1997). Hence, when measuring MEL, the bound and free hormone have to be considered separately. MEL measured in saliva, which is approximately 70% lower than in plasma, apparently reflects the unbound form (Nowak et al. 1987; Kennaway and Voultsios 1998).

Circa 70–80% of the circulating MEL is *metabolized* in the liver, first to 6-hydroxymelatonin and then to 6-hydroxymelatonin sulfate (6-sulfatoxymelatonin) (Jones et al. 1969; Young et al. 1985) followed by excretion through the urine, the remainder being excreted unmetabolized (Kopin et al. 1961). Glucuronide conjugation of 6-hydroxymelatonin plays a minor role.

The *half-life* of MEL in serum has been calculated to be less than 30 min (Weitzman et al. 1978; Brown et al. 1997), 43.6 min (Iguchi et al. 1982), and 57 ± 34 min (Claustrat et al. 1986), respectively. For further details on MEL pharmacokinetics see Mallo et al. (1990).

2.3.3 Interindividual Variation of MEL Secretion

There is relatively strong interindividual variation of MEL secretion (Arendt et al. 1979). Within individuals the rhythm is rather constant (Bojkowski et al. 1987b). For interindividual differences, coefficients of variation of 25% have been calculated, whereas 8% was calculated for intraindividual differences (Lerchl and Partsch 1994). Low and high MEL secretors can be distinguished. Circa 5% of the general population are thought to not synthesize appreciable amounts of MEL (Langer et al. 1997). In low MEL secretors, the peak serum levels at night lay between 18 and 40 pg/ml, whereas in high MEL secretors they were between 54 and 75 pg/ml and occasionally even higher (Arendt et al. 1979; Laakso et al. 1990). With respect to nocturnal MEL excretion, the mean cut-off point between low and high secretors was calculated to be 0.25 nmol/l (Bergiannaki et al. 1995).

2.3.4 Day/Night Rhythm of MEL Formation in Humans

Like in animal species, there is a pronounced day < night rhythm of serum MEL levels in humans, the day/night increase being up to 30-fold.

The rhythm was apparently first described by Pelham and colleagues (Pelham et al. 1973) and amply confirmed using different techniques, including bioassay (Lynch et al. 1975a, b; Vaughan et al. 1976), radioimmunoassay (RIA) (Arendt et al. 1975, 1977, 1978; Lynch et al. 1978b; Weitzman et al. 1978; Fevre-Montange et al. 1983a, b; Touitou et al. 1984; Bartsch et al. 1985, 1992b; Sieghart et al. 1987; Di et al. 1998), gas chromatography mass spectrometry (Wilson et al. 1977, 1978), the highly specific gas chromatography negative chemical ionization mass spectrometry (Lewy and Markey 1978), as well as high-performance liquid chromatography (Vieira et al. 1992; Peniston Bird et al. 1993). The day < night rhythm has also been demonstrated in postmortem pineal glands (Stanley and Brown 1988; Schmid et al. 1994; Hofman et al. 1995), saliva (Miles et al. 1985a, b, 1987; McIntyre et al. 1987; Nowak et al. 1987; Nowak 1988; Laakso et al. 1990, 1994; Nickelsen et al. 1991; Demel 1992; Kennaway and Voultsios 1998), milk (Illnerova et al. 1993b), amniotic fluid (Kivelä et al. 1989), and urine, here mostly in the form of the metabolic product of MEL, 6-sulfatoxymelatonin (Bartsch et al. 1992b).

The shape of the rhythm is roughly sinusoidal (Fig. 2.3). In the normal population, daytime MEL secretion is invariably very low. With the most sensitive assay methods, plasma/serum MEL levels have been found to lie in the region of 1.5 pg/ml (Lewy and Markey 1978). The *nocturnal increase* is slow. Statistically significant day/night differences occur between 2100 hours (Arendt et al. 1977; Arendt et al. 1982; Brzezinski et al. 1988) and 2300 hours (Oxenkrug et al. 1988), the time point of increase being season dependent (see below). With the magnitude of the RIA error taken into consideration, nocturnal MEL secretion onset was calculated to occur at 2206 hours (1940–0029 hours) (Brown et al. 1997). This wide range is understandable as the phase of the rhythm depends on the season (see below) and whether the subjects studied are morning- or evening-type people, the latter showing a phase delay compared to the former (Duffy et al. 1999). Nocturnal MEL secretion usually peaks around 0200 hours (Arendt et al. 1977, 1982; Smith et al. 1977; Wetterberg 1978; Bojkowski et al. 1987b; Brzezinski et al. 1987; Webley and Lenton 1987; Oxenkrug et al. 1988; Schober et al. 1989), but peaks appearing between 0200 and 0500 hours have also been described

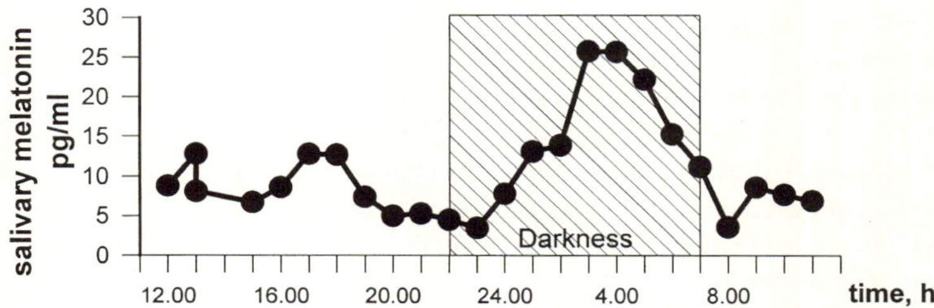

Fig. 2.3. Salivary melatonin levels of a single male adult, measured at 1-h intervals during a single normal-routine day by means of radioimmunoassay. The curve obtained is roughly sinusoidal, with low levels during daytime, a slow increase beginning at 2300 hours and a distinct peak from 0300 to 0400 hours. Note that the two small peaks during daytime are unusual

(Touitou et al. 1981, 1984; Arendt et al. 1982). The nocturnal maximum lies around 42 pg/ml (Lewy and Markey 1978), but, as pointed out above, in high MEL secretors it may exceed 75 pg/ml. MEL secretion decreases in the morning, reaching daytime levels around 0900 (Arendt et al. 1982) or 1100 hours (Brzezinski et al. 1988). More recently, offset times of 0621 hours (median range 0246–0817 hours) have been calculated (Brown et al. 1997). The total amount of MEL synthesized during a 24-h period has been calculated to be 30 µg (Fellenberg et al. 1980), 80% of which is formed during nighttime (Fellenberg et al. 1982).

Methods of Analyzing the Day/Night Rhythm of MEL Synthesis. A frequently applied technique to characterize the rhythm is the cosinor method (Nelson et al. 1979; Lerchl and Partsch 1994), giving details such as acrophase (calculated peak time), mesor (calculated 24-h mean concentration), and amplitude. To determine the exact circadian phase position, MEL secretion onset in the evening is determined in subjects exposed to dim light (Lewy and Sack 1989). Brown and co-workers (Brown et al. 1997) have applied the recently developed Bayesian statistical procedures to incorporate the magnitude of the immunoassay error into the analysis of rhythm parameters. Deconvolution analysis has also been used (Geoffriau et al. 1999).

2.3.4.1 The Day/Night Rhythm of MEL Secretion Is Basically a Circadian Rhythm

That the day/night rhythm of MEL secretion is basically endogenous and circadian in nature, i.e., with a period slightly longer than 24 h, has been demonstrated in a large proportion of blind people without conscious light perception (Lewy and Newsome 1983; Lewy et al. 1985; Sack et al. 1991, 1992; Nakagawa et al. 1992; Sack and Lewy 1993, 1997; Lockley et al. 1997; Klerman et al. 1998; Skene et al. 1999), in people living in the Antarctic in winter (Broadway and Arendt 1986; Broadway et al. 1987; Kennaway and Van Dorp 1991), and in people exposed to constant dim light (Middleton et al. 1997). Interestingly, free-running MEL rhythms have been described in *sighted* people (Hoban et al. 1989; Hashimoto et al. 1997a). Period length of the free-running MEL rhythm was calculated to lie between 24.2 and 24.79 h (Lewy and Newsome 1983; Lockley et al. 1997; Middleton et al. 1997; Sack and Lewy 1997; Hashimoto et al. 1997a). In one case, a period of 25.1 h has been reported (Hoban et al. 1989). Complete inversion of the MEL rhythm takes about 5–7 (Lynch et al. 1978a; Nowak 1988) or even 11 days (Fevre-Montange et al. 1981).

Under normal conditions, the circadian rhythm is entrained by environmental light and darkness. In people living in the Antarctic, the MEL rhythm is entrained to clock time as long as the sun shines, becomes free-running in winter and re-entrains to daylight in spring (Kennaway and Van Dorp 1991).

2.3.4.2 Ontogenesis of the Circadian Rhythm

MEL synthesis is demonstrable as early as in neonates. Yet there is no day/night rhythm (Kennaway et al. 1992; Munoz Hoyos et al. 1993), as the circadian rhythm-generating system is not mature at birth, but matures progressively (Rivkees 1997). The day < night difference present in blood taken from the umbilical vessels (Munoz

Hoyos et al. 1992) is of maternal and not infant origin (Munoz Hoyos et al. 1993). It takes a relatively long time for the day/night rhythm of MEL secretion to develop. As revealed by studying the development of 6-sulfatoxymelatonin excretion in urine (Kennaway et al. 1992), a day/night rhythm becomes apparent between 9 and 12 weeks of age in full-term infants and 2–3 weeks later in preterm infants. At 24 weeks of age, total 6-sulfatoxymelatonin excretion was 25% that of adult levels. The low mean night-time serum MEL concentrations increased to a peak value at 1–3 years of age, being considerably lower in individuals aged 15–20 years, when correlated with body weight (Waldhauser et al. 1988).

2.3.5 Influence of Light and Dark on MEL Secretion

In the early days of MEL research in humans, there was no evidence that acute exposure to light or darkness had any effect on MEL secretion. The day/night rhythm was found to continue normally in blindfolded people (Vaughan et al. 1979c) and nocturnal MEL secretion could not be inhibited by light (Jimerson et al. 1977; Wetterberg 1978; Vaughan et al. 1979b). However, there is now ample evidence that acute *light* exposure at night affects the pineal gland. There are two effects: it acutely suppresses MEL synthesis and may shift the circadian oscillator leading to a phase shift of the MEL rhythm.

2.3.5.1 *Suppression of MEL Synthesis by Light*

That light suppresses MEL synthesis in humans was first unequivocally shown by Lewy and colleagues (Lewy et al. 1980). In subjects exposed to 2500 lx from 0200 to 0400 hours, plasma MEL concentrations dropped after 10 to 20 min from 50 pg/ml to below 20 pg/ml and increased again to nocturnal levels at 0430 hours; 1500 lx was less suppressive and 500 lx had no effect. Later it was shown that even 200–500 lx suffices to suppress MEL levels (Bojkowski et al. 1987a; Strassman et al. 1987; Brainard et al. 1997), 100 lx being without effect (Strassman et al. 1987), and that there is a clear dose–response relationship of light and MEL suppression (Brainard et al. 1988; McIntyre et al. 1989b, c; Dollins et al. 1993b; Mayeda et al. 1998). The minimum intensities/duration of light to demonstrate the inhibitory effects on MEL in saliva were 285 lx/120 min, 339 lx/90 min, 366 lx/60 min, and 393 lx/30 min (Aoki et al. 1998). The wider the pupils, the more effective the light stimulus (Gaddy et al. 1993). In view of the slow time-dependent increase of MEL secretion at night, it is not surprising that light given early at night may yield different results than when given later (McIntyre et al. 1989a; Petterborg et al. 1991). Like in several animal species, white light and green light are most effective in suppressing MEL secretion (Horne et al. 1991; Morita et al. 1997). In color-vision-deficient humans, the depressive effect of light on MEL synthesis is not altered compared to normal controls (Ruberg et al. 1996). The depressive effect of bright light on MEL levels is accentuated by caffeine (Wright et al. 1997) and can be blunted by psoralen (Souetre et al. 1987, 1988). In women, light has been reported to suppress MEL secretion more strongly (by 40%) than in men (Monteleone et al. 1995). In unipolar depressed patients, light did not affect MEL secretion differently than in normal controls (Cummings et al. 1989).

2.3.5.2 *Light Phase Shifts the MEL Rhythm*

In addition to suppressing MEL secretion, light exposure may also phase shift the MEL rhythm. Depending on the time of day light is given it may lead to a phase advance or a phase delay. Phase advances were achieved by exposing subjects to bright light for 3 h in the morning on 3–6 consecutive days (Lemmer et al. 1994; Dijk and Cajochen 1997), but even 1-day exposure suffices (Buresova et al. 1991; Samkova et al. 1997), showing that the circadian pacemaker can be rapidly affected by light. Even room light (180 lx) is effective (Boivin and Czeisler 1998). The phase advance is apparently more pronounced in the morning (1.2–2.6 h) than in the evening (0.6–2.2 h) (Buresova et al. 1991), but it may also be that only the time point of the evening MEL rise is phase advanced (Vondrasova Jelinkova et al. 1999). Phase advances have also been observed when light is given for 3 h both in the morning and in the evening. In winter, a phase advance to summer conditions was achieved within 3 days, the nocturnal MEL signal being 3 h shorter than in the winter condition (Buresova et al. 1992). The phase shift persisted for 3 days after withdrawal of bright light (Illnerova et al. 1993a). Also in the Antarctic, 1 h of bright light both in the morning and in the evening phase advanced the MEL rhythm (Broadway et al. 1987).

It is not quite clear yet whether light exposure alone is effective or whether additional factors are involved. The latter assumption is supported by the fact that a phase advance of 1 h occurred when subjects, instead of being awakened at 0600 hours and exposed to bright light from 0600–0900 hours, received dim light during this time span and were awake (Samkova et al. 1997). A role of the quality of light and/or season is suggested by the fact that in summer, 16 h of natural sunlight phase advances the MEL rhythm and shortens the nocturnal MEL signal by 2 h compared to subjects exposed in winter to the same time span of artificial + natural daylight (Vondrasova et al. 1997). In the case of a free-running MEL rhythm in a young man, a forced fixed sleep schedule did not act as a zeitgeber for the MEL rhythm (Hashimoto et al. 1997a).

The question on whether acute exposure to *darkness* influences the MEL rhythm has been little examined. In one study, exposure to *darkness and sleep* in the afternoon rapidly phase advanced MEL onset by 2 h and the offset by 1 h (Van Cauter et al. 1998), leaving the question on whether darkness alone is effective unanswered.

Whether *phase delays* can be achieved by giving light early at night has been little studied and the results obtained are controversial, perhaps because of the different exposure times used. Thus, in one study, exposure to 500 lx from 2300 to 2400 hours phase delayed the MEL rhythm (Laakso et al. 1993), whereas 2500 lx given from 1800 to 2100 hours had no effect (Lemmer et al. 1994).

2.3.6 Seasonal Differences in MEL Secretion

As light has profound effects on MEL secretion (see above) and as the durations of light and darkness vary considerably in the temperate zones between seasons, seasonal differences in MEL secretion can be expected to exist. However, as due to understandable reasons, long-term studies in individuals and populations are difficult to perform, the data are relatively scarce. Moreover, the data obtained cannot be easily compared as often different aspects have been examined. The general impression gained is that seasonal differences do exist, despite the fact that in some studies (Griffiths et al. 1986), or in certain parts of a study (Bojkowski and Arendt 1988), no such differences could be detected.

In several studies, *phase shifts* of the MEL rhythm have been described. Thus, the *nocturnal increase* (Matthews et al. 1991) as well as the *nocturnal peaks* occur about 1.5 h earlier in summer than in winter (Illnerova et al. 1985; Bojkowski and Arendt 1988; Honma et al. 1992a). In contrast, remarkably stable nocturnal acrophases have also been noted, occurring around 0300 hours throughout the seasons (Touitou et al. 1984). There is also evidence for a bimodal annual rhythm of plasma MEL levels with peaks in January and July and nadirs in May and October (Arendt et al. 1977, 1979).

Whereas some studies point in the direction that plasma MEL levels are higher in winter than in summer (Kauppila et al. 1987; Kivelä et al. 1988), the reverse has been found in young men (Touitou et al. 1984).

Whether the *duration* of enhanced nocturnal MEL synthesis is season-dependent is not entirely clear. While in some studies no differences were noted (Illnerova et al. 1985; Matthews et al. 1991), more recently it was found that the duration was 2–3 h longer in winter than in summer conditions (Wehr 1991; Buresova et al. 1992; Vondrasova et al. 1997).

Total 24-h 6-sulfatoxymelatonin excretion did not reveal seasonal differences, but the amount excreted between 1200–1800 hours showed a statistically significant seasonal rhythm, with peaks in December/January and in July (Bojkowski and Arendt 1988).

Data on pinealocyte receptors in relation to season also exist (Luboshitzky et al. 1997).

2.3.7 MEL Secretion in Old Age and Pineal Calcification

There is clear evidence that MEL secretion decreases with age (Brown et al. 1979; Touitou et al. 1981, 1984; Sack et al. 1986; Bojkowski and Arendt 1990). On average, the 24-h mean plasma levels of MEL in the elderly are half of those in the young. This is not to say that old people have invariably low MEL levels. In 70–90-year-old people, 33% had normal daytime MEL levels, 53% exhibited decreased levels (0.17 nmol/l), and the remainder were found to have increased levels (0.43 nmol/l or higher) (Touitou et al. 1985). Interestingly, daytime MEL levels exhibited a sex difference, being higher in elderly women than in elderly men. Usually, the circadian rhythm of MEL secretion persists in old age (Touitou et al. 1981, 1984). A lack of circadian rhythmicity was demonstrated in 1 of 13 elderly subjects (Ohashi et al. 1997). Whether the morning decline of MEL levels is generally faster in the elderly (Ohashi et al. 1997) remains to be established.

An important question is whether or not the decrease of MEL secretion in old age is related to the common presence in the pineal gland of mulberry-shaped calcareous concretions also termed acervulus, corpora arenacea, or brain sand (Ganapathy and Kalyanaraman 1978; Vollrath 1981; Krstic 1986). The concretions consist mainly of calcium and phosphorus in the form of hydroxyapatite, like in bone, and variable amounts of Mg and trace elements as well as organic substances (Krstic 1976; Michotte et al. 1977; Galliani et al. 1989, 1990; Galliani and Mongiorgi 1991; Bocchi and Valdre 1993; Kodaka et al. 1994; Schmid et al. 1994; Nakamura et al. 1995; Schmid and Raykhtsaum 1995). Individual concretions may measure up to 3000 µm in diameter (Allen et al. 1981). The center of calcification is usually within 2 mm of the midline of the gland, occasionally 2.6 mm from the midline (Pilling and Hawkins 1977). In a rare case an asymmetrical calcification of the pineal gland was present, giving the false impression of a shift of the gland from the mid-sagittal plane of the brain (Ketonen et al. 1980).

Concretions increase in number and size with age (Vollrath 1981). When averaged over all age groups, in the Western world, ca. 50% of the pineal glands are calcified. Pineal calcification is apparently much less frequent in India where concretions were present in 16.7% of the patients examined (Kohli et al. 1992). In the Western world, pineal concretion occur very early in life. Already in the first year of life, about 3% of the children studied roentgenologically and by means of computed tomography revealed concretions, increasing up to 7.1% in 10-year-old children (Helmke and Winkler 1986; Winkler and Helmke 1987). There are apparently no sex differences in their incidence. This may be different in India (Kohli et al. 1992). Although pineal concretions greatly increase in frequency and size with age, they may be absent even in people over 80 years of age (Hasegawa et al. 1987; Galliani et al. 1989).

The functional significance of the concretions is obscure. Experimental studies in Mongolian gerbils have suggested that their occurrence is related to the secretory activity of the gland because they are much rarer in animals in which MEL synthesis is depressed due to pineal sympathetic denervation (Reiter et al. 1976; Welsh et al. 1979; Champney et al. 1985) or to chronic administration of the β-adrenoceptor blocker propranolol (Vaughan et al. 1986b). Also in humans, the formation of concretions has been related to the secretory activity of the pineal gland (Galliani et al. 1990) or to an alteration in calcium homeostasis in the pinealocytes (Schmid et al. 1994). Which consequences does pineal calcification have in humans? As judged from 6-sufatoxy-melatonin excretion, it does not appear to affect MEL secretion (Bojkowski and Arendt 1990). It has been shown to impair the sense of direction, perhaps by altering the intracranial electromagnetic environment (Bayliss et al. 1985) and to lead to daytime tiredness and sleep disturbance (Kunz et al. 1998).

2.3.8 Various Factors Influencing MEL Secretion

A large number of studies have been performed to examine which factors affect MEL secretion, both in health and disease. In normal *healthy* subjects, usually no *sex differences* in MEL secretion are demonstrable (Arendt et al. 1979; Beck Friis et al. 1984; Sack et al. 1986; Bojkowski and Arendt 1990), but in the elderly, MEL levels were higher in women than men (Touitou et al. 1985). In different phases of the *menstrual cycle,* usually no differences in MEL secretion are seen (Fellenberg et al. 1982; Brzezinski et al. 1988; Kivelä et al. 1988; Berga and Yen 1990). In one study it was higher in the luteal than in the follicular phase (Brun et al. 1987), whereas in another study morning serum MEL levels decreased significantly during the luteal phase (Fernandez et al. 1990). During parturition MEL levels do not change (Kivelä et al. 1989). After 2 days of *fasting,* a 19% decrease of MEL levels was noted, which did not occur when glucose supplements were given (Rojdmark and Wetterberg 1989). In another study, after 72-h fasting, daytime MEL levels increased, but nocturnal levels were unchanged (Meyer et al. 1990).

There are some studies that suggest that *physical exercise* and *stress* affect MEL levels. Physical exercise has been shown to increase daytime plasma MEL levels in women, but this effect diminished with increasing intensities of exercise (Skrinar et al.

1989). More detailed studies revealed that MEL levels may change within minutes, and that the onset of MEL secretion at night may be affected 12–24 h later, the changes depending on the duration, intensity, type, and time point of exercise. No effect was noted when physical exercise was performed early in the morning or during the day. Late evening exercise during the rising phase of MEL secretion blunted the increase. High-intensity exercise performed when nocturnal MEL levels were already high led to a 50% increase of MEL levels. Low- and high-intensity exercise during the night led to a phase delay of MEL onset the next night (Buxton et al. 1997). No or little effect has also been described (David et al. 1987; Monteleone et al. 1992, 1993). There is evidence that *posture* affects nocturnal MEL levels. They increase when moving from a supine to a standing position and decrease when the positions are reversed (Deacon and Arendt 1994; Nathan et al. 1998). They are also higher in sitting than in lying subjects, but there was no increase when subjects changed from sitting to standing (Nathan et al. 1998). *Sleep deprivation* increased MEL levels in the first night in one study (Salin Pascual et al. 1988), but in another study, 36 h of sleep deprivation had no effect (von Treuer et al. 1996).

MEL changes have been reported in several *diseases*, but only a few will be mentioned here. In *cluster headache*, the nocturnal MEL levels were clearly depressed (Chazot et al. 1984, 1987; Claustrat et al. 1989). In *coronary heart disease*, strongly depressed nocturnal MEL levels were found (Brugger et al. 1995). In *orthostatic hypotension*, depression or a strong increase of urinary MEL excretion was observed (Tetsuo et al. 1981). In *hypothalamic amenorrhea*, MEL levels were higher than in normal controls (Berga et al. 1988; Brzezinski et al. 1988). In *Turner's syndrome*, the day/night rhythm was unchanged and estrogen administration did not have an effect on MEL levels (Schober et al. 1989). In *anorexia nervosa* and *bulimia nervosa*, MEL levels were not different from controls, but when the eating disorder was combined with depression, MEL levels were decreased (Kennedy et al. 1989). On the other hand, it was observed that in anorectic people, nocturnal MEL (Arendt et al. 1992) and mean 24-h MEL secretion (Brambilla et al. 1988) were higher than in healthy people, and that the MEL rhythm was abnormal with peaks during the day and lack of peaks at night (Brambilla et al. 1988). In *depressed patients*, MEL levels were found to be unaffected (Thompson et al. 1988) or decreased (Claustrat et al. 1984; Beck Friis et al. 1985; Wetterberg 1985; Almay et al. 1987). Under *antidepressant medication*, MEL levels decreased after imipramine and mianserin treatment and increased slightly after phenelzine (Stewart and Halbreich 1989). In healthy volunteers the antidepressant desipramine increased evening production of MEL (Franey et al. 1986). In *schizophrenia* (Fanget et al. 1989), *chronic pain* (Almay et al. 1987), and *Alzheimer's disease* (Uchida et al. 1996; Liu et al. 1999; Mishima et al. 1999), lower than normal MEL levels were found and occasionally also rhythm irregularities. Pineal glands of Alzheimer's patients contain abnormal, swollen noradrenergic axons (Jengeleski et al. 1989).

2.3.9 MEL and Seasonal Affective Disorder

In view of the fact that MEL secretion is light/dark dependent, considerable effort has been devoted to studying which role MEL plays in *seasonal affective disorder* (SAD), especially as the symptoms of winter depression can be successfully treated with bright light (Rosenthal et al. 1986; Terman et al. 1988; Winton et al. 1989; Wehr 1992;

Wirz Justice et al. 1993). Usually 2500–5000 lx are used, but 500 and 1000 lx are sufficient to depress MEL levels in SAD patients (Gaddy et al. 1990). Usually, the light is administered in the morning for several hours, but the time of day of the treatment does not seem to be important (Wehr et al. 1986; Wirz Justice et al. 1993). On the other hand, when light was given either in the morning or in the evening, evening light was less antidepressant in one study (Lewy et al. 1998), whereas in another trial it was more effective (Wirz Justice et al. 1993). As pointed out (Rosenthal and Wehr 1987), there are no plausible comprehensive theories of the mechanism of phototherapy. Two main theories have been suggested: first, that it acts by modifying MEL secretion (Rosenthal et al. 1988); and second that it acts by correcting abnormally timed circadian rhythms (Lewy et al. 1987 a, b). Although there is no doubt that light depresses MEL secretion and sets the circadian rhythm clock (see above), the precise role of MEL in SAD is unclear. MEL suppression by light as well as retinal contrast sensitivity and visual-evoked EEG responses are not different in SAD patients compared to healthy controls (Murphy et al. 1993; Partonen et al. 1997), although in one study SAD patients showed a significant seasonal variation in sensitivity to MEL suppression by light, with supersensitivity in the winter and subsensitivity in the summer (Thompson et al. 1990).

Compared to healthy controls, patients suffering from winter depression exhibited phase delays of the MEL rhythm (Terman et al. 1987, 1988; Dahl et al. 1993; Wirz Justice et al. 1993), which in other respects was normal (Checkley et al. 1993; Partonen et al. 1996). MEL levels were unchanged (Partonen et al. 1996) or higher during daytime (Danilenko et al. 1994).

After a successful antidepressive light therapy, the MEL rhythm was phase advanced (Lewy et al. 1987 a, 1998; Terman et al. 1987; Danilenko et al. 1994; Rice et al. 1995), MEL levels increased (Terman et al. 1987; Salinas et al. 1992; Partonen et al. 1995; Rice et al. 1995) or were unchanged (Rao et al. 1992) (nonseasonal depression!).

Apparently, the therapeutic effect of light in SAD is not mediated by phase shifts in melatonin secretion (Checkley et al. 1993) and it is doubtful whether the phase of the MEL rhythm plays a causative role in SAD. Phase position is apparently not correlated with depth of depression or with a preferential response to morning or evening light (Wirz Justice et al. 1993). When responders and nonresponders to the light therapy were compared, the time of onset of the nocturnal MEL increase did not differ (Rice et al. 1995). The fact that under two different light regimes in which MEL levels were equally suppressed the antidepressive effects of the two light treatments differed (Winton et al. 1989) speaks against a decisive role of MEL levels. MEL administration had no effect on the depressive symptoms, neither when given in the morning nor in the evening (Wirz Justice et al. 1990). Also, the antidepressive symptoms did not differ after atenolol, which suppresses MEL synthesis and placebo (Rosenthal et al. 1986, 1988). In some patients, the antidepressive effect of the light therapy could be partially reversed by oral administration of MEL (Rosenthal et al. 1986).

2.3.10 Electromagnetic and Magnetic Fields

The question of whether or not electromagnetic or magnetic fields affect MEL secretion/excretion in humans has not yet been definitively answered.

Exposure to electromagnetic fields (50–60 Hz, 20 µT, continuous or intermittent) were found to be without effects in some studies (Graham et al. 1996, 1997; Selmaoui et al. 1996 a, b). The finding in one experiment that men with preexisting low levels of MEL showed significantly greater suppression of MEL when they were exposed to 200 mG magnetic-field condition could not be repeated (Graham et al. 1996, 1997). Wilson and colleagues (Wilson et al. 1990) noted that exposure to conventional electric blankets did not have an effect on 6-sulfatoxymelatonin excretion; however 7 of 28 volunteers exposed to continuous polymer wire blankets which switched on and off twice as often and produced 50% stronger magnetic fields than the ordinary blankets, showed an increased 6-sulfatoxymelatonin excretion. After exposure to 20 µT (50 Hz, circularly polarized), a roughly 30-min delay in MEL secretion onset time with indications of a marginal reduction in maximum MEL levels was found in about 20% of the test persons, magnetic fields generated by square-wave currents being more effective compared to sinusoidal waveforms (Wood et al. 1998). In electric utility workers exposed to 60 Hz, a reduced 6-sulfatoxymelatonin excretion was noted (Burch et al. 1998, 1999). A significant depression in nocturnal MEL rise was observed in patients suffering from low back pain syndrome after long-term intermittent exposure to a very low-frequency magnetic field (2.9 mT, 40 Hz, square wave, bipolar) (Karasek et al. 1998).

Little work has been published on possible effects on MEL secretion of *high-frequency* (900 MHz, 1800 MHz) electromagnetic fields as used in cellular phones. Both 900 MHz (Mann et al. 1998; de Seze et al. 1999) and 1800 MHz (de Seze et al. 1999) were found to be without effect.

Strong, static magnetic fields as used in NMR imaging, likewise did not suppress nocturnal MEL levels (Prato et al. 1989; Schiffman et al. 1994).

2.3.11 Function of MEL in Humans

As with other hormones, the function of MEL depends on the presence of receptors. However, receptor-independent functions, such as scavenging of radicals etc., have to be considered as well (see below). In contrast to animal species, information on MEL receptors in humans is scarce.

2.3.11.1 MEL Receptors (Brydon et al. 1999)

There are principally three high-affinity MEL receptor subtypes, MEL1a, MEL1b, and MEL1c, but only the first two are found in humans. The receptors are all G-protein-coupled. In humans, MEL1a receptors were first detected in the hypothalamic SCN (Reppert et al. 1988; Weaver et al. 1993). In the SCN, also the MEL1a receptor mRNA was demonstrable, but not the mRNA of the MEL1b receptor (Weaver and Reppert 1996). The gene location for the MEL1a receptor is 4q35.1 (Slaugenhaupt et al. 1995). The MEL1b receptor maps to 11q21–22, is present in retina and brain, and inhibits adenylyl cylase (Reppert et al. 1995). It is now thought that the effect of MEL on the circadian system is mediated via the MEL1a receptor. It is not yet clear in which parts of the human body MEL receptors are well represented. The MEL1a receptor is apparently also present in kidney, intestine (Lee and Pang 1993; Song et al. 1997), and melanoma cells (Ying et al. 1993).

2.3.11.2 Routes of MEL Administration

In humans, MEL is usually administered orally. In exceptional cases, the dosages were as high as 6.6 g daily for 35 days or 1.35 g daily for 51 days (Lerner and Nordlund 1978). Currently, doses of not more than 5 mg are usually given. To obtain plasma MEL levels comparable to normal nocturnal levels, ca. 0.3 mg of MEL has to be given orally (Zhdanova et al. 1996).

MEL has also been administered intravenously (Lerner and Nordlund 1978), intranasally (Vollrath et al. 1981), transdermally (Lee et al. 1994; Bangha et al. 1997a, b; Benes et al. 1997), or oral-transmucosally (Benes et al. 1997). Except for transdermally, where it takes 2–4 h before plasma MEL concentrations increased above baseline (Lee et al. 1994), MEL is rapidly absorbed. When given orally, plasma MEL levels increase strikingly after 30–150 min, with elimination half-lives between 0.5 and 0.8 h (Waldhauser et al. 1984; Aldhous et al. 1985). There is clear intersubject variation in MEL absorption (Waldhauser et al. 1984; Lee et al. 1994; Benes et al. 1997), varying 25-fold among subjects (Waldhauser et al. 1984). A melatonin preparation with a pulsatile liberation pattern has been described (Hoffmann et al. 1999).

Although MEL administration does not appear to have side effects (Lerner and Nordlund 1978; Waldhauser and Wurtman 1983; Palm et al. 1997; Avery et al. 1998; McArthur and Budden 1998; Suhner et al. 1998b), caution has been expressed against regular use because of a lack of controlled long-term studies and in particular when unlicensed preparations are taken (Arendt 1997; Arendt and Deacon 1997; Cupp 1997). No data on the toxicology of MEL in humans are available (Guardiola Lemaitre 1997). In view of the fact that in lower vertebrates MEL affects melanophores responding with pigment aggregation (Lerner et al. 1958), it is relevant to note that long-term administration of MEL did not affect skin color in humans, as measured by reflectometry (McElhinney et al. 1994). The pharmacokinetics of MEL in humans (Mallo et al. 1990) and the toxicity of MEL in rats and mice (Sugden 1983) are dealt with in the indicated publications.

2.3.11.3 Effects of MEL Administration

Principally, two groups of effects of MEL have to be considered in humans: *acute effects,* possibly not involving the circadian oscillator, and *delayed effects* mediated by the SCN becoming apparent in the circadian organization of the body. Although, in practice, the two groups of effects may intricately interact, they are considered separately to get a clearer picture. *Long-term* effects of MEL which are of utmost importance in photoperiodic mammals and account for seasonal changes are apparently not present in humans. When considering the effects of MEL administration it should be borne in mind that most often pharmacological and not physiological dosages were applied. The most frequently documented effects of MEL administration relate to sedation and body temperature regulation.

Sedative Effects

Numerous studies have demonstrated *sleep-promoting effects* of MEL treatment in humans, as evidenced by subjects' self-reports, polysomnographic recordings, and continuous actigraphic registration of motor activity (Dawson and Encel 1993; Arendt et al. 1997; Zhdanova et al. 1997; Defrance and Quera Salva 1998; Geoffriau et al. 1998). However, reports showing no effect have also been published (James et al. 1987). Already in the 1970s and 1980s, *sedation* and *sleep induction* were clearly documented following MEL administration (Anton Tay et al. 1971, 1974; Cramer et al. 1974, 1976; Vollrath et al. 1981; Arendt et al. 1984, 1985a; Lieberman et al. 1984; Borbely 1986; Nickelsen et al. 1989), and also more recent studies support the early findings (Palm et al. 1997; Jean Louis et al. 1998; McArthur and Budden 1998; Sack et al. 1998; Skene et al. 1999; Zisapel 1999). As little as 0.3 mg of MEL given at different time points during the day or evening produces a sedative effect, usually within 1 h (Dollins et al. 1993a, 1994; Zhdanova et al. 1995, 1996). Recently, 10 µg MEL administered intravenously was found to be effective (van den Heuvel et al. 1999). The latency to maximum effect varied linearly from 3 h 40 min at 1200 hours to 1 h at 2100 hours (Tzischinsky and Lavie 1994). There is a decrease of latency to sleep onset and to stage 2 sleep, usually without altering the sleep architecture (Reid et al. 1996; Gilbert et al. 1999). Also, actual sleep time and sleep efficiency have been described to increase (Tzischinsky and Lavie 1994; Nave et al. 1995, 1996; Attenburrow et al. 1996; McArthur and Budden 1998). The duration of REM sleep is apparently not affected by MEL (Dijk et al. 1995). With respect to non-REM sleep, different results have been obtained. An increase (Attenburrow et al. 1996) and no effect were noted (Dijk et al. 1995). However, in the latter study, MEL enhanced EEG power density in non-REM sleep in the 13.75–14.0 Hz bin range (i.e., within the frequency range of sleep spindles). MEL significantly increased sleep propensity, the spectral power in the theta, delta and spindles bands, and significantly decreased the power in the alpha and beta bands (Attenburrow et al. 1996). In the EEG, sedative effects of MEL can be demonstrated rather quickly, even before any subjective soporific effects are recognized (Cajochen et al. 1997; Milstein et al. 1998). These findings indicate that MEL possesses a time-dependent hypnotic effect and suggest that endogenous MEL may participate in sleep–wake regulation. MEL administered *perioperatively* was sedative as well (Naguib and Samarkandi 1999).

In view of these findings, many attempts have been made to treat sleep disorders with MEL. MEL works very well in people with sleep disorders due to total blindness (Folkard et al. 1990; Palm et al. 1991, 1997; Tzischinsky et al. 1992), old age (Garfinkel et al. 1995; Hughes et al. 1998; Brusco et al. 1999), jet lag (Suhner et al. 1998a), and a number of neurological diseases such as Alzheimer's disease (Brusco et al. 1999) and tuberous sclerosis (O'Callaghan et al. 1999). In *chronic insomniacs,* MEL does not seem to be very effective (James et al. 1990; Ellis et al. 1996). In one study, MEL was found to reinforce REM sleep (Kunz and Bes 1997). A beneficial effect has also been reported in a study in which very high dosages (75 mg for 14 days, at 2200 hours) had been used (MacFarlane et al. 1991). *Sleep loss* and exacerbation of symptoms of dysphoria following MEL administration were noted in moderately to severely depressed patients and in patients with Huntington's chorea (Carman et al. 1976).

In view of the sedative effects of MEL, it is not surprising that also changes in *neurobehavioral performance* were noted, such as a decrease of number of correct responses in auditory vigilance, response latency in reaction time and self-reported vigor as well as increases in self-reported fatigue, confusion, and sleepiness (Dollins et al. 1993a). Decrements in alertness (Arendt and Deacon 1997), in performance on tracking tasks, and on response and reaction time scores for visual choice and extended two-choice visual tasks were also observed (Rogers et al. 1998).

Body Temperature

There is clear evidence that MEL depresses body temperature (BT) (Carman et al. 1976; Cagnacci et al. 1992, 1994; Dollins et al. 1993a, 1994; Tzischinsky and Lavie 1994; Reid et al. 1996; Arendt and Deacon 1997; Cagnacci et al. 1997; Gilbert et al. 1999). MEL can accelerate the evening decline in BT (Krauchi et al. 1997) and MEL and BT are inversely correlated, MEL accounting for ca. 40% of body temperature amplitude (Cagnacci et al. 1992). Thus, suppression of MEL secretion during the night increases BT (Cagnacci et al. 1997). Interestingly, MEL's hypothermic effect could not be demonstrated in women during the luteal phase and in elderly women. While the lack of effect in elderly women has been interpreted to represent a result of ageing, that of the luteal phase could be meaningful to avoid possible interference with embryo implantation (Cagnacci et al. 1997).

MEL's Function As Zeitgeber for the Circadian Oscillator

As pointed out above, the MEL rhythm is generated by neurons lying in the SCN which are equipped with MEL1a receptors. It is therefore most probable that also in humans, like in other mammalian species, MEL provides feedback to the SCN (Stehle et al. 1989) and shifts the pacemaker. As a result, not only the sleep–wake cycle and the core body temperature rhythm are affected (Middleton et al. 1997), but also that of MEL itself. The chronobiotic properties of MEL have recently been reviewed (Arendt and Deacon 1997; Arendt et al. 1997).

Depending on the time of day MEL is given, it delays or phase advances its own rhythm. A *phase delay* occurs after administration in the morning (Lewy et al. 1992), a *phase advance* results from MEL treatment in the evening (Mallo et al. 1988; Lewy et al. 1992; Attenburrow et al. 1995; Middleton et al. 1997). The phase shift can occur within 1 day, even with physiological doses of MEL (Lewy and Sack 1997). Data on a phase response curve following MEL administration are also available (Lewy et al. 1996). MEL administration can alleviate jet lag and tiredness after long-haul flights (Petrie et al. 1989, 1993) and can accelerate resynchronization of the MEL rhythm (Samel et al. 1991). There is evidence that the MEL rhythm can be dissociated from the sleep–wake rhythm (Arendt et al. 1986; Hashimoto et al. 1997). In a non-24-h sleep–wake syndrome with free-running MEL rhythm, MEL administration partly entrained the sleep–wake cycle and MEL rhythm (Hashimoto et al. 1998).

MEL's Effect on the Endocrine System and Relation to Puberty in Humans

As mentioned above, MEL has profound effects on the endocrine and reproductive systems in seasonally breeding mammals. In humans, the endocrine system is relatively little affected by MEL.

Among the hormones that do respond to MEL are prolactin, which tends to increase (Arendt et al. 1985a; Webley and Lenton 1987; Webley et al. 1988). Neither the various hormones themselves, such as cortisol, growth hormone, luteinizing hormone, thyroxine, testosterone etc., are affected (Weinberg et al. 1981; Arendt et al. 1985a) nor their responses to gonadotropin-releasing hormone (GnRH), thyrotropin-releasing hormone

(TRH), and adrenocorticotropic hormone (ACTH) (Weinberg et al. 1980; Paccotti et al. 1987). Nevertheless, numerous publications have pointed out that hormone-related diseases are accompanied by changes in MEL secretion, the general opinion being that there are no causal relationships between MEL and the diseases.

Ever since the report by Heubner (Heubner 1898) that a 4.5-year-old boy with a tumor in the pineal region developed precocious puberty, numerous attempts have been made to elucidate the role of the pineal gland and its hormone MEL in relation to puberty. The data obtained show that young children have higher MEL secretion rates relative to body size than young adults and that around puberty there is a fall in MEL secretion. Apparently, the decrease in MEL levels is the result of an increase of body mass and a causal relationship does not seem to exist between MEL decrease and the occurrence of puberty (Waldhauser and Dietzel 1985; Waldhauser and Steger 1986; Cavallo 1993, 1997; Langer et al. 1997).

MEL and the Immune System

It is beyond the scope of this review to deal with the interactions of MEL and the immune system. Suffice it to say that many authors are of the opinion that interactions do exist and that MEL exerts a stimulatory role on the immune system (Lissoni et al. 1989, 1994; Withyachumnarnkul et al. 1990; Guerrero and Reiter 1992; Maestroni 1993; Wajs et al. 1995; Liebmann et al. 1997; Rogers et al. 1997; Fraschini et al. 1998; Martins et al. 1998; Garcia Maurino et al. 1999).

MEL and Cancer

There is a considerable amount of data pointing in the direction of interrelationships between the pineal gland, its hormone MEL, and cancer. There is clear evidence that in patients suffering from certain cancers, MEL secretion/excretion is depressed (Bartsch et al. 1981, 1992a; Tamarkin et al. 1982), the decrease not being the result of increased peripheral metabolism (Bartsch et al. 1991). Moreover, MEL has oncostatic properties (Blask and Hill 1986; Blask et al. 1986, 1997, 1999; Ying et al. 1993). As several factors are involved in the genesis of malignancies, it is relevant that MEL is a strong scavenger of radicals (Poeggeler et al. 1994; Reiter et al. 1994, 1999; Hardeland et al. 1996a; Reiter 1996) and appears to affect the immune system (see above) and the cytoskeleton (Benitez King et al. 1990; Huerto Delgadillo et al. 1994; Matsui and Machado Santelli 1997). Finally, intrinsic pineal peptides with antitumor effects are of interest (Bartsch et al. 1987, 1990, 1992c). No doubt, these aspects will be dealt with exhaustively in the contributions to follow.

2.4 Conclusions

The present review has shown that a wealth of data exist on MEL secretion and its regulation in humans. Together with data obtained in different mammalian species, we appear to have a fairly complete picture of how the pineal gland functions. MEL secretion is basically controlled by the circadian oscillator lying in the hypothalamus in the suprachiasmatic nucleus; however, environmental lighting conditions affect MEL secretion as well. The circadian oscillator and the neurons linking the oscillator with the pineal gland lead to very low MEL synthesis and release during daytime, the reverse occurring at night. The pineal gland "measures" day/night length and adjusts its MEL synthesis accordingly. An important point is that MEL has different functions in mammals. In photoperiodic mammals the duration of MEL secretion at night induces seasonal changes, e.g., inhibition or stimulation of reproduction, change in fur colour, etc. In humans, MEL has no such functions. Here MEL appears to be play an important role as a component of the circadian timing system of the body. It has sedative and hypothermic effects. With respect to cancer, it is relevant that MEL secretion is depressed in patients suffering from certain cancer types, and that MEL and intrinsic pineal peptides have oncostatic properties.

References

Aldhous M, Franey C, Wright J, Arendt J (1985) Plasma concentrations of melatonin in man following oral absorption of different preparations. Br J Clin Pharmacol 19:517–521

Allen DJ, Allen JS, Didio LJ, McGrath JA (1981) Scanning electron microscopy and x-ray microanalysis of the human pineal body with emphasis on calcareous concretions. J Submicrosc Cytol 13:675–695

Almay BG, von Knorring L, Wetterberg L (1987) Melatonin in serum and urine in patients with idiopathic pain syndromes. Psychiatry Res 22:179–191

Amir S, Robinson B (1995) Ultraviolet light entrains rodent suprachiasmatic nucleus pacemaker. Neuroscience 69:1005–1011

Anton Tay F (1974) Melatonin: effects on brain function. Adv Biochem Psychopharmacol 11:315–324

Anton Tay F, Diaz JL, Fernandez Guardiola A (1971) On the effect of melatonin upon human brain. Its possible therapeutic implications. Life Sci I 10:841–850

Aoki H, Yamada N, Ozeki Y, Yamane H, Kato N (1998) Minimum light intensity required to suppress nocturnal melatonin concentration in human saliva. Neurosci Lett 252:91–94

Arendt J (1978) Melatonin assays in body fluids. J Neural Transm Suppl 13:265–278

Arendt J (1995) Melatonin and the mammalian pineal gland. Chapman & Hall, London

Arendt J (1997) Safety of melatonin in long-term use. J Biol Rhythms 12:673–681

Arendt J, Deacon S (1997) Treatment of circadian rhythm disorders–melatonin. Chronobiol Int 14:185–204

Arendt J, Paunier L, Sizonenko PC (1975) Melatonin radioimmunoassay. J Clin Endocrinol Metab 40:347–350

Arendt J, Wetterberg L, Heyden T, Sizonenko PC, Paunier L (1977) Radioimmunoassay of melatonin: human serum and cerebrospinal fluid. Horm Res 8:65–75

Arendt J, Wirz Justice A, Bradtke J, Kornemark M (1979) Long-term studies on immunoreactive human melatonin. Ann Clin Biochem 16:307–312

Arendt J, Hampton S, English J, Kwasowski P, Marks V (1982) 24-hour profiles of melatonin, cortisol, insulin, C-peptide and GIP following a meal and subsequent fasting. Clin Endocrinol Oxf 16:89–95

Arendt J, Symons AM, Laud CA, Pryde SJ (1983) Melatonin can induce early onset of the breeding season in ewes. J Endocrinol 97:395–400

Arendt J, Borbely AA, Franey C, Wright J (1984) The effects of chronic, small doses of melatonin given in the late afternoon on fatigue in man: a preliminary study. Neurosci Lett 45:317–321

Arendt J, Bojkowski C, Folkard S, Franey C, Marks V, Minors D, Waterhouse J, Wever RA, Wildgruber C, Wright J (1985 a) Some effects of melatonin and the control of its secretion in humans. Ciba Found Symp 117:266–283

Arendt J, Bojkowski C, Franey C, Wright J, Marks V (1985 b) Immunoassay of 6-hydroxymelatonin sulfate in human plasma and urine: abolition of the urinary 24-hour rhythm with atenolol. J Clin Endocrinol Metab 60:1166–1173

Arendt J, Aldhous M, Marks V (1986) Alleviation of jet lag by melatonin: preliminary results of controlled double blind trial. Br Med J Clin Res Ed 292:1170

Arendt J, Bhanji S, Franey C, Mattingly D (1992) Plasma melatonin levels in anorexia nervosa. Br J Psychiatry 161:361–364

Arendt J, Skene DJ, Middleton B, Lockley SW, Deacon S (1997) Efficacy of melatonin treatment in jet lag, shift work, and blindness. J Biol Rhythms 12:604–617

Attenburrow ME, Dowling BA, Sargent PA, Sharpley AL, Cowen PJ (1995) Melatonin phase advances circadian rhythm. Psychopharmacology Berl 121:503–505

Attenburrow ME, Cowen PJ, Sharpley AL (1996) Low dose melatonin improves sleep in healthy middle-aged subjects. Psychopharmacology Berl 126:179–181

Avery D, Lenz M, Landis C (1998) Guidelines for prescribing melatonin. Ann Med 30:122–130

Bakos J, Nagy N, Thuroczy G, Szabo LD (1995) Sinusoidal 50 Hz, 500 microT magnetic field has no acute effect on urinary 6-sulphatoxymelatonin in Wistar rats. Bioelectromagnetics 16:377–380

Bangha E, Elsner P, Kistler GS (1997a) Suppression of UV-induced erythema by topical treatment with melatonin (N-acetyl-5-methoxytryptamine). Influence of the application time point. Dermatology 195:248–252

Bangha E, Lauth D, Kistler GS, Elsner P (1997b) Daytime serum levels of melatonin after topical application onto the human skin. Skin Pharmacol 10:298–302

Bartness TJ, Powers JB, Hastings MH, Bittman EL, Goldman BD (1993) The timed infusion paradigm for melatonin delivery: what has it taught us about the melatonin signal, its reception, and the photoperiodic control of seasonal responses? J Pineal Res 15:161–190

Bartsch C, Bartsch H, Jain AK, Laumas KR, Wetterberg L (1981) Urinary melatonin levels in human breast cancer patients. J Neural Transm 52:281–294

Bartsch C, Bartsch H, Fluchter SH, Attanasio A, Gupta D (1985) Evidence for modulation of melatonin secretion in men with benign and malignant tumors of the prostate: relationship with the pituitary hormones. J Pineal Res 2:121–132

Bartsch H, Bartsch C, Noteborn HP, Flehmig B, Ebels I, Salemink CA (1987) Growth-inhibiting effect of crude pineal extracts on human melanoma cells in vitro is different from that of known synthetic pineal substances. J Neural Transm 69:299–311

Bartsch H, Bartsch C, Gupta D (1990) Tumor-inhibiting activity in the rat pineal gland displays a circannual rhythm. J Pineal Res 9:171–178

Bartsch C, Bartsch H, Bellmann O, Lippert TH (1991) Depression of serum melatonin in patients with primary breast cancer is not due to an increased peripheral metabolism. Cancer 67:1681–1684

Bartsch C, Bartsch H, Lippert TH (1992a) The pineal gland and cancer: facts, hypotheses and perspectives. The Cancer Journal 5:194–199

Bartsch C, Bartsch H, Schmidt A, Ilg S, Bichler KH, Fluchter SH (1992b) Melatonin and 6-sulfatoxymelatonin circadian rhythms in serum and urine of primary prostate cancer patients: evidence for reduced pineal activity and relevance of urinary determinations. Clin Chim Acta 209:153–167

Bartsch H, Bartsch C, Simon WE, Flehmig B, Ebels I, Lippert TH (1992c) Antitumor activity of the pineal gland: effect of unidentified substances versus the effect of melatonin. Oncology 49:27–30

Bayliss CR, Bishop NL, Fowler RC (1985) Pineal gland calcification and defective sense of direction. Br Med J Clin Res Ed 291:1758–1759

Beck O, Borg S, Lundman A (1982) Concentration of 5-methoxyindoles in the human pineal gland. J Neural Transm 54:111–116

Beck Friis J, von Rosen D, Kjellman BF, Ljunggren JG, Wetterberg L (1984) Melatonin in relation to body measures, sex, age, season and the use of drugs in patients with major affective disorders and healthy subjects. Psychoneuroendocrinology 9:261–277

Beck Friis J, Kjellman BF, Aperia B, Unden F, von Rosen D, Ljunggren JG, Wetterberg L (1985) Serum melatonin in relation to clinical variables in patients with major depressive disorder and a hypothesis of a low melatonin syndrome. Acta Psychiatr Scand 71:319–330

Benes L, Claustrat B, Horriere F, Geoffriau M, Konsil J, Parrott KA, DeGrande G, McQuinn RL, Ayres JW (1997) Transmucosal, oral controlled-release, and transdermal drug administration in human subjects: a crossover study with melatonin. J Pharm Sci 86:1115–1119

Benitez King G, Huerto Delgadillo L, Anton Tay F (1990) Melatonin effects on the cytoskeletal organization of MDCK and neuroblastoma N1E-115 cells. J Pineal Res 9:209–220

Benshoff HM, Brainard GC, Rollag MD, Lynch GR (1987) Suppression of pineal melatonin in Peromyscus leucopus by different monochromatic wavelengths of visible and near-ultraviolet light (UV-A). Brain Res 420:397–402

Benson B, Ebels I (1978) Pineal peptides. J Neural Transm Suppl 13:157–173

Berga SL, Yen SS (1990) Circadian pattern of plasma melatonin concentrations during four phases of the human menstrual cycle. Neuroendocrinology 51:606–612

Berga SL, Mortola JF, Yen SS (1988) Amplification of nocturnal melatonin secretion in women with functional hypothalamic amenorrhea. J Clin Endocrinol Metab 66:242–244

Bergianaki JD, Soldatos CR, Paparrigopoulos TJ, Syrengelas M, Stefanis CN (1995) Low and high melatonin excretors among healthy individuals. J Pineal Res 18:159–164

Bergstrom WH, Hakanson DO (1998) Melatonin: the dark force. Adv Pediatr 45:91–106

Bernard M, Donohue SJ, Klein DC (1995) Human hydroxyindole-O-methyltransferase in pineal gland, retina and Y79 retinoblastoma cells. Brain Res 696:37–48

Bhatnagar KP (1994) Synaptic ribbons of the mammalian pineal gland : enigmatic organelles of poorly understood function. Advances in Structural Biology 3:47–94

Bieck PR, Antonin KH, Balon R, Oxenkrug G (1988) Effect of brofaromine and pargyline on human plasma melatonin concentrations. Prog Neuropsychopharmacol Biol Psychiatry 12:93–101

Biesalski HK, Welker HA, Thalmann R, Vollrath L (1988) Melatonin and other serotonin derivatives in the guinea pig membranous cochlea. Neurosci Lett 91:41–46

Blask DE, Hill SM (1986) Effects of melatonin on cancer: studies on MCF-7 human breast cancer cells in culture. J Neural Transm Suppl 21:433–449

Blask DE, Hill SM, Orstead KM, Massa JS (1986) Inhibitory effects of the pineal hormone melatonin and underfeeding during the promotional phase of 7,12-dimethylbenzanthracene-(DMBA)-induced mammary tumorigenesis. J Neural Transm 67:125–138

Blask DE, Wilson ST, Zalatan F (1997) Physiological melatonin inhibition of human breast cancer cell growth in vitro: evidence for a glutathione-mediated pathway. Cancer Res 57:1909–1914

Blask DE, Sauer LA, Dauchy R, Holowachuk EW, Ruhoff MS (1999) New actions of melatonin on tumor metabolism and growth. Biol Signals Recept 8:49–55

Bocchi G, Valdre G (1993) Physical, chemical, and mineralogical characterization of carbonate-hydroxyapatite concretions of the human pineal gland. J Inorg Biochem 49:209–220

Boivin DB, Czeisler CA (1998) Resetting of circadian melatonin and cortisol rhythms in humans by ordinary room light. Neuroreport 9:779–782

Bojkowski CJ, Arendt J (1988) Annual changes in 6-sulphatoxymelatonin excretion in man. Acta Endocrinol Copenh 117:470–476

Bojkowski CJ, Arendt J (1990) Factors influencing urinary 6-sulphatoxymelatonin, a major melatonin metabolite, in normal human subjects. Clin Endocrinol Oxf 33:435–444

Bojkowski CJ, Aldhous ME, English J, Franey C, Poulton AL, Skene DJ, Arendt J (1987a) Suppression of nocturnal plasma melatonin and 6-sulphatoxymelatonin by bright and dim light in man. Horm Metab Res 19:437–440

Bojkowski CJ, Arendt J, Shih MC, Markey SP (1987b) Melatonin secretion in humans assessed by measuring its metabolite, 6-sulfatoxymelatonin. Clin Chem 33:1343–1348

Borbely AA (1986) Endogenous sleep-substances and sleep regulation. J Neural Transm Suppl 21:243–254

Brainard GC, Richardson BA, King TS, Reiter RJ (1984) The influence of different light spectra on the suppression of pineal melatonin content in the Syrian hamster. Brain Res 294:333–339

Brainard GC, Podolin PL, Leivy SW, Rollag MD, Cole C, Barker FM (1986) Near-ultraviolet radiation suppresses pineal melatonin content. Endocrinology 119:2201–2205

Brainard GC, Lewy AJ, Menaker M, Fredrickson RH, Miller LS, Weleber RG, Cassone V, Hudson D (1988) Dose-response relationship between light irradiance and the suppression of plasma melatonin in human volunteers. Brain Res 454:212–218

Brainard GC, Rollag MD, Hanifin JP (1997) Photic regulation of melatonin in humans: ocular and neural signal transduction. J Biol Rhythms 12:537–546

Brambilla F, Fraschini F, Esposti G, Bossolo PA, Marelli G, Ferrari E (1988) Melatonin circadian rhythm in anorexia nervosa and obesity. Psychiatry Res 23:267–276

Broadway J, Arendt J (1986) Delayed recovery of sleep and melatonin rhythms after nightshift work in Antarctic winter [letter]. Lancet 2:813–814

Broadway J, Arendt J, Folkard S (1987) Bright light phase shifts the human melatonin rhythm during the Antarctic winter. Neurosci Lett 79:185–189

Brown EN, Choe Y, Shanahan TL, Czeisler CA (1997) A mathematical model of diurnal variations in human plasma melatonin levels. Am J Physiol 272:E506–516

Brown GM, Young SN, Gauthier S, Tsui H, Grota LJ (1979) Melatonin in human cerebrospinal fluid in daytime; its origin and variation with age. Life Sci 25:929–936

Bruce J, Tamarkin L, Riedel C, Markey S, Oldfield E (1991) Sequential cerebrospinal fluid and plasma sampling in humans: 24-hour melatonin measurements in normal subjects and after peripheral sympathectomy. J Clin Endocrinol Metab 72:819–823

Brugger P, Marktl W, Herold M (1995) Impaired nocturnal secretion of melatonin in coronary heart disease. Lancet 345:1408

Brun J, Claustrat B, David M (1987) Urinary melatonin, LH, oestradiol, progesterone excretion during the menstrual cycle or in women taking oral contraceptives. Acta Endocrinol Copenh 116:145–149

Brusco LI, Fainstein I, Marquez M, Cardinali DP (1999) Effect of melatonin in selected populations of sleep-disturbed patients. Biol Signals Recept 8:126–131

Brydon L, Petit L, de Coppet P, Barrett P, Morgan PJ, Strosberg AD, Jockers R (1999) Polymorphism and signalling of melatonin receptors. Reprod Nutrit Devel 39:315–324

Brzezinski A, Lynch HJ, Wurtman RJ, Seibel MM (1987) Possible contribution of melatonin to the timing of the luteinizing hormone surge [letter]. N Engl J Med 316:1550–1551

Brzezinski A, Lynch HJ, Seibel MM, Deng MH, Nader TM, Wurtman RJ (1988) The circadian rhythm of plasma melatonin during the normal menstrual cycle and in amenorrheic women. J Clin Endocrinol Metab 66:891–895

Burch JB, Reif JS, Yost MG, Keefe TJ, Pitrat CA (1998) Nocturnal excretion of a urinary melatonin metabolite among electric utility workers. Scand J Work Environ Health 24:183–189

Burch JB, Reif JS, Yost MG, Keefe TJ, Pitrat CA (1999) Reduced excretion of a melatonin metabolite in workers exposed to 60 Hz magnetic fields. Am J Epidemiol 150:27–36

Buresova M, Dvorakova M, Zvolsky P, Illnerova H (1991) Early morning bright light phase advances the human circadian pacemaker within one day. Neurosci Lett 121:47–50

Buresova M, Dvorakova M, Zvolsky P, Illnerova H (1992) Human circadian rhythm in serum melatonin in short winter days and in simulated artificial long days. Neurosci Lett 136:173–176

Buxton OM, L'Hermite Baleriaux M, Hirschfeld U, Cauter E (1997) Acute and delayed effects of exercise on human melatonin secretion. J Biol Rhythms 12:568–574

Cagnacci A, Elliott JA, Yen SS (1992) Melatonin: a major regulator of the circadian rhythm of core temperature in humans. J Clin Endocrinol Metab 75:447–452

Cagnacci A, Soldani R, Romagnolo C, Yen SS (1994) Melatonin-induced decrease of body temperature in women: a threshold event. Neuroendocrinology 60:549–552

Cagnacci A, Krauchi K, Wirz Justice A, Volpe A (1997) Homeostatic versus circadian effects of melatonin on core body temperature in humans. J Biol Rhythms 12:509–517

Cajochen C, Krauchi K, Wirz Justice A (1997) The acute soporific action of daytime melatonin administration: effects on the EEG during wakefulness and subjective alertness. J Biol Rhythms 12:636–643

Cardinali DP, Lynch HJ, Wurtman RJ (1972) Binding of melatonin to human and rat plasma proteins. Endocrinology 91:1213–1218

Carman JS, Post RM, Buswell R, Goodwin FK (1976) Negative effects of melatonin on depression. Am J Psychiatry 133:1181–1186

Cassone VM, Chesworth MJ, Armstrong SM (1986) Entrainment of rat circadian rhythms by daily injection of melatonin depends upon the hypothalamic suprachiasmatic nuclei. Physiol Behav 36:1111–1121

Cassone VM, Roberts MH, Moore RY (1987) Melatonin inhibits metabolic activity in the rat suprachiasmatic nuclei. Neurosci Lett 81:29–34

Cavallo A (1993) Melatonin and human puberty: current perspectives. J Pineal Res 15:115–121

Cavallo A (1997) Is there a role for melatonin in children? In: Webb SM, Puig-Domingo M, Moller M, and Pevet P (eds) Pineal Update. PJD Publications Ltd, Westbury, NY. pp 235–244

Champney TH, Joshi BN, Vaughan MK, Reiter RJ (1985) Superior cervical ganglionectomy results in the loss of pineal concretions in the adult male gerbil (Meriones unguiculatus). Anat Rec 211:465–468

Chazot G, Claustrat B, Brun J, Jordan D, Sassolas G, Schott B (1984) A chronobiological study of melatonin, cortisol growth hormone and prolactin secretion in cluster headache. Cephalalgia 4:213–220

Chazot G, Claustrat B, Brun J, Zaidan R (1987) Effects on the patterns of melatonin and cortisol in cluster headache of a single administration of lithium at 7.00 p.m. daily over one week: a preliminary report. Pharmacopsychiatry 20:222–223

Checkley SA, Murphy DG, Abbas M, Marks M, Winton F, Palazidou E, Murphy DM, Franey C, Arendt J (1993) Melatonin rhythms in seasonal affective disorder. Br J Psychiatry 163:332–337

Claustrat B, Chazot G, Brun J, Jordan D, Sassolas G (1984) A chronobiological study of melatonin and cortisol secretion in depressed subjects: plasma melatonin, a biochemical marker in major depression. Biol Psychiatry 19:1215–1228

Claustrat B, Brun J, Garry P, Roussel B, Sassolas G (1986) A once-repeated study of nocturnal plasma melatonin patterns and sleep recordings in six normal young men. J Pineal Res 3:301–310

Claustrat B, Loisy C, Brun J, Beorchia S, Arnaud JL, Chazot G (1989) Nocturnal plasma melatonin levels in migraine: a preliminary report. Headache 29:242–245

Collin J-P (1971) Differentiation and regression of the cells of the sensory line in the epiphysis cerebri. In: Wolstenholme GEW and Knight J (eds) The Pineal Gland. Churchill Livingstone, Edinburgh London. pp 79–125

Coon SL, Mazuruk K, Bernard M, Roseboom PH, Klein DC, Rodriguez IR (1996) The human serotonin N-acetyltransferase (EC 2.3.1.87) gene (AANAT): structure, chromosomal localization, and tissue expression. Genomics 34:76–84

Cowen PJ, Fraser S, Sammons R, Green AR (1983) Atenolol reduces plasma melatonin concentration in man. Br J Clin Pharmacol 15:579–581

Cramer H, Rudolph J, Consbruch U, Kendel K (1974) On the effects of melatonin on sleep and behavior in man. Adv Biochem Psychopharmacol 11:187–191

Cramer H, Bohme W, Kendel K, Donnadieu M (1976) Freisetzung von Wachstumshormon und von Melanozyten stimulierendem Hormon im durch Melatonin gebahnten Schlaf beim Menschen. Arzneimittelforschung 26:1076–1078

Cremer Bartels G, Krause K, Küchle HJ (1983) Influence of low magnetic-field-strength variations on the retina and pineal gland of quail and humans. Graefes Arch Clin Exp Ophthalmol 220:248–252

Cremer Bartels G, Krause K, Mitoskas G, Brodersen D (1984) Magnetic field of the earth as additional zeitgeber for endogenous rhythms? Naturwissenschaften 71:567–574

Cummings MA, Berga SL, Cummings KL, Kripke DF, Haviland MG, Golshan S, Gillin JC (1989) Light suppression of melatonin in unipolar depressed patients. Psychiatry Res 27:351–355

Cupp MJ (1997) Melatonin. Am Fam Physician 56:1421–1425, 1428

Dahl K, Avery DH, Lewy AJ, Savage MV, Brengelmann GL, Larsen LH, Vitiello MV, Prinz PN (1993) Dim light melatonin onset and circadian temperature during a constant routine in hypersomnic winter depression. Acta Psychiatr Scand 88:60–66

Danilenko KV, Putilov AA, Russkikh GS, Duffy LK, Ebbesson SO (1994) Diurnal and seasonal variations of melatonin and serotonin in women with seasonal affective disorder. Arctic Med Res 53:137–145

David M, Harthe C, Brun J, Mouren P, Veillas G, Claustrat B (1987) Plasma catecholamines and melatonin levels during acute exercise in cardiac patients. Neuroendocrinol Lett 9:267–272

Dawson D, Encel N (1993) Melatonin and sleep in humans. J Pineal Res 15:1–12

de Seze R, Ayoub J, Peray P, Miro L, Touitou Y (1999) Evaluation in humans of the effects of radiocellular telephones on the circadian patterns of melatonin secretion, a chronobiological rhythm marker. J Pineal Res 27:237–242

Deacon S, Arendt J (1994) Posture influences melatonin concentrations in plasma and saliva in humans. Neurosci Lett 167:191–194

Deacon S, English J, Tate J, Arendt J (1998) Atenolol facilitates light-induced phase shifts in humans. Neurosci Lett 242:53–56

Defrance R, Quera Salva MA (1998) Therapeutic applications of melatonin and related compounds. Horm Res 49:142–146

Deguchi T (1975) Ontogenesis of a biological clock for serotonin: acetyl coenzyme A N-acetyltransferase in pineal gland of rat. Proc Natl Acad Sci USA 72:2814–2818

Deguchi T (1978) Ontogenesis of circadian rhythm of melatonin synthesis in pineal gland of rat. J Neural Transm Suppl 13:115–128

Delfs TM, Baars S, Fock C, Schumacher M, Olcese J, Zimmermann RC (1994) Sex steroids do not alter melatonin secretion in the human. Hum Reprod 9:49–54

Demel AW (1992) Radioimmunologische Untersuchung der Zirkadianrhythmen von Cortisol und Melatonin im Speichel. Wien Klin Wochenschr 104:423–425

Di WL, Kadva A, Djahanbakhch O, Silman R (1998) Radioimmunoassay of bound and free melatonin in plasma. Clin Chem 44:304–310

Dijk DJ, Cajochen C (1997) Melatonin and the circadian regulation of sleep initiation, consolidation, structure, and the sleep EEG. J Biol Rhythms 12:627–635

Dijk DJ, Roth C, Landolt HP, Werth E, Aeppli M, Achermann P, Borbely AA (1995) Melatonin effect on daytime sleep in men: suppression of EEG low frequency activity and enhancement of spindle frequency activity. Neurosci Lett 201:13–16

Dollins AB, Lynch HJ, Wurtman RJ, Deng MH, Kischka KU, Gleason RE, Lieberman HR (1993a) Effect of pharmacological daytime doses of melatonin on human mood and performance. Psychopharmacology Berl 112:490–496

Dollins AB, Lynch HJ, Wurtman RJ, Deng MH, Lieberman HR (1993b) Effects of illumination on human nocturnal serum melatonin levels and performance. Physiol Behav 53:153–160

Dollins AB, Zhdanova IV, Wurtman RJ, Lynch HJ, Deng MH (1994) Effect of inducing nocturnal serum melatonin concentrations in daytime on sleep, mood, body temperature, and performance. Proc Natl Acad Sci USA 91:1824–1828

Donohue SJ, Roseboom PH, Illnerova H, Weller JL, Klein DC (1993) Human hydroxyindole-O-methyltransferase: presence of LINE-1 fragment in a cDNA clone and pineal mRNA. DNA Cell Biol 12:715–727

Drijfhout WJ, van der Linde AG, Kooi SE, Grol CJ, Westerink BH (1996a) Norepinephrine release in the rat pineal gland: the input from the biological clock measured by in vivo microdialysis. J Neurochem 66:748–755

Drijfhout WJ, van der Linde AG, de Vries JB, Grol CJ, Westerink BH (1996b) Microdialysis reveals dynamics of coupling between noradrenaline release and melatonin secretion in conscious rats. Neurosci Lett 202:185–188

Drijfhout WJ, Grol CJ, Westerink BH (1996c) Parasympathetic inhibition of pineal indole metabolism by prejunctional modulation of noradrenaline release. Eur J Pharmacol 308:117–124

Drijfhout WJ, Brons HF, Oakley N, Hagan RM, Grol CJ, Westerink BH (1997) A microdialysis study on pineal melatonin rhythms in rats after an 8-h phase advance: new characteristics of the underlying pacemaker. Neuroscience 80:233–239

Dubbels R, Reiter RJ, Klenke E, Goebel A, Schnakenberg E, Ehlers C, Schiwara HW, Schloot W (1995) Melatonin in edible plants identified by radioimmunoassay and by high performance liquid chromatography-mass spectrometry. J Pineal Res 18:28–31

Duffy JF, Dijk DJ, Hall EF, Czeisler CA (1999) Relationship of endogenous circadian melatonin and temperature rhythms to self-reported preference for morning or evening activity in young and older people. J Investig Med 47:141–150

Ebihara S, Marks T, Hudson DJ, Menaker M (1986) Genetic control of melatonin synthesis in the pineal gland of the mouse. Science 231:491–493

Ebihara S, Hudson DJ, Marks T, Menaker M (1987) Pineal indole metabolism in the mouse. Brain Res 416:136–140

Ehret M, Pevet P, Maitre M (1991) Tryptophan hydroxylase synthesis is induced by 3′,5′-cyclic adenosine monophosphate during circadian rhythm in the rat pineal gland. J Neurochem 57: 1516–1521

Ellis CM, Lemmens G, Parkes JD (1996) Melatonin and insomnia. J Sleep Res 5:61–65

Fanget F, Claustrat B, Dalery J, Brun J, Terra JL, Marie Cardine M, Guyotat J (1989) Nocturnal plasma melatonin levels in schizophrenic patients. Biol Psychiatry 25:499–501

Fellenberg AJ, Phillipou G, Seamark RF (1980) Measurement of urinary production rates of melatonin as an index of human pineal function. Endocr Res Commun 7:167–175

Fellenberg AJ, Phillipou G, Seamark RF (1982) Urinary 6-sulphatoxy melatonin excretion during the human menstrual cycle. Clin Endocrinol Oxf 17:71–75

Fernandez B, Malde JL, Montero A, Acuna D (1990) Relationship between adenohypophyseal and steroid hormones and variations in serum and urinary melatonin levels during the ovarian cycle, perimenopause and menopause in healthy women. J Steroid Biochem 35:257–262

Fevre-Montange M, Van Cauter E, Refetoff S, Desir D, Tourniaire J, Copinschi G (1981) Effects of "jet lag" on hormonal patterns. II. Adaptation of melatonin circadian periodicity. J Clin Endocrinol Metab 52:642–649

Fevre-Montange M, Estour B, Abou Samra AB, Bajard L, Tourniaire J (1983a) Twenty-four hour melatonin secretory pattern in men with idiopathic hemochromatosis. Psychoneuroendocrinology 8:321–326

Fevre-Montange M, Tourniaire J, Estour B, Bajard L (1983b) 24 hour melatonin secretory pattern in Cushing's syndrome. Clin Endocrinol Oxf 19:175–181

Finocchiaro L, Callebert J, Launay JM, Jallon JM (1988) Melatonin biosynthesis in *Drosophila*: its nature and its effects. J Neurochem 50:382–387

Folkard S, Arendt J, Aldhous M, Kennett H (1990) Melatonin stabilises sleep onset time in a blind man without entrainment of cortisol or temperature rhythms. Neurosci Lett 113:193–198

Follenius M, Weibel L, Brandenberger G (1995) Distinct modes of melatonin secretion in normal men. J Pineal Res 18:135–140

Franey C, Aldhous M, Burton S, Checkley S, Arendt J (1986) Acute treatment with desipramine stimulates melatonin and 6-sulphatoxy melatonin production in man. Br J Clin Pharmacol 22: 73–79

Fraschini F, Demartini G, Esposti D, Scaglione F (1998) Melatonin involvement in immunity and cancer. Biol Signals Recept 7:61–72

Freund D, Arendt J, Vollrath L (1977) Tentative immunohistochemical demonstration of melatonin in the rat pineal gland. Cell Tissue Res 181:239–244

Gaddy JR, Stewart KT, Byrne B, Doghramji K, Rollag MD, Brainard GC (1990) Light-induced plasma melatonin suppression in seasonal affective disorder. Prog Neuropsychopharmacol Biol Psychiatry 14:563–568

Gaddy JR, Rollag MD, Brainard GC (1993) Pupil size regulation of threshold of light-induced melatonin suppression. J Clin Endocrinol Metab 77:1398–1401

Galliani I, Mongiorgi R (1991) Further observations on pineal brain sand formation and evolution in man. Boll Soc Ital Biol Sper 67:837–844

Galliani I, Frank F, Gobbi P, Giangaspero F, Falcieri E (1989) Histochemical and ultrastructural study of the human pineal gland in the course of aging. J Submicrosc Cytol Pathol 21: 571–578

Galliani I, Falcieri E, Giangaspero F, Valdre G, Mongiorgi R (1990) A preliminary study of human pineal gland concretions: structural and chemical analysis. Boll Soc Ital Biol Sper 66:615–622

Ganapathy K, Kalyanaraman S (1978) The calcified pineal in plain X-rays of the skull. Neurol India 26:131–134

Garcia Maurino S, Pozo D, Carrillo Vico A, Calvo JR, Guerrero JM (1999) Melatonin activates Th1 lymphocytes by increasing IL-12 production. Life Sci 65:2143–2150

Garfinkel D, Laudon M, Nof D, Zisapel N (1995) Improvement of sleep quality in elderly people by controlled-release melatonin. Lancet 346:541–544

Gastel JA, Roseboom PH, Rinaldi PA, Weller JL, Klein DC (1998) Melatonin production: proteasomal proteolysis in serotonin N-acetyltransferase regulation. Science 279:1358–1360

Geoffriau M, Brun J, Chazot G, Claustrat B (1998) The physiology and pharmacology of melatonin in humans. Horm Res 49:136–141

Geoffriau M, Claustrat B, Veldhuis J (1999) Estimation of frequently sampled nocturnal melatonin production in humans by deconvolution analysis: Evidence for episodic or ultradian secretion. J Pineal Res 27:139–144

Gilbert SS, van den Heuvel CJ, Dawson D (1999) Daytime melatonin and temazepam in young adult humans: equivalent effects on sleep latency and body temperatures. J Physiol Lond 514:905–914

Goridis C, Neff NH (1972) Evidence for specific monoamine oxidases in human sympathetic nerve and pineal gland. Proc Soc Exp Biol Med 140:573–574

Goto M, Oshima I, Tomita T, Ebihara S (1989) Melatonin content of the pineal gland in different mouse strains. J Pineal Res 7:195–204

Govitrapong P, Phansuwan Pujito P, Ebadi M (1989) Studies on the properties of muscarinic cholinergic receptor sites in bovine pineal gland. Comp Biochem Physiol C 94:159–164

Graham C, Cook MR, Riffle DW, Gerkovich MM, Cohen HD (1996) Nocturnal melatonin levels in human volunteers exposed to intermittent 60 Hz magnetic fields. Bioelectromagnetics 17:263–273

Graham C, Cook MR, Riffle DW (1997) Human melatonin during continuous magnetic field exposure. Bioelectromagnetics 18:166–171

Griffiths PA, Folkard S, Bojkowski C, English J, Arendt J (1986) Persistent 24-h variations of urinary 6-hydroxy melatonin sulphate and cortisol in Antarctica. Experientia 42:430–432

Grota LJ, Reiter RJ, Keng P, Michaelson S (1994) Electric field exposure alters serum melatonin but not pineal melatonin synthesis in male rats. Bioelectromagnetics 15:427–437

Guardiola Lemaitre B (1997) Toxicology of melatonin. J Biol Rhythms 12:697–706

Guerrero JM, Reiter RJ (1992) A brief survey of pineal gland-immune system interrelationships. Endocr Res 18:91–113

Hardeland R, Behrmann G, Fuhrberg B, Poeggeler B, Burkhardt S, Uria H, Obst B (1996a) Evolutionary aspects of indoleamines as radical scavengers. Presence and photocatalytic turnover of indoleamines in a unicell, *Gonyaulax polyedra*. Adv Exp Med Biol 398:279–284

Hardeland R, Fuhrberg B, Uria H, Behrmann G, Meyer TJ, Burkhardt S, Poeggeler B (1996b) Chronobiology of indoleamines in the dinoflagellate *Gonyaulax polyedra*: metabolism and effects related to circadian rhythmicity and photoperiodism. Braz J Med Biol Res 29:119–123

Hasegawa A, Ohtsubo K, Mori W (1987) Pineal gland in old age; quantitative and qualitative morphological study of 168 human autopsy cases. Brain Res 409:343–349

Hashimoto S, Nakamura K, Honma S, Honma K (1997) Free-running circadian rhythm of melatonin in a sighted man despite a 24-hour sleep pattern: a non-24-hour circadian syndrome. Psychiatry Clin Neurosci 51:109–114

Hashimoto S, Nakamura K, Honma S, Honma K (1998) Free-running of plasma melatonin rhythm prior to full manifestation of a non-24 hour sleep-wake syndrome. Psychiatry Clin Neurosci 52:264–265

Helmke K, Winkler P (1986) Die Häufigkeit von Pinealisverkalkungen in den ersten 18 Lebensjahren. ROFO Fortschr Geb Rontgenstr Nuklearmed 144:221–226

Heubner, No initial (1898) No title. Dtsch Med. Wschr 29:214–215

Hoban TM, Sack RL, Lewy AJ, Miller LS, Singer CM (1989) Entrainment of a free-running human with bright light? Chronobiol Int 6:347–353

Hoffmann A, Farker K, Dittgen M, Hoffmann H (1999) A melatonin preparation with a pulsatile liberation pattern: a new form of melatonin in replacement therapy. Biol Signals Recept 8:96–104

Hofman MA, Skene DJ, Swaab DF (1995) Effect of photoperiod on the diurnal melatonin and 5-methoxytryptophol rhythms in the human pineal gland. Brain Res 671:254–260

Honma K, Honma S, Kohsaka M, Fukuda N (1992a) Seasonal variation in the human circadian rhythm: dissociation between sleep and temperature rhythm. Am J Physiol 262:R885–891

Honma S, Kanematsu N, Katsuno Y, Honma K (1992b) Light suppression of nocturnal pineal and plasma melatonin in rats depends on wavelength and time of day. Neurosci Lett 147:201–204

Horne JA, Donlon J, Arendt J (1991) Green light attenuates melatonin output and sleepiness during sleep deprivation. Sleep 14:233–240

Horton TH, Ray SL, Stetson MH (1989) Maternal transfer of photoperiodic information in Siberian hamsters. III. Melatonin injections program postnatal reproductive development expressed in constant light. Biol Reprod 41:34–39

Huang SK, Klein DC, Korf HW (1992) Immunocytochemical demonstration of rod-opsin, S-antigen, and neuron-specific proteins in the human pineal gland. Cell Tissue Res 267:493–498

Huerto Delgadillo L, Anton Tay F, Benitez King G (1994) Effects of melatonin on microtubule assembly depend on hormone concentration: role of melatonin as a calmodulin antagonist. J Pineal Res 17:55–62

Hughes RJ, Sack RL, Lewy AJ (1998) The role of melatonin and circadian phase in age-related sleep-maintenance insomnia: assessment in a clinical trial of melatonin replacement. Sleep 21:52–68

Iguchi H, Kato KI, Ibayashi H (1982) Melatonin serum levels and metabolic clearance rate in patients with liver cirrhosis. J Clin Endocrinol Metab 54:1025–1027

Illnerova H, Vanecek J (1979) Response of rat pineal serotonin N-acetyltransferase to one min light pulse at different night times. Brain Res 167:431–434

Illnerova H, Zvolsky P, Vanecek J (1985) The circadian rhythm in plasma melatonin concentration of the urbanized man: the effect of summer and winter time. Brain Res 328:186–189

Illnerova H, Buresova M, Nedvidkova J, Dvorakova M, Zvolsky P (1993a) Maintenance of a circadian phase adjustment of the human melatonin rhythm following artificial long days. Brain Res 626:322–326

Illnerova H, Buresova M, Presl J (1993b) Melatonin rhythm in human milk. J Clin Endocrinol Metab 77:838–841

Jagota A, Olcese J, Harinarayana Rao S, Gupta PD (1999) Pineal rhythms are synchronized to light–dark cycles in congenitally anophthalmic mutant rats. Brain Res 825:95–103

James SP, Mendelson WB, Sack DA, Rosenthal NE, Wehr TA (1987) The effect of melatonin on normal sleep. Neuropsychopharmacology 1:41–44

James SP, Sack DA, Rosenthal NE, Mendelson WB (1990) Melatonin administration in insomnia. Neuropsychopharmacology 3:19–23

Janavs JL, Pierce ME, Takahashi JS (1991) N-acetyltransferase and protein synthesis modulate melatonin production by Y79 human retinoblastoma cells. Brain Res 540:138–144

Jastrow H, von Mach MA, Vollrath L (1997) The shape of synaptic ribbons in the rat pineal gland. Cell Tissue Res 287:255–261

Jean Louis G, von Gizycki H, Zizi F (1998) Melatonin effects on sleep, mood, and cognition in elderly with mild cognitive impairment. J Pineal Res 25:177–183

Jengeleski CA, Powers RE, O'Connor DT, Price DL (1989) Noradrenergic innervation of human pineal gland: abnormalities in aging and Alzheimer's disease. Brain Res 481:378–382

Jentsch A, Lehmann M, Schone E, Thoss F, Zimmermann G (1993) Weak magnetic fields change extinction of a conditioned reaction and daytime melatonin levels in the rat. Neurosci Lett 157:79–82

Jimerson DC, Lynch HJ, Post RM, Wurtman RJ, Bunney WE, Jr. (1977) Urinary melatonin rhythms during sleep deprivation in depressed patients and normals. Life Sci 20:1501–1508

John TM, Liu GY, Brown GM (1998) 60 Hz magnetic field exposure and urinary 6-sulphatoxy-melatonin levels in the rat. Bioelectromagnetics 19:172–180

Jones RL, McGeer PL, Greiner AC (1969) Metabolism of exogenous melatonin in schizophrenic and non-schizophrenic volunteers. Clin Chim Acta 26:281–285

Kanematsu N, Honma S, Katsuno Y, Honma K (1994) Immediate response to light of rat pineal melatonin rhythm: analysis by in vivo microdialysis. Am J Physiol 266:R1849–1855

Karasek M, Lewinski A, Vollrath L (1988) Precise annual changes in the numbers of "synaptic" ribbons and spherules in the rat pineal gland. J Biol Rhythms 3:41–48

Karasek M, Woldanska Okonska M, Czernicki J, Zylinska K, Swietoslawski J (1998) Chronic exposure to 2.9 mT, 40 Hz magnetic field reduces melatonin concentrations in humans. J Pineal Res 25:240–244

Kato M, Honma K, Shigemitsu T, Shiga Y (1993) Effects of exposure to a circularly polarized 50-Hz magnetic field on plasma and pineal melatonin levels in rats. Bioelectromagnetics 14:97–106

Kato M, Honma K, Shigemitsu T, Shiga Y (1994) Circularly polarized 50-Hz magnetic field exposure reduces pineal gland and blood melatonin concentrations of Long-Evans rats. Neurosci Lett 166:59–62

Kauppila A, Kivela A, Pakarinen A, Vakkuri O (1987) Inverse seasonal relationship between melatonin and ovarian activity in humans in a region with a strong seasonal contrast in luminosity. J Clin Endocrinol Metab 65:823–828

Kendler BS (1997) Melatonin: media hype or therapeutic breakthrough? Nurse Pract 22:66–67, 71–62, 77

Kennaway DJ, Van Dorp CF (1991) Free-running rhythms of melatonin, cortisol, electrolytes, and sleep in humans in Antarctica. Am J Physiol 260:R1137–1144

Kennaway DJ, Voultsios A (1998) Circadian rhythm of free melatonin in human plasma. J Clin Endocrinol Metab 83:1013–1015

Kennaway DJ, Stamp GE, Goble FC (1992) Development of melatonin production in infants and the impact of prematurity. J Clin Endocrinol Metab 75:367–369

Kennedy SH, Garfinkel PE, Parienti V, Costa D, Brown GM (1989) Changes in melatonin levels but not cortisol levels are associated with depression in patients with eating disorders. Arch Gen Psychiatry 46:73–78

Kennedy SH, Gnam W, Ralevski E, Brown GM (1995) Melatonin responses to clonidine and yohimbine challenges. J Psychiatry Neurosci 20:297–304

Ketonen L, Tallroth K, Taavitsainen M, Borowski MM (1980) Asymmetrische Verkalkung des Corpus pineale, welche eine Verlagerung der Glandula pinealis vortäuscht. ROFO Fortschr Geb Rontgenstr Nuklearmed 133:105–106

Khoory R (1987) Compensation of the natural magnetic field does not alter N-acetyltransferase activity and melatonin content of rat pineal gland. Neurosci Lett 76:215–220

Kivelä A, Kauppila A, Ylostalo P, Vakkuri O, Leppaluoto J (1988) Seasonal, menstrual and circadian secretions of melatonin, gonadotropins and prolactin in women. Acta Physiol Scand 132: 321–327

Kivelä A, Kauppila A, Leppaluoto J, Vakkuri O (1989) Serum and amniotic fluid melatonin during human labor. J Clin Endocrinol Metab 69:1065–1068

Klein DC, Weller JL (1972a) A rapid light-induced decrease in pineal serotonin N-acetyltransferase activity. Science 177:532–533

Klein DC, Weller JL (1972b) Rapid light-induced decrease in pineal serotonin N-acetyltransferase activity. Science 177:532–533

Klein DC, Buda MJ, Kapoor CL, Krishna G (1978) Pineal serotonin N-acetyltransferase activity: abrupt decrease in adenosine 3′,5′-monophosphate may be signal for "turnoff". Science 199: 309–311

Klerman EB, Rimmer DW, Dijk DJ, Kronauer RE, Rizzo JF, 3rd, Czeisler CA (1998) Nonphotic entrainment of the human circadian pacemaker. Am J Physiol 274:R991–996

Kneisley LW, Moskowitz MA, Lynch HG (1978) Cervical spinal cord lesions disrupt the rhythm in human melatonin excretion. J Neural Transm Suppl 13:311–323

Kodaka T, Mori R, Debari K, Yamada M (1994) Scanning electron microscopy and electron probe microanalysis studies of human pineal concretions. J Electron Microsc Tokyo 43:307–317

Kohli N, Rastogi H, Bhadury S, Tandon VK (1992) Computed tomographic evaluation of pineal calcification. Indian J Med Res 96:139–142

Kopin IJ, Pare CMB, Axelrod J, Weissbach H (1961) The fate of melatonin in animals. J Biol Chem 236:3072–3075

Kopp N, Claustrat B, Tappaz M (1980) Evidence for the presence of melatonin in the human brain. Neurosci Lett 19:237–242

Korf HW (1996) Innervation of the pineal gland. In: Unsicker K (ed) Autonomic-endocrine interactions. Harwood, Amsterdam. pp 129–180

Korf H-W, Moller M (1984) The innervation of the mammalian pineal gland with special reference to central pinealopetal projections, Alan R. Liss, New York

Korf HW, Foster RG, Ekstrom P, Schalken JJ (1985a) Opsin-like immunoreaction in the retinae and pineal organs of four mammalian species. Cell Tissue Res 242:645–648

Korf HW, Moller M, Gery I, Zigler JS, Klein DC (1985b) Immunocytochemical demonstration of retinal S-antigen in the pineal organ of four mammalian species. Cell Tissue Res 239: 81–85

Korf HW, Klein DC, Zigler JS, Gery I, Schachenmayr W (1986a) S-antigen-like immunoreactivity in a human pineocytoma. Acta Neuropathol Berl 69:165–167

Korf HW, Oksche A, Ekstrom P, Gery I, Zigler JS, Jr., Klein DC (1986b) Pinealocyte projections into the mammalian brain revealed with S-antigen antiserum. Science 231:735–737

Korf HW, Gotz W, Herken R, Theuring F, Gruss P, Schachenmayr W (1990) S-antigen and rod-opsin immunoreactions in midline brain neoplasms of transgenic mice: similarities to pineal cell tumors and certain medulloblastomas in man. J Neuropathol Exp Neurol 49:424–437

Korf HW, White BH, Schaad NC, Klein DC (1992) Recoverin in pineal organs and retinae of various vertebrate species including man. Brain Res 595:57–66

Korf HW, Schomerus C, Maronde E, Stehle JH (1996) Signal transduction molecules in the rat pineal organ: Ca2+, pCREB, and ICER. Naturwissenschaften 83:535–543

Korf HW, Schomerus C, Stehle JH (1998) The pineal organ, its hormone melatonin, and the photo-neuroendocrine system. Adv Anat Embryol Cell Biol 146:1–100

Krauchi K, Cajochen C, Danilenko KV, Wirz Justice A (1997) The hypothermic effect of late evening melatonin does not block the phase delay induced by concurrent bright light in human subjects. Neurosci Lett 232:57–61

Krstic R (1976) A combined scanning and transmission electron microscopic study and electron probe microanalysis of human pineal acervuli. Cell Tissue Res 174:129–137

Krstic R (1986) Pineal calcification: its mechanism and significance. J Neural Transm Suppl 21: 415–432

Kunz D, Bes F (1997) Melatonin effects in a patient with severe REM sleep behavior disorder: case report and theoretical considerations. Neuropsychobiology 36:211–214

Kunz D, Bes F, Schlattmann P, Herrmann WM (1998) On pineal calcification and its relation to subjective sleep perception: a hypothesis-driven pilot study. Psychiatry Res 82:187–191

Laakso ML, Porkka Heiskanen T, Alila A, Stenberg D, Johansson G (1990) Correlation between salivary and serum melatonin: dependence on serum melatonin levels. J Pineal Res 9:39–50

Laakso ML, Hatonen T, Stenberg D, Alila A, Smith S (1993) One-hour exposure to moderate illuminance (500 lux) shifts the human melatonin rhythm. J Pineal Res 15:21–26

Laakso ML, Porkka Heiskanen T, Alila A, Stenberg D, Johansson G (1994) Twenty-four-hour rhythms in relation to the natural photoperiod: a field study in humans. J Biol Rhythms 9:283–293

Langer M, Hartmann J, Turkof H, Waldhauser F (1997) Melatonin beim Menschen – ein Überblick. Wien Klin Wochenschr 109:707–713

LaPorte R, Kus L, Wisniewski RA, Prechel MM, Azar Kia B, McNulty JA (1990) Magnetic resonance imaging (MRI) effects on rat pineal neuroendocrine function. Brain Res 506:294–296

Lee BJ, Parrott KA, Ayres JW, Sack RL (1994) Preliminary evaluation of transdermal delivery of melatonin in human subjects. Res Commun Mol Pathol Pharmacol 85:337–346

Lee JM, Jr., Stormshak F, Thompson JM, Thinesen P, Painter LJ, Olenchek EG, Hess DL, Forbes R, Foster DL (1993) Melatonin secretion and puberty in female lambs exposed to environmental electric and magnetic fields. Biol Reprod 49:857–864

Lee JM, Jr., Stormshak F, Thompson JM, Hess DL, Foster DL (1995) Melatonin and puberty in female lambs exposed to EMF: a replicate study. Bioelectromagnetics 16:119–123

Lee PP, Pang SF (1993) Melatonin and its receptors in the gastrointestinal tract. Biol Signals 2:181–193

Lemmer B, Bruhl T, Witte K, Pflug B, Kohler W, Touitou Y (1994) Effects of bright light on circadian patterns of cyclic adenosine monophosphate, melatonin and cortisol in healthy subjects. Eur J Endocrinol 130:472–477

Lerchl A, Partsch CJ (1994) Reliable analysis of individual human melatonin profiles by complex cosinor analysis. J Pineal Res 16:85–90

Lerchl A, Nonaka KO, Stokkan KA, Reiter RJ (1990) Marked rapid alterations in nocturnal pineal serotonin metabolism in mice and rats exposed to weak intermittent magnetic fields. Biochem Biophys Res Commun 169:102–108

Lerchl A, Nonaka KO, Reiter RJ (1991) Pineal gland "magnetosensitivity" to static magnetic fields is a consequence of induced electric currents (eddy currents). J Pineal Res 10:109–116

Lerchl A, Reiter RJ, Howes KA, Nonaka KO, Stokkan KA (1991b) Evidence that extremely low frequency Ca^{2+}-cyclotron resonance depresses pineal melatonin synthesis in vitro. Neurosci Lett 124:213–215

Lerchl A, Nordhoff V, Gerding H (1998a) Expression of the gene for the retinal protein peripherin in the pineal gland of humans and Djungarian hamsters (*Phodopus sungorus*). Neurosci Lett 258:187–189

Lerchl A, Zachmann A, Ali MA, Reiter RJ (1998b) The effects of pulsing magnetic fields on pineal melatonin synthesis in a teleost fish (brook trout, *Salvelinus fontinalis*). Neurosci Lett 256:171–173

Lerner AB, Case JD, Takahashi Y, Lee Y, Mori W (1958) Isolation of melatonin, the pineal gland factor that lightens melanocytes. J Amer Chem Soc 80:2587

Lerner AB, Case JD, Heinzelman RV (1959) Structure of melatonin. J Amer Chem Soc 81:6084–6085

Lerner AB, Nordlund JJ (1978) Melatonin: clinical pharmacology. J Neural Transm Suppl 13:339–347

Letz B, Schomerus C, Maronde E, Korf HW, Korbmacher C (1997) Stimulation of a nicotinic ACh receptor causes depolarization and activation of L-type Ca2+ channels in rat pinealocytes. J Physiol Lond 499:329–340

Lewy AJ, Markey SP (1978) Analysis of melatonin in human plasma by gas chromatography negative chemical ionization mass spectrometry. Science 201:741–743

Lewy AJ, Newsome DA (1983) Different types of melatonin circadian secretory rhythms in some blind subjects. J Clin Endocrinol Metab 56:1103–1107

Lewy AJ, Sack RL (1989) The dim light melatonin onset as a marker for circadian phase position. Chronobiol Int 6:93–102

Lewy AJ, Sack RL (1997) Exogenous melatonin's phase-shifting effects on the endogenous melatonin profile in sighted humans: a brief review and critique of the literature. J Biol Rhythms 12:588–594

Lewy AJ, Wehr TA, Goodwin FK, Newsome DA, Markey SP (1980) Light suppresses melatonin secretion in humans. Science 210:1267–1269

Lewy AJ, Sack RL, Singer CM (1985) Melatonin, light and chronobiological disorders. Ciba Found Symp 117:231–252

Lewy AJ, Siever LJ, Uhde TW, Markey SP (1986) Clonidine reduces plasma melatonin levels. J Pharm Pharmacol 38:555–556

Lewy AJ, Sack RL, Miller LS, Hoban TM (1987a) Antidepressant and circadian phase-shifting effects of light. Science 235:352–354

Lewy AJ, Sack RL, Singer CM, White DM (1987b) The phase shift hypothesis for bright light's therapeutic mechanism of action: theoretical considerations and experimental evidence. Psychopharmacol Bull 23:349–353

Lewy AJ, Ahmed S, Jackson JM, Sack RL (1992) Melatonin shifts human circadian rhythms according to a phase-response curve. Chronobiol Int 9:380–392

Lewy AJ, Ahmed S, Sack RL (1996) Phase shifting the human circadian clock using melatonin. Behav Brain Res 73:131–134

Lewy AJ, Bauer VK, Cutler NL, Sack RL, Ahmed S, Thomas KH, Blood ML, Jackson JM (1998) Morning vs evening light treatment of patients with winter depression. Arch Gen Psychiatry 55:890–896

Li Y, Jiang DH, Wang ML, Jiao DR, Pang SF (1989) Rhythms of serum melatonin in patients with spinal lesions at the cervical, thoracic or lumbar region. Clin Endocrinol Oxf 30:47–56

Lieberman HR, Waldhauser F, Garfield G, Lynch HJ, Wurtman RJ (1984) Effects of melatonin on human mood and performance. Brain Res 323:201–207

Liebmann PM, Wolfler A, Felsner P, Hofer D, Schauenstein K (1997) Melatonin and the immune system. Int Arch Allergy Immunol 112:203–211

Lissoni P, Tancini G, Barni S, Viviani S, Archili C, Cattaneo G, Fiorelli G (1989) Alterations of pineal gland and of T lymphocyte subsets in metastatic cancer patients: preliminary results. J Biol Regul Homeost Agents 3:181–183

Lissoni P, Barni S, Tancini G, Ardizzoia A, Ricci G, Aldeghi R, Brivio F, Tisi E, Rovelli F, Rescaldani R, et al. (1994) A randomised study with subcutaneous low-dose interleukin 2 alone vs interleukin 2 plus the pineal neurohormone melatonin in advanced solid neoplasms other than renal cancer and melanoma. Br J Cancer 69:196–199

Little KY, Kirkman JA, Duncan GE (1996) Beta-adrenergic receptor subtypes in human pineal gland. J Pineal Res 20:15–20

Liu RY, Zhou JN, van Heerikhuize J, Hofman MA, Swaab DF (1999) Decreased melatonin levels in postmortem cerebrospinal fluid in relation to aging, Alzheimer's disease, and apolipoprotein E-epsilon4/4 genotype. J Clin Endocrinol Metab 84:323–327

Lockley SW, Skene DJ, Tabandeh H, Bird AC, Defrance R, Arendt J (1997) Relationship between napping and melatonin in the blind. J Biol Rhythms 12:16–25

Löscher W, Mevissen M, Lerchl A (1998) Exposure of female rats to a 100-microT 50 Hz magnetic field does not induce consistent changes in nocturnal levels of melatonin. Radiat Res 150:557–567

Luboshitzky R, Dharan M, Goldman D, Herer P, Hiss Y, Lavie P (1997) Seasonal variation of gonadotropins and gonadal steroids receptors in the human pineal gland. Brain Res Bull 44:665–670

Lucas RJ, Freedman MS, Munoz M, Garcia Fernandez JM, Foster RG (1999) Regulation of the mammalian pineal by non-rod, non-cone, ocular photoreceptors. Science 284:505–507

Lynch HJ, Ozaki Y, Shakal D, Wurtman RJ (1975a) Melatonin excretion of man and rats: effect of time of day, sleep, pinealectomy and food consumption. Int J Biometeorol 19:267–279

Lynch HJ, Wurtman RJ, Moskowitz MA, Archer MC, Ho MH (1975b) Daily rhythm in human urinary melatonin. Science 187:169–171

Lynch HJ, Jimerson DC, Ozaki Y, Post RM, Bunney WE, Jr., Wurtman RJ (1978a) Entrainment of rhythmic melatonin secretion in man to a 12-hour phase shift in the light/dark cycle. Life Sci 23:1557–1563

Lynch HJ, Ozaki Y, Wurtman RJ (1978b) The measurement of melatonin in mammalian tissues and body fluids. J Neural Transm Suppl 13:251–264

MacFarlane JG, Cleghorn JM, Brown GM, Streiner DL (1991) The effects of exogenous melatonin on the total sleep time and daytime alertness of chronic insomniacs: a preliminary study. Biol Psychiatry 30:371–376

Maestroni GJ (1993) The immunoneuroendocrine role of melatonin. J Pineal Res 14:1–10

Mallo C, Zaidan R, Faure A, Brun J, Chazot G, Claustrat B (1988) Effects of a four-day nocturnal melatonin treatment on the 24 h plasma melatonin, cortisol and prolactin profiles in humans. Acta Endocrinol Copenh 119:474–480

Mallo C, Zaidan R, Galy G, Vermeulen E, Brun J, Chazot G, Claustrat B (1990) Pharmacokinetics of melatonin in man after intravenous infusion and bolus injection. Eur J Clin Pharmacol 38:297–301

Mann K, Wagner P, Brunn G, Hassan F, Hiemke C, Roschke J (1998) Effects of pulsed high-frequency electromagnetic fields on the neuroendocrine system. Neuroendocrinology 67:139–144

Martin XD, Malina HZ, Brennan MC, Hendrickson PH, Lichter PR (1992) The ciliary body – the third organ found to synthesize indoleamines in humans. Eur J Ophthalmol 2:67–72

Martins E, Jr., Fernandes LC, Bartol I, Cipolla Neto J, Costa Rosa LF (1998) The effect of melatonin chronic treatment upon macrophage and lymphocyte metabolism and function in Walker-256 tumour-bearing rats. J Neuroimmunol 82:81–89

Mason R, Brooks A (1988) The electrophysiological effects of melatonin and a putative melatonin antagonist (N-acetyltryptamine) on rat suprachiasmatic neurones in vitro. Neurosci Lett 95:296–301

Matsui DH, Machado Santelli GM (1997) Alterations in F-actin distribution in cells treated with melatonin. J Pineal Res 23:169–175

Matthews CD, Guerin MV, Wang X (1991) Human plasma melatonin and urinary 6-sulphatoxy melatonin: studies in natural annual photoperiod and in extended darkness. Clin Endocrinol Oxf 35:21–27

Mayeda A, Mannon S, Hofstetter J, Adkins M, Baker R, Hu K, Nurnberger J, Jr. (1998) Effects of indirect light and propranolol on melatonin levels in normal human subjects. Psychiatry Res 81:9–17

McArthur AJ, Budden SS (1998) Sleep dysfunction in Rett syndrome: a trial of exogenous melatonin treatment. Dev Med Child Neurol 40:186–192

McArthur AJ, Gillette MU, Prosser RA (1991) Melatonin directly resets the rat suprachiasmatic circadian clock in vitro. Brain Res 565:158–161

McCord CP, Allen FB (1917) Evidences associating pineal gland function with alterations in pigmentation. J Exp Zool 23:207–224

McElhinney DB, Hoffman SJ, Robinson WA, Ferguson J (1994) Effect of melatonin on human skin color. J Invest Dermatol 102:258–259

McIntyre IM, Norman TR, Burrows GD, Armstrong SM (1987) Melatonin rhythm in human plasma and saliva. J Pineal Res 4:177–183

McIntyre IM, Burrows GD, Norman TR (1988) Suppression of plasma melatonin by a single dose of the benzodiazepine alprazolam in humans. Biol Psychiatry 24:108–112

McIntyre IM, Norman TR, Burrows GD, Armstrong SM (1989a) Human melatonin response to light at different times of the night. Psychoneuroendocrinology 14:187–193

McIntyre IM, Norman TR, Burrows GD, Armstrong SM (1989b) Human melatonin suppression by light is intensity dependent. J Pineal Res 6:149–156

McIntyre IM, Norman TR, Burrows GD, Armstrong SM (1989c) Quantal melatonin suppression by exposure to low intensity light in man. Life Sci 45:327–332

McIntyre IM, Norman TR, Burrows GD, Armstrong SM (1993) Alterations to plasma melatonin and cortisol after evening alprazolam administration in humans. Chronobiol Int 10:205–213

McMillen IC, Houghton DC, Young IR (1995) Melatonin and the development of circadian and seasonal rhythmicity. J Reprod Fertil Suppl 49:137–146

McNulty JA, Fox LM (1992) Pinealocyte synaptic ribbons and neuroendocrine function. Microsc Res Tech 21:175–187

McNulty JA, Fox LM, Spurrier WA (1990) A circannual cycle in pinealocyte synaptic ribbons in the hibernating and seasonally reproductive 13-lined ground squirrel (*Spermophilus tridecemlineatus*). Neurosci Lett 119:237–240

Mevissen M, Lerchl A, Löscher W (1996) Study on pineal function and DMBA-induced breast cancer formation in rats during exposure to a 100-mG, 50 Hz magnetic field. J Toxicol Environ Health 48:169–185

Meyer BJ, Steyn ME, Moncrieff J (1990) Effects of fasting on plasma melatonin levels. Suid-Afrikaanse Tydskrif vir Wetenskap 86:145–147

Michotte Y, Lowenthal A, Knaepen L, Collard M, Massart DL (1977) A morphological and chemical study of calcification of the pineal gland. J Neurol 215:209–219

Middleton B, Arendt J, Stone BM (1997) Complex effects of melatonin on human circadian rhythms in constant dim light. J Biol Rhythms 12:467–477

Miles A, Philbrick D, Tidmarsh SF, Shaw DM (1985a) Direct radioimmunoassay of melatonin in saliva [letter]. Clin Chem 31:1412–1413

Miles A, Philbrick DR, Shaw DM, Tidmarsh SF, Pugh AJ (1985b) Salivary melatonin estimation in clinical research [letter]. Clin Chem 31:2041–2042

Miles A, Thomas DR, Grey JE, Pugh AJ (1987) Salivary melatonin assay in laboratory medicine–longitudinal profiles of secretion in healthy men [letter]. Clin Chem 33:1957–1959

Milstein V, Small JG, Spencer DW (1998) Melatonin for sleep EEG. Clin Electroencephalogr 29:49–53

Mishima K, Tozawa T, Satoh K, Matsumoto Y, Hishikawa Y, Okawa M (1999) Melatonin secretion rhythm disorders in patients with senile dementia of Alzheimer's type with disturbed sleep-waking. Biol Psychiatry 45:417–421

Moller M (1976) The ultrastructure of the human fetal pineal gland. II. Innervation and cell junctions. Cell Tissue Res 169:7–21

Monteleone P, Maj M, Fuschino A, Kemali D (1992) Physical stress in the middle of the dark phase does not affect light-depressed plasma melatonin levels in humans. Neuroendocrinology 55:367–371

Monteleone P, Maj M, Franza F, Fusco R, Kemali D (1993) The human pineal gland responds to stress-induced sympathetic activation in the second half of the dark phase: preliminary evidence. J Neural Transm Gen Sect 92:25–32

Monteleone P, Esposito G, La Rocca A, Maj M (1995) Does bright light suppress nocturnal melatonin secretion more in women than men? J Neural Transm Gen Sect 102:75–80

Moore RY, Heller A, Wurtman RJ, Axelrod J (1967) Visual pathway mediating pineal response to environmental light. Science 155:220–223

Morin D, Simon N, Depres Brummer P, Levi F, Tillement JP, Urien S (1997) Melatonin high-affinity binding to alpha-1-acid glycoprotein in human serum. Pharmacology 54:271–275

Morita T, Tokura H, Wakamura T, Park SJ, Teramoto Y (1997) Effects of the morning irradiation of light with different wavelengths on the behavior of core temperature and melatonin in humans. Appl Human Sci 16:103–105

Moriyama Y, Yamamoto A (1995) Microvesicles isolated from bovine pineal gland specifically accumulate L-glutamate. FEBS Lett 367:233–236

Munoz Hoyos A, Rodriguez Cabezas T, Molina Carballo A, Martinez Sempere JJ, Ruiz Cosano C, Acuna Castroviejo D (1992) Melatonin concentration in the umbilical artery and vein in human preterm and term neonates and neonates with acute fetal distress. J Pineal Res 13:184–191

Munoz Hoyos A, Jaldo Alba F, Molina Carballo A, Rodriguez Cabezas T, Molina Font JA, Acuna Castroviejo D (1993) Absence of plasma melatonin circadian rhythm during the first 72 hours of life in human infants. J Clin Endocrinol Metab 77:699–703

Murata J, Sawamura Y, Ikeda J, Hashimoto S, Honma K (1998) Twenty-four hour rhythm of melatonin in patients with a history of pineal and/or hypothalamo-neurohypophyseal germinoma. J Pineal Res 25:159–166

Murphy DG, Murphy DM, Abbas M, Palazidou E, Binnie C, Arendt J, Campos Costa D, Checkley SA (1993) Seasonal affective disorder: response to light as measured by electroencephalogram, melatonin suppression, and cerebral blood flow [see comments]. Br J Psychiatry 163:327–331, 335–327

Murphy DL, Tamarkin L, Sunderland T, Garrick NA, Cohen RM (1986) Human plasma melatonin is elevated during treatment with the monoamine oxidase inhibitors clorgyline and tranylcypromine but not deprenyl. Psychiatry Res 17:119–127

Naguib M, Samarkandi AH (1999) Premedication with melatonin: a double-blind, placebo-controlled comparison with midazolam. Br J Anaesth 82:875–880

Nakagawa H, Isaki K, Sack RL, Lewy AJ (1992) Free-running melatonin, sleep propensity, cortisol and temperature rhythms in a totally blind person. Jpn J Psychiatry Neurol 46:210–212

Nakamura KT, Nakahara H, Nakamura M, Tokioka T, Kiyomura H (1995) Ultrastructure and x-ray microanalytical study of human pineal concretions. Anat Anz 177:413–419

Nathan PJ, Jeyaseelan AS, Burrows GD, Norman TR (1998) Modulation of plasma melatonin concentrations by changes in posture. J Pineal Res 24:219–223

Nave R, Peled R, Lavie P (1995) Melatonin improves evening napping. Eur J Pharmacol 275:213–216

Nave R, Herer P, Haimov I, Shlitner A, Lavie P (1996) Hypnotic and hypothermic effects of melatonin on daytime sleep in humans: lack of antagonism by flumazenil. Neurosci Lett 214:123–126

Nelson W, Tong YL, Lee JK, Halberg F (1979) Methods for cosinor-rhythmometry. Chronobiologia 6:305–323

Neuwelt EA, Lewy AJ (1983) Disappearance of plasma melatonin after removal of a neoplastic pineal gland. N Engl J Med 308:1132–1135

Nickelsen T, Demisch L, Demisch K, Radermacher B, Schoffling K (1989) Influence of subchronic intake of melatonin at various times of the day on fatigue and hormonal levels: a placebo-controlled, double-blind trial. J Pineal Res 6:325–334

Nickelsen T, Samel A, Maass H, Vejvoda M, Wegmann H, Schoffling K (1991) Circadian patterns of salivary melatonin and urinary 6-sulfatoxymelatonin before and after a 9-hour time-shift. Adv Exp Med Biol 294:493–496

Novier A, Nicolas D, Krstic R (1996) Calretinin immunoreactivity in pineal gland of different mammals including man. J Pineal Res 21:121–130

Nowak R (1988) The salivary melatonin diurnal rhythm may be abolished after transmeridian flights in an eastward but not a westward direction [letter]. Med J Aust 149:340–341

Nowak R, McMillen IC, Redman J, Short RV (1987) The correlation between serum and salivary melatonin concentrations and urinary 6-hydroxymelatonin sulphate excretion rates: two non-invasive techniques for monitoring human circadian rhythmicity. Clin Endocrinol Oxf 27:445–452

O'Callaghan FJ, Clarke AA, Hancock E, Hunt A, Osborne JP (1999) Use of melatonin to treat sleep disorders in tuberous sclerosis. Dev Med Child Neurol 41:123–126

Ohashi Y, Okamoto N, Uchida K, Iyo M, Mori N, Morita Y (1997) Differential pattern of the circadian rhythm of serum melatonin in young and elderly healthy subjects. Biol Signals 6:301–306

Oksche A (1971) Sensory and glandular elements of the pineal organ. In: Wolstenholme GEW and Knight J (eds) The Pineal Gland. Churchill Livingstone, Edinburgh London. pp 127–146

Olcese J, Hurlbut E (1989) Comparative studies on the retinal dopamine response to altered magnetic fields in rodents. Brain Res 498:145–148

Olcese J, Reuss S (1986) Magnetic field effects on pineal gland melatonin synthesis: comparative studies on albino and pigmented rodents. Brain Res 369:365–368

Olcese J, Reuss S, Vollrath L (1985) Evidence for the involvement of the visual system in mediating magnetic field effects on pineal melatonin synthesis in the rat. Brain Res 333:382–384

Olcese J, Reuss S, Stehle J, Steinlechner S, Vollrath L (1988) Responses of the mammalian retina to experimental alteration of the ambient magnetic field. Brain Res 448:325–330

Oxenkrug GF, McIntyre IM, Balon R, Jain AK, Appel D, McCauley RB (1986) Single dose of tranylcypromine increases human plasma melatonin. Biol Psychiatry 21:1085–1089

Oxenkrug GF, Balon R, Jain AK, McIntyre IM (1988) Melatonin plasma response to MAO inhibitor: influence on human pineal activity? Acta Psychiatr Scand 77:160–162

Paccotti P, Terzolo M, Torta M, Vignani A, Schena M, Piovesan A, Angeli A (1987) Acute administration of melatonin at two opposite circadian stages does not change responses to gonadotropin releasing hormone, thyrotropin releasing hormone and ACTH in healthy adult males. J Endocrinol Invest 10:471–477

Palm L, Blennow G, Wetterberg L (1991) Correction of non-24-hour sleep/wake cycle by melatonin in a blind retarded boy. Ann Neurol 29:336–339

Palm L, Blennow G, Wetterberg L (1997) Long-term melatonin treatment in blind children and young adults with circadian sleep-wake disturbances. Dev Med Child Neurol 39:319–325

Pangerl B, Pangerl A, Reiter RJ (1990) Circadian variations of adrenergic receptors in the mammalian pineal gland: a review. J Neural Transm Gen Sect 81:17–29

Partonen T, Vakkuri O, Lamberg Allardt C (1995) Effects of exposure to morning bright light in the blind and sighted controls. Clin Physiol 15:637–646

Partonen T, Vakkuri O, Lamberg Allardt C, Lonnqvist J (1996) Effects of bright light on sleepiness, melatonin, and 25-hydroxyvitamin D(3) in winter seasonal affective disorder. Biol Psychiatry 39:865–872

Partonen T, Vakkuri O, Lonnqvist J (1997) Suppression of melatonin secretion by bright light in seasonal affective disorder. Biol Psychiatry 42:509–513

Pelham RW, Vaughan GM, Sandock KL, Vaughan MK (1973) Twenty-four-hour cycle of a melatonin-like substance in the plasma of human males. J. Clin. Endocrinol. Metab 37:341–344

Peniston Bird JF, Di WL, Street CA, Kadva A, Stalteri MA, Silman RE (1993) HPLC assay of melatonin in plasma with fluorescence detection. Clin Chem 39:2242–2247

Petrie K, Conaglen JV, Thompson L, Chamberlain K (1989) Effect of melatonin on jet lag after long haul flights. Br Med J 298:705–707

Petrie K, Dawson AG, Thompson L, Brook R (1993) A double-blind trial of melatonin as a treatment for jet lag in international cabin crew. Biol Psychiatry 33:526–530

Petterborg LJ, Kjellman BF, Thalen BE, Wetterberg L (1991) Effect of a 15 minute light pulse on nocturnal serum melatonin levels in human volunteers. J Pineal Res 10:9–13

Pfeffer M, Kuhn R, Krug L, Korf HW, Stehle JH (1998) Rhythmic variation in β_1-adrenergic receptor mRNA levels in the rat pineal gland: circadian and developmental regulation. Eur J Neurosci 10:2896–2904

Pilling JR, Hawkins TD (1977) Distribution of calcification within the pineal gland. Br J Radiol 50:796–798

Poeggeler B, Saarela S, Reiter RJ, Tan DX, Chen LD, Manchester LC, Barlow Walden LR (1994) Melatonin–a highly potent endogenous radical scavenger and electron donor: new aspects of the oxidation chemistry of this indole accessed in vitro. Ann NY Acad Sci 738:419–420

Prato FS, Ossenkopp K-P, Kavaliers M, Uksik P, Nicholson RL, Drost D, Sestini EA (1989) Effects of exposure to magnetic resonance imaging on nocturnal serum melatonin and other hormone levels in adult males: preliminary findings. Journal of Bioelectricity 7:169–180

Quay W (1974) Pineal Chemistry, Charles C Thomas, Springfield, Illinois

Rao ML, Muller Oerlinghausen B, Mackert A, Strebel B, Stieglitz RD, Volz HP (1992) Blood serotonin, serum melatonin and light therapy in healthy subjects and in patients with nonseasonal depression. Acta Psychiatr Scand 86:127–132

Redecker P (1995) The ras-like rab3 A protein is present in pinealocytes of the gerbil pineal gland. Neurosci Lett 184:117–120

Redecker P (1996) Synaptotagmin I, synaptobrevin II, and syntaxin I are coexpressed in rat and gerbil pinealocytes. Cell Tissue Res 283:443–454

Redecker P, Veh RW (1994) Glutamate immunoreactivity is enriched over pinealocytes of the gerbil pineal gland. Cell Tissue Res 278:579–588

Redecker P, Pabst H, Gebert A, Steinlechner S (1997) Expression of synaptic membrane proteins in gerbil pinealocytes in primary culture. J Neurosci Res 47:509–520

Reid K, Van den Heuvel C, Dawson D (1996) Day-time melatonin administration: effects on core temperature and sleep onset latency. J Sleep Res 5:150–154

Reiter RJ. (1981). The Pineal Gland. Volumes I to III. CRC Press, Boca Raton, Florida.

Reiter RJ (1982) Neuroendocrine effects of the pineal gland and of melatonin. In: Ganong WF and Martini L (eds) Frontiers in Neuroendocrinology. Raven Press, New York. pp 287–316

Reiter RJ (1985) Action spectra, dose-response relationships, and temporal aspects of light's effects on the pineal gland. Ann N Y Acad Sci 453:215–230

Reiter RJ (1986) Normal patterns of melatonin levels in the pineal gland and body fluids of humans and experimental animals. J Neural Transm Suppl 21:35–54

Reiter RJ (1996) The indoleamine melatonin as a free radical scavenger, electron donor, and anti-oxidant. In vitro and in vivo studies. Adv Exp Med Biol 398:307–313

Reiter RJ, Welsh MG, Vaughan MK (1976) Age-related changes in the intact and sympathetically denervated gerbil pineal gland. Am J Anat 146:427–431

Reiter RJ, Steinlechner S, Richardson BA, King TS (1983) Differential response of pineal melatonin levels to light at night in laboratory-raised and wild-captured 13-lined ground squirrels (*Spermophilus tridecemlineatus*). Life Sci 32:2625–2629

Reiter RJ, Tan DX, Poeggeler B, Menendez Pelaez A, Chen LD, Saarela S (1994) Melatonin as a free radical scavenger: implications for aging and age-related diseases. Ann N Y Acad Sci 719:1–12

Reiter RJ, Tan DX, Poeggeler B, Kavet R (1998) Inconsistent suppression of nocturnal pineal melatonin synthesis and serum melatonin levels in rats exposed to pulsed DC magnetic fields. Bioelectromagnetics 19:318–329

Reiter RJ, Tan DX, Cabrera J, D'Arpa D, Sainz RM, Mayo JC, Ramos S (1999) The oxidant/antioxidant network: role of melatonin. Biol Signals Recept 8:56–63

Reppert SM, Weaver DR (1995) Melatonin madness. Cell 83:1059–1062

Reppert SM, Weaver DR, Rivkees SA, Stopa EG (1988) Putative melatonin receptors in a human biological clock. Science 242:78–81

Reppert SM, Weaver DR, Cassone VM, Godson C, Kolakowski LF, Jr. (1995) Melatonin receptors are for the birds: molecular analysis of two receptor subtypes differentially expressed in chick brain. Neuron 15:1003–1015

Reuss S (1987) Electrical activity of the mammalian pineal gland. In: Reiter RJ (ed) Pineal Research Reviews. Alan R. Liss, Inc., New York. pp 153–189

Reuss S (1996) Components and connections of the circadian timing system in mammals. Cell Tissue Res 285:353–378

Reuss S (1999) Trigeminal innervation of the mammalian pineal gland. Microscopy Resarch & Technique 46:305–309

Reuss S, Olcese J (1986) Magnetic field effects on the rat pineal gland: role of retinal activation by light. Neurosci Lett 64:97–101

Reuss S, Semm P, Vollrath L (1983) Different types of magnetically sensitive cells in the rat pineal gland. Neurosci Lett 40:23–26

Reuss S, Olcese J, Vollrath L, Skalej M, Meves M (1985) Lack of effect of NMRI-strength magnetic fields on rat pineal melatonin synthesis. IRCS Med Sci 13:471

Reuss S, Schröder B, Schröder H, Maelicke A (1992) Nicotinic cholinoceptors in the rat pineal gland as analyzed by western blot, light- and electron microscopy. Brain Res 573:114–118

Rice J, Mayor J, Tucker HA, Bielski RJ (1995) Effect of light therapy on salivary melatonin in seasonal affective disorder. Psychiatry Res 56:221–228

Richardson BA, Yaga K, Reiter RJ, Morton DJ (1992) Pulsed static magnetic field effects on in-vitro pineal indoleamine metabolism. Biochim Biophys Acta 1137:59–64

Richardson BA, Yaga K, Reiter RJ, Hoover P (1993) Suppression of nocturnal pineal N-acetyltransferase activity and melatonin content by inverted magnetic fields and induced eddy currents. Int J Neurosci 69:149–155

Rivkees SA (1997) Developing circadian rhythmicity. Basic and clinical aspects. Pediatr Clin North Am 44:467–487

Rodriguez IR, Mazuruk K, Schoen TJ, Chader GJ (1994) Structural analysis of the human hydroxy-indole-O-methyltransferase gene. Presence of two distinct promoters. J Biol Chem 269:31969–31977

Rogers N, van den Heuvel C, Dawson D (1997) Effect of melatonin and corticosteroid on in vitro cellular immune function in humans. J Pineal Res 22:75–80

Rogers NL, Phan O, Kennaway DJ, Dawson D (1998) Effect of daytime oral melatonin administration on neurobehavioral performance in humans. J Pineal Res 25:47–53

Rogers WR, Reiter RJ, Barlow Walden L, Smith HD, Orr JL (1995a) Regularly scheduled, day-time, slow-onset 60 Hz electric and magnetic field exposure does not depress serum melatonin concentration in nonhuman primates. Bioelectromagnetics Suppl 3:111–118

Rogers WR, Reiter RJ, Smith HD, Barlow Walden L (1995b) Rapid-onset/offset, variably scheduled 60 Hz electric and magnetic field exposure reduces nocturnal serum melatonin concentration in nonhuman primates. Bioelectromagnetics Suppl 3:119–122

Rojdmark S, Wetterberg L (1989) Short-term fasting inhibits the nocturnal melatonin secretion in healthy man. Clin Endocrinol Oxf 30:451–457

Rommel T, Demisch L (1994) Influence of chronic beta-adrenoreceptor blocker treatment on melatonin secretion and sleep quality in patients with essential hypertension. J Neural Transm Gen Sect 95:39–48

Roseboom PH, Coon SL, Baler R, McCune SK, Weller JL, Klein DC (1996) Melatonin synthesis: analysis of the more than 150-fold nocturnal increase in serotonin N-acetyltransferase messenger ribonucleic acid in the rat pineal gland. Endocrinology 137:3033–3045

Rosen LA, Barber I, Lyle DB (1998) A 0.5 G, 60 Hz magnetic field suppresses melatonin production in pinealocytes. Bioelectromagnetics 19:123–127

Rosenthal NE, Wehr TA (1987) Seasonal affective disorders. Psychiatric Annals 17:670–674

Rosenthal NE, Sack DA, Jacobsen FM, James SP, Parry BL, Arendt J, Tamarkin L, Wehr TA (1986) Melatonin in seasonal affective disorder and phototherapy. J Neural Transm Suppl 21:257–267

Rosenthal NE, Jacobsen FM, Sack DA, Arendt J, James SP, Parry BL, Wehr TA (1988) Atenolol in seasonal affective disorder: a test of the melatonin hypothesis. Am J Psychiatry 145:52–56

Rousseau A, Petrén S, Planntin J, Eklundh T, Nordin C (1999) Serum and cerebrospinal fluid concentrations of melatonin: a pilot study in healthy male volunteers. J. Neural Transm. 106: 883–888

Ruberg FL, Skene DJ, Hanifin JP, Rollag MD, English J, Arendt J, Brainard GC (1996) Melatonin regulation in humans with color vision deficiencies. J Clin Endocrinol Metab 81:2980–2985

Rudolph K, Wirz Justice A, Krauchi K, Feer H (1988) Static magnetic fields decrease nocturnal pineal cAMP in the rat. Brain Res 446:159–160

Sack RL, Lewy AJ (1993) Human circadian rhythms: lessons from the blind. Ann Med 25:303–305

Sack RL, Lewy AJ (1997) Melatonin as a chronobiotic: treatment of circadian desynchrony in night workers and the blind. J Biol Rhythms 12:595–603

Sack RL, Lewy AJ, Erb DL, Vollmer WM, Singer CM (1986) Human melatonin production decreases with age. J Pineal Res 3:379–388

Sack RL, Lewy AJ, Blood ML, Stevenson J, Keith LD (1991) Melatonin administration to blind people: phase advances and entrainment. J Biol Rhythms 6:249–261

Sack RL, Lewy AJ, Blood ML, Keith LD, Nakagawa H (1992) Circadian rhythm abnormalities in totally blind people: incidence and clinical significance. J Clin Endocrinol Metab 75:127–134

Sack RL, Lewy AJ, Hughes RJ (1998) Use of melatonin for sleep and circadian rhythm disorders. Ann Med 30:115–121

Salin Pascual RJ, Ortega Soto H, Huerto Delgadillo L, Camacho Arroyo I, Roldan Roldan G, Tamarkin L (1988) The effect of total sleep deprivation on plasma melatonin and cortisol in healthy human volunteers. Sleep 11:362–369

Salinas EO, Hakim Kreis CM, Piketty ML, Dardennes RM, Musa CZ (1992) Hypersecretion of melatonin following diurnal exposure to bright light in seasonal affective disorder: preliminary results. Biol Psychiatry 32:387–398

Samel A, Wegmann HM, Vejvoda M, Maass H, Gundel A, Schutz M (1991) Influence of melatonin treatment on human circadian rhythmicity before and after a simulated 9-hr time shift. J Biol Rhythms 6:235–248

Samkova L, Vondrasova D, Hajek I, Illnerova H (1997) A fixed morning awakening coupled with a low intensity light maintains a phase advance of the human circadian system. Neurosci Lett 224:21–24

Schenda J, Vollrath L (1998a) Demonstration of action-potential-producing cells in the rat pineal gland in vitro and their regulation by norepinephrine and nitric oxide. J Comp Physiol A 183: 573–581

Schenda J, Vollrath L (1998b) Modulatory role of acetylcholine in the rat pineal gland. Neuroscience Letters 254:13–16

Schenda J, Vollrath L (1999) An intrinsic neuronal-like network in the rat pineal gland. Brain Res 823:231–233

Schiffman JS, Lasch HM, Rollag MD, Flanders AE, Brainard GC, Burk DL, Jr. (1994) Effect of MR imaging on the normal human pineal body: measurement of plasma melatonin levels. J Magn Reson Imaging 4:7–11

Schmid HA, Raykhtsaum G (1995) Age-related differences in the structure of human pineal calcium deposits: results of transmission electron microscopy and mineralographic microanalysis. J Pineal Res 18:12–20

Schmid HA, Requintina PJ, Oxenkrug GF, Sturner W (1994) Calcium, calcification, and melatonin biosynthesis in the human pineal gland: a postmortem study into age-related factors. J Pineal Res 16:178–183

Schneider T, Thalau HP, Semm P (1994) Effects of light or different earth-strength magnetic fields on the nocturnal melatonin concentration in a migratory bird. Neurosci Lett 168:73–75

Schober E, Waldhauser F, Frisch H, Schuster E, Bieglmayer C (1989) Melatonin secretion in Turner's syndrome: lack of effect of oestrogen administration. Clin Endocrinol Oxf 31:475–479

Schomerus C, Ruth P, Korf HW (1994) Photoreceptor-specific proteins in the mammalian pineal organ: immunocytochemical data and functional considerations. Acta Neurobiol Exp Warsz 54: 9–17

Schomerus C, Laedtke E, Korf HW (1995) Calcium responses of isolated, immunocytochemically identified rat pinealocytes to noradrenergic, cholinergic and vasopressinergic stimulations. Neurochem Int 27:163–175

Schröder H, Weihe E, Nohr D, Vollrath L (1988) Immunohistochemical evidence for the presence of peptides derived from proenkephalin, prodynorphin and pro-opiomelanocortin in the guinea pig pineal gland. Histochemistry 88:333–341

Schröder H, Bendig A, Dahl D, Gröschel Stewart U, Vollrath L (1990) Neuronal markers in the rodent pineal gland – an immunohistochemical investigation. Histochemistry 94:309–314

Selmaoui B, Touitou Y (1995) Sinusoidal 50-Hz magnetic fields depress rat pineal NAT activity and serum melatonin. Role of duration and intensity of exposure. Life Sci 57:1351–1358

Selmaoui B, Bogdan A, Auzeby A, Lambrozo J, Touitou Y (1996a) Acute exposure to 50 Hz magnetic field does not affect hematologic or immunologic functions in healthy young men: a circadian study. Bioelectromagnetics 17:364–372

Selmaoui B, Lambrozo J, Touitou Y (1996b) Magnetic fields and pineal function in humans: evaluation of nocturnal acute exposure to extremely low frequency magnetic fields on serum melatonin and urinary 6-sulfatoxymelatonin circadian rhythms. Life Sci 58:1539–1549

Semm P, Schneider T, Vollrath L (1980) Effects of an earth-strength magnetic field on electrical activity of pineal cells. Nature 288:607–608

Shaw D, Goldman BD (1995) Influence of prenatal photoperiods on postnatal reproductive responses to daily infusions of melatonin in the Siberian hamster (*Phodopus sungorus*). Endocrinology 136:4231–4236

Sieghart W, Ronca E, Drexler G, Karall S (1987) Improved radioimmunoassay of melatonin in serum. Clin Chem 33:604–605

Skene DJ, Lockley SW, Arendt J (1999) Melatonin in circadian sleep disorders in the blind. Biol Signals Recept 8:90–95

Skrinar GS, Bullen BA, Reppert SM, Peachey SE, Turnbull BA, McArthur JW (1989) Melatonin response to exercise training in women. J Pineal Res 7:185–194

Slaugenhaupt SA, Roca AL, Liebert CB, Altherr MR, Gusella JF, Reppert SM (1995) Mapping of the gene for the Mel1a-melatonin receptor to human chromosome 4 (MTNR1 A) and mouse chromosome 8 (Mtnr1a). Genomics 27:355–357

Smith JA, Padwick D, Mee TJ, Minneman KP, Bird ED (1977) Synchronous nyctohemeral rhythms in human blood melatonin and in human post-mortem pineal enzyme. Clin Endocrinol Oxf 6:219–225

Smith JA, Barnes JL, Mee TJ (1979) The effect of neuroleptic drugs on serum and cerebrospinal fluid melatonin concentrations in psychiatric subjects. J Pharm Pharmacol 31:246–248

Smith JA, Mee TJ, Padwick DJ, Spokes EG (1981) Human post-mortem pineal enzyme activity. Clin Endocrinol Oxf 14:75–81

Song Y, Chan CW, Brown GM, Pang SF, Silverman M (1997) Studies of the renal action of melatonin: evidence that the effects are mediated by 37 kDa receptors of the Mel1a subtype localized primarily to the basolateral membrane of the proximal tubule. FASEB J 11:93–100

Souetre E, Salvati E, Belugou JL, de Galeani B, Krebs B, Ortonne JP, Darcourt G (1987) 5-Methoxypsoralen increases the plasma melatonin levels in humans. J Invest Dermatol 89:152–155

Souetre E, Salvati E, Lacour JC, Belugou JL, Krebs B, Ortonne JP, Darcourt G (1988) Psoralen and the suppression of melatonin secretion by bright light. Photochem Photobiol 47:579–581

Sparks DL (1998) Anatomy of a new paired tract of the pineal gland in humans. Neurosci Lett 248:179–182

Stankov B, Cimino M, Marini P, Lucini V, Fraschini F, Clementi F (1993) Identification and functional significance of nicotinic cholinergic receptors in the rat pineal gland. Neurosci Lett 156:131–134

Stanley M, Brown GM (1988) Melatonin levels are reduced in the pineal glands of suicide victims. Psychopharmacol Bull 24:484–488

Stehle JH (1995) Pineal gene expression: dawn in a dark matter. J Pineal Res 18:179–190

Stehle J, Reuss S, Schröder H, Henschel M, Vollrath L (1988) Magnetic field effects on pineal N-acetyltransferase activity and melatonin content in the gerbil–role of pigmentation and sex. Physiol Behav 44:91–94

Stehle J, Vanecek J, Vollrath L (1989) Effects of melatonin on spontaneous electrical activity of neurons in rat suprachiasmatic nuclei: an in vitro iontophoretic study. J Neural Transm 78:173–177

Stehle JH, Foulkes NS, Molina CA, Simonneaux V, Pevet P, Sassone Corsi P (1993) Adrenergic signals direct rhythmic expression of transcriptional repressor CREM in the pineal gland. Nature 365:314–320

Stewart JW, Halbreich U (1989) Plasma melatonin levels in depressed patients before and after treatment with antidepressant medication. Biol Psychiatry 25:33–38

Strassman RJ, Peake GT, Qualls CR, Lisansky EJ (1987) A model for the study of the acute effects of melatonin in man. J Clin Endocrinol Metab 65:847–852

Sugden D (1983) Psychopharmacological effects of melatonin in mouse and rat. J Pharmacol Exp Ther 227:587–591

Suhner A, Schlagenhauf P, Johnson R, Tschopp A, Steffen R (1998a) Comparative study to determine the optimal melatonin dosage form for the alleviation of jet lag. Chronobiol Int 15:655–666

Suhner A, Schlagenhauf P, Tschopp A, Hauri Bionda R, Friedrich Koch A, Steffen R (1998b) Impact of melatonin on driving performance. J Travel Med 5:7–13

Sun JH, Yaga K, Reiter RJ, Garza M, Manchester LC, Tan DX, Poeggeler B (1993) Reduction in pineal N-acetyltransferase activity and pineal and serum melatonin levels in rats after their exposure to red light at night. Neurosci Lett 149:56–58

Suranyi Cadotte B, Lal S, Nair NP, Lafaille F, Quirion R (1987) Coexistence of central and peripheral benzodiazepine binding sites in the human pineal gland. Life Sci 40:1537–1543

Surrall K, Smith JA, Bird H, Okala B, Othman H, Padwick DJ (1987) Effect of ibuprofen and indomethacin on human plasma melatonin. J Pharm Pharmacol 39:840–843

Takahashi JS (1994) Circadian rhythms. ICER is nicer at night (sir!). Curr Biol 4:165–168

Takahashi K, Deguchi T (1983) Entrainment of the circadian rhythms of blinded infant rats by nursing mothers. Physiol Behav 31:373–378

Tamarkin L, Danforth D, Lichter A, DeMoss E, Cohen M, Chabner B, Lippman M (1982) Decreased nocturnal plasma melatonin peak in patients with estrogen receptor positive breast cancer. Science 216:1003–1005

Tan CH, Khoo JC (1981) Melatonin concentrations in human serum, ventricular and lumbar cerebrospinal fluids as an index of the secretory pathway of the pineal gland. Horm Res 14:224–233

Taugner R, Schiller A, Rix E (1981) Gap junctions between pinealocytes. A freeze-fracture study of the pineal gland in rats. Cell Tissue Res 218:303–314

Taylor RL, Albuquerque ML, Burt DR (1980) Muscarinic receptors in pineal. Life Sci 26:2195–2200

Terman M, Quitkin FM, Terman JS, Stewart JW, McGrath PJ (1987) The timing of phototherapy: effects on clinical response and the melatonin cycle. Psychopharmacol Bull 23:354–357

Terman M, Terman JS, Quitkin FM, Cooper TB, Lo ES, Gorman JM, Stewart JW, McGrath PJ (1988) Response of the melatonin cycle to phototherapy for Seasonal Affective Disorder. Short note. J Neural Transm 72:147–165

Tetsuo M, Polinsky RJ, Markey SP, Kopin IJ (1981) Urinary 6-hydroxymelatonin excretion in patients with orthostatic hypotension. J Clin Endocrinol Metab 53:607–610

Thiele G, Meissl H (1987) Action spectra of the lateral eyes recorded from mammalian pineal glands. Brain Res 424:10–16

Thompson C, Franey C, Arendt J, Checkley SA (1988) A comparison of melatonin secretion in depressed patients and normal subjects. Br J Psychiatry 152:260–265

Thompson C, Stinson D, Smith A (1990) Seasonal affective disorder and season-dependent abnormalities of melatonin suppression by light. Lancet 336:703–706

Tosini G, Menaker M (1996) Circadian rhythms in cultured mammalian retina [see comments]. Science 272:419–421

Tosini G, Menaker M (1998) The clock in the mouse retina: melatonin synthesis and photoreceptor degeneration. Brain Res 789:221–228

Touitou Y, Fevre M, Lagoguey M, Carayon A, Bogdan A, Reinberg A, Beck H, Cesselin F, Touitou C (1981) Age- and mental health-related circadian rhythms of plasma levels of melatonin, prolactin, luteinizing hormone and follicle-stimulating hormone in man. J Endocrinol 91:467–475

Touitou Y, Fevre M, Bogdan A, Reinberg A, De Prins J, Beck H, Touitou C (1984) Patterns of plasma melatonin with ageing and mental condition: stability of nyctohemeral rhythms and differences in seasonal variations. Acta Endocrinol Copenh 106:145–151

Touitou Y, Fevre Montange M, Proust J, Klinger E, Nakache JP (1985) Age- and sex-associated modification of plasma melatonin concentrations in man. Relationship to pathology, malignant or not, and autopsy findings. Acta Endocrinol Copenh 108:135–144

Trinchard Lugan I, Waldhauser F (1989) The short term secretion pattern of human serum melatonin indicates apulsatile hormone release. J Clin Endocrinol Metab 69:663–669

Truong H, Yellon SM (1997) Effect of various acute 60 Hz magnetic field exposures on the nocturnal melatonin rise in the adult Djungarian hamster. J Pineal Res 22:177–183

Turek FW (1996) Melatonin hype hard to swallow. Nature 379:295–296

Tzischinsky O, Lavie P (1994) Melatonin possesses time-dependent hypnotic effects. Sleep 17:638–645

Tzischinsky O, Pal I, Epstein R, Dagan Y, Lavie P (1992) The importance of timing in melatonin administration in a blind man. J Pineal Res 12:105–108

Uchida K, Okamoto N, Ohara K, Morita Y (1996) Daily rhythm of serum melatonin in patients with dementia of the degenerate type. Brain Res 717:154–159

Van Cauter E, Moreno Reyes R, Akseki E, L'Hermite Baleriaux M, Hirschfeld U, Leproult R, Copinschi G (1998) Rapid phase advance of the 24-h melatonin profile in response to afternoon dark exposure. Am J Physiol 275:E48–54

van den Heuvel CJ, Kennaway DJ, Dawson D (1999) Thermoregulatory and soporific effects of very low dose melatonin injection. Am J Physiol 276:E249–254

Vaughan GM, Pelham RW, Pang SF, Loughlin LL, Wilson KM, Sandock KL, Vaughan MK, Koslow SH, Reiter RJ (1976) Nocturnal elevation of plasma melatonin and urinary 5-hydroxyindoleacetic acid in young men: attempts at modification by brief changes in environmental lighting and sleep and by autonomic drugs. J Clin Endocrinol Metab 42:752–764

Vaughan GM, McDonald SD, Jordan RM, Allen JP, Bohmfalk GL, Abou Samra M, Story JL (1978) Melatonin concentration in human blood and cerebrospinal fluid: relationship to stress. J Clin Endocrinol Metab 47:220–223

Vaughan GM, Allen JP, de la Pena A (1979a) Rapid melatonin transients. Waking Sleeping 3:169–173

Vaughan GM, Bell R, de la Pena A (1979b) Nocturnal plasma melatonin in humans: episodic pattern and influence of light. Neurosci Lett 14:81–84

Vaughan GM, McDonald SD, Jordan RM, Allen JP, Bell R, Stevens EA (1979c) Melatonin, pituitary function and stress in humans. Psychoneuroendocrinology 4:351–362

Vaughan GM, Mason AD, Jr., Reiter RJ (1986a) Serum melatonin after a single aqueous subcutaneous injection in Syrian hamsters. Neuroendocrinology 42:124–127

Vaughan MK, Joshi BN, Reiter RJ (1986b) Daily propranolol administration reduces pineal concretion formation in the Mongolian gerbil. Proc Soc Exp Biol Med 182:372–374

Vieira R, Miguez J, Lema M, Aldegunde M (1992) Pineal and plasma melatonin as determined by high-performance liquid chromatography with electrochemical detection. Anal Biochem 205: 300–305

Vivien Roels B, Pevet P, Beck O, Fevre Montange M (1984) Identification of melatonin in the compound eyes of an insect, the locust (*Locusta migratoria*), by radioimmunoassay and gas chromatography-mass spectrometry. Neurosci Lett 49:153–157

Vollrath L (1973) Synaptic ribbons of a mammalian pineal gland: circadian changes. Z Zellforsch Mikrosk Anat 145:171–183

Vollrath L (1981) The Pineal Organ. Springer-Verlag, Berlin Heidelberg New York

Vollrath L, Schröder H (1987) Neuronal properties of mammalian pinealocytes? In: Trentini GP, de Gaetani C, and Pévet P (eds) Fundamentals and clinics in pineal research. Raven Press, New York. pp 13–23

Vollrath L, Semm P, Gammel G (1981) Sleep induction by intranasal application of melatonin. In: Birau N and Schloot W (eds) Melatonin – Current Status and Perspectives. Pergamon Press, Oxford and New York. pp 129–133

Vollrath L, Seidel A, Huesgen A, Manz B, Pollow K, Leiderer P (1989) One millisecond of light suffices to suppress nighttime pineal melatonin synthesis in rats. Neurosci Lett 98:297–298

Vollrath L, Spessert R, Kratzsch T, Keiner M, Hollmann H (1997) No short-term effects of high-frequency electromagnetic fields on the mammalian pineal gland. Bioelectromagnetics 18: 376–387

von Treuer K, Norman TR, Armstrong SM (1996) Overnight human plasma melatonin, cortisol, prolactin, TSH, under conditions of normal sleep, sleep deprivation, and sleep recovery. J Pineal Res 20:7–14

Vondrasova D, Hajek I, Illnerova H (1997) Exposure to long summer days affects the human melatonin and cortisol rhythms. Brain Res 759:166–170

Vondrasova Jelinkova D, Hajek I, Illnerova H (1999) Adjustment of the human melatonin and cortisol rhythms to shortening of the natural summer photoperiod. Brain Res 816:249–253

Wagner HJ (1997) Presynaptic bodies ("ribbons"): from ultrastructural observations to molecular perspectives. Cell Tissue Res 287:435–446

Wajs E, Kutoh E, Gupta D (1995) Melatonin affects pro-opiomelanocortin gene expression in the immune organs of the rat. Eur J Endocrinol 133:754–760

Waldhauser F, Dietzel M (1985) Daily and annual rhythms in human melatonin secretion: role in puberty control. Ann N Y Acad Sci 453:205–214

Waldhauser F, Steger H (1986) Changes in melatonin secretion with age and pubescence. J Neural Transm Suppl 21:183–197

Waldhauser F, Wurtman RJ (1983) The secretion and actions of melatonin. In: Litwak G (ed) Biochemical actions of hormones, volume X. Academic Press, New York. pp 187–225

Waldhauser F, Waldhauser M, Lieberman HR, Deng MH, Lynch HJ, Wurtman RJ (1984) Bioavailability of oral melatonin in humans. Neuroendocrinology 39:307–313

Waldhauser F, Weiszenbacher G, Tatzer E, Gisinger B, Waldhauser M, Schemper M, Frisch H (1988) Alterations in nocturnal serum melatonin levels in humans with growth and aging. J Clin Endocrinol Metab 66:648–652

Wallace RB, Altman J, Das GD (1969) An autoradiographic and morphological investigation of the postnatal development of the pineal body. Am J Anat 126:175–183

Weaver DR (1997) Reproductive safety of melatonin: a "wonder drug" to wonder about. J Biol Rhythms 12:682–689

Weaver DR, Reppert SM (1986) Maternal melatonin communicates daylength to the fetus in Djungarian hamsters. Endocrinology 119:2861–2863

Weaver DR, Reppert SM (1996) The Mel1a melatonin receptor gene is expressed in human suprachiasmatic nuclei. Neuroreport 8:109–112

Weaver DR, Stehle JH, Stopa EG, Reppert SM (1993) Melatonin receptors in human hypothalamus and pituitary: implications for circadian and reproductive responses to melatonin. J Clin Endocrinol Metab 76:295–301

Webb SM, Champney TH, Lewinski AK, Reiter RJ (1985) Photoreceptor damage and eye pigmentation: influence on the sensitivity of rat pineal N-acetyltransferase activity and melatonin levels to light at night. Neuroendocrinology 40:205–209

Webley GE, Lenton EA (1987) The temporal relationship between melatonin and prolactin in women. Fertil Steril 48:218–222

Webley GE, Bohle A, Leidenberger FA (1988) Positive relationship between the nocturnal concentrations of melatonin and prolactin, and a stimulation of prolactin after melatonin administration in young men. J Pineal Res 5:19–33

Wehr TA (1991) The durations of human melatonin secretion and sleep respond to changes in daylength (photoperiod). J Clin Endocrinol Metab 73:1276–1280

Wehr TA (1992) Seasonal vulnerability to depression. Implications for etiology and treatment. Encephale 18 Spec No 4:479–483

Wehr TA, Jacobsen FM, Sack DA, Arendt J, Tamarkin L, Rosenthal NE (1986) Phototherapy of seasonal affective disorder. Time of day and suppression of melatonin are not critical for antidepressant effects. Arch Gen Psychiatry 43:870–875

Weinberg U, D'Eletto RD, Weitzman ED, Erlich S, Hollander CS (1979) Circulating melatonin in man: episodic secretion throughout the light-dark cycle. J Clin Endocrinol Metab 48:114–118

Weinberg U, Weitzman ED, Fukushima DK, Cancel GF, Rosenfeld RS (1980) Melatonin does not suppress the pituitary luteinizing hormone response to luteinizing hormone-releasing hormone in men. J Clin Endocrinol Metab 51:161–162

Weinberg U, Weitzman ED, Horowitz ZD, Burg AC (1981) Lack of an effect of melatonin on the basal and L-dopa stimulated growth hormone secretion in men. J Neural Transm 52:117–121

Weitzman ED, Weinberg U, D'Eletto R, Lynch H, Wurtman RJ, Czeisler C, Erlich S (1978) Studies of the 24 hour rhythm of melatonin in man. J Neural Transm Suppl 13:325–337

Welker HA, Semm P, Willig RP, Commentz JC, Wiltschko W, Vollrath L (1983) Effects of an artificial magnetic field on serotonin N-acetyltransferase activity and melatonin content of the rat pineal gland. Exp Brain Res 50:426–432

Welsh MG, Hansen JT, Reiter RJ (1979) The pineal gland of the gerbil, Meriones unguiculatus. III. Morphometric analysis and fluorescence histochemistry in the intact and sympathetically denervated pineal gland. Cell Tissue Res 204:111–125

Wessler I, Reinheimer T, Bittinger F, Kirkpatrick CJ, Schenda J, Vollrath L (1997) Day-night rhythm of acetylcholine in the rat pineal gland. Neurosci Lett 224:173–176

Wetterberg L (1978) Melatonin in humans – physiological and clinical studies. J Neural Transm Suppl 13:289–310

Wetterberg L (1985) Melatonin and affective disorders. Ciba Found Symp 117:253–265

Wetterberg L, Hayes DK, Halberg F (1987) Circadian rhythm of melatonin in the brain of the face fly, *Musca autumnalis* De Geer. Chronobiologia 14:377–381

Wilson BW (1978) The application of mass spectrometry to the study of the pineal gland. J Neural Transm Suppl 13:279–288

Wilson BW, Snedden W, Silman RE, Smith I, Mullen P (1977) A gas chromatography-mass spectrometry method for the quantitative analysis of melatonin in plasma and cerebrospinal fluid. Anal Biochem 81:283–291

Wilson BW, Anderson LE, Hilton DI, Phillips RD (1981) Chronic exposure to 60-Hz electric fields: effects on pineal function in the rat. Bioelectromagnetics 2:371–380

Wilson BW, Chess EK, Anderson LE (1986) 60-Hz electric-field effects on pineal melatonin rhythms: time course for onset and recovery. Bioelectromagnetics 7:239–242

Wilson BW, Wright CW, Morris JE, Buschbom RL, Brown DP, Miller DL, Sommers Flannigan R, Anderson LE (1990) Evidence for an effect of ELF electromagnetic fields on human pineal gland function. J Pineal Res 9:259–269

Winkler P, Helmke K (1987) Age-related incidence of pineal gland calcification in children: a roentgenological study of 1,044 skull films and a review of the literature. J Pineal Res 4:247–252

Winton F, Corn T, Huson LW, Franey C, Arendt J, Checkley SA (1989) Effects of light treatment upon mood and melatonin in patients with seasonal affective disorder. Psychol Med 19:585–590

Wirz Justice A (1996) Melatonin: ein neues Wundermittel? Schweiz Rundsch Med Prax 85:1332–1336

Wirz Justice A, Graw P, Krauchi K, Gisin B, Arendt J, Aldhous M, Poldinger W (1990) Morning or night-time melatonin is ineffective in seasonal affective disorder. J Psychiatr Res 24:129–137

Wirz Justice A, Graw P, Krauchi K, Gisin B, Jochum A, Arendt J, Fisch HU, Buddeberg C, Poldinger W (1993) Light therapy in seasonal affective disorder is independent of time of day or circadian phase. Arch Gen Psychiatry 50:929–937

Withyachumnarnkul B, Nonaka KO, Santana C, Attia AM, Reiter RJ (1990) Interferon-gamma modulates melatonin production in rat pineal glands in organ culture. J Interferon Res 10: 403–411

Wood AW, Armstrong SM, Sait ML, Devine L, Martin MJ (1998) Changes in human plasma melatonin profiles in response to 50 Hz magnetic field exposure. J Pineal Res 25:116–127

Wright KP, Jr., Badia P, Myers BL, Plenzler SC, Hakel M (1997) Caffeine and light effects on nighttime melatonin and temperature levels in sleep-deprived humans. Brain Res 747:78–84

Wurtman RJ, Anton Tay F (1969) The mammalian pineal as a neuroendocrine transducer. Recent Prog Horm Res 25:493–522

Wurtman RJ, Axelrod J (1965) The pineal gland. Scientific American 213:50–60

Wurtman RJ, Axelrod J, Kelly DE (1968) The Pineal. Academic Press, New York and London

Yamada H, Yamaguchi A, Moriyama Y (1997) L-aspartate-evoked inhibition of melatonin production in rat pineal glands. Neurosci Lett 228:103–106

Yamada H, Ogura A, Koizumi S, Yamaguchi A, Moriyama Y (1998a) Acetylcholine triggers L-glutamate exocytosis via nicotinic receptors and inhibits melatonin synthesis in rat pinealocytes. J Neurosci 18:4946–4952

Yamada H, Yatsushiro S, Ishio S, Hayashi M, Nishi T, Yamamoto A, Futai M, Yamaguchi A, Moriyama Y (1998b) Metabotropic glutamate receptors negatively regulate melatonin synthesis in rat pinealocytes. J Neurosci 18:2056–2062

Yatsushiro S, Yamada H, Kozaki S, Kumon H, Michibata H, Yamamoto A, Moriyama Y (1997) L-aspartate but not the D form is secreted through microvesicle-mediated exocytosis and is sequestered through Na+-dependent transporter in rat pinealocytes. J Neurochem 69:340–347

Yellon SM (1994) Acute 60 Hz magnetic field exposure effects on the melatonin rhythm in the pineal gland and circulation of the adult Djungarian hamster. J Pineal Res 16:136–144

Yellon SM, Longo LD (1987) Melatonin rhythms in fetal and maternal circulation during pregnancy in sheep. Am J Physiol 252:E799–802

Yellon SM, Longo LD (1988) Effect of maternal pinealectomy and reverse photoperiod on the circadian melatonin rhythm in the sheep and fetus during the last trimester of pregnancy. Biol Reprod 39:1093–1099

Ying SW, Niles LP, Crocker C (1993) Human malignant melanoma cells express high-affinity receptors for melatonin: antiproliferative effects of melatonin and 6-chloromelatonin. Eur J Pharmacol 246:89–96

Young IM, Leone RM, Francis P, Stovell P, Silman RE (1985) Melatonin is metabolized to N-acetyl serotonin and 6-hydroxymelatonin in man. J Clin Endocrinol Metab 60:114–119

Young SN, Gauthier S, Kiely ME, Lal S, Brown GM (1984) Effect of oral melatonin administration on melatonin, 5-hydroxyindoleacetic acid, indoleacetic acid, and cyclic nucleotides in human cerebrospinal fluid. Neuroendocrinology 39:87–92

Zawilska JB, Jarmak A, Woldan Tambor A, Nowak JZ (1995) Light-induced suppression of nocturnal serotonin N-acetyltransferase activity in chick pineal gland and retina: a wavelength comparison. J Pineal Res 19:87–92

Zawilska JB, Rosiak J, Nowak JZ (1998) Effects of near-ultraviolet light on the nocturnal serotonin N-acetyltransferase activity of rat pineal gland. Neurosci Lett 243:49–52

Zecca L, Mantegazza C, Margonato V, Cerretelli P, Caniatti M, Piva F, Dondi D, Hagino N (1998) Biological effects of prolonged exposure to ELF electromagnetic fields in rats: III. 50 Hz electromagnetic fields. Bioelectromagnetics 19:57–66

Zemdegs IZ, McMillen IC, Walker DW, Thorburn GD, Nowak R (1988) Diurnal rhythms in plasma melatonin concentrations in the fetal sheep and pregnant ewe during late gestation. Endocrinology 123:284–289

Zhdanova IV, Wurtman RJ, Lynch HJ, Ives JR, Dollins AB, Morabito C, Matheson JK, Schomer DL (1995) Sleep-inducing effects of low doses of melatonin ingested in the evening. Clin Pharmacol Ther 57:552–558

Zhdanova IV, Wurtman RJ, Morabito C, Piotrovska VR, Lynch HJ (1996) Effects of low oral doses of melatonin, given 2–4 hours before habitual bedtime, on sleep in normal young humans. Sleep 19:423–431

Zhdanova IV, Lynch HJ, Wurtman RJ (1997) Melatonin: a sleep-promoting hormone. Sleep 20: 899–907

Zisapel N (1999) The use of melatonin for the treatment of insomnia. Biol Signals Recept 8:84–89

3 The Role of Melatonin in the Neuroendocrine System: Multiplicity of Sites and Mechanisms of Action

Daniel P. Cardinali, Rodolfo A. Cutrera, Luis I. Brusco, Ana I. Esquifino

Abstract

Melatonin is synthesized and secreted during the dark period of the light-dark cycle, a rhythm generated by a circadian clock located within the suprachiasmatic nuclei (SCN) of the hypothalamus and entrained to a 24-h period by the light–dark cycle. Direct effects of melatonin on the SCN, perhaps affecting γ-aminobutyric acid (GABA)-containing neurons, can explain melatonin's effects on the circadian system. In addition, the periodic secretion of melatonin is a potential circadian signal to several non-neural cells that can "read" the message. Duration of melatonin nocturnal secretion is proportional to the length of the night and has experimentally been demonstrated to be an integral part of photoperiodic mechanisms. The intracellular sites and mechanisms of action of melatonin to affect circadian and photoperiodic responses are far from being elucidated. Although action through specific membrane receptor sites has been identified as important, interactions with specific intracellular proteins like calmodulin or tubulin, or with nuclear receptor sites should also be considered in view of the lipophilic nature of the melatonin molecule. Melatonin is also a potent free radical scavenger and pharmacologic effects of melatonin could be explained through direct scavenging of free radicals, or through induction of enzymes that improve total antioxidative defense capacity. Melatonin has a significant immunomodulatory activity, playing a major role in the organism's defense. Melatonin seems to act as an arm of the circadian clock, giving a time-related signal to a number of restorative body functions.

3.1 Melatonin Is a Universal Time-Related Signal

Research in the last 20 years has clearly defined two central, and presumably distinct, effects of melatonin, namely, photoperiodic time measurement to induce seasonality, and circadian entrainment of a number of 24-h physiological rhythms (Pévet and Pitrosky 1997). Melatonin plays an established role in controlling seasonal reproduction in a number of mammalian and non-mammalian species (Reiter 1991). It has also a major influence on the circadian organization of vertebrates including human beings (Dawson and Armstrong 1996), as well as exerting diverse behavioral effects (Golombek et al. 1996).

A feasible nature of the effect of the melatonin signal on the circadian system is to influence the entrainment pathways to the major circadian oscillator, the suprachiasmatic nuclei (SCN) of the hypothalamus (Arendt 1995). In the fetus, strong evidence exists for a physiological role of the maternal melatonin signal as an internal synchronizer.

It is also clear that melatonin has a number of functions that far transcend its actions at the SCN level. In this respect, melatonin is a ubiquitous signal potentially available to every cell in the organism that can "read" the message. It is also a free radical scavenger and has indirect antioxidant effects (Reiter et al. 1997), as well as a strong immuno-modulatory activity (Liebmann et al. 1997; Cardinali et al. 1997a).

How melatonin acts at the cellular level to modify expression of the cellular clock remains a mystery. Although it is widely held that melatonin acts on specific receptors in cell membranes, an interaction of melatonin with nuclear receptor sites and with intracellular proteins, like calmodulin or tubulin-associated proteins, as well as the direct antioxidant effects of melatonin, can explain many general functions of the hormone (Cardinali et al. 1997c).

This chapter deals with a number of aspects of receptor- and non-receptor-mediated effects of melatonin. Its aim is to set a wider scope on the subject of melatonin's mechanism of action than what is currently considered.

3.2 Sites of Melatonin Action Are Multiple

Photoperiodic time measurement in adult and fetal mammals is critically dependent upon the melatonin signal (Arendt 1995). The mechanisms sensitive to melatonin, which mediate the clock message sent by the pineal gland in the form of a melatonin cycle, may reside in the brain, in the biological clock itself, namely, the SCN.

There are also data indicating that the melatonin receptors located in different organs may also convey circadian-meaningful information to every cell in the organism, i.e., melatonin can play the role of an internal synchronizer (Dawson and Armstrong 1996). Less is known about the melatonin-sensitive mechanisms that mediate seasonal changes in reproductive physiology, although several data suggest that they may be located in the hypophyseal pars tuberalis (Masson-Pévet et al. 1994).

The existence of saturable melatonin binding sites in the brain and rapid exchange of melatonin between cerebrospinal fluid (CSF) and vascular compartments was suggested in the early 1970s by kinetics studies employing [^{3}H]melatonin (for references, see Cardinali 1981). Less than 1% of injected melatonin remained after 20 min of injection, indicating the rapid exchange between CSF and vascular compartments.

The first experiments on brain melatonin receptor sites were carried out in the late 1970s by employing [^{3}H]melatonin as a ligand and indicated the existence of melatonin acceptor sites in bovine hypothalamic, cerebral cortex, and cerebellar membranes (Cardinali 1981). The introduction of the 2-[^{125}I]iodomelatonin analog allowed detection of melatonin binding in several brain areas, the choroid plexus, and in some brain arteries as well as in peripheral organs, like the harderian glands, primary and secondary lymphoid organs, adrenal glands, heart and lungs, gastrointestinal tract, mammary glands, kidney, and male reproductive organs. Indeed, the picture that has emerged indicates that almost every tissue and organ in the body could "decode" information carried by the melatonin signal (Masson-Pévet et al. 1994; Chow et al. 1996; Pang et al. 1996).

A first classification of putative melatonin receptor sites into ML1 and ML2 was based on kinetic and pharmacological differences of 2-[^{125}I]iodomelatonin binding. A nomenclature of melatonin receptors has recently been proposed by the International Union of Pharmacology (IUPHAR) (Dubocovich et al. 1998). Of the two subtypes cloned in mammals, mt_1 and mt_2 (formerly called Mel_{1a} and Mel_{1b}), the mt_1 receptor, whose messenger RNA is expressed in the SCN and pars tuberalis, was supposed to mediate the circadian and reproductive effects of the hormone, respectively (Reppert 1997); however, conclusive evidence on this is yet to be provided.

Considering the lipophilic nature of melatonin, it is also probable that intracellular sites of action exist. Although the idea that melatonin is a natural ligand for the orphan nuclear hormone receptor superfamily RZR/ROR is now being contested, it should not be taken to negate the possibility of a nuclear site of action of melatonin (Cardinali et al. 1997c). Nuclear melatonin receptors may mediate melatonin immunomodulatory phenomena, as indicated by the finding that thiazolidine diones, specific ligands for RZR/ROR receptors, exhibited a potent antiarthritic activity in the rat adjuvant arthritis model (Missbach et al. 1996). This could explain the improvement of circadian rhythmicity of immune cell proliferation in rats injected with Freund's adjuvant and given melatonin (Cardinali et al. 1996a, b).

3.3 Melatonin Promotes GABAergic Responses in SCN and the Brain

Among the various neurotransmitter systems that could mediate melatonin action in the circadian system, γ-aminobutyric acid (GABA) has emerged as important. GABA is implicated in the regulation of a variety of behavioral functions, including biological rhythms. Indeed, GABA is considered as the principal neurotransmitter of the mammalian circadian system, being present in every cell of the SCN (for references, see Cardinali and Golombek 1998).

In principle, to demonstrate that a neurotransmitter system is involved in the mediation of a melatonin effect, two requirements should be fulfilled: (a) the neurotransmitter system should show dynamic changes as a consequence of melatonin injection; (b) functional obliteration of the neurotransmitter system should modify the melatonin effect. A demonstration of this sort implies either that the neurotransmitter system is an integral part of the sequence of events triggered by melatonin action, or that it plays a permissive role in the examined effect.

It is interesting that a number of monoamine pathways within the brain are not important for melatonin entrainment of circadian rhythmicity. For example, the intraventricular injection of 6-hydroxydopamine and 5,7-dihydroxytryptamine, which deplete catecholamines and indolamines, failed to alter melatonin-induced entrainment (Cassone et al. 1986).

We previously reported daily rhythms of the GABA turnover rate in the preoptic-medial basal hypothalamus (containing the SCN), cerebral cortex, and pineal gland of Syrian hamsters (Kanterewicz et al. 1993). In the four tissues examined, nocturnal acrophases were apparent at about 4 h after lights off. In agreement with this, earlier studies indicated a diurnal rhythm in the number of rat brain high-affinity $GABA_A$ and benzodiazepine (BZP) binding sites, with nocturnal and diurnal maxima, respectively (Rosenstein and Cardinali 1990). Not only in the CNS was the BZP binding rhythm circadian, but it was also abolished by ablation of the circadian clock, as well as by removal of the pineal gland.

Rhythmic fluctuations in the association constant of low-affinity GABA binding sites were also reported, with maximal affinity in the middle of the night (Kanterewicz et al. 1995). Daily changes in affinity correlated with variations in a major marker of GABA postsynaptic activity, namely, the uptake of Cl^- by synaptoneurosomes, which showed a nocturnal peak. Further supporting the concept that the brain GABAergic system displays a rhythmic pattern with a nocturnal maximum, a mid-late night peak of [^{3}H]GABA release from preoptic-medial basal hypothalamic explants (containing the SCN) was observed in hamsters (Yannielli et al. 1996). Behavioral correlates of GABA neurochemical rhythms, e. g., GABA-BZP-related activities like anxiety, as assessed in hamsters in an elevated plus-maze, showed distinct maxima (i. e., in anxiolysis and locomotor activity) during the night (Yannielli et al. 1996). This rhythm persisted under constant conditions (Kanterewicz et al. 1995), indicating its circadian nature. Therefore, a number of neurochemical changes in brain GABA activity and behavioral phenomena linked to $GABA_A$ receptors strongly correlate with exhibiting maxima during the night.

Pinealectomy disrupted diurnal rhythmicity of $GABA_A$ and BZP high-affinity binding in rat brain, an effect counteracted by melatonin replacement. A single melatonin injection (25–100 µg/kg) augmented significantly GABA turnover and synthesis in the hypothalamus, cerebral cortex, or cerebellum of rats (Rosenstein et al. 1990). Nanomolar concentrations of melatonin augmented [^{36}Cl$^-$]influx in vitro by potentiating a GABA-induced increase of chloride ion uptake, evening activity of melatonin being detectable at concentrations an order of magnitude greater than those effective during daylight. Therefore, melatonin is an effective therapeutic agent to promote brain GABAergic activity. In experiments on the spontaneous firing activity of single neurons in parietal cortex of anesthetized rabbits, Stankov et al. (1992) showed that melatonin and 2-iodomelatonin had GABA-like effects and were able alone, in nanomolar concentrations, to significantly slow the neuronal firing activity, as well as to potentiate the effect of GABA.

A significant correlation between brain GABAergic activity and melatonin effects on behavioral parameters like hot-plate anesthesia, locomotor behavior, anxiolysis, or 3-mercaptopropionic acid-induced seizures was found (Golombek et al. 1996). These effects of melatonin, as well as melatonin-induced reentrainment of circadian rhythms in hamsters subjected to a jet-lag paradigm, were inhibited by co-injection with the central-type BZP antagonist flumazenil, indicating the existence of significant interactions between melatonin and BZP-sensitive mechanisms in rodents. Indeed, melatonin enhanced the anxiolytic actions of diazepam in several behavioral tests in mice in accordance with results that link melatonin with GABAergic activity (Guardiola-Lemaitre et al. 1992). However, a clinical study designed to investigate whether flumazenil blocks the hypnotic and hypothermic effects of melatonin in humans indicated occurrence of melatonin effects independently of central BZP receptor blockade (Nave et al. 1996).

Central-type BZP receptors appear to mediate the antidopaminergic effect of melatonin (as well as clonazepam) in 6-hydroxydopamine-lesioned rats (Tenn and Niles 1995). In addition, Tenn and Niles (1997) reported that the antidopaminergic action of the melatonin analog S-20098 is mediated by BZP/$GABA_A$ receptors in the striatum. Direct inhibitory effects of BZP on melatonin binding were reported (Anis et al. 1992; Atsmon et al. 1996).

Clinical correlates of these experimental observations deserve to be mentioned. Melatonin has been reported as a useful anticonvulsant treatment in a number of situations (Champney et al. 1996; Molina Carballo et al. 1997). Melatonin enhanced EEG power density in non-REM sleep and reduced activity in the 15.25–16.5 Hz band, these changes in the EEG being to some extent similar to those induced by BZP hypnotics (Dijk et al. 1995). Improvement of sleep quality by controlled-release of melatonin in BZP-treated elderly insomniacs was shown by Garfinkel et al. (1997). Melatonin treatment significantly increased sleep efficiency and total sleep time and decreased wake after sleep onset, sleep latency, number of awakenings, and fragmental index, as compared to placebo. Concomitant treatment with melatonin and BZP enabled patients to completely cease any BZP use, with an improvement in sleep quality and no side effects (Dagan et al. 1997). This suggested that some of the individuals suffering from insomnia and addicted to BZP may successfully undergo withdrawal from these drugs and improve their sleep by means of treatment with melatonin. Indeed, we recently demonstrated that 8 out of 13 patients with primary insomnia taking melatonin reduced or abandoned concomitant BZP use (Fainstein et al. 1997).

3.4 Melatonin Acts on Cells Through cAMP- and Non-cAMP-Signal Transduction Pathways

Among presumptive second messengers for melatonin action, the cAMP generating system has received paramount attention. The main signal transduction pathway of high-affinity mt_1 receptors in both neuronal and non-neuronal tissues is the inhibition of cAMP formation through a pertussis toxin-sensitive inhibitory G protein (Reppert 1997).

Coupling of the high-affinity melatonin receptors to other signaling pathways has also been reported. Melatonin augmented cGMP levels in rat MBH (Vacas et al. 1981), rabbit aorta (Satake et al. 1991), hamster retina (Faillace et al. 1996), murine mammary glands (Cardinali et al. 1992), and human lymphocytes (López González et al. 1992). Early studies also indicated that melatonin decreased depolarization-induced $^{45}Ca^{2+}$ uptake in brain cellular and subcellular preparations, suggesting a modulatory activity of the hormone on voltage-controlled calcium channels (Zisapel 1988; Cardinali et al. 1991). Melatonin inhibited gonadotropin releasing hormone-induced $(Ca^{2+})_i$ increase and luteinizing hormone (LH) release in neonatal pituitary cells; the effect of melatonin on LH was markedly reduced in a low Ca^{2+} medium, indicating that melatonin may act by inhibiting Ca^{2+} influx (Vanecek and Klein 1995).

Nanomolar concentrations of melatonin decreased rat MBH prostaglandin (PG) E_2 release (Cardinali 1981), as well as arachidonic acid (AA) conversion to cyclooxygenase derivatives (Franchi et al. 1987). PGE_2 levels and release were also prevented by melatonin injections.

Eicosanoid involvement in melatonin action was also indicated by studies on human platelets (for references, see Cardinali et al. 1993). Melatonin inhibits platelet aggregation and eicosanoid synthesis, exhibiting a diurnal variation with a greater activity of melatonin in the evening in healthy men. Melatonin binding sites are detectable in human platelet membranes. Studies on labeled AA metabolism suggested that melatonin acts upstream in the AA metabolic pathway, presumably at the level of cyclooxygenase or higher. Such a direct effect of melatonin on cyclooxygenase was also observed in seminal vesicle microsomes (Cardinali et al. 1993).

Therefore, it appears that melatonin action in tissues is presumably much more complicated than a single inhibitory effect on cAMP synthesis. Even in the case of the single melatonin acceptor site best known, the mt_1 receptor, studies on the cloned receptor indicated that it is coupled to parallel signal transduction pathways involving inhibition of adenyl cyclase and potentiation of phospholipase activation in target cells (Godson and Reppert 1997). Direct effects of melatonin on calmodulin and other intracellular proteins, nuclear receptor activity for melatonin, and the free radical scavenging properties of melatonin should also be considered.

3.5 Melatonin Interacts with Cytoskeletal Proteins

In the early 1970s, a hypothesis was put forth that melatonin affects microtubule- or microfilament-dependent processes in living organisms. This hypothesis was based on observations indicating that melatonin inhibited colchicine effects on pigment granules in amphibian melanocytes, competed with colchicine to inhibit regeneration in Stentor coeruleus, and caused mitotic arrest in onion root tips by acting on the mitotic spindle (for references, see Cardinali 1981). To explain all these effects, achieved in vitro at pharmacological concentrations of melatonin, binding of melatonin to tubulin at the colchicine binding sites was postulated.

Although such a binding could not be eventually demonstrated, a number of observations supported the view that melatonin affects contractile protein-dependent processes in cells. Melatonin locally applied to retinal cells or to sciatic nerve fibers impaired fast axonal transport of proteins, a microtubule-dependent process (Cardinali 1981). Likewise, electron microscopic studies of MBH of rats treated with melatonin, indicated the existence of ultrastructural changes in the median eminence compatible with modified axoplasmic transport.

MDCK and neuroblastoma N1 E-115 cells cultured with physiological concentrations of melatonin showed an increase in the number of elongations and in the incidence of contacts with neighboring cells, as well as in neurite processes (Benitez-King et al. 1990). Actin antibody stain showed the appearance of thicker fluorescent fibers beneath the cell membrane and over the nucleus in the melatonin-treated cells. In addition, an increase of microtubules was found in neuroblastoma cells cultured with nanomolar concentrations of melatonin (Melendez et al. 1996). This effect was due to an increase in the polymerization state of tubulin, without significant changes in the levels of β-tubulin or its mRNA. These findings supported the idea that the tubulin polymerization process is one of the targets of melatonin action.

Anton-Tay and co-workers (1998) demonstrated changes in calmodulin levels in MDCK and N1 E-115 cells cultured with nanomolar concentrations of melatonin, inhibition of calmodulin-dependent phosphodiesterase, and specific binding of [^{3}H]melatonin to calmodulin. Melatonin inhibited the activity of Ca^{2+}/calmodulin-dependent protein kinase II (CaM-kinase II), one of the principal calmodulin-dependent enzymes in brain, as well as inhibiting enzyme autophosphorylation (Benitez-King et al. 1996). Thus, melatonin may act as a universal calmodulin antagonist in cells.

As far as other cytoskeletal components are concerned, in vivo melatonin injections inhibited the expression of the mRNA for actin in the hypothalamus (Iovanna et al.

1990a). The synthesis of actin exhibited, in the rat hypothalamus, a diurnal variation with a maximum during morning hours (Iovanna et al. 1990b). Melatonin treatment induced a significant depression in the incorporation of [^{35}S]methionine into 43 kDa actin. Hypothalamic actin mRNA levels decreased after the injection of 100 mg/kg melatonin for 10 days. At this dosage, melatonin did not modify hypothalamic somatostatin or H-*ras* mRNA concentration (Iovanna et al. 1990a).

Collectively, the results suggest that cytoskeletal components like tubulin and actin can be significantly affected by melatonin. Melatonin may act as a calmodulin antagonist and its effects in vivo and in vitro on tubulin polymerization and cytoskeletal organization in cells can be related to changes in Ca^{2+}-calmodulin activity. At higher, pharmacological, concentrations a non-specific inhibitory effect on tubulin can occur. Cellular functions may be rhythmically regulated by melatonin modulation of calmodulin-dependent protein phosphorylation and gene expression (e.g., actin mRNA).

3.6 Melatonin Exerts Antioxidant Effects

The existence of universal melatonin functions in the body is supported by the ease with which melatonin crosses physiological barriers like the blood–brain barrier, or enters cells and subcellular compartments. One of these universal functions can be its antioxidative activity.

It is still an open question whether the antioxidant effect of melatonin is only pharmacological or arises at physiological concentrations as well. Relevant to this, an important question in melatonin physiology, not yet satisfactorily answered, concerns the actual concentration of melatonin reaching the target sites. It must be noted that the levels of a lipophilic substance like melatonin entering cell membranes under physiological or pharmacological conditions can be considerably higher than the circulating hormone concentration. For example, in early studies measuring radioimmunoassayable hypothalamic melatonin levels in rats, values of about 1–4 pg/mg hypothalamus were reported (Pang and Brown 1983; Catala et al. 1987). By employing high-pressure liquid chromatography, rat MBH melatonin levels were about 10 pg/mg hypothalamus at night and about 2 pg/mg hypothalamus during daylight (Cardinali et al. 1991). These melatonin levels are well within a high nanomolar range if one assumes a homogeneous distribution of the compound, and possibly in a substantially greater concentration if a preferential distribution of melatonin in the membrane's lipid moiety occurs. Thus, melatonin could be present even at micromolar concentrations in the close proximity to cell membrane lipids, major targets for free radicals.

In recent years, several reports have demonstrated that melatonin is an efficient free radical scavenger and a general antioxidant at micromolar concentrations (for references, see Reiter et al. 1997). In vitro, melatonin scavenges the toxic hydroxyl radical, on equimolar basis being significantly more efficient than mannitol or glutathione. Melatonin also scavenges the peroxyl radical generated during lipid peroxidation with an activity similar to vitamin E in several but not all models (Longoni et al. 1998). Besides its direct antioxidant activity, melatonin was reported to alter the activities and to increase the levels of mRNA of enzymes that improve the total antioxidative defense capacity, e.g., glutathione peroxidase, copper-zinc superoxide dismutase, or manganese superoxide dismutase (Kotler et al. 1998).

In several animal models, pharmacological concentrations of melatonin prevented oxidative damage (Reiter et al. 1997). These pathological situations include nuclear DNA damage caused by the injection of the chemical carcinogen safrole, damage to DNA in human lymphocytes due to ionizing radiation, and cataracts in glutathione-depleted newborn rats. Other sequela of oxidative damage like ischemia reperfusion injury (Bertuglia et al. 1996) and those that follow lipopolysaccharide (LPS) administration, e.g., hepatotoxicity (Sewerynek et al. 1995), DNA damage (Sewerynek et al. 1996), and death (Maestroni 1996), are prevented or greatly reduced by melatonin injection. Melatonin treatment curtails seizures induced by cyanide- and L-cysteine, paraquat-induced damage of lung and liver, neuronal apoptosis, mortality following septic and hemorrhagic shock, and UV-induced erythema in humans (for references, see Reiter et al. 1997).

Melatonin also blocks the acute gastric lesions induced by stress, ethanol, ischemia, or aspirin (Brzozowski et al. 1997) and the gastric ischemia reperfusion injury in rats (De la Lastra et al. 1997). It protects primary cultures of rat cortical neurons from NMDA excitotoxicity and hypoxia/reoxygenation (Cazevieille et al. 1997) and maintains glutathione homeostasis in kainic acid-exposed rat brain tissue (Floreani et al. 1997). Acutely administered melatonin reduces oxidative damage in lung and brain induced by hyperbaric oxygen (Pablos et al. 1997), and in vitro prevents oxidative damage of adult rat hepatocytes in culture (Kojima et al. (1997).

Melatonin prevented striatal and hippocampal lesions induced by MPTP administration (Acuña-Castroviejo et al. 1996). In accordance with this, melatonin rescued dopamine neurons from cell death in several tissue culture models of oxidative stress (Iacovitti et al. 1997). Brain oxidative stress, as well as hyperlipidemic nephropathy, induced by Adriamycin was effectively inhibited by melatonin (Montilla et al. 1997a, b).

It is interesting to note that for some of these effects, removal of the major endogenous source of melatonin by pinealectomy augments the damage produced, thus indicating that melatonin activity can be seen even at physiological (i.e., low nanomolar) concentrations (Reiter et al. 1997).

Melatonin not only neutralizes oxygen-derived free radicals but can also scavenge species of other types such as peroxynitrite, a toxic oxidant formed from the reaction of superoxide and nitric oxide under conditions of inflammation and oxidant stress (Gilad et al. 1997). Another important oxidative damage prevented by melatonin is that arising from carbon-centered free radicals, like those presumably involved in Alzheimer's disease (AD). Indeed, in recent years the hypothesis of a possible therapeutic relevant effect of melatonin in AD has been entertained. Support for this hypothesis derived from in vitro data indicating that melatonin protects neurons against β-amyloid toxicity and inhibits amyloid formation (Pappolla et al. 1998), prevents β-amyloid-induced lipid peroxidation (Daniels et al. 1998), alters the metabolism of the β-amyloid precursor protein (Song and Lahiri 1997), and prevents the oxidative damage by β-amyloid of mitochondrial DNA (Bozner et al. 1997).

Recently, we reported that short-term treatment of dementia patients having sleep disorders with melatonin (3 mg p.o.) improved significantly "sundowning", i.e., the episodes of agitated behavior that were more severe at night (Fainstein et al. 1997). Moreover, in a case report of two monozygotic twins suffering from AD of 8 years of evolution, the disease was halted in the melatonin-treated subject (6 mg p.o. for 36 months), as indicated by a stable, milder impairment of mnemic function and a substantial improvement of sleep quality and reduction of sundowning (Brusco et al.

1998 a). Indeed, the probability of an absent melatonin rhythm is higher in dementia patients as compared with subjects without dementia (Uchida et al. 1996). In a recent open pilot study (Brusco et al. 1998 b) on the efficacy of melatonin in treatment of sleep and cognitive disorders in 14 AD patients, treated daily with 9 mg melatonin for 1–3 years, we observed that a significant improvement of sleep quality was found in all cases, with a halted evolution of the cognitive and behavioral signs of the disease during the time they received melatonin. A substantial improvement of sleep quality and reduction of sundowning was found in all patients studied. Thus melatonin treatment at relatively moderate pharmacological doses can be useful to ameliorate behavior disorders and to halt cognitive impairment in elderly patients with dementia. To what extent this is due to an effect of melatonin on β-amyloid-related processes or is the consequence of an improvement in sleep quality and reduction of sundowning agitation awaits further investigation.

3.7 Melatonin Restores Amplitude of Circadian Rhythm in Chronic Inflammation or Aging

Significant impairment or abolition of circadian rhythms is observed as a consequence of several chronic diseases. This is an effect named as "masking" in chronobiology jargon. Another situation in which significant distortion of circadian rhythmicity is found is aging. Old age has been associated with changes in the amplitude and period of circadian rhythms in several species, including humans. In rodents, reported age-related changes in circadian rhythmicity include alterations in entrainment to the light–dark cycle and reduced amplitude of rhythms. Neuronal degeneration at the SCN and in the amplitude of melatonin rhythm have been reported as a consequence of aging (for references, see Golombek et al. 1997; Cardinali et al. 1998 a).

Mycobacterial-induced experimental arthritis in rodents is a model of heuristic value to assess the chronobiological consequences of chronic inflammation. Adjuvant-induced arthritis represents a T-lymphocyte-dependent, strain-specific autoimmune disease (Whitehouse 1988). Although the clinical onset of the inflammatory disease of the joints occurs on days 12–15, a number of neuroendocrine sequela become apparent at an early phase of arthritis development. For example, on days 3–5 after injection, increases of pituitary PRL (PRL) mRNA and decreases of growth hormone (GH) mRNA were found, coincident with augmented circulating PRL and decreased GH levels (Neidhart 1996).

We have verified the existence of significant effects of immune-mediated inflammation on ACTH, GH, PRL, and TSH release at an early phase (day 3) after Freund's adjuvant injection into rats (Selgas et al. 1997, 1998). After mycobacterial Freund's adjuvant injection, 24-h variations of TSH levels became blunted while 24-h variation of PRL and ACTH persisted. Freund's adjuvant injection increased serum ACTH and PRL levels and decreased GH and TSH levels. Injection of Freund's adjuvant reduced the amplitude of the daily variation of norepinephrine content, shifted the maximum of dopamine turnover, and blunted the daily variation in serotonin turnover in several hypothalamic areas, including the preoptic-anterior hypothalamic area and the SCN.

Since immunosuppression by cyclosporine restored rhythmicity of several of the neuroendocrine parameters examined, the immune-related nature of the studied phenomena seemed to be warranted. Immunocompetent cells affect neural processes by paracrine means through the release of cytokines influencing the nerves situated in close proximity or conveyed to the central nervous system via the bloodstream. There is information on changes in neurotransmitter dynamics in several brain regions, as well as in the peripheral autonomic nervous system, in the course of an immune reaction or as a consequence of cytokine administration (Besedovsky and Del Rey 1996; Cardinali and Esquifino 1998).

More recently, we examined the effect of melatonin injection on Freund's adjuvant-induced changes in levels and 24-h rhythms of circulating ACTH, PRL, and LH (Esquifino et al. 1999). Animals received s.c. injections of melatonin (30 µg) or vehicle, 1 h before lights off for 12 days, and 10 days after melatonin treatment they were injected with Freund's complete adjuvant or its vehicle. In melatonin-treated rats, an augmentation of the amplitude of serum ACTH rhythm could be detected. Melatonin effectively counteracted the decrease of circulating LH and restored its 24-h rhythmicity in immunized rats. As far as PRL is concerned, melatonin injection produced a significant diurnal variation in immunized rats only. Therefore, several early changes in levels and 24-h rhythms of circulating ACTH, PRL, and LH in Freund's adjuvant-injected rats were sensitive to treatment with pharmacological amounts of melatonin.

Another aspect significantly modified by progression of adjuvant's arthritis is the circadian organization of the immune response. Two circadian systems have been isolated in both experimental animals and man: (1) the circulation of T, B, or NK lymphocyte subsets in peripheral blood, and (2) the density of epitope molecules (e. g., $CD4^+$, $CD8^+$) at their surface, which may be related to the cellular reaction to antigen exposure. Changes in lymphocyte subset populations can be attributed to time of day-associated changes in cell proliferation in immunocompetent organs as well as to nyctohemeral modifications in lymphocyte release and traffic among lymphoid organs (for references, see Cardinali et al. 1996c; Cardinali and Esquifino 1998).

During the last few years, we have examined the regulation of circadian rhythmicity of cell proliferation and mitogenic responses in rat submaxillary lymph nodes and spleen during the immune reaction. In both immunized and nonimmunized rats a significant diurnal variation of lymph cell proliferation (as measured by ornithine decarboxylase, ODC) displaying maximal activity in the afternoon was found, coincident with maximal lipopolysaccharide- and concanavalin A-induced cell proliferation. Splenic cell proliferation was maximal at morning hours, which coincided temporally with changes in some immune-related parameters (i. e., mitogenic activity) (Esquifino et al. 1996).

Theoretically, two possible pathways through which the SCN can modulate the circadian organization of the immune response should be considered. One is purely neuroendocrine and involves the strong circadian profile of secretion of hormones like melatonin. Another includes the direct circadian control of immune reactivity through the autonomic nervous system innervating the lymphoid organs (Esquifino and Cardinali 1994). In sympathetically denervated plus parasympathetically decentralized submaxillary lymph nodes the normal diurnal variation in cell proliferation, with an afternoon peak, was blunted. It was thus concluded that an important regulation of circadian rhythms in cell proliferation in lymph nodes is derived from the autonomic nerves.

The involvement of melatonin in maintaining a normal circadian structure of the immune response was also examined. In an experiment assessing the effect of pineal suppression on diurnal variations in lymph node and splenic cell proliferation, ODC activity in pinealectomized rats decreased by about half while still exhibiting a low amplitude, significant diurnal variation with a maximum during the afternoon. Administration of melatonin at late evening restored ODC levels and the amplitude of diurnal rhythmicity (Cardinali et al. 1996a; 1997b). The results were compatible with the view that the pineal gland plays a role in circadian changes of immune responsiveness in lymph nodes and spleen via an immunopotentiating effect of melatonin on cell proliferation.

Indeed, the functional relationship between the pineal gland and the immune system has been known, and was initially indicated by the disorganization of thymic structures after pinealectomy of newborn rats. Pinealectomy or any other experimental procedure that inhibits melatonin synthesis and secretion induces a state of immunodepression, counteracted by melatonin in several species (for references, see Maestroni 1995; Liebmann et al. 1997). Under in vivo conditions, melatonin displays an immunoenhancing effect, particularly apparent in immunodepressive states. Melatonin administration augments antibody responses, such as plate-forming cell activity, cytotoxic T-cell responses, and antibody-dependent cellular cytotoxicity. It corrects immunodeficiency that follows stress, immunosuppressant drugs, or viral infections, and prevents apoptosis in hematopoietic cells and thymocytes. Melatonin, in physiological concentrations, stimulated activated $CD4^+$ T cells in vitro to release opioid agonist(s) with immunoenhancing and antistress properties (Maestroni 1995).

The nature of the mechanisms involved in the immunomodulatory activity of melatonin remains unsettled. There is evidence suggesting the existence of membrane-specific binding sites for melatonin in immune cells and the binding of labeled melatonin to peripheral blood T cells is markedly higher than that observed in B lymphocytes. Melatonin binding sites were mostly observed in $CD4^+$ cells rather than in $CD8^+$ cells, suggesting that $CD4^+$ lymphocytes are the targets of melatonin among human lymphocyte subpopulations. In human lymphocytes, melatonin augmented cyclic GMP production but was ineffective in modifying cyclic AMP production. mRNA expression of mt_1 receptors is detectable in lymphocytes. Since drugs that bind to RZR/ROR receptors are active in experimental models of autoimmune diseases, the possible nuclear effect of melatonin should be taken into account (Missbach et al. 1996).

The interaction between the CNS and the immune system is a bi-directional process: immunocompetent cells affect local neural processes by paracrine means through the release of cytokines and by endocrine means on the CNS via the bloodstream. During the immune reaction there is an increase of sympathetic and parasympathetic activity in local nerves. Submaxillary lymph node and splenic sympathetic activity attained their maximum at early night, while cholinergic activity in lymph nodes peaked during the afternoon. After pineal removal, a significant decrease of the amplitude of rhythmic presynaptic activity in lymphoid organs was found. These sequela of pinealectomy were counteracted by evening administration of melatonin (Cardinali et al. 1996a, 1997b).

Mycobacterial-induced experimental arthritis in rodents also demonstrated to be a useful animal model to examine the chronobiological activity of melatonin in aging (Cardinali et al. 1997a; 1998a, b). In a study carried out in young (50 days old) and old (18 months old) rats, 24-h rhythms of splenic and lymph node ODC and neuronal activity exhibited progressively smaller amplitudes of the daily variations as arthritis developed. In every case, mesor and amplitude of ODC activity were lowest in old rats. Melatonin administration was very effective in reversing age-related changes of immune responsiveness. Moreover, old rats exhibited lower melatonin content than young animals.

It must be noted that the effects of melatonin are not always beneficial. Melatonin administration to young rats after mycobacterium or collagen injection induced a more severe arthritis than that expected in control animals (Cardinali et al. 1998a). Therefore, pharmacological levels of melatonin in young animals may overstimulate the immune system causing exacerbation of both autoimmune processes.

3.8 Concluding Remarks

In mammals, pineal melatonin is produced rhythmically and functions as a photo-periodic signal. Its role as a circadian mediator, neglected for a long time, appears also to be important. Through the several mechanisms reviewed herein, melatonin can play the role of a universal synchronizer within the body. Accordingly, the melatonin rhythm may be one of the most important efferent SCN signals for imparting synchronicity to the rest of the organism.

Most biological changes after melatonin treatment require chronic (i.e., from several days to weeks) administration schedules. It seems feasible that the time-related effects, if they become repetitive with time, allow the animal's physiology to adapt to a new associative relationship. This process is not dissimilar from that involved in associative learning. The brain, as well as the peripheral cells, may be able to encode an accumulated memory of melatonin signals and thereby define time intervals around the 24-h cycle and the annual scale. It is within the framework of this concept that the effect of melatonin on sleep disturbances in aged people has to be considered, i.e., melatonin treatment will reinstall the circadian organization of the sleep – wake cycle.

References

Acuña Castroviejo D, Cotomontes A, Monti MG, Ortiz GG, Reiter RJ (1996) Melatonin is protective against MPTP-induced striatal and hippocampal lesions. Life Sci 60:PL23–PL29

Anis Y, Nir I, Schmidt U, Zisapel N (1992) Modification by oxazepam of the diurnal variations in brain ^{125}I-melatonin binding sites in sham-operated and pinealectomized rats. J Neural Transm 89:155–166

Anton-Tay F, Martínez I, Tovar R, Benítez-King G (1998) Modulation of the subcellular distribution of calmodulin by melatonin in MDCK cells. J Pineal Res 24:35–42

Arendt J (1995) Melatonin and the Mammalian Pineal Gland. Chapman & Hall, London

Atsmon J, Oaknin S, Laudon M, Laschiner S, Gavish M, Dagan Y, Zisapel N (1996) Reciprocal effects of chronic diazepam and melatonin on brain melatonin and benzodiazepine binding sites. J Pineal Res 20:65–71

Benítez-King G, Huerto Delgadillo L, Anton-Tay F (1990) Melatonin effects on the cytoskeletal organization of MDCK and neuroblastoma N1E-115 cells. J Pineal Res 9:209–220

Benítez-King G, Ríos A, Martínez A, Anton-Tay F (1996) In vitro inhibition of Ca^{2+}/calmodulin-dependent kinase II activity by melatonin. Biochim Biophys Acta 1290:191–196

Bertuglia S, Marchiafava PL, Colantuoni A (1996) Melatonin prevents ischemia reperfusion injury in hamster cheek pouch microcirculation. Cardiovasc Res 31:947–952

Besedovsky HO, Del Rey A (1996) Immune-neuro-endocrine interactions: facts and hypotheses. Endocr Rev 17:64–102

Bozner P, Grishko V, Ledoux SP, Wilson GL, Chyan YC, Pappolla MA (1997) The amyloid beta protein induces oxidative damage of mitochondrial DNA. J Neuropathol Exp Neurol 56:1356–1362

Brusco LI, Marquez M, Cardinali DP (1998a) Monozygotic twins with Alzheimer's disease treated with melatonin. Case report. J Pineal Res 25:260–263

Brusco LI, Marquez M, Cardinali DP (1998b) Melatonin treatment stabilizes chronobiologic and cognitive symptoms in Alzheimer's disease. Neuroendocrinol Lett 19:111–115

Brzozowski T, Konturek PC, Konturek SJ, Pajdo R, Bielanski W, Brzozowska I, Stachura J, Hahn EG (1997) The role of melatonin and L-tryptophan in prevention of acute gastric lesions induced by stress, ethanol, ischemia, and aspirin. J Pineal Res 23:79–89

Cardinali DP (1981) Melatonin. A mammalian pineal hormone. Endocr Rev 2:327–346

Cardinali DP, Esquifino AI (1998) Neuroimmunoendocrinology of the cervical autonomic nervous system. Biomed Rev, in press

Cardinali DP, Golombek DA (1998) The rhythmic GABAergic system. Neurochem Res 23:607–614

Cardinali DP, Rosenstein RE, Golombek DA, Chuluyan HE, Kanterewicz B, Del Zar MM, Vacas MI (1991) Melatonin binding sites in brain: Single or multiple? Adv Pineal Res 5:159–165

Cardinali DP, Bonanni Rey RA, Mediavilla MD, Sánchez Barceló E (1992). Diurnal changes in cyclic nucleotide response to pineal indoles in murine mammary glands. J Pineal Res 13:111–116

Cardinali DP, Del Zar MM, Vacas MI (1993) The effects of melatonin on human platelets. Acta Physiol Pharmacol Ther Latinoamer 43:1–13

Cardinali DP, Cutrera RA, Castrillon P, Esquifino AI (1996a) Diurnal rhythms in ornithine decarboxylase activity and norepinephrine and acetylcholine synthesis of rat submaxillary lymph nodes: Effect of pinealectomy, superior cervical ganglionectomy and melatonin replacement. Neuroimmunomodulation 3:102–111

Cardinali DP, Cutrera RA, Castrillon P, Garcia Bonacho M, Selgas L, Esquifino AI (1996b) Circadian organization of immune response in rat submaxillary lymph nodes: A site for melatonin action. In: Tang PL, Pang SF, Reiter RJ (eds) Melatonin: A Universal Photoperiodic Signal with Diverse Actions. Karger, Basel, pp 60–70

Cardinali DP, Della Maggiore V, Selgas L, Esquifino AI (1996c) Diurnal rhythm in ornithine decarboxylase activity and noradrenergic and cholinergic markers in rat submaxillary lymph nodes. Brain Res 711:153–162

Cardinali DP, Brusco LI, Selgas L, Esquifino AI (1997a) Melatonin: A synchronizing signal for the immune system. Neuroendocrinol Lett 18:73–84

Cardinali DP, Cutrera RA, Garcia Bonacho M, Esquifino AI (1997b) Effect of pinealectomy, superior cervical ganglionectomy and melatonin treatment on 24-hour rhythms in ornithine decarboxylase and tyrosine hydroxylase activities of rat spleen. J Pineal Res 22:210–220

Cardinali DP, Golombek DA, Rosenstein RE, Cutrera RA, Esquifino AI (1997c) Melatonin site and mechanism of action: Single or multiple? J Pineal Res 23:32–39

Cardinali DP, Brusco LI, Garcia Bonacho M, Esquifino AI (1998a) Effect of melatonin on 24-hour rhythms of ornithine decarboxylase activity and norepinephrine and acetylcholine synthesis in submaxillary lymph nodes and spleen of young and aged rats. Neuroendocrinology 67:349–362

Cardinali DP, Brusco LI, Selgas L, Esquifino AI (1998b) Diurnal rhythms in ornithine decarboxylase activity and norepinephrine and acetylcholine synthesis in submaxillary lymph nodes and spleen of young and aged rats during Freund's adjuvant-induced arthritis. Brain Res 789:283–292

Cassone VM, Chesworth MJ, Armstrong SM (1986) Entrainment of rat circadian rhythms by daily injection of melatonin depends upon the hypothalamic suprachiasmatic nuclei. Physiol Behav 36:1111–1121

Catala MD, Quay WB, Timiras PS (1987) Developmental changes in hypothalamic melatonin levels of male rats. Int J Dev Neurosci 5:313–318

Cazevieille C, Osborne NN (1997) Retinal neurones containing kainate receptors are influenced by exogenous kainate and ischaemia while neurones lacking these receptors are not – Melatonin counteracts the effect of ischaemia and kainate. Brain Res 755:91–100

Champney TH, Hanneman WH, Legare ME, Appel K (1996) Acute and chronic effects of melatonin as an anticonvulsant in male gerbils. J Pineal Res 20:79–83

Chow PH, Lee PN, Poon AMS, Shiu SYM, Pang SF (1996) The gastrointestinal tract: A site of melatonin paracrine action. In: Tang PL, Pang SF, Reiter RJ (eds) Melatonin: A Universal Photoperiodic Signal with Diverse Actions. Karger, Basel, pp 123–132

Dagan Y, Zisapel N, Nof D, Laudon M, Atsmon J (1997) Rapid reversal of tolerance to benzodiazepine hypnotics by treatment with oral melatonin: a case report. Eur Neuropsychopharmacol 7:157–160

Daniels WMU, Vanrensburg SJ, Vanzyl JM, Taljaard JJF (1998) Melatonin prevents beta-amyloid-induced lipid peroxidation. J Pineal Res 24:78–82

Dawson D, Armstrong SM (1996) Chronobiotics-drugs that shift rhythms. Pharmacol Ther 69:15–36

De la Lastra CA, Cabeza J, Motilva V, Martin MJ (1997) Melatonin protects against gastric ischemia-reperfusion injury in rats. J Pineal Res 23:47–52

Dijk DJ, Roth C, Landolt HP, Werth E, Aeppli M, Achermann P, Borbely AA (1995) Melatonin effect on daytime sleep in men: Suppression of EEG low frequency activity and enhancement of spindle frequency activity. Neuroscience Lett 201:13–16

Dubocovich ML, Cardinali DP, Guardiola-Lemaitre B, Hagan RM, Krause DN, Sugden D, Vanhoutte PM, Yocca FD (1998) Melatonin receptors. In: The IUPHAR Compendium of Receptor Characterization and Classification. IUPHAR Media, London, in press

Esquifino AI, Selgas L, Arce A, Della Maggiore V, Cardinali DP (1996) Twenty four hour rhythms in immune responses in rat submaxillary lymph nodes and spleen. Effect of cyclosporine. Brain Behav Immunity 10:92–102

Esquifino AI, Cardinali DP (1994) Local regulation of the immune response by the autonomic nervous system. Neuroimmunomodulation 1:265–273

Esquifino AI, Castrillón P, Garcia-Bonacho M, Vara E, Carcinali DP (199) Effect of melatonin treatment on 24-hour rhythms of serum ACTH, growth hormone, prolactin, luteinizing hormone and insulin in rats injected with Freund's adjuvant. J Pineal Res 27:15–23

Faillace MP, Keller Sarmiento MI, Rosenstein RE (1996) Melatonin effect on the cyclic GMP system in the golden hamster retina. Brain Res 711:112–117

Fainstein I, Bonetto A, Brusco LI, Cardinali DP (1997) Effects of melatonin in elderly patients with sleep disturbance. A pilot study. Curr Ther Res 58:990–1000

Floreani M, Skaper SD, Facci L, Lipartiti M, Giusti P (1997) Melatonin maintains glutathione homeostasis in kainic acid-exposed rat brain tissues. FASEB J 11:1309–1315

Franchi AM, Gimeno MF, Cardinali DP, Vacas MI (1987) Melatonin, 5-methoxytryptamine and some of their analogs as cyclooxygenase inhibitors in rat medial basal hypothalamus. Brain Res 405:384–388

Garfinkel D, Laudon M, Zisapel N (1997) Improvement of sleep quality by controlled-release melatonin in benzodiazepine-treated elderly insomniacs. Arch Gerontol Geriatr 24:223–231

Gilad E, Cuzzocrea S, Zingarelli B, Salzman AL, Szabo C (1997) Melatonin is a scavenger of peroxynitrite. Life Sci 60:PL169–PL174

Gimeno MF, Landa ME, Speziale NS, Cardinali DP, Gimeno AL (1980) Melatonin blocks in vitro generation of prostaglandins by the uterus and hypothalamus. Eur J Pharmacol 62:309–317

Godson C, Reppert SM (1997) The Mel (1a) melatonin receptor is coupled to parallel signal transduction pathways. Endocrinology 138:397–404

Golombek DA, Pévet P, Cardinali DP (1996) Melatonin effect on behavior: Possible mediation by the central GABAergic system. Neurosci Biobehav Rev 20:403–412

Golombek DA, Rosenstein RE, Yannielli P, Sarmiento M, Cardinali DP (1997) Aging attenuates diurnal variation in hamster locomotion, anxiolysis and GABA turnover. Neurosci Lett 233:9–12

Guardiola Lemaitre B, Lenegre A, Porsolt RD (1992) Combined effects of diazepam and melatonin in two tests for anxiolytic activity in the mouse. Pharmacol Biochem Behav 41:405–408

Iacovitti L, Stull ND, Johnston K (1997) Melatonin rescues dopamine neurons from cell death in tissue culture models of oxidative stress. Brain Res 768:317–326

Iovanna J, Dusetti N, Cadenas B, Cardinali DP (1990a) Time-dependent effect of melatonin on actin mRNA levels and incorporation of ^{35}S-methionine into actin and proteins by the rat hypothalamus. J Pineal Res 9:51–63

Iovanna J, Dusetti N, Calvo E, Cardinali DP (1990b) Diurnal changes in actin mRNA levels and incorporation of ^{35}S-methionine into actin in the rat hypothalamus. Cell Mol Neurobiol 10:207–216

Kanterewicz B, Golombek DA, Rosenstein RE, Cardinali DP (1993) Diurnal changes of GABA turnover rate in brain and pineal gland of Syrian hamsters. Brain Res Bull 31:661–666

Kanterewicz B, Rosenstein RE, Golombek DA, Yannielli P, Cardinali DP (1995) Daily variations in GABA receptor function in Syrian hamster cerebral cortex. Neurosci Lett 200:211–213

Kelly RW, Amato F, Seamark RF (1984) N-acetyl-5-methoxy kynurenamine, a brain metabolite of melatonin, is a potent inhibitor of prostaglandin biosynthesis. Biochem Biophys Res Commun 121:372–379

Kojima T, Mochizuki C, Mitaka T, Mochizuki Y (1997) Effects of melatonin on proliferation, oxidative stress and Cx32 gap junction protein expression in primary cultures of adult rat hepatocytes. Cell Structure and Function 22:347–356

Kotler M, Rodriguez C, Sainz RM, Antolin I, Menendez Pelaez A (1998) Melatonin increases gene expression for antioxidant enzymes in rat brain cortex. J Pineal Res 24:83–89

Liebmann PM, Wolfer A, Felsner P, Hofer D, Schauenstein K (1997) Melatonin and the immune system. Int Arch Allergy Immunol 112:203–211

Longoni B, Salgo MG, Pryor WA, Marchiafava PL (1998) Effects of melatonin on lipid peroxidation induced by oxygen radicals. Life Sci 62:853–859

Lopez Gonzalez MA, Calvo JR, Osuna C, Guerrero JM (1992) Interaction of melatonin with human lymphocytes: evidence for binding sites coupled to potentiation of cyclic AMP stimulated by vasoactive intestinal peptide and activation of cyclic GMP. J Pineal Res 12:97–104

Maestroni GJM (1995) T-helper-2 lymphocytes as a peripheral target of melatonin. J Pineal Res 18:84–89

Maestroni GJM (1996) Melatonin as a therapeutic agent in experimental endotoxic shock. J Pineal Res 20:84–89

Masson-Pevet M, George D, Kalsbeek A, Saboureau M, Lakdhar-Ghazal N, Pevet P (1994) An attempt to correlate brain areas containing melatonin binding sites with rhythmic functions: A study in five hibernator species. Cell Tiss Res 278:97–106

Melendez J, Maldonado V, Ortega A (1996) Effect of melatonin on beta-tubulin and MAP2 expression in NIE-115 cells. Neurochem Res 21:653–658

Missbach M, Jagher B, Sigg I, Nayeri S, Carlberg C, Wiesenberg I (1996) Thiazolidine diones, specific ligands of the nuclear receptor retinoid Z receptor retinoid acid receptor-related orphan receptor alpha with potent antiarthritic activity. J Biol Chem 271:13515–13522

Molina Carballo A, Muñoz Hoyos A, Reiter RJ, Sanchez Forte M, Moreno Madrid F, Rufo Campos M, Molina Font JA, Acuña Castroviejo D (1997) Utility of high doses of melatonin as adjunctive anticonvulsant therapy in a child with severe myoclonic epilepsy: two years' experience. J Pineal Res 23:97–105

Montilla P, Tunez I, Munoz MC, Lopez A, Soria JV (1997a) Hyperlipidemic nephropathy induced by adriamycin: effect of melatonin administration. Nephron 76:345–350

Montilla P, Tunez I, Munoz MC, Soria JV, Lopez A (1997b) Antioxidative effect of melatonin in rat brain oxidative stress induced by adriamycin. J Physiol Biochem 53:301–305

Nave R, Herer P, Haimov I, Shlitner A, Lavie P (1996) Hypnotic and hypothermic effects of melatonin on daytime sleep in humans: Lack of antagonism by flumazenil. Neurosci Lett 214:123–126

Neidhart M (1996) Effect of Freund's complete adjuvant on the diurnal rhythms of neuroendocrine processes and ornithine decarboxylase activity in various tissues of male rats. Experientia 52:900–908

Pablos MI, Reiter RJ, Chuang JI, Ortiz GG, Guerrero JM, Sewerynek E, Agapito MT, Melchiorri D, Lawrence R, Deneke SM (1997) Acutely administered melatonin reduces oxidative damage in lung and brain induced by hyperbaric oxygen. J Appl Physiol 83:354–358

Pang CS, Tang PL, Song Y, Pang SF, Ng KW, Guardiola Lemaitre B, Delagrange P, Brown GM (1997) Differential inhibitory effects of melatonin analogs and three naphthalenic ligands on 2-[I-^{125}]iodo-melatonin binding to chicken tissues. J Pineal Res 23:148–155

Pang SF, Brown GM (1983) Regional concentrations of melatonin in the rat brain in the light and dark period. Life Sci 33:1199–1204

Pappolla MA, Sos M, Omar RA, Bick RJ, Hickson Bick DL, Reiter RJ, Efthimiopoulos S, Robakis NK (1997) Melatonin prevents death of neuroblastoma cells exposed to the Alzheimer amyloid peptide. J Neurosci 17:1683–1690

Pévet P, Pitrosky B (1997) The nocturnal melatonin peak and the photoperiodic response. In: Maestroni GJM, Conti A, Reiter RJ (eds) Therapeutic Potential of Melatonin. Front Horm Res, vol 23 Karger, Basel, pp 14–24

Prevedello MR, Ritta MN, Cardinali DP (1979) Fast axonal transport in rat sciatic nerve. Inhibition by pineal indoles. Neuroscience Lett 13:29–34

Reiter RJ (1991) Pineal melatonin: Cell biology of its synthesis and of its physiological interactions. Endocr Rev 12:151–180

Reiter RJ, Tang L, Garcia JJ, Munozhoyos A (1997) Pharmacological actions of melatonin in oxygen radical pathophysiology. Life Sci 60:2255–2271

Reppert SM (1997) Melatonin receptors: Molecular biology of a new family of G protein-coupled receptors. J Biol Rhyth 12:528–531

Rosenstein RE, Cardinali DP (1990) Central GABAergic mechanisms as target for melatonin activity. Neurochem Int 17:373–379

Satake N, Oe H, Shibata S (1991) Vasorelaxing action of melatonin in rat isolated aorta; possible endothelium dependent relaxation. Gen Pharmacol 22:1127–1133

Selgas L, Arce A, Esquifino AI, Cardinali DP (1997) Twenty-hour hour rhythms of serum ACTH, prolactin, growth hormone and thyroid-stimulating hormone, and of median eminence norepinephrine, dopamine and serotonin, in rats injected with Freund's adjuvant. Chronobiol Int 14:253–265

Selgas L, Pazo D, Arce A, Esquifino AI, Cardinali DP (1998) Circadian rhythms in adenohypophysial hormone levels and hypothalamic monoamine turnover in mycobacterial adjuvant-injected rats. Biol Signals Recept 7:15–24

Sewerynek E, Abe M, Reiter RJ, Barlow-Walden LR, Chen L, McCabe TJ, Roman LJ, Diaz Lopez B (1995) Melatonin administration prevents lipopolysaccharide- induced oxidative damage in phenobarbital-treated animals. J Cell Biochem 58:436–444

Sewerynek E, Reiter RJ, Melchiorri D, Ortiz GG, Lewinski A (1996) Oxidative damage in the liver induced by ischemia-reperfusion: protection by melatonin. Hepatogastroenterology 43: 898–905

Song WH, Lahiri DK (1997) Melatonin alters the metabolism of the beta-amyloid precursor protein in the neuroendocrine cell line PC12. J Mol Neurosci 9:75–92

Stankov B, Biella G, Panara C, Lucini V, Capsoni S, Fauteck J, Cozzi B, Fraschini F (1992) Melatonin signal transduction and mechanism of action in the central nervous system: using the rabbit cortex as a model. Endocrinology 130:2152–2159

Tenn CC, Niles LP (1995) Central-type benzodiazepine receptors mediate the antidopaminergic effect of clonazepam and melatonin in 6-hydroxydopamine lesioned rats: involvement of a GABAergic mechanism. J Pharmacol Exp Ther 274:84–89

Tenn CC, Niles LP (1997) The antidopaminergic action of S-20098 is mediated by benzodiazepine/ $GABA_A$ receptors in the striatum. Brain Res 756:293–296

Uchida K, Okamoto N, Ohara K, Morita Y (1996) Daily rhythm of serum melatonin in patients with dementia of the degenerate type. Brain Res 717:154–159

Vacas MI, Keller Sarmiento MI, Cardinali DP (1981) Melatonin increases cGMP and decreases cAMP levels in rat medial basal hypothalamus in vitro. Brain Res 225:207–211

Vanecek J, Klein DC (1995) Melatonin inhibition of GnRH-induced LH release from neonatal rat gonadotroph: involvement of Ca^{2+} not cAMP. Am J Physiol 269:E85–E90

Whitehouse MW (1988) Adjuvant-induced polyarthritis in rats. In: Greenwald RA, Diamad HS (eds) Handbook of Animal Models for the Rheumatic Diseases. CRC Press, New York, pp 3–16

Yannielli P, Kanterewicz B, Cardinali DP (1996) Circadian changes in anxiolysis-related behavior of Syrian hamsters. Correlation with hypothalamic GABA release. Biol Rhythm Res 27:363–373

Zisapel N (1988) Melatonin receptors revisited. J Neural Transm 73:1–5

4 The Pineal Gland and Chronobiologic History: Mind and Spirit as Feedsidewards in Time Structures for Prehabilitation*

Franz Halberg, Germaine Cornélissen, Ario Conti,
Georges J. M. Maestroni, Cristina Maggioni, Federico Perfetto, Roberto Salti,
Roberto Tarquini, George S. Katinas, Othild Schwartzkopff

Abstract

Not only circadian rhythms – recurring patterns with a period of about 24 h (in the range of 20–28 h) – but also ultradian and infradian rhythms (with periods shorter than 20 h and longer than 28 h, respectively), characterize melatonin in humans, whether it is measured in blood, saliva, or urine. Among infradians, the about-yearly (circannual) and half-yearly (circasemiannual) components are noteworthy. At mid-latitude, circannuals may predominate in circulating melatonin during the daytime, whereas circasemiannuals may become more prominent during the nighttime. A stable half-yearly component also prominently characterizes the geomagnetic disturbance index Kp. Support for the hypothesis that Kp may influence human melatonin is provided by the fact that closer to the pole, at 65°N in Oulu, Finland, geomagnetic effects are stronger. There, circulating melatonin, measured around noon, exhibits a clear circasemiannual variation. Circaseptans and circasemiseptans, with periods of about a week and half a week, are found ubiquitously in relation to the pineal gland. In the case of melatonin secreted into the superfusion fluid by the pike pineal in vitro, kept at constant temperature in continuous darkness, the circaseptan component has an amplitude larger than that of the circadian rhythm. Circaseptans are also observed in the mouse pineal gland in vivo, wherein the presence of melatonin has been questioned, yet established by three independent groups of investigators who all documented a circadian variation peaking during the dark (rest) span.

In vitro intermodulations among the pineal, pituitary, and adrenal glands were found to involve a rhythmically recurring and thus predictable sequence of opposite effects, i.e., a chronomodulation. Attenuation, no-effect, and amplification alternate circadian-dependently in the effect exerted by aqueous pineal extract or melatonin on the adrenal corticosterone response to pituitary incubation fluid or to an

---

* Dedicated on his birthday, 27 September, to Brian Brockway, who built Data sciences for physiologic telemetry on laboratory animals, and thus to the challenge to enable, by the development of instrumentation, a change from a single-sample-based health care into one of chronomes for prehabilitation to reduce or eliminate, whenever possible, the need for rehabilitation.
Support: U.S. Public Health Service (GM-13981), National Heart, Lung, and Blood Institute, National Institutes of Health (HL-40650), University of Minnesota Supercomputer Institute, Dr. h.c. Dr. h.c. Earl Bakken Fund and Dr. Betty Sullivan Fund, and Mr. Lynn Peterson, United Business Machines, Fridley, MN.

adrenocorticotropic hormone (ACTH) analog. This predictable sequence of effects is also seen for melatonin acting directly upon the adrenal gland. These results represent a challenge to those interested in integrative functions in health and in major diseases. Rather than remaining a confusing source of variability, the assessment of broad time structures (chronomes) yields new end points that can serve for treatment optimization by timing, and for risk assessment and management through the institution of preventive action as soon as risk elevation is detected. Several interdigitating and partly interwoven fields are centered around the pineal gland, where interactions between life and the cosmos may well lie. One of these fields concerns the rules of intermodulations among rhythms far beyond the circadian system. Another field consists of individuals' adjustments and adaptations of species as features of the integration of organisms, as open systems with their ever-changing, predictable insofar as cyclic, environment. Life has coded time structure in genes, but organisms still respond to remote drummers. The pineal gland may play a critical role as the window not only for photic and non-photic effects of the sun, but also for other galactic effects. Barriers to these fields exist only in the mind as deeply rooted conventions, such as homeostasis. The third field, chronobioethics, pertinent to pinealists' concerns about the soul, whether they come from philosophy or molecular biology, is sketched only in a historical context here.

4.1 Introduction

Many barriers exist only in the mind. Once we realize that it is possible to walk through walls, because the walls are only there by our own deeply rooted conventions, we arrive

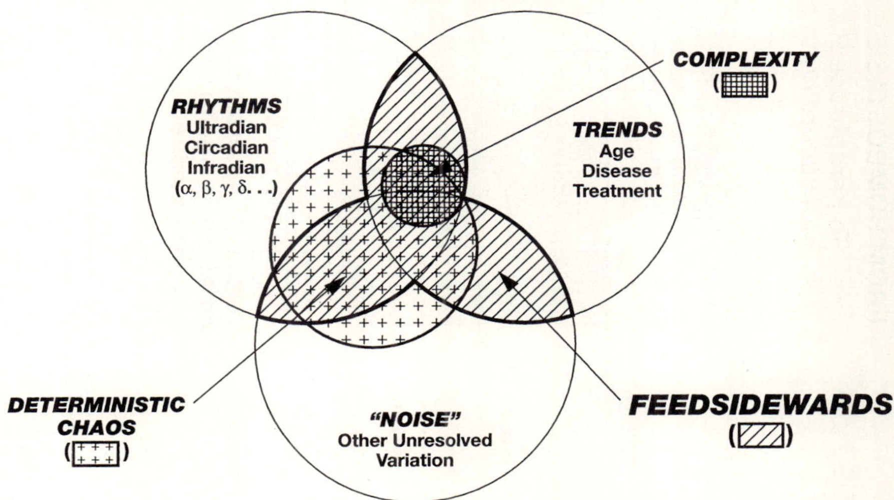

Fig. 4.1. Chronomes consist of multifrequency rhythms, elements of chaos, trends in chaotic and rhythmic end points, and other, as yet unresolved variability ("noise"). (Copyright Halberg)

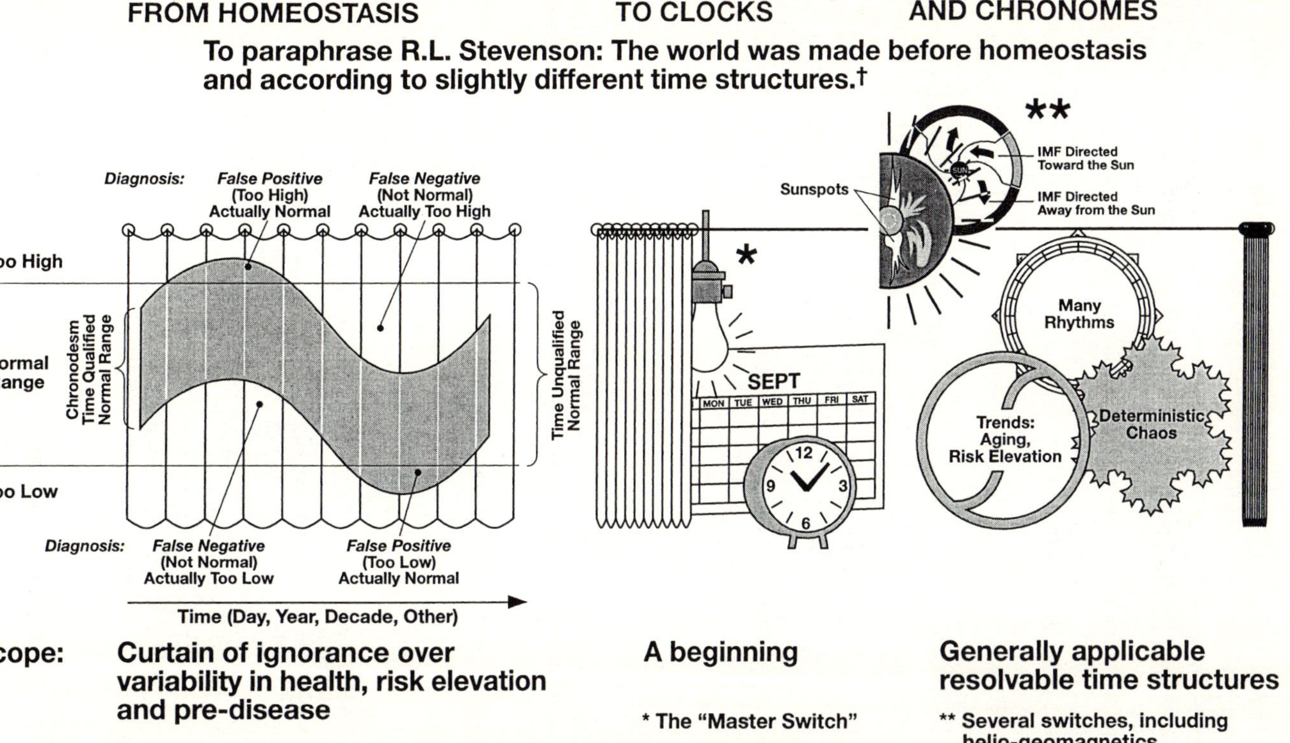

Fig. 4.2. The chronome in us, that came about as a function of chronomes around us, to be eventually coded genetically, awaits further exploration in health care, notably for stroke prevention. (Copyright Halberg)

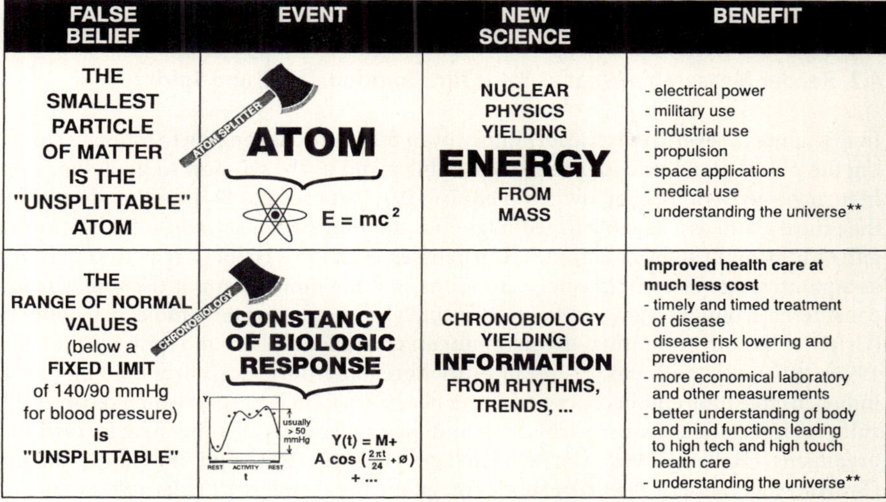

Fig. 4.3. By understanding biological activity as a function of time, notably in medicine, chronobiology can be aligned in importance with the splitting of the atom (*). Understanding of the origin of the universe by nuclear physics is matched by a greater understanding via chronobiology of theoretical, experimental and applied biology as a whole, with applications to veterinary sciences, nutrition, animal husbandry, pest control, and other aspects of agriculture, including the concerns for the broadest environmental integrity, beyond chronobiology's major promise of cost-effective health care (**). (Copyright Halberg)

at the verge of making a contribution to progress in science and ethics, and thus in their concomitant interdigitated pursuit.[1] This recognition may serve pinealists, whose historical background is the search for the soul. To vitalists who believe that the soul leaves the body with its last breath, the soul is the anima. Cicero's *animus mundi* may be used as a preferred alternative in focusing on relations among cycles and events viewed as time structures (Fig. 4.1) in and around us (Fig. 4.2). To resolve these chronomes (portmanteau'd from *chronos,* time, and *nomos,* rule), we must abandon the current general position with respect to physiological variation in the normal range, which corresponds to that in preatomic physics. It was then believed that the "a-tom" was the smallest particle that could not be further split (Fig. 4.3). Breaking the atom opened the door to a new universe of particles governed by new forces and physical laws. The field of nuclear physics evolved and brought with it new knowledge, a new energy source, and a wealth of practical applications (Fig. 4.3) (Cornélissen and Halberg 1994). Indeed, the analogy applies to splitting of the normal range into the time structures of everyday physiology (Table 4.A1). From picking

---

[1] We may wish to first tear down the barrier between science and spirituality and approach both objectively. Next, we may tear down the barriers within the sciences, such as those among disciplines as different as physics and biology. Barriers between integrationists and reductionists then fade when the pineal gland is at issue.

different times of day and seasons for study, a transdisciplinary science, chrono-biology, emerged to search objectively for the physiology of cognition, if not of the *animus mundi.*

4.2 Render Measurable What as Yet Is Not: Emotion, Mind, and Spirit

In this context, several inextricably interwoven fields are challenging to scholars study-ing the pineal gland, whether they regard this gland as the window to light, the latter in turn viewed as a "master" switch (Pennisi 1997; Plautz et al. 1997), and/or also accept the pineal gland as a keyhole to geomagnetics, another switch at noon at high latitudes and only by night at lower latitudes (Cornélissen et al. 1997; Halberg et al. 1997), and/or as a putative site of consciousness, emotions, and the mind, if not of the soul, already considered from the viewpoints of (eventually we hope chrono-) molecular biology (Crick 1994) or from that of an eminent crystallographer and physicist (Tiller 1997). Whatever the focus, or for all of the foregoing purposes, chronobiology is an indispensable tool, and in order to use it, a necessary field is needed which resolves the rules of intermodulations of rhythms and other elements of time structure within organisms (Halberg 1969; Cornélissen and Halberg 1994) and the subtle factors (Brown 1960; Halberg 1969a) or subtle (magnetic field) energies (Tiller 1997) that may be involved with life and the environment.

This endeavor leads to a related, interdigitating field, chronoastrobiology (Halberg et al. 1991a; Cornélissen et al. 1999), which strives to elucidate the past and present integration of organisms into their largely wavy environment (Fig. 4.4). In the course of evolution, organisms may have coded many of the cycles and events of their proximal and distant habitat in their genes, as Fig. 4.5 suggests for the circadian change of human heart rate (Halberg 1983; Hanson et al. 1984). From this viewpoint, life on earth is a living fossil, telling us perhaps about the cosmos of the past (Halberg 1969; Halberg et al. 1991a; Breus et al. 1995; Cornélissen et al. 1999). On the other hand, the biosphere is still responsive today to the very cycles and events that also characterize its current environment (Halberg et al. 1998; Cornélissen et al. 1999). Facts in these overlapping fields are summarized in Tables 4.A1 and 4.A2.

A third field, chronobioethics (Cornélissen et al. 1994a, b; Halberg et al. 1994), is most pertinent to those concerned about the mind and spirit, if not soul, whether they come from philosophy, molecular biology, or integrative physics. Tiller (1997, 1999) and Tiller, Dibble, and Kohane (1999) provide a dual four-space frame, complementing the physical four dimensions of space-time with a mathematically underpinned reciprocal four-space which they embedded into emotions (a nine-space), that in turn is embedded in the frame of mind (a ten-space), while the embedding frame for all this is the frame of spirit (Scheme 1, p. 73). The task of chronobiology is to test this concept in inferential statistical terms. Our goal consists of measuring what is measurable and rendering measurable what as yet is not, in time and meaningfully *(Omnia metire quæcumque licet et immensa ad mensuram tempestive et ergo significative redige).* We sketch approaches to the pineal gland first in a historical context, thereafter turning to our main task, which is physiological.

Fig. 4.4. Chronoastrobiology strives to elucidate the past and present integration of organisms into their largely wavy environment. (Copyright Halberg)

WE ARE THERE* AND HERE AND

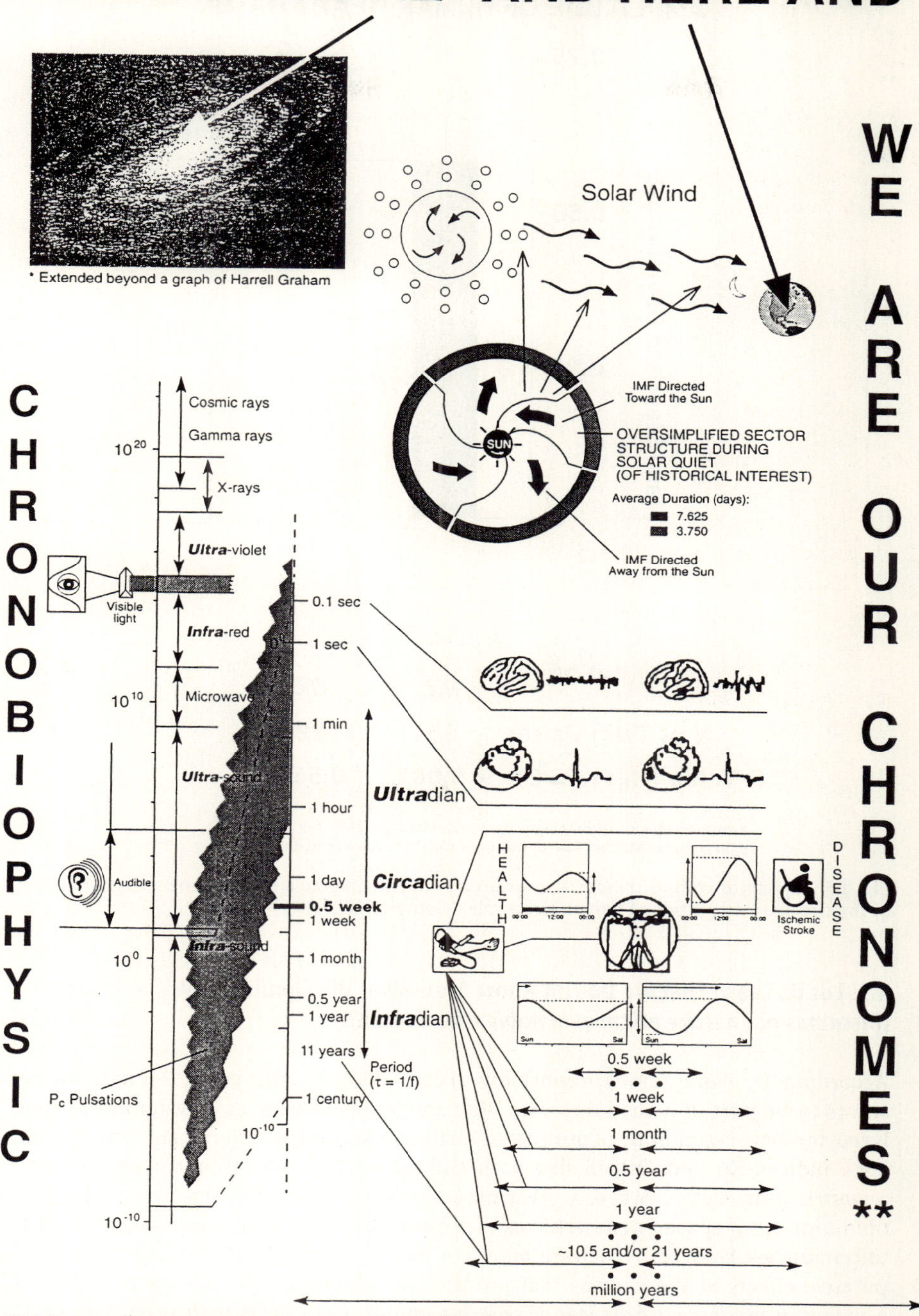

**** Which further comprise age and other trends, including adaptive, integrative and cultural evolution toward a chrononoosphere, topics of chronobiology broadly.**

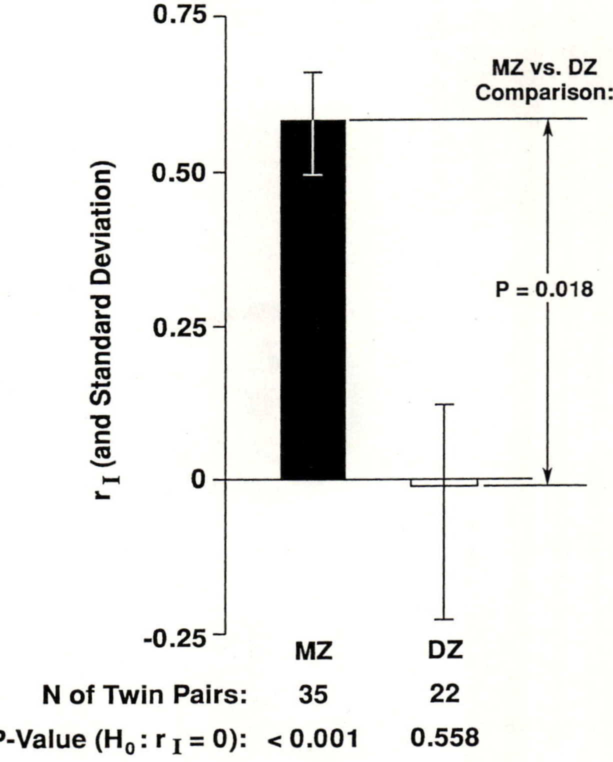

Fig. 4.5. Groups of around-the-clock records on twin pairs reared apart document the heritability of the circadian amplitude of human heart rate. (Copyright Halberg)

4.3 Let Us Learn That We Do Not Know Even What We Should Know
(Discamus nos nescire etiam quid nobis sciendam sit)

According to Tiller (1997), reportedly (Pecci 1997) "…the entire human species seems to be part of a vast organism". We single out instead, as a system to be analyzed, the even vaster body of our planet earth, by using its planetary magnetic disturbance indices Kp and Dst, in line with Gilberd's (1600) *magnum magnes est orbis terrestris* (our earthly orb is a great magnet). We look further into the problem of obtaining from space-borne vehicles measures of the interplanetary magnetic field, to paraphrase Gilberd by writing *magnum magnes est orbis solaris*, only to find the putative effects of the cosmos that prompt us to view our physiology as reflecting a *magnum magnis est orbis cosmi* or *orbis mundi* if not *orbis multiversi*. Table 4.A2 in this context provides a view of statistically validated dividends, and Scheme 1 is

Scheme 1*
Tiller's (1997) dynamic equations of nature (1 to 3)

1. Function $\Leftrightarrow$ Structure $\Leftrightarrow$ Chemistry
2. Function $\Leftrightarrow$ Structure $\Leftrightarrow$ Chemistry $\Leftrightarrow$ (Electromagnetic energy fields)
3. Function $\Leftrightarrow$ Structure $\Leftrightarrow$ Chemistry $\Leftrightarrow$ (Electromagnetic energy fields)
($\Leftrightarrow$ Emotions)*$\Leftrightarrow$ Mind ($\Leftrightarrow$ Spirit)*

lead to

4. a biological radiation detector

and the BIOCOS project (Biosphere anD Cosmos) viewing
life as a genetic-memory endowed electromagnetic field and radiation detector
(I to VII)

Approach		Biosphere (indicator of physical process)		Cosmos
I.	Ind	decrease in HRV	$\Leftrightarrow$	increase in Kp
II.	Ind	decrease in ~ 46 sec spectral power of RR intervals	$\Leftrightarrow$	increase in Kp
III.	Pop	decrease in HRV	$\Leftrightarrow$	increase in Kp
IV.	Pop	increase in incidence of MI	$\Leftrightarrow$	north-south turn of vertical component of interplanetary magnetic field (Bz)
V.	Pop	increase in incidence of MI	$\Leftrightarrow$	Forbush decrease in intensity of galactic cosmic rays (FD)
VI.	Pop	increase in incidence of stroke	$\Leftrightarrow$	FD but not Bz turn
VII.	Pop	decrease in bacterial sectoring	$\Leftrightarrow$	decrease in Dst below – 100 nT

* Perspective is complementary as reciprocal to that in Table 4.A2. A long-term memory as genes prompts resonance at selected frequencies that are detected as cross-spectral coherence (Table 4.A2). In approach 1, the "altered" time structure of the electrocardiogram (ECG), i.a., the presence of a very prominent circaseptan component and a greatly reduced circadian, desynchronized from 24 hours, prompted the search for some underlying environmental event, and led to the record showing a magnetic storm in the timespan with the altered ECG. Actually, the circadian changes observed during the storm in this healthy subject resembled earlier findings in rabbits. Approaches on humans (I-VI) focusing on the individual (Ind) as well as population (Pop) are complemented by those on bacteria (VII). Increases in Kp and decreases in Dst, both geomagnetic disturbance indices may be complemented by biomarkers of magnetic storms and/or other (galactic ?) events. Our interpretation of Tiller's (1997) view expressed in equation 3 on p. 4. This can prompt objective tests, e.g., of emotionality, like grief, or mind, like conflict, by sensitive chronobiological indices. Particularly pertinent are gross alterations of the environment, such as a precise anthropogenic week in magnetic disturbance when the natural physical planetary geomagnetic index shows a component differing from a precise week. For intentionality, see also Jahn (1995, 1996)
HRV = heart rate variability; MI = myocardial infarction.
The frequencies, and at each frequency, further parameters remain to be specified.

a challenge to biologists and physicists to use the organisms themselves as environmental gauges.

Those who doubt the desirability of viewing the organism as an open system in the context of its environment should try to hold their breath; that should promptly convince them that they need the air they breathe. This should have been clear to the first human. Equally early, humans and other animals may have realized that every so often they must rest or sleep. The extent to which we depend on more subtle effects impinging upon us may have been considered by those who coined the concept of the *animus mundi* over two millennia ago. In the context of the pineal gland, the concept of the *anima orbis* ("world soul") has been cited as a dynamic substrate for the vital principle associated with the body in life and perhaps departing from it at death. It may be more than chance that in the form of *animus* (rarely *anima*), the term was also used for the mind as a rational human principle. Tiller (1993, 1997) has approached the emotions, mind, and spirit in thermodynamic terms, and dimensions embedding the classical four time-space dimensions and their four reciprocals. We propose Cicero's *animus mundi,* or if there is a need to refer to more than one universe, *animus multiversi,* to conclude as Tiller did (1997), that organisms are (biological) detectors of photic and non-photic effects from the sun and from beyond. On the applied side, there is the precedent of distinguishing between grief and conflict in objective chronobiological terms (Halberg et al. 1996).

4.4 Tarquini and Pineal Gland History

Late in 1998, unfortunately without bibliographical references, with an outline presented orally, the late Brunetto Tarquini, professor and head of a department of medicine in Florence, traced interest in the pineal gland as the anatomical substrate of the soul, from Chinese medicine to the Cartesian "mind" (Halberg et al. 1999). Interest in the pineal gland as the "house" of the soul (animus/anima) started, according to Tarquini, presumably with the legendary Yellow Emperor (said to have reigned from 2697–2597 BC) (Huang Ti Nei Ching Su Wen 1949). Other scholars of this time have failed to list the gland in their index (e. g. Huang Ti Nei Ching Su Wen 1949). In the Vedas, the religious teachings of India, the pineal gland was one of the chakras, the centers of vital energy. This chakra was the "door to perfect peace and harmony." In the Vedic pantheon, one of the most potent deities was Varona, the god of the nocturnal sky, which directs the course of the sun.

Herophilus of Alexandria (325–280 BC), who taught anatomy and physiology during the reign of the first two Ptolemys, mentioned the pineal gland as an entity that is functionally involved in the coordination of the flow of thought (psychikon): as a sphincter between the ventricles of the brain and the nerves; so did Galen by indicating that the soul is in the center of the brain, in the pineal gland, rather than in the heart as Aristotle claimed. Herophilus's lasting fame stems first and foremost more generally from studies on the brain, the eyes, and the circulation. He described the nerves, the calamus scriptorius (or calamus herophilii), the occipital bone, the retina, and the duodenum. Herophilus may have been first to count the beats of the pulse, with an ingenious water clock, and thereby to diagnose even amorous passion, if not the healing, subtle energies of love, interpreted by a psychiatrist (Pecci 1997) in the light of subtle electromagnetic energy fields.

In the Confessions of St. Augustine (354–430 AD), a broad concept of the soul, *anima,* was equated with the Greek "psyche" and Hebrew "nephes": the latter means "life", including everything that contributes to life. Tarquini speculated whether the teachings of the Vedas became known in Europe through the Templars, the monk-soldiers of St. Bernard, eventually mostly annihilated by Philip "the Beautiful" of France, except for those retained by the initiators of Freemasonry (Barber 1994; Demurger 1989; Partner 1982; Seward 1995). From the Templars may stem the kiss of the behind, which should excite or awaken the serpent Kundalini, a cosmic force which resides in the roots of the dorsal spine and in the sex glands, and which once awakened reaches the pineal gland, an explicit reference relating this gland to the cosmos. The Templars are credited with the recognition of the "third eye" in the pineal gland, providing a direct vision of time and space.

In the seventeenth century, Rene Descartes (1596–1650) wrote: "There is a small gland in the brain, the pineal, in which the soul exercises its function in a more peculiar way than in all other parts." (Descartes 1664/1972). He thus maintained, interpretations to the contrary notwithstanding, that spirituality (Tiller's 11th dimension, the spirit) cannot be localized in any one part of the body; but he considered the pineal gland as the organ where the mind (Tiller's 10th dimension) or anima *(res cogitans)* meets the material, the fleeting mortal body *(res extensa).* With Tiller's view of humanity as a vast body, *on revient toujours à ses anciens amours.*

4.5 Anatomical–Clinical Associations

Against this mythical historical background, the development of less philosophical, albeit homeostatic studies, was delayed until the middle of the eighteenth century, when some investigations contested the Cartesian view. Around the middle of the seventeenth century, Giovanni Battista Morgagni (1682–1771), the father of pathological anatomy, in "De sedibus", with his thorough postmortem examinations, provides case histories for patients with "disturbances of the mind" as having anomalous, atrophic, or calcified pineal glands. The pineal gland has a particular propensity to calcify, as now documented with modern methods of imaging (Schorner et al. 1991; Sandyk and Awerbuch 1991; Kohli et al. 1992; Sandyk 1993). The calcification begins with the deposition of hydroxylapatite in pineal cells. The concretions, acerbuli, or sand of the brain are eventually expelled into intercellular space where they form aggregates. The calcification of the pineal gland has had different interpretations. The then-leading endocrinologist Nicola Pende, following up on Morgagni, maintained correlations of the pineal gland with "psychosexual alterations", which he defined as including sadism, masochism, homosexuality, and satyriasis. In the following century, other scholars, like Thomas Arnold (1795–1842), made the same observation as Morgagni and diffused the conviction that madness can be tied to the presence of an abnormal pineal gland.

In the last years of the nineteenth century, the first clinical cases of affections of pineal origin were reported. The anatomists of this time regarded the pineal gland as a vestigial organ, biologically redundant, as did the clinicians based on morphologic and comparative studies. Some of those who more recently assessed the degree of calcification in the pineal gland by computerized tomography have correlated this degree of calcification with a reduced production of melatonin, with alterations of sleep – wakefulness and tiredness during the day; but this was not found for 6-sulfat-oxymelatonin in a small sample (Bojkowski and Arendt 1990). The pineal gland has been studied for possible involvement in: multiple sclerosis (Sandyk and Awerbuch 1992, 1994), a point of interest from a geophysical viewpoint (Resch 1995); schizophrenia onset (Sandyk 1992a); and more generally (Sandyk 1992b, 1993), bipolar emotional disease (Sandyk and Pardeshi 1990) and psychiatric illness (Sandyk and Awerbuch 1993).

4.6 The Homeostatic Melatonin Era

Lerner discovered melatonin, an indole derived from 5-hydroxytryptamine, as the major but not sole secretion of the pineal gland, as recently as 1958. For the subsequent 20 years, the pineal gland remained in obscurity (its natural environment!). It was catalogued as a neurochemical fossil, only to gain worldwide fame almost immediately thereafter, not so much because of the progress made in the discovery of the growing importance of the photofraction in agriculture but as a public response to the many hypotheses in experimental physiopathology about the critical properties of melatonin. From the end of the 1970s to the first years of the 1990s, the pineal gland came to be resuscitated as an omnipotent, oncostatic organ, a proposition that had been made in Vienna in the pre-melatonin era (Engel and Bergmann 1952). Melatonin also became the pill for eternal youth, even if today aging, rather than representing a hope, constitutes more of a menace (except for good wine which improves with age!).

Embryologically derived from the ependyma, which covers the ceiling of the third ventricle, the pineal gland or epiphysis, also called the penis of the brain, remains an ontogenetically very old organ, a member in the family of circumventricular secretory organs which include the subcommissural organ (SCO), the subfornical organ (SFO), the organum vasculosum of the lamina terminalis (OVLT), the ependyma of the median eminence, and the area postrema at the posterior border of the ceiling of the fourth ventricle. These organs, described also as "windows of the brain", with the exception of the SCO, are areas which lack a blood–brain barrier. Their functions are not known with certainty. Along an evolutionary scale, these organs could be involved in the metabolism of fluids, osmolality, and electrolyte transport. Melatonin is indeed found in insects (Wetterberg et al. 1987) and unicells (Balzer and Fuhrberg 1996).

In the 1960s, Farrell postulated a hormone of pineal origin that coordinates the glomerulosa of the adrenal cortex, which produces aldosterone – adrenocortical glomerulotropin or glomerulotropin. The biosynthesis of aldosterone in the glomerulosa is influenced by a number of factors, notably by the octapeptide angiotensin II (Aguilera 1993; cf. also Franchimont 1964; Haulica et al. 1981; Damian 1989). In some amphibians and fishes, pineal cells have characteristic photoreceptors with photosensitivity and electrical activity suggestive of a "third eye". In birds and reptiles, in addition to photoreception, there is a secretory function which becomes more pronounced in higher vertebrates. The vascularization of the pineal gland is rich when viewed in relation to the amount of tissue. After the kidney, Brunetto regarded the pineal gland as the second most irrigated organ. This certainly constitutes a biological paradox, if the pineal gland is a redundant vestigial organ.

The pineal gland is innervated by postganglionic fibers that derive from the sympathetic ganglion and are found along the great vein of Galen. The sympathetic nervous input into the pineal gland is coordinated by impulses from the suprachias-matic nuclei, which in turn are innervated directly from the retinal-hypothalamic tract. Today, many investigators explore the potential role of melatonin, besides that in onco-logy and geriatrics, in the treatment of many psychiatric conditions, including depres-sion, bipolar manic-depressive illness, schizophrenia, and Seasonal Affective Disorder (SAD). With SAD and its geographic distribution, the foregoing broad perspective of the pineal gland from thousands of years ago led to data that recognized the half-yearly signature of geo- and heliomagnetics in the human circulating melatonin concentra-tion (Tarquini et al. 1997b; Maggioni et al. 1999).

For basic science, the data document the master switch of helio- and geomagne-tism, complementing that of visible light. This invisible switch acts at night in the absence of sunlight and heat; its effects may be less tangible at middle than at high latitudes and await study as an oncological risk indicator for humans wishing to venture into space (Halberg and Cornélissen 1998). Before such studies are designed, the dynamics of melatonin should be explored.

4.7 Sampling and Assessing Chronomes

Each variable's time structure, its chronome (Fig. 4.1), constitutes the desirable, albeit often at best only partially obtainable, control information and should replace an imaginary "baseline", as outlined in Table 4.A1 (Halberg et al. 1991b; cf. Cornélissen and Halberg 1994; Macey 1994). The predictable element of chronomes is a broad spectrum of rhythms (Halberg 1969, 1980; Cornélissen and Halberg 1994) in which about 24-h patterns, circadians, are prominent and represent, for certain variables, an often indispensable dimension to be assessed or at least considered in time-specified spot-checks. But the relative prominence of patterns differs among variables, even in the same system. The blood pressure of a 73-year-old man may show a prominent about-yearly pattern, while in the same circulation, heart rate may show no circannual component while the third harmonic of the year stands out in the spectrum (Katinas et al. 1998). In the same circulation of seven clinically healthy medical students, cortisol may be prominently circadian, while endothelin-1 (ET-1) may be about 8-hourly (Herold et al. 1998) and about 84-hourly periodic (Tarquini et al. 1997a). These same circaoctohorans and circasemiseptans, superficially orphan rhythms (as long as they have no generally known matching environmental counterparts), were also found in murine pinnal dermis in the population densities of endotheliocytes, the cells producing ET-1, with the circadians in the same population densities appearing only as a transient response to trauma (Katinas unpublished). Here, we show further, as illustrated by studies of melatonin and circulatory variables, that certain generally neglected patterns with frequencies other than circadian need to be considered, along with other aspects of the chronome, such as chaos and trends. In the 1960s, an about 24-h (circadian) rhythm in the melatonin content of pineal glands from rats was documented by Quay (1964) and subsequently confirmed by others (Axelrod 1974; Reiter 1980). Arendt et al (1977; see also 1984, 1985) documented a circannual rhythm (see also Arendt 1986, Arendt and Broadway 1986).

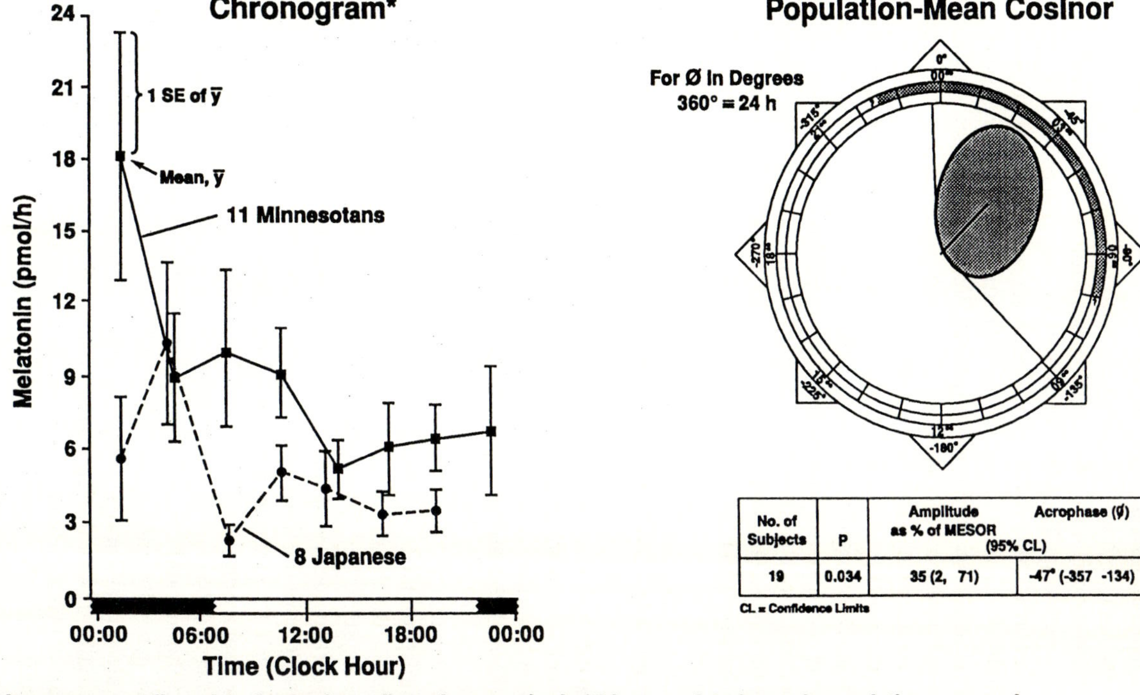

Fig. 4.6. Circadian rhythm of urinary melatonin. Note differences in pattern. (Copyright Halberg)

4.8 Subtle Human Melatonin Rhythms

A circadian rhythm in human melatonin secretion and excretion was mapped by Lynch et al. (1975). Richard J. Wurtman's group showed that very small doses of exogenous melatonin, which raised blood concentrations to those present at night, accelerated sleep onset and sustained sleep when given at 1145 hours, as a "sleep hormone" might do (Dollins et al. 1994). Figure 4.6 shows circadian rhythms and models fitted to them to derive new end points for quantifying dynamics, such as amplitudes and acrophases, as measures of the extent and timing of change. Figure 4.7 shows two ultradian components in the circulating melatonin of children, with periods of 3.4 and 1.5 h, which may be related to the REM cycle (Salti et al. 1999). These two components are visualized in this figure, obtained by stacking the data of eight boys and eight girls over an idealized cycle. The original data from each child were expressed as a percentage of the corresponding time series mean value in order to eliminate interindividual differences in nocturnal melatonin concentration. Thus, common features in the pattern of samples obtained every 30 min from 1900 to 0700 hours could be viewed more easily. These summaries are complemented by a one-way analysis of variance testing for the equality of all class means. Even though this approach is less powerful than the cosinor which consistently yields statistical significance, an analysis of variance also validates the 3.4-h component with statistical significance and the 1.5-h component with borderline statistical significance.

Figure 4.8 shows within-week as well as within-day variations (Herold et al. 1999). While these are readily seen by the naked eye, it should be recognized that the ordinate

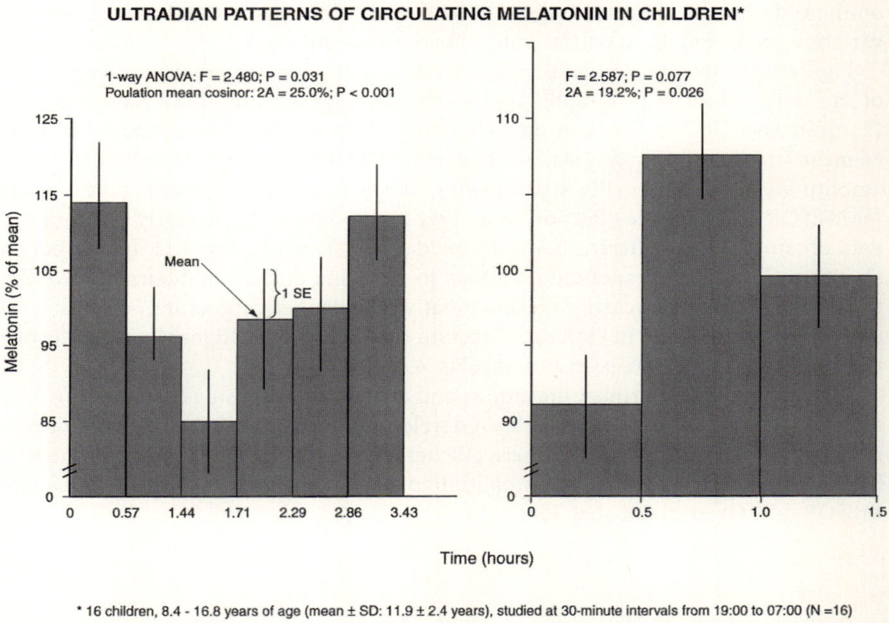

Fig. 4.7. Ultradian components characterizing nocturnal circulating melatonin concentration in children. (Salti et al. in press) (Copyright Halberg)

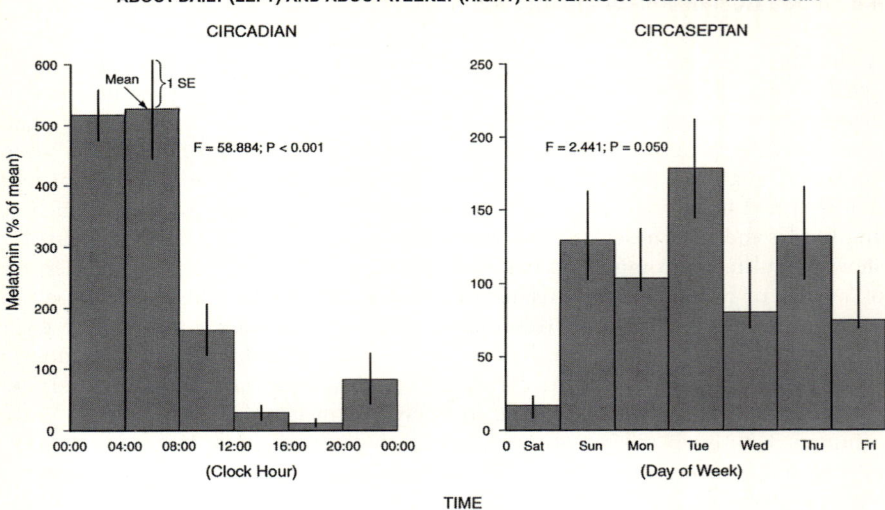

Fig. 4.8. Circadian and circaseptan changes in salivary melatonin concentrations. (Herold et al. in press) (Copyright Halberg)

on the right for the pattern along the 1-week scale differs from that for the circadian variation on the left. Both patterns are statistically significant.

Figure 4.9 shows, with serially independent sampling, the about-yearly pattern of melatonin at 43°N geographic and 43.98 geomagnetic latitude in Florence, Italy (Tarquini et al. 1997b). The same data are summarized as changes in chronome-adjusted mean [midline-estimating statistic of rhythm (MESOR)] on the left and as circadian amplitudes on the right in Fig. 4.10, showing an opposite time course along the seasons for MESOR vs amplitude (Tarquini et al. 1997b). Profiles of about-yearly variations in data obtained at six different hours of the day are shown in Fig. 4.11 and reveal an about-yearly pattern, responding perhaps to sunshine during the hours of daylight, whereas an about-half-yearly pattern, a putative geomagnetic signature, is seen at 0000 and 0400 hours by night in Florence (Tarquini et al. 1997b). At a higher latitude, the two peaks during the year are seen at noon (Fig. 4.12).

The possible use of urinary melatonin, and of the ratio of plasma cortisol to urinary melatonin for assessment of the risks of developing cardiovascular disease or emotional depression is discussed elsewhere (Wetterberg et al. 1986). For the assessment of breast cancer risk, the circadian amplitude of melatonin may serve if the circadian MESOR fails (Wetterberg et al. 1986).

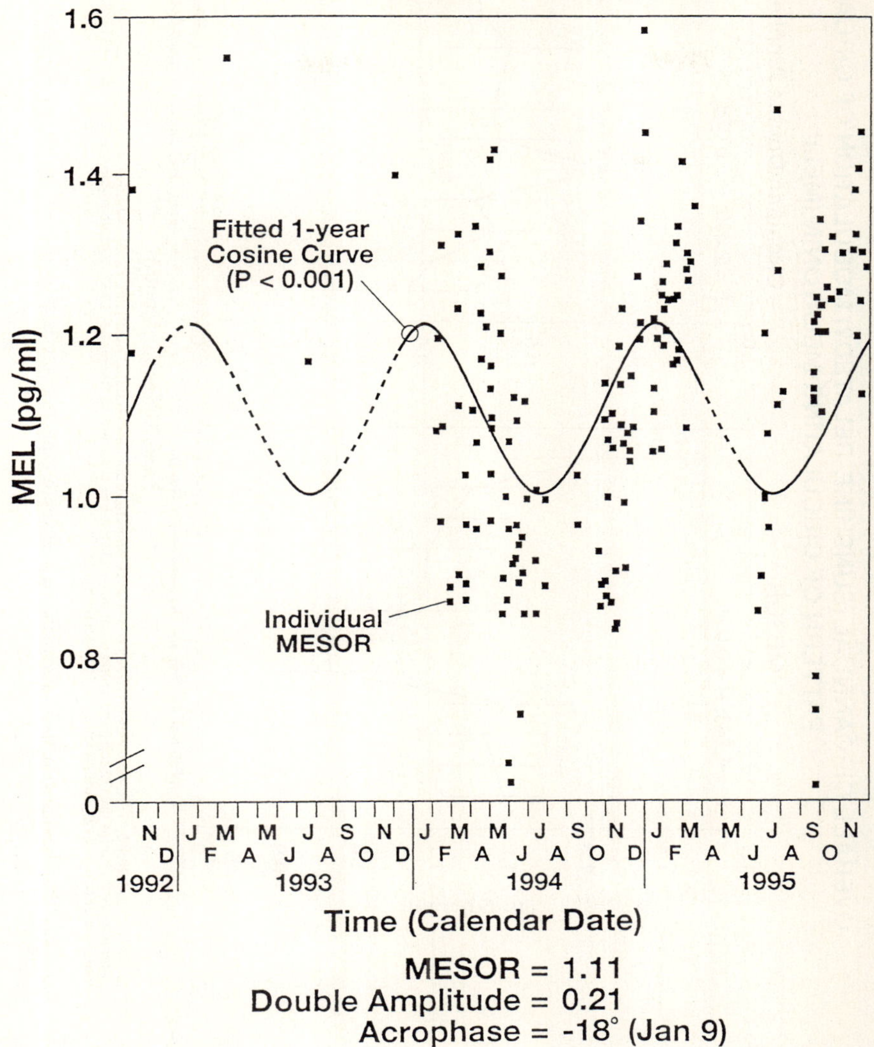

* Single cosinor analysis on $\log_{10}$-transformed data from 172 subjects; squares are individual MESORs estimated from 4-hourly sampling for 24 hours, in Florence, Italy, 43.47 °N.

Fig. 4.9. With serially independent sampling, an about-yearly variation is demonstrated for human circulating melatonin. (Copyright Halberg)

OVERALL CIRCANNUAL (SUNSHINE-RELATED?) MODULATION OF CIRCADIAN PATTERN OF CIRCULATING MELATONIN (MEL)*

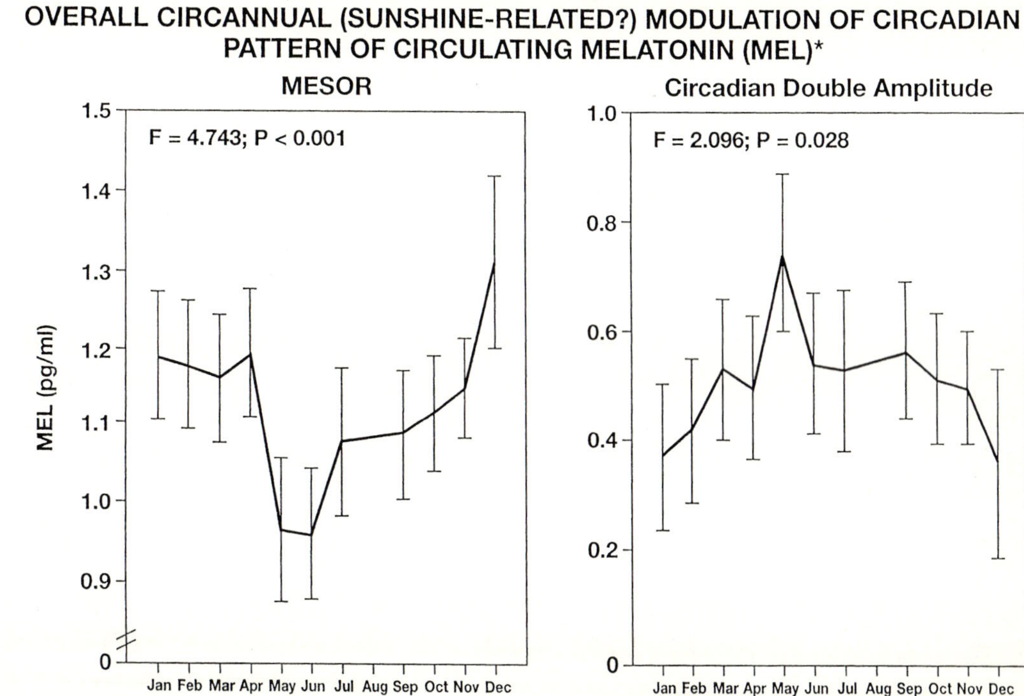

* In Florence, Italy, 43.47 °N; one-way analyses of variance on log_{10}-transformed data: means and 95% confidence limits; data collected at 4-hour intervals for 24 hours on 172 subjects studied between 13 October 1992 and 15 December 1995. Result on the MESOR qualified by differing patterns between day- (light) and night- (dark) time hours.

Fig. 4.10. The circannual pattern of circulating melatonin exhibits opposite time courses for the MESOR and circadian amplitude. (Copyright Halberg)

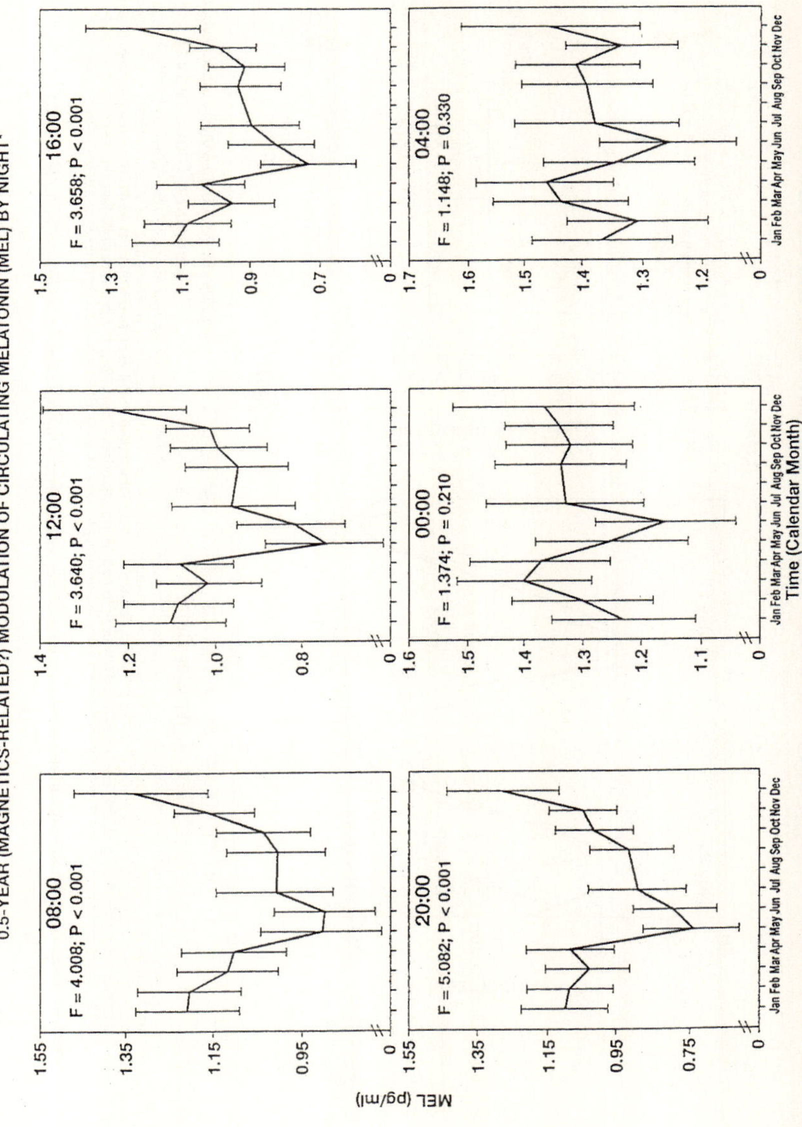

* During daylight (08:00–20:00) versus nightly darkness (00:00; 04:00) hours, about-yearly (sunshine-related?) and about-half-yearly (magnetics-related?) effects are seen; one-way analyses of variance on $\log_{10}$-transformed data: means and 95% confidence limits; data collected at 4-hour intervals for 24 hours on 172 subjects studied between 13 October 1992 and 15 December 1995, in Florence, Italy, 43.47 °N.

Fig. 4.11. At 43.47 °N latitude, circulating melatonin may undergo an about-yearly rhythm during the daylight hours and an about half-yearly rhythm during the hours of darkness. (Copyright Halberg)

ALONG THE SCALE OF THE YEAR, CIRCULATING MELATONIN REVEALS LATITUDE AND CIRCADIAN STAGE-DEPENDENT EFFECTS*

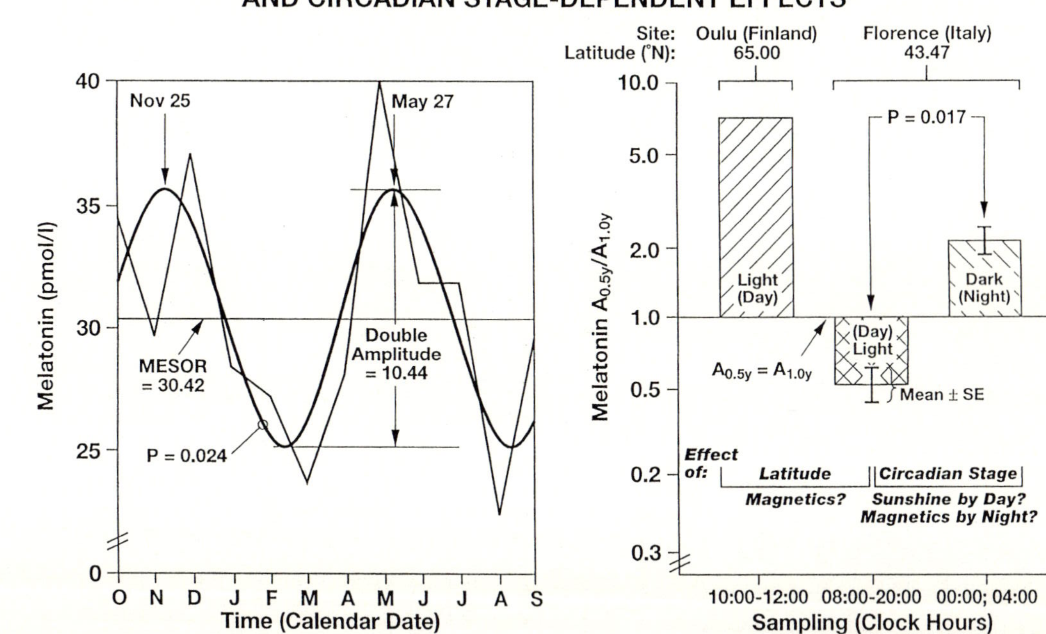

* Half-yearly (possibly magnetics-related) response pattern is seen at 65.00 °N in Oulu, Finland (left half of graph) when measured around noon; accordingly, the log-ratio of the half-yearly to yearly amplitudes exceeds unity (first bar in right half of graph) whereas this ratio (original data not shown) is much smaller than unity (second bar in right half of graph) during waking hours (including noon) in Florence, Italy, at 43.47 °N, where the amplitude ratio exceeds unity only during nightly darkness (third bar in right half of graph); original data from 172 patients of B. Tarquini, each sampled every 4 hours for 24 hours; data in Oulu are from 11 healthy men sampled monthly for one year (Martikainen et al., Acta Endocr. 109:446-450, 1985).

Fig. 4.12. At 65.00 °N, in Oulu, Finland, human circulating melatonin is characterized by a circasemiannual pattern. (Copyright Halberg)

4.9 Pineal Gland and Adrenal Cortex, Feedsidewards

Any pineal study profits from including the adrenal cortex, a known mechanism of rhythmicity, indispensable for the maintenance of the rhythm in circulating eosinophils of humans (Halberg et al. 1951; Halberg 1953; Kaine et al. 1955) and of mice (Halberg et al. 1953). In the 1950s, we also learned that corticosterone in mouse blood was circadian periodic (Halberg et al. 1958a) and by the 1960s we had reported that corticosterone production by the adrenal gland in response to ACTH was clearly time-dependent (Ungar and Halberg 1962; Halberg 1965), as was pituitary ACTH activity (Ungar and Halberg 1963). In a survey in October 1982, we studied the difference at each time point between the corticosterone produced in response to the stimulating effect of ACTH 1-17 alone, i.e., without the addition of pineal homogenate, vs the response to ACTH in the presence of pineal homogenate. The differences at the six test times were drastically different, changing sign at two of the six test times so that addition of pineal homogenate enhanced corticosterone production at two test times and reduced it at four other test times, revealing the rhythmically changing interdependency of the three hormones (Brown et al. 1983) (Table 4.1).

Accordingly, we refrained from using the feedback concept describing time-unqualified interactions between two interdependent endocrine glands, e.g., the hypophysis and adrenal (Table 4.A1), replacing it with the "feed-sideward", a concept based on time-qualified facts. Feed-sideward was coined (Macey 1994; cf. Halberg 1983; Sánchez de la Peña 1993) to describe a changing modulating relation between three or more a priori periodic entities, such as the pineal, pituitary, and adrenal glands that led to a rhythmic and to that extent predictable change from amplification over no-effect to damping by the same entity, the modulator (the pineal gland), of the interaction of two other periodic entities, the pituitary and adrenal glands (Table 4.1). In a large series of such experimental studies, a model of circadian rhythmically recurring changes, including opposite effects of melatonin was defined in extenso in 1983 as the feedsideward (Halberg 1983; Sánchez de la Peña 1993). The first evidence was a three-

Table 4.1. First report of a circadian rhythm in pineal melatonin (top) and of a pineal feedsideward* (bottom). (From Brown et al. 1983)

Experiment	Melatonin (APH) (pg/pineal)					
HALO[§]	02	06	10	14	18	22
10/11/82	<50.0	<50.0	51.5	170.0	<50.0	<50.0
10/31/82	<50.0	<50.0	51.5	2007.2	200.0	<50.0

APH effect upon stimulation by ACTH 1–17 of adrenal corticosterone (difference in ng/ml per hour between ACTH 1–17 alone or with APH, *$P < 0.05$ by t test)

Date HALO[#]	02	06	10	14	18	22
10/31/82	−34.5*	−34.5	−35.5*	+40.9*	+85.5*	−47.8*

B6D2F$_1$ female mice, like rats, show dramatic circadian variation in the melatonin content of an aqueous pineal homogenate (APH). After 4 weeks at 24 ± 1 °C and ~68% relative humidity on six LD12:12 regimens staggered by 4 h each, with Purina Chow and water freely available, 13–20 mice from each regimen were killed at the same clock-hour to sample six circadian stages. Pineal and adrenal glands were removed, pineal glands pooled in a conical glass tube, homogenized with 9% NaCl solution (1 pineal/1 ml) and APH partly used in adrenal incubations and partly stored at −90 °C until melatonin RIA.
* Cyclic factors other than melatonin are also likely to contribute to this feedsideward.
HALO, hours after light onset; §, pineal harvest time; #, adrenal harvest time.

way interaction with aqueous pineal homogenate or melatonin being incubated with pituitary glands and the incubation fluid then being incubated with adrenal glands to measure corticosterone production by the latter (Table 4.1). The same change in sign of response, a rhythmically alternating stimulation and inhibition in vitro by melatonin incubated with the isolated adrenal or pituitary gland, was also shown as the time-varying response of these glands directly to melatonin without any mediation by the pituitary gland (Brown et al. 1983; Sánchez de la Peña et al. 1988).

The differences in adrenal corticosterone production in vitro between the effect of ACTH alone or with pineal homogenate, as profiles each consisting of results at six test times, were replicated in our laboratory in a series of studies. The studies were carried out isophasically, with the adrenal, pituitary, and pineal gland at each test time from the same circadian stage, and heterophasically, with the three glands from different circadian stages (Sánchez de la Peña, 1993). In each of the profiles, the result changed in sign, so that the addition of pineal homogenate enhanced corticosterone production at some test times and reduced it at other test times, revealing the rhythmically changing interdependency of the three hormones. The mean of five isophasic studies, summarized in Fig. 4.13, describes the changing relation between three a priori periodic entities; the pineal, pituitary, and adrenal glands. One finds a rhythmic and to that extent predictable change from amplification over no-effect to damping by the same entity, the modulator (the pineal gland), of the interaction of two other periodic entities, the pituitary and adrenal gland, a sequence of effects also reproduced as a direct effect of melatonin upon the production of corticosterone by the adrenal gland in vitro (Sanchez de la Peña et al. 1986; cf. Halberg 1983, Sánchez de la Peña 1993,

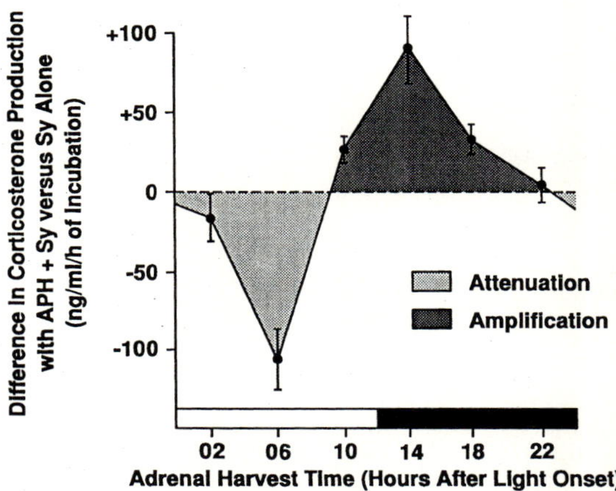

Lack of effect, attenuation or amplification by Aqueous Pineal Homogenate (APH) of corticosterone production by bisected adrenals in response to Sy; mean of 5 isophasic studies

Fig. 4.13. The pineal modulates, in a predictable since rhythmic fashion, the effect of the pituitary upon the adrenal gland, insofar as corticosterone production by bisected adrenal glands, stimulated by ACTH 1–17, is concerned. (Copyright Halberg)

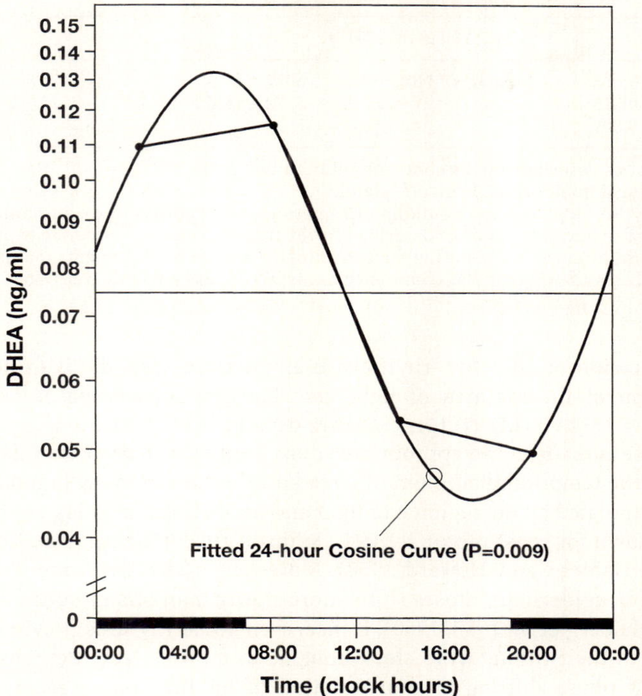

Fig. 4.14. The production of DHEA stimulated by melatonin shows quantitative differences in the expression of an adrenal chronosensitivity. (Copyright Halberg)

Macey 1994). In contrast, in a meta-analysis of data published as failing to show a time effect in light of an analysis of variance (Fig. 4.14) (Haus et al. 1996), a meta-analysis by cosinor found quantitative but not qualitative differences in the production of dehydroepiandrosterone (DHEA).

4.10 Circaseptans

The biological week was well known to Hippocrates, Galen, and Avicenna (Hildebrandt and Bandt-Reges 1992) and in later centuries, long before ours (Hufeland 1797). In 1986, an about-weekly (circaseptan)-circadian intermodulation in the melatonin content of the rat pineal gland was also found (Sánchez de la Peña et al. 1986). Table 4.2 shows both circadian and circaseptan rhythmicity in rat pineal melatonin. These periodicities were anticipated and validated cost-effectively in a chronobiologic pilot design (Sánchez de la Peña et al. 1983a). In such a design, the feasibility of shifting

Table 4.2. Periodic components found in chronobiologic pilot study of rat pineal melatonin[a]

Period (τ) fitted in h	P value (Ho: A = 0)	MESOR (M) ± SE	Amplitude (A) ± SE	Acrophase (ϕ) (95% CL)
		(pg/100 µl of APH[b])		
168	0.006	8652 ± 1340	6622 ± 1824	−142° (−108°, −175°)
24	0.013		5772 ± 1887	−277° (−240°, −314°)
Overall	0.001			

[a] Characteristics described by the concomitant fit of two periods.
[b] Aqueous pineal homogenate from one gland.
MESOR, a rhythm-adjusted mean (midline-estimating statistic of rhythm); amplitude, one half of the measure of the extent of change described by the fitted curve; acrophase, lag from start of light span of peak in the curve representing the concomitant fit of two components (Sánchez de la Peña et al. 1988); SE, standard error; CL, confidence limits; acrophase (ϕ) in degrees; 360°, τ.

the time location of circadian rhythms to any desired schedule is exploited by the manipulation of the regimen of light and darkness alternating at 12-h intervals [light:dark = 12:12 h (LD 12:12)] (Sánchez de la Peña et al. 1983a, b).

Allowance is made for an appropriate adjustment time. Earlier work had shown, as noted, that the temporal placement of circadian rhythms at many organization levels can be manipulated by the regimen of light and darkness alternating at 12-h intervals. This was shown for gross motor activity (Johnson 1926; Halberg et al. 1957, 1958a, b, 1959, 1960; Halberg and Howard 1958). Moreover, under the same light intensity, the rhythm in epidermal mitoses shifts more slowly than others, such as that in liver glycogen (Halberg et al. 1957, 1960). Hence, even some rhythms previously believed to be fixed in their timing as to clock hour, as in mitoses, were demonstrated to be amenable to phase shifting by manipulation of lighting, once several weeks were allowed for their adjustment to the new regimen (Halberg et al. 1957; cf. also Sánchez de la Peña et al. 1983a). The circadian rhythm in seizure susceptibility shifted relatively rapidly in rodents (Halberg et al. 1958a) and humans (Halberg et al. 1959).

The analyses of the Fig. 4.15 data in Table 4.2 reveal that both the 24-h and 168-h cosine curves have a non-zero amplitude, whereby the assumptions of "no-circadian pattern" and „no-circaseptan pattern" are both rejected. The double amplitude in Table 4.2, expressed as a percentage of MESOR, indicates the large extent of change in the about 7-day cycle and expresses further that the circaseptan double amplitude can be larger than the circadian one. The concomitant assessment of at least two components is a step toward changing from focus upon a circadian clock to the spectral element of multifrequency rhythms in the chronome (Fig. 4.2) (Halberg et al. 1991b; Cornélissen and Halberg 1992, 1994; Macey 1994). For the assessment of certain other rhythms, such as ultradians, and the chaotic element of the chronome, the data at 4-h intervals are not dense enough. For assessing yet another element, trends (in end points of rhythms and chaos), the time series here analyzed covering only 1 week, are not long enough and limitations as to density and length restrict the search for any other unresolved variation (Fig. 4.1).

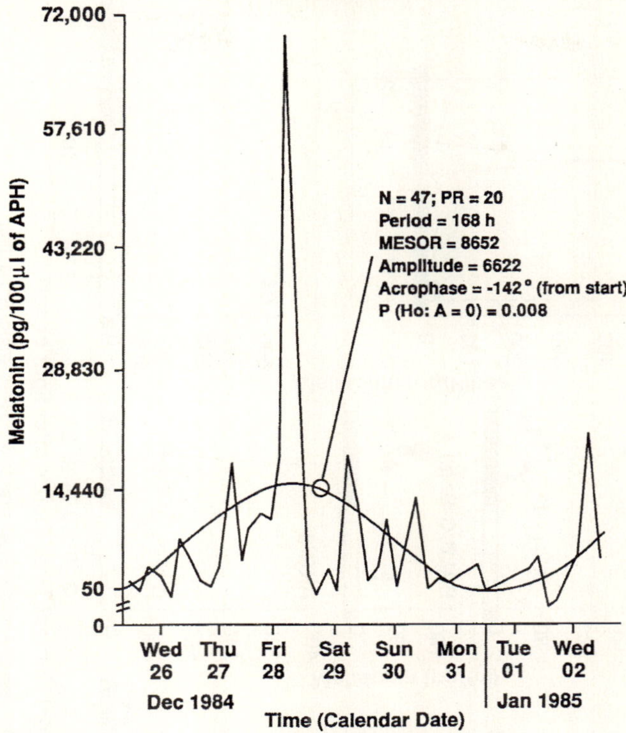

Fig. 4.15. Circadian and circaseptan patterns resolved in chronobiologic pilot study of rat pineal melatonin. (Copyright Halberg)

4.11 The Need for Dense Sampling Before the Ruling Out of Rhythms and Presence of Melatonin

Melatonin in the mouse pineal gland and a rhythm in this hormone were reported by us in 1983 (Brown et al. 1983) (Fig. 4.16), with new data appearing a few years later (Brown et al. 1986; cf. Sánchez de la Peña et al. 1988). The first two profiles, published by Brown et al. in 1983 (Table 4.1), were carried out about 3 weeks apart, on 11 and 31 October, as specified by Sánchez de la Peña et al. (1988). We used animals of the same sex, about 11 and 14 weeks (±3 days) of age, respectively; the mice were from the same

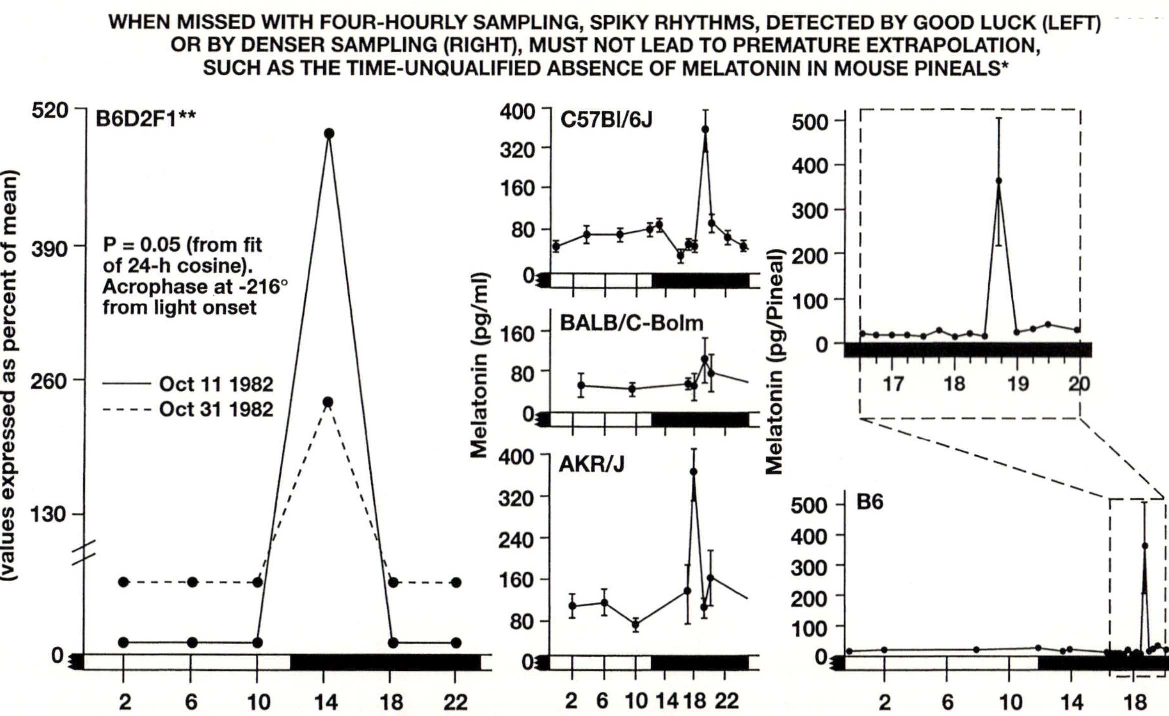

* Brown G.M., Grota L.J., Sánchez de la Peña S., Halberg F., Halberg E. Abstracts of papers, Minn. Acad. Sci. 51st Annual Spring Meeting. University of Minnesota-Duluth, April 29-30, 1983, p. 12. Maestroni G.J.M., Conti A., Pierpaoli W. Clin. exp. Immunol. 68: 384-391, 1987. Conti A., Maestroni G.J.M. J. Pineal Res. 20: 138-144, 1996.

** First generation hybrids of subline 6 of C57 Black strain (B6) and subline 2 of Dilute Brown strain (D2).

Fig. 4.16. The "spikiness" of the circadian variation in murine melatonin may be responsible for the failure of others to find melatonin in some strains of mice. (Copyright Halberg)

shipment provided by an extramural supplier. They were presumably of a comparable genetic background, i.e., first generation hybrids (F_1) of two inbred strains: the C57 Black, subline 6 (B6), and the Dilute Brown, subline 2 (D2). The parent strains had been brother-to-sister mated for nearly half a century. Six groups, each consisting of 13–20 $B6D2F_1$ female mice were kept in six separate chambers on six different LD 12:12 regimens staggered by 4 h. In each chamber, 12 h of light alternated with 12 h of darkness. The daily light span began in chambers 1–6 at 0800, 1200, 1600, 2000, 0000, and 0400 hours, respectively. Hence six different circadian stages could be tested at a single clock-hour. The pineal glands were analyzed in Dr. Gregory Brown's laboratory by the same radioimmunoassay technique, nearly at the same time.

In the first experiment, radioimmunoassayed melatonin was found in aqueous pineal homogenate from glands removed at 14 h after light onset, whereas at five other test times none was detected (Fig. 4.17, top left). In the replication, melatonin was found only at 14 and 18 h after light onset. Sharp peaks were found in each experiment at 14 h after light onset. The timing of the peaks was similar, as could be anticipated from studies carried out only 3 weeks apart, presumably at roughly comparable stages of any about-weekly and about-yearly rhythms. The height of the peaks, however, differed drastically in the two studies. The data of Fig. 4.17 were included in a review in 1988, with the following comment: "The extremely sharp spikiness of the peaks in both studies prompts the question whether the very top of a true peak was actually sampled in one study but not in the other. These circumstances as well as infradian changes may contribute to a stunning interstudy difference in MESORs." (Sánchez de la Peña et al. 1988).

In 1986, we confirmed our earlier finding on the spikiness of pineal melatonin once more (Brown et al. 1986). The finding of spikiness in the melatonin of mice was extended at about the same time to melatonin in the circulation of three strains of mice by Maestroni et al. (1986, 1987), who demonstrated a circadian rhythm of plasma melatonin in 3–4-month-old inbred C57Bl/6J BALB/c Bohn and AKR/J female mice, (Fig. 4.17). In the C57Bl/6J mice, the mean values at 14 time points were less than 100 pg/ml, while at one time point during the dark span the average was 360 pg/ml.

4.12 A Controversy Resolved

In Fig. 4.17, it is easy to see the circadian rhythm by the unaided eye. The zero-amplitude (no rhythm) assumption, however, is rejected only by fitting a 24-h cosine curve to the $\log_{10}$-transformed data. The data of Figs. 4.16 and 4.17 were also presented at a Gordon Conference in the early 1980s, where a lack of melatonin in mouse pineal glands was sweepingly described by a scholar in the field. His statement was immediately rectified with the demonstration of the Fig. 4.16 data (Brown et al. 1983). Nonetheless, without reference to our report, the same scholar subsequently restricted his claim at the next Gordon Conference to the statement that the pineal gland only of the domesticated mouse does not contain melatonin. A subsequent paper from the same laboratory, again more cautiously, concluded that melatonin is absent from the pineal gland of *some* domesticated mice, such as C57Bl mice, while it was present in the pineal of wild-derived mice. This paper in *Science* (Ebihara et al. 1986) was at variance with the original work (Brown et al. 1983; Halberg 1983) and with its extension to the finding of the presence and spikiness of pineal melatonin also in the plasma of the C57Bl

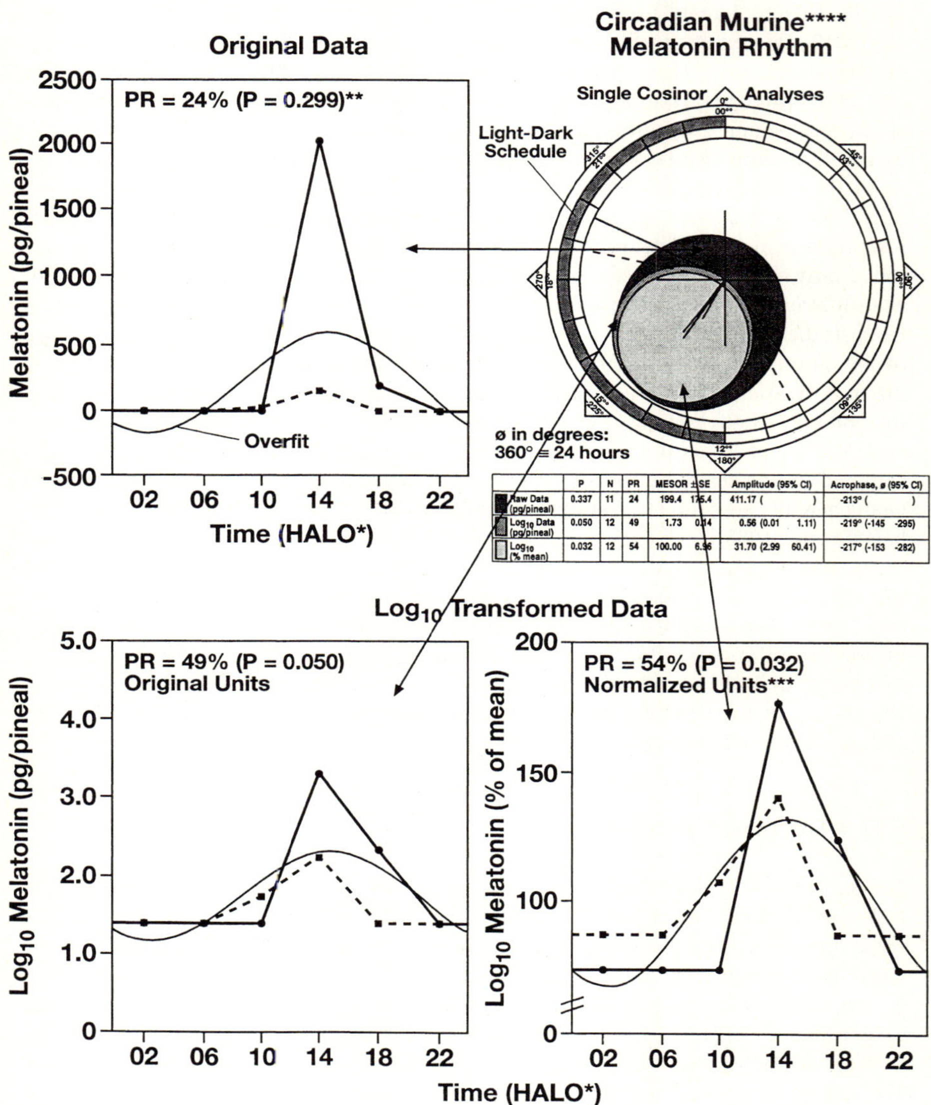

	P	N	PR	MESOR ±SE		Amplitude (95% CI)	Acrophase, ø (95% CI)
Raw Data (pg/pineal)	0.337	11	24	199.4	175.4	411.17 ( )	-213° ( )
Log10 Data (pg/pineal)	0.050	12	49	1.73	0.14	0.56 (0.01 1.11)	-219° (-145 -295)
Log10 (% mean)	0.032	12	54	100.00	6.96	31.70 (2.99 60.41)	-217° (-153 -282)

* HALO: <u>H</u>ours <u>A</u>fter <u>L</u>ight <u>O</u>nset; ** P-values for ordering only in the want of a better generally applicable model; *** By eliminating inter-series difference in average; **** Female $B_6D_2F_1$ mice.

Fig. 4.17. Demonstration of a circadian rhythm of melatonin in the mouse pineal gland; methodological considerations for its assessment in view of its spiky waveform. (Copyright Halberg)

strain, for which the lack of pineal melatonin was reported (Ebihara et al. 1986). Eventually, the above-discussed reports based on radioimmunoassays were validated by high performance liquid chromatography (HPLC) (Conti and Maestroni 1996), but the paper in *Science* deserves more detailed methodological comment.

Ebihara et al. (1986) were unable to find pineal melatonin in certain strains of mice which had been inbred in the laboratory for more than 40 years. In contrast, in pineal glands of strains which were inbred in the laboratory for less than 30 years, they could eventually demonstrate the presence of melatonin, which earlier had been stated by one of the authors to be categorically absent in the pineal gland of mice, without any qualification whatever. In a genetic analysis, the authors postulated that two independently segregating autosomal recessive mutant genes are probably responsible for the lack of enzymes NAT (*N*-acetyltransferase) and HIOMT (hydroxyindole-*O*-methyltransferase), which are necessary to convert serotonin into melatonin and which they were also unable to find in their experiments, using radioimmunoassays for the determination of melatonin and the transferase system. Serotonin, measured by HPLC with electrochemical determination (HPLC-EC), was found to be normal in the pineal glands of the C57Bl/6J strain. Of course, it could well be that after 40 years of inbreeding in the laboratory some mutation of the transferase genes could have occurred, were it not for the earlier studies on mice in Minnesota in 1982 and 1986 that had already shown the presence of melatonin in the pineal gland of mice in hybrids of strains inbred for over 40 years. The added original Minnesotan observation that melatonin could only be found in the pineal gland during a very limited time span, peaking only quickly, could have been discussed as a minimum, even if material presented at conferences is not to be cited. It was clear from the original work that the presence of melatonin would only be evident if densely timed test results were available.

At this point, Conti and Maestroni (1996) discussed the hypothesis of Ebihara et al. (1986) about a genetic defect accounting for the putative deficiency of melatonin in certain laboratory inbred strains of mice. Conti and Maestroni presumably used the same inbred laboratory strain C57Bl/6 of mice and tested their pineal melatonin in a much more densely timed fashion in three sets:

1. The animals were killed and their pineal gland tested for melatonin at 0000, 0400, 0800, 1200, 1300, and 1330 hours, and every 15 min between 1630 and 1830 Hours After Light Onset (HALO)
2. Every 30 min from 1330 to 2000 HALO
3. Every 30 min during dark hours only between 1630 and 2000 HALO

In all three sets of experiments, Conti and Maestroni were able to demonstrate the occurrence of melatonin in the pineal hydrolysate using the HPLS-EC method. They found a definite circadian rhythm with a very narrow peak during the dark hours (Fig. 4.16, middle and right). These authors found that the retention time for melatonin is extremely short, only of 10 min and 45 s, which easily can account for the lack of melatonin in less densely timed studies, as was already pointed out earlier.

To find proof for the presence or to confirm the absence of NAT and HIOMT, tests with a similar dense sampling would be necessary. Conti and Maestroni (1996) further reported evidence of a melatonin metabolite in the urine: 6-hydroxy-melatonin-sulphate. The possibility that melatonin may be produced at other sites, e.g., the harderian gland, retina, or enterochromaffin cells of the intestinal tract, cannot be excluded (Wetterberg et al. 1990). A circadian rhythm is demonstrated with differences

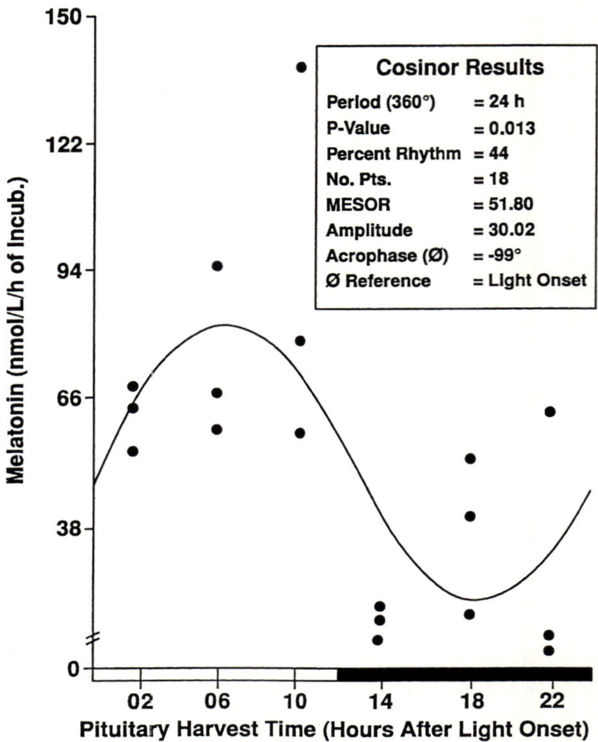

Fig. 4.18. Circadian stage-dependence of melatonin produced by the anterior pituitary gland. (Copyright Halberg)

in timing for melatonin release from the anterior pituitary gland and the hypothalamus of spontaneously hypertensive stroke-prone rats in Figs. 4.18 and 4.19, and a cosinor summary of phase relations to the pineal gland is presented in Fig. 4.20.

In a different study, Conti and Maestroni discussed the effect of melatonin on the immune system studying its role in the development of autoimmune diabetes in non-obese diabetic mice, counteracting the immune suppressive effect by, e.g., corticosteroid treatment, triggering the synthesis and/or release of opioid peptides from activated CD4 T lymphocytes. A feedsideward documented in 1983 at the Minnesota Academy of Sciences and later again discussed by Brown et al. (1986) is pertinent. When we study the effect of melatonin, using, e.g., the aqueous pineal homogenate from test animals to examine any effect on the hypophysis, we have to account for feedsidewards along several time scales (Cecchettin et al. 1986).

Furthermore, for a chronome approach, the assessment of age effects (Fig. 4.21) is essential. In this context, it will be important not to be satisfied by detecting a statistically significant change during several years, say 3 years, whatever its statistical significance. Thus, over 3 years, there can be a decrease followed by an increase, both

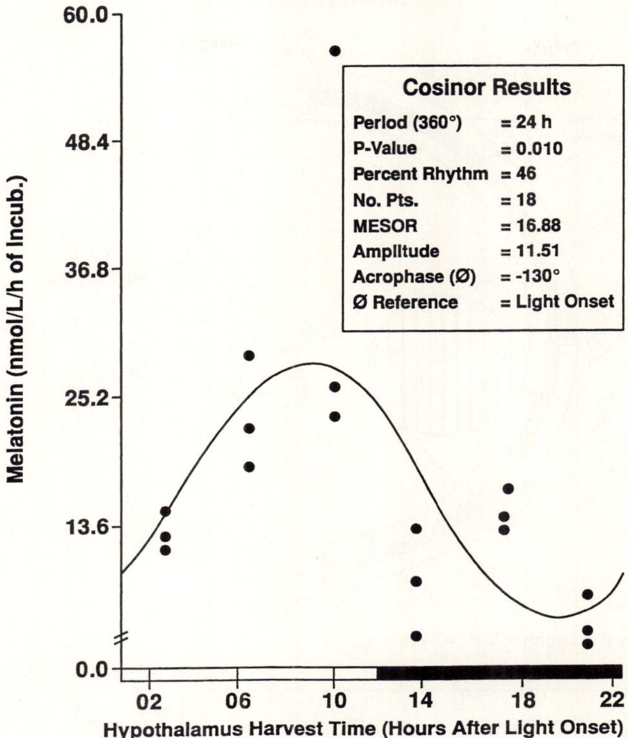

Fig. 4.19. Circadian stage-dependence of melatonin produced by the hypothalamus. (Copyright Halberg)

spurious since a variable such as human 17-ketosteroid excretion can be characterized by an about 10-year cycle. It seems possible that melatonin may also undergo a similar rhythm.

4.13 Summary

A series of studies in vitro all support the tenet that a broader-than-circadian focus by pinealists is overdue. The evidence from different latitudes is in keeping with a role for melatonin as a mediator of the photic circannual and non-photic circasemiannual effects of the sun. As a minimum, the about-weekly change in pineal function should be assessed. One argument stems from the fact, shown in detail in Fig. 4.22, that under controlled conditions, the extent of change in a circaseptan cycle is more prominent than the circadian change in the melatonin secreted by the superfused isolated pineal gland of the pike, kept in continuous darkness at a constant temperature (Cornélissen et al. 1995;

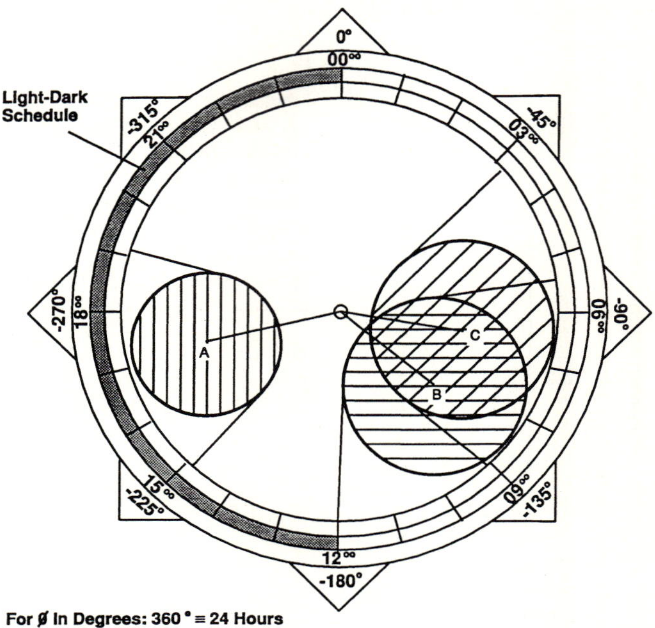

Key to Ellipses	P	No. Obs.	PR	MESOR ± SE		Amplitude° (95% CL)	Acrophase ($\emptyset$)
▥ A Pineal Glands	0.001	12	77.0	82	7.9	61.0 (28.7 94.0)	-255° (-223 -287)
▤ B Hypothalamus	0.010	18	45.6	17	2.3	11.5 (2.7 20.3)	-130° (-81 -180)
▨ C Anterior Pituitary	0.013	18	44.0	52	6.2	30.0 (6.3 54.0)	-99° (-47 -152)

Units: nmol/L/h of Inc.; CL = confidence limits

Fig. 4.20. Cosinor summary indicating the phase relations between the circadian variation of melatonin produced by the pineal gland, the anterior pituitary gland, and the hypothalamus. (Copyright Halberg)

Falcón et al. 1996), as is also the case in the rat pineal gland in situ. The opposite relation, namely a more prominent circadian rhythm, characterizes saliva in healthy humans (Fig. 4.8). It seems reasonable to advocate the assessment of both about-daily and about-weekly rhythms when the circadian and circaseptan patterns of drug administration can contribute to the difference between the inhibition and stimulation of a malignancy (Halberg and Halberg 1980). Certainly, neglect of considering daytime values may dilute a latitude (Wetterberg et al. 1999) or obscure an age effect (Zeitzer et al. 1999). Thus, the use of a sensitive radioimmunoassay on new, as yet unpublished data validates what cosinor end points on log-transformed data published by others (Zeitzer et al. 1999) reveal in Fig. 4.23. When the circadian amplitude of circulating melatonin concentrations takes into account the daytime data, this index decreases with age in adulthood.

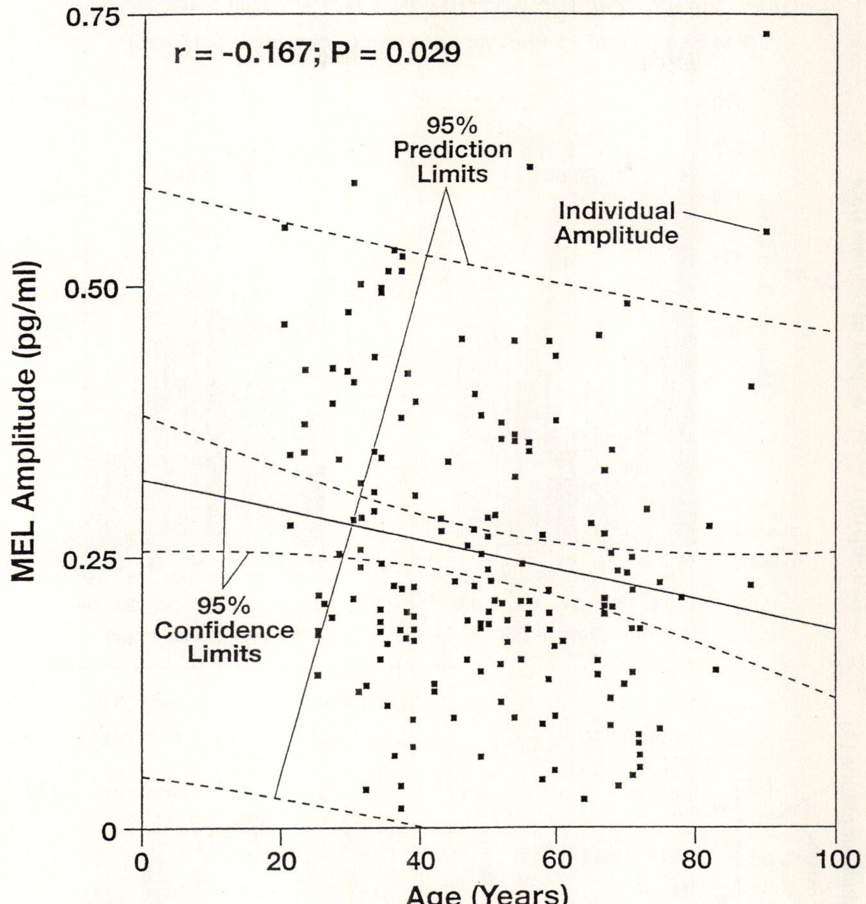

* Analysis of $\log_{10}$-transformed data from 172 subjects; squares are individual amplitudes estimated from 4-hourly sampling for 24 hours.

Fig. 4.21. Changes with age in the prominence of the circadian rhythm in human circulating melatonin concentration. (Copyright Halberg)

The task of taking the entire time structure of melatonin into account, including its broad spectrum of rhythms, seems complex and utopian at first, like the building of roads in an unknown terrain or the suggestion to take to the air. But in building highways or airplanes to bridge distances, it must be kept in mind that once the roads or the planes are available, they need not be built over and over again for each trip. Users may pay taxes, pay a toll, or buy a ticket. This is what the BIOCOS project (Halberg et al. 1998), collecting reference standards that are complex and costly, is all about. It is hoped that the reference database will be augmented for further work and used in the interim.

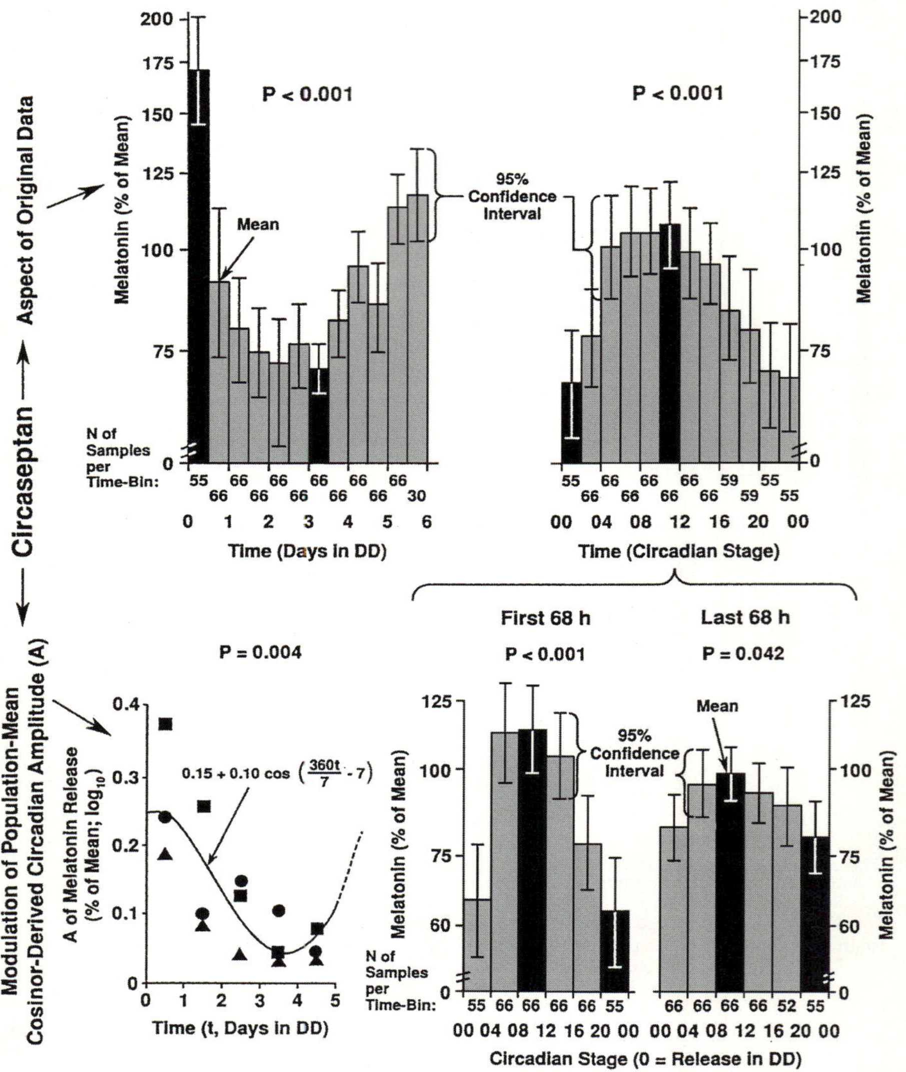

Fig. 4.22. Results of a meta-analysis presented herein show on the *left* that the variation in pineal secretion of melatonin along the scale of days can be larger than the variation along the scale of hours. These results are to be interpreted in the light of follow-up series each covering several weeks, showing that the about-weekly change in the melatonin secreted by the pineal gland into a superfusate can be larger in amplitude than the about-daily change in melatonin secretion. (Copyright Halberg)

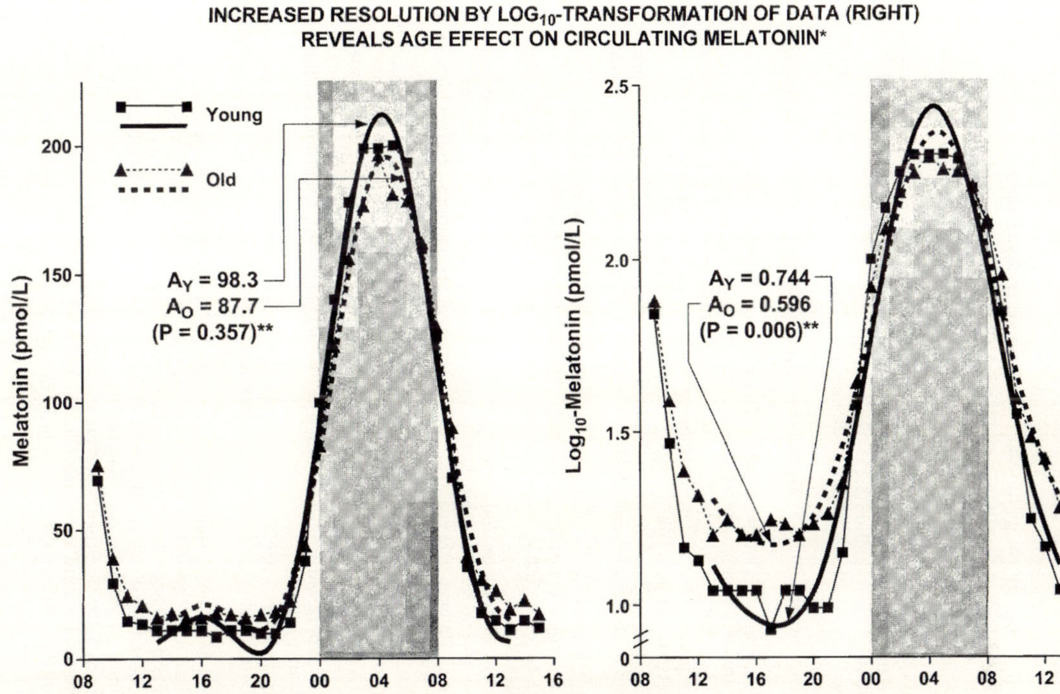

* Circadian waveform of circulating melatonin in two age groups approximated by model consisting of cosine curves
 with periods of 24 and 12 hours.
**P-value from test of equality of 24-hour amplitudes of young (A_Y) and old (A_O) subjects ($H_0 : A_Y = A_O$).

Fig. 4.23. The circadian amplitude of melatonin, reported as age-invariant by Zeitzer et al. (1999) (*left*), decreases with age (*right*), as also established earlier (Tarquini et al. 1997b) and by as yet unpublished data. (Copyright Halberg)

References

Aguilera G (1993) Factors controlling steroid biosynthesis in the zona glomerulosa of the adrenal. J Steroid Biochem Molec Biol 45:147–151

Arendt J (1986) Role of the pineal gland and melatonin in seasonal reproductive function in mammals. Oxford Reviews of Reproductive Biology 8:266–320

Arendt J (1997) Safety of melatonin in long-term use (?). J Biol Rhythms 12:707–708

Arendt J (1998a) Biological rhythms: the science of chronobiology. J Roy Coll Physicians Lond 32:27–35

Arendt J (1998b) Complex effects of melatonin. Therapie 53:479–488

Arendt J (1998c) Melatonin and the pineal gland: influence on mammalian seasonal and circadian physiology. Reviews of Reproduction 3:13–22

Arendt J, Broadway J (1986) Phase response of human melatonin rhythms to bright light in Antarctica. J Physiol 377:68

Arendt J, Wirz-Justice A, Bradtke J (1978) Annual rhythm of serum melatonin in man. Neurosci Lett 7:327–330

Arendt J, Borbely AA, Franey C, Wright J (1984) The effect of chronic, small doses of melatonin given in the late afternoon on fatigue in man: a preliminary study. Neurosci Lett 45:317–321

Arendt J, Bojkowski C, Franey C, Wright J, Marks V (1985) Immunoassay of 6-hydroxymelatonin sulfate in human plasma and urine: abolition of the urinary 24-hour rhythm with atenolol. J Clin Endocr Metab 60:1166–1173

Axelrod J (1974) The pineal gland: a neurochemical transducer. Chemical signals from nerves regulate synthesis of melatonin and convey information about internal clocks. Science 184:1341–1348

Balzer I, Fuhrberg B (1996) Presence of melatonin in the unicell *Euglena gracilis.* 5th Mtg. Society for Research on Biological Rhythms, Jacksonville, Florida, May 8–12, 1996 (p 82)

Barber M (1994) The New Knighthood: a history of the order of the temple. Cambridge University Press, Cambridge, UK

Bojkowski CJ, Arendt J (1990) Factors influencing urinary 6-sulphatoxymelatonin, a major melatonin metabolite, in normal human subjects. Clin Endocrinol 33:435–444

Breus T, Cornélissen G, Halberg F, Levitin AE (1995) Temporal associations of life with solar and geophysical activity. Annales Geophysicæ 13:1211–1222

Brown FA Jr (1960) Response to pervasive geophysical factors and the biological clock problem. Cold Spr Harb Symp Quant Biol 25:57–71

Brown GM, Grota LJ, Sánchez de la Peña S, Halberg F, Halberg E (1983) Circadian melatonin rhythm: part of stimulatory feed-sideward in pineal modulation of adrenal response to ACTH 1–17? Abstracts of papers, Minn Acad Sci 51st Annual Spring Meeting, University of Minnesota-Duluth, April 29–30, 1983, p 12

Brown GM, Sánchez de la Peña S, Marques N, Grota LJ, Ungar F, Halberg F (1986) More on the circadian melatonin rhythm in pineals from domesticated $B_6D_2F_1$ mice. Chronobiologia 13:335–338

Cecchettin M, Sánchez de la Peña S, Halberg F, Cornélissen G, Ungar F, Brown G, Mantero F, Vecsei P, Simmons D, Walker RF, Scheving LE (1986) Reproducibility of pineal feed-sideward upon the ACTH 1–17 effect on adrenocortical function. Chronobiologia 13:71–73

Conti A, Maestroni GJM (1996) HPLC validation of a circadian melatonin rhythm in the pineal gland of inbred mice. J Pineal Res 20:138–144

Cornélissen G, Halberg F (1992) Broadly pertinent chronobiology methods quantify phosphate dynamics (chronome) in blood and urine. Clin Chem 38:329–333

Cornélissen G, Halberg F (1994) Introduction to Chronobiology. Medtronic Chronobiology Seminar #7, April 1994 (Library of Congress Catalog Card #94–060580; URL http://revilla.mac.cie.uva.es/chrono)

Cornélissen G, Bingham C, Siegelová J, Fiser B, Dusek J, Prikryl P, Sonkowsky RP, Halberg F (1994a) Cardiovascular disease risk monitoring in the light of chronobioethics. Chronobiologia 21:321–325

Cornélissen G, Zaslavskaya RM, Kumagai Y, Romanov Y, Halberg F (1994b) Chronopharmacologic issues in space. J. Clin Pharmacol 34:543–551

Cornélissen G, Portela A, Halberg F, Bolliet V, Falcón J (1995) Toward a chronome of superfused pike pineals: about-weekly (circaseptan) modulation of circadian melatonin release. In vivo 9:323–329

Cornélissen G, Tarquini B, Syutkina EV, Siegelova J, Halberg F (1997) *Cherchez le* feedsideward: contradictory results versus predictable outcomes in infradian, notably circaseptan, meta-chronanalytic perspective. Cancer Biotherapy and Radiopharmaceuticals 12:419–420

Cornélissen G, Halberg F, Schwartzkopff O, Delmore P, Katinas G, Hunter D, Tarquini B, Tarquini R, Perfetto F, Watanabe Y, Otsuka K (1999) Chronomes, time structures, for chronobioengineering for "a full life". Biomedical Instrumentation & Technology 33:152–187

Crick F (1994) The Astonishing Hypothesis: the scientific search for the soul. Scribners, New York

Damian E (1989) The antisteroid, antigonadotropic and metabolism-regulation activity of the pineal gland. Endocrinologie 27:57–64

Demurger A (1989) Vie et mort de l'ordre du Temple. Seuil, Paris

Descartes R (1972) Traite de l'homme (Treatise of Man, Hall T.S. trans.). Harvard University Press, Cambridge, Mass. Original work published 1664

Dollins AB, Zhdanova IV, Wurtman RJ, Lynch HJ, Deng MH (1994) Effect of inducing nocturnal serum melatonin concentrations in daytime on sleep, mood, body temperature, and performance. Proc Natl Acad Sci USA 91:1824–1828

Ebihara S, Marks T, Hudson DJ, Menaker M (1986) Genetic control of melatonin synthesis in the pineal gland of the mouse. Science 231:491–493

Engel P, Bergmann W (1952) Die physiologische Funktion der Zirbeldrüse und ihre therapeutische Anwendung. Z Vitamin- Hormon- Fermentforsch 4:564–594

Falcón J, Ravault J-P, Cornélissen G, Halberg F (1996) Circaseptan-modulated circadian rhythmic melatonin release from isolated pike pineal in alternating light and darkness. 4° Convegno Nazionale, Società Italiana di Cronobiologia, Gubbio (Perugia), Italy, June 1–2, 1996, pp 39–40

Franchimont P (1964) [Action of melatonin and glomerulotropin on the genital and adrenal functions of the rat.] Arch Int Physiol Biochim 72:735–740

Gilberd W (1600) De magnete, magneticisque corporibus, et de magno magnete tellure; physiologia noua, plurimis et argumentis, et experimentis demonstrata [On magnetism, magnetic bodies, and the great magnet Earth]. London, Short, pp 204

Halberg E, Halberg F (1980) Chronobiologic study design in everyday life, clinic and laboratory. Chronobiologia 7:95–120

Halberg F (1953) Some physiological and clinical aspects of 24-hour periodicity. J Lancet (USA) 73:20–32

Halberg F (1965) Organisms as circadian systems; temporal analysis of their physiologic and pathologic responses, including injury and death. In: Symposium on Medical Aspects of Stress in the Military Climate, Walter Reed Army Institute of Research (Col. William D. Tigertt, Medical Corps, Director and Commandant), Walter Reed Army Medical Center (Maj. Gen. A.L. Tynes, Medical Corps, Commanding), Washington DC, 22–24 April 1964. US Government Printing Office: 1965– 778–714, pp 1–36

Halberg F (1969) Chronobiology. Ann Rev Physiol 31:675–725

Halberg F (1969a) Chronobiologie; rythmes et physiologie statistique. In: Marois M (ed) Theoretical Physics and Biology. North-Holland, Amsterdam, pp 347–393. Discussions pp 339–344 and 394–411

Halberg F (1980) Chronobiology: methodological problems. Acta Med Rom 18:399–440

Halberg F (1983) *Quo vadis* basic and clinical chronobiology: promise for health maintenance. Am J Anat 168:543–594

Halberg F, Cornélissen G (1998) Chronoastrobiology. In: Proc. 3rd International Symposium of Chronobiology and Chronomedicine, Kunming, China, October 7–12, 1998 (pp 128–142)

Halberg F, Howard RB (1958) 24-hour periodicity and experimental medicine. Example and interpretations. Postgrad Med 24:349–358

Halberg F, Visscher MB, Flink EB, Berge K, Bock F (1951) Diurnal rhythmic changes in blood eosinophil levels in health and in certain diseases. J Lancet (USA) 71:312–319

Halberg F, Engel R, Treloar AE, Gully RJ (1953) Endogenous eosinopenia in institutionalized patients with mental deficiency. AMA Arch Neurol Psychiatr 69:462–469

Halberg F, Bittner JJ, Smith D (1957) Belichtungswechsel und 24-Stundenperiodik von Mitosen im Hautepithel der Maus. Z Vitamin- Hormon- u Fermentforsch 9:69–73

Halberg F, Barnum CP, Silber RH, Bittner JJ (1958a) 24-hour rhythms at several levels of integration in mice on different lighting regimens. Proc Soc Exp Biol (NY) 97:897–900

Halberg F, Jacobson E, Wadsworth G, Bittner JJ (1958b) Audiogenic abnormality spectra, 24-hour periodicity and lighting. Science 128:657–658

Halberg F, Halberg E, Barnum CP, Bittner JJ (1959) Physiologic 24-hour periodicity in human beings and mice, the lighting regimen and daily routine. In: Withrow RB (ed) Photoperiodism and Related Phenomena in Plants and Animals, Ed Publ No 55, Am Assn Adv Sci, Washington DC, pp 803–878

Halberg F, Albrecht PG, Barnum CP Jr (1960) Phase shifting of liver-glycogen rhythm in intact mice. Amer J Physiol 199:400–402

Halberg F, Breus TK, Cornélissen G, Bingham C, Hillman DC, Rigatuso J, Delmore P, Bakken E, International Womb-to-Tomb Chronome Initiative Group (1991a) Chronobiology in space. Keynote, 37th Ann Mtg Japan Soc for Aerospace and Environmental Medicine, Nagoya, Japan, November 8–9, 1991. University of Minnesota/Medtronic Chronobiology Seminar Series, #1, December 1991, 21 pp of text, 70 figures

Halberg F, Cornélissen G, Carandente F (1991b) Chronobiology leads toward preventive health care for all: cost reduction with quality improvement. A challenge to education and technology *via* chronobiology. Chronobiologia 18:187–193

Halberg F, Bingham C, Siegelová J, Fiser B, Dusek J, Prikryl P, Sonkowsky RP, Cornélissen G (1994) "Cancer marker chronomes" assessed in the light of chronobioethics. Chronobiologia 21:327–330

Halberg F, Cornélissen G, Halpin C, Burchell H, Watanabe Y, Kumagai Y, Otsuka K, Zaslavskaya R (1996) Fleeting "monitor-", "conflict-" or "grief-associated" blood pressure disorders: MESOR-hypertension and circadian hyperamplitudetension (CHAT). EuroRehab 6:225–240

Halberg F, Cornélissen G, Tarquini B, Grafe A, Syutkina EV, Otsuka K, Watanabe Y, Siegelova J, Sanchez de la Pena S, Carandente F, Schwartzkopff O (1997) Pineals, cancers and feedsidewards in the Biosphere and the Cosmoi (BIOCOS). Cancer Biotherapy and Radiopharmaceuticals 12:421–422

Halberg F, Syutkina EV, Cornélissen G (1998) Chronomes render predictable the otherwise-neglected human "physiological range": position paper of BIOCOS project. Human Physiology 24:14–21

Halberg F, Cornélissen G, Schwartzkopff O, Cagnoni M, Perfetto F, Tarquini R (1999) Pineal mythology and chronorisk: the swan song of Brunetto Tarquini (* February 8, 1938 – † December 10, 1998). Neuroendocrinol Lett 20:91–100

Hanson BR, Halberg F, Tuna N, Bouchard TJ Jr, Lykken DT, Cornélissen G, Heston LL (1984) Rhythmometry reveals heritability of circadian characteristics of heart rate of human twins reared apart. Cardiologia 29:267–282

Haulica I, Coculescu M, Petrescu G, Stratone A, Cringu A, Rosca V (1981) Possible hormonal role of the angiotensin-like peptides in mammals' pineal gland. Physiologie 18:155–165

Haus E, Nicolau GY, Ghinea E, Dumitriu L, Petrescu E, Sackett-Lundeen L (1996) Stimulation of the secretion of dehydroepiandrosterone by melatonin in mouse adrenals in vitro. Life Sci 14:PL263–PL267

Herold M, Cornélissen G, Loeckinger A, Koeberle D, Koenig P, Halberg F (1998) About 8-hourly variation of circulating human endothelin-1 (ET-1) in clinical health. Peptides 19:821–825

Herold M, Cornélissen G, Loeckinger A, Klotz W, Rawson MJ, Katinas G, Alinder C, Bratteli C, Cohn J, Halberg F (1999) About-daily (circadian) and about-weekly (circaseptan) patterns of human salivary melatonin. Abstract, Serono Int. Symp. Endocrinology of Aging, Tempe, Arizona, Oct. 27–30, 1999, p 32

Herold M, Cornélissen G, Rawson MJ, Katinas GS, Alinder C, Bratteli C, Gubin D, Halberg F (in press). About-daily (circadian) and about-weekly (circaseptan) patterns of human salivary melatonin. J Anti-Aging Med

Hildebrandt G, Bandt-Reges I (1992) Chronobiologie in der Naturheilkunde: Grundlagen der Circaseptanperiodik. Haug, Heidelberg

Huang Ti Nei Ching Su Wen [The Yellow Emperor's Classic of Internal Medicine] (Veith I, trans.) (1949) University of California Press, Berkeley

Hufeland CW (1797) Macrobiotics: the art of prolonging life. 2nd English translation, printed for J. Bell, London

Jahn RG (1995) Out of this aboriginal sensible muchness: consciousness, information, and human health. J Am Soc Psychical Res 89:301–312

Jahn RG (1996) Information, consciousness, and health. Alternative Therapies in Health and Medicine 2:32–38

Johnson MS (1926) Activity and distribution of certain wild mice in relation to biotic communities. J Mammalogy 7:215–277

Kaine HD, Seltzer HS, Conn JW (1955) Mechanism of diurnal eosinophil rhythm in man. J Lab Clin Med 45:247–252

Katinas GS, Cornélissen G, Homans D, Rhodus N, Siegelova J, Machat R, Halberg F (1998) Individualized combination chronotherapy of coexisting CHAT and MESOR-hypertension including diltiazem HCl. Abstract 8, Neinvazivni metody v kardiovaskularnim vyzkumu, 6th International Fair of Medical Technology and Pharmacy, MEFA Congress, Brno, Czech Republic, November 3–4, 1998

Kohli N, Rastogi H, Bhadury S, Tandon VK (1992) Computed tomographic evaluation of pineal calcification. Indian J Med Res 96:139–142

Lerner AB, Case JD, Takahashi Y, Lee TH, Mori W (1958) Isolation of melatonin, the pineal gland factor that lightens melanocytes. J Am Chem Soc 80:2587

Lynch HJ, Wurtman R, Moskowitz MA, Archer MC, Ho MH (1975) Daily rhythm in human urinary melatonin. Science 187:169–171

Macey SL (ed) (1994) Encyclopedia of Time. Garland Publishing, New York

Maestroni GJM, Conti A, Pierpaoli W (1986) Role of the pineal gland in immunity: Circadian synthesis and release of melatonin modulates the antibody response and antagonizes the immunosuppressive effect of corticosterone. J Neuroimmunol 13:19–30

Maestroni GJM, Conti A, Pierpaoli W (1987) Role of the pineal gland in immunity: II. Melatonin enhances the antibody response via an opiatergic mechanism. Clin Exp Immunol 68:384–391

Maggioni C, Cornélissen G, Antinozzi R, Ferrario M, Grafe A, Halberg F (1999) A half-yearly aspect of circulating melatonin in pregnancies complicated by intrauterine growth retardation. Neuroendocrinol Lett 20:55–68

Partner P (1982) The murdered magicians: the templars and their myth. Oxford University Press, Oxford

Pecci EF (1997) Forward. In: Tiller WA: Science and Human Transformation: Subtle Energies, Intentionality and Consciousness. Pavior, Walnut Creek, California, pp xiii–xix

Pennisi E (1997) Multiple clocks keep time in fruit fly tissues. Science 278:1560–1561

Plautz JD, Kaneko M, Hall JC, Kay SA (1997) Independent photoreceptive circadian clocks throughout *Drosophila*. Science 278:1632–1635

Quay WB (1964) General biochemistry of the pineal gland of mammals. In: Reiter RJ (ed) The Pineal Gland. CRC Press, Boca Raton, Florida, pp 173–198

Reiter RJ (1980) The pineal and its hormones in the control of reproduction in mammals. Endocrinol Rev 1:109–131

Resch J (1995) Geographische Verteilung der Multiplen Sklerose und Vergleich mit geophysikalischen Groessen. Soz Praeventivmed 40:161–171

Salti R, Galluzzi F, Bindi G, Perfetto F, Tarquini R, Halberg F, Cornélissen G (1999) Ultradian structure of circulating melatonin by night in children. Abstract, Serono Int. Symp. Endocrinology of Aging, Tempe, Arizona, Oct. 27–30, 1999, p 33

Salti R, Galluzzi F, Bindi G, Perfetto F, Tarquini R, Halberg F, Cornélissen G (in press). Nocturnal melatonin patterns in children. J Clin Endocr Metab

Sánchez de la Peña S (1993) The feedsideward of cephalo-adrenal immune interactions. Chronobiologia 20:1–52

Sánchez de la Peña S, Halberg F, Halberg E, Ungar F, Cornélissen G, Sánchez E, Brown G, Scheving LE, Yunis EG, Vecsei P (1983a) Pineal modulation of ACTH 1–17 effect upon murine corticosterone production. Brain Res Bull 11:117–125

Sánchez de la Peña S, Halberg F, Ungar F, Haus E, Lakatua D, Scheving LE, Sánchez E, Vecsei P (1983b) Circadian pineal modulation of pituitary effect on murine corticosterone in vitro. Brain Res Bull 10:559–565

Sánchez de la Peña S, Brown GM, Ungar F, Marques N, Scheving LE, Grota LJ, Halberg F (1986) Chronobiologic lead study cost-effectively assesses circadian-circaseptan intermodulation in murine pineal melatonin content. Chronobiologia 13:329–333

Sánchez de la Peña S, Halberg F, Ungar F, Lakatua D (1988) Ex vivo hierarchy of circadian-infradian rhythmic pineal-pituitary-adrenal intermodulations in rodents. In: Pancheri P, Zichella L (eds) Biorhythms and Stress in the Physiopathology of Reproduction. Hemisphere, New York, pp 177–214

Sandyk R (1992a) The pineal gland and the mode of onset of schizophrenia. Int J Neurosci 67:9–17

Sandyk R (1992b) Pineal and habenula calcification in schizophrenia. Int J Neurosci 67:19–30

Sandyk R (1993) The relationship of thought disorder to third ventricle width and calcification of the pineal gland in chronic schizophrenia. Int J Neurosci 68:53–59

Sandyk R, Awerbuch GI (1991) The pineal gland in multiple sclerosis. Int J Neurosci 61:61–67

Sandyk R, Awerbuch GI (1992) Nocturnal plasma melatonin and alpha-melanocyte stimulating hormone levels during exacerbation of multiple sclerosis. Int J Neurosci 67:173–186

Sandyk R, Awerbuch GI (1993) Multiple sclerosis: the role of the pineal gland in its timing of onset and risk of psychiatric illness. Int J Neurosci 72:95–106

Sandyk R, Awerbuch GI (1994) Pineal calcification and its relationship to the fatigue of multiple sclerosis. Int J Neurosci 74:95–103

Sandyk R, Pardeshi R (1990) The relationship between ECT nonresponsiveness and calcification of the pineal gland in bipolar patients. Int J Neurosci 54:301–306

Schorner W, Kunz D, Henkes H, Sander B, Schmidt D, Felix R (1991) [The demonstration of calcifications in magnetic resonance tomography (MRT). The effect of different parameters on the MRT imaging of cerebral calcifications.] ROFO: Fortschritte auf dem Gebiete der Röntgenstrahlen und der Neuen Bildgebenden Verfahren 154:430–437

Seward D (1995) The Monks of War: The Military Religious Orders. Penguin, London (revised ed.)

Tarquini B, Cornélissen G, Perfetto F, Tarquini R, Halberg F (1997a) About-half-weekly (circasemiseptan) component of the endothelin-1 (ET-1) chronome and vascular disease risk. Peptides 18:1237–1241

Tarquini B, Cornélissen G, Perfetto F, Tarquini R, Halberg F (1997b) Chronome assessment of circulating melatonin in humans. In vivo 11:473–484

Tiller WA (1997) Science and Human Transformation: Subtle Energies, Intentionality and Consciousness. Pavior, Walnut Creek, California

Tiller WA (1999) Towards a predictive model of subtle domain connections to the physical domain aspect of reality: the origins of wave-particle duality, electric-magnetic monopoles and the mirror principle. J Scientific Exploration 13:41–67

Tiller WA, Dibble WE Jr, Kohane MJ (1999) Towards objectifying intention via electronic devices. Subtle Energies & Energy Medicine 8:103–123

Ungar F, Halberg F (1962) Circadian rhythm in the in vitro response of mouse adrenal to adrenocorticotropic hormone. Science 137:1058–1060

Ungar F, Halberg F (1963) In vitro exploration of a circadian rhythm in adrenocorticotropic activity of C mouse hypophysis. Experientia (Basel) 19:158–159

Wetterberg L, Halberg F, Halberg E, Haus E, Kawasaki T, Ueno M, Uezono K, Cornélissen G, Matsuoka M, Omae T (1986) Circadian characteristics of urinary melatonin from clinically healthy women at different civilization disease risk. Acta Med Scand 220:71–81

Wetterberg L, Hayes DK, Halberg F (1987) Circadian rhythms of melatonin in the brain of the face fly, *Musca autumnalis* De Geer. Chronobiologia 14:377–381

Wetterberg L, Sánchez de la Peña S, Halberg F (1990) Circadian rhythm of melatonin release from pineal, hypothalamus and pituitary in hypertensive rats. Progress in Clinical and Biological Research 341B:145–153

Wetterberg L, Bratlid T, Knorring L, Eberhard G, Yuwiler A (1999) A multinational study of the relationships between nighttime urinary melatonin production, age, gender, body size, and latitude. Eur Arch Psychiatr Neurosci 249:256–262

Zeitzer JM, Daniels JE, Duffy JF, Klerman EB, Shanahan TL, Dijk DJ, Czeisler CA (1999) Do plasma melatonin concentrations decline with age? Am J Med 107:432–436

Appendix

Table 4.A1. Chronobiologic concepts, tools, and long-term goals for pinealists

View of:	I. Homeostatic (1–8) Response Physiology	II. Chronome (6–42) Physiology	Utility of II
1. Definition of normalcy, e.g., health	Negative: absence of abnormality, e.g., of disease[a]	Positive: parametric and nonparametric assessment	Time structure: control in whatever we do
2. Quantification of normalcy, e.g., health	Population-based: percent abnormality, e.g., morbidity and mortality	Individualized: P values for statistical significance and for scientific (e.g., clinical) signification	Recognizing risk of abnormality before the fait accompli of catastrophe
3. Interpretation of reality	Putative (imaginary) set points	Chronomes: consisting of (a) rhythms, (b) trends, (c) deterministic and other chaos, (d) any residuals, and (e) interactions among a, b, c, and d	Chronorisk syndromes: (a) circadian over-swinging of blood pressure (15) or (b) chronome alteration with heart rate jitter deficit (16) or (c) circadian rhythm alteration (17) or (d) altered about-yearly rhythms in circulating prolactin and TSH signaling breast and prostatic cancer risk elevation (14, 18) maintaining normal dynamics
4. Variability	Confounder (foe)	Of interest in its own right (friend)	As a tool and source of information [1]

Footnotes see pp. 107, 108.

Table 4.A1 (continued)

View of:	I. Homeostatic (1–8) Response Physiology	II. Chronome (6–42) Physiology	Utility of II
5. Biosystems' behavior if perturbed (6)	Settling down to a steady state (constancy) or limited random "hunting", e. g., as (mistakenly anticipated) when a single blood pressure is taken after some (30) minutes of rest (7, 8)	Dynamic chronomes (13) that characterize health within chronobiologic limits set by the intermodulation of the chronomes' α-, β-, γ-, and δ- (spontaneous, reactive, and modulating) rhythms, e. g., a large circadian change in blood pressure during 24-h bed rest (7, 8)	Positive individualized quantification of health (13, 14)
6. Analogy	Thermostats with "hunting" noise	Pendulums in resolvable chronomes	Prediction
7. Physiologic or normal ranges of variation	Broad, random, indivisible; equated to noise in current standard for diagnosis and treatment	Structured, predictable [2]; resolved into reference ranges (chronodesms) for chronomes	Circadian blood pressure amplitude (BP-A) or circadian standard deviation (SD) for detecting effect of in utero exposure to betamimetics (19)
8. Action?	Confounder elimination (20) incompatible with detection of circadian blood pressure disorder (15)	Monitoring and as-one-goes analyses and, on this basis, action	Detects treatable over-swinging of BP-A, which carries a 720% increase in risk of ischemic stroke (7); improves cancer treatment (21, 22)
9. End points	Original values: casual measurements at times of convenience, not necessarily of pertinence (e. g., of "the" blood pressure with >40% uncertainty in diagnosis in cases of borderline hypertension) (15) Time-unspecified: mean; standard error	Time-specified chrones in chronomes: time-coded: original values; standard deviations (e. g., 6-h, 24-h); MESOR(s); periods; amplitude(s); waveform(s); acrophase(s); trends; chaotic dimensions; residuals	Chronobiologic software: provides information, e. g., on c and d above; guides timed treatment that has greatly prolonged the survival of cancer patients (21–24)
10. Sources of variation	Exogenous responses to stimuli from proximity mostly from the habitat niche	Endogenous and exogenous: responses to stimuli from near and far, including cosmos (13, 32–37)	Resolution of impact of storms in space on myocardial infarctions on earth: space weather report? [3]
11. Mechanism	Feedbacks along axes: unstructured "modulation" like the deus ex machina in a physiological tragedy since outcomes may be unpredictable	Feedsidewards in networks with alternating outcomes: predictable (insofar as rhythmic) as a chronomodulation (13, 14, 25–27)	Predictable since rhythmic neuro-endo-crino-vascular inter-modulations (13) can account for outcomes that may be as different as stimulation vs inhibition of immunity
12. Hierarchy	Up/down	Collateral: alternating primacy among inter-modulating multifrequency rhythms in chronomes	Focusing on selected tasks at different times

Footnotes see pp. 107, 108.

Table 4.A1 (continued)

View of:	I. Homeostatic (1–8) Response Physiology	II. Chronome (6–42) Physiology	Utility of II
13. Teleonomy	Righting and regulation	Anticipatory, preparatory coordination	Greater flexibility
14. Simplified analogy	Thermostat	Pendulum	–
15. Biologic evolution	Darwinian, externally adaptive	More and more internal and integrative while externally adaptive to both nature and nurture	Instrumented self-help
16. Health and environmental care	Medical treatment often limited and late, given mostly after the diagnosis of overt disease [4]	Optimization according to marker chronomes (of interventions by drugs and/or devices, e.g., pacemakers, with diagnosis and treatment refined by narrowed reference range and assessment within that range of chronorisk leading to preventive treatment timed by marker rhythms, that also serve to validate effect) (13, 28–31)	e.g., Catastrophic and iatrogenic disease prevention (13, 19)
17. Animal husbandry, apiculture, aquaculture, and economic entomology	Convenience	Chronome-based [5]	Optimization: greater efficacy; fewer undesirable effects
18. Value	Often wasteful	Cost-effective	Waste reduced
19. Seeking inanimate and animate origins	Stratigraphy for identifying, in geologically analyzed space, sequences in time; radiocarbon dating	Additional tracing of chronomo-ontogeny and chronomo-phylogeny [6] in the context of glimpses of cycles in corresponding spans of a figurative cosmo-ontogeny (32–36)	Adds to knowledge of the past to better optimize the future
20. Life in the scheme of physical and cultural things	Survival of the fittest with humans dominating food chains viewed in the perspective of bio-energetics in a mostly terrestrial ecology	Physically and socially chronomodulating and thus informatively and integratively evolving biota molded by human culture; homo not only faber but cosmoinformans and chronomodulans in a budding broad chronocosmoecology (13, 32–37) [7]	Humans safeguard the integrity of the biosphere as it extends into the cosmos and as we speculatively yet [by joining the approaches by (36), ablations superimposed epochs (32–34) and resonance tests (37)] concomitantly explore the temporal aspects of our origins, possibly represented by our chronomes that in turn may reflect a long-past environment

Footnotes see pp. 107, 108.

Table 4.A1 (continued)

View of:	I. Homeostatic (1–8) Response Physiology	II. Chronome (6–42) Physiology	Utility of II
21. Investigator satisfaction	Frustrating work when (without specification of chronobiologic timing, even at the same clock-hours) one gets confusing and/or obscuring, even oppositeresults from the same intervention	Sheer fun: long-standing controversy is resolved by accounting for both the genetic and broadly environmental bases of the feedsidewards among inanimate and animate cycles that constitute life; disease risk recognition promises to lead to the prevention or timed treatment of catastrophic diseases such as stroke, cancer or sudden death	Increased productivity

Just as contemporary physics, by fission and fusion, gathers more and more energy by splitting the atom, biomedicine gathers more and more information by splitting the normal value range into time structures, thereby resolving, e.g., rhythms (fission) and looking at their feedsideward interrelations (fusion) for a better understanding of an interdigitated, indivisible Janus-faced inseparable soma and psyche.

[a] Health promotion is a step in the right direction, by its recommendations of attention to diet, exercise, or relaxation, as long as it is then followed by a chronobiologic assessment of the effect of recommended procedures, rather than merely by the old reliance of ruling out the occurrence of values outside the normal range.

[b] Location and dispersion indices include the determination from histogram of values, of means (arithmetic, geometric, harmonic), median, mode, minimum, maximum, 100%, and 90% ranges, interquartile range, standard deviation, standard error. These end points are computed from time-unspecified single values in the context of the homeostatic approach, whereas in the chronobiologic framework the location and dispersion indices are used on time-specified samples and on time series-derived parameters, i.e., on each of the end points (chrones: $M, A, \phi, [A_n, \phi_n]$, etc.) of the chronomes.

[1] An international womb-to-tomb chronome initiative with aims primarily at stroke and other catastrophic vascular disease prevention. It focuses on chronocardiology in general and blood pressure and heart rate dynamics in particular. Those interested may consult the chronobiology home page at http://revilla.mac.uva.es/chrono

[2] Information from the physiologic range for prevention, diagnosis, or treatment is much refined when this range is individualized and interpreted in the light of a personalized background as well as in the context of gender-, age-, ethnicity-, and chronome stage-specification.

[3] The need for forecasting storms in space should be explored further on the basis of systematic studies aligning physiological lifetime monitoring and clinical and archival statistical studies with ongoing physical data collection near and far, both for ascertaining effects and in studying countermeasures (32–37). Blood pressure, heart rate, and other physiological and psychological monitoring would also provide basic information on any cross-spectral and other associations (feedsidewards) within and among biological and environmental chronomes while further providing reference values of medical interest.

[4] Even if some preventive measures have also been long implemented, e.g., by vaccination, and even if recently hygienic measures (such as exercise and caloric, fat, and sodium restriction) are also popular, all can be greatly improved by timing designed with chronobiologic individualization. The alternative, current action based on group results, its unquestionable overall merits notwithstanding, fails to recognize, for instance, that the blood pressure response to salt may differ as a function of circadian stage (38), and there are indeed individuals in whom the addition of salt lowers rather than raises blood pressure (8, 39).

[5] Even after the death of a cockroach, when bacteria take over, periodicities (e.g., in oxygen consumption) may not be "eliminated" but continue with increased amplitude. Critical information may be lost by filtering variation deemed to be undesirable since it lies beyond one's conventional scope.

[6] Development from the egg of rhythms (some may be much older than shards) and of other constituents of chronomes to trace their homeo- or heterochronically roughly "recapitulatory" development across species, with both ontogeny and phylogeny, perhaps tracing in their turn the concomitant development of the geocosmic environment (32–37). This distant basic goal can be pursued with the immediate reward of obtaining indispensable reference values for the diagnosis of two chronobiologic risk syndromes, circadian hyperamplitude tension (CHAT) and chronome alterations of heart rate variability (CAHRVs), e. g., an extreme deficit in heart rate jitter (16, 40) – associated with an increase in the risk of ischemic stroke or of a myocardial infarction of 720 and 550 %, respectively.

References for Table 4.A1

1. Bernard C (1885) Leçons sur les phénomènes de la vie communs aux animaux et aux végétaux. JB Baillière, Paris
2. Bernard C (1865) De la diversité des animaux soumis à l'expérimentation. De la variabilité des conditions organiques dans lesquells ils s'offrent à l'expérimentateur. J de l'Anatomie et de la Physiologie normales et pathologiques de l'homme et des animaux 2:497–506
3. Cannon WB (1932) The Wisdom of the Body. WW Norton, New York
4. Guyton AC (1996) Textbook of Medical Physiology, 9th edn. W.B. Saunders, Philadelphia
5. Field J (Editor-in-Chief) (1959) Handbook of Physiology. American Physiological Society, Washington D.C.
6. Goldberger AL (1991) Is the normal heartbeat chaotic or homeostatic? News in Physiologic Science (Int Union Physiol Sci/Am Physiol Soc) 6:87–91
7. Halberg F, Cornélissen G, Bakken E (1990) Caregiving merged with chronobiologic outcome assessment, research and education in health maintenance organizations (HMOs). Prog Clin Biol Res 341B:491–549
8. Halberg F, Cornélissen G, Halberg E, Halberg J, Delmore P, Shinoda M, Bakken E (1988) Chronobiology of human blood pressure, 4th edn. Medtronic Continuing Medical Education Seminars
9. Halberg F (1969) Chronobiology. Ann Rev Physiol 31:675–725
10. Reinberg A, Smolensky MH (1983) Biological rhythms and medicine. Cellular, metabolic, physiopathologic, and pharmacologic aspects. Springer, New York
11. Komarov FI (ed) (1989) Chronobiology and chronomedicine. Medicina, Moscow
12. Touitou Y, Haus E (eds) (1992) Biological Rhythms in Clinical and Laboratory Medicine. Springer, Berlin
13. Cornélissen G, Halberg F (1994) Introduction to Chronobiology. Medtronic Chronobiology Seminar #7, April 1994
14. Halberg F (1983) *Quo vadis* basic and clinical chronobiology: promise for health maintenance. Am J Anat 168:543–594
15. Halberg F, Cornélissen G (1995) International Womb-to-Tomb Chronome Initiative Group: Resolution from a meeting of the International Society for Research on Civilization Diseases and the Environment (New SIRMCE Confederation), Brussels, Belgium, March 17–18, 1995: Fairy tale or reality? Medtronic Chronobiology Seminar #8, April 1995. URL http://revilla.mac.cie.uva.es/chrono. Cf. also Halberg F, Cornélissen G, Siegelova J, Fiser B. Rezoluce S.I.R.M.C.E. AmireporT 2:105–106 (In Czech)
16. Otsuka K, Cornélissen G, Halberg F (1997) Circadian rhythmic fractal scaling of heart rate variability in health and coronary artery disease. Clin Cardiol 20:631–638
17. Cornélissen G, Bakken E, Delmore P, Orth-Gomér K, Åkerstedt T, Carandente O, Carandente F, Halberg F (1990) From various kinds of heart rate variability to chronocardiology. Am J Cardiol 66:863–868
18. Halberg F, Cornélissen G, Sothern RB, Wallach LA, Halberg E, Ahlgren A, Kuzel M, Radke A, Barbosa J, Goetz F, Buckley J, Mandel J, Schuman L, Haus E, Lakatua D, Sackett L, Berg H, Wendt HW, Kawasaki T, Ueno M, Uezono K, Matsuoka M, Omae T, Tarquini B, Cagnoni M, Garcia Sainz M, Perez Vega E, Wilson D, Griffiths K, Donati L, Tatti P, Vasta M, Locatelli I, Camagna A, Lauro R, Tritsch G, Wetterberg L (1981) International geographic studies of oncological interest on chronobiological variables. In Kaiser H (ed) Neoplasms – Comparative Pathology of Growth in Animals, Plants and Man. Williams and Wilkins, Baltimore
19. Syutkina EV, Cornélissen G, Halberg F, Grigoriev AE, Abramian AS, Yatsyk GV, Morozova NA, Ivanov AP, Shevchenko PV, Polyakov YA, Bunin AT, Safin SR, Maggioni C, Alvarez M, Fernandez O, Tarquini B, Mainardi G, Bingham C, Kopher R, Vernier R, Rigatuso J, Johnson D (1995) Effects lasting into adolescence of exposure to betamimetics in utero. Clinical Drug Investigation 9:354–362
20. Sheps SG, Canzanello VJ (1994) Current role of automated ambulatory blood pressure and self-measured blood pressure determinations in clinical practice. Mayo Clin Proc 69:1000–1005

21. Halberg F (1995) The week in phylogeny and ontogeny: opportunities for oncology. In Vivo 9:269–278
22. Halberg Francine, Halberg J, Halberg E, Halberg Franz (1989) Chronobiology, radiobiology and steps toward the timing of cancer radiotherapy. In: Kaiser H (series ed), Goldson AL (volume ed) Cancer Growth and Progression. (Vol 9) Kluwer Academic Publ, Dordrecht
23. Hrushesky WJM (1985) Circadian timing of cancer chemotherapy. Science 228:73–75
24. Halberg F, Cornélissen G, Bingham C, Fujii S, Halberg E (1992) From experimental units to unique experiments: chronobiologic pilots complement large trials. In Vivo 6:403–428
25. Halberg E, Halberg F (1980) Chronobiologic study design in everyday life, clinic and laboratory. Chronobiologia 7:95–120
26. Sánchez de la Peña S (1993) The feedsideward of cephalo-adrenal immune interactions. Chronobiologia 20:1–52
27. Halberg F, Guillaume F, Sánchez de la Peña S, Cavallini M, Cornélissen G (1986) Cephaloadrenal interactions in the broader context of pragmatic and theoretical rhythm models. Chronobiologia 13:137–154
28. Zaslavskaya RM (1993) Chronodiagnosis and chronotherapy of cardiovascular diseases. 2nd edn. Translation into English from Russian. Moscow: Medicina
29. Halberg F, Watanabe H (eds) (1992) Workshop on Computer Methods on Chronobiology and Chronomedicine. 20th Int. Cong. Neurovegetative Research, Tokyo, Sept. 10–14, 1990. Medical Review, Tokyo
30. Otsuka K, Cornélissen G, Halberg F (eds) (1993) Chronocardiology and Chronomedicine: Humans in Time and Cosmos. Life Science Publishing, Tokyo
31. Cornélissen G, Halberg E, Bakken E, Delmore P, Halberg F (eds) (1993) Toward phase zero pre-clinical and clinical trials: chronobiologic designs and illustrative applications. (2nd extended edn.) University of Minnesota Medtronic Chronobiology Seminar Series, #6
32. Halberg F, Breus TK, Cornélissen G, Bingham C, Hillman DC, Rigatuso J, Delmore P, Bakken E (1991) International womb-to-tomb chronome initiative group: chronobiology in space. Keynote, 37th Ann. Mtg. Japan Soc. for Aerospace and Environmental Medicine, Nagoya, Japan, November 8–9, 1991. University of Minnesota Medtronic Chronobiology Seminar Series, #1
33. Breus T, Cornélissen G, Halberg F, Levitin AE (1995) Temporal associations of life with solar and geophysical activity. Annales Geophysicæ 13:1211–1222
34. Roederer JG (1995) Are magnetic storms hazardous to your health? Eos, Transactions, American Geophysical Union 76:441, 444–445
35. Halberg F, Wendt H, Cornélissen G, Hawkins D, Sothern RB, Haus E, Garcia Alonso L, Portela A, Syutkina EV, Breus TK, Vernova Ye S, Kleitman E, Kleitman N, Stebbings JH, Johnson D (1998) Chronobiologic monitoring of health and environmental integrity (in English; from Fiziologiya Cheloveka 24:84–90, 1998, in Russian). Human Physiology 24:728–733
36. Cornélissen G, Halberg F, Wendt HW, Bingham C, Sothern RB, Haus E, Kleitman E, Kleitman N, Revilla MA, Revilla M Jr, Breus TK, Pimenov K, Grigoriev AE, Mitish MD, Yatsyk GV, Syutkina EV (1996) Resonance of about-weekly human heart rate rhythm with solar activity change. Biologia (Bratislava) 51:749–756
37. Halberg F, Cornélissen G, Hillman DC, Bingham C, Halberg E, Guillaume F, Barnwell F, Wu JY, Wang ZR, Halberg FE, Holte J, Schmitt OH, Kellogg PJ, Luyten W, Breus TK, Komarov FI, Mikulecky M, Garcia L, Lodeiro C, Iglesias T, Quadens O, Muller C, Kaada B, Miles L, Hayes DK (1992) Chronobiology in a moon-based chemical analysis and physiologic monitoring laboratory. In: Ponnamperuma C, Gehrke CE (eds) A Lunar-Based Chemical Laboratory (LBCAL), A Deepak Publishing, Hampton, Virginia
38. Itoh K, Kawasaki T, Cugini P (1996) Effects of timing of salt intake to 24-hour blood pressure and its circadian rhythm. Ann N Y Acad Sci 783:324–325
39. Bittle CC Jr, Molina DJ, Bartter FC (1985) Salt sensitivity in essential hypertension as determined by the cosinor method. Hypertension 7:989–994
40. Baevsky RM, Petrov VM, Cornélissen G, Halberg F, Orth-Gomér K, Åkerstedt T, Otsuka K, Breus T, Siegelova J, Dusek J, Fiser B (1997) Meta-analyzed heart rate variability, exposure to geomagnetic storms, and the risk of ischemic heart disease. Scripta Medica 70:199–204
41. Cornélissen G, Halberg F (1996) Impeachment of casual blood pressure measurements and the fixed limits for their interpretation and chronobiologic recommendations. In: Portaluppi F, Smolensky MH (eds) Time-dependent structure and control of arterial blood pressure. Ann NY Acad Sci 783:24–46
42. Halberg F, Cornélissen G, Haus E, Northrup G, Portela A, Wendt H, Otsuka K, Kumagai Y, Watanabe Y, Zaslavskaya R (1996) Clinical relevance of about-yearly changes in blood pressure and the environment. Int J Biometeorol 39:161–175

Table 4.A2. Biomedicine, solar activity, and terrestrial magnetism: inferential statistical analyses[a]

Study (reference) Design* Variable(s)**	Location (latitude [geographic/ geomagnetic])	Population Sample Size Age (years)	Sampling Span	Interval	Results* Spectral	Cross-spectral coherence components (frequency) [period]
I. Data longitudinally covering up to 10 Schwabe (about 10.5-year) cycles on height and other morphology[b]						
1. Weber et al. (1) T BH	Austria (49.02°N/48.57 [northernmost]- 46.27°N/46.08 [southernmost])	Male recruits 507,125 18	10 years	Monthly	After detrending (linearly): about 10-yearly component modulating circannual variation	With Kp: 0.813 (2.20 year^{-1}) [5.45 months] With sunshine: 0.545 (1.90 year^{-1}) [6.32 months] 0.963 (0.90 year^{-1}) [13.33 months] (using 22 degrees of freedom)
2. Weber et al. (2) T BH	Austria (49.02°N/48.57 [northernmost]- 46.27°N/46.08 [southernmost])	Male recruits 713,162 18	14 years	Yearly	After detrending (linearly): common about 9.25-year component consistent among recruits from eight separate socioeconomic strata	
3. Nikityuk (u.p.†) T i. BH; BW; HC ii. CC; AC	Moscow, Russia (55.45°N/50.76)	Russian babies 25 – 150/year Birth	i. 112 years ii. 41 years	Yearly	About 10.5- and/or 21-year cycles	BH with Kp: 0.819 (0.143 year^{-1}) [6.99 years] BW with Kp: 0.867 (0.143 year^{-1}) [6.99 years] (using 10 degrees of freedom)
4. Nikityuk (u.p.†) T BH; BW; HC; CC; AC	Alma-Ata, Kazakstan (43.19°N/33.67)	Russian and Kazak babies 25 – 150/year Birth	40 years	Yearly	About 10.5- and 21-year cycles with common characteristics for different end points	
5. Cornélissen (u.p.†) T BW	Minnesota (45.00°N/55.00)	Newborns 2,150,122 Birth	33 years	Yearly	About 21-year cycle	
6. Otto et al. (3) T BH; BW	Germany: Berlin (52.31°N/52.06) Leipzig (51.20°N/51.19)	Newborns 574,600 Birth	1959; 1961; 1963; 1964	Monthly	About-yearly and half-yearly changes	

7. Henneberg and Louw (4) T BH; BW (z-score)	South Africa (33.56°S/−32.70)	Impoverished rural schoolchildren 1,522 6–18	1 year	Monthly	About-yearly change in BW and half-yearly variation in BH	
8. Garcia et al. (5) H BH; BW	La Coruña, Spain (43.22°N/47.40)	Newborns 674 Birth	0–16 months	About-monthly	About-yearly and half-yearly components, more prominent by reference to the time of birth, i.e., as a partly endogenous function of age, than by reference to a fixed calendar date, i.e., as a function of exogenous factors such as sunlight, temperature, and nutrition	

II. Supportive evidence on human pathology

9. Cornélissen et al. (6) T Incidence	Worldwide	Morbidity 6,304,025 (largest only) Various ages	Meta-analysis of 47 studies		About-weekly and half-weekly patterns of incidence of cardiovascular morbid events	
10. Halberg et al. (7) T Incidence	Moscow, Russia (55.45°N/50.76)	Morbidity 6,304,025 Various ages	3 years	Daily	About-yearly incidence of cardio-vascular morbid events and half-yearly pattern of incidence of epileptic attacks	MI with Kp: $\underline{0.51}$ (0.315 day^{-1}) [3.17 days] MI with Bz: $\underline{0.58}$ (0.315 day^{-1}) [3.17 days] (using 25 degrees of freedom)
11. Düll and Düll (8) T Morbid events	Copenhagen, Denmark (55.43°N/55.19)	Adults 36,000 Various ages	5 years	Daily	Maximal cross-correlation between magnetic storms and mortality at 1-day lag on data summarized by superimposed epochs in relation to peaks in geomagnetic disturbance ("electron invasion")	
12. Faraone et al. (9) T Sectoring in colonies of microorganisms	Milan & Rome, Italy (45.28°N/46.31 & 41.53°N/41.89) 200–250/day	Colonies of air bacteria and *Staphylococcus aureus*	12.5 (air) and 7 (staph) years	Daily	Prominent components resembling time structure of Kp and WN, notably with periods of about 5 and 0.5 years	Air bacteria with Dst: $\underline{0.647}$ (0.054 day^{-1} [18.6 days]) Staph with Kp: $\underline{0.700}$ (2.262 year^{-1} [0.442 year]) and $\underline{0.660}$ (0.303 day^{-1} [3.3 days], using 20 degrees of freedom)

Footnotes see p. 115.

Table 4.A2 (continued)

Study (reference) Design* Variable(s)**	Location (latitude [geographic/ geomagnetic])	Population Sample Size Age (years)	Sampling Span	Interval	Results* Spectral	Cross-spectral coherence components (frequency) [period]
III. Supportive evidence on human physiology						
12. Portela et al. (10) L BP	USA (45.00°N/55.00; 45.33°N/51.42; 42.19°N/53.20)	Adult males 3 20; 65; 71	14; 15; 26 years	5–6/day; 2/day; about-weekly	About 11-year cycles paralleling solar activity	
13. Cornélissen et al. (11) H HR	USA (45.00°N/55.00 [2]; 41.18°N/52.19; 34.00°N/41.01; 47.17°N; 47.53)	Adults 5 28–81	Five selected spans between Aug 1967 and Apr 1975	1–6/day (N: 382–2840)	About 11.6-year cycle; remove-and-replace approach showing resonance of about 7-day component with corresponding variation in solar activity. When the sun "removes" an about 7-day component from the spectrum of its velocity changes, the circaseptan HR amplitude is smaller	
14. Baevsky et al. (12) H HR	Space (Soyuz spacecraft)	Russian cosmonauts 49 25–50	2–15 min	Beat-to-beat	About 30% reduction in HR variability (gauged by standard deviation of R-R intervals) during magnetic storms vs quiet days in extraterrestrial space	
15. Syutkina et al. (13) H BP; HR	Moscow, Russia (55.45°N/50.76)	Newborns 32 First month of life	Up to 20 days	15 min	Correlation between nonlinearly-determined period of about 7-day component of BP and HR vs. local magnetic disturbance (K)	With K: $\geq$0.70 (around 0.3 day^{-1} and around 0.14 hour^{-1}) [3.33 days and 8.77 hours] (using 10 degrees of freedom)
16. Halberg et al. (7) L BP; HR; blood pH	Minneapolis, Minnesota, USA (44.59°N/54.60)	Premature baby 1 Birth	Up to 26 months	Up to 1–5/day; denser at outset	Near-match of some spectral peaks of BP and HR vs. Kp	DBP with Bz: 0.74 (2.0 week^{-1}) [3.5 days] (using 14 degrees of freedom)

17. Halberg et al. (14; u.p.†) L BP; HR	Ancona, Italy (43.37°N/43.48)	Adult woman 1 28	1 year (267 days in isolation from society)	15–30 min	Closeness of nonlinearly-determined period of about-half-weekly component of HR vs Kp during isolation	HR: with Kp: <u>0.558</u> (0.045 hour^{-1}) [22.22 hours] with CR: <u>0.524</u> (0.138 hour^{-1}) [7.25 hours] with 3hCR-SD: <u>0.546</u> (0.153 hour^{-1}) [6.54 hours] (using 22 degrees of freedom)
18. Watanabe et al. (15) L BP; HR	Tokyo, Japan (35.42°N/25.75)	Adult man 1 35 (at start)	3 years	15–30 min		With Kp: <u>>0.5</u> (0.036 day^{-1}) [27.8 days] (using 26 degrees of freedom)
19. Watanabe (u.p.†) L BP; HR	Tokyo, Japan (35.42°N/25.75)	Adult man 1 35 (at start)	11 years min	15–30 monthly summary	About 10.5-year component in mean and SD of HR and in SD of SBP	HR with WN: <u>0.664</u> (1.636 year^{-1}) [7.33 months] (using 14 degrees of freedom)
20. Halberg et al. (7) L BP; HR	Minneapolis, Minnesota, USA (44.59°N/54.60)	Adult man 1 68 (at start)	4 years (with interruptions)	15–30 min	About 5% increase in HR during magnetic storm and in BP on day preceding a magnetic storm on earth	
21. Sothern (RBS) (u.p.†) L BW; SBP; DBP; HR; RR; PEF; TE; EH; mood; vigor	St. Paul, Minnesota, USA (45.00°N/55.00)	Minnesotan clinically healthy man 1 20.5 (at start)	30.8 years	1–6/day (N > 50,000/ variable; total N».5 million values)	About 10.5- and 21-year as well as yearly ($P < 0.001$) cycles	With Kp: <u>0.740</u> (2.03 year^{-1}) [5.91 months] (using 22 degrees of freedom) [Cross-correlation with Kp of BP and HR maximal near lag 0]
22. Halberg et al. (16) L Urine volume; urinary 17-ketosteroid excretion	Copenhagen, Denmark (55.43°N/55.19)	Clinically healthy man 1 43–58 years	15 years	Daily	About 9.28 years (17-KS) or 4.18 years (UV), 1.0 year, 7 days and 3.5 days components, the latter two free-running from social schedule for 17-KS during last 3 years when subject self- administered testosterone	17-KS with Kp: <u>0.588</u> (11.98 year^{-1} [4.36 weeks] using 20 degrees of freedom)

Footnotes see p. 115.

Table 4.A2 (continued)

Study (reference) Design* Variable(s)**	Location (latitude [geographic/ geomagnetic])	Population Sample Size Age (years)	Sampling Span	Interval	Results* Spectral	Cross-spectral coherence components (frequency) [period]
IV. Modulating role of melatonin?						
23. Tarquini et al. (17) H melatonin (MEL)	Florence, Italy (43.78°N/44.26)	Adults 172 20–90	3 years	4-hourly for 24 h	About-yearly variation in MEL during daytime but half-yearly changes during nighttime at latitude of 43.47 °N; half-yearly variation at 65.00 °N around noon	
24. Maggioni et al. (18) H Melatonin	Milan, Italy (45.28°N/46.31)	Women in 3rd trimester of pregnancy (14 healthy and 11 IUGR)	1 year	4-hourly for 24 h	About half-yearly variation found only in IUGR group, with 1.0 year A and 0.5 year (A,ϕ) difference between the two groups	
V. Theoretical Computations						
25. Ulmer et al. (19)	For a field of about 1 nT typical of a magnetic field associated with the human circulatory system and with the interplanetary magnetic field, the oscillating period of some ions found in cells (Na^+, K^+, Ca^{++}, Mg^{++}) is about 1 week and that of some proteins (albumin, hemoglobin) is about 1 month, both periods corresponding to prominent components in the time structure of human physiology (computations based on Earth's magnetic field of 0.5×10^{-4} T typically found at mid-latitudes for ions or molecules in a vacuum)					

Footnotes see p. 115.

Footnotes for Table 4.A2

Cross-spectral coherence coefficients have in common with correlation coefficients that they describe the relation between two variables. Cross-spectral coherence coefficients are less unspecific in that they describe the relationship at a specific frequency. In order to avoid listing spurious associations, only cross-spectral coherence coefficients away from spectral peaks are listed herein. *Abbreviations:* L, longitudinal; T, transverse; H, hybrid (linked cross-sectional): BH, body height; BW, body weight; HC, head circumference; CC, chest circumference; AC, abdomen circumference; BP, blood pressure (S, systolic; D, diastolic); HR, heart rate; RR, respiratory rate; PEF, peak expiratory flow; TE, 1-min time estimation; EH, eye–hand coordination; MI, myocardial infarctions; MEL, circulating melatonin; Kp, geomagnetic disturbance index; WN, Wolf number of solar activity; Bz, vertical component of interplanetary magnetic field; CR, cosmic ray intensity; 3hCR-SD, 3-h standard deviation of CR; SD, standard deviation: † u.p., unpublished.

[a] Some analyses or meta-analyses in the Chronobiology Laboratories of the University of Minnesota in Minneapolis, Minnesota; a review of the vast literature is beyond the scope of this table.

[b] In dealing with biomedical equivalents of the about 21-year Hale cycle (and there are components with even lower frequencies and even larger amplitudes in our anthropometric data), one cannot collect too many cycles over a single lifetime and turns for replications to populations; in the case of each of the population rhythms, one deals with findings on many individuals' cycles, as in the case of findings covering 112 years of population anthropometry. In each of the two studies by Weber et al.[1,2], over half a million individuals must be sufficiently concordant to allow a demonstration of the 10-year population rhythm. The 30.8 years of self-measurements of over 10 variables around the clock for a total of over half a million values provide a longitudinal validation check, supporting other evidence collected transversely, by virtue of (bio)ergodicity properties (that is the consistency of findings made transversely on populations and longitudinally on individuals, notably in relation to temporal characteristics of a process, assumed to be [bio]stationary in the sense of reproducibility of some of its characteristics).

[*] Strengthening and broadening the scope as to mechanisms of the propositions of Weber et al.[1] by cross-spectral coherence, superimposed epochs, and remove-and-replace approach, among other analyses on even larger and more diverse data sets. Results thus obtained reveal not only about-yearly rhythms but also other components with periods ranging from half a week to about 21 years that are not shared with sunshine. Overall, the evidence points to mechanisms complementing sunshine effects, including geomagnetic disturbance effects that may be mediated via intermodulations involving melatonin produced by the pineal gland.

References for Table 4.A2

1. Weber GW, Prossinger H, Seidler H (1998) Height depends on month of birth. Nature 391: 754–755
2. Weber GW, Seidler H, Wilfing H, Hauser G (1995) Secular change in height in Austria: an effect of population stratification? Ann Hum Biol 22:277–288
3. Otto W, Reissig G (1963) Zur Anthropologie der Neugeborenen. 4. Mitteilung. Laenge und Gewicht der Neugeborenen in den verschiedenen Monaten. Monatsberichte der Deutschen Akademie der Wissenschaften zu Berlin 5:549–559
4. Henneberg M, Louw GJ (1993) Further studies on the month-of-birth effect on body size: rural schoolchildren and an animal model. Am J Phys Anthropol 91:235–244
5. Garcia Alonso L, Hillman D, Cornelissen G, Garcia Penalta X, Wang ZR, Halberg F (1993) In: Otsuka K, Cornélissen G, Halberg F (eds) Chronocardiology and Chronomedicine: Humans in Time and Cosmos. Life Science Publishing, Tokyo, pp 71–75
6. Cornélissen G, Breus TK, Bingham C, Zaslavskaya R, Varshitsky M, Mirsky B, Teibloom M, Tarquini B, Bakken E, Halberg F, International Womb-to-Womb Initiative Group (1993) Beyond circadian chronorisk: worldwide circaseptan-circasemiseptan patterns of myocardial infarctions, other vascular events, and emergencies. Chronobiologia 20:87–115
7. Halberg F, Breus TK, Cornelissen G, Bingham C, Hillman DC, Rigatuso J, Delmore P, Bakken E, International Womb-to-Womb Initiative Group (1991) Chronobiology in space. Keynote, 37th Ann. Mtg. Japan Soc. for Aerospace and Environmental Medicine, Nagoya, Japan, November 8–9, 1991. University of Minnesota/Medtronic Chronobiology Seminar Series, #1
8. Düll T, Düll B (1935) Zusammenhänge zwischen Störungen des Erdmagnetismus und Häufungen von Todesfällen. Deutsch Med Wschr 61:95–97
9. Faraone P, Cornelissen G, Katinas GS, Halberg F, Siegelova J (1999) Astrophysical influences on sectoring in colonies of microorganisms. Abstract 17, MEFA 1999, Brno, Czech Republic, November 3–6, 1999

10. Portela A, Northrup G, Halberg F, Cornelissen G, Wendt H, Melby JC, Haus E (1996) Changes in human blood pressure with season, age and solar cycles: a 26-year record. Int J Biometeorol 39:176–181
11. Cornélissen G, Halberg F, Wendt HW, Bingham C, Sothern RB, Haus E, Kleitman E, Kleitman N, Revilla MA, Revilla M Jr, Breus TK, Pimenoy K, Grogoriev AE, Mitish MD, Yatsyk GV, Syutkina EV (1996) Resonance of about-weekly human heart rate rhythm with solar activity change. Biologia (Bratislava) 51:749–756
12. Baevsky RM, Petrov VM, COrnelissen G, Halberg F, Orth-Gomer K, Akerstedt T, Otsuka K, Breus T, Siegelova J, Dusek J, Fiser B (1997) Meta-analyzed heart rate variability, exposure to geomagnetic storms, and the risk of ischemic heart disease. Scripta Medica 70:199–204
13. Syutkina EV, Cornelissen G, Sonkowsky RP, Lanzoni C, Galvango A, Montalbini M, Swartzkopff O (1997) Neonatal intensive care may consider associations of cardiovascular rhythms with local magnetic disturbance. Scripta Medica 70:217–226
14. Halberg F, Cornelissen G, Sonkowsky RP, Lanzoni C, Galvango A, Montalbini M, Swartzkopff O (1998) Chrononursing (chronutrics), psychiatry and language. New Trends in Experimental and Clinical Psychiatry 14:15–26
15. Watanabe Y, Hillman DC, Otsuka K, Binham C, Breus TK, Cornelissen G, Halberg F (1994) Cross-spectral coherence between geomagnetic disturbance and human cardiovascular variables at non-societal frequencies. Chronobiologia 21:265–272
16. Halberg F, Engeli M, Hamburger C, Hillman D (1965) Spectral resolution of low-frequency, small-amplitude rhythms in excreted 17-ketosteroid; probable androgen induced circaseptan desychronization. Acta Endocrinol (Kbh) Suppl 103:5–54
17. Tarquini B, Cornélissen G, Perfetto F, Tarquini R, Halberg F (1997) Chronome assessment of circulating melatonin in humans. In Vivo 11:473–484
18. Maggioni C, Cornelissen G, Antinozzi R, Ferrario M, Grafe A, Halberg F (1999) A half-yearly aspect of circulating melatonin in pregnancies complicated by intrauterine growth retardation. Neuroendocrinol Lett 20:55–68
19. Ulmer W, Cornélissen G, Halberg F (1995) Physical chemistry and the biologic week in the perspective of chrono-oncology. In Vivo 9:363–374

5 Melatonin Involvement in Cancer: Methodological Considerations*

Germaine Cornélissen, Franz Halberg, Federico Perfetto, Roberto Tarquini, Cristina Maggioni, Lennart Wetterberg

Abstract

On the one hand, melatonin has been determined in different body fluids to seek markers of a heightened risk of cancer development. On the other hand, it has been used as an oncostatic drug in the treatment plan of cancer patients. For both applications, timing has proved to be critical, not only along the circadian scale but in relation to a number of other multifrequency rhythms as well. Like other immunomodulators, melatonin has been shown in the experimental laboratory to inhibit cancer growth when administered at one circadian stage but to enhance it when given at another. Such circadian stage-dependence of melatonin effects was observed in a mouse leukemia model, in a mouse sarcoma model, and in spontaneous mammary carcinogenesis. In the latter model, circannual and/or other secular changes were also noted. "First do no harm" is hence a particularly restrictive possibility in the clinical use of immunomodulators such as melatonin. Marker rhythmometry is a useful tool offered by chronobiology to guide treatment timing while, if specific, it can also serve for the assessment of the patient's response to treatment.

In relation to cancer chronorisk, a prominent circadian rhythm was found to characterize urinary melatonin in groups of clinically healthy women at high or low risk of developing breast cancer. The major difference between women at high vs low breast cancer risk was in the circadian amplitude, the rhythm being more

* Dedicated to the memory of Lawrence E. Scheving and also to Sergio Cardoso and John Pauly. With all three, in a series of laboratory experiments, we demonstrated the merits of timing antileukemic treatment. This work is further dedicated to the memory of John Raidel, whose legs were amputated above the knees but who was agile on a scooter, showing the will to live and love by remembering earliest friendships. He fostered them by attempting to reunite even his classmates. He passed away due to leukemia, without benefiting from the results of extensive laboratory experimentation that first demonstrated the doubling of survival time by the chronobiologic timing of treatment and then resulted in cures not obtained by a (homeostatic) treatment schedule specified as the best by the U.S. National Institutes of Health. Timing is most important in the case of immunomodulators, such as pineal agents, perhaps other than melatonin, and preferably for prehabilitation, that is for action at the time of an elevated disease risk, rather than waiting for the fait accompli of disease.

Support: U.S. Public Health Service (GM-13981), National Heart, Lung, and Blood Institute, National Institutes of Health (HL-40650), University of Minnesota Supercomputer Institute, Dr. h.c. Dr. h.c. Earl Bakken Fund and Dr. Betty Sullivan Fund, and Mr. Lynn Peterson, United Business Machines, Fridley, MN.

prominent in the high risk women. In the presence of actual disease, however, the circadian amplitude of circulating melatonin is statistically significantly reduced in comparison with healthy women. In view of the absence of a difference in mean value, it thus becomes critical to assess the circadian variation since a difference can only be validated at some, but not all circadian stages. This is particularly true since the validated difference can be in either direction. Doing no harm, in this case and many others, gene therapy included, is a critical commandment, when harm can be avoided by scheduling according to body time. Individual differences in the timing of the acrophase of the circadian melatonin rhythm further speak against reliance on spot-checks, even if they are clock time-specified.

Whether focus is placed on chronorisk assessment or on chronotherapy, the multifrequency rhythms of a broad time structure (chronome) should be kept in mind. Central to the chronomes are feedsideward mechanisms, that is interactions among multiple rhythmic entities that result in predictable sequences of attenuation, no effect, and amplification of the effect of one entity (the actor) upon another entity (the reactor) as it may be modified by a third entity (the modulator). The pineal gland, whose major product is melatonin, has been documented to be primarily involved in feedsidewards, which may account for seemingly contradictory phenomena that become predictable once the data are properly time-specified and analyzed for rhythms. Apart from circadian behavior, the pineal gland is also eminently circaseptan-periodic and is known to be sensitive to changes in magnetic field, aspects awaiting exploitation in the clinic, where circaseptan scheduling may be more practical to optimize than the circadian routine; but both these and other roles of timing can be assessed by marker rhythmometry. Variability, a terrible foe when it remains unassessed, can become a new resolving tool, particularly useful in the study of neuroimmunomodulation.

5.1 Introduction

The problem of the pineal gland and cancer, viewed chronobiologically, starts with primum nil nocere, insofar as treatment is concerned. Toxic drugs given at the wrong time can harm (Halberg et al. 1973; Halberg Francine et al. 1989; Halberg J et al. 1989). Moreover, one can speed up a cancer rather than inhibit it with agents such as melatonin that seem harmless (Halberg E and Halberg F 1980; Bartsch and Bartsch 1981; Bartsch et al. 1990a, b, 1994; Sánchez 1993; Tarquini et al. 1997). Such empirical results on chronotherapy, based on circaseptan and circannual as well as circadian rhythms, should be placed on a biochemically mapped basis. A first step in this endeavor was completed before the end of the 1950s. A circadian cell cycle was first described in growing mouse liver (Fig. 5.1) (Halberg et al. 1959), and later approximated in a unicell (Edmunds and Halberg 1981). The demonstration of the hours of changing resistance followed (Fig. 5.2) (Halberg et al. 1955; Halberg 1960). Growth hormone increas-

Fig. 5.1. Demonstration of a circadian cell cycle by the changes along the 24-h scale of glycogen, ▶
mitosis, phospholipid labeling, and RNA and DNA formation. The sequence of RNA formation
preceding that of DNA, first demonstrated in growing mouse liver, is also approximated in a unicell.
(From Edmunds and Halberg 1981, copyright Halberg)

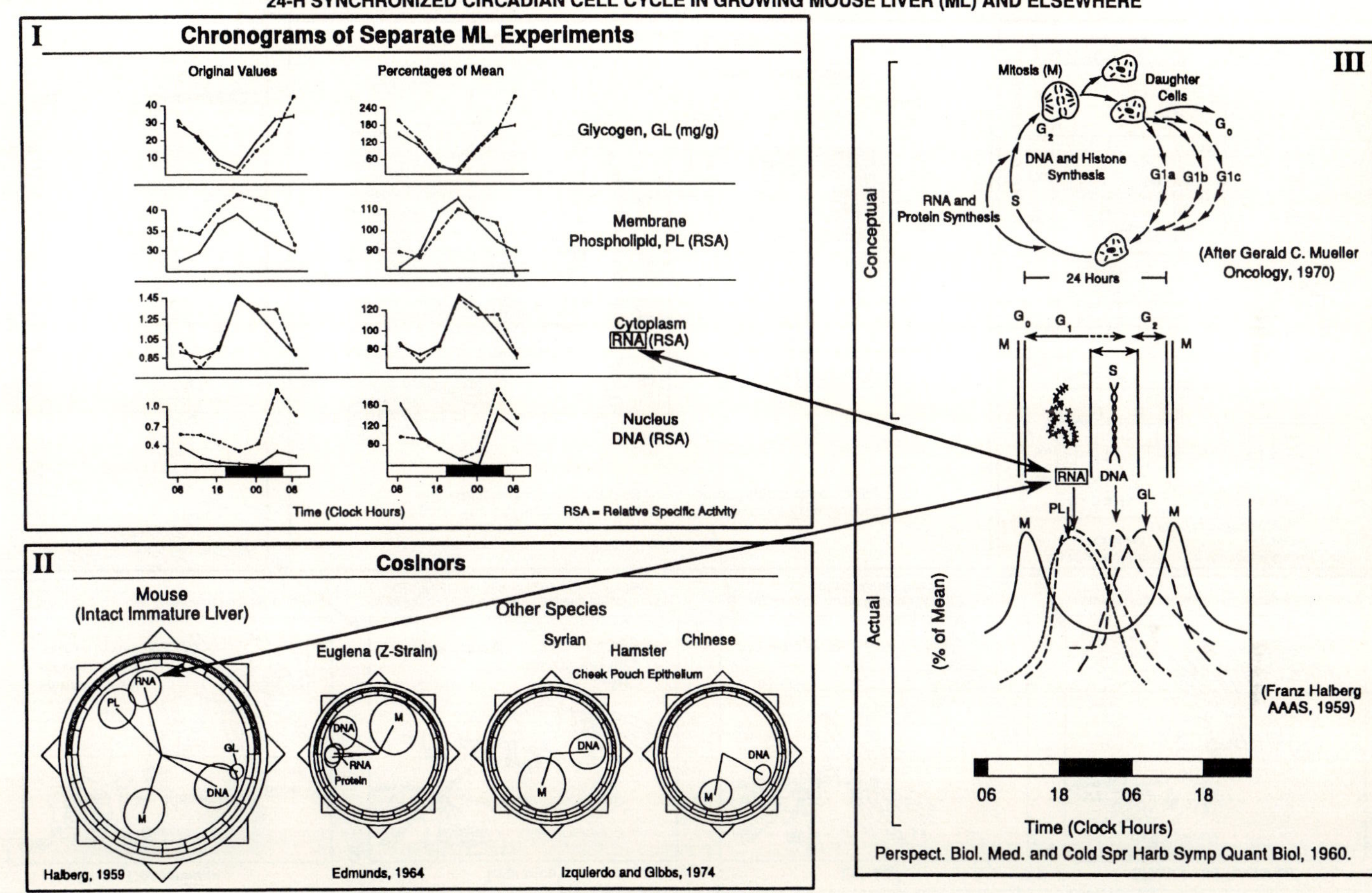

Fig. 5.1. Legend see p. 118

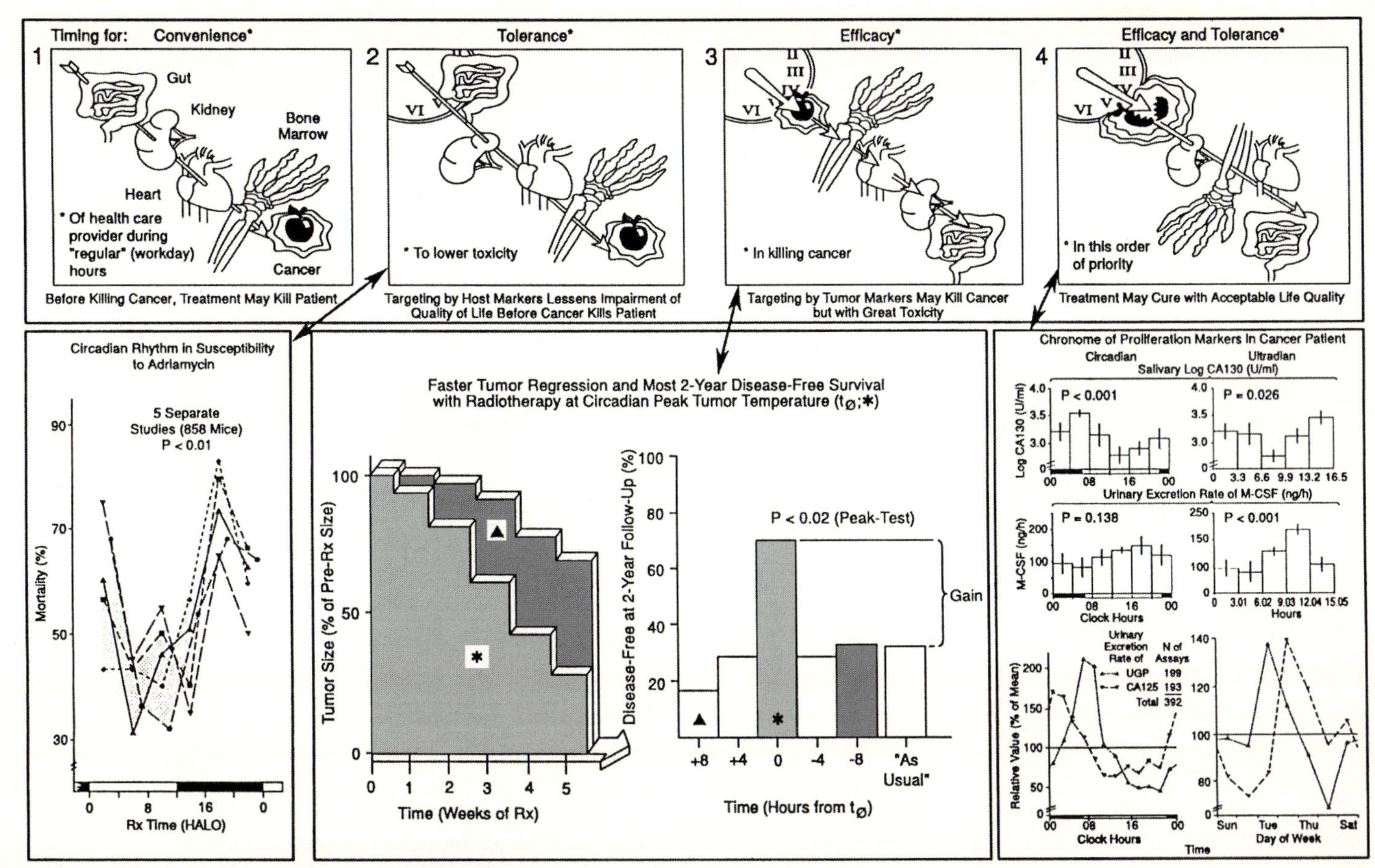

Fig. 5.2. Legend see p. 121

◄ **Fig. 5.2.** The times of changing resistance are illustrated in the context of cancer chronotherapy. Whereas major focus has been placed on chronotherapeutic trials aimed at increasing tolerance by minimizing the side effects of cytotoxic drugs, the availability of tumor markers renders it now possible to emphasize the optimization of treatment efficacy. Using tumor temperature as a marker for timing treatment, it was possible to double the 2-year disease-free survival of patients with perioral cancer. (Copyright Halberg)

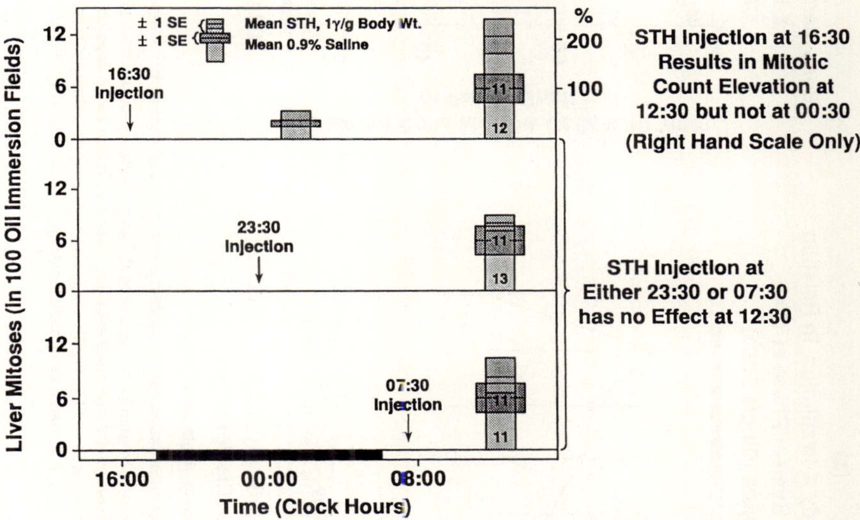

Fig. 5.3. Growth hormone increases mitotic activity in murine hepatic parenchyma when it is given before the circadian acrophase of RNA synthesis, but not thereafter. (Copyright Halberg)

ed mitoses in hepatic parenchyma when it was given at a circadian stage before RNA synthesis but not thereafter (Fig. 5.3) (Halberg et al. 1973); the detection of the effect in itself had already been demonstrated earlier to be circadian stage-dependent (Litman et al. 1958).

Thus, it eventually became clear that the same stimulus, applied in the same dose in the case of radiation (Fig. 5.4) or a drug (Fig. 5.5), is harmful, even lethal, or is harmless, solely as a function of timing (Halberg 1960, 1962; Haus 1964; Haus et al. 1973; Scheving 1976; Reinberg and Halberg 1979; Bartsch and Bartsch 1981). We also found that the toxicity of a number of chemical agents, including eventually a number of different anticancer drugs (Fig. 5.6), and also the LD_{50} in response to whole-body X-ray irradiation (Fig. 5.4), were dependent on timing along the 24-h scale (see discussion of LD_{50} in Halberg 1960; cf. also Haus 1964; Haus et al. 1972; Halberg et al. 1973).

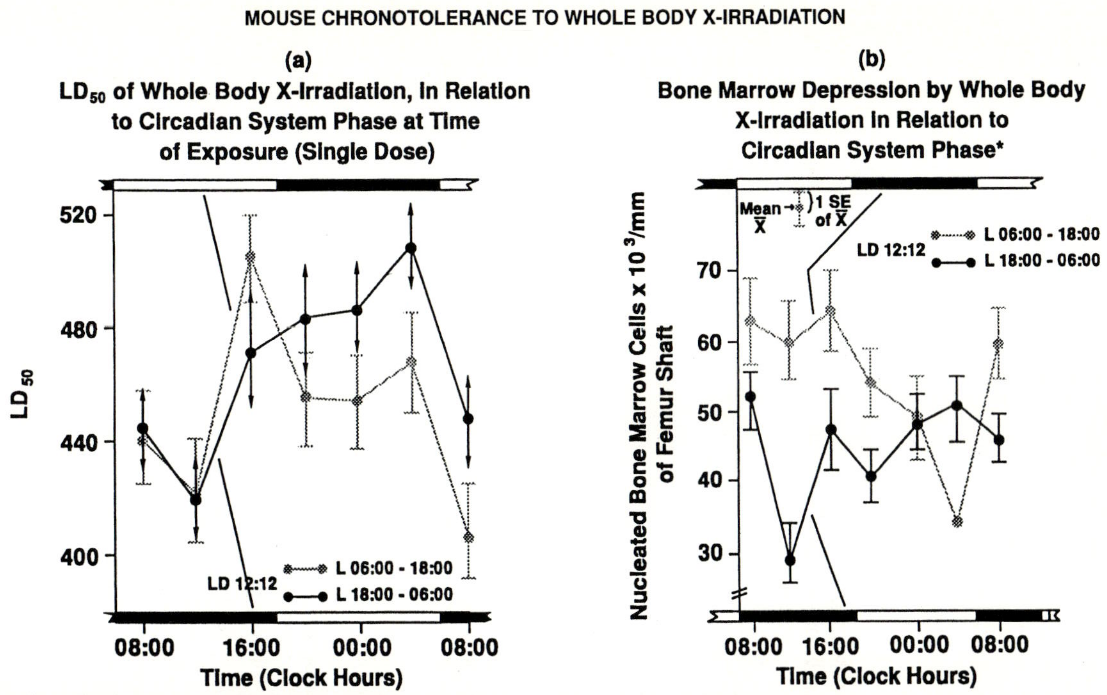

* At time of exposure of adult male C-mice to single dose (350 R)

(a) LD_{50} of whole body X-ray irradiation of mice and circadian system phase at time of exposure to single dose

(b) Bone marrow depression by whole body X-irradiation in relation to circadian system phase at time of exposure of adult male C-mice to single dose (350 r). Circadian radiosensitivity rhythm of mice after standardization for 14 days on 2 lighting regimens differing 180° in phase. LD_{50} in roentgens. Data obtained in 1958 in the Chronobiology Laboratories, University of Minnesota. Circadian rhythm in LD_{50} (a) and bone marrow depression (b). A remarkable circadian change in radiosensitivity also characterizes the Chinese hamster

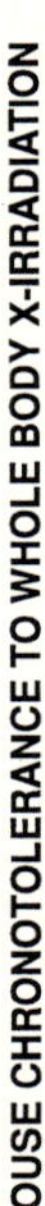

Fig. 5.4. Demonstration of a circadian rhythm in susceptibility to whole-body X-irradiation in the mouse. The LD_{50} from exposure to a single dose is circadian periodic, with an acrophase depending on the temporal placement of the spans of light and darkness of the environmental lighting regimen (From Halberg 1960, copyright Halberg)

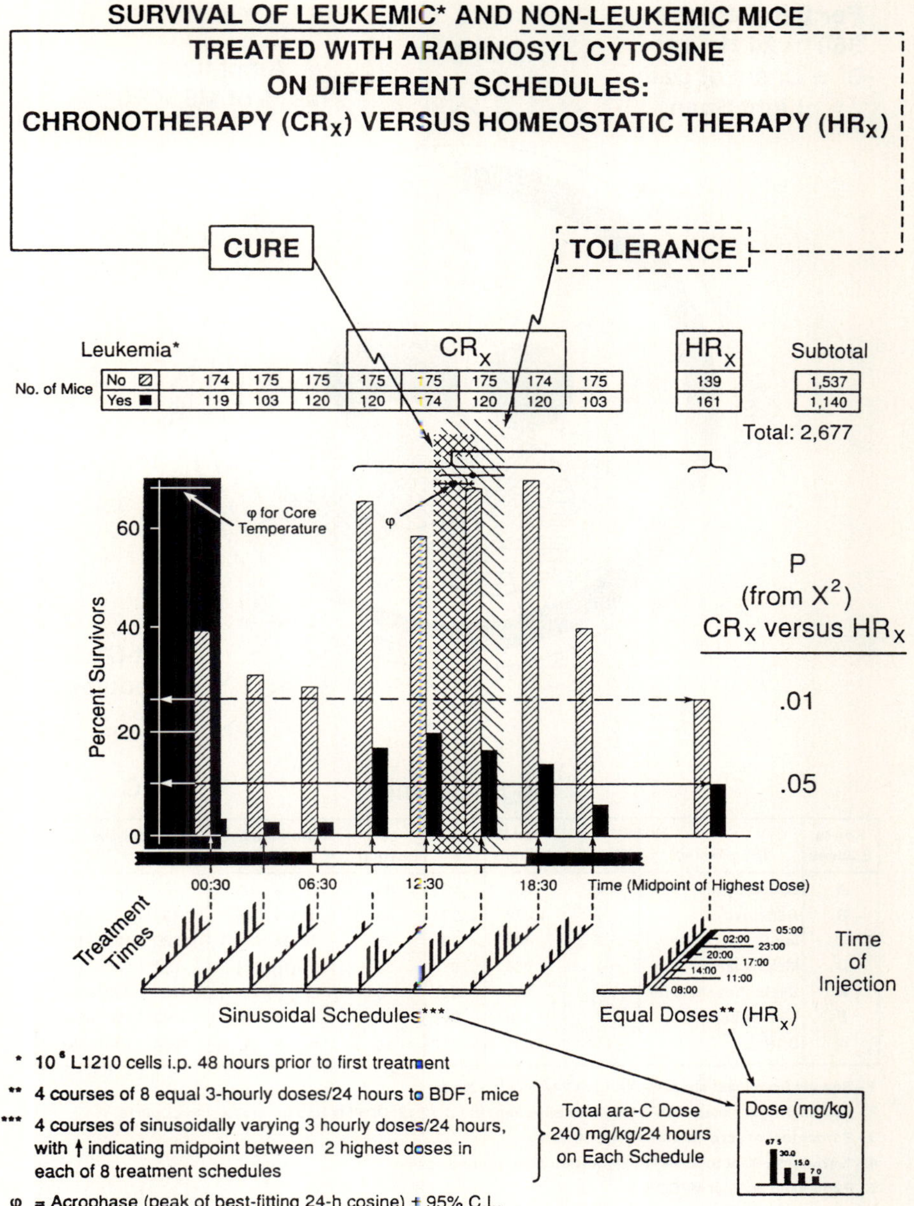

Fig. 5.5. The tolerance and efficacy of drugs depends critically on the timing of their administration in relation to the circadian (and other) bodily rhythms. This is demonstrated herein for the case of the anti-cancer drug ara-C, given in either equal or 24-h modulated 3-hourly doses. Similar results are obtained when drugs are given in single daily doses, even in the case of 24-h formulations. (Copyright Halberg)

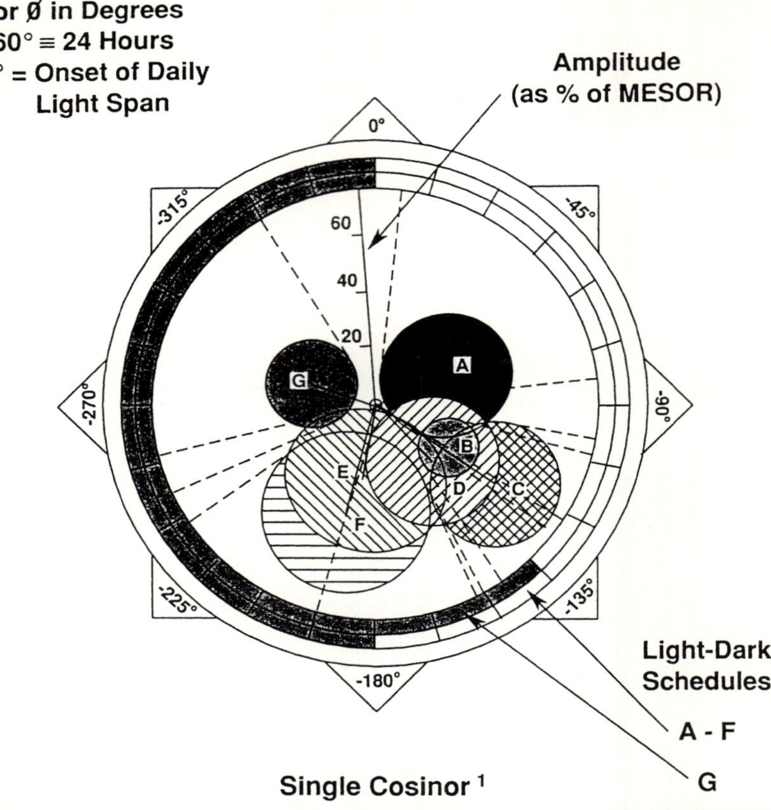

Single Cosinor [1]

Key to Ellipses	Drug Tested[2]	N Studies	Total N Animals	P[3]	% Rhythm[4]	Amplitude[5] (95% CL)	Acrophase (Ø) (95% CL)
A	Daunomycin	4	690	.021	31	29 (4 54)	-67° (-8 -127)
B	Adriamycin	6	1,072	< .001	60	33 (21 45)	-124° (-103 -146)
C	ara-C	2	480	< .001	76	56 (32 80)	-126° (-100 -151)
D	Melphalan	4	456	.015	33	31 (18 57)	-138° (-83 -193)
E	Cyclophosphamide	6	826	.027	18	32 (3 60)	-185° (-123 -255)
F	Vincristine	2	239	.007	67	46 (15 76)	-193° (-150 -235)
G	DDPt	11	1,503	.002	18	24 (8 39)	-289° (-246 -331)

1 - Results from least-squares fitting of 24-h cosine curve

2 - I.p., All drugs, except DDPt, tested in mice kept in LD 12:12; DDPt tested in rats kept in LD; 8:16

3 - P from test of zero-amplitude hypothesis

4 - % rhythm = % of total variability attributable to fitted cosine

5 - Expressed as % of MESOR

Fig. 5.6. The toxicity of a number of chemical agents, including various anticancer drugs, depends on the timing of their administration along the 24-h scale. (Copyright Halberg)

5.2 Chronoradiotherapy

During the Eisenhower administration, the United States sent much grain to fight hunger in India. The rupees paid to the United States for the grain under US Public Law 480 became available for research in India and were to be used to benefit both countries. In charge of the United States section of an international biological program, one of us (FH) was to design research that met the aim of mutual benefit for the two countries. These funds were used solely for planning: using only his own resources for the work, Dr. B. D. Gupta, head of radiotherapy at the Postgraduate Institute for Medical Education and Research in Le Corbusier's city of Chandigarh, realized a chronotherapeutic dream when he showed that in patients with very advanced perioral cancers (now only of historical interest since such cancers are presently treated long before they become an impediment to speech and food intake), radiation treatment doubled the two-year disease-free survival rate when given at peak tumor temperature, this variable being used as a marker rhythm (Halberg 1977; Halberg et al. 1977).

5.3 Historical Perspective

Those involved then assumed that chronotherapy was here to stay. But a quarter-century later, at the turn of the millennium, there is still reluctance to accept timing, which too many investigators regard as a "complication", just as scrubbing before surgery was long avoided (Halberg 1974; Halberg et al. 2000).

Before 1843, in Cordova, Spain, in the twelfth century CE, Averroës (1126–1198), the pantheistic Islamic scholar of law and philosophy, also learned in medicine, forced into hiding for some time on suspicion of heresy, apparently insisted on washing hands before surgery. Other Arab physicians of Cordova followed the practice. When hand washing was transferred to surgery is unknown, but the merits of cleanliness as such had been recognized in priestly Jewish documents dating at least 500 and perhaps 1300 years before the Common Era (Exodus 30: 17–21) (Rabbi Leonard A. Schoolman, personal communication). The washing of hands before meals and before any ritual act may have eventually led to the requirement of cleanliness before surgery by the time of Averröes, if not earlier. To some extent, the atmosphere eventually recognizing the need for scrubbing was prepared for thousands of years, even if the resolution of bacteria by the microscope was to give it a rational basis. Today, scrubbing and hygiene in general involve many aspects of medicine.

But this does not apply to timing as yet. At a session of the Gynecologic Oncology Group of the United States, a statistician at the US National Institutes of Health (NIH) was publicly against chronotherapy. His argument sounds simplistic: "Another sheet of paper would have to be filled out and would need analysis". We can understand him and his colleagues, who believe that they are underpaid and overworked and who, in any event, often confront messy, occasionally undecipherable records kept by officious if not efficient health care providers. For many more reasons, most of them equally untenable, the timing of cancer treatment, with notable exceptions, remains mostly neglected even today, as it was several decades ago at the start of work in the field which developed into chronobiology. While circadian optimization may find some obstacles from a practical viewpoint, optimization along the scale of the week may be easier (Cornélissen et al. 1999 a), and certainly deserves testing, notably in view of prominent circaseptan rhythms in tumor growth that do not require scheduling during odd hours, a major obstacle (Ulmer 1999; Ulmer et al. 1999).

5.4 Chronochemotherapy

Before the first funding for cancer chronotherapy was provided by the NIH, chronobiologists had to show that they could do better than the best treatment of an experimental leukemia then available. We did so on the experimental animal model specified by the NIH (using an L1210 leukemia advocated by the chemotherapy section head): We doubled the survival time of mice with this malignancy in three consecutive studies simply by replacing the eight equal doses of ara-C per day with a sinusoidally varying treatment totalling the same dose per day, thereby taking into account the net effect of periodicities in the host and in the tumor, as published in *Science* and elsewhere; eventually cures could be had when none were otherwise available (Cardoso et al. 1970; Haus et al. 1972; Halberg et al. 1973, 1979; Scheving et al. 1976). Despite accumulating laboratory evidence in Minnesota and Arkansas (USA) and clinical findings in Chandigarh, in Minnesota, and eventually in Europe, there is an enormous resistance by radiation and drug therapists to use chronobiology.

5.5 Need for Marker Rhythmometry

Chronobiology with marker rhythmometry (Cornélissen and Halberg 1989, 1992a) involves added work and may require activity outside clinic hours. Furthermore, the recommended use of marker rhythms can add cost and complexity. Complexity

Fig. 5.7. Rhythms with a very wide range of frequencies, organizing chaotic changes, and trends ▶
with growth, development, and aging– the chronome – involve change that can be of very large extent. Stages of high and low susceptibility will occur as a function of age and multifrequency rhythms, as risk stages built into us, but interacting with the environment, the socioecologic niche, and the even broader cosmos. The rhythms' periods can be viewed as adaptations to our environment, refined by internal integrative interactions among body parts, in an open dynamic nonlinear system as integrations occurring concomitantly among the organism's chronomes and those of its environment. Chronorisks as they affect the outcome of human health or disease – victory or defeat – are depicted in two cabbage-like graphs, containing the organism in their middle. They suggest that a cell or human body as a whole can live with often-unattended open neuroendocrine and immune gates of susceptibility. It is much too costly and hardly necessary to defend all the body's weak spots all the time. Instead, each cell or organism coordinates its chronome. In the two *"cabbages"*, each set of *wavy circles* represents the different time scales of the periods of the chronome, labeled for the larger *cabbage*. The *triangles, quadrangles*, and *closed* or *open circles* represent low points (bathyphases of rhythms with different frequencies). Vulnerability will most likely be highest at the beginning and near the end of the individual's life, but critical spans for one or the other challenge occur in utero as a feature of development, modulated already by circadian changes in susceptibility, with the latter recurring throughout life. The gates of susceptibility in circannual, circadecennian and other infradian systems must not be disregarded, as the host cell (*in the middle*) is approached, even if it seems easiest to study circadians and ultradians. The risk of cancer, whether posed by a virus, a chemical, radiation, or some other factor, shown by *arrows*, may be stopped at any one level of organization by the rhythms of defense in a broad sense, changing with one of the many intermodulating frequencies. The organism wins the day if the first line of defense, the critical stages of the life span, and of circadecennian, circannual, and other rhythms are defended. The risk dependent upon the stages of rhythms with several frequencies, called chronorisk, is at its highest when a potentially harmful stimulus finds several or all of the gates open concurrently, as shown in the graph at the *bottom left*. An agent even of a relatively low intensity then readily defeats the host. (Copyright Halberg)

CHRONORISK

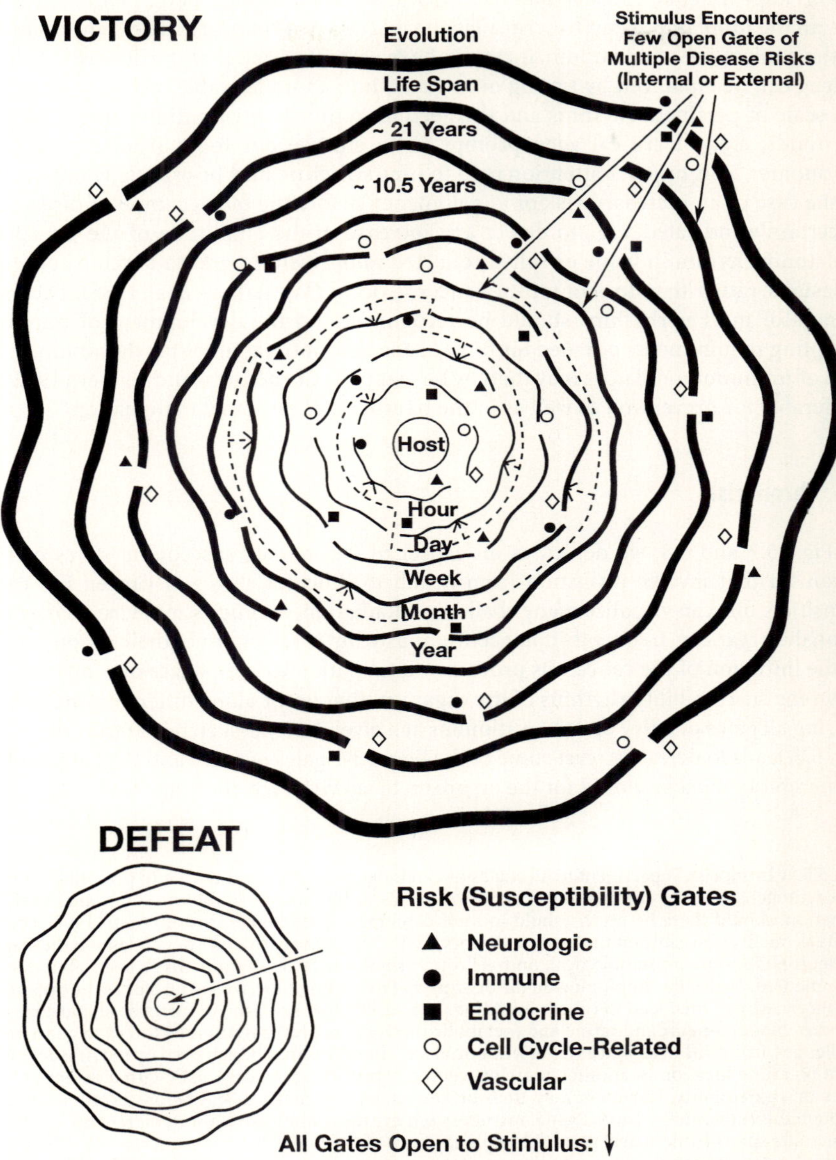

Fig. 5.7. Legend see p. 126

applies in particular to the pineal gland, since melatonin can stimulate or inhibit a cancer as a function of when it is administered (Bartsch and Bartsch 1981; Levi et al. 1981; Langevin et al. 1983; Wrba et al. 1985, 1986, 1989; Bartsch et al. 1990a, b, 1994). It is all the more important to scrutinize the relevance of time structures (chronomes) to the clinical use of melatonin. It cannot be overemphasized that the desired effect of a drug can be enhanced by timing or reversed into an undesirable one. Timing along the scale of pertinent rhythms and the need for a timely intervention along the scale of trends, e.g., of critical ages, prompt us to enlarge our focus from rhythms to chronomes. This broader attention paid to time structure may be especially important in the case of an increased risk of a malignancy before the fait accompli of disease. It is certainly indicated, e.g., to detect any increase in the amplitude of the circadian melatonin rhythm in urine or saliva collected noninvasively, and to develop counter-measures, rather than to wait for the cancer to occur (Wetterberg et al. 1986). The usefulness of marker rhythms should be recognized and the development of minimal sampling requirements pursued for their use so that optimal times for the administration of immunomodulators may thereby be mapped (Cornélissen and Halberg 1992b), preferably for prevention as well as in the treatment of an actual malignancy.

5.6 Chronorisk

In Figs. 5.7 and 5.8, we depict, as an aspect of chronorisk, susceptible states of the organism that involve transition from health to different stages of cancer. For each transition, one can visualize many challenges. Radiation, chemicals, and viruses may act upon the organism, frequently if not continuously, but whether or not their action results in the initiation of the cancerous process or a transition to later stages depends largely upon the susceptibility rhythms of the organism that occur along different time scales. Having all gates of entry open to a stimulus at a given time, as sketched at the bottom of Fig. 5.7, leads to defeat. At least some of the figurative gates of entry and thus of possible vulnerability must be closed for the organism to survive each challenge to it.

Fig. 5.8. Chronorisk is pertinent to all stages of carcinogenesis. Ideally, one should prevent exposure ▶ to carcinogens, but the elimination of exposure seems to be utopian in the case of stimuli such as radiation, should there be no threshold to their carcinogenic potential. One alternative is hence to focus as much as possible on the earliest chronorisk, that is on the early risks dependent on the stages of neuroendocrine, immunological, and cell cycle rhythms. Such rhythms have been abundantly documented. A stimulus impinging on the organism at the wrong time, wrong insofar as the organism is concerned, can then lead to cancer initiation. In a successful organism, in turn, the signal corps of physico-chemico-neuroendocrine and metabolic messengers prompts immune forces to go to meet challenges, from within as well as from without, at any time. This may be achieved by moving defenses from one time location to another, thereby eventually providing the organism with a defense of all gates of vulnerability. Intruders may then be caught for the first cell kill before or after passage through the vulnerability gates. Even if intruders achieve the initiation of cancer, there remain critical vulnerable spans (time-gates) at all subsequent stages of carcinogenesis. Since it may be difficult to close all gates of susceptibility, a complementary strategy may be to move the time location of the rhythms along the different pertinent cycle lengths shown in the figure. This may be done systematically and/or randomly by manipulating one's routine, or by chronobiologic response modifiers such as melatonin, aimed at cancer prevention as well as treatment. Before or after the initiation of a cancer (during or after cancer promotion or during progression), victory depends on whether cancer risks have been successfully met by the organism's strategic defense initiative, because defense forces are at the gate of vulnerability or because the cancer (chrono)risk happened to coincide with peaks of resistance rather than susceptibility. (Copyright Halberg)

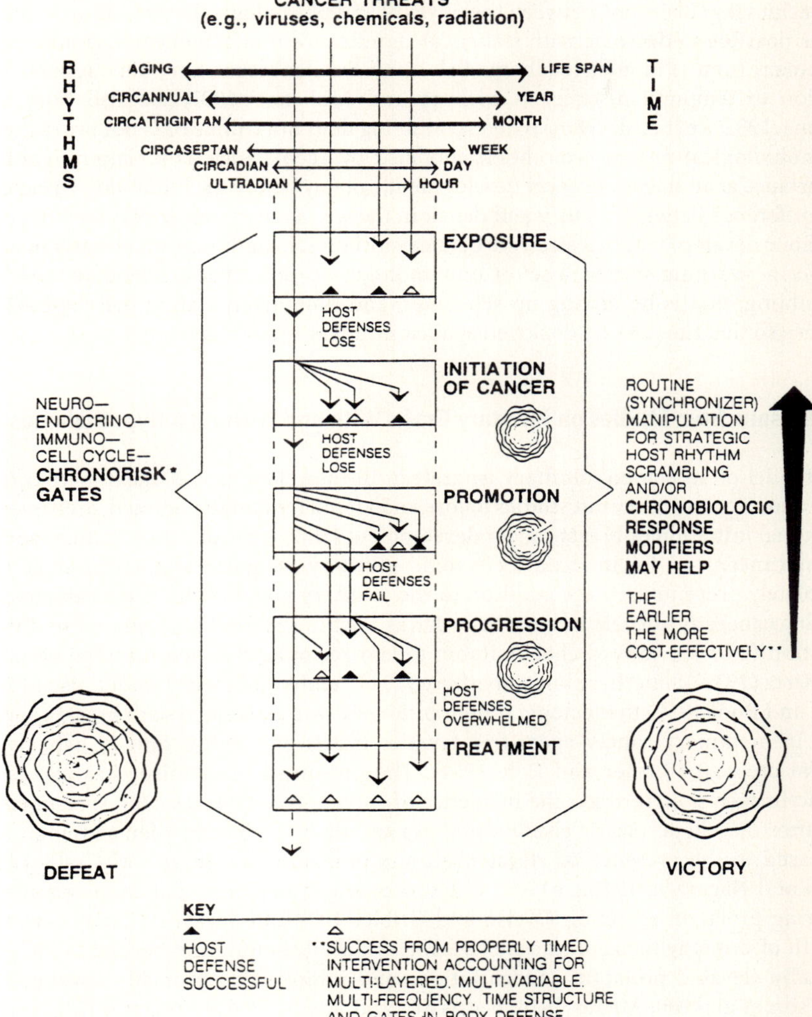

Fig. 5.8. Legend see p. 128

5.7 Rhythm Scrambling

By moving defensive missiles in space to prevent their destruction, one might imitate what the organism has to do in time by intentional rhythm scrambling. With repeated shifts of the lighting regimen, while keeping the time of food availability the same, thus achieving "disorder" between the physical and physiological circadian variations, it was possible to decrease with statistical significance the incidence of spontaneous mammary tumors in mice (Halberg et al. 1986). Similar results have been reported in relation to temporal disorder (Carlebach and Ashkenazi 1987; see also Kort and Weijma, 1982; Kort et al. 1986). When scrambling does not suffice or is not practicable, chronobiological response modifiers modeled by agents such as melatonin gain in importance at all stages of cancer development, notably before the initiation of a cancer. The difference between victory and defeat in the war against cancer may well depend upon our development of a strategic defense initiative, including, with an early neuro-endocrine warning system, a set of immunologic weapons that are rendered safe by scrambling, that is by mixing up schedules, and thus manipulating the rhythms of defense so that they can be deployed against invaders at any time.

5.8 Lessons from Studies on Pituitary Grafts With and Without a Hypothalamus

The model of an ectopic pituitary isograft in Bittner (1936, 1952) agent (virus)-free mice allows one to carry out studies before and after a temporally defined, presumably hormonal initiation of breast cancer development (Halberg et al. 1994). In this model, breast cancer can be induced by ectopic pituitary isografts and reduced, if not completely prevented, by the addition to the pituitary gland of the hypothalamus. In viewing cancer as a developmental disorder, as a feature of the disappearance of differentiation (tropism toward chaos), Rubin (1985) reemphasized the findings by Loeb and Kirtz (1939), Silberberg and Silberberg (1949), and Mühlbock (1958; cf. also Mühlbock and Boot 1959) that ectopic rodent pituitaries are associated with breast cancer, even in mammary cancer agent-free females (C3Hf and Balb/c and F_1 hybrids of several crosses) (Bittner and Cole 1961). The presumed mechanism was excessive prolactin secretion. In mice, the incidence of spontaneously developing (presumably precancerous) hyperplastic alveolar nodules and the incidence of adenocarcinoma is increased in the presence of elevated plasma prolactin (Nandi and McGrath 1973; Yanai and Nagasawa 1972a, b) and reduced by ergot alkaloids (dopamine agonists) lowering prolactin secretion (Welsh and Gribler 1979). In Sprague-Dawley rats, the growth of dimethylbenzanthracene (DMBA)-induced tumors is enhanced by experimentally elevated prolactin concentrations and inhibited by prolactin suppression (Pearson et al. 1969; Meites 1974; Nagasawa and Meites 1970). Prolactin thus seems to be essential for the promotion of both spontaneous and carcinogen-induced mammary tumors in rodents.

Blask (1984) and Blask and Hill (1986) reported that pineal activation, in the model of the anosmic rat, can lead to an earlier development of prolactin-dependent mammary tumors, but also to an increased extent of tumor regression, and that pinealectomy negates these effects and additionally stimulates the growth of more tumors per rat. The authors postulate that pineal-induced changes in prolactin secretion may account for the altered tumor development and growth. Since the tumors are prolactin-dependent, the alterations appear to be both undesirable (earlier tumor development) and desirable (increased regression). Studies in the laboratory are overdue on the prolactin and melatonin chronomes, notably since in the case of prolactin in humans at personal and familial risk of developing breast cancer, the alteration detected is in the circannual rhythm, as discussed below. Accordingly, when El-Domeiri and Das Gupta (1976) report that melatonin may not have a direct effect on tumor growth, changes produced by melatonin deficiency may perhaps be due to a complex reaction involving centers in the hypothalamus and brain stem. Studies examining the effects of pinealectomy upon the rhythmic element in the chronomes of marker variables are lacking.

5.9 Methodological Considerations

On the one hand, melatonin, a main hormone produced by the pineal gland and other sites (Wetterberg et al. 1990), is thought to be involved in the etiology of cancer, notably of breast cancer (Tamarkin et al. 1982; see also Halberg et al. 1988; Cornélissen and Halberg 1992a; Ronco and Halberg 1996). On the other hand, melatonin, and the pineal gland more generally, have long been thought to be useful for oncotherapy. To some extent, however, a breakthrough has been elusive. Below, we review some methodological aspects of pineal research, which may be responsible for the status quo.

5.9.1 Specificity

Melatonin, while being the primary product of the pineal gland, is also produced elsewhere in organisms (Wetterberg et al. 1990; Plautz et al. 1997), including unicells (Balzer and Fuhrberg 1996). Moreover, the pineal gland is affected externally by light (Reiter 1978) and other non-photic variables of the physical environment (see below), and internally by several hormones (Cardinali and Vacas 1976) and sympathetic nervous input (Ueck 1979). It is classically viewed as coordinating influences on several aspects of endocrine, immune, and metabolic physiology (Blask 1984; Maestroni et al. 1988), as a neuroendocrine transducer, with its hormonal secretion coordinated by nervous signals generated by environmental lighting and transmitted to the pineal gland by sympathetic fibers (Wurtman et al. 1968). As such, in defined serum-free medium, an effect of melatonin may depend on the dose, addition of serum, and a complex interaction with hormones. Candidates are estradiol and/or prolactin (Blask and Hill 1986). The important considerations of timing according to a host of predictable rhythmic components and age trends are reviewed below. It is hence critical to specify the exact experimental conditions in the reporting of results.

Bartsch et al. (1992b) and Bartsch and Bartsch (1997) also showed evidence that the antineoplastic properties of the pineal gland can only partially be attributed to its hormone melatonin. They postulated the existence of an as yet unidentified pineal compound (Catrina et al. 1999) showing antitumor activity with considerable therapeutic potential. In vitro work on human tumor cell lines indicates that one of the primary metabolites of melatonin, namely 6-hydroxymelatonin, appears to exert significantly greater cytotoxicity than melatonin itself (Shellard et al. 1989). The high degree of complexity of mechanisms involved in the interactions between tumor growth and the immunoneuroendocrine system was further illustrated by Bartsch et al. (1999) in their serial transplants of DMBA-induced mammary tumors in rats, serving as a model for human breast cancer.

The pineal gland and melatonin seem to be important modulators of the immune system (Ronco and Halberg 1996). The gonadal inhibition usually displayed by melatonin reportedly disappears with pineal hypofunction, an observation thought to be relevant to the etiology of breast cancer (Cohen et al. 1978). Melatonin is also thought to antagonize the immunosuppressor effects of corticosteroids (Maestroni et al. 1986), to increase the cytotoxic activity of natural killer cells (Angeli et al. 1988), and to interact with β-endorphins. Melatonin may stimulate the lymphocytic T proliferative response by action on the opioid receptors in the lymphocytes (Sandyck and Mukherjee 1989).

A link between melatonin and cellular immunity was suggested from studies in experimental animals (Bartsch et al. 1995), with the further observation that there may be stimulation of the immune system and the pineal gland at early stages but inhibition at advanced stages of cancer. The pineal gland has been hypothesized to be "hyperactive" at the outset, as a break to carcinogenesis, and to be "exhausted" thereafter, as the tumor grows (Relkin 1976). It is possible that the circadian amplitude is a gauge of hyperactivity when it is too large and of exhaustion when it is too small, and that the amplitude increase may precede the onset of carcinogenesis. If so, the results of Bartsch et al. (1985, 1992a) and of Tarquini et al. (1999) fit exhaustion, and our earlier finding of an increased circadian amplitude in women at an elevated personal and familial risk of breast cancer, presumably in the absence of an overt cancer (Wetterberg et al. 1986), fits hyperactivity. These tempting philosophical ideas may be too sweeping and may be resolved as feedsideward interactions in chronomes.

5.9.2 Risk Versus Disease

With respect to breast cancer risk, clinically healthy women in Minnesota (USA) and Kyushu (Japan) were sampled around the clock, once in one to four seasons. Possible differences that could reflect the large difference in breast cancer incidence in these two geographic locations were investigated (Halberg et al. 1981). In particular, any involvement of the pineal gland was examined (Wetterberg et al. 1986). A prominent circadian rhythm was found to characterize urinary melatonin in both populations, peaking in the middle of the night (Fig. 5.9). The American women who were at a higher epidemiological risk of developing breast cancer exhibited a larger circadian amplitude of urinary melatonin, as compared to women at low breast cancer risk ($P < 0.025$). A circannual rhythm was also apparent in the North American women, but not in the Japanese women, complementing other associations between the circannual amplitude of circulating prolactin and thyroid-stimulating hormone (TSH) and breast cancer risk (Halberg et al. 1981).

In the presence of actual disease, Tamarkin et al. (1982) and Bartsch et al. (1995) reported that the nocturnal peak in circulating melatonin may be decreased at certain stages and in certain kinds of cancer. Moreover, Bartsch et al. (1989) reported a progressive reduction of the nocturnal peak with increasing tumor size in patients with primary breast cancer. Bartsch et al. (1997a) also reported a decrease in nocturnal urinary 6-sulfatoxymelatonin excretion in patients with primary breast cancer as compared to age-matched healthy peers, as well as a negative correlation with tumor size. Moreover, Bartsch et al. (1997b) found that nocturnal urinary 6-sulfatoxymelatonin

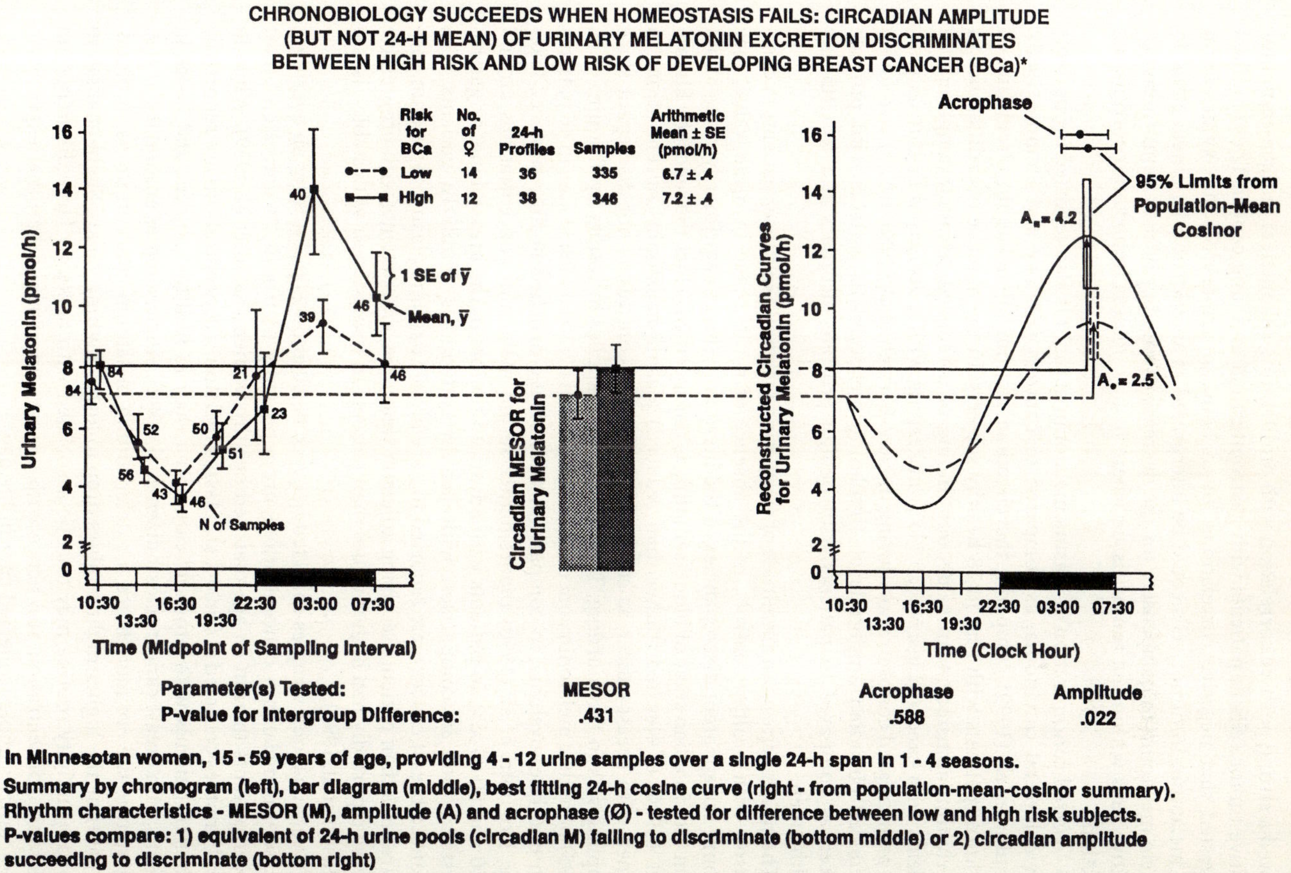

Fig. 5.9. Least-squares fit of 24-h cosine curve to urinary melatonin excretion documents circadian variation and assesses rhythm characteristics as point-and-interval estimates amenable to testing by inferential statistical methods. (Copyright Halberg)

showed a strong positive correlation with proliferating cell nuclear antigen-immuno-positive tumor cells in patients with gastrointestinal and lung cancer. These authors suggested that negative correlations found between urinary 6-sulfatoxymelatonin and melatonin in tumor cells could be interpreted as an effort of the pineal gland to secrete melatonin to compensate for the decrease in number of melatonin-immuno-positive cells within the tumor tissue where it may possess important coordinating functions.

A lowered concentration of plasma melatonin at night is also reported for patients with colorectal carcinoma, as compared to controls (Khoory and Stemme 1988). Tarquini et al. (1999) compared the circadian rhythm of melatonin circulating in 39 cancer patients with that in 28 healthy subjects. In the absence of a difference in the rhythm-adjusted mean value, our finding of a reduced circadian amplitude in patients with cancer (Figs. 5.10, 5.11) is in keeping with a decreased nocturnal peak and an increased concentration during the day. Bartsch et al. (1985, 1992a) also found a reduced circadian amplitude of circulating melatonin in patients with prostate cancer, as compared to patients with benign prostate disorders and to clinically healthy men.

The smaller circadian amplitude of circulating melatonin in cancer patients in comparison with healthy controls, in the absence of a difference in mean value, may account for apparent discrepancies in the literature. This may primarily come about when the circadian stage of blood sampling is not specified. Thus, the report by Lissoni et al. (1990) of elevated (rather than reduced) plasma melatonin concentrations in women with breast cancer, as compared to healthy controls, is not discrepant since sampling was in the morning. The same remark applies to results by Tarquini et al. (1995) of higher melatonin concentrations between 07:30 and 09:30 A.M. in 46 patients with multiple myeloma, in comparison with 31 age-matched healthy subjects. Like-wise, Dogliotti et al. (1990) report higher serum melatonin concentrations around 08:00 and 00:00 A.M. in patients with advanced tumors of the breast, lung, or gastro-intestinal tract in comparison with healthy individuals. Similarly, patients with advanced breast cancer were reported to have higher daytime plasma melatonin con-centrations than patients receiving adjuvant treatment, and patients with progressive disease reportedly had higher values than patients in remission or with stable disease (Falkson et al. 1990).

In venous blood samples collected between 08:00 and 09:00 A.M. after an overnight fast, serum melatonin concentrations are reportedly higher in 74 untreated breast cancer patients than in 46 age-matched healthy women (Barni et al. 1989). While in blood samples collected at 08:00 A.M., plasma melatonin concentrations are reportedly higher in patients with prostatic carcinoma than in healthy subjects, and lower in patients with breast cancer (Oosthuizen et al. 1989). Single samples around 08:00 A.M. may not be the optimal choice for determining melatonin concentrations in blood. At that time the higher nightly values drop to lower daytime values. Slight changes in the daily activity routine may thus be associated with relatively large differences in melatonin concentration and hence with a larger uncertainty than that expected to characterize samples collected later in the day.

It is thus important, as a minimum, to specify when samples have been collected with as much information as possible concerning the subject's daily routine. Preferably, samples should be obtained at different times around the clock so that the circadian variation can be assessed under conditions standardized as much as possible.

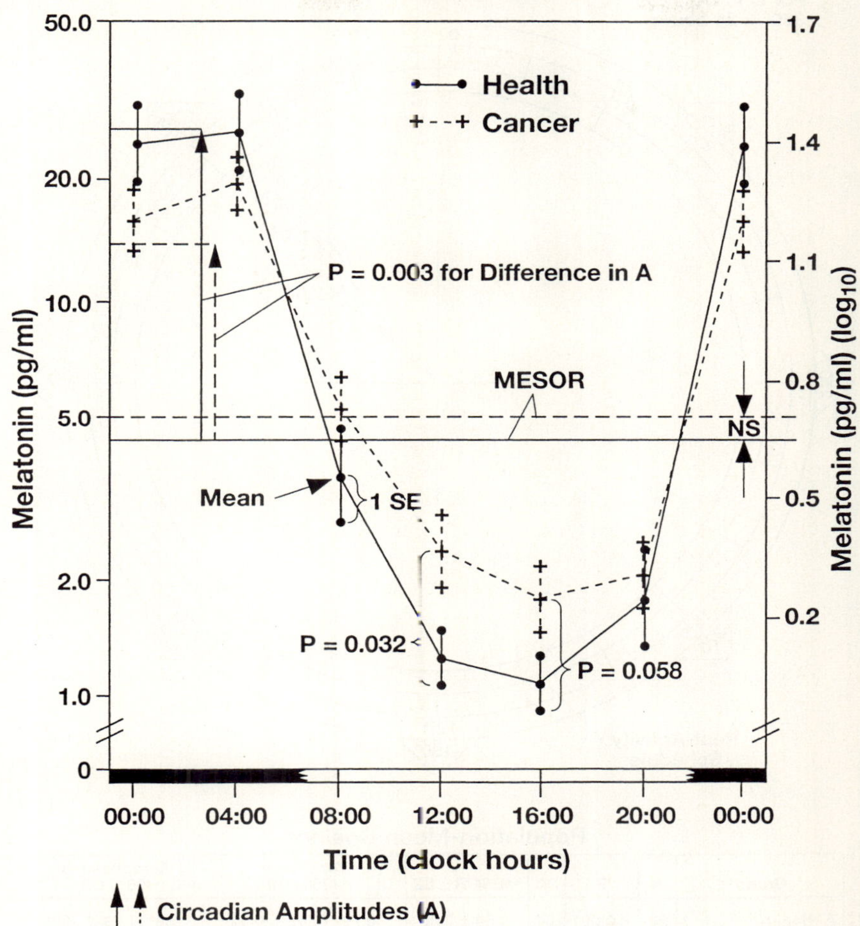

Fig. 5.10. In the absence of a difference in the rhythm-adjusted mean value of circulating melatonin, a difference in circadian amplitude is found between the group of 39 cancer patients and that of 28 healthy subjects. In view of this difference in circadian amplitude, in comparison with healthy subjects, cancer patients have numerically lower circulating melatonin concentrations during the rest span but statistically significantly higher melatonin concentrations during the active span (around noon) (Tarquini et al. 1999). (Copyright Halberg)

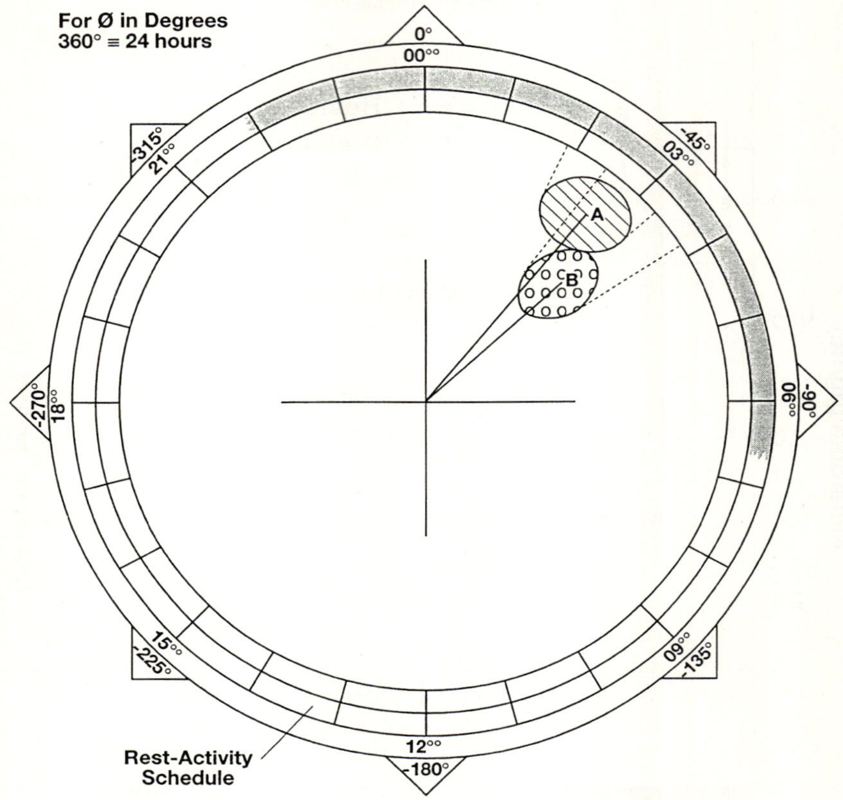

Population-Mean Cosinor

Group	N	P	PR	MESOR ± SE	Amplitude (95% CI)			Acrophase, ø (95% CI)		
A Health	28	< 0.001	84	0.64 ± 0.08	0.79	(0.67	0.92)	-39°	(-29	-50)
B Cancer	39	< 0.001	77	0.70 ± 0.06	0.57	(0.43	0.70)	-48°	(-30	-58)
					P = 0.003					

Analyses on log$_{10}$-transformed data.

Fig. 5.11. Cosinor display of the circadian rhythm in circulating melatonin concentration in cancer patients in comparison with healthy subjects. The smaller amplitude characterizing the cancer patients as compared with the healthy subjects is statistically significant ($P = 0.003$) (Tarquini et al. 1999). (Copyright Halberg)

5.9.3 Melatonin as an Oncostatic Drug: Importance of Timing

Melatonin, at physiological concentrations, was reported to exert a direct but reversible, antiproliferative effect on human breast cancer cell (MCF-7) growth in culture (Hill and Blask 1988). According to these authors, the reported antiproliferative effect of melatonin is associated with striking changes in the ultrastructural features of these cells, suggestive of a sublethal but reversible cellular injury. In vivo work by Blask et al. (1991) on the *N*-methyl-*N*-nitrosurea (NMU) model of hormone-responsive carcinogenesis in the rat breast suggests that melatonin inhibits tumorigenesis by acting as an anti-promoting rather than an anti-initiating agent. Given in the afternoon or evening, alone or in combination with interleukin (IL)-2, melatonin has shown promise in the treatment of cancer patients who do not respond to standard antitumor therapies (Lissoni et al. 1989, 1991 a, b, 1992 a–c, 1993, 1994).

The role of melatonin as a sensitizer of standard chemotherapeutic modalities awaits future study. There is a precedent with camptothecins used as radiation sensitizers in vivo for which the optimal timing of administration was investigated (Kirichenko and Rich 1999). Specifically, the effects of irradiation combined with 9-aminocamptothecin (9AC) on a mouse mammary cancer and the gastrointestinal tract were examined, as was the circadian dependency for cytotoxicity, radiation sensitization, and acute toxicity after single doses of 9AC given at six different times over 24 h (Kirichenko and Rich 1999). A reanalysis of the survival of C3Hf/Kam mice treated with 9AC, administered i. m. either in a 2 mg/kg dosage twice a week for 2 weeks or as a single injection of 4 mg/kg, revealed a statistically significant circadian rhythm ($P = 0.025$) with a double amplitude of 68.6%, attesting to its prominence, and with least toxic effects found when 9AC is administered in the middle of the light (rest) span. When combined with radiation, this timing also corresponded to the longest tumor regrowth delay (Kirichenko and Rich 1999).

Differences in the time of melatonin administration may account for many contradictory results of its anticancer effects. Bartsch and Bartsch (1981) have shown that melatonin, presumably a major component of aqueous pineal homogenate, inhibits, stimulates, or has no effect upon tumor growth and survival as a function of the timing of its administration. There may also be changes in the response to melatonin administration as a function of the time elapsed since the institution of treatment. For instance, this is the case for the response of natural killer cell activity to treatment with melatonin (vs saline) (administered at 3:00 P.M.) observed in clinically healthy subjects: whereas melatonin stimulated killer cell activity 2 and 4 h after treatment, at 6 h post-treatment, the killer cell activity of subjects receiving melatonin was considerably decreased (Lissoni et al. 1986). A similar finding conclusively determined the importance of timing the administration of growth hormone in relation to the acrophase in RNA synthesis (Halberg et al. 1973; cf. Litman et al. 1958).

The circadian stage-dependent effects of melatonin were investigated in Minneapolis on 115 CD2F$_1$ female mice, housed four per cage on a staggered schedule of light and darkness alternating at 12-h intervals. All animals were injected intraperitoneally with 5000 live L1210 leukemia cells. Starting 72 h later, until death, 1 mg/kg of melatonin or saline was injected subcutaneously daily at one of six circadian stages. A third group of mice was handled for weighing only every second day, rather than daily, as were the other two groups. The effect of melatonin was shown to be circadian stage-dependent (Langevin et al. 1983; Halberg et al. 1988). On the average, survival time is shortened as an overall effect of the (handling for) daily vehicle (saline) injection.

The question whether melatonin effects are also time-dependent in the case of mouse breast cancer was investigated with us in Vienna on a model of breast cancer in which most animals spontaneously develop the cancer (Wrba et al. 1985, 1990). A circadian chronomodulation by melatonin upon mammary carcinogenesis was demonstrated (Wrba et al. 1985). Seven groups, each of 30 $C3CF_1$ mice, were kept at $25 \pm 1\,°C$, on six staggered light:dark = 12:12 h (LD12:12) regimens. One group was left untreated. Two sets of six other groups received the vehicle or 10 mg/kg of melatonin every day at one of six circadian times, 4 h apart, starting before the appearance of any tumors. In comparison to tumor incidence in the untreated groups, the six sub-groups of mice given daily injections of saline (as a control procedure) all had a higher early average subgroup breast cancer rate ($P < 0.05$), an effect possibly reduced by melatonin ($\chi^2 = 3.48$; $P < 0.10$). At 71 weeks of age, over 75% of these mice had breast cancer. The 10 mg/kg of melatonin, given daily to six groups of mice kept on six staggered LD12:12 regimens, affect carcinogenesis in a circadian stage-dependent fashion. The differences in mean times to tumor appearance between saline- and melatonin-injected groups of mice reveal a beneficial effect of melatonin between 07 and 19 hours after light onset (HALO) and a detrimental effect at 23 and 03 HALO (Wrba et al. 1985, 1990). Using the timing of the temperature acrophase for the untreated, unshifted mice as the target phase for temperature on each LD12:12 regimen, some phases were slightly delayed by $1-3$ h if treatment was given at 03 or 23 HALO, while their timing was unchanged or advanced by as much as 4 h at other times (in zero-amplitude test; $P = 0.003$ by fit of 24- and 12-h cosine model). The temperature acrophase may be pulled toward the time of daily injections of melatonin or placebo (Wrba et al. 1985).

The fact that circaseptan and circannual rhythm stages may also be important, primarily in the case of the pineal gland which is eminently periodic along these time scales, is suggested by another study involving 375 mice (15 mice per group) similar in design to the preceding one but with two additional sets of six groups each of 15 mice receiving 1.0 or 0.1 mg/kg bw of melatonin. In the face of a confounding age effect, and beyond circadian variation in melatonin effect, inter-study differences were found in carcinogenesis for comparisons of untreated, placebo-treated, and melatonin-treated groups (Wrba et al. 1990). Circannual changes in endogenous defense mechanisms against cancer have also been reported by Bartsch et al. (1990a). The extent to which differences in non-photic solar effects between the two studies of Wrba et al. played a role remains a task for further study.

The change in sign of the effect of melatonin in carcinogenesis should be viewed in the light of feedsideward mechanisms (Halberg 1983; Sánchez et al. 1988). In any study of hormonal interactions, hormones change dynamically; they have to interact in response to changing conditions, which therefore add a statistical element. It seems reasonable that even complex neuroendocrine interactions are neither entirely random nor strictly deterministic. These interactions cannot be completely explained by time-unqualified feedbacks and feedforwards along classical axes. The result of an oversimplified search for "inhibition" vs "stimulation" is the status quo, yielding much controversy. The alternative is the study of multiple rhythmic interactions in the cephalo-adrenal neuroendocrine network.

A pineal feedsideward constitutes an interactive rhythm involving three (or more) entities, one of them modifying the relation between the other two in a predictable (chronomodulatory) way. The approach to the study of multiple time-dependent interactions as feedsidewards provides a framework whereby seemingly contradictory phenomena can be properly time-specified and thereafter analyzed experimentally for underlying mechanisms. Usually, pineal modulation involves amplification at the time of a high response of the adrenal to the pituitary gland, and attenuation at the time of a low adrenal response to the pituitary gland (Halberg 1983; Sánchez 1993). Thus, it is critical to record the time of treatment administration as well as the time elapsed from the time of treatment. Since treatment may affect not only the mean value, but also the dynamic characteristics of change such as the amplitude, acrophase, and/or waveform of multifrequency rhythms, it is recommended to sample around the clock before and after treatment for as complete an assessment of treatment effects as possible. The choice of marker rhythms is particularly important in this case and is best complemented by actual long-term outcomes whenever possible.

5.9.4 Coordination Via Feedsidewards: Importance of Circaseptans

The pineal gland and melatonin may be critical in terms of the coordination of a broad time structure. This is partly reflected by the fact that melatonin has been associated not only with breast cancer risk but also with cardiovascular disease risk and emotional depression (Wetterberg et al. 1986), and in part by the prominence of rhythms with other frequencies, circaseptans in particular.

Prominent circaseptan changes were observed in superfused pike pineals kept in continuous darkness and constant temperature (Cornélissen et al. 1995b; Falcon et al. 1996). A circaseptan modulation of the circadian amplitude of the melatonin content of pineal glands has also been shown in the case of female Lewis/S rats (Sánchez et al. 1986). A circaseptan modulation of the amplitude of a free-running circadian rhythm in RIA-assayable melatonin from the isolated pineal gland of lizards kept in continuous darkness was found by analyses of published, albeit limited, data (Halberg 1983). A time-macroscopic infradian modulation is shown for melatonin production by the chicken pineal gland in superfusion, studied by Leung et al. (1990) in LD12:12. Microscopic analyses of these data reveal a 121.4-h amplitude modulation of a free-running circadian (26-h) rhythm (Leung et al. 1992). Circaseptans were also found to be critical in a longitudinal study of a patient with an advanced adenocarcinoma of the ovary, shedding new light on the involvement of the pineal gland in carcinogenesis. The changes in the circadian MESOR (midline estimating statistic of rhythm; a rhythm-adjusted mean) of 6-sulfatoxymelatonin following a course of chemotherapy differed in relation to the success or failure of treatment, but the MESOR did not correlate with tumor burden, assessed by circulating CA 125. In contrast, the ratio of circaseptan-to-circadian amplitudes involving two chronome components correlated with the cancer marker (Fig. 5.12) (Cornélissen et al. 1995a). To that extent, the study revealed a critical circaseptan aspect of pineal gland involvement in cancer progression. A circaseptan component was also found to characterize other tumor markers assessed longitudinally in this patient (Figs. 5.13–15). Implications for treatment include the possible optimization of treatment efficacy on the basis of rhythmic components other than circadian. Until all treatment is implemented with programmable drug delivery devices that allow the timed release of medications in keeping with all pertinent chronome components, anything that has to be done during clinic hours (usually limited during the daytime) is more readily optimized by timing according to the day of the week than by the time of day.

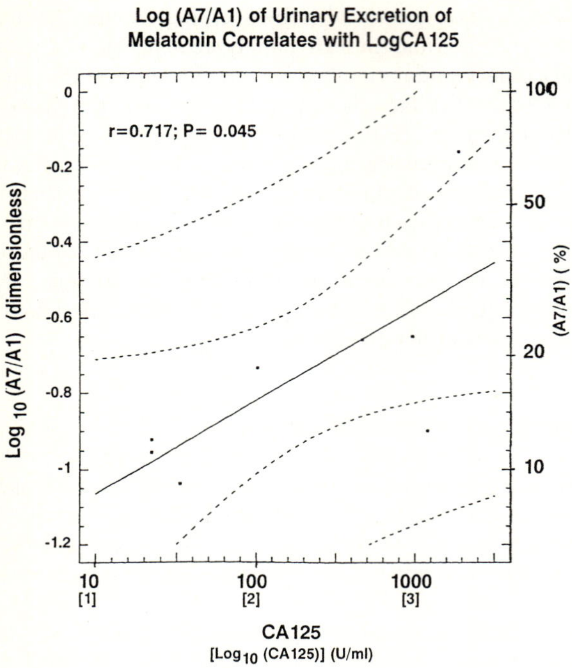

**Log (A7/A1) of Urinary Excretion of
Melatonin Correlates with LogCA125**

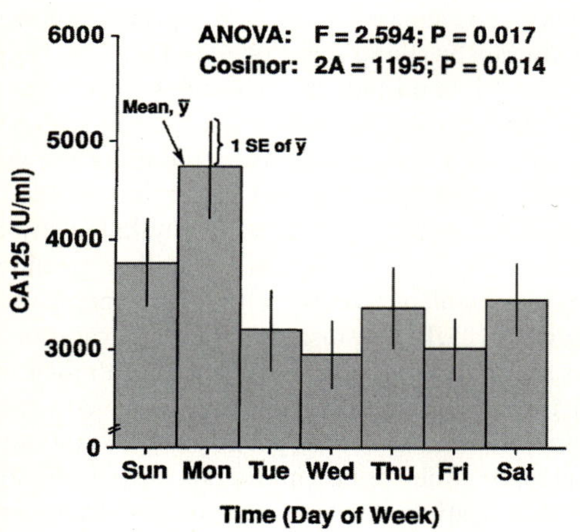

Fig. 5.12. In the absence of any correlation of the rhythm-adjusted mean value, the circaseptan-to-circadian amplitude ratio of 6-sulfoxy-melatonin correlates positively with serum CA 125 concentration used as a gauge of tumor burden. The result implies not only the involvement of the pineal gland, but of the circaseptan aspect of the pineal gland in cancerous growth. Regression line shown with 95% confidence and 95% prediction limits. (Copyright Halberg)

**CIRCASEPTAN PATTERN IN
SALIVARY CA125 (EH, 72y)**

Fig. 5.13. In addition to a prominent circadian variation (not shown), the ovarian tumor marker CA 125, assessed noninvasively in saliva, is circaseptan periodic. The about-weekly rhythm is assessed longitudinally in a 72-year-old woman who collected saliva samples several times a day during several months while receiving treatment with different oncostatic drugs. (Copyright Halberg)

Fig. 5.14. A circaseptan component also characterizes the urinary excretion of CA 125, an ovarian tumor marker determined in samples collected around the clock by a 72-year-old woman with advanced ovarian carcinoma. (Copyright Halberg)

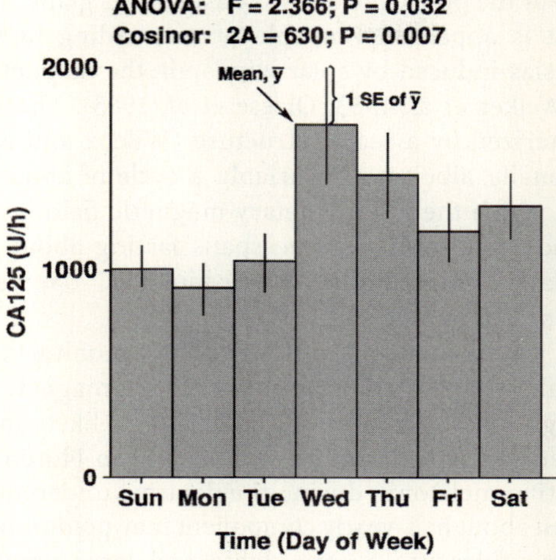

Fig. 5.15. In addition to a prominent circadian rhythm (not shown), a weekly variation is observed further for urinary gonadotropic peptide, used as an ovarian tumor marker in samples collected longitudinally several times a day by a 72-year-old woman diagnosed with an advanced ovarian carcinoma. (Copyright Halberg)

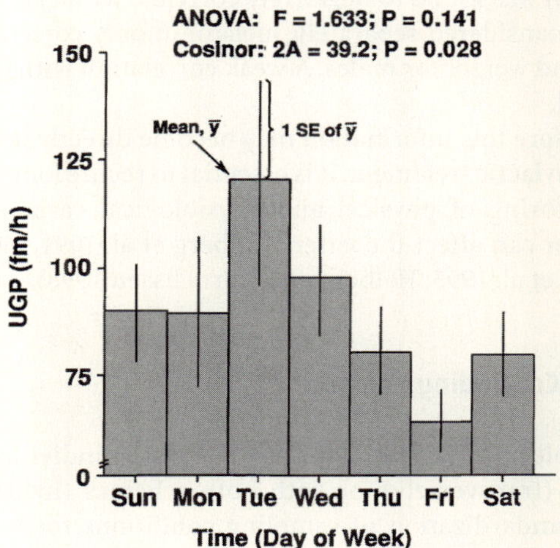

5.9.5 Environmental Effects from Near and Far

In terms of mechanisms through which prominent circaseptans constitute a major feature of the pineal gland. We consider this gland's sensitivity to changes in magnetic field; it is apparently capable of responding to variations of the order of a few nanoteslas induced by solar winds via the magnetosphere (Reuss et al. 1983; Semm 1983; Welker et al. 1983; Olcese et al. 1985). The latter has been determined to be characterized by a sector structure (Wilcox and Ness 1965; Wilcox 1969) involving occasionally, albeit not invariably, a cycle of about 27 days consisting of four spans during which the interplanetary magnetic field is directed in alternation toward the sun and away from it, three spans lasting about 7.625 days and the fourth about 3.750 days. The integrative genetic biologic 7-day week may thus also have an adaptive component (Halberg et al. 1991).

The future concomitant longitudinal monitoring of circulating melatonin and of physical variables, such as the local geomagnetic disturbance index K, may also shed light on any non-photic solar effects likely mediated via the pineal gland. One pertinent study by Tarquini et al. (1997) in Florence, Italy (43.47 °N), suggested that circulating melatonin during the daytime undergoes circannual changes, whereas by night, an about half-yearly component may predominate. In this respect it is noteworthy that Kp undergoes a very stable and large amplitude circasemiannual variation, peaking at the equinoxes, in keeping with the patterns seen for nocturnal melatonin concentrations. Based on data from Martikainen et al. (1985), Randall (1990) emphasized that at high latitudes (Oulu, Finland; 65.00 °N), prominent circasemiannual components are found in plasma melatonin. He aligned these observations with the magnetic disturbance induced by the solar wind, which is most pronounced in the auroral zones, where the corpuscular radiation enters the atmosphere.

In another study (Wetterberg et al. 1999), overnight urine was collected each month for 12–16 months from 321 healthy subjects at 19 medical centers in 14 countries distributed on 5 continents at latitudes from 31.01 °S – 77.00 °N. Mean melatonin concentration was found to negatively correlate with age, weight, and height. When the sexes were considered separately, melatonin only correlated with age for females and with age and weight for males. A weak correlation with latitude, but not longitude, was also found.

Before this information may become directly useful to optimize curative as well as prophylactic treatment, it is essential to secure longitudinal records of the concomitant monitoring of physical and physiological variables, to better understand how the former can affect the latter (Halberg et al. 1991, 1998; Cornélissen et al. 1994, 1999b; Breus et al. 1995; Halberg and Cornélissen 1998).

5.10 Concluding Remarks

Reliable spectra of biological rhythms on individuals (longitudinally), across individuals (transversely), or with both schemes (hybrid design) require provisions for: (a) standardization of sampling conditions for biophysical, biochemical, and other behavioral observations; (b) repetition and replication of sets of observations; (c) marker rhythm monitoring with sufficient density and for a span of sufficient length, in order to derive (d) inferential statistical point-and-interval estimates of rhythm characteristics (e) for the assessment of modulation, in the strict (mathematical) sense and (f) for the assessment of any rhythm alteration for groups and for the individual by means of parameter tests (Bingham et al. 1982) and control charts (Cornélissen et al. 1997). Specification of the stages of several built-in but environmentally synchronized rhythms is also indicated. Thus, a calendar date and clock time with information regarding light–dark schedules for laboratory animals, and also rest–activity, vacation, travel, etc., for humans, is required in order to approximate along with the season the day of the week and the circadian stage.

It should further be realized that once variability is assessed and replications are planned, it is usually cost-effective to assign, from a set of any six experimental units (e.g., patients), one to each of six different rhythm stages (e.g., 4 h apart for a circadian rhythm assessment) rather than testing all six units at the same time point (Gunther et al. 1980; Cornélissen et al. 1991, 1992; Bingham et al. 1993; Halberg et al. 1993). In addition to obtaining a mean value that is both more accurate and more precise, dynamic characteristics of change can then be derived by curve-fitting with the precautions of the cosinor. These parameters can be useful in their own right, not only for the diagnosis of abnormality but also for assessing health and, what is most important, for the recognition of earliest risk elevation, as documented for the personal and familial risk of human breast cancer in Fig. 5.10. The assessment of chronomes also serves as a sensitive gauge of any response to intervention such as chronotherapy, implemented on an individualized basis and reported in inferential statistical terms (Halberg et al. 1992). Variability, a terrible foe, then becomes a new resolving tool of particular use in the study of pineal and other neuroimmunomodulation.

References

Angeli A, Gatti G, Sartori ML, Del Ponte D, Carignola R (1988) Effects of exogenous melatonin on human natural killer (NK) cell activity. An approach to the immunomodulatory role of the pineal gland. In: Gupta D, Attanasio A, Reiter R (eds), The Pineal Gland and Cancer, Brain Research Promotion, London-Tübingen, pp 145–156

Balzer I, Fuhrberg B (1996) Presence of melatonin in the unicell *Euglena gracilis*. 5th Mtg. Society for Research on Biological Rhythms, Jacksonville, Florida, May 8–12, 1996 (p 82)

Barni S, Lissoni P, Sormani A, Pelizzoni F, Brivio F, Crispino S, Tancini G (1989) The pineal gland and breast cancer: serum levels of melatonin in patients with mammary tumors and their relation to clinical characteristics. Int J Biological Markers 4:157–162

Bartsch H, Bartsch C (1981) Effect of melatonin on experimental tumors under different photoperiods and times of administration. J Neural Transm 52:269–279

Bartsch C, Bartsch H (1997) Significance of melatonin in malignant diseases. Wiener Klinische Wochenschrift 109:722–729

Bartsch C, Bartsch H, Flüchter S, Attanasio A, Gupta D (1985) Evidence for modulation of melatonin secretion in men with benign and malignant tumors of the prostate. Relationship with the pituitary hormones. J Pineal Res 2:121–132

Bartsch C, Bartsch H, Fuchs U, Lippert TH, Bellmann O, Gupta D (1989) Stage-dependent depression of melatonin in patients with primary breast cancer. Correlation with prolactin, thyroid-stimulating hormone, and steroid receptors. Cancer 64:426–433

Bartsch H, Bartsch C, Gupta D (1990a) Seasonal variations of endogenous defence mechanisms against cancer. In: Gupta D, Wollman HA, Ranke MB (eds) Neuroendocrinology: New Frontiers. Brain Research Promotion, Tübingen, pp 333–339

Bartsch H, Bartsch C, Gupta D (1990b) Tumor-inhibiting activity in the rat pineal gland displays a circannual rhythm. J Pineal Res 9:171–178

Bartsch C, Bartsch H, Schmidt A, Ilg S, Bichler KH, Fluchter SH (1992a) Melatonin and 6-sulfatoxymelatonin circadian rhythms in serum and urine of primary prostate cancer patients: evidence for reduced pineal activity and relevance of urinary determinations. Clin Chim Acta 209:153–167

Bartsch H, Bartsch C, Simon WE, Fleming B, Ebels I, Lippert TH (1992b) Antitumor activity of the pineal gland: effect of unidentified substances vs. the effect of melatonin. Oncology 49:27–30

Bartsch H, Bartsch C, Mecke D, Lippert TH (1994) Seasonality of pineal melatonin production in the rat: possible synchronization by the geomagnetic field. Chronobiology Int 11:21–26

Bartsch C, Bartsch H, Buchberger A, Rokos H, Mecke D, Lippert TH (1995) Serial transplants of DMBA-induced mammary tumors in Fischer rats as model system for human breast cancer. IV. Parallel changes of biopterin and melatonin indicate interactions between the pineal gland and cellular immunity in malignancy. Oncology 52:278–283

Bartsch C, Bartsch H, Karenovics A, Franz H, Peiker G, Mecke D (1997a) Nocturnal urinary 6-sulfatoxymelatonin excretion is decreased in primary breast cancer patients compared to age-matched controls and shows negative correlation with tumor size. J Pineal Research 23:53–58

Bartsch C, Kvetnoy I, Kvetnaia T, Bartsch H, Molotkov A, Franz H, Raikhlin N, Mecke D (1997b). Nocturnal urinary 6-sulfatoxymelatonin and proliferating cell nuclear antigen-immunopositive tumor cells show strong positive correlations in patients with gastrointestinal and lung cancer. J Pineal Res 23:90–96

Bartsch C, Bartsch H, Buchberger A, Stieglitz A, Effenberger-Klein A, Kruse-Jarres JD, Besenthal I, Rokos H, Mecke D (1999) Serial transplants of DMBA-induced mammary tumors in Fischer rats as a model system for human breast cancer. VI. The role of different forms of tumor-associated stress for the regulation of pineal melatonin secretion. Oncology 56:169–176

Bingham C, Arbogast B, Cornélissen Guillaume G, Lee JK, Halberg F (1982) Inferential statistical methods for estimating and comparing cosinor parameters. Chronobiologia 9:397–439

Bingham C, Cornélissen G, Halberg F (1993) Power of "Phase 0" chronobiologic trials at different signal-to-noise ratios and sample sizes. Chronobiologia 20:179–190

Bittner JJ (1936) Some possible effects of nursing on the mammary gland tumor incidence in mice: preliminary report. Science 84:162

Bittner JJ (1952) The genesis of breast cancer in mice. Texas Rep Biol Med 10:160–166

Bittner JJ, Cole HL (1961) Induction of mammary cancer in agent-free mice bearing pituitary isografts correlated with inherited hormonal mechanisms. J Nat Cancer Inst 27:1273–1284

Blask DE (1984) The pineal: an oncostatic gland? In: Reiter RJ (ed), The Pineal Gland, Raven Press, New York, pp 253–284

Blask DE, Hill SM (1986) Effects of melatonin on cancer: studied on MCF-7 human breast cancer cells in culture. J Neural Transm 21:433–439

Blask DE, Pelletier DB, Hill SM, Lemus-Wilson A, Grosso DS, Wilson ST, Wise ME (1991) Pineal melatonin inhibition of tumor promotion in the N-nitroso-N-methylurea model of mammary carcinogenesis: potential involvement of antiestrogenic mechanisms in vivo. J Cancer Res Clin Oncol 117:526–532

Breus T, Cornélissen G, Halberg F, Levitin AE (1995) Temporal associations of life with solar and geophysical activity. Annales Geophysicæ 13:1211–1222

Cardinali DP, Vacas MI (1976) Mechanisms underlying hormone effects on pineal function: a model for the study of integrative neuroendocrine processes. J Endocrinol Invest 1:89–96

Cardoso SS, Scheving LE, Halberg F (1970) Mortality of mice as influenced by the hour of the day of drug (ara-C) administration. Pharmacologist 12:302 (abstract)

Carlebach R, Ashkenazi IE (1987) Temporal disorder: therapeutic considerations. Chronobiologia 14:162

Catrina SB, Catrina AI, Sirzen F, Griffiths W, Bergman T, Biberfeld P, Coculescu M, Mutt V (1999) A cytotoxic, apoptotic, low-molecular weight factor from pineal gland. Life Sci 65:1047–1057

Cohen M, Lippman M, Chabner B (1978) Role of pineal gland in aetiology and treatment of breast cancer. Lancet 2:814–816

Cornélissen G, Halberg F (1989) The chronobiologic pilot study with special reference to cancer research: Is chronobiology or, rather, its neglect wasteful? In: Goldson AL (ed), Kaiser H (series ed), Cancer Growth and Progression, vol 9, ch 9, Dordrecht: Kluwer Academic Publ, pp 103–133

Cornélissen G, Halberg F, Prikryl P, Dankova E, Siegelova J, Dusek J, International Womb-to-Tomb Chronome Study Group (1991) Prophylactic aspirin treatment: the merits of timing. JAMA 266:3128–3129

Cornélissen G, Bingham C, Wilson D, Halberg F (1992) Illustrating power of cost-effective "Phase 0" chronobiologic trials in endocrinology, psychiatry and oncology. Plenary lecture, 13th meeting Int. Soc. Clinical Biostatistics, Copenhagen, Denmark, August 17–21, 1992. In: Cornélissen G, Halberg E, Bakken E, Delmore P, Halberg F (eds): Toward phase zero preclinical and clinical trials: chronobiologic designs and illustrative applications. University of Minnesota Medtronic Chronobiology Seminar Series, #6, September 1992, pp 138–181

Cornélissen G, Halberg F (1992a) Chronobiologic response modifiers and breast cancer development: classical background and chronobiologic tasks remaining. In vivo 6:387–402

Cornélissen G, Halberg F (1992b) Toward a „chron-sensus" on neuroimmunomodulation, with "modulation" operationally and inferentially defined. In: Fabris N, Jankovic BD, Markovic BM, Spector NH (eds), Ontogenetic and Phylogenetic Mechanisms of Neuroimmunomodulation: From molecular biology to psychosocial sciences, Ann NY Acad Med Sci 650:60–67

Cornélissen G, Wendt HW, Guillaume F, Bingham C, Halberg F, Breus TK, Rapoport S, Komarov F (1994) Disturbances of the interplanetary magnetic field and human pathology. Chronobiologia 21:151–154

Cornélissen G, Berg H, Haus E, Halberg F (1995a) Chronome of urinary 6-sulfoxy-melatonin excretion, circulating CA125, cancer progression and therapeutic response. In vivo 9:375–378

Cornélissen G, Portela A, Halberg F, Bolliet V, Falcón J (1995b) Toward a chronome of superfused pike pineals: about-weekly (circaseptan) modulation of circadian melatonin release. In vivo 9:323–329

Cornélissen G, Halberg F, Hawkins D, Otsuka K, Henke W (1997) Individual assessment of antihypertensive response by self-starting cumulative sums. J. Medical Engineering & Technology 21:111–120

Cornélissen G, Gubin D, Halberg Francine, Milano G, Halberg Franz (1999a) Chronomedical aspects of oncology and geriatrics (review). in vivo 13:77–82

Cornélissen G, Halberg F, Schwartzkopff O, Delmore P, Katinas G, Hunter D, Tarquini B, Tarquini R, Perfetto F, Watanabe Y, Otsuka K (1999b) Chronomes, time structures, for chronobioengineering for "a full life". Biomedical Instrumentation & Technology 33:152–187

Dogliotti L, Berruti A, Buniva T, Torta M, Bottini A, Tampellini M, Terzolo M, Faggiuolo R, Angeli A (1990) Melatonin and human cancer. J Steroid Biochem Molec Biol 37:983–987

Edmunds LN, Halberg F (1981) Circadian time structure of *Euglena:* a model system amenable to quantification. In: Kaiser H (ed), Neoplasms: Comparative Pathology of Growth in Animals, Plants and Man, Williams and Wilkins, Baltimore, pp 105–134

El-Domeiri AA, Das Gupta TK (1976) The influence of pineal ablation and administration of melatonin on growth and spread of hamster melanoma. J Surg Oncol 8:197–205

Falcón J, Ravault J-P, Cornélissen G, Halberg F (1996) Circaseptan-modulated circadian rhythmic melatonin release from isolated pike pineal in alternating light and darkness. 4° Convegno Nazionale, Società Italiana di Cronobiologia, Gubbio (Perugia), Italy, June 1–2, 1996, pp 39–40

Falkson G, Falkson HC, Steyn ME, Rapoport BL, Meyer BJ (1990) Plasma melatonin in patients with breast cancer. Oncology 47:401–405

Günther R, Herold M, Halberg E, Halberg F (1980) Circadian placebo and ACTH effects on urinary cortisol in arthritics. Peptides 1:387–390

Halberg E, Halberg F (1980) Chronobiologic study design in everyday life, clinic and laboratory. Chronobiologia 7:95–120

Halberg Francine, Halberg J, Halberg E, Halberg Franz (1989) Chronobiology, radiobiology and steps toward the timing of cancer radiotherapy. In: Goldson AL (ed), Kaiser H (series ed), Cancer Growth and Progression, vol. 9, ch. 19, Dordrecht: Kluwer Academic Publ, pp 227–253

Halberg F (1960) Temporal coordination of physiologic function. Cold Spr Harb Symp Quant Biol 25:289–310; see also discussion for LD50 of whole body X-irradiation

Halberg F (1962) Physiologic 24-hour rhythms: A determinant of response to environmental agents. In: Schaefer KE (ed) Man's Dependence on the Earthly Atmosphere. Macmillan, New York, pp 48–98

Halberg F (1974) Protection by timing treatment according to bodily rhythms: an analogy to protection by scrubbing before surgery. Chronobiologia 1 (Suppl. 1):27–68

Halberg F (1977) Biological as well as physical parameters relate to radiology. Guest Lecture, Proc. 30th Ann. Cong. Rad, January 1977, Post-Graduate Institute of Medical Education and Research, Chandigarh, India, p 8

Halberg F (1983) *Quo vadis* basic and clinical chronobiology: promise for health maintenance. Am J Anat 168:543–594

Halberg F, Bittner JJ, Gully RJ, Albrecht PG, Brackney EL (1955) 24-hour periodicity and audiogenic convulsions in I mice of various ages. Proc Soc Exp Biol (NY) 88:169–173

Halberg F, Cornélissen G (1998) Chronoastrobiology. In: Proc. 3rd International Symposium of Chronobiology and Chronomedicine, Kunming, China, October 7–12, 1998, pp 128–142

Halberg F, Halberg E, Barnum CP, Bittner JJ (1959) Physiologic 24-hour periodicity in human beings and mice, the lighting regimen and daily routine. In: Withrow RB (ed) Photoperiodism and Related Phenomena in Plants and Animals, Ed Publ No 55, Washington, DC: Am Assn Adv Sci, pp 803–878

Halberg F, Haus E, Cardoso SS, Scheving LE, Kühl JFW, Shiotsuka R, Rosene G, Pauly JE, Runge W, Spalding JF, Lee JK, Good RA (1973) Toward a chronotherapy of neoplasia: tolerance of treatment depends upon host rhythms. Experientia (Basel) 29:909–934

Halberg F, Gupta BD, Haus E, Halberg E, Deka AC, Nelson W, Sothern RB, Cornélissen G, Lee JK, Lakatua DJ, Scheving LE, Burns ER (1977) Steps toward a cancer chronopolytherapy. In: Proc. XIV International Congress of Therapeutics, Montpellier, France: L'Expansion Scientifique Française, pp 151–196

Halberg F, Nelson W, Cornélissen G, Haus E, Scheving LE, Good RA (1979) On methods for testing and achieving cancer chronotherapy. Cancer Treatment Rep 63:1428–1430

Halberg F, Cornélissen G, Sothern RB, Wallach LA, Halberg E, Ahlgren A, Kuzel M, Radke A, Barbosa J, Goetz F, Buckley J, Mandel J, Schuman L, Haus E, Lakatua D, Sackett L, Berg H, Wendt HW, Kawasaki T, Ueno M, Uezono K, Matsuoka M, Omae T, Tarquini B, Cagnoni M, Garcia Sainz M, Perez Vega E, Wilson D, Griffiths K, Donati L, Tatti P, Vasta M, Locatelli I, Camagna A, Lauro R, Tritsch G, Wetterberg L (1981) International geographic studies of oncological interest on chronobiological variables. In: Kaiser H (ed) Neoplasms– Comparative Pathology of Growth in Animals, Plants and Man. Williams and Wilkins, Baltimore, pp 553–596

Halberg F, Sánchez de la Peña S, Nelson W (1986) Rhythm scrambling and tumorigenesis in CD2F$_1$ mice. III Int. Symposium, Social Diseases and Chronobiology, Florence, Nov. 29, 1986, B. Tarquini, R. Vergassola (eds) pp 59–61

Halberg F, Sánchez de la Peña S, Wetterberg L, Halberg J, Halberg Francine, Wrba H, Dutter A, Hermida Dominguez RC (1988) Chronobiology as a tool for research, notably on melatonin and tumor development. In: Pancheri P, Zichella L (eds) Biorhythms and Stress in the Physiopathology of Reproduction Hemisphere. New York, pp 131–175

Halberg F, Breus TK, Cornélissen G, Bingham C, Hillman DC, Rigatuso J, Delmore P, Bakken E, International Womb-to-Tomb Chronome Initiative Group (1991) Chronobiology in space. Keynote, 37th Ann. Mtg. Japan Soc. for Aerospace and Environmental Medicine, Nagoya, Japan, November 8–9, 1991. University of Minnesota/Medtronic Chronobiology Seminar Series, #1, December 1991, 21 pp. of text, 70 figures

Halberg F, Cornélissen G, Bingham C, Fujii S, Halberg E (1992) From experimental units to unique experiments: chronobiologic pilots complement large trials. In vivo 6:403–428

Halberg F, Bingham C, Cornélissen G (1993) Clinical trials: the larger the better? Chronobiologia 20:193–212

Halberg F, Cornélissen G, Fujii S, Takagi M, Haus E, Lakatua D, Ikonomov O, Stoynev A, Portela A, Illera JC, Illera M (1994) Hypothalamic transplantation for cancer prevention: a thought experiment with old and new data and tools: marker chronomes. Chronobiologia 21:140–142

Halberg F, Syutkina EV, Cornélissen G (1998) Chronomes render predictable the otherwise-neglected human "physiological range": position paper of BIOCOS project. Human Physiology 24:14–21

Halberg F, Smith HN, Cornélissen G, Delmore P, Schwartzkopff O, International BIOCOS Group (2000) Hurdles to asepsis, universal literacy, and chronobiology – all to be overcome. Neuroendocrinol Lett 21:145–160

Halberg J, Cornélissen G, Halberg F, Halberg Francine, Halberg E (1989) The sphygmochron for chronobiologic blood pressure and heart rate assessment in cancer patients. In: Woolley PV (ed), Kaiser H (series ed) Cancer Growth and Progression. vol. 10, ch. 18, Kluwer Academic Publ, Dordrecht, pp 189–195

Haus E (1964) Periodicity in response and susceptibility to environmental stimuli. Ann NY Acad Sci 117:281–291

Haus E, Halberg F, Scheving L, Pauly JE, Cardoso S, Kühl JFW, Sothern R, Shiotsuka RN, Hwang DS (1972) Increased tolerance of leukemic mice to arabinosyl cytosine given on schedule adjusted to circadian system. Science 177:80–82

Haus E, Halberg F, Loken MK, Kim US (1973) Circadian rhythmometry of mammalian radiosensitivity. In: Tobias A, Todd P (eds) Space Radiation Biology. Academic Press, New York, pp 435–474

Hill SM, Blask DE (1988) Effects of the pineal hormone melatonin on the proliferation and morphological characteristics of human breast cancer cells (MCF-7) in culture. Cancer Res. 48:6121–6126

Khoory R, Stemme D (1988) Plasma melatonin levels in patients suffering from colorectal carcinoma. J Pineal Res 5:251–258

Kirichenko AV, Rich TA (1999) Radiation enhancement by 9-aminocamptothecin: the effect of fractionation and timing of administration. Int J Radiation Oncology Biol Phys 44:659–664

Kort WJ, Weijma JM (1982) Effect of chronic light-dark shift stress on the immune response of the rat. Physiol Behav 29:1083–1087

Kort WJ, Zondervan PE, Hulsman LOM, Weijma JM, Westbroeck DL (1986) Light-dark shift shown, with special reference to spontaneous tumor incidence in female BN rats. JNCI 76:439–446

Langevin T, Hrushesky W, Sánchez S, Halberg F (1983) Melatonin (M) modulates survival of CD$_2$F$_1$ mice with L1210 leukemia. Chronobiologia 10:173–174

Leung B, Cornélissen G, Hillman D, Wang ZR, Binkley S, Bingham C, Halberg F (1992) Halting steps toward a circadian-infradian pineal melatonin chronome. In: Halberg F, Watanabe H (eds) Proc. Workshop on Computer Methods on Chronobiology and Chronomedicine, Tokyo, Sept. 13, 1990, Medical Review, Tokyo, pp 263–285

Leung FC, Poindexter CA, Miller DL, Anderson LE (1990) Biochemical and hormonal evaluation of pineal gland in vitro. Abstract, 12th Ann. Mtg. Bioelectromagnetics Society, San Antonio, TX, June 10–14, 1990; cf. also Technical Report: Biochemical and Hormonal Evaluation of Pinealocytes Exposed in vitro to Electric and Magnetic Fields, submitted to Electric Power Research Institute, Palo Alto, CA 94303, September 1990, by Battelle, Pacific Northwest Laboratories, P. O. Box 999, Richland, WA 99352

Levi F, Halberg F, Nesbit M, Haus E, Levine H (1981) Chrono-oncology. In: Kaiser H (ed), Neoplasms – Comparative Pathology of Growth in Animals, Plants and Man, Williams and Wilkins, Baltimore, pp 267–316

Lissoni P, Marelli O, Mauri R, Resentini M, Franco P, Esposti D, Esposti G, Fraschini F, Halberg F, Sothern RB, Cornélissen G (1986) Ultradian chronomodulation by melatonin of a placebo effect upon human killer cell activity. Chronobiologia 13:339–343

Lissoni P, Barni S, Tancini G, Crispino S, Paolorossi F, Cattaneo G, Lucini V, Mariani M, Esposti D, Esposti G, Fraschini F (1988) Relation between lymphocyte subpopulations and pineal function in patients with early or metastatic cancer. Ann NY Acad Sci 521:290–299

Lissoni P, Barni S, Crispino S, Tancini G, Fraschini F (1989) Endocrine and immune effects of melatonin therapy in metastatic cancer patients. Eur J Cancer Clin Oncol 25:789–795

Lissoni P, Crispino S, Barni S, Sormani A, Brivio F, Pelizzoni F, Brenna A, Bratina G, Tancini G (1990) Pineal gland and tumor cell kinetics: serum levels of melatonin in relation to Ki-67 labeling rate in breast cancer. Oncology 47:275–277

Lissoni P, Barni S, Cattaneo G, Tancini G, Esposti G, Esposti D, Fraschini F (1991a) Clinical results with the pineal hormone melatonin in advanced cancer resistant to standard antitumor therapies. Oncology 48:448–450

Lissoni P, Tisi E, Brivio F, Ardizzoia A, Crispino S, Barni S, Tancini G, Conti A, Maestroni GJM (1991b) Modulation of interleukin-2-induced macrophage activation in cancer patients by the pineal hormone melatonin. J Biol Regul & Homeostatic Agents 5:154–156

Lissoni P, Barni S, Ardizzoia A, Brivio F, Tancini G, Conti A, Maestroni GJM (1992a) Immunological effects of a single evening subcutaneous injection of low-dose interleukin-2 in association with the pineal hormone melatonin in advanced cancer patients. J Biol Regul & Homeostatic Agents 6:132–136

Lissoni P, Barni S, Ardizzoia A, Paolorossi F, Crispino S, Tancini G, Tisi E, Archili C, De Toma D, Pipino G, Conti A, Maestroni GJM (1992b) Randomized study with the pineal hormone melatonin versus supportive care alone in advanced nonsmall cell lung cancer resistant to first-line chemotherapy containing cisplatin. Oncology 49:336–339

Lissoni P, Tisi E, Barni S, Ardizzoia A, Rovelli F, Rescaldani R, Ballabio D, Benenti C, Angeli M, Tancini G, Conti A, Maestroni GJM (1992c) Biological and clinical results of a neuroimmunotherapy with interleukin-2 and the pineal hormone melatonin as a first line treatment in advanced non-small cell lung cancer. Br J Cancer 66:155–158

Lissoni P, Barni S, Rovelli F, Brivio F, Ardizzoia A, Tancini G, Conti A, Maestroni GJM (1993) Neuroimmunotherapy of advanced solid neoplasma with single evening subcutaneous injection of low-dose interleukin-2 and melatonin: preliminary results. Eur J Cancer 29 A:185–189

Lissoni P, Barni S, Tancini G, Ardizzoia A, Ricci G, Aldeghi R, Brivio F, Tisi E, Rovelli F, Rescaldani R, Quadro G, Maestroni G (1994) A randomised study with subcutaneous low-dose interleukin 2 alone vs interleukin 2 plus the pineal neurohormone melatonin in advanced solid neoplasma other than renal cancer and melanoma. Br J Cancer 69:196–199

Litman T, Halberg F, Ellis S, Bittner JJ (1958) Pituitary growth hormone and mitoses in immature mouse liver. Endocrinology 62:361–364

Loeb L, Kirtz MM (1939) The effect of transplants of anterior lobes of the hypophysis on the growth of the mammary gland and on the development of mammary gland carcinoma in various strains of mice. Am J Cancer 36:56–82

Maestroni GJM, Conti A, Pierpaoli W (1986) Role of the pineal gland in immunity: circadian synthesis and release of melatonin modulates the antibody response and antagonists immunosuppressive effect of corticosterone. J Neuroimmunol 13:19–30

Maestroni GJM, Conti A, Pierpaoli W (1988) Pineal melatonin, its fundamental immunoregulatory role in aging and cancer. In: Pierpaoli W, Spector NH (eds) Neuroimmunomodulation: interventions in Aging and Cancer. Ann NY Acad Sci 521:140–148

Martikainen H, Tapanainen J, Vakkuri O, Leppaluoto J, Huhtaniemi I (1985) Circannual concentrations of melatonin, gonadotrophins, prolactin and gonadal steroids in males in a geographical area with a large annual variation in daylight. Acta Endocrinol (Copenhagen) 109:446–450

Meites J (1974) Relation of prolactin to mammary tumorigenesis and growth in rats. In: Boyns AR, Griffiths K (eds) Prolactin and Carcinogenesis, 4th Tenovus Workshop, Alpha Omega Alpha, Cardiff, p 54

Mühlbock O (1958) The hormonal genesis of mammary cancer. (Proc Soc Endocrinol) J Endocrinol 17:vii–xv

Mühlbock O, Boot LM (1959) Induction of mammary cancer in mice without the mammary tumor agent by isografts of hypophyses. Cancer Res 19:402–412

Nagasawa H, Meites J (1970) Suppression by ergocornine and iprioazine of carcinogen-induced mammary tumors in rats: effects upon serum and urinary prolactin levels. Proc Soc exp Biol Med 135:469

Nandi S, McGrath CN (1973) Mammary neoplasia in mice. Adv Cancer Res 17:353

Olcese J, Reuss S, Vollrath L (1985) Evidence for the involvement of the visual system in mediating magnetic field effects on pineal melatonin synthesis in the rat. Brain Res 333:382–384

Oosthuizen JM, Bornman MS, Barnard HC, Schulenburg GW, Boomker D, Reif S (1989) Melatonin and steroid-dependent carcinoma. Andrologia 21:429–431

Pearson OH, Llerena O, Llerena L, Molina A, Butler T (1969) Prolactin-dependent rat mammary cancer: a model for man? Trans Ass Am Physicians 82:225–238

Plautz JD, Kaneko M, Hall JC, Kay SA (1997) Independent photoreceptive circadian clocks throughout *Drosophila*. Science 278:1632–1635

Randall W (1990) The solar wind and human birth rate: a possible relationship due to magnetic disturbances. Int J Biometeorol 34:42–48

Reinberg A, Halberg F (eds) (1979) Chronopharmacology, Proc. Satellite Symp. 7th Int. Cong. Pharmacol, Paris 1978, Pergamon Press, Oxford/New York

Reiter RJ (1978) Interaction of photoperiod, pineal and seasonal reproduction as exemplified by findings in the hamster. In: Reiter RJ (ed) Progress in Reproductive Biology, Karger, Basel, v 4, pp 169–190

Relkin R (1976) The pineal and human disease. In: Relkin R (ed) Annual Research Reviews.: Lundsdale House, Hornby, vol 1: The Pineal, pp 76–79

Reuss S, Semm P, Vollrath L (1983) Different types of magnetically sensitive cells in the rat pineal gland. Neurosci Lett 40:23–26

Ronco A, Halberg F (1996) The pineal gland and cancer. Anticancer Res 16:2033–2040

Rubin H (1985) Cancer as a dynamic developmental disorder. Cancer Res 45:2935–2942

Sánchez de la Peña S (1993) The feedsideward of cephalo-adrenal immune interactions. Chronobiologia 20:1–52

Sánchez de la Peña S, Marques N, Halberg F (1986) Infradian modulation of corticosterone production by hamster adrenals in vitro. In: Halberg F, Reale L, Tarquini B (eds) Proc. 2nd Int. Conf. Medico-Social Aspects of Chronobiology, Florence, Oct. 2, 1984, Istituto Italiano di Medicina Sociale, Rome, pp 771–775

Sánchez de la Peña S, Halberg F, Ungar F, Lakatua D (1988) Ex vivo hierarchy of circadian-infradian rhythmic pineal-pituitary-adrenal intermodulations in rodents. In: Pancheri P, Zichella L (eds) Biorhythms and Stress in the Physiopathology of Reproduction. Hemisphere, New York, pp 177–214

Sandyck R, Mukherjee S (1989) Attenuation of response-induced catalepsy by melatonin and the role of the opioid system. Int J Neurosci 48:297–301

Scheving LE (1976) The dimension of time in biology and medicine: chronobiology. Endeavour 35:66–72

Scheving LE, Haus E, Kühl JFW, Pauly JE, Halberg F, Cardoso S (1976) Close reproduction by different laboratories of characteristics of circadian rhythm in 1-β-D-arabinofuranosylcytosine tolerance by mice. Cancer Res 36:1133–1137

Semm P (1983) Neurobiological investigations on the magnetic sensitivity of the pineal gland in rodents and pigeons. Comp Biochem Physiol 76:683–689

Shellard SA, Whelan RDH, Hill BT (1989) Growth inhibitory and cytotoxic effects of melatonin and its metabolites on human tumor cell lines in vitro. Brit J Cancer 60:288–290

Silberberg R, Silberberg M (1949) Mammary cancer in castrate male mice receiving ovarian and hypophyseal grafts at various ages. Proc Soc exp Biol Med 70:510–513

Tamarkin L, Danforth D, Lichter A, DeMoss E, Cohen M, Chabner B, Lippman M (1982) Decreased nocturnal plasma melatonin peak in patients with estrogen receptor positive breast cancer. Science 216:1003–1005

Tarquini R, Perfetto F, Zoccolante A, Salti F, Piluso A, De Leonardis V, Lombardi V, Guidi G, Tarquini B (1995) Serum melatonin in multiple myeloma: natural brake or epiphenomenon? Anticancer Research 15:2633–2637

Tarquini B, Cornélissen G, Perfetto F, Tarquini R, Halberg F (1997) Chronome assessment of circulating melatonin in humans. In vivo 11:473–484

Tarquini B, Cornélissen G, Tarquini R, Perfetto F, Halberg F (1999) General and unspecific damping by malignancy of the circadian amplitude of circulating human melatonin? Neuroendocrinology Letters 20:25–28

Ueck M (1979) Innervation of the vertebrate pineal. In: Kappers JA, Pevet P (eds) Progress in Brain Research, Elsevier/North-Holland, Amsterdam, Vol 52, pp 45–88

Ulmer W (1999) Experimentelle und theoretische Untersuchungen zum Wachstum und ATP-Metabolismus von Sphäroiden und ihre Bedeutung für die Computersimulation, Regelungstheorie und Radioonkologie. Festschrift zur Würdigung der 27-jährigen wissenschaftlichen Arbeit v. Univ.-Prof. Dr.-Ing. Werner Düchting am Institut für Regelungs- und Steuerungstechnik der Universität-GH Siegen aus Anlaß seiner Emeritierung zum 28.02.1999, pp 77–93

Ulmer W, Cornélissen G, Revilla M, Halberg F (1999) Circadian and circaseptan dependence of the beta-ATP peak of four different cancer cell cultures: implications for chronoradiotherapy. Abstract 13, MEFA, Brno, Czech Rep., Nov 3–6, 1999

Welker HA, Semm P, Willig RP, Commentz JC, Witschko W, Vollrath L (1983) Effects of an artificial magnetic field on serotonin N-acetyltransferase activity and melatonin content of the rat pineal gland. Exp Brain Res 50:426–432

Welsch CW, Gribler C (1979) Prophylaxis of spontaneously developing mammary carcinoma in C_3H–HeJ female mice by suppression of prolactin. Cancer Res 33:2939

Wetterberg L, Halberg F, Halberg E, Haus E, Kawasaki T, Ueno M, Uezono K, Cornélissen G, Matsuoka M, Omae T (1986) Circadian characteristics of urinary melatonin from clinically healthy women at different civilization disease risk. Acta Med Scand 220:71–81

Wetterberg L, Sánchez de la Peña S, Halberg F (1990) Circadian rhythm of melatonin release from pineal, hypothalamus and pituitary in hypertensive rats. Progress in Clinical and Biological Research 341B:145–153

Wetterberg L, Bratlid T, Knorring L v, Eberhard G, Yuwiler A (1999) A multinational study of the relationships between nighttime urinary melatonin production, age, gender, body size, and latitude. Eur Arch Psychiatr Neurosci 249 256–262

Wilcox JM (1969) The interplanetary magnetic field: solar origin and terrestrial effects. Space Sci Rev 8:258

Wilcox JM, Ness NF (1965) Quasi-stationary corotating structure in the interplanetary medium. J Geophys Res 70:5793–5805

Wrba H, Halberg F, Dutter A (1985) Melatonin circadian-stage-dependently delays breast cancer development in mice injected daily for several months. In: Neuroimmunomodulation, Proc. 1st Int. Workshop on NIM, Bethesda, MD, Nov. 27–30, 1984, IWGN, Bethesda, pp 258–261

Wrba H, Halberg F, Dutter A (1986) Melatonin circadian-stage-dependently delays breast tumor development in mice injected daily for several months. Chronobiologia 13:123–128

Wrba H, Dutter A, Sánchez de la Peña S, Wu J, Carandente F, Halberg F (1989) Secular or circannual immunomodulation of placebo (Pl) and melatonin (Mt) effects on murine breast cancer? Chronobiologia 16:197

Wrba H, Dutter A, Sánchez de la Peña S, Cornélissen G, Halberg F (1990) Secular or circadian effects of placebo and melatonin on murine breast cancer? Progress in Clinical and Biological Research 341 A:31–40

Wurtman RJ, Axelrod J, Kelly DE (1968) The Pineal. Academic Press, New York, p 108

Yanai R, Nagasawa H (1972a) Inhibition of mammary tumorigenesis by ergot alkaloids and promotion of mammary tumorigenesis by pituitary isografts in adreno-ovariectomized mice. J Nat Cancer Inst 48:715

Yanai R, Nagasawa H (1972b) Enhancement by pituitary isografts of mammary hyperplastic nodules in adreno-ovariectomized mice. J Nat Cancer Inst 48:115

Section II
Effect of Tumor Growth on the Production and Secretion of Pineal Melatonin

6 Analysis of Melatonin in Patients with Cancer of the Reproductive System

Christian Bartsch, Hella Bartsch, Dieter Mecke

Abstract

This review summarizes findings relating to the concentrations of circulating melatonin as well as of the urinary excretion of 6-sulfatoxymelatonin (aMT6s, main peripheral metabolite of melatonin) in patients with different types of cancer of the reproductive system. According to these clinical studies, nocturnal circulating melatonin is diminished in patients with localized primary tumors of the mammary gland and endometrium as well as of the prostate gland. The most prominent depletions are found among patients with advanced localized primary tumors leading to a negative correlation between circulating melatonin and tumor size in patients with either mammary or prostate cancer. The phenomenon of a reduction of circulating melatonin appears to be transient since patients with recurrences show a normalization or even an elevation of melatonin. Since the surgical removal of the primary tumor does not lead to a normalization of depressed melatonin it is assumed that complex regulatory mechanisms are involved in the down-regulation of melatonin. It is unclear whether circulating melatonin is depleted in cancer patients due to a reduced production by the pineal gland or due to unknown peripheral metabolic processes. According to our investigations, an enhanced hepatic degradation to aMT6s can be ruled out in patients with mammary or prostate cancer. The depletion of circulating melatonin was found to be accompanied by neuroendocrine disturbances affecting the circadian secretion of the adenohypophyseal hormones prolactin, somatotropin and thyroid-stimulating hormone (TSH). It is conceivable that either the depletion of melatonin leads to these disturbances or that primary malignant tumor growth affects widespread systemic changes which in parallel disturb the pineal gland as well as the hypothalamo-adenohypophyseal system. As opposed to patients with mammary, endometrial or prostate cancer, extreme interindividual variations are found for unknown reasons among patients with ovarian cancer, some of them showing exceedingly highly levels of melatonin. The possible significance of the above-described findings in patients with cancers of the reproductive system is discussed with a view to achieve a better understanding of the modulation of signaling processes within the neuroimmunoendocrine network due to neoplasia.

6.1 Introduction

Engel (1935) and Bergmann (Bergmann and Engel 1950) as well as Hofstätter (1959) demonstrated that pineal gland extracts inhibit experimental tumors and are even able to control human cancers. Subsequently, it was found that removal of the pineal gland (pinealectomy) stimulated the growth of experimental primary tumors and their metastases (Barone and Das Gupta 1970; Lapin 1974). All these findings point to a potentially important role of the pineal gland in the control of malignant tumors. A central question is whether these effects can be attributed essentially to melatonin, the main pineal hormone, or not (Bartsch and Bartsch 1981). Experimental in vivo studies showed that mainly hormone-responsive tumors can be inhibited by melatonin (Tamarkin et al. 1981) whereas fast-growing and undifferentiated tumors are not influenced by this compound (Bartsch H et al. 1994). Lissoni and co-workers reported on the administration of melatonin in cancer patients and indicated that the pineal hormone may possess the potential to be a favorable palliative therapeutic measure when all standard forms of treatment failed (Barni et al. 1990) or could help to alleviate the side effects as well as to improve the therapeutic efficacy of immunotherapies using interleukin-2 (Lissoni et al. 1990; Lissoni 1997). If melatonin indeed has anti-tumor activity, the question arises whether its synthesis, secretion, and circulating levels may be reduced in cancer patients. The following chapters address this problem both from a clinical and experimental point of view.

6.2 Methodological Considerations and Problems Encountered in Estimating the Function of the Pineal Gland in Cancer Patients

6.2.1 Analytical Methods

Clinical studies aiming at estimating the function of the pineal gland with respect to the secretion of melatonin have to consider the regulatory mechanisms involved in its production since melatonin is secreted upon production and does not seem to be further regulated. Among all investigated mammalian species including humans, pineal melatonin biosynthesis displays a 24-h rhythm with a clear peak occurring around the middle of the scotoperiod (i.e., the period of darkness) irrespective whether the species is diurnal (day-active), such as sheep (Ravault et al. 1996), monkey (Guerin and Matthews 1990), and human (Lynch et al 1975), or nocturnal (night-active) such as mice (Vivien-Roels et al. 1998) or rats (Ralph et al. 1971). Therefore it is also the foremost methodological requirement for every clinical study dealing with the analysis of melatonin to secure that samples are collected repeatedly during a 24-h cycle to reliably monitor the profile of melatonin secretion (Arendt 1978). This can be achieved either by determining the concentration of melatonin in blood, saliva or urine or by measuring the urinary excretion of 6-sulfatoxymelatonin (Aldhous and Arendt 1988), the main peripheral metabolite of the pineal hormone (Jones et al. 1969). The question that arises is how frequently samples have to be collected. Blood sampling with the help of an indwelling catheter essentially avoids disturbances of the patients and thus enhances compliance so that the resolution of temporal fine structures including pulsatile patterns superimposed upon the basal daily profile of mela-tonin secretion is possible (Follenius et al. 1995; Luboshitzky et al. 1996, 1997). In our own clinical studies, we usually collect blood samples for the measurement of endocrine parameters including melatonin at 4-hourly intervals. This is sufficient to perform statistical analyses according to the cosinor method (Nelson et al. 1979) testing for the presence of daily rhythms, including the respective rhythm parameters such as the amplitude and the acrophase (time of the peak during the 24-h cycle) with the paradigm of a simple sinusoidal variation due to the sleep–wake cycle during the 24-hourly observation period. Recently, a related but more complex statistical analysis called COSIFIT was developed (Teicher et al. 1990), using multiple sinusoidal components for curve fitting which, however, like other statistical methods applied for rhythm analysis, requires more frequent blood sampling. This is difficult to realize under most clinical conditions. For this reason, sample collections at 4- to 2-hourly intervals have been used to determine the daily fluctuations of a number of physiological parameters including the components of the hematopoietic and hemostatic system as well as the temporal modulation of the therapeutic efficacy of drugs and their toxicity (Redfern and Lemmer 1997).

Also in the case of urine collections, we found 4-hourly sample intervals to be adequate to monitor the daily variations of melatonin or of its main metabolite 6-sulfatoxymelatonin (aMT6s). Since it is advisable not to disturb sleep, however, we prefer to collect only one 8-hourly sample, unless patients are equipped with a catheter. Melatonin- or aMT6s-determinations from urine samples appear to be of special relevance from a clinical point since they allow to monitor pineal function in a noninvasive fashion.

The formation of aMT6s in the liver is a two-step process in which melatonin is first converted to 6-hydroxymelatonin by a microsomal monooxygenase of the cytochrome P-450 system and subsequently conjugated predominantly by a cytosolic sulfotransferase to aMT6s (70–80%; Jones et al. 1969; Kopin et al. 1961). According to Arendt et al. (1985), the urinary excretion of aMT6s reliably reflects circulating melatonin in healthy subjects and according to our own investigations this applies to untreated cancer patients as well (Bartsch et al. 1992). The excretion of unmetabolized melatonin only occurs in traces constituting around 1% of the total production by the pineal gland. The urinary concentration of melatonin, however, correlates well with the corresponding concentrations of circulating melatonin and can therefore be used to estimate the melatonin secretory activity of the pineal gland (Bartsch et al. 1992). This phenomenon is principally due to passive tubular reabsorption during the renal processes: melatonin, like other low-molecular weight substances (Rocci et al. 1981), passes into the primary filtrate after glomerular filtration but subsequently diffuses back into renal venous blood until an equilibrium is attained between both fluids. Wetterberg (1978) reported an excellent correlation between plasma melatonin concentrations ($r = 0.89$) in first morning urine and 0200 hours serum melatonin concentrations in healthy humans, and subsequently Lang et al. (1981) detected a good correlation ($r = 0.74$) between plasma concentrations (2400 hours) and nocturnal urinary melatonin (2100–0700 hours). In our investigations on healthy men as well as those suffering from unoperated and untreated prostate tumors, a strong correlation ($r = 0.772$) was found between the concentration of serum melatonin at 0300 hours and the nocturnal urinary concentration (2300–0700 hours; Bartsch et al. 1992). Recently, Graham et al. (1998) reported that both melatonin and aMT6s determinations from morning urine samples, representing urine production between 2300 and 0700 hours, reflect the area under the curve for plasma melatonin determined during this period at hourly intervals. These studies emphasize that determinations of melatonin as well as of aMT6s from nocturnal or early morning urine samples are well-suited to monitor the daily variations of melatonin and represent simple but adequate tools to determine pineal function in cancer patients. These noninvasive techniques ensure good compliance and avoid potential disturbances that might affect pineal melatonin production. The collection of naturally passed nocturnal urine samples will not disturb sleep. Evidence exists that sleep disturbances may affect the physiological production and secretion of melatonin (Hajak et al. 1995; Suzuki et al. 1993). It has repeatedly been reported that it is feasible to measure melatonin in saliva which also reflects the day–night variations in blood (Laakso et al. 1994). The collection of saliva, however, is an active process and in our opinion is not compatible with undisturbed sleep.

The above considerations lead to the conclusion that determinations of melatonin from either blood or urine, as well as of aMT6s from urine, render relevant information concerning the day–night variations of pineal melatonin if samples are collected at suitable intervals over a complete 24-h cycle. If it can be assumed that no phase-shifts occur and/or emphasis is laid primarily on the estimation of the nocturnal surge of production, measurements of melatonin or aMT6s from overnight urine samples will suffice, as is the case in certain types of clinical studies dealing with patients suffering from cancer. A word of caution, however, should be expressed at this point concerning studies in which circulating melatonin is measured in morning or even noon samples. Due to the clear day–night rhythm of pineal melatonin secretion, the levels of melatonin in serum or plasma are, according to our experience, low and are sometimes near or even below the assayable level and should therefore be interpreted very cautiously.

6.2.2 Parameters Interfering with the Estimation of Pineal Function in Patients

Photic retinal impulses synchronize the circadian oscillation of the suprachiasmatic nuclei to the environmental light–dark cycle and thus the rhythm of pineal melatonin production (Moore and Klein 1974). In this process, light at night is known to inhibit the secretion of melatonin. This phenomenon is dependent on the quality and quantity of light as well as on the duration of light exposure and can vary considerably among species. In the rat, 1 min of 150 lx white light leads to a rapid depression of the activity of the rate-limiting enzyme of pineal melatonin biosynthesis, the serotonin-N-acetyl-transferase (SNAT), as well as to a drastic reduction of plasma melatonin to daytime concentrations (Illnerova et al. 1979). As opposed to that, a pronounced suppression of nocturnal melatonin secretion in humans to basal daytime-like levels is elicited only by very bright light, such as 2500 lx for 2 h (Lewy et al. 1980). Light of lower intensity, e. g., 200–300 lx, only leads to a partial suppression of melatonin secretion (Bojkowski et al. 1987) emphasizing the existence of a dose–response curve for the suppressive effect of light on nocturnal melatonin. With respect to wavelength, it appears that light of 509 nm seems to possess the strongest inhibitory potential in humans whereas light beyond 600 nm exerts practically no suppressive effect (Brainard et al. 1985). For these reasons it is important for clinical studies to verify that light at night during sample collection does not disturb the physiological secretion of the pineal hormone. During sample collection at night it should therefore be seen that only dim and if possible red light is used. In this context it may also be mentioned that there are some hints that the circadian profile of melatonin secretion in humans may exhibit phase-shifts among different seasons (Illnerova et al. 1985) due to different light–dark cycles. For well-controlled clinical studies it would therefore appear desirable to perform them within comparable seasons. Since the daily melatonin rhythm is so strictly controlled by the light–dark cycle, it is also important to study exclusively well-entrained individuals who recently have not performed shift-work. The inclusion of blind patients into clinical-oncological studies aiming at the measurement of pineal function should also be avoided since these individuals very often possess a free-running melatonin rhythm with peaks occurring at unpredictable times during the day or night (Lewy and Newsome 1983).

From these findings and considerations it is clear that light has to be considered as a "drug" for the secretion of melatonin, but there are a number of other true pharmacological agents which interfere with the biosynthesis of melatonin. These interactions are immediate consequences of the mechanisms involved in the regulation of pineal melatonin biosynthesis. Pineal melatonin secretion is strictly controlled by nerves (Klein 1985) which enter the pineal gland as postganglionic sympathetic fibers originating from the superior cervical ganglia and control the activity of the rate-limiting step of the biosynthesis of melatonin, SNAT, via a cascade of intracellular events involving adenylate cyclase, cAMP as a second messenger, and the activation of specific protein kinases which ultimately induce the activation of SNAT. Due to this regulatory mechanism, the administration of β- or α-blockers has to be strictly avoided in studies dealing with the effect of oncological diseases on pineal function (Arendt et al. 1985) so that the physiological course of melatonin secretion can be monitored correctly.

Benzodiazepine derivatives frequently used as preoperative sedatives can also negatively interfere with melatonin secretion (Kabuto et al. 1986; Monteleone et al. 1989), although the detailed mechanisms involved in this phenomenon are not exactly understood. Lowenstein et al. (1984) and Suranyi-Cadotte et al. (1987) reported that both central and peripheral benzodiazepine binding sites are found in human pineal glands.

According to in vivo studies with rat pineal glands, it appears that the inhibitory action of diazepam on melatonin secretion is elicited via a reduced activity of the rate-limiting step of the biosynthesis of melatonin (SNAT) (Wakabayashi et al. 1991). The studies of Winters et al. (1991), however, lead to the conclusion that the in vivo inhibition of melatonin production by benzodiazepines does not involve SNAT, but rather a step before or after this enzymatic reaction including the hydroxyindole-O-methyl-transferase (HIOMT).

According to our own experience, the administration of insulin seems to negatively affect melatonin secretion whereas oral antidiabetic substances, sulfonylurea derivatives, do not affect melatonin (C. Bartsch et al., unpublished observations). Pharmacological elevations of melatonin are elicited by a number of drugs acting on the central nervous system such as the neuroleptic chlorpromazine (Smith et al. 1979). In this case, the stimulatory action on melatonin, however, does not seem to be exerted at the level of the pineal but rather via a reduced peripheral metabolization of melatonin in the liver. The MAO-A inhibitors clorgyline and tranylcypromine but not the MAO-B inhibitor deprenyl were found to lead to an elevation of plasma melatonin (Murphy et al. 1986) prompting these authors to conclude that mainly an increased availability of serotonin, the precursor of the pineal hormone, and an enhanced noradrenergic function due to a higher availability of norepinephrine appear to be responsible for the observed elevation of melatonin. Fluvoxamine, an antidepressant with a selective inhibitory effect on the uptake of serotonin, was also found to elevate circulating melatonin as well as desipramine, a noradrenaline uptake inhibitor (Skene et al. 1994). The naturally occurring peptide destyrosine-γ-endorphin, possessing an antidepressant action according to Chazot et al. (1985), leads to an enhanced secretion of melatonin. Tetrahydrocannabinol increases circulating melatonin (Lissoni et al. 1986). The list of drugs affecting melatonin is far from complete but underlines that the secretion of the pineal hormone may easily be misinterpreted due to the concomitant administration of pharmacological agents. It is therefore indispensable to perform a meticulous process of selection of those patients who should become part of clinical studies to analyze possible effects of tumor growth on pineal function.

Melatonin secretion, however, is not only regulated by external factors, such as light and drugs, but the pineal gland as an integrative part of the neuroendocrine system (Cardinali, 1984) is also influenced by endogenous factors such as hormones and, according to recent findings (Maestroni 1993), by immune processes. This implies that whenever pineal gland function is to be monitored in cancer patients they have to be otherwise free from endocrine and immune diseases. Clinical studies on pineal gland function are furthermore complicated by the fact that the secretion of melatonin declines with age (Touitou et al. 1981), which is of particular interest for the proper choice of nonmalignant controls. Finally, it has to be stressed that these patients have to be carefully selected and characterized with respect to their disease, i.e., they should have tumors of a comparable histopathological appearance and the malignant disease should be at a defined stage. In our studies we focussed our attention on cases who had developed a tumor for the first time, being otherwise healthy, and who had not been treated in any way thus avoiding possible pharmacological influences on the synthesis and metabolism of melatonin.

These considerations demonstrate methodological limitations for relevant studies on melatonin secretion in cancer patients and illustrate why melatonin will not easily meet the requirements of a feasible tumor marker as may have been hoped by clinicians. This fact, however, by no means lowers the potential importance of this hormone in the physiological control of malignant tumors exerted by the different systems of the body. It is rather felt that well-controlled clinical studies estimating pineal function in cancer patients will contribute to a better understanding of how the neuroimmunoendocrine network responds to different forms and stages of malignant disease.

6.3 Studies on Patients with Breast Cancer

Interest in a possible role of the pineal gland and its hormone melatonin in the etiology of breast cancer was stimulated in the late 1970s and early 1980s by the so-called Cohen hypothesis published in *The Lancet* in 1978 and the results of a subsequent study published by this American group. They hypothesized that a diminished function of the pineal gland may promote the development of breast cancer since it is associated with estrogen excess and that, on the other hand, melatonin inhibits ovarian estrogen production, gonadotrophin production, sexual development, as well as maturation.

They further argued that pineal calcification, possibly connected with a lowered secretion of melatonin, is commonest in those countries with high rates of breast cancer, whereas the incidences of pineal calcification and of breast cancer are moderate among the black population in the United States. If, on the other hand, pineal function is impaired, they argued it may trigger puberty which may be the cause of early menarche being a risk factor for breast cancer. Finally, they assumed, although at that time information was still scarce, that the pineal gland and melatonin may influence tumor induction and growth in experimental animals (Anisimov et al. 1973; Karmali et al. 1978). Evidence for this assumption was soon found in several studies dealing with experimental breast cancer using different in vivo model systems. Tamarkin et al. (1982) and Shah et al. (1984) found that administration of melatonin inhibits DMBA-induced mammary cancer. Blask et al. (1991) observed NMU-induced breast cancer occurrence in rats, and Subramanian and Kothari (1991) detected that spontaneous mammary cancers in C3H/Jax mice are reduced by melatonin. Later it was found that human mammary cancer cell lines can be inhibited by physiological concentrations of melatonin in vitro (Blask et al. 1997). These data support the concept of an involvement of endogenous melatonin in the etiology of human breast cancer.

Parallel to Cohen and collaborators, we formulated a similar hypothesis and started designing clinical studies on the measurement of melatonin in human breast cancer patients. In our opinion, the main reason for linking the pineal gland with breast cancer as a hormone-dependent type of malignancy were the findings that pinealectomy leads to a stimulation of experimental tumor growth affecting both primary growth as well as formation of metastases (Barone and Das Gupta 1970), and the early observations of Bergmann and Engel (1950) in Vienna in the 1930s that the administration of pineal extracts inhibited the growth of benzopyrene-induced tumors in mice. We assumed a particularly close link between melatonin and hormone-dependent cancers, such as breast and prostate cancer, since melatonin was reported to affect pubertal development as well as seasonal reproduction in animals via central neuroendocrine mechanisms. A lowered secretion of melatonin, as an agent controlling normal mammary development, may thus also promote the formation of breast cancer. The actual functional relationship between the endogenous production of melatonin and the risk to develop breast cancer, however, appears to be considerably more complex than expected due to nonlinear and dynamic changes occurring in the levels of circulating melatonin during different phases of tumor development and the fact that early sub-clinical stages of tumor development are almost impossible to be studied. Indications for complex interactions between pineal gland activity and different phases of mammary tumor development also stem from our analytical experimental studies on tumor-bearing rats and will be discussed in detail in another chapter of this volume.

Wetterberg et al. (1979) were the first to publish analytical results on melatonin with respect to the breast cancer problem. They compared the circadian variation of urinary melatonin in clinically healthy women in Japan and the United States of America since the incidence of mammary cancer is very low in Japan compared to the United States. They found a lower excretion of melatonin in Japanese than in North-American white women, indicating that the risk to develop breast cancer is positively correlated with a high pineal secretory activity. They interpreted this observation as an effort of the body's systems to control the process of tumor development. These findings were subsequently confirmed by them in another study (Wetterberg et al.

1986) in which, in addition, a circannual rhythm of melatonin was found in North-American but not in Japanese women.

Cohen and colleagues published the results of their first clinical study in 1982 (Tamarkin et al. 1982) analyzing the circadian profile of plasma melatonin in 20 women with unoperated primary breast cancer at the early clinical stages I and II. Among the ten patients with estrogen receptor-positive tumors, they detected a significant depression of nocturnal melatonin. This finding was subsequently confirmed by the same group in another study (Danforth et al. 1985) in which only patients with estrogen receptor-positive tumors showed lowered melatonin compared to healthy controls. It is interesting to note that in a third group of women with a high risk to develop breast cancer, melatonin did not differ from the controls, in contrast to Wetterberg et al. (1979).

In our initial study performed at the All-India Institute of Medical Sciences, New Delhi, in December 1979, we collected samples from cancer patients and suitable controls under conditions which met those stringent methodological criteria mentioned above. Urinary melatonin was analyzed in its circadian profile in untreated postmenopausal breast cancer patients as well as in age-matched controls free of cancer (Bartsch et al. 1981). Both groups of patients were studied within a single month and the collection of urine samples (4-hourly intervals during the day and one 8-hourly interval at night) was performed over two to three consecutive days. The nine controls had uterovaginal prolapse and were free from endocrine and immunological disease. Out of the ten breast cancer patients analyzed, nine had an unoperated primary tumor (stage I, $n = 1$; stage II, $n = 1$; stage III, $n = 6$; stage IV, $n = 1$) and one patient showed a local recurrence. The overall group of breast cancer patients revealed a 30 % depression of the 24-h excretion of melatonin which was accompanied by a phase delay of the circadian peak of urinary melatonin excretion, showing higher levels in the morning than at night (Bartsch et al. 1981). An unusual feature of these two groups of Indian women was that melatonin showed a high excretion during the afternoon (1400–1800 hours) which was even higher than the night surge (2200–600 hours). The excretion of melatonin in the cancer patients was around 50 % lower during the afternoon (1400–1800 hours) and night interval (2200–0600 hours) than in the controls, and was even more pronounced and became statistically significant when cancer patients with an advanced primary tumor (> stage II) were considered.

We performed another study on breast cancer patients in Germany during the 1980s in which the circadian rhythm of serum melatonin was monitored using 4-hourly sampling intervals (Bartsch et al. 1989). Twenty-three patients with unoperated primary tumors and 12 patients with recurrences were analyzed. Twenty-eight patients with different types of benign breast disease served as controls. Since some individuals among all groups of patients had additional diseases and received different types of pharmacological treatments, the data analysis was performed in two ways. Initially, all patients in the different groups irrespective of the presence of other pathological conditions or treatments were analyzed forming the so-called clinical cross-section. Afterwards, only those patients within each group who were practically healthy and suffered only from either benign or malignant diseases of the mammary gland were selected resulting in the so-called central groups. In addition, the central group of controls was subdivided into younger and older patients so that the latter could be age-matched with the central group of unoperated primary breast cancer patients. These 12 patients showed a significant 67 % depression of the circadian amplitude of melatonin, compared to the eight age-matched controls suffering from benign breast diseases such as fibroadenoma or fibrocystic mastopathy. A similar significant depression of 56 % was found when the 23 patients of the clinical cross-section of patients with primary breast cancer were compared to their respective age-matched controls ($n = 15$). In contrast, patients with secondary breast tumors which appeared after surgical removal of the primary tumor, either locally at the site of operation or within the contralateral breast in some cases forming distant metastases, did not give any indication of a depression of circulating melatonin. Patients with recurrences under antiestrogen therapy (tamoxifen) even showed slightly higher amplitudes than their age-matched controls (Bartsch et al. 1989). These results indicate that the levels of circulating melatonin may be transiently depressed due to primary tumor growth but return to normal after operation and particularly when secondary tumors arise. A functional role for the presence of primary tumors in the suppression of melatonin is also suggested by the fact that a subdivision of primary breast cancer patients according to tumor size (stage T1, < 2 cm in diameter; T2, > 2 cm, < 5 cm; T3, > 5 cm) showed an increasing depression of the nocturnal surge parallel to tumor size: –27 % at T1, –53 % at T2, and –73 % at T3 (Bartsch et al. 1989). The relevance of these findings was verified again by excluding effects of age or pharmacological treatments. As opposed to the studies of Tamarkin et al. (1982) and Danforth et al. (1985), in our study (Bartsch et al. 1989) no correlation between melatonin and the levels of estrogen receptors within the tumors was detectable. The discrepancies between these studies may have been due to the fact that Tamarkin (1982) as well as Danforth (1985) analyzed exclusively patients at the initial clinical stages I and II, whereas we studied a number of patients at the advanced stage III, who may tend to show a lower expression of estrogen receptors within tumor tissue.

Concerning the possible mechanisms involved in the depression of circulating melatonin, we assumed, at this stage of our investigations, that the pineal secretory activity may be inhibited due to primary breast tumor growth. In order to test this hypothesis indirectly, the main peripheral metabolite of melatonin, aMT6s, was quantified in the sera of breast cancer patients. Parallel circadian profiles of aMT6s and melatonin in serum were found in all investigated groups including patients with primary breast cancer (Bartsch et al. 1991), which confirmed previous observations of Markey et al. (1985) in healthy subjects and indicated that the depression of circulating melatonin in cancer patients is apparently not due to changes in the hepatic metabolism of the pineal hormone. In addition, these results paved the way for future clinical studies with the help of measurements of aMT6s. The stability of this compound and its excretion into urine allowed the design of new studies using the noninvasive collection of nocturnal urinary samples for the estimation of pineal secretory activity.

The aim of our subsequent studies was to confirm with this method our previous findings of a depression of circulating melatonin in primary mammary cancer patients. Therefore, urine samples were collected from a larger population of breast cancer patients using more than 100 hospitalized women with different types of benign gynecological complaints as controls (Bartsch et al. 1997). Prior to the final evaluation when detailed anamneses had become available, patients with breast cancer as well as their controls were thoroughly screened and only those patients who had no other serious pathological conditions, received no pharmacological treatments, and were within a comparable age range were selected. Among the group of primary mammary cancer patients, 17 remained. The group of controls allowed strict age-matching in the range of +1–3 years in a ratio of two per each breast cancer patient. Among these oncological patients, the nocturnal urinary excretion of aMT6s was significantly depressed by 48% confirming the observations of our two preceding studies. When breast tumor patients were subdivided according to the stage of their tumor, an inverse correlation was found between aMT6s and tumor size: patients with T2 tumors showed a 40% depression and those with T3 tumors a 71% significant depression (Bartsch et al. 1997), which rendered support for the concept that primary mammary tumor growth is actively involved in a depletion of circulating melatonin. Hoffmann et al. (1996) also observed a negative correlation between primary tumor size and the concentrations of nocturnal melatonin in breast cancer patients further supporting the correctness of the concept. Since the size of the primary tumor is apparently actively involved in lowering circulating melatonin, it is logical to conclude that the average melatonin value of a group of cancer patients will only differ significantly from age-matched controls in cases where adequate numbers of patients with larger tumors are present. This may also explain why Skene et al. (1990) were not able to detect changes of the urinary excretion of aMT6s in their breast cancer patients compared to age-matched normal controls. They, however, unfortunately give no clinical details concerning the size and stage of the tumors of their patients.

Dogliotti et al. (1990) also addressed the question of circulating melatonin in patients with breast cancer by measuring the pineal hormone at midnight as well as in the morning at 0800 hours. They analyzed 30 cancer patients with stage I–II as well as 45 patients with stage III–IV. Melatonin did not differ significantly between the 26 controls and stage I–II the patients. In contrast, patients with the advanced stages III–IV showed highly significantly elevated night as well as morning melatonin concentrations in serum compared to controls and patients with stages I–II. These findings are in obvious contradiction to those of Tamarkin et al. (1982) as well as our own results (Bartsch et al. 1981, 1989, 1997). A possible explanation could be that Dogliotti et al. (1990) studied a number of breast cancer patients with stage IV and distant metastases, since, according to our findings, it appears that circulating melatonin declines parallel to the growth of the primary tumor as long as it is localized but again normalizes and can even be elevated when distant metastases arise (Bartsch et al. 1989). We assume that a recognition of distant metastases by components of the immune system may be involved in this phenomenon.

Oosthuizen et al. (1989) compared the levels of plasma melatonin collected at 0800 hours in 18 patients with breast cancer and compared them to 18 controls. They found statistically significant lower levels in patients with breast cancer. The authors, however, give no further clinical details about the patients and their controls, making it impossible to compare their results with those of other studies. Furthermore, it is unclear what these results of single time point determinations of morning melatonin in blood may imply for the shape of the circadian rhythm of melatonin in these patients: is the amplitude lowered or does perhaps a phase advance of their acrophase occur? Also Barni et al. (1989) exclusively studied the concentration of melatonin in morning serum samples analyzing 74 untreated breast cancer patients of the stages T1–3N0–2M0 and compared them to 46 age-matched healthy women as controls. They found that mean serum melatonin levels were significantly higher among the breast cancer patients than in the controls and that melatonin was highest in those patients with the best prognosis, i.e., estrogen receptor-positive and node-negative cases. In continuation of this study, the group of Lissoni (1990a) analyzed the levels of melatonin at 0800 hours in the serum of 25 untreated breast cancer patients with a locally limited disease and estimated cell proliferation in the corresponding tumors with the help of Ki-67 labeling measurements. Forty-six healthy women were used as controls. Breast cancer patients, as in their previous study, showed significantly higher mean melatonin levels than the controls. Patients with negative Ki-67 labeling rate in their tumors, having a low rate of cellular proliferation, exhibited significantly higher levels of the pineal hormone than those with a positive Ki-67 rate. The authors interpreted these results in a similar way to their previous data, namely that an elevation of melatonin represents a favorable sign for the prognosis of the malignant disease. In another study, Lissoni et al. (1987) addressed the question whether radical mastectomy may affect the levels of serum melatonin in the morning. They studied 24 patients with T1–2N0–2M0 breast cancer representing relatively early and localized stages of the disease and compared the preoperative levels in these patients to 14 women undergoing surgery for reasons other than neoplastic disease. Mean melatonin in patients with breast cancer before surgery was again significantly higher than in the unoperated controls. Melatonin showed divergent changes 15 days after radical mastectomy as compared to the preoperative values: it was not affected in 13 patients, whereas it was enhanced in 5 women and showed a decline in 6 cases. This study again confirmed the previous trend of the results of Lissoni's group, namely that early stages of cancer with a good prognosis are apparently connected with high morning melatonin levels although the observed incongruent changes after mastectomy are of still unknown clinical significance.

A group in South Africa (Falkson et al. 1990) also analyzed melatonin in breast cancer patients with the help of blood samples collected in the morning (0730–830 hours). Eighty-six patients were studied from whom 280 assays were performed and compared with the clinical status of the patients. According to this study, patients in the advanced disease group had significantly higher levels than those in the adjuvant treatment group, and patients with progressive disease had significantly higher values than those in remission or with stable disease. Multiple-regression tests showed a significant inverse correlation between survival and melatonin values. These results appear to be in obvious contradiction to those of Lissoni and co-workers, who found a positive correlation between high melatonin levels and a favorable prognosis. It has to be pointed out, however, that the types of breast cancer patients analyzed by the two groups were completely different. Lissoni et al. (1990a) studied untreated patients at early stages of the disease free from distant metastases, while the South African group also analyzed patients undergoing anticancer therapy who in several cases had distant metastases. Since these authors mention that they could not observe an effect of the presence of metastases on the levels of melatonin, it appears that there are unexplainable contradictions between the results of these two groups of researchers. However, it has to be stressed that these contradictions are found in studies dealing exclusively with determinations of melatonin in early morning samples and only render limited information with respect to the complex circadian profile of melatonin. It is conceivable that additional phenomena such as seasonal changes may distort findings in relation to the secretory activity of the pineal gland in breast cancer patients.

Holdaway et al. (1991) determined the circadian profile of melatonin in women with previous breast cancer at the summer and winter solstices. Although the overall melatonin secretion assessed by either the amplitude of the nocturnal melatonin pulse or the area under the 24-h melatonin curve (AUC) was not different between the 20 premenopausal women with previous breast cancer and the nine controls, both the amplitude and the AUC significantly declined in breast cancer patients in winter. In a more detailed analysis, this depletion was apparently confined to those patients whose tumor was detected in winter. Since, according to another study of the same group (Mason et al. 1990), breast cancer detection in winter is connected with a poorer prognosis than in spring/summer, these results may indicate that the abnormal reduction of melatonin in winter may be involved in tumor growth promotion during this season and could thus contribute to a decreased survival time in these patients.

Recently, Maestroni and Conti (1996) addressed the important question of the levels of melatonin within breast cancer tissue. They analyzed the concentration of melatonin in 15 human neoplastic and adipose tissues sampled during mastectomy or tumorectomy with high-performance liquid chromatography. They found melatonin to be three orders of magnitude more concentrated in these tissues than in sera from the corresponding cancer patients or from healthy controls. Melatonin was also measured in a limited number of normal breast tissues ($n = 3$) where it showed equally elevated concentrations. The authors further determined the nuclear grade of the malignant tissue specimens and correlated this parameter with the corresponding melatonin concentrations and detected a clear inverse relationship. This means that tumors with a high nuclear grade, connected with a poor prognosis of the patients concerned, show relatively low amounts of intratumoral melatonin. Contrary to that, a positive correlation was found between melatonin and the estrogen receptor status, high levels of estrogen receptors being a favorable prognostic marker. Due to the exceedingly high levels of melatonin in breast specimens, Maestroni and Conti (1996) hypothesize that melatonin may be synthesized by normal and malignant epithelial mammary cells and that an adequate amount of melatonin produced by these cells might constitute a protective factor against malignant processes. There is clear evidence that melatonin is also produced by extra-pineal tissues including the retina (Pevet et al. 1980), parts of the gastrointestinal tract (Huether 1993), as well as the ovaries (Itoh et al. 1999). It therefore could indeed be possible that the mammary gland produces melatonin although it cannot be completely ruled out that, in addition, circulating melatonin may be concentrated and metabolized by mammary gland tissue as is known for sex steroids (Soderqvist 1998). An interplay may even exist between melatonin and sex steroids which are fundamentally involved in the endocrine control of the components of the reproductive tract thus contributing to the local homeostasis, as it has been demonstrated for the human prostate (Gilad et al. 1997) and for malignant human mammary cells under in vitro conditions (Molis et al. 1995). It is therefore important to clarify by further clinical as well as biochemical studies which detailed mechanisms are involved in the modulation of circulating melatonin in patients with breast cancer at different stages of their disease and to resolve whether primarily a variation in pineal secretory activity, peripheral processes at the level of the tumor, or complex systemic changes including the neuroimmunoendocrine network are involved.

6.4 Studies on Patients with Gynecological Cancers

6.4.1 Endometrial Cancer

In 1992, Sandyk published a hypothesis in which he linked the pineal gland with the pathogenesis of endometrial carcinoma, also termed carcinoma of the uterine corpus. In Western industrialized nations, endometrial carcinoma is one of the most common neoplasms of the female genital tract (Silverberg at al. 1990). The exact pathogenesis of this type of cancer is unknown although there are a number of risk factors known, namely obesity, nulliparity, diabetes mellitus, and hypertension (Bardawil 1987). Patients with a long-lasting history of anovulatory cycles and those with "delayed climacterium" are apparently at higher risk (Bardawil 1987). All these risk factors have in common a connection with increased or extended estrogenic exposure which is unopposed by progesterone. During the normal menstrual cycle, estrogens are known to stimulate the endometrium whereas the luteal production of progesterone leads to transformation into the so-called secretory stage of the endometrium facilitating implantation of a fertilized ovum. Chronic estrogen exposure unopposed by gestagens predisposes the endometrium in susceptible women to malignant transformation. Such chronic estrogen exposure may stem either from endogenous sources, such as estrogen-secreting ovarian tumors and chronic anovulation, or from exogenous sources such as contraceptive pills as well as stilbestrol (Bardawil 1987). High circulating levels of androstenedione and estrone have also been associated with an elevated risk of endometrial cancer (Potischman et al. 1996), since androstenedione is aromatized to estrone in adipose tissue (Nimrod and Ryan 1975) which is further converted to estrogens. For this reason, obese women possess an elevated production of estrogen compared to slimmer women of comparable age. Sandyk (1992) in his subsequent line of thought argues that a deficient secretion of melatonin should be considered as an additional risk factor for the development of endometrial carcinoma because melatonin possesses antiestrogenic properties and declines in plasma during the menopause when the risk to develop endometrial cancer increases. Similar to the Cohen hypothesis on the involvement of melatonin and the pineal gland in the etiology of breast cancer (Cohen et al. 1978), Sandyk (1992) assumes a possible involvement of increased pineal calcification (connected with reduced melatonin secretion) in the pathogenesis of endometrial cancer. Sandyk (1992) further argues that pineal calcification is extremely low among black Africans (Daramola and Olowu 1972) and is also lower in black Americans than in the white population of the United States (Adeloye and Felson 1974), which shows a higher incidence of endometrial cancer than black women (Plaxe and Saltzstein 1997). According to the findings of Bojkowski and Arendt (1990), however, it is doubtful whether an actual correlation exists between the age-related decline of melatonin production and the amount of corresponding pineal calcification.

Sandyk (1992) predicted a depression of circulating melatonin in patients with endometrial cancer compared to age-matched healthy controls. The current clinical studies dealing with the analysis of melatonin in patients with endometrial cancer seem to support this assumption. Karasek et al. (1996) studied ten untreated patients with adenocarcinoma of the uterine corpus and they found a significant depletion of the nocturnal surge of melatonin by more than 50% compared to age-matched controls. Grin and Grünberger (1998), analyzing the circadian rhythm of melatonin in

68 patients with confirmed endometrial cancer in comparison to 70 controls, also observed a drastic depletion of melatonin by about 80%.

The mechanisms involved in the depletion of melatonin in patients with endometrial cancer as well as the functional significance of these findings for the prognosis of this malignancy still await elucidation. In this context it is worth mentioning that in recent years we performed two major experiments to test whether melatonin administration or physiological pinealectomy due to exposure to constant light may affect the survival of virgin BDII/Han rats, an inbred strain that develops spontaneous endometrial cancers in more than 90% of all animals. Life-long administration of the pineal hormone beginning from day 30 of life led to a small but statistically significant increase of the median survival time compared to the respective controls. However, no effect of this treatment was detectable if melatonin was given from day 50 or day 180 of life onwards until the end of life. The effect of physiological pinealectomy was critically dependent on the time of the beginning of treatment: only if constant light was applied from day 30 of life onwards was a pronounced shortening of the median survival time noticed, but no effect was noticed if the treatment was begun on day 50 of life (Deerberg et al. 1997; C. Bartsch, unpublished results). These findings clearly show that in this experimental model system for endometrial cancer administration of melatonin or a deficiency of the pineal hormone under constant light does not acutely influence the course of the disease (once it is established) but that the pineal gland seems to exert an indirect effect via a modification of developmental processes within the reproductive tract. This illustrates and emphasizes the complexity of the mechanisms involved in the neuroendocrine control of the development of endometrial cancer under participation of the pineal gland which seems to include modulatory effects on pubertal processes. Evidence for an inhibitory influence of melatonin on puberty in the rat has been documented (Rivest et al. 1986) and is also discussed for humans (Waldhauser et al. 1993). A delay of puberty due to melatonin would lead to a reduction of the overall amount of estrogen to which the system is exposed throughout life and the risk of estrogen-sensitive tissues, such as the endometrium, to undergo malignant transformation would thus be reduced. These considerations underline that it is indispensable to differentiate between the role of the pineal hormone in neuroendocrine events which may affect the development of endometrial carcinomas indirectly via developmental processes and acute effects during the growth and progression of malignancy. In both cases, however, a central question is to which extent melatonin may affect estrogen and progesterone production and the balance between the two hormones which seems to be an important factor for the development of endometrial cancer.

With respect to the control of estrogens by melatonin, there is increasing evidence that the pineal hormone is able to exert neuroendocrine control over the production of ovarian estrogen and to interfere with the action of the steroid at the level of target tissues. Control over the production of ovarian estrogen during the female cycle of seasonally breeding animals is a well-accepted fact and seems to occur either via central structures affecting luteinizing hormone-releasing hormone (LHRH) and subsequently luteinizing hormone (LH) (Skinner and Robinson 1997; although the detailed mechanisms with respect to the distribution of melatonin receptors in the relevant brain areas remain elusive) or directly at the level of the ovary (Tamura et al. 1998). It is also known that melatonin modulates the local action of estrogen at the level of target tissues by affecting the transcription of the estrogen receptor gene expression

(Molis et al. 1994), receptor-binding activity (Danforth et al. 1983) and the activation of the estrogen receptor for DNA-binding (Rato et al. 1999). A body of evidence exists that melatonin exerts a stimulatory action on the formation of progesterone (Durotoye et al. 1997; Webley et al. 1988) so that melatonin may in fact control the estrogen/progesterone balance, being a critical determinant for the development of endometrial cancer. Since there are currently no solid data to support a role for melatonin in the control of ovarian steroid production in women, future studies should define further to what extent this assumption may apply to the human situation so that possible endocrine consequences of a depression of circulating melatonin could be predicted adequately.

6.4.2 Ovarian Cancer

Ovarian cancer constitutes the most frequent cause of death among patients affected by gynecologic malignancies (Parkin et al. 1988). Although the etiology of ovarian cancer is still unknown, a number of epidemiological risk factors have been identified that are accompanied by a history of incessant ovulation, e.g., due to nulliparity, early menarche, late menopause, and not having taken oral contraceptives (Fathalla 1971). The mechanisms involved in the development of ovarian cancer are assumed to be connected to normal physiological processes which occur during ovulation, namely the rupturing of the ovarian surface epithelium and subsequent proliferation to heal the wound gene-rated within this tissue. Since epithelial ovarian carcinomas exclusively develop from the superficial ovarian epithelium it appears that during the above-described proliferative process mutations occur for unknown reasons leading to uninterrupted autocrine and paracrine growth stimulation of these cells (Godwin et al. 1992, 1993). It appears that ovarian cancer constitutes the end point of a multi-step carcinogenic process in which two oncogenes are predominantly involved, namely *Her-2/neu* (Ross et al. 1999) as well as mutations of the tumor suppressor gene p53 (Diebold 1999).

Together with I.M. Kvetnoy and T.V. Kvetnaia (Obninsk, Russia), we analyzed the nocturnal urinary excretion of aMT6s in 119 Russian patients with ovarian carci-nomas comparing them to 27 age-matched controls. Among the patients, a high variability of the measured aMT6s values was detected with some of them showing very low levels whereas others exhibited extremely high values. A similar observation was made by us in a follow-up study of a woman with an inoperable ovarian can-cer showing extremely high values of nocturnal aMT6s excretion (A. Karenovics, C. Bartsch, H. Bartsch, unpublished results). Since these abnormal values are found irrespective of age, we assume that they could be connected to certain, yet unknown, parameters involved in the growth of ovarian tumors such as oncogene-encoded growth factors secreted by these tumors which could exert a stimulatory action on pineal melatonin secretion. Another possibility is that melatonin produced by ovarian tumors may reach the circulation and thus lead to the observed high concentrations of melatonin. Melatonin has been found to be higher in human preovulatory follicular fluid than in the circulation (Brzezinski et al. 1987). Itoh (1997) obtained evidence for the production of melatonin by ovarian tissue in the rat. It seems that the function of ovarian melatonin, according to the study of Yie (1995) and as discussed before, could be to regulate ovarian steroidogenesis thereby inhibiting estrogen and stimulating progesterone production.

6.5 Studies on Patients with Prostate Cancer

The prostate as an accessory sex organ is functionally dependent on the secretion of gonadal testosterone which is converted within this gland to the potent androgen dihydrotestosterone (DHT) via the enzyme 5α-reductase (Bruchovsky and Wilson 1968). Androgens support both the fibromuscular as well as the epithelial cell elements of the prostate and support their proliferative as well as secretory activity (Aumüller 1983). Estrogen, which is also formed to a minor extent by the testes (Payne et al. 1987) and the prostate (West et al. 1988), suppresses testosterone uptake into the prostate epithelial cell and reduces 5α-reductase activity (Lee et al. 1973). Thus, estrogen at pharmacological doses affects atrophy of the epithelial cell elements of the prostate and in case of a chronic administration leads to squamous hyperplasia and metaplasia. The development of benign prostatic hypertrophy (BPH), which is connected with an increase of the stromal cell elements in aging men, is thought to occur due to a shift in the androgen/estrogen balance accompanied by lowered testosterone and elevated estrogen production (Habenicht and el Etreby 1991). For the etiology of prostatic cancer (PC) on the other hand, it would appear logical to assume that an elevation of testosterone exerting a stimulatory action on epithelial cell elements should be involved. Scientific evidence for such an assumption had been lacking and, only recently, Gann et al. (1996) reported the results of a major prospective study showing that an elevation of the testosterone/estrogen ratio precedes the development of PC. For this reason, it would also appear necessary to investigate what type of neuroendocrine mechanisms may be involved in the dysregulation of men who develop PC and which affect the hypothalamic-pituitary-gonadal axis as well as melatonin. The pineal hormone exerts multiple control over testosterone by inhibiting its secretion at the gonadal level (Valenti et al. 1995) as well as by controlling its production via an inhibition of LHRH and LH secretion (Rasmussen 1993). In addition, melatonin controls the action of testosterone on the prostate, modifying its metabolism so that less DHT is formed (Horst and Adam 1982). The presence of specific melatonin receptors constitutes the molecular basis for the action of the pineal hormone on the prostate (Gilad et al. 1998). Melatonin receptors have also been found in human malignant prostatic cells (Gilad et al. 1999), rendering an explanation on why the androgen-sensitive human prostatic cancer cell line LNCaP (Lupowitz and Zisapel 1999) is inhibited by this compound. Melatonin has also been reported to inhibit the growth of the hormone-sensitive Dunning prostatic adenocarcinoma R3327H in the rat (Philo and Berkowitz, 1988). It is interesting to note that in contrast to that, the androgen-insensitive Dunning prostatic adenocarcinoma, R3327 HIF (Buzzell 1988) is stimulated by melatonin, illustrating that this hormone exerts a differential effect on hormone-sensitive or insensitive prostate cancer cells and may thus play a decisive role in the regulation of this type of malignancy.

Our clinical findings concerning the levels of circulating melatonin in patients with prostate cancer were obtained with protocols and analytical methods very similar to those described above for our other clinical studies. The main aim of these investigations was to compare the circadian profiles of serum melatonin in patients with primary unoperated prostate cancer (PC) with those in patients of similar age suffering from benign prostatic hyperplasia (BPH). In an additional group, young men (YM) working in the ward or hospitalized for minor urological complaints were studied to document possible age-related changes. The evaluation of results focussed on essentially healthy subjects of the so-called central groups who received no drugs that might have interfered with pineal melatonin secretion. The central groups were formed from the overall so-called clinical cross-section of all analyzed patients.

The first study comprised 63 patients of the clinical cross-section: 14 YM, 25 patients with BPH, and 24 PC patients. Our initial evaluation of these groups revealed that 50% of all PC patients showed very low levels of circulating melatonin at night (< 0.05 pmol/ml), whereas only 20% of all BPH patients showed this phenomenon (Bartsch et al. 1982). PC patients with low melatonin possessed tumors of a low degree of histological differentiation whereas patients with highly differentiated malignant tumors exhibited very pronounced nocturnal melatonin peaks which are usually found among young men only. The subsequent selection of untreated patients from the above-described clinical cross-section yielded the following central groups: ten YM (average age 33 years), 13 elderly patients with BPH, and nine elderly patients with PC (average ages 68–70 years), as well as five patients with incidental carcinomas (PCi) who were diagnosed within BPH tissue during the histopathological examination consisting of small foci of highly differentiated malignant cells. Patients with unoperated primary PC exhibited extremely low levels of nocturnal serum melatonin leading to the loss of a statistically detectable circadian rhythmicity whereas patients with BPH showed significant circadian rhythms of melatonin being only 36% lower than in YM. Patients with PCi had melatonin profiles comparable to YM (Bartsch et al. 1983, 1985). The question arises whether high levels of melatonin in patients with PCi are a direct consequence of the presence of controlled malignant prostate cancer cells or not. A considerable number of elderly men are known to die with such type of dormant malignant cells and the incidence does not seem to differ widely among countries even for those which show either high or low prevalence of PC such as in the United States and Japan. It is assumed that additional environmental and socioeconomic factors are involved in the transformation of dormant to progressing PC cells. Possible tumor promoters for PC are components of animal fat, making this type of cancer the most frequent malignant tumor among men in Western industrialized nations (Rowley and Mason 1997).

In order to verify the results of the first study, further investigations were initiated using the same protocol as well as analytical methods. In this case the total clinical cross-section consisted of 43 patients: ten YM, 18 unoperated patients with BPH and 15 patients with PC without metastases; the central groups had seven patients with BPH and nine patients with PC (average ages 65–69 years) as well as eight YM (average age 27 years). A comparison of the central groups showed that patients with PC suffered a significant depletion of the circadian melatonin amplitude (–65%; Bartsch et al. 1992) thus confirming the results of our first study. Since the first and second study were performed within the same department using identical protocols and analytical techniques, data could be pooled and an overall evaluation of both studies was performed. Among the 18 PC patients of the combined central groups, the circadian amplitude of melatonin was significantly depressed by 71% as compared to the 20 patients with BPH (Bartsch et al. 1998). It could well be that this result (in spite of a relatively small number of patients analyzed to this point) may prove to be representative for patients with PC, since the individuals of the central group were healthy and untreated so that a true effect of tumor growth upon pineal melatonin secretion could be observed. A further evaluation of the combined data of our two studies indicates that tumor growth seems to actively interfere with the levels of circulating melatonin. When patients with PC and BPH were subdivided according to the size of their tumor, namely into small (BPH+/T1), medium-sized (BPH++/T2) and big (BPH+++/T3,4), patients with T1 tumors showed a slight depression of their melatonin amplitude by 28% as compared to BPH+, whereas patients with T2 and T3,4 tumors exhibited both a significant depletion by 78% in relation to BPH++ and BPH+++, respectively (Bartsch et al. 1998). Since the amplitude of melatonin remained stable among patients with BPH of different sizes and the age-ranges for patients of the matched groups were comparable, it can be concluded that the growth of localized primary prostate cancer negatively affects the concentration of circulating melatonin at night.

The mechanisms involved in this phenomenon are at present unknown but a change of the main hepatic pathway of metabolism of melatonin to aMT6s can be excluded. Parallel measurements of aMT6s in serum in the first study showed a depression in PC very similar to that of melatonin (Bartsch and Bartsch 1994). Also in the second study, comparable amplitude depressions of aMT6s were detected in serum and urine of patients with PC, very similar to those detected for melatonin in serum and urine (Bartsch et al. 1992). In an attempt to test a possible direct functional involvement of prostate cancer growth in the depression of circulating melatonin, a limited number of patients with low nocturnal melatonin were studied before and after the surgical removal of an advanced tumor. However, no normalization of melatonin occurred after surgery, and only if patients were treated with high doses of estrogen after surgery was melatonin found to increase in some of the cases (Bartsch et al. 1986). An endogenous normalization of circulating melatonin seems to take place in PC patients with distant metastases showing very pronounced circadian amplitudes as compared to those with lymph node metastases (Bartsch and Bartsch 1994). These different findings underline that the presence of an advanced, though still localized, primary tumor of the prostate is obviously not directly responsible for the depression of circulating melatonin, but rather indicate that other systemic changes (within the endocrine and immune system) occurring in response to malignant growth may be involved. In patients of the first and second prostate cancer study a number of additional hormones in their circadian profile, which included LH, follicle-stimulating hormone (FSH), testosterone, thyroid-stimulating hormone (TSH), thyroxine, prolactin, somatotropin, and cortisol were analyzed. Surprisingly, hormones of the adenohypophyseal-gonadal axis (LH, FSH, and testosterone) which exert control over the function of the prostate were not found to show any specific changes in patients with PC. In contrast to that, the circadian profiles of prolactin and somatotropin were seriously affected, showing a transition to ultradian rhythmicity (having a shortened phase-length of 12 h instead of 24). Changes occurred also in the adenohypophyseal-thyroidal axis: TSH was abnormally low in spite of thyroxine being within normal limits (Bartsch et al. 1985, 1994, 1998). These observations are supported by similar findings in breast cancer patients (Bartsch et al. 1989) and support the assumption that (neuro)endocrine imbalances occur in cancer patients. It would be important to understand whether these hormonal changes are a direct consequence of tumor growth and which mechanisms are involved. It could be hypothesized that the depression of melatonin plays a functional role in this phenomenon affecting central parts of the endocrine system since the pineal hormone is thought to control the adenohypophyseal secretion via hypothalamic nuclei (Morgan et al. 1994). An alternative possibility is that the secretion of melatonin and of other hormones is disturbed because of a negative effect of tumor growth upon those parts of the autonomic nervous system in the brain that control both the pineal gland and the hypothalamo-adenohypophyseal unit.

References

Adeloye A, Felson B (1974) Incidence of normal pineal gland calcification in skull roentgenograms of black and white Americans. Am J Roentgenol Radium Ther Nucl Med 122:503–507

Aldhous ME, Arendt J (1988) Radioimmunoassay for 6–sulphatoxymelatonin in urine using an iodinated tracer. Ann Clin Biochem 25:298–303

Anisimov VN, Morozov VG, Khavinson Vkh, Dilman VM (1973) Comparison of the anti-tumour activity of extracts of the epiphysis and hypothalamus, melatonin and sigetin in mice with transplanted mammary gland cancer. Vopr Onkol 19:99–101

Arendt J (1978) Melatonin assays in body fluids. J Neural Transm Suppl 13:265–278

Arendt J, Bojkowski C, Franey C, Wright J, Marks V (1985) Immunoassay of 6-hydroxymelatonin sulfate in human plasma and urine: abolition of the urinary 24-hour rhythm with atenolol. J Clin Endocrinol Metab 60:1166–1173

Aumüller G (1983) Morphologic and endocrine aspects of prostatic function. Prostate 4:195–214

Bardawil WA (1987) Endometrium. In: Gold JJ, Josimovich JB (eds) Gynecologic endocrinology. Plenum Medical Book, New York, pp 185–245

Barni S, Lissoni P, Sormani A, Pelizzoni F, Brivio F, Crispino S, Tancini G (1989) The pineal gland and breast cancer: serum levels of melatonin in patients with mammary tumours and their relation to clinical characteristics. Int J Biol Markers 4:157–162

Barni S, Lissoni P, Paolorossi F, Crispino S, Archili C, Tancini G (1990) A study of the pineal hormone melatonin as a second line therapy in metastatic colorectal cancer resistant to fluorouracil plus folates. Tumori 76:58–60

Barone RM, Das Gupta TK (1970) Role of pinealectomy on Walker 256 carcinoma in rats. J Surg Oncol 2:313–322

Bartsch C, Bartsch H (1994) Melatonin secretion in oncological patients: current results and methodological considerations. In: Maestroni GJM, Conti A, Reiter RJ (eds) Advances in Pineal Research, Vol 7. John Libbey, London, pp 283–301

Bartsch C, Bartsch H, Jain AK, Laumas KR, Wetterberg L (1981) Urinary melatonin levels in human breast cancer patients. J Neural Transm 52:281–294

Bartsch C, Bartsch H, Flüchter SH, Harzmann R, Bichler K-H, Gupta D (1982) Serum melatonin and cortisol rhythms of patients with benign and malignant tumours of the prostate. Neuroendocrinol Lett 4:176 (Abstract)

Bartsch C, Bartsch H, Flüchter SH, Harzmann R, Attanasio A, Bichler K-H, Gupta D (1983) Circadian rhythms of serum melatonin, prolactin and growth hormone in patients with benign and malignant tumours of the prostate and in non–tumour controls. Neuroendocrinol Lett 5:377–386

Bartsch C, Bartsch H, Flüchter SH, Attanasio A, Gupta D (1985) Evidence for modulation of melatonin secretion in men with benign and malignant tumours of the prostate: relationship with the pituitary hormones. J Pineal Res 2:121–132

Bartsch C, Bartsch H, Flüchter SH, Attanasio A., Gupta D (1986) Melatonin rhythms in prostate cancer patients: effect of operation and hormone treatment. J Neural Transm Suppl 21:491 (Abstract)

Bartsch C, Bartsch H, Fuchs U, Lippert TH, Bellmann O, Gupta D (1989) Stage-dependent depression of melatonin in patients with primary breast cancer. Correlation with prolactin, thyroid stimulating hormone, and steroid receptors. Cancer 64:426–433

Bartsch C, Bartsch H, Bellmann O, Lippert TH (1991) Depression of serum melatonin in patients with primary breast cancer is not due to an increased peripheral metabolism. Cancer 67:1681–1684

Bartsch C, Bartsch H, Schmidt A, Ilg S, Bichler K-H, Flüchter SH (1992) Melatonin and 6-sulfatoxymelatonin circadian rhythms in serum and urine of primary prostate cancer patients: evidence for reduced pineal activity and relevance of urinary determinations. Clin Chim Acta 209:153–167

Bartsch C, Bartsch H, Flüchter SH, Mecke D, Lippert TH (1994) Diminished pineal function coincides with disturbed circadian endocrine rhythmicity in untreated primary cancer patients. Consequence of premature aging or of tumour growth? Ann N Y Acad Sci 719:502–525

Bartsch C, Bartsch H, Karenovics A, Franz H, Peiker G, Mecke D (1997) Nocturnal urinary 6-sulphatoxymelatonin excretion is decreased in primary breast cancer patients compared to age-matched controls and shows negative correlation with tumour-size. J Pineal Res 23:53–58

Bartsch C, Bartsch H, Bichler K-H, Flüchter SH (1998) Prostate cancer and tumour stage–dependent circadian neuroendocrine disturbances. Aging Male 1:188–199

Bartsch H, Bartsch C (1981) Effect of melatonin on experimental tumours under different photoperiods and times of administration. J Neural Transm 52:269–279

Bartsch H, Bartsch C, Mecke D, Lippert TH (1994) Differential effect of melatonin on early and advanced passages of a DMBA-induced mammary carcinoma in the female rat. In: Maestroni GJM, Conti A, Reiter RJ (eds) Advances in Pineal Research, Vol 7. John Libbey, London, pp 247–252

Bergmann W, Engel P (1950) Über den Einfluß von Zirbelextrakten auf Tumoren bei weißen Mäusen und bei Menschen. Wien Klin Wochenschr 62:79–82

Blask DE, Pelletier DB, Hill SM, Lemus-Wilson A, Grosso DS, Wilson ST, Wise ME (1991) Pineal melatonin inhibition of tumour promotion in the N-nitroso-N-methylurea model of mammary carcinogenesis: potential involvement of antiestrogenic mechanisms in vivo. J Cancer Res Clin Oncol 117:526–532

Blask DE, Wilson ST, Zalatan F (1997) Physiological melatonin inhibition of human breast cancer cell growth in vitro: evidence for a glutathione-mediated pathway. Cancer Res 57:1909–1914

Bojkowski CJ, Aldhous ME, English J, Franey C, Poulton AL, Skene DJ, Arendt J (1987) Suppression of nocturnal plasma melatonin and 6-sulphatoxymelatonin by bright light and dim light in man. Hormon Metab Res 19:437–440

Bojkowski CJ, Arendt J (1990) Factors influencing urinary 6-sulphatoxymelatonin, a major melatonin metabolite, in normal human subjects. Clin Endocrinol (Oxf) 33:435–444

Brainard GC, Lewy AJ, Menaker M, Fredrickson RM, Miller LS, Weleber RG, Cassone V, Hudson D (1985) Effect of light wavelength on the suppression of nocturnal plasma melatonin in normal volunteers. Ann NY Acad Sci 453:376–378

Bruchovsky N, Wilson JD (1968) The conversion of testosterone to 5-alpha-androstan-17-beta-ol-3-one by rat prostate in vivo and in vitro. J Biol Chem 243:2012–2021

Brzezinski A, Seibel MM, Lynch HJ, Deng MH, Wurtman RJ (1987) Melatonin in human preovulatory follicular fluid. J Clin Endocrinol Metab 64:865–867

Buzzell GR (1988) Studies on the effects of the pineal hormone melatonin on an androgen-insensitive rat prostatic adenocarcinoma, the Dunning R 3327 HIF tumour. J Neural Transm 72:131–140

Cardinali DP (1984) Neural–hormonal integrative mechanisms in the pineal gland and superior cervical ganglia. In: Reiter RJ (ed) The Pineal Gland. Raven Press, New York, pp 83–107

Chazot G, Claustrat B, Brun J, Olivier M (1985) Rapid antidepressant activity of destyr-gamma-endorphin: correlation with urinary melatonin. Biol Psychiatry 20:1026–1030

Cohen M, Lippman M, Chabner B (1978) Role of the pineal gland in aetiology and treatment of breast cancer. Lancet 2(8094):814–816

Danforth DN Jr, Tamarkin L, Do R, Lippman ME (1983) Melatonin-induced increase in cytoplasmic estrogen receptor activity in hamster uteri. Endocrinology 113:81–85

Danforth DN Jr, Tamarkin L, Mulvihill JJ, Bagley CS, Lippman ME (1985) Plasma melatonin and the hormone-dependency of human breast cancer. J Clin Oncol 3:941–948

Daramola GF, Olowu AO (1972) Physiological and radiological implications of a low incidence of pineal calcification in Nigeria. Neuroendocrinology 9:41–57

Deerberg F, Bartsch C, Pohlmeyer G, Bartsch H (1997) Effect of melatonin and physiological pinealectomy on the development of spontaneous endometrial carcinoma in BDII/Han rats. Cancer Biotherapy 12:420 (Abstract)

Diebold J (1999) Molecular genetics of ovarian carcinomas. Histol Histopathol 14:269–277

Dogliotti L, Berruti A, Buniva T, Torta M, Bottini A, Tampellini M, Terzolo M, Faggiuolo R, Angeli A (1990) Melatonin and human cancer. J Steroid Biochem Mol Biol 37:983–987

Durotoye LA, Webley GE, Rodway RG (1997) Stimulation of the production of progesterone by the corpus luteum of the ewe by the perfusion of melatonin in vivo and by treatment of granulosa cells with melatonin in vitro. Res Vet Sci 62:87–91

Engel P (1935) Wachstumsbeeinflussende Hormone und Tumorwachstum. Z Krebsforsch 41:488–496

Falkson G, Falkson HC, Steyn ME, Rapoport BL, Meyer BJ (1990) Plasma melatonin in patients with breast cancer. Oncology 47:401–405

Fathalla MF (1971) Incessant ovulation – a factor in ovarian neoplasia? Lancet 2(7716):163

Follenius M, Weibel L, Brandenberger G (1995) Distinct modes of melatonin secretion in normal men. J Pineal Res 18:135–140

Gann PH, Hennekens CH, Ma J, Longcope C, Stampfer MJ (1996) Prospective study of sex hormone levels and risk of prostate cancer. J Natl Cancer Inst 88:1118–1126

Gilad E, Matzkin H, Zisapel N (1997) Interplay between sex steroids and melatonin in regulation of human benign prostate epithelial cell growth. J Clin Endocrinol Metab 82:2535–2541

Gilad E, Pick E, Matzkin H, Zisapel N (1998) Melatonin receptors in benign prostate epithelial cells: evidence for the involvement of cholera and pertussis toxins-sensitive G proteins in their signal transduction pathways. Prostate 35:27–34

Gilad E, Laufer M, Matzkin H, Zisapel N (1999) Melatonin receptors in PC3 human prostate tumor cells. J Pineal Res 26:211–220

Godwin AK, Testa JR, Handel LM, Liu Z, Vanderveer LA, Tracey PA, Hamilton TC (1992) Spontaneous transformation of rat ovarian surface epithelial cells: association with cytogenetic changes and implications of repeated ovulation in the etiology of ovarian cancer. J Natl Cancer Inst 84:592–601

Godwin AK, Testa JR, Hamilton TC (1993) The biology of ovarian cancer development. Cancer 71 (2 Suppl):530–536

Graham C, Cook MR, Kavet R, Sastre A, Smith DK (1998) Prediction of nocturnal plasma melatonin from morning urinary measures. J Pineal Res 24:230–238

Grin W, Grünberger W (1998) A significant correlation between melatonin deficiency and endometrial cancer. Gynecol Obstet Invest 45:62–65

Guerin MV, Matthews CDJ (1990) Plasma melatonin exhibits a diurnal secretion in the common marmoset (*Callithrix jacchus*): relationship to the rest–activity cycle. Pineal Res 8:237–244

Habenicht UF, el Etreby MF (1991) Rationale for using aromatase inhibitors to manage benign prostatic hyperplasia. Experimental studies. J Androl 12:395–402

Hajak G, Rodenbeck A, Staedt J, Bandelow B, Huether G, Ruther E (1995) Nocturnal plasma melatonin levels in patients suffering from chronic primary insomnia. J Pineal Res 19:116–122

Hoffmann G, Pollow K, Kreienberg R, Hütz R, Schaffrath M, Strittmatter, H-J, Weickel W, Pollow B (1996) 24-h-Melatonin-Serumprofile bei Mammakarzinom-Patientinnen im Vergleich zu einer tumorfreien Kontrollgruppe. Geburtshilfe Frauenheilkd 56:307–316

Hofstätter R (1959) Versuche der postoperativen Krebsbehandlung mit Zirbelstoffen. Krebsarzt 14:307–316

Holdaway IM, Mason BH, Gibbs EE, Rajasoorya C, Hopkins KD (1991) Seasonal changes in serum melatonin in women with previous breast cancer. Br J Cancer 64:149–153

Horst HJ, Adam KU (1982) Influence of melatonin on the human prostatic androgen metabolism in vitro. Horm Metabol Res 14:54

Huether G (1993) The contribution of extrapineal sites of melatonin synthesis to circulating melatonin levels in higher vertebrates. Experientia 49:665–670

Illnerova H, Vanecek J, Krecek J, Wetterberg L, Saaf J (1979) Effect of one minute exposure to light at night on rat pineal serotonin *N*-acetyltransferase and melatonin. J Neurochem 32:673–675

Illnerova H, Zvolsky P, Vanecek J (1985) The circadian rhythm in plasma melatonin concentration of the urbanized man: the effect of summer and winter time. Brain Res 25:186–189

Itoh MT, Ishizuka B, Kudo Y, Fusama S, Amemiya A, Sumi Y (1997) Detection of melatonin and serotonin *N*-acetyltransferase and hydroxyindole-*O*-methyltransferase activities in rat ovary. Mol Cell Endocrinol 136:7–13

Itoh MT, Ishizuka B, Kuribayashi Y, Amemiya A, Sumi Y (1999) Melatonin, its precursors, and synthesizing enzyme activities in the human ovary. Mol Hum Reprod 5:402–408

Jones RL, McGeer PL, Greiner AC (1969) Metabolism of exogenous melatonin in schizophrenic and non-schizophrenic volunteers. Clin Chim Acta 26:281–285

Kabuto M, Namura I, Saitoh Y (1986) Nocturnal enhancement of plasma melatonin could be suppressed by benzodiazepines in humans. Endocrinol Jpn 33:405–414

Karmali RA, Horrobin DF, Ghayur T (1978) Role of pineal gland in aetiology and treatment of breast cancer. Lancet II:1002

Karasek M, Dec W, Kowalski AJ, Bartsch H, Bartsch C (1996) Serum melatonin circadian profile in women with adenocarcinoma of uterine corpus. Int J Thymology 4 (Suppl1):80–83

Klein DC (1985) Photoneural regulation of the mammalian pineal gland. In: Evered D, Clark S (eds) Photoperiodism, melatonin and the pineal. Ciba Foundation Symposium 117, Pitman, London, pp 38–51

Kopin IJ, Pare CMB, Axelrod J, Weissbach H (1969) The fate of melatonin in animals. J Biol Chem 236:3072–3075

Laakso ML, Porkka Heiskanen T, Alila A, Stenberg D, Johansson G (1994) Twenty-four-hour rhythms in relation to the natural photoperiod: a field study in humans. J Biol Rhythms 9:283–293

Lang U, Kornemark M, Aubert ML, Paunier L, Sizonenko PC (1981) Radioimmunological determination of urinary melatonin in humans: correlation with plasma levels and typical 24-hour rhythmicity. J Clin Endocrinol Metab 53:645–650

Lapin V (1974) Influence of simultaneous pinealectomy and thymectomy on the growth and formation of metastases of the Yoshida sarcoma in rats. Exp Pathol Jena 9:108–112

Lee DK, Bird CE, Clark AF (1973) In vitro effects of estrogens on rat prostatic 5 alpha-reduction of testosterone. Steroids 22:677–685

Lewy AJ, Wehr TA, Goodwin FK, Newsome DA, Markey SP (1980) Light suppresses melatonin secretion in humans. Science 210:1267–1269

Lewy AJ, Newsome DA (1983) Different types of melatonin circadian secretory rhythms in some blind subjects. J Clin Endocrinol Metab 56:1103–1107

Lissoni P (1997) Effects of low–dose recombinant interleukin–2 in human malignancies. Cancer J Sci Am 3, Suppl 1:S115–S120

Lissoni P, Resentini M, Mauri R, Esposti G, Rossi D, Legname G, Fraschini F (1986) Effects of tetrahydrocannabinol on melatonin secretion in man. Horm Metab Res 18:77–78

Lissoni P, Paolorossi F, Barni S, Tancini G, Crispino S, Rovelli F, Ferri L, Esposti G, Esposti D, Fraschini F (1987) Correlation between changes in prolactin and melatonin serum levels after radical mastectomy. Tumori 73:263–267

Lissoni P, Brivio F, Barni S, Tancini G, Cattaneo G, Archili C, Conti A, Maestroni GJ (1990) Neuro-immunotherapy of human cancer with interleukin-2 and the neurohormone melatonin: its efficacy in preventing hypotension. Anticancer Res 10:1759–1761

Lissoni P, Crispino S, Barni S, Sormani A, Brivio F, Pelizzoni F, Brenna A, Bratina G, Tancini G (1990a) Pineal gland and tumour cell kinetics: serum levels of melatonin in relation to Ki-67 labeling rate in breast cancer. Oncology 47:275–277

Lowenstein PR, Rosenstein R, Caputti E, Cardinali DP (1984) Benzodiazepine binding sites in human pineal gland. Eur J Pharmacol 106:399–403

Luboshitzky R, Lavi S, Thuma I, Lavie P (1996) Nocturnal melatonin and luteinizing hormone rhythms in women with hyperprolactinemic amenorrhea. J Pineal Res 20:72–78

Luboshitzky R, Herer P, Lavie P (1997) Pulsatile patterns of melatonin secretion in patients with gonadotropin-releasing hormone deficiency: effects of testosterone treatment. J Pineal Res 22:95–101

Lupowitz Z, Zisapel N (1999) Hormonal interactions in human prostate tumour LNCaP cells. J Steroid Biochem Mol Biol 68:83–88

Lynch HJ, Wurtman RJ, Moskowitz MA, Archer MC, Ho MH (1975) Daily rhythm in human urinary melatonin. Science 187:169–171

Maestroni GJM (1993) The immunoneuroendocrine role of melatonin. J Pineal Res 14:1–10

Maestroni GJM, Conti A (1996) Melatonin in human breast cancer tissue: association with nuclear grade and estrogen receptor status. Lab Invest 75:557–561

Markey SP, Higa S, Shih M, Danforth DN, Tamarkin L (1985) The correlation between human plasma melatonin levels and urinary 6-hydroxymelatonin excretion. Clin Chim Acta 150:221–225

Mason BH, Holdaway IM, Stewart AW, Neave LM, Kay RG (1990) Season of initial discovery of tumour as an independent variable predicting survival in breast cancer. Br J Cancer 61:137–141

Molis TM, Spriggs LL, Hill SM (1994) Modulation of estrogen receptor mRNA expression by melatonin in MCF-7 human breast cancer cells. Mol Endocrinol 8:1681–1690

Molis TM, Spriggs LL, Jupiter Y, Hill SM (1995) Melatonin modulation of estrogen-regulated proteins, growth factors, and proto-oncogenes in human breast cancer. J Pineal Res 18:93–103

Monteleone P, Forziati D, Orazzo C, Maj M (1989) Preliminary observations on the suppression of nocturnal plasma melatonin levels by short-term administration of diazepam in humans. J Pineal Res 6:253–258

Moore RY, Klein DC (1974) Visual pathways and the central neural control of a circadian rhythm in pineal serotonin N-acetyltransferase activity. Brain Res 71:17–33

Morgan PJ, Barrett P, Howell HE, Helliwell R (1994) Melatonin receptors: localization, molecular pharmacology and physiological significance. Neurochem Int 24:101–146

Murphy DL, Tamarkin L, Sunderland T, Garrick NA, Cohen RM (1986) Human plasma melatonin is elevated during treatment with the monoamine oxidase inhibitors clorgyline and tranylcypromine but not deprenyl. Psychiatry Res 17:119–127

Nelson W, Tong YL, Lee JK, Halberg F (1979) Methods for cosinor-rhythmometry. Chronobiologia 6:305–323

Nimrod A, Ryan KJ (1975) Aromatization of androgens by human abdominal and breast fat tissue. J Clin Endocrinol Metab 40:367–372

Oosthuizen JM, Bornman MS, Barnard HC, Schulenburg GW, Boomker D, Reif S (1989) Melatonin and steroid-dependent carcinomas. Andrologia 21:429–431

Parkin DM, Laara E, Muir CS (1988) Estimates of the worldwide frequency of sixteen major cancers in 1980. Int J Cancer 41:184–197

Payne AH, Perkins LM, Georgiou M, Quinn PG (1987) Intratesticular site of aromatase activity and possible function of testicular estradiol. Steroids 50:435–448

Pevet P, Balemans MG, Legerstee WC, Vivien-Roels B (1980) Circadian rhythmicity of the activity of hydroxyindole-O-methyl transferase (HIOMT) in the formation of melatonin and 5-methoxytryptophol in the pineal, retina, and harderian gland of the golden hamster. J Neural Transm 49:229–245

Philo R, Berkowitz AS (1988) Inhibition of Dunning tumour growth by melatonin. J Urol 139:1099–1102

Plaxe SC, Saltzstein SL (1997) Impact of ethnicity on the incidence of high-risk endometrial carcinoma. Gynecol Oncol 65:8–12

Potischman N, Hoover RN, Brinton LA, Siiteri P, Dorgan JF, Swanson CA, Berman ML, Mortel R, Twiggs LB, Barrett RJ, Wilbanks GD, Persky V, Lurain JR (1996) Case-control study of endogenous steroid hormones and endometrial cancer. J Natl Cancer Inst 88:1127–1135

Ralph CL, Mull D, Lynch HJ, Hedlund L (1971) A melatonin rhythm persists in rat pineals in darkness. Endocrinology 89:1361–1366

Rasmussen DD (1993) Diurnal modulation of rat hypothalamic gonadotropin-releasing hormone release by melatonin in vitro. J Endocrinol Invest 16:1–7

Rato AG, Pedrero JG, Martinez MA, del Rio B, Lazo PS, Ramos S (1999) Melatonin blocks the activation of estrogen receptor for DNA binding. FASEB J 13:857–868

Ravault JP, Chesneau D, Ouvray M, Locatelli A (1996) Pineal microdialysis of the melatonin in conscious sheep: methodology, application to a diurnal rhythm and effect of isoproterenol. J Neuroendocrinol 1996 8:387–394

Redfern PH, Lemmer B (1997) Physiology and pharmacology of biological rhythms. Handbook of Experimental Pharmacology, Vol 125. Springer, Berlin

Rivest RW, Aubert ML, Lang U, Sizonenko PC (1986) Puberty in the rat: modulation by melatonin and light. J Neural Transm Suppl 21:81–108

Rocci ML Jr, Szefler SJ, Acara M, Jusko WJ (1981) Prednisolone metabolism and excretion in the isolated perfused rat kidney. Drug Metab Dispos 9:177–182

Ross JS, Yang F, Kallakury BV, Sheehan CE, Ambros RA, Muraca PJ (1999) HER-2/neu oncogene amplification by fluorescence in situ hybridization in epithelial tumours of the ovary. Am J Clin Pathol 111:311–316

Rowley KH, Mason MD (1997) The aetiology and pathogenesis of prostate cancer. Clin Oncol R Coll Radiol 9:213–218

Sandyk R, Anastasiadis PG, Anninos PA, Tsagas N (1992) Is the pineal gland involved in the pathogenesis of endometrial carcinoma. Int J Neurosci 62:89–96

Shah PN, Mhatre MC, Kothari LS (1984) Effect of melatonin on mammary carcinogenesis in intact and pinealectomized rats in varying photoperiods. Cancer Res 44:3403–3407

Silverberg E, Boring CC, Squires BA (1990) Cancer Statistics 1990. CA Cancer J Clin 40:9–26

Skene DJ, Bojkowski CJ, Currie JE, Wright J, Boulter PS, Arendt J (1990) 6–sulphatoxymelatonin production in breast cancer patients. J Pineal Res 8:269–276

Skene DJ, Bojkowski CJ, Arendt J (1994) Comparison of the effects of acute fluvoxamine and desipramine administration on melatonin and cortisol production in humans. Br J Clin Pharmacol 37:181–186

Skinner DC, Robinson JE (1997) Luteinising hormone secretion from the perifused ovine pars tuberalis and pars distalis: effects of gonadotropin-releasing hormone and melatonin. Neuroendocrinology 66:263–270

Smith JA, Barnes JL, Mee TJ (1979) The effect of neuroleptic drugs on serum and cerebrospinal fluid concentrations in psychiatric subjects. J Pharm Pharmacol 31:246–248

Soderqvist G (1998) Effects of sex steroids on proliferation in normal mammary tissue. Ann Med 30:511–524

Subramanian A, Kothari L (1991) Melatonin, a suppressor of spontaneous murine mammary tumours. J Pineal Res 10:136–140

Suranyi Cadotte B, Lal S, Nair NP, Lafaille F, Quirion R (1987) Coexistence of central and peripheral benzodiazepine binding sites in the human pineal gland. Life Sci 40:1537–1543

Suzuki S, Dennerstein L, Greenwood KM, Armstrong SM, Sano T, Satohisa E (1993) Melatonin and hormonal changes in disturbed sleep during late pregnancy. J Pineal Res 15:191–198

Tamarkin L, Cohen M, Roselle D, Reichert C, Lippman M, Chabner B (1981) Melatonin inhibition and pinealectomy enhancement of 7,12-dimethylbenz(a)anthracene-induced mammary tumours in the rat. Cancer Res 41:4432–4436

Tamarkin L, Danforth D, Lichter A, DeMoss E, Cohen M, Chabner B, Lippman M (1982) Decreased nocturnal plasma melatonin peak in patients with estrogen receptor positive breast cancer. Science 216:1003–1005

Tamura H, Nakamura Y, Takiguchi S, Kashida S, Yamagata Y, Sugino N, Kato H (1998) Melatonin directly suppresses steroid production by preovulatory follicles in the cyclic hamster. J Pineal Res 25:135–141

Teicher MH, Barber NI (1990) COSIFIT: an interactive program for simultaneous multioscillator cosinor analysis of time-series data. Comput Biomed Res 23:283–295

Touitou Y, Fevre M, Lagoguey M, Carayon A, Bogdan A, Reinberg A, Beck H, Cesselin F, Touitou C (1981) Age- and mental health-related circadian rhythms of plasma levels of melatonin, prolactin, luteinizing hormone and follicle-stimulating hormone in man. J Endocrinol 91:467–475

Valenti S, Guido R, Giusti M, Giordano G (1995) In vitro acute and prolonged effects of melatonin on purified rat Leydig cell steroidogenesis and adenosine 3′,5′-monophosphate production. Endocrinology 136:5357–5362

Vivien-Roels B, Malan A, Rettori MC, Delagrange P, Jeanniot JP, Pevet P (1998) Daily variations in pineal melatonin concentrations in inbred and outbred mice. J Biol Rhythms 13:403–439

Wakabayashi H, Shimada K, Satoh T (1991) Effects of diazepam administration on melatonin synthesis in the rat pineal gland in vivo. Chem Pharm Bull (Tokyo) 39:2674–2676

Waldhauser F, Ehrhart B, Forster E (1993) Clinical aspects of the melatonin action: impact of development, aging, and puberty, involvement of melatonin in psychiatric disease and importance of neuroimmunoendocrine interactions. Experientia 49:671–681

Webley GE, Luck MR, Hearn JP (1988) Stimulation of progesterone secretion by cultured human granulosa cells with melatonin and catecholamines. J Reprod Fertil 84:669–677

West NB, Roselli CE, Resko JA, Greene GL, Brenner RM (1988) Estrogen and progestin receptors and aromatase activity in rhesus monkey prostate. Endocrinology 123:2312–2322

Wetterberg L (1978) Melatonin in humans: physiological and clinical studies. J Neural Transm 13 (suppl):289–310

Wetterberg L, Halberg F, Tarquini B, Cagnoni M, Haus E, Griffith K, Kawasaki T, Wallach LA, Ueno M, Uezo K, Matsuoka M, Kuzel M, Halberg E, Omae T (1979) Circadian variation in urinary melatonin in clinically healthy women in Japan and the United States of America. Experientia 35: 416–419

Wetterberg L, Halberg F, Halberg E, Haus E, Kawasaki T, Ueno M, Uezono K, Cornelissen G, Matsuoka M, Omae T (1986) Circadian characteristics of urinary melatonin from clinically healthy young women at different civilization disease risks. Acta Med Scand 220:71–81

Winters WD, Yuwiler A, Oxenkrug GF (1991) The effects of benzodiazepines on basal and isoproterenol-stimulated N-acetyltransferase activity by the rat pineal gland, in vivo and in vitro. J Pineal Res 10:151–158

Yie SM, Brown GM, Liu GY, Collins JA, Daya S, Hughes EG, Foster WG, Young Lai, EV (1995) Melatonin and steroids in human pre-ovulatory follicular fluid: seasonal variations and granulosa cell steroid production. Hum Reprod 10:50–55

7 Melatonin in Patients with Cancer of Extra-Reproductive Location

Tatiana V. Kvetnaia, *Igor M. Kvetnoy*, Hella Bartsch,
Christian Bartsch, Dieter Mecke

Abstract

A review is given on the analysis of melatonin (MT) and of its main peripheral metabolite 6-sulfatoxymelatonin (aMT6s) in patients with different types of primary unoperated cancers outside of the reproductive system. A very low production of melatonin (as estimated by the nocturnal urinary excretion of aMT6s) was found in male patients with lung or stomach cancer compared to aged-matched controls as well as in female patients with thyroid cancer. The levels of aMT6s in these women, however, did not differ from female patients with benign thyroid diseases indicating a general suppressive effect of thyroid disease on the pineal gland. A similar but opposite phenomenon was observed in male patients with primary unoperated colorectal cancer who showed an elevated production of melatonin as estimated by urinary aMT6s when compared to healthy men but not when compared to patients with colitis ulcerosa. The mechanisms involved in these phenomena are poorly understood and seem to include central as well as peripheral components. This view is supported by the finding that in spite of varying urinary aMT6s excretion measured in patients with different types of tumors, aMT6s shows comparable positive correlations with the degree of tumor cell proliferation (as estimated by the number of PCNA-immunopositive cells). Therefore the amount of aMT6s excreted (as well as the corresponding concentration of circulating MT) has to be understood as the net result of a number of different effects exerted by the tumor on the system which includes the secretion of melatonin by the pineal gland, its peripheral metabolism, and possibly also its production and release by tumor cells.

7.1 Introduction

Melatonin (MT), the 5-methoxy-*N*-acetylated derivative of serotonin (5-hydroxytryptamine) was first discovered in the bovine pineal gland by Lerner et al. (1958) and for a number of years the pineal gland was considered to be the exclusive site of its biosynthesis (Reiter 1991). The wide application of immunocytochemical methods for the study of the cellular localization and metabolism of many biochemical substances, including hormones, has allowed the acquisition of evidence that MT-immunopositive cells are detectable in a number of extra-pineal tissues including the retina, Harderian gland (Bubenik et al. 1978), mucosa of the gut (Bubenik et al. 1977), respiratory tract, liver, kidney, adrenal glands, thymus, placenta, endometrium, endothelium, cerebellum, and neuronal ganglia (Raikhlin and Kvetnoy 1994; Kvetnoy and Yuzhakov 1993, 1994). The discovery of the key enzymes of the MT biosynthetic pathway, namely of N-acetyl-transferase and hydroxyindole-O-methyltransferase, in the retina, gut, as well as in other tissues mentioned above (Hickman et al. 1998; Quay and Ma 1976; Slominski et al. 1996) gave evidence for extra-pineal sites of production thus leading to a revision of the view that the pineal gland is the only source of MT (Raikhlin and Kvetnoy 1994, Kvetnoy 1999). Today it is well established that MT possesses a wide spectrum of biological activities playing a key role as a neuroendocrine signal molecule for general and local coordination including the synchronization of biological rhythms (Kvetnoy et al. 1997), thus fundamentally affecting the nervous, endocrine, and immune systems as well as the organism as a whole. One of the most important properties of MT probably is its influence on cell division. MT is able to inhibit cell mitosis, with a partial delay occurring at the metaphase stage (Banerjee and Margulis 1973). Several reports have been published concerning an inhibitory action of MT on the growth of sex steroid hormone-dependent cells (Cos et al. 1996; Hill et al. 1992) as well as on the development of experimental malignant tumors both under in vivo and in vitro conditions (Blask 1984; Blask et al. 1994). In view of these findings, it was recently proposed that MT is a naturally occurring oncostatic neurohormone for the prevention of neoplastic growth (Blask et al. 1994). Therefore it appears logical to assume that clinical studies dealing with the estimation of MT secretion in cancer patients would be of fundamental theoretical importance for a better understanding of the endogenous mechanisms involved in the process of neoplastic transformation, and may also be used as a diagnostic and prognostic tool for malignant diseases.

Determinations of MT in cancer patients have been performed for the last 20 years. Initially, the content of serum MT in cancer patients was studied with the help of gas chromatography combined with mass spectrometry (Pico et al. 1979) and the levels of MT were found to be lower in these patients compared with healthy controls. Later on, a decrease of nocturnal MT was detected in breast cancer patients (Tamarkin et al. 1982). Touitou et al. (1985) analyzed the daily excretion of MT in cancer patients and showed an increase of MT in patients with cancer of the uterus and ovary, and a decrease in patients affected with cancer of the liver, kidney, lung, and skin. Numerous studies followed and those dealing with the determination of MT in patients with cancer of the reproductive organs are surveyed in the preceding chapter of this book.

This chapter focuses on MT in patients with tumors originating in organs outside of the reproductive system. In these studies, the nocturnal production of MT was estimated noninvasively with the help of 6-sulfatoxymelatonin (aMT6s) measurements in urine. aMT6s is the major peripheral metabolite of MT (Kopin et al. 1961), and its measurement became possible after the development of a simple and reliable radioimmunoassay by Arendt et al. (1985). The nocturnal excretion of aMT6s in urine was found to reflect the nocturnal secretion of MT in blood in healthy subjects as well as in cancer patients (Bartsch. et al. 1992). The second part of this chapter deals with the analysis of correlations between the nocturnal production of MT, as estimated by aMT6s measurements, with intratumoral parameters at the cellular level in order to better understand the possible mechanisms involved in the pineal–cancer crosstalk. These parameters include the proliferative potential of tumor cells studied immunocytochemically using antibodies to proliferating cell nuclear antigen (PCNA) as well as the number of MT-immunopositive cells within tumors.

7.2 Patients and Methods

Clinical Material. Between 1991 and 1996, we studied more than 200 female and male cancer patients who were hospitalized at the Medical Radiological Research Center, Obninsk, the Oncological Research Center, Moscow, or the Medical University Hospital, Saratov. The patients were suffering from cancer of the thyroid, larynx, lung, stomach, colon, and rectum, as well as of the urinary bladder. The patients were studied before surgical removal of their primary tumor and received neither chemo- nor radiotherapy or other types of treatment that were known to affect the synthesis and secretion of MT.

The female and male control groups consisted of patients suffering from different types of nonmalignant diseases, i.e., ischemic heart disease, chronic gastritis, choleli- thiasis, rheumatoid arthritis, and deforming arthrosis, as well as of healthy volunteers. In addition, two separate control groups were studied with colitis ulcerosa or with benign thyroid diseases to allow comparison with the corresponding malignant diseases, namely colorectal and thyroid cancer. To facilitate relevant comparisons among the different groups of patients, care was taken that patients with similar age-ranges were selected out of the overall populations in each group. For further clinical details of the groups see Table 7.1.

Urine was collected from each patient or control subject during the night be- tween 11 P.M. and 7 A.M. and immediately after collection it was aliquoted and frozen at $-20\,^{\circ}$C and transported to Germany in liquid nitrogen where assays were perform- ed. Specimens from different parts of the respective tumors were collected during operation and were immediately fixed in 10% phosphate-buffered formaldehyde and in phosphate-buffered Bouin's fluid.

Measurement of aMT6s. The radioimmunoassay of aMT6s was carried out in Germany from the above-described frozen urine aliquots using specific antibodies for aMT6s, standards and iodinated tracer of Stockgrand Ltd. (Guildford, Surrey, UK)

Table 7.1. Clinical details of the groups studied

Group	n	Average age (years $\pm$ SEM)	TNM staging
1. Men			
Healthy	21	56 ± 1	
Chronic colitis ulcerosa	6	53 ± 4	
Colorectal cancer	15	60 ± 2	$T_{2-4}N_{0-4}M_{0-2}$
Larynx cancer	25	56 ± 2	$T_{2-4}N_{0-2}M_{0-1}$
Lung cancer	31	58 ± 1	$T_{1-4}N_{0-3}M_{0-1}$
Stomach cancer	8	65 ± 3	$T_{3-4}N_{X-2}M_0$
Cancer of urinary bladder	7	55 ± 4	$T_{2-4}N_{X-1}M_0$
2. Women			
Healthy	27	55 ± 2	
Benign thyroid disease	12	57 ± 2	
Thyroid cancer	6	55 ± 2	$T_{3-4}N_{0-2}M_{0-1}$

applying the protocol of Arendt et al. (1985). The inter-assay coefficient of variation was 7% and the intra-assay coefficient of variation was 4%. The results were expressed as ng aMT6s/h.

Histopathological Examination of the Tumors. From the formalin-fixed tumor specimens, paraffin sections were prepared and stained with hematoxylin/eosin according to routine histopathological methods. Tumors were examined by a histopathologist and classified according to the WHO standard (Hermanek and Sobin 1992).

Immunocytochemical Study of Proliferating Cell Nuclear Antigen (PCNA) and MT Within Tumors. Specimens from the zone of active proliferation without necrosis were fixed in 10% phosphate-buffered formaldehyde of pH 7.2 for the detection of PCNA and in phosphate-buffered Bouin's fluid of pH 7.2 for the detection of MT. Immunocytochemical reactions were performed with the mouse monoclonal anti-PCNA antibody PC-10 (Dako, Glostrup, Denmark) and with the ultra-specific rabbit polyclonal anti-MT antiserum (URMA) of CIDtech Research Inc., Mississauga, Canada, according to established protocols using biotinylated rabbit anti-mouse IgG (for PC-10), biotinylated goat anti-rabbit IgG (for URMA) as secondary antibodies, and peroxidase-conjugated avidin and 3,3-diaminobenzidine tetrahydrochloride for staining (all reagents from Dako).

Morphometric Analysis of PCNA- and MT-Positive Cells. Nuclei staining for PCNA and cytoplasm staining for MT were accepted as positive. The quantification of PCNA-positive nuclei and of MT-positive cells was performed with the computer image analysis system MORPHOSTAR with the application software module COLQUANT (Imstar S.A., Paris, France). PCNA immunopositive nuclei of tumor cells, and MT-positive cells were counted in 50 randomly selected visual fields, each covering 0.785 mm^2, at $\times 200$ original magnification ($\times 20$ objective and $\times 10$ ocular). At least five sections were counted from each tumor. The results were expressed as the mean number of nuclei or cells (or as PCNA index) per one visual field at $\times 200$ original magnification. The PCNA index was calculated as the number of PCNA positive nuclei/total number of nuclei.

Statistical Analysis. The data were analyzed with the program STATISTICA of Statsoft (Tulsa, OK, USA) using non-parametric tests exclusively, the ANOVA of Kruskal-Wallis and the Spearman rank order correlation test.

7.3 Results

7.3.1 Nocturnal Urinary aMT6s Excretion in Cancer Patients

Thyroid Cancer Patients. The six female patients with large primary carcinomas of the thyroid studied before operation (see Table 7.1) showed a 56% lower nocturnal excretion of aMT6s (539 ± 104 ng/h) than the 27 female controls of comparable age (55 ± 2 years) not affected by thyroid disease (1223 ± 176 ng/h). The 12 controls affected by different types of benign thyroid enlargement (11 simple goiters as well as adenomas and one Hashimoto thyroiditis) with an average age of 57 ± 2 years showed a significant 60% depletion of aMT6s (492 ± 89; $P < 0.05$) compared with controls without thyroid disease. Six to eight days after surgery of a malignant thyroid tumor (subtotal resection or thyroidectomy) the nocturnal excretion of aMT6s appeared to be decreasing further from 539 ± 104 to 348 ± 83 ng/h.

Laryngeal Cancer Patients. Male patients with primary laryngeal cancers ($n = 25$) which were histologically characterized as squamous cell carcinomas of grades G1–3 and of different clinical stages (T2–4N0–2M0–1), showed a marginal increase of the average nocturnal excretion of aMT6s by 12% (1126 ± 191 ng/h) compared with controls (1004 ± 146 ng/h). Both groups had the same average age of 56 years (see Table 7.1). A subdivision of the group of patients with laryngeal cancer according to the size of their primary tumor reveals that patients with relatively small tumors at the T2 stage ($n = 4$) exhibit a 125% increase of aMT6s (2262 ± 690 ng/h) compared with controls, whereas patients with T4 tumors ($n = 5$) show a 62% decrease (386 ± 132 ng/h). Patients with T3 tumors ($n = 16$) are more or less at the level of controls (1074 ± 183 ng/h). Since the average age of the patients in subgroups with different tumor sizes is comparable (55–57 years), it appears that the size of a primary laryngeal carcinoma exerts a modulatory effect on the nocturnal production of MT. In the presence of smaller tumors (T2) the production is elevated whereas an inhibition seems to occur due to large tumors (T4). A subdivision according to the histological grade of the tumor indicates that aMT6s excretion is elevated in the presence of highly differentiated G1 tumors ($n = 9$), being on average 52% (1524 ± 405 ng) higher than in controls, whereas the two patients with undifferentiated G3 tumors had extremely low aMT6s (114 and 384 ng/h). Patients with poorly differentiated G2 tumors were found to have an average aMT6s excretion which was practically indistinguishable from controls (996 ± 195 ng/h).

Lung Cancer Patients. The average nocturnal aMT6s excretion of the 31 male patients affected by unoperated primary lung cancer was found to be highly significantly diminished by 59% ($P = 0.0019$; 408 ± 63 ng/h) compared with tumor-free controls of similar average age (see Table 7.1). Patients with cancer had tumors of clinical stages T1–4N0–3M0–1, the majority being non-small-cell lung cancer, characterized histologically as squamous cell carcinomas and adenocarcinomas. In addition, one patient with a small-cell carcinoma was analyzed. A division of patients according to the size of their primary tumors did not give any clear indication for an effect on the nocturnal excretion of aMT6s since the suppression was comparable among the different subgroups being in the range of 36–50% (T2, $n = 11$, 501 ± 141 ng/h; T3, $n = 13$, 378 ± 91 ng/h; T4, $n = 4$, 362 ± 29 ng/h).

Stomach Cancer Patients. The nocturnal excretion of aMT6s showed a significant 59% ($P = 0.0023$) decrease in the eight male patients studied with primary unoperated stomach cancer (410 ± 75 ng/h) compared with controls. These patients had large tumors without distant metastases (T3–4NX–2M0) histologically characterized as adenocarcinomas.

Colorectal Cancer Patients. The 15 male patients studied had mostly large tumors of stages T3 and T4 which in some cases showed pronounced lymph node infiltration (N2–4) and in a few cases distant metastases. Tumors were classified histologically as adenocarcinomas. The average nocturnal urinary aMT6s excretion was 1449 ± 309 ng/h, thus being 44 % higher than in tumor-free controls of comparable average age and not affected with diseases of the colon. The excretion of aMT6s was also found to be high in six male patients of similar average age suffering from chronic colitis ulcerosa (1909 ± 553 ng/h).

Patients with Urinary Bladder Cancer. Among the seven male patients with primary unoperated cancers of the urinary bladder the average nocturnal excretion of aMT6s was marginally lower (−15 %; 858 ± 132 ng/h) than in tumor-free controls of the same average age. Tumors were mostly large (T3–4) but localized and without distant metastases (see Table 7.1). Six of them were classified histologically as transitional cell carcinomas showing different degrees of differentiation (G1–3), and one was an adenocarcinoma.

7.3.2 Correlations of Nocturnal Urinary aMT6s Excretion with Immunocytochemical Parameters Measured in Gastrointestinal and Lung Cancer

In 26 male patients with unoperated primary colorectal, stomach, or lung cancer without distant metastases, the number of PCNA-, and MT-positive cells were quantitated morphometrically and correlated with the corresponding urinary aMT6s excretion.

Urinary aMT6s and PCNA-Positive Tumor Cells. A very strong positive correlation was found between the urinary aMT6s excretion and the number of PCNA-positive cells in tumor tissue for all analyzed patients (Spearman rank test $r_S = +0.915$; $P < 0.0001$). A subdivision of patients according to their tumor type also revealed strong positive correlations ($r_S = 0.736 – 1.000$; $P < 0.01 – 0.001$), particularly in patients with colorectal cancer who exhibited a much higher relative number of PCNA-positive tumor cells (64 ± 9) than patients with stomach (36 ± 5) or lung cancer (39 ± 5).

Urinary aMT6s and Number of MT-Positive Tumor Cells. The urinary excretion of aMT6s showed clear negative correlations with the intratumoral MT content not only among the merged group of all 26 analyzed cancer patients ($r_S = −0.846$; $P < 0.0001$) but also within the three subgroups of patients suffering from colorectal ($n = 6$; $r_S = −0.928$; $P < 0.01$), stomach ($n = 8$; $r_S = −0.905$; $P < 0.0025$), and lung cancer ($n = 12$; $r_S = −0.735$; $P < 0.01$). The average relative number of MT-positive tumor cells was lower in colorectal (12±3) than in stomach or lung cancer (17 ± 4) indicating that fast-proliferating tumors show a lower degree of MT-positive cells.

7.3.3 Correlations Among Intratumoral Parameters in Gastrointestinal Cancers

PCNA and MT. Since the average relative number of MT-positive tumor cells was lower in colorectal than in stomach or lung cancer with a higher proliferative activity as estimated by PCNA, it is not surprising that clear negative correlations were found between the individual PCNA- and MT-values of the different tumors. This applied to the overall group of patients analyzed ($n = 26$; $r_S = +0.924$; $P < 0.001$) as well as to the subgroups with different types of tumors ($r_S = +0.886 – 0.928$; $P < 0.01 – 0.0001$).

7.4 Discussion

7.4.1 Melatonin Secretion in Human Cancer Patients

Thyroid Cancer. The limited number of thyroid cancer patients studied indicates that the nocturnal production of MT as estimated by the urinary excretion of aMT6s is lower as compared to controls not affected by thyroid disease. The inhibitory effect of thyroid cancer upon pineal function, however, does not seem to be a specific phenomenon confined to the presence of malignant thyroid cells, since patients with a benign enlargement of the thyroid including simple goiter, adenomas as well as Hashimoto thyroiditis show a statistically significant 60% depletion of urinary excretion of aMT6s ($P < 0.05$) compared with controls without thyroid enlargement. These differences are not due to age since patients of the different groups were of comparable age. It has to be concluded that thyroid enlargement in general, irrespective of benign or malignant pathology, appears to inhibit pineal MT production. It is conceivable that the depletion of circulating MT may have a negative effect for the patients' prognosis since MT has been reported to control thyroid function including the mitotic activity of thyroid follicular cells (Lewinski and Sewerynek 1986; Wajs and Lewinski 1992) and [^{3}H]thymidine incorporation into the DNA of thyroid lobes (Wajs and Lewinski 1991), whereas pinealectomy was found to stimulate thyroid growth (Lewinski et al. 1992; Wajs and Lewinski 1992). Our findings of a decline of pineal MT secretory activity are supported by the observation of a depressed circadian amplitude of plasma MT in female patients with recurrent struma (Kuzdak and Komorowski 1995). Rojdmark et al. (1991) detected a higher peak of MT in patients with hypothyroidism and a phase-advanced peak in thyrotoxic patients. It is unclear whether changed thyroid hormone secretion alone accounts for the observed depletion of MT secretion. It is conceivable that central changes within the hypothalamo-pituitary-thyroidal axis which is modulated by MT (Vriend 1983) may reciprocally affect the pineal gland, or that thyroid growth factors secreted (Lewinski et al. 1993) by the respective tumors inhibit MT production. Another conceivable mechanism could be that homeostatic changes within benign or malignant thyroid glands negatively affect the superior cervical ganglia which exert sympathetic control over both the thyroid and the pineal gland (Cardinali et al. 1983).

Larynx Cancer. The male larynx is an androgen-dependent structure and possesses respective sex steroid receptors (Beckford et al. 1985). Laryngeal carcinomas have been reported to contain both testosterone and estrogen receptors (Marsigliante et al. 1992; Resta et al. 1994). Therefore, the normal and neoplastic larynx are endocrine-dependent structures which are affected by changes within the hypothalamo-pituitary-gonadal axis as well as by the neuroendocrine hormone MT. Although alcohol abuse and tobacco smoking appear to be the main causative factors for the development of laryngeal cancer (Seitz et al. 1998), changes of the levels of circulating testosterone due to aging as well as a result of pronounced ethanol consumption (Shi et al. 1998) and tobacco smoking (Ochedalski et al. 1994) appear to be important co-factors which determine the susceptibility of laryngeal cells towards malignant transformation by the primary noxae (Kambic et al. 1984). The exact role of androgens is, however, still unclear mainly due to the paradox that the highest incidence of laryngeal carcinomas occurs in the presence of reduced androgen levels connected with aging, although these androgens exert a stimulatory action on laryngeal mucosa and their tumors (Kleemann and Kunkel 1996). A similar paradox is known for prostate cancer which almost exclusively develops in elderly men and in many cases is androgen-dependent. An overproduction of autocrine and paracrine cytokines due to elevated oncogenic expression may be regarded as a compensatory mechanism towards declining circulating androgen levels normally controlling prostatic development and function in order to foster the initial and autonomous growth of prostatic cancer cells (Chan et al. 1998). Due to the presence of androgen receptors, these cells are also stimulated in their growth by androgens. Similar mechanisms may also apply to the development of laryngeal cancer in elderly men (Whisler et al. 1998) particularly because of the puzzling observation of elevated circulating androgens in these patients (Kambic et al. 1984). The documented negative effects of alcohol consumption on the levels of circulating MT (Wetterberg et al. 1992) may aggravate endocrine disturbances in patients with laryngeal cancer, such as an elevation of circulating testosterone due to an inhibitory action of MT on the reproductive system (Pang et al. 1998). A decline of circulating MT by alcohol abuse may also directly stimulate the development of laryngeal cancer because of its diminished availability as a free radical scavenger molecule (Reiter et al. 1999) to neutralize metabolites of ethanol and particularly of tobacco constituents (often consumed together with alcohol) possessing a high mutagenic potential upon laryngeal tissue (Phillips 1996).

The main result of our current study on male patients with unoperated primary laryngeal cancer is that there is a striking tumor size dependent effect upon the nocturnal production of MT with a clear elevation in the presence of relatively small tumors (T2) and a depression due to large ones (T4). MT production was also found to depend upon the histological grade of differentiation being high in case of G1 and very low in G3 tumors. These pronounced tumor-size- and differentiation-dependent effects are the reason why the production of MT in the overall group of patients with laryngeal cancer does not differ from controls. Similar tumor-size- as well as differentiation-dependent changes have been reported for circulating testosterone in patients with laryngeal cancer (Szeja et al. 1996) and it is conceivable that these may affect pineal MT secretory activity via androgen receptors present in this gland (Luboshitzky et al. 1997). The detailed mechanisms involved in these divergent effects of tumor size on the nocturnal production of MT, however, are still unclear but similar phenomena have been observed in patients with prostate or breast cancer (Bartsch C et al. 1989; Bartsch C and Bartsch H 1994; Bartsch C et al. 1997). Also in animals with different serial transplants derived from a chemically-induced mammary cancer, differential effects have been observed on pineal MT, namely a stimulation during the growth of a relatively well-differentiated early passage and an inhibition in the presence of an undifferentiated late passage (Bartsch C et al. 1995). The stimulation due to early passages seems to involve a recognition by the cell-mediated immune system, whereas the inhibition of circulating MT may involve a degradation of the pineal hormone by tumorous tissue (Bartsch C et al. 1999). Similar mechanisms may apply to the tumor-stage-dependent modulation of MT in patients with different types of cancer, since MT is high in patients with laryngeal as well as prostate cancer of high differentiation and low if poorly differentiated tumors are present (Bartsch C and Bartsch H 1997; Bartsch C et al. 1998). Evidence for an activation of cellular immunity has also been obtained in patients with primary laryngeal cancer (Cozzolino et al. 1986). These observations underline the necessity to strictly differentiate analytical biochemical observations in cancer patients according to the stage of their disease and illustrate the complexity of neuroimmunoendocrine interactions involved in cancer.

Lung Cancer. Male patients with primary unoperated lung cancer of different clinical stages mostly being non-small-cell lung cancers (squamous carcinomas and adeno-carcinomas) showed statistically significantly lower nocturnal urinary aMT6s excretion values than tumor-free controls (-59%, $P = 0.0019$), no differences were detected between patients with different tumor sizes. Viviani et al. (1992) reported obliterated day/night variations of circulating MT in patients with non-small-cell lung cancer which could be restored by the administration of interleukin (IL)-2. Dogliotti et al. (1990), in a mixed group of male and female patients, reported an overall elevation of MT in case of stage III–IV lung cancer, with very high early morning and nighttime MT levels due to non-small-cell lung cancer and very low concentrations in the presence of small-cell lung cancer. The discrepancies among these studies may be due to incongruent distributions of the clinical stages analyzed by the different authors.

Concerning the mechanisms involved in the clear depression of MT in our as well as Viviani et al.'s (1992) investigation, it can be hypothesized that either central or peripheral processes are involved. Central mechanisms could include an inhibition of the biosynthesis and secretion of MT which invariably would be connected with a depressed hepatic formation of aMT6s. Peripheral mechanisms may consist of either a lowered formation of aMT6s by the liver or an enhanced degradation of MT to metabolites other than aMT6s. It is presently unknown which mechanism may apply to patients with lung cancer but different alternatives can be discussed on theoretical grounds. A central inhibition of the formation of MT by the pineal gland could occur in lung cancer patients due to the formation of hormones or growth factors (Rozengurt 1999) produced by the tumor. A considerable proportion of bronchial cancers, particularly of the small-cell type, are known to produce hormones ectopically, e.g., ACTH (Wajchenberg et al. 1995) as well as neurohypophyseal peptides (Moses and Scheinman 1991). The peripheral mechanisms leading to the observed depletion of the urinary aMT6s excretion could include a potential inhibition of the hepatic formation of this metabolite due to malignancy, but neither in breast (Bartsch C et al. 1991) nor in prostate cancer (Bartsch C et al. 1992) has evidence for such changes been found. It is more likely that the enzyme indolamine 2,3-dioxygenase (IDO, E.C. 1.13.11.17) may be involved, which has been reported to be expressed in extra-hepatic normal tissues, including lung, as well as in lung cancer (Yoshida and Hayaishi 1984; Yasui et al. 1986). IDO, among others, is controlled in its activity by interferon-γ (Taylor and Feng 1991). The metabolization of MT in the brain occurs primarily via the kynurenine pathway (Hirata et al. 1974) and highest IDO activity is found in the choroid plexus as well as in the pineal gland (Fujiwara et al. 1978). The presence of a further elevated IDO activity in the brain of lung cancer patients would open up the possibility that the depression of aMT6s might even be caused by an enhanced central degradation of the pineal hormone.

The depression of MT is not the only drastic endocrine change observed in lung cancer patients. It has been repeatedly reported that the pituitary-gonadal axis is disturbed in these patients characterized by a depression of circulating testosterone in the absence of any changes of luteinizing hormone (LH) (Blackman et al. 1988; Taggart et al. 1993). Although the mechanisms involved in this phenomenon are as yet unknown, the depression of testosterone correlates with poor performance of the patients concerned (Aasebo et al. 1991).

A central question is whether a substitutional therapy with MT in patients with a pronounced depression of the pineal hormone in their circulation may perhaps be able to favorably influence the course of this disease. According to experimental studies on animals, the primary growth of aggressive lung cancers does not appear to be controlled by MT effectively (Lapin and Ebels 1974) since fast-growing and undifferentiated tumors usually respond poorly to MT (Bartsch H et al. 1986); although according to a recent study the formation of metastases is inhibited (Mocchegiani et al. 1999). The clinical studies of Lissoni and co-workers, however, indicate that MT in conjunction with other therapies may help to improve therapeutic results (Lissoni et al. 1998) as well as to diminish the myelotoxic side effects of chemotherapy (Lissoni et al. 1997a); although the latter was not confirmed by a subsequent double-blind study (Ghielmini et al. 1999).

If, however, MT will prove to be a physiologically relevant endogenous free radical scavenger molecule as is discussed in numerous experimental studies (Reiter et al. 1999), it could well be that the circulating levels of the pineal hormone would reduce the susceptibility of bronchial tissue towards the mutagenic action of highly reactive metabolites of tobacco constituents due to oxidative damage (Park et al. 1998), partially because specific MT receptors were identified in lung tissue (Paul et al. 1999). Changes of MT levels in blood with aging or in the course of other diseases may thus elevate the risk to develop lung cancer which could be reduced by the administration of MT.

Stomach Cancer. In the current study, a statistically significantly depressed nocturnal excretion of aMT6s was found in men with primary unoperated stomach cancer (-59%, $P = 0.0023$) compared with tumor-free controls of comparable age. Colombo et al. (1991), exclusively analyzing morning MT in blood, reported depressed levels in patients with unoperated stomach cancer which, however, due to the circadian rhythm of MT secreted by the pineal gland render only limited information. In our previous study of stomach cancer patients, the day/night variation of urinary MT was monitored and the excretion of MT during the day was found to be higher than at night, whereas the total 24-h excretion remained unchanged compared with healthy controls as well as patients with benign tumors of the stomach (adenomatous polyps) who showed the usual nocturnal surge of MT (Kvetnoy and Levin 1987). This observation indicates the presence of a phase shift of the urinary excretion of MT in stomach cancer patients. A similar observation was made for the urinary excretion of MT in postmenopausal Indian breast cancer patients (Bartsch C et al. 1981), which was interpreted to facilitate the progression of the disease due to the circadian-stage-dependent effect of MT on tumor growth characterized by an inhibition during the early night and a stimulation or no effect during the day (Bartsch H and Bartsch C 1981). In further studies, parallel determinations of MT in blood and of aMT6s in urine throughout day and night should clarify whether such acrophase changes can be confirmed or not. Such studies would also help to find answers to the central question of whether the secretion of MT by the pineal or the hepatic metabolism of MT to aMT6s is inhibited in stomach cancer patients. An additional possibility to explain the observed depletion of nocturnal aMT6s in stomach cancer would be to consider an involvement of the degradation of MT via the enzyme IDO leading to the corresponding kynurenine derivatives, which has been reported to occur in normal stomach tissue (Hayaishi 1976).

In further clinical studies, it should be analyzed whether the levels of circulating MT are affected by the presence of precancerous conditions of this type of malignancy, e. g., chronic gastritis, gastric adenoma, chronic peptic ulcer as well as pernicious anemia (Bajtai and Hidvegi 1998), since MT, under experimental conditions, possesses a protective action towards both stress- (Khan et al. 1990; Kato et al. 1998) and indomethacin-induced gastric injuries (Alarcon de la Lastra 1999). According to Konturel et al. (1997), the gastroprotective action of MT against stress- and ischemia-induced lesions is mediated via the scavenging of oxygen radicals. Hirokawa et al. (1998) state that such highly reactive compounds are apparently also generated during the metabolization of ethanol in gastric mucosal cells. This would mean that MT could lower the documented risk for the development of stomach cancer due to alcohol consumption (de Stefani et al. 1998) via its free radical scavenging property. In the same way, other risks for stomach cancer such as smoking (Chow et al. 1999) could be reduced by adequate levels of circulating MT. This view is supported by the observation of Lewinski et al. (1991) concerning a certain anti-mitotic effect of MT on epithelial cells of the gastric mucosa.

Data concerning the effect of MT administration to patients with gastric cancer are sparse: Lissoni et al. (1993, 1993a, 1994) applied in their preliminary studies low doses of IL-2 in combination with MT to 16 advanced stomach cancer patients and concluded that this type of combination therapy may be promising and that such investigations should therefore be extended.

Colorectal Cancer Patients. Among the male patients studied with primary unoperated colorectal adenocarcinomas, the nocturnal urinary aMT6s excretion showed a trend towards elevation compared with tumor-free controls of comparable age. The enhanced formation of aMT6s, however, does not appear to be a cancer-specific phenomenon since in patients affected by chronic colitis ulcerosa similar changes were observed. In our previous study, the daily urinary MT excretion was found to be elevated to nocturnal levels (Kvetnoi and Levin 1987) indicating a potential acrophase shift of the circadian pattern of pineal MT secretion. Lissoni et al. (1988) reported an increase of morning MT in blood in those colorectal cancer patients who responded favorably to chemotherapy. Khoory and Stemme (1988), analyzing the circadian variations of MT in plasma of primary unoperated colorectal carcinoma patients with and without distant metastases, observed a pronounced depletion of nocturnal MT. It is currently difficult to explain the apparent discrepancies between our findings and those of Khoory and Stemme (1988). It may be conceivable that the peripheral metabolism of MT in the liver towards aMT6s is enhanced in colorectal cancer patients. This assumption, however, is not supported by clinical studies on breast (Bartsch C et al. 1991) and prostate cancer patients (Bartsch C et al. 1992). In order to test this hypothesis, parallel determinations of MT in blood and of aMT6s in blood or urine should be performed over a complete 24-h cycle in patients with malignant and benign colorectal diseases. Such a study will also help to test whether acrophase changes of circulating MT occur and would allow first insights into other potential ways of how MT is cleared from the circulation or is secreted into the same. A reduction of circulating MT could be elicited via IDO, an enzyme degrading MT to he respective kynurenine derivative (Hayaishi 1976), as well as by a modification of the observed pronounced retention of MT in gut tissue (Messner et al. 1998). Possible elevations of circulating MT may not only be affected by an enhanced secretion by the pineal gland in response to certain benign or malignant pathologies of the colon and rectum, but also by a potential release from these tissues where a considerable amount of extra-pineal MT biosynthesis occurs (Huether 1993). MT seems to serve in gastrointestinal physiology and homeostasis by being involved in intestinal motility (Harlow and Weekley 1986) and in the feedback system with serotonin which regulates appetite and digestive processes both in the gastrointestinal tract and the brain (Bubenik and Pang 1994). MT also controls the mitotic activity within colonic mucosa epithelial cells (Lewinski et al. 1991), and therefore it is not surprising that exogenous MT is able to inhibit both in vivo (Anisimov et al. 1997) and in vitro experimental colon cancer (Melen-Mucha et al. 1998). Preliminary evidence exists that human colorectal carcinomas can be inhibited by the administration of MT in combination with low doses of IL-2 (Barni et al. 1992, 1995; Lissoni et al. 1993, 1993a, 1997). MT may not only be useful in the treatment of established colorectal cancers but may also possess a preventive action against pre-neoplastic conditions such as severe colitis (Bubenik et al. 1998), which is supported by studies on experimental dextran-induced colitis (Pentney and Bubenik 1995). In this connection, details of the interplay between pineal and gastrointestinal MT require to be clarified as does the question of which MT may primarily contribute to the protection of the colon and rectum towards benign and malignant lesions. The enhanced production of aMT6s observed in the present study in patients with colitis ulcerosa and with colon cancer can be viewed as an effort of the pineal gland to inhibit these proliferative processes within the gastrointestinal tract, an effect facilitated via the presence of specific MT receptors within the human gastrointestinal tract (Poon et al. 1996).

Patients with Other Types of Malignancies. In our own study, no obvious change in the urinary excretion of aMT6s was found in patients with primary unoperated *cancer of the urinary bladder* and no other reports have, to our knowledge, been published on this type of cancer. Panzer and Viljoen (1998) reported divergent changes in the 24-h urinary excretion of aMT6s among 10 patients with *osteosarcoma*: in eight patients lower values were found than in the controls, whereas in the remaining two patients, aMT6s was dramatically higher for no obvious reason. In some of his female patients with *Hodgkin's sarcoma*, Lissoni et al. (1986) reported clearly elevated levels of nocturnal MT. Pfletschinger (1994) studied daytime values of plasma MT (8–10 A.M.) in children with neuroblastoma or with acute *lymphoblastic leukemia (ALL)* and detected a statistically significant depletion in the presence of neuroblastomas compared with children with ALL. These results were, however, not compared with data from healthy age-matched controls of same sex and it is questionable in which way these morning values correspond to the nocturnal peak concentrations. Tarquini et al. (1995) reported that morning MT in plasma (7.30–9.30 A.M.) of patients with *multiple myeloma* was significantly elevated compared with controls. More extensive analytical clinical investigations of MT in patients with hematopoietic neoplasias are urgently required to better understand the role of the pineal gland in these types of malignancies where circulating MT may show changes quite different from those observed in patients with solid tumors. This assumption is derived from findings in experimental therapeutic studies on the effect of MT in animals with leukemias where the pineal hormone stimulated the malignant process whereas pinealectomy was inhibitory (Conti et al. 1992).

7.4.2 Correlation Between MT Production and Proliferative Activity in Tumor Cells

The 26 male patients with colorectal, stomach, or lung cancer showed strong positive correlations between the expression of PCNA in tumors and the nocturnal urinary aMT6s excretion, in comparison with the combined group of all cancer patients as well as the subgroups with different types of cancers (Bartsch C et al. 1997a). Similar strong positive correlations were observed in patients with ovarian cancer between nocturnal urinary aMT6s excretion and the expression of PCNA in the corresponding tumor tissue (I. M. Kvetnoy and C. Bartsch, unpublished results). The mechanisms involved in the interaction between proliferation within tumor tissue and pineal secretory activity are not understood, but the observed correlations may point to a feedback between tumor cells and the pineal gland, which exerts an inhibitory effect on the proliferation of both normal (Bindoni 1971) and malignant cells (Barone and Das Gupta 1970). A central question is in which way tumor cells may signal information about their proliferative state to the pineal gland. It is conceivable that oncogeneticly encoded growth factors are secreted by tumor cells and reach the pineal gland via the circulation and thus affect pineal MT biosynthesis. Another possibility is that growth factors secreted by tumor cells affect parts of the sympathetic nervous system which are involved in the neural stimulation of pineal MT biosynthesis, i.e., the postganglionic sympathetic noradrenergic nerves innervating the pineal and regulating the activity of the rate-limiting enzyme, serotonin-N-acetyltransferase.

The current findings of a close correlation between PCNA and nocturnal urinary aMT6s supports the idea that the proliferative activity of tumor cells could be monitored in a noninvasive fashion for both diagnostic as well as prospective purposes (since the proliferative activity of tumor cells plays a key role for invasiveness and formation of metastases). PCNA is regarded by some authors as one of the most important proliferation marker (Yu et al. 1992). PCNA is a non-histone protein with a molecular weight of 36 kDa which functions as an auxiliary protein of DNA-polymerase delta (Prelich et al. 1987) and its expression increases drastically during the S phase of the cell cycle (Zuber et al. 1989). With respect to the prognostic relevance of PCNA determinations in malignant tissue, it appears that an unfavorable prognosis is connected with high PCNA expression as was reported for patients with carcinoma of the thyroid, stomach (Maeda et al. 1994), or breast (Cummings et al. 1993), with non-Hodgkin's lymphoma (Klemi et al. 1992), with soft-tissue sarcomas, and with malignant fibrous histiocytomas (Choong et al. 1995).

7.4.3 Correlation Among Tumoral Parameters

The negative correlation observed between PCNA- and MT-immunopositive tumor cells indicates that the binding or uptake of circulating MT by tumor cells and/or the biosynthesis of MT within tumor cells declines if proliferation increases. This confirms previous findings in some types of tumors with a low degree of differentiation (Raikhlin and Kvetnoy 1994). Poon et al. (1996) detected specific MT receptors within the gastrointestinal tract and MT is synthesized by extra-pineal tissues (Huether 1993). The loss of immunocytochemically detectable MT in fast-proliferating tumors can be viewed as a sign of diminished cellular differentiation, which according to a number of studies is connected with a lowered sensitivity towards the inhibitory action of MT (Bartsch H et al. 1986; Bartsch C and Bartsch H 1997). A central question is whether MT-immunopositive cells as detected in different types of carcinomas in humans (Raikhlin and Kvetnoy 1994; Bartsch C et al. 1997a) result from local biosynthesis or whether MT secreted by the pineal gland is bound and/or taken up by these malignant tissues. Subsequent studies should therefore test whether MT-immunopositive cells are identical with MT-receptor positive cells or not, and whether MT is indeed produced within tumors. Irrespective of the exact origin of intratumoral MT, the localization of this hormone within malignant tissue can be regarded to be of considerable importance for growth and invasiveness due to its effects on cellular proliferation and differentiation, modulation of immune reactions (Maestroni et al. 1994) and of radiosensitivity (Kvetnoy et al. 1994). The presence of MT within a tumor is likely to be of diagnostic and prognostic importance as has been demonstrated for other biologically active substances including hormones, neuropeptides and biogenic amines. If MT is indeed synthesized within carcinomatous tumor tissue, it would render further support to the concept that neuroendocrine substances are not produced exclusively in classical neuroendocrine tumors but also in different types of carcinomas (Bonkhoff et al. 1995; Bosman 1984; Maluf and Koerner 1994). We assume that about 25–30% of all epithelial malignant tumors are able to produce biogenic amines or peptide hormones (Raikhlin and Kvetnoy 1994) which could lead not only to a re-consideration of the concept on how neoplasias develop, but would also have practical applications for both diagnostic and prognostic purposes.

7.5 Conclusion

The current analytical results in patients with cancers outside of the reproductive system render further support to the concept of a close link between MT and cancer (Blask 1993), as evidenced by previous experimental studies which showed that pinealectomy stimulates experimental tumor growth whereas aqueous pineal extracts and MT are inhibitory (Blask 1984; Bartsch H et al. 1987; Blask et al. 1991).

Patients with different malignancies outside of the reproductive system show varying changes of the urinary excretion of aMT6s, which depend on both the histological type of the tumor and the stage of the disease. The mechanisms involved in these phenomena are poorly understood and seem to include central as well as peripheral components which may be superimposed. This view is supported by the observation that in spite of a varying nocturnal urinary excretion of aMT6s in patients with different types of tumors, e.g., stomach, lung, and colorectal cancer, aMT6s shows comparable positive correlations with the degree of tumor cell proliferation as estimated by the number of PCNA-immunopositive cells. This means that the amount of aMT6s excretion (as well as the corresponding concentration of circulating MT) can be regarded as the net result of a number of different effects exerted by the tumor on the system, e.g., the secretion of pineal MT, its peripheral metabolism as well as the production and release by tumor cells. It is felt that a deeper understanding of these processes will help to better understand the complex systemic effects of different types and stages of tumor growth upon the neuroimmunoendocrine network, and might in future even be applied for both diagnostic as well as prognostic purposes.

Acknowledgements. We are grateful to our colleagues Drs. N. Raikhlin, V. Medvedev, O. Karjakin, P. Garbuzov, A. Brodsky, Yu. Romanko, M. Davydov, I. Stilidi, Y. Timofeev, A. Perevoshikov, D. Chuvashkin, I. Kozlova and V. Minaev for the assistance in collecting urine samples and tumor specimens. This work was supported by grants of Deutsche Forschungsgemeinschaft (436 RUS 17/129/1995) as well as of the Russian Foundation of Fundamental Research (RFFR: 96–04–49273 and 98–04–48763).

References

Aasebo U, Bremnes RM, Aakvaag A, Slordal L (1991) Gonadal endocrine dysfunction in patients with lung cancer: relation to responsiveness to chemotherapy, respiratory function and performance status. J Steroid Biochem Mol Biol 39:375–80

Anisimov VN, Popovich IG, Zabezhinski MA (1997) Melatonin and colon carcinogenesis: I. Inhibitory effect of melatonin on development of intestinal tumours induced by 1,2-dimethylhydrazine in rats. Carcinogenesis 18:1549–1453

Alarcon de la Lastra C, Motilva V, Martin MJ, Nieto A, Barranco MD, Cabeza J, Herrerias JM (1999) Protective effect of melatonin on indomethacin-induced gastric injury in rats. J Pineal Res 26:101–107

Arendt J, Bojkowski C, Franey C, Wright J, Marks V (1985) Immunoassay of 6-hydroxymelatonin sulfate in human plasma and urine: abolition of the urinary 24-hour rhythm with atenolol. J Clin Endocrinol Metab 60:1166–1173

Bajtai A, Hidvegi J (1998) The role of gastric mucosal dysplasia in the development of gastric carcinoma. Pathol Oncol Res 4:297–300

Banerjee S, Margulis L (1973) Mitotic arrest by melatonin. Exp. Cell Res 78:314–318

Barni S, Lissoni P, Cazzaniga M, Ardizzoia A, Paolorossi F, Brivio F, Perego M, Tancini G, Conti A, Maestroni G (1992) Neuroimmunotherapy with subcutaneous low-dose interleukin-2 and the pineal hormone melatonin as a second-line treatment in metastatic colorectal carcinoma. Tumori 78:383–387

Barni S, Lissoni P, Cazzaniga M, Ardizzoia A, Meregalli S, Fossati V, Fumagalli L, Brivio F, Tancini G (1995) A randomized study of low-dose subcutaneous interleukin-2 plus melatonin versus supportive care alone in metastatic colorectal cancer patients progressing under 5-fluorouracil and folates. Oncology 52:243–245

Barone RM, Das Gupta TK (1970) Role of pinealectomy on Walker 256 carcinoma in rats. J Surg Oncol 2:313–322

Bartsch C, Bartsch H (1994) Melatonin secretion in oncological patients: current results and methodological considerations. In: Maestroni GJM, Conti A, Reiter RJ (eds) Advances in pineal research: 7. John Libbey and Comp., London, pp 283–301

Bartsch C, Bartsch H (1997) Significance of melatonin in malignant diseases (in German). Wien Klin Wochenschr 109:722–729

Bartsch C, Bartsch H, Jain AK, Laumas KR, Wetterberg L (1981) Urinary melatonin levels in human breast cancer patients. J Neural Transm 52:281–294

Bartsch C, Bartsch H, Fuchs U, Lippert TH, Bellmann O, Gupta D (1989) Stage-dependent depression of melatonin in patients with primary breast cancer: correlation with prolactin, thyroid stimulating hormone and steroid receptors. Cancer 64:426–433

Bartsch C, Bartsch H, Bellmann O, Lippert TH (1991) Depression of serum melatonin in patients with primary breast cancer is not due to an increased peripheral metabolism. Cancer 67:1681–1684

Bartsch C, Bartsch H, Schmidt A, Ilg S, Bichler KH, Flüchter SH (1992) Melatonin and 6-sulfatoxymelatonin circadian rhythms in serum and urine of primary prostate cancer patients: evidence for reduced pineal activity and relevance of urinary determinations. Clin Chim Acta 209:153–167

Bartsch C, Bartsch H, Buchberger A, Rokos H, Mecke D, Lippert TH (1995) Serial transplants of DMBA-induced mammary tumours in Fischer rats as model system for human breast cancer. IV. Parallel changes of biopterin and melatonin indicate interactions between the pineal gland and cellular immunity in malignancy. Oncology 52:278–283

Bartsch C, Bartsch H, Karenovics A, Franz H, Peiker G, Mecke D (1997) Nocturnal urinary 6-sulphatoxymelatonin excretion is decreased in primary breast cancer patients compared to age-matched controls and shows negative correlation with tumour-size. J Pineal Res 23:53–58

Bartsch C, Kvetnoy IM, Kvetnaia TV, Bartsch, H, Molotkov AO, Franz H, Raikhlin NT, Mecke D (1997a) Nocturnal urinary 6-sulphatoxymelatonin and proliferating cell nuclear antigen-immunopositive tumour cells show strong positive correlations in patients with gastrointestinal and lung cancer. J Pineal Res 23:90–96

Bartsch C, Bartsch H, Bichler K-H, Flüchter S-H (1998) Prostate cancer and tumour stage-dependent circadian neuroendocrine disturbances. The Aging Male 1:188–199

Bartsch C, Bartsch H, Buchberger A, Stieglitz A, Effenberger-Klein A, Kruse-Jarres JD, Besenthal I, Rokos H, Mecke D (1999) Serial transplants of DMBA-induced mammary tumours in Fischer rats as a model system for human breast cancer. VI. The role of different forms of tumour-associated stress for the regulation of pineal melatonin secretion. Oncology 56:169–176

Bartsch H, Bartsch C (1981) Effect of melatonin on experimental tumours under different photoperiods and times of administration. J Neural Transm 52:269–279

Bartsch H, Bartsch C, Flehmig B. (1986) Differential effect of melatonin on slow and fast growing passages of a human melanoma cell line. Neuroendocrinol Lett 8:289–293

Bartsch H, Bartsch C, Noteborn HPJM, Flehmig B, Ebels I, Salemink CA (1987) Growth-inhibiting effect of crude pineal extracts on human melanoma cells in vitro is different from that of known synthetic pineal substances. J Neural Transm 69:299–311

Beckford NS, Rood SR, Schaid D, Schanbacher B (1985) Androgen stimulation and laryngeal development. Ann Otol Rhinol Laryngol 94(6 Pt 1):634–640

Bindoni M (1971) Relationships between the pineal gland and the mitotic activity of some tissues. Arch Sci Biol (Bologna) 55:3–21

Blackman MR, Weintraub BD, Rosen SW, Harman SM (1988) Comparison of the effects of lung cancer, benign lung disease, and normal aging on pituitary-gonadal function in men. J Clin Endocrinol Metab 66:88–95

Blask DE (1984) The pineal: an oncostatic gland? In: Reiter RJ (ed) The pineal gland. Raven Press, New York, pp 253–284

Blask DE (1993) Melatonin in oncology. In: Yu HS, Reiter RJ (eds) Melatonin biosynthesis, physiological effects, and clinical applications. CRC Press, Boca Raton, pp 447–475

Blask DE Cos S, Hill SM, Burns DM (1991) Melatonin action on oncogenesis. In: Fraschini F, Reiter RJ (eds) Role of melatonin and pineal peptides in neuroimmunomodulation. Plenum Press, New York, pp 233–240

Blask DE, Wilson ST, Lemus-Wilson AM (1994) The oncostatic and oncomodulatory role of the pineal gland and melatonin. In: Maestroni GJM, Conti A, Reiter RJ (eds) Advances in pineal research: 7. John Libbey and Comp., London, pp 235–241

Bonkhoff H, Stein U, Remberger K (1995) Endocrine-paracrine cell types in the prostate and prostatic adenocarcinoma are postmitotic cells. Hum Pathol 26:167–170

Bosman F (1984) Neuroendocrine cells in non-endocrine tumours. In: Falkmer S, Hakanson R, Sundler F (eds) Evolution and tumour pathology of the neuroendocrine system. Elsevier Science Publishers B. V., Amsterdam, pp 519–543

Bubenik GA, Brown GM, Grota LJ (1977) Immunohistological localization of melatonin in the rat digestive system. Experientia 33:662–663

Bubenik GA, Purtill RA, Brown GM, Grota LJ (1978) Melatonin in the retina and the Harderian gland. Ontogeny, diurnal variations and melatonin treatment. Exp Eye Res 27:323–333

Bubenik GA, Pang SF (1994) The role of serotonin and melatonin in gastrointestinal physiology: ontogeny, regulation of food intake, and mutual serotonin-melatonin feedback. J Pineal Res 16:91–99

Bubenik GA, Blask DE, Brown GM, Maestroni GJ, Pang SF, Reiter RJ, Viswanathan M, Zisapel N (1998) Prospects of the clinical utilization of melatonin. Biol Signals Recept 7:195–219

Choong PF, Akerman M, Willem H, Anderson C, Gustafson P, Alvegard T, Rydholm A (1995) Expression of proliferating cell nuclear antigen (PCNA) and Ki-67 in soft tissue sarcoma: is prognostic significance histotype-specific? APMIS 103:797–805

Cardinali DP, Vacas MI, Gejman PV (1981) The sympathetic superior cervical ganglia as peripheral neuroendocrine centers. J Neural Transm 52:1–21

Chan JM, Stampfer MJ, Giovannucci EL (1998) What causes prostate cancer? A brief summary of the epidemiology. Semin Cancer Biol 8:263–273

Chow WH, Swanson CA, Lissowska J, Groves FD, Sobin LH, Nasierowska-Guttmejer A, Radziszewski J, Regula J, Hsing AW, Jagannatha S, Zatonski W, Blot WJ (1999) Risk of stomach cancer in relation to consumption of cigarettes, alcohol, tea and coffee in Warsaw, Poland. Int J Cancer 81:871–876

Colombo F, Iannotta F, Fachinetti A, Giuliani F, Cornaggia M, Finzi G, Mantero G, Fraschini F, Malesci A, Bersani M, Pigni E, Bonato C, de Martini G, Calati AM, Scotti S (1991) Changes in hormonal and biochemical parameters in gastric adenocarcinoma (in Italian). Minerva Endocrinol 16:127–139

Conti A, Haran-Ghera N, Maestroni GJ (1992) Role of pineal melatonin and melatonin-induced-immuno-opioids in murine leukemogenesis. Med Oncol Tumour Pharmacother 9:87–92

Cos S, Fernandez F, Sanchez-Barcelo EJ (1996) Melatonin inhibits DNA synthesis in MCF-7 human breast cancer cells in vitro. Life Sci 58:2447–2453

Cozzolino F, Torcia M, Castigli E, Selli C, Giordani R, Carossino AM, Squadrelli M, Cagnoni M, Pistoia V, Ferrarini M (1986) Presence of activated T-cells with a T8+ M1+ Leu 7+ surface phenotype in invaded lymph nodes from patients with solid tumours. J Natl Cancer Inst 77:637–641

Cummings MC, Furnival CM, Parsons PG, Townsend E (1993) PCNA immunostaining in breast cancer. Aust N Z J Surg 63:630–636

Dogliotti L, Berruti A, Buniva T, Torta M, Bottini A, Tampellini M, Terzolo M, Faggiuolo R, Angeli A (1990) Melatonin and human cancer. J Steroid Biochem Mol Biol 37:983–987

Fujiwara M, Shibata M, Watanabe Y, Nukiwa T, Hirata F, Mizuno N, Hayaishi O (1978) Indoleamine 2,3-dioxygenase. Formation of L-kynurenine from L-tryptophan in cultured rabbit pineal gland. J Biol Chem 253:6081–6085

Ghielmini M, Pagani O, de Jong J, Pampallona S, Conti A, Maestroni G, Sessa C, Cavalli F (1999) Double-blind randomized study on the myeloprotective effect of melatonin in combination with carboplatin and etoposide in advanced lung cancer. Br J Cancer 80:1058–1061

Harlow HJ, Weekley BL (1986) Effect of melatonin on the force of spontaneous contractions of in vitro rat small and large intestine. J Pineal Res 3:277–284

Hayaishi O (1976) Properties and function of indoleamine 2,3-dioxygenase. J Biochem (Tokyo) 79:13P–21P

Hermanek P, Sobin LH (1992) TNM: Classification of malignant tumours. Springer-Verlag, Berlin

Hickman D, Pope J, Patil SD, Fakis G, Smelt V, Stanley LA, Payton M, Unadkat JD, Sim E (1998) Expression of arylamine N-acetyltransferase in human intestine. Gut 42:402–409

Hill SM, Spriggs LL, Simon MA, Muraoka H, Blask DE (1992) The growth inhibitory action of melatonin on human breast cancer cells is linked to the estrogen response system. Cancer Lett 64:249–256

Hirata F, Hayaishi O, Tokuyama T, Seno S (1974) In vitro and in vivo formation of two new metabolites of melatonin. J Biol Chem 249:1311–1313

Hirokawa M, Miura S, Yoshida H, Kurose I, Shigematsu T, Hokari R, Higuchi H, Watanabe N, Yokoyama Y, Kimura H, Kato S, Ishii H (1998) Oxidative stress and mitochondrial damage precedes gastric mucosal cell death induced by ethanol administration. Alcohol Clin Exp Res 22(3 Suppl):111S–114 S

Huether G (1993) The contribution of extrapineal sites of melatonin synthesis to circulating melatonin levels in higher vertebrates. Experientia 49:665–670

Kambic V, Radsel Z, Prezelj J, Zargi M (1984) The role of testosterone in laryngeal carcinogenesis. Am J Otolaryngol 5:344–349

Kato K, Murai I, Asai S, Matsuno Y, Komuro S, Kaneda N, Iwasaki A, Ishikawa K, Nakagawa S, Arakawa Y, Kuwayama H (1998) Protective role of melatonin and the pineal gland in modulating water immersion restraint stress ulcer in rats. J Clin Gastroenterol 27 Suppl 1:S110–115

Khan R, Burton S, Morley S, Daya S, Potgieter B (1990) The effect of melatonin on the formation of gastric stress lesions in rats. Experientia 46:88–89

Khoory R, Stemme D (1988) Plasma melatonin levels in patients suffering from colorectal carcinoma. J Pineal Res 5:251–258

Kleemann D, Kunkel S (1996) Serum dihydrotestosterone versus total testosterone values of patients with laryngeal carcinomas and chronic laryngitis (in German). Laryngorhinootologie 75:351–355

Klemi PJ, Alanen K, Jalkanen S, Joensuu H (1992) Proliferating cell nuclear antigen (PCNA) as a prognostic factor in non-Hodgkin's lymphoma. Brit J Cancer 66:739–743

Kopin IJ, Pare CMB, Axelrod J, Weissbach H (1961) The fate of melatonin in animals. J Biol Chem 236:3072–3075

Kuzduk K, Komorowski J (1995) Nocturnal rhythm of melatonin secretion in strumectomized patients with normal thyrotropin blood levels and recurrent non-toxic benign thyroid nodules. Neuroendocrinol Lett 17:237–243

Kvetnoy IM (1999) Extrapineal melatonin: location and role within diffuse neuroendocrine system. Histochem J 31:1–12

Kvetnoi IM, Levin IM (1987) Diurnal excretion of melatonin in cancer of the stomach and large intestine. Vopr Oncol 33:29–32

Kvetnoy IM, Yuzhakov VV (1993) Extrapineal melatonin: advances in microscopical identification of hormones in endocrine and non-endocrine cells. Microscopy and Analysis 21:27–29

Kvetnoy IM, Yuzhakov VV (1994) Extrapineal melatonin: non-traditional localization and possible significance for oncology. In: Maestroni GJM, Conti A, Reiter RJ (eds) Advances in pineal research: 7. John Libbey and Comp., London, pp 197–210

Kvetnoy IM, Kvetnaia TV, Konopljannikov AG, Tsyb AF, Yuzhakov VV (1994) Melatonin and tumour growth: from experiments to clinical application. In: Kvetnoy IM, Reiter RJ (eds) Melatonin: general biological and oncoradiological aspects. MRRC Press, Obninsk, pp 17–23

Kvetnoy IM, Sandvik AK, Waldum HL (1997) The diffuse neuroendocrine system and extrapineal melatonin. J Mol. Endocrinol 18:1–3

Lapin V, Ebels I (1976) Effects of some low molecular weight sheep pineal fractions and melatonin on different tumours in rats and mice. Oncology 33:110–113

Lerner AB, Case JD, Takahashi Y, Lee TH, Mori W (1958) Isolation of melatonin, the pineal gland factor that lightens melanocytes. J Am Chem Soc 80:2587

Lewinski A, Sewerynek E (1986) Melatonin inhibits the basal and TSH-stimulated mitotic activity of thyroid follicular cells in vivo and in organ culture. J Pineal Res 3:291–299

Lewinski A, Rybicka I, Wajs E, Szkudlinski M, Pawlikowski M (1991) Influence of pineal indoleamines on the mitotic activity of gastric and colonic mucosa epithelial cells in the rat: interaction with omeprazole. J Pineal Res 10:104–108

Lewinski A, Wajs E, Karbownik M (1992) Effects of pineal-derived indolic compounds and of certain neuropeptides on the growth processes in the thyroid gland. Thyroidology 4:11–15

Lewinski A, Pawlikowski M, Cardinali DP (1993) Thyroid growth-stimulating and growth-inhibiting factors. Biol Signals 2:313–351

Lissoni P, Viviani S, Bajetta E, Buzzoni R, Barreca A, Mauri R, Resentini M, Morabito F, Esposti D, Esposti G, Fraschini F (1986) A clinical study of the pineal gland activity in oncologic patients. Cancer 57:837–842

Lissoni P, Tancini G, Barni S, Crispino S, Paolorossi F, Rovelli F, Cattaneo G, Fraschini F (1988) Melatonin increase as predictor for tumour objective response to chemotherapy in advanced cancer patients. Tumori 74:339–345

Lissoni P, Barni S, Tancini G, Ardizzoia A, Rovelli F, Cazzaniga M, Brivio F, Piperno A, Aldeghi R, Fossati D, Characiejus D, Kothari L, Conti A, Maestroni GJM (1993) Immunotherapy with subcutaneous low-dose interleukin-2 and the pineal indole melatonin as a new effective therapy in advanced cancers of the digestive tract. Br J Cancer 67:1404–1407

Lissoni P, Brivio F, Ardizzoia A, Tancini G, Barni S (1993a) Subcutaneous therapy with low-dose interleukin-2 plus the neurohormone melatonin in metastatic gastric cancer patients with low performance status. Tumori 79:401–404

Lissoni P, Barni S, Cazzaniga M, Ardizzoia A, Rovelli F, Brivio F, Tancini G (1994) Efficacy of the concomitant administration of the pineal hormone melatonin in cancer immunotherapy with low-dose IL-2 in patients with advanced solid tumours who had progressed on IL-2 alone. Oncology 51:344–347

Lissoni P, Fumagalli L, Paolorossi F, Rovelli F, Roselli MG, Maestroni GJ (1997) Anticancer neuroimmunomodulation by pineal hormones other than melatonin: preliminary phase II study of the pineal indole 5-methoxytryptophol in association with low-dose IL-2 and melatonin. J Biol Regul Homeost Agents 11:119–122

Lissoni P, Paolorossi F, Ardizzoia A, Barni S, Chilelli M, Mancuso M, Tancini G, Conti A, Maestroni GJ (1997a) A randomized study of chemotherapy with cisplatin plus etoposide versus chemo-endocrine therapy with cisplatin, etoposide and the pineal hormone melatonin as a first-line treatment of advanced non-small-cell lung cancer patients in a poor clinical state. J Pineal Res 23:15–19

Lissoni P, Giani L, Zerbini S, Trabattoni P, Rovelli F (1998) Biotherapy with the pineal immunomodulating hormone melatonin versus melatonin plus aloe vera in untreatable advanced solid neoplasms. Nat Immun 16:27–33

Luboshitzky R, Dharan M, Goldman D, Hiss Y, Herer P, Lavie P (1997) Immunohistochemical localization of gonadotropin and gonadal steroid receptors in human pineal glands. J Clin Endocrinol Metab 82:977–981

Maeda K, Chung YS, Onoda N, Kato Y, Nitta A, Arimoto Y, Yamada N, Kondo Y, Sowa M (1994) Proliferating cell nuclear antigen labeling index of preoperative biopsy specimens in gastric carcinoma with special reference to prognosis. Cancer 73:528–533

Maestroni GJM, Conti A, Covacci V (1994) Melatonin-induced immuno-opioids: fundamentals and clinical perspectives. In: Maestroni GJM, Conti A, Reiter RJ (eds) Advances in Pineal Research: 7. John Libbey, London, pp 73–81

Maluf HM, Koerner FC (1994) Carcinomas of the breast with endocrine differentiation: a review. Virch Arch 425:449–457

Marsigliante S, Leo G, Resta L, Storelli C (1992) Steroid receptor status in malignant and non-malignant larynx. Ann Oncol 3:387–392

Melen-Mucha G, Winczyk K, Pawlikowski M (1998) Somatostatin analogue octreotide and melatonin inhibit bromodeoxyuridine incorporation into cell nuclei and enhance apoptosis in the transplantable murine colon 38 cancer. Anticancer Res 18:3615–3619

Messner M, Hardeland R, Rodenbeck A, Huether G (1998) Tissue retention and subcellular distribution of continuously infused melatonin in rats under near physiological conditions. J Pineal Res 25:251–259

Mocchegiani E, Perissin L, Santarelli L, Tibaldi A, Zorzet S, Rapozzi V, Giacconi R, Bulian D, Giraldi T (1999) Melatonin administration in tumour-bearing mice (intact and pinealectomized) in relation to stress, zinc, thymulin and IL-2. Int J Immunopharmacol 21:27–46

Moses AM, Scheinman SJ (1991) Ectopic secretion of neurohypophyseal peptides in patients with malignancy. Endocrinol Metab Clin North Am 20:489–506

Ochedalski T, Lachowicz-Ochedalska A, Dec W, Czechowski B (1994) Examining the effects of tobacco smoking on levels of certain hormones in serum of young men (in Polish). Ginekol Pol 65:87–93

Pang SF, Li L, Ayre EA, Pang CS, Lee PP, Xu RK, Chow PH, Yu ZH, Shiu SY (1998) Neuroendocrinology of melatonin in reproduction: recent developments. J Chem Neuroanat 14:157–166

Panzer A, Viljoen M (1998) Urinary 6-sulfatoxymelatonin levels in osteosarcoma: a case-control study. Med Sci Res 26:43–45

Park EM, Park YM, Gwak YS (1998) Oxidative damage in tissues of rats exposed to cigarette smoke. Free Radic Biol Med 25:79–86

Paul P, Lahaye C, Delagrange P, Nicolas JP, Canet E, Boutin JA (1999) Characterization of 2-[125I]iodomelatonin binding sites in Syrian hamster peripheral organs. J Pharmacol Exp Ther 290:334–340

Pentney PT, Bubenik GA (1995) Melatonin reduces the severity of dextran-induced colitis in mice. J Pineal Res 19:31–39

Pfletschinger U (1994) Melatonin and neuropeptides in children with acute lymphoblastic leukemia and neuroblastoma (in German). Medical dissertation at the University of Tübingen

Phillips DH (1996) DNA adducts in human tissues: biomarkers of exposure to carcinogens in tobacco smoke. Environ Health Perspect 104(Suppl 3):453–458

Pico JL, Mathe G, Young IM, Leone RM, Hooper J, Silman RE (1979) Role of hormones in the aetiology of human cancer. Pineal indole hormones and cancer. Cancer Treatment Reports 63:1204 (Abstract).

Poon AM, Mak AS, Luk HAT (1996) Melatonin and 2[125I]iodomelatonin binding sites in the human colon. Endocr Res 22:77–94

Prelich G, Tan CK, Kostura M, Mathews MB, So AG, Downey KM, Stillman B (1987) Functional identity of proliferating cell nuclear antigen and a DNA polymerase-delta auxiliary protein. Nature 326:517–520

Quay WB, Ma YH (1976) Demonstration of gastrointestinal hydroxyindole-O-methyltransferase. IRCS Med Sci 4:563

Raikhlin NT, Kvetnoy IM (1994) The APUD system (diffuse endocrine system) in normal and pathological states. Physiol Gen Biol Rev 8, part 4:1–44

Reiter RJ (1991) Melatonin: that ubiquitously acting pineal hormone. News Physiol Sci 6:223–228

Reiter RJ, Tan DX, Cabrera J, D'Arpa D, Sainz RM, Mayo JC, Ramos S (1999) The oxidant/antioxidant network: role of melatonin. Biol Signals Recept 8:56–63

Resta L, Fiorella R, Marsigliante S, Leo G, Di Nicola V, Marzullo A, Botticella MA (1994) The status of receptors in laryngeal carcinoma: a study of receptors for estrogen, progesterone, androgens and glucocorticoids (in Italian). Acta Otorhinolaryngol Ital 14:385–392

Rojdmark S, Berg A, Rossner S, Wetterberg L (1991) Nocturnal melatonin secretion in thyroid disease and in obesity. Clin Endocrinol (Oxf) 35:61–65

Rozengurt E (1999) Autocrine loops, signal transduction, and cell cycle abnormalities in the molecular biology of lung cancer. Curr Opin Oncol 11:116–122

Seitz HK, Poschl G, Simanowski UA (1998) Alcohol and cancer. Recent Dev Alcohol 14:67–95

Shi Q, Hales DB, Emanuele NV, Emanuele MA (1998) Interaction of ethanol and nitric oxide in the hypothalamic-pituitary-gonadal axis in the male rat. Alcohol Clin Exp Res 22:1754–1762

Slominski A, Baker J, Rosano TG, Guisti LW, Ermak G, Grande M, Gaudet SJ (1996) Metabolism of serotonin to N-acetylserotonin, melatonin, and 5-methoxytryptamine in hamster skin culture. J Biol Chem 271:12281–12286

Szmeja Z, Kruk-Zagajewska A, Kozak W, Szyfter W, Wojtowicz J (1996) Correlation between the clinical stage of larynx cancer in male and female and levels of steroid hormones in serum (in Polish). Otolaryngol Pol 50:21–31

De Stefani E, Boffetta P, Carzoglio J, Mendilaharsu S, Deneo-Pellegrini H (1998) Tobacco smoking and alcohol drinking as risk factors for stomach cancer: a case-control study in Uruguay. Cancer Causes Control 9:321–329

Taggart DP, Gray CE, Bowman A, Faichney A, Davidson KG (1993) Serum androgens and gonadotrophins in bronchial carcinoma. Respir Med 87:455–460

Tamarkin L, Danforth D, Lichter A, Demoss E, Cohen M, Chabner B, Lippman M (1982) Decreased nocturnal plasma melatonin peak in patients with estrogen receptor positive breast cancer. Science 216:1003–1005

Tarquini R, Perfetto F, Zoccolante A, Salti F, Piluso A, De Leonardis V, Lombardi V, Guidi G, Tarquini B (1995) Serum melatonin in multiple myeloma: natural brake or epiphenomenon? Anticancer Res 15:2633–2637

Taylor MW, Feng GS (1991) Relationship between interferon-gamma, indoleamine 2,3-dioxygenase, and tryptophan catabolism. FASEB J 5:2516–2522

Touitou Y, Fevre-Montange M, Proust J, Klinger E, Nakache JP (1985) Age- and sex-associated modification of plasma melatonin concentrations in man. Relationship to pathology, malignant or not, and autopsy findings. Acta Endocrinol (Copenh) 108:135–144

Viviani S, Bidoli P, Spinazze S, Rovelli F, Lissoni P (1992) Normalization of the light/dark rhythm of melatonin after prolonged subcutaneous administration of interleukin-2 in advanced small cell lung cancer patients. J Pineal Res 12:114–117

Vriend J (1983) Evidence for pineal gland modulation of the neuroendocrine-thyroid axis. Neuroendocrinology 36:68–78

Wajchenberg BL, Mendonca B, Liberman B, Adelaide M, Pereira A, Kirschner MA (1995) Ectopic ACTH syndrome. J Steroid Biochem Mol Biol 53:139–151

Wajs E, Lewinski A (1991) Effects of melatonin on [3H]-thymidine incorporation into DNA of rat thyroid lobes in vitro. Biochem Biophys Res Commun 181:1187–1191

Wajs E, Lewinski A (1992) Inhibitory influence of late afternoon melatonin injections and the counter-inhibitory action of melatonin-containing pellets on thyroid growth process in male Wistar rats: comparison with effects of other indole substances. J Pineal Res 13:158–166

Wetterberg L, Aperia B, Gorelick DA, Gwirtzman HE, McGuire MT, Serafetinides EA, Yuwiler A (1992) Age, alcoholism and depression are associated with low levels of urinary melatonin. J Psychiatry Neurosci 17:215–224

Whisler LC, Wood NB, Caldarelli DD, Hutchinson JC, Panje WR, Friedman M, Preisler HD, Leurgans S, Nowak J, Coon JS (1998) Regulators of proliferation and apoptosis in carcinoma of the larynx. Laryngoscope 108:630–638

Yasui H, Takai K, Yoshida R, Hayaishi O (1986) Interferon enhances tryptophan metabolism by inducing pulmonary indoleamine 2,3-dioxygenase: its possible occurrence in cancer patients. Proc Natl Acad Sci U S A 83:6622–6626

Yoshida R, Hayaishi O (1987) Indoleamine 2,3-dioxygenase. Methods Enzymol 142:188–195

Yu CC, Woods AL, Levison DA (1992) The assessment of cellular proliferation by immunohistochemistry: a review of currently available methods and their application. Histochem J 24:121–131

Zuber M, Tan EM, Ryoji M (1989) Involvement of proliferating cell nuclear antigen (cyclin) in DNA replication in living cells Mol Cell Biol 9:57–66

8 The Modulation of Melatonin in Tumor-Bearing Animals: Underlying Mechanisms and Possible Significance for Prognosis

Hella Bartsch, Christian Bartsch, Dieter Mecke

Abstract

Earlier clinical studies showed that circulating melatonin is depressed in patients with primary tumors of different histological types including endocrine-dependent (mammary, endometrial, prostate) as well as endocrine-independent (lung, gastric) malignancies. The depression of circulating melatonin is most pronounced in patients with advanced localized primary tumors, where a negative correlation with the size of the tumor is found. In contrast, patients with a high risk to develop breast cancer or with early stages of prostate cancer show a very pronounced secretion of melatonin. Also a considerable number of patients with ovarian cancer exhibit a high melatonin production. The underlying mechanisms involved in the modulation of circulating melatonin in cancer patients is poorly understood and therefore studies on experimental tumor-bearing animals were performed in order to better understand this phenomenon.

Most studies have been carried out in relation to breast cancer using 7,12-dimethylbenz[a]anthracene (DMBA)-induced mammary cancers as well as serial transplants derived thereof. These experiments demonstrate that nocturnal circulating melatonin is modulated due to the presence of different types and stages of these experimental tumors, as has been observed for the respective human tumors underlining the relevance of these studies. Rats with chemically induced mammary tumors and slow-growing as well as well-differentiated serial transplants containing epithelial cell elements (adenocarcinomas and carcinosarcomas) have an enhanced production of melatonin involving activation of the rate-limiting enzyme of pineal melatonin biosynthesis (serotonin-N-acetyltransferase). This is probably due to an elevation of the sympathetic tone in response to a stimulation of the cellular immune system during malignant growth. In contrast, nocturnal circulating melatonin is depleted in animals with fast-growing mammary tumor transplants when myoepithelial–mesenchymal conversion leads to pure sarcomas. The depletion of melatonin appears to be due to either a reduced availability of the precursor amino acid tryptophan because of a glucocorticoid-induced activation of the hepatic enzyme tryptophan 2,3-dioxygenase or a direct peripheral degradation of melatonin via indolamine 2,3-dioxygenase expressed in tumor and/or other tissues. The significance of these findings is discussed in terms of a possible extrapolation to the clinical situation as well as to increase our theoretical knowledge concerning the mechanisms involved in complex host–tumor interactions within the framework of the neuroimmunoendocrine network.

8.1 Introduction

In the preceding chapters, findings relating to the measurement of circulating melatonin as well as the excretion of melatonin and of its main metabolite, 6-sulfatoxymelatonin (aMT6s), in urine of patients with different types and stages of cancer are reviewed. It has become obvious that melatonin and aMT6s are modulated in different ways by the same type of malignancy depending on the histological type as well as different stage of the tumor. The underlying mechanisms, however, remain unclear and may be difficult to understand solely on the basis of clinical studies because of self-evident ethical restrictions. For this reason it is indispensable to complement such diagnostic clinical studies with related experimental studies on tumor-bearing animals. Clinical studies on patients with breast and prostate cancer with the help of parallel determinations of melatonin and aMT6s indicate that the depression of circulating melatonin is not due to a simple modification of the peripheral conversion of melatonin to aMT6s in the liver (Bartsch et al. 1991, 1992). Therefore it was supposed that the pineal melatonin biosynthetic activity may be reduced due to tumor growth as indicated by the initial investigations of Lapin and Frowein (1981), who detected an inverse correlation between the pineal content of melatonin and increasing size of transplantable Yoshida sarcomas in Wistar rats and interpreted this finding as evidence for an involvement of melatonin in the stimulatory effect of pinealectomy on malignant growth (Lapin 1974). Further analytical experimental studies until now have mainly focused on female rats bearing chemically induced mammary tumors as well as serial transplants derived therefrom.

8.2 Studies with Chemically-Induced Mammary Cancers

Mammary cancer in female Sprague-Dawley rats induced chemically with 7,12-dimethylbenz[a]anthracene (DMBA) (Huggins et al. 1959) represents one of the most commonly used experimental model systems for hormone-dependent human breast cancer (Welsch 1985; Vorherr 1987). DMBA-induced mammary tumors are regulated neuroendocrinologically mainly via the hypothalamo-adenohypophyseal-gonadal axis (Meites 1980) and prolactin (Meites 1980). In addition, the thyroid gland (Chen et al. 1977; Goodman et al. 1980), adrenal cortex (Hallberg 1990), and β-endorphin (Tejwani et al. 1991) are involved. Melatonin administration, probably because of its neuroendocrine effects (Blask 1987; Chap. 15 of this volume), inhibits these types of tumors whereas pinealectomy stimulates them (Tamarkin et al. 1981; Shah et al. 1984). Therefore it appears to be of interest to study the levels of circulating melatonin as well as to analyze components of the pineal melatonin biosynthetic pathway in order to better understand the possible feedback of tumor growth on the pineal gland, an organ which may possess a central role for the temporal organization of the endocrine system (Bartsch et al. 1998). In addition, studies on DMBA-generated mammary cancers allow analysis of the direct effect of this tumor-inducing chemical on circulating hormones such as melatonin and other neuroendocrine parameters.

8.2.1 Acute Effects of DMBA-Administration on Circulating Melatonin: Evidence for an Induction of the Hepatic Degradation of the Hormone

DMBA is metabolically activated by the liver as well as mammary tissue to its ultimate carcinogen, 3,4-dihydrodiol-1,2-epoxide (DiGiovanni and Juchau 1980) by the action of cytochrome P450 isozymes (CYP1A1, CYP1A2, and CYP1B1; Christou et al. 1995; Shou et al. 1996) which are induced by the mammary carcinogen. The administration of DMBA has been reported to lead to a number of endocrine changes which include modification of the timing of the luteinizing hormone (LH) peak (Kerdelhue and Peck 1981) as well as of prolactin secretion (Kerdelhue and El Abed 1979), which appear to take place due to estrogen-like effects of DMBA modifying the regulation of the hypothalamo-pituitary axis in the female rat (Pasqualini et al. 1990). These hormonal changes are supposed to regulate the sensitivity of the developing mammary gland towards the carcinogen, which is at its maximum when the animals are 50–55 days old. In addition, there are inter-strain differences, Sprague-Dawley rats (el Abed et al. 1987) being the most sensitive animals. The DMBA model system therefore allows study of the interplay between a metabolically activated carcinogen and concomitant changes within the neuroendocrine milieu, both of which are essential and interdependent determinants for tumor initiation.

In our experiment, 16 female Sprague-Dawley rats were treated on day 51 of life with a single intragastric dose of DMBA in 1 ml of peanut oil (10 mg/100 g body weight), and, on the same day, 16 animals of the same age and strain received 1 ml of peanut oil only (Huggins et al. 1959). Eight animals from each group were sacrificed 2 days later when they were 53 days old (between 1 and 2 A.M. under very dim red light, $\lambda > 665$ nm) and plasma and pineal glands were collected. The remaining eight animals in each group were sacrificed 5 days later in the same way when animals were 58 days old. Nocturnal plasma melatonin was found to be significantly depressed both 2 and 7 days after DMBA-administration (–37% and –31%, $P < 0.05$), whereas the main peripheral metabolite of melatonin, 6-sulfatoxymelatonin (aMT6s), did not differ compared to controls, indicating an increased degradation of the pineal hormone due to DMBA (Bartsch et al. 1990, 1990a). The content of pineal melatonin was not affected by DMBA treatment but 2 days after carcinogen treatment, serotonin (+31%, $P < 0.025$) and N-acetylserotonin (+25%, $P < 0.05$) were significantly elevated (Bartsch et al. 1990, 1990a). Further evidence for an enhanced peripheral degradation of melatonin due to DMBA was obtained in a subsequent study (Praast et al. 1995) in which it was found that the hepatic microsomal monooxygenases of the cytochrome P450 system catalyzing the 6-hydroxylation of melatonin, though primarily inducible by phenobarbital, are also enhanced by DMBA treatment, probably because of an involvement of CYP1A2.

The observed depression of melatonin 2 and 7 days after DMBA treatment could have a direct functional significance for experimental mammary cancer initiation. Melatonin, on one hand, possesses free radical scavenging properties (Reiter et al. 1999) which could thus reduce the metabolic activation of DMBA via radical intermediary products and, on the other hand, exerts pronounced regulatory functions within the neuroendocrine system by controlling prolactin (Esquifino et al. 1997) and the hypothalamo-pituitary-gonadal axis including the timing of the LH surge (Chiba et al. 1994; Evans et al. 1999), thereby counteracting the above-described array of endocrine imbalances generated by the administration of DMBA (Kerdelhue and

El Abed 1979; Kerdelhue et al. 1981; Pasqualini et al. 1990). The observed depression of nocturnal circulating melatonin may therefore be regarded as an event favoring the process of mammary cancer initiation, an assumption supported by the observation that melatonin administration during the initiation phase of DMBA-induced mammary cancer development inhibits these tumors (Subramaniam and Kothari 1991) and modifies the metabolism of this carcinogen (Kothari and Subramaniam 1992).

8.2.2 The Effects of DMBA-Induced Mammary Tumor Growth on Circulating Melatonin and the Pineal Melatonin Biosynthetic Pathway

Available data concerning the levels of circulating hormones in animals with DMBA-induced mammary cancers are sparse. According to the results of Burenin and Piskareva (1984), the growth of DMBA-induced breast cancer results in an increase of circulating thyrotropin-releasing hormone (TRH), somatotropin, and prolactin, whereas the basal levels of adrenocorticotropic hormone (ACTH), LH, and luteinizing hormone-releasing hormone (LHRH) remain unchanged. With respect to the peripheral hormones, the concentrations of estradiol, corticosterone, and insulin are enhanced and those of testosterone, progesterone, and thyroxine lowered. In the following section, our data on the levels of circulating melatonin and on the melatonin biosynthetic pathway in DMBA-treated and control animals are described.

8.2.2.1 Experiment with Sprague-Dawley Rats

DMBA was administered in 1 ml peanut oil (10 mg/100 g body weight) to 22 female Sprague-Dawley rats on day 51 of life. Twenty-one animals serving as controls received 1 ml of peanut oil as vehicle. Approximately 130 days after DMBA treatment, when 73% of all animals had developed palpable tumors, they were sacrificed at night between 2–3 A.M. under dim red light ($\lambda > 665$ nm). The histopathological investigations proved all tumors to be adenocarcinomas according to Prof. Dr. E. Kaiserling (Institute of Pathology, Tübingen) with very pronounced concentrations of cytosolic estrogen receptors (Bartsch 1988). Tumors were of different size and in many cases individual animals had more than one. In order to allow a better comparison between the animals, the total tumor load for each animal was calculated. Melatonin was quantitated in both plasma and pineal gland. In addition, different steps of the pineal melatonin biosynthetic pathway were analyzed including the rate-limiting enzyme serotonin-N-acetyltransferase (SNAT), hydroxyindole-O-methyltransferase (HIOMT), and the intermediary products serotonin and N-acetylserotonin.

In animals with mammary cancers, no changes of the average values of any of the quantitated parameters were detected, but among the six tumor-free animals treated with DMBA, a nominal increase of SNAT activity was found (+31%, Bartsch 1988). Plots of the corresponding total tumor burden of each animal versus the corresponding plasma melatonin concentrations, as well as versus the different components of the pineal melatonin biosynthetic pathway, showed hyperbolic relationships for pineal and plasma melatonin as well as for SNAT and N-acetylserotonin with an initial increase parallel to growing tumor size of up to 10 cm^3 and a subsequent decline when tumor load increased further (Bartsch 1988; Bartsch et al. 1990a). This indicates that the effect of malignant growth upon the pineal gland critically depends on the tumor burden. Since DMBA-induced mammary tumors in Sprague-Dawley rats develop at different intervals and in varying numbers, it is impossible to form groups of adequate number to allow quantitative analytical endocrine studies with this model system of human breast cancer.

8.2.2.2 Experiment with Inbred F344 Fischer Rats

Because of the above-described limitations of the DMBA model system in Sprague-Dawley rats, we decided to switch to another experimental paradigm, namely serial transplants derived from a DMBA-induced mammary cancer in inbred female F344 Fischer rats (Lee et al. 1981), which will be dealt with in detail later. The availability of a specific radioimmunoassay for 6-sulfatoxymelatonin (aMT6s) (Aldhous and Arendt 1988) as well as a set of metabolic cages enabled us to monitor the nocturnal production of melatonin during different phases of mammary tumor development. Since female F344 Fischer rats, as well as most other rat strains, are less susceptible to the mammary cancer-inducing action of DMBA than Sprague-Dawley rats (Moore et al. 1988), tumors developed more slowly in these animals and a considerable number of mammary tumors showed benign histology.

As in the previous DMBA experiments, the carcinogen was administered once intragastrically at 10 mg/100 g body weight in 1 ml of peanut oil on the 51st day of life. Nocturnal urine (11 P.M.–7 A.M.) was collected in metabolic cages once before the beginning of the experiment in June and after DMBA- or vehicle-treatment at weekly to monthly intervals for more than a year. It was possible to analyze nocturnal aMT6s from these samples in seven vehicle-treated tumor-free controls as well as in seven animals each with a single malignant (adenocarcinoma) or a single benign mammary tumor (fibroadenoma), allowing a specific analysis of the effect of different types of tumor growth on the production of pineal melatonin.

The tumor-free controls showed a significant circannual rhythm of aMT6s production characterized by two summer peaks and a winter trough (Bartsch H et al. 1994) in spite of controlled photoperiods (light:dark = 12 h:12 h), air-conditioned rooms and food ad libitum. This led to the hypothesis of a possible functional involvement of seasonal variations of the horizontal component H of the geomagnetic field for the generation of this phenomenon, since pineal gland activity had indeed been shown to be influenced by variations of H under experimental conditions (Olcese et al. 1988). In animals with malignant mammary tumors, no circannual rhythm of aMT6s was detectable due to the absence of a clear winter decline when tumors grew in size and because of a flattened rise during the second summer (Bartsch H et al. 1993). These observations seem to support our previous results on the analysis of melatonin and the pineal melatonin biosynthetic pathway in Sprague-Dawley rats with DMBA-induced tumors (Bartsch et al. 1990a), in which indications for a stimulation of the pineal melatonin biosynthetic activity were found during the growth of small malignant mammary tumors, and an inhibition was observed when tumors became big. Rats with benign tumors showed a flattened, though statistically significant, circannual rhythm with relatively high aMT6s values in winter and a surge during the second summer being even more pronounced than in controls (Bartsch H et al. 1993). These findings indicate that a stimulation of pineal melatonin production occurs due to the growth of both benign and malignant DMBA-induced mammary tumors.

When animals were sacrificed after more than 1 year of observation, the average values of plasma melatonin as well as of the parameters of the pineal melatonin biosynthetic pathway did not differ between the tumor-free controls and the two tumor groups, which was in good agreement with the results of our previous study on Sprague-Dawley rats (Bartsch C et al. 1990) where antagonistic effects by tumors of different size led to the absence of any overall group changes. Due to the small number of Fischer rats in each group, non-linear regression analyses between the corresponding tumor sizes and the measured parameters could not be calculated.

8.3 Studies with Serial Transplants Derived from DMBA-Induced Mammary Cancers

In order to initiate the process of serial passaging, a cell suspension of one histologically confirmed mammary adenocarcinoma induced in inbred F344 Fischer rats with DMBA was injected subcutaneously into young female rats of the same strain according to Lee et al. (1981). When these tumors became well-palpable they were checked for their histology using frozen sections and, if characterized as malignant (adenocarcinoma), the tumor was transplanted again. It took two attempts for such transplants to grow successfully at the first passage. Difficulties to establish serial transplants derived from DMBA-induced mammary tumors even in syngeneic rats have been reported by others as well (Welsch 1985).

8.3.1 Early Passage: Localized Carcinosarcoma

Nocturnal urine was collected from nine animals bearing passage 2 tumors as well as from six tumor-free controls with the help of a set of metabolic cages at weekly to fortnightly intervals until an average tumor volume of 25 cm^3 was attained after 16 weeks. The animals were then sacrificed at night under dim red light (as described above) at the time of the nocturnal peak of pineal melatonin production (2–3 A.M.) and plasma, pineal glands as well as tumor tissues were collected.

The tumors were characterized histopathologically as carcinosarcomas and they showed localized growth. It was found that nocturnal urinary melatonin increased by 50% shortly after successful transplantation, remaining elevated throughout the subsequent growth of the tumor compared to tumor-free controls. Plasma melatonin showed a 42% increase when animals were sacrificed (Bartsch C et al. 1995). The analysis of the pineal melatonin biosynthetic pathway revealed a 45% increase of the rate-limiting enzyme, SNAT, accompanied by a significant 62% elevation of N-acetyl-serotonin ($P < 0.05$). Since the activity of SNAT is under control of adrenergic nerves (Klein 1985), it was assumed that an activation of the sympathetic tone of the autonomic nervous system may be involved in the observed elevated formation of N-acetyl-serotonin via SNAT. Noradrenaline was found to be significantly augmented by 243% ($P < 0.005$) in nocturnal urine collected 1 week before the animals were sacrificed, whereas adrenaline remained unchanged (Bartsch et al. 1999). This supports the assumption that the growth of tumors of passage 2 led to an elevation of the sympathetic tone. It is conceivable that this phenomenon is due to an activation of the cellular immune system in response to the growth of these tumor transplants, since urinary biopterin steadily rose as the carcinosarcomas increased in size and was significantly augmented by 214% ($P < 0.005$) towards the end of the experiment (Bartsch et al. 1995). Biopterin is formed by murine macrophages under the control of interferon-γ for which a 12-fold rise was detected (Bartsch et al. 1999). These findings may be viewed as an effort of the body to control malignant growth, probably via an immuno-logical rejection of the tumor. We observed an inhibitory effect of melatonin on tumors of passage 2 (Bartsch H et al. 1994a) and it is conceivable that this phenomenon is primarily due to the immunostimulatory action of the pineal hormone (Maestroni 1993) rather than to its neuroendocrine effects on prolactin and, to a lesser extent, gonadal steroids (passage 2 tumors showed no biochemically detectable estrogen and progesterone receptors).

Immunological mechanisms may in fact be of considerable general importance for this tumor transplantation system and could even be responsible for the histological changes taking place during serial passaging. The original DMBA-induced adeno-carcinoma of the mammary gland changed to a carcinosarcoma at passage 2 consist-ing of transformed epithelial as well as mesenchymal components (Bartsch et al., in press) due to the replacement of epithelial tissue elements by sarcomatous tissue, which led to a complete loss of epithelial cells and the development of a pure sarcoma at passages 4–6 when transplanted further (Bartsch et al. 2000). The mechanisms involved in this process of epithelial–mesenchymal tissue remodeling (Guarino 1995, Birchmeier et al. 1996) are complex but may in part consist of immunological pro-cesses leading to a rejection of epithelial cells. The observed elevation of macrophage-derived biopterin as well as interferon-γ support this view. The progressive growth of abnormal mesenchymal cells could be viewed as a compensatory mechanism to main-tain the overall tissue mass of the tumor or as a misdirected wound-healing process within the tumor tissue which may be partially destroyed by immune processes.

8.3.2 Late Passages: Metastasizing Sarcoma

An experiment comparable to the one described for passage 2 was performed when serial transplantation had reached the 12th passage, being a pure sarcoma with a tendency to develop pulmonary metastases. Twelve animals were used as tumor recipients and 12 controls received vehicle injections. Nocturnal urine samples (11 P.M.–7 A.M.) were collected from all animals in the same way as described above. Seven weeks after transplantation when tumors had reached a volume of 25 cm^3, animals were sacrificed at night (2–3 A.M.) and plasma, pineal glands, as well as tumor tissues were collected.

Nocturnal urinary melatonin was found to be significantly depressed by 22% ($P < 0.025$) towards the end of the experiment and plasma melatonin was 56% lower ($P < 0.01$) than in controls (Bartsch et al. 1995, 1999). The analysis of the pineal mela-tonin biosynthetic pathway, however, did not reveal any concomitant changes of SNAT or HIOMT, but a statistically significant reduction of pineal serotonin ($-24\%, P < 0.05$) was detected. In addition, the concentration of plasma tryptophan, the precursor of serotonin and melatonin, was significantly reduced ($-34\%, P < 0.0001$). It was there-fore assumed that the observed depression of circulating melatonin in tumor-bearing animals of this passage may be due to a reduced availability of this precursor amino acid (Bartsch et al. 1999) since acute tryptophan depletions are known to negatively affect nocturnal melatonin (Zimmermann et al. 1993), which was confirmed in our laboratory under in vitro conditions analyzing the secretion of perifused pineal glands exposed to a tryptophan-free medium (Schmidt 1996).

The depletion of tryptophan in tumor-bearing animals may result from an activation of the hypothalamo-pituitary-adrenal axis accompanied by an enhanced secretion of corticosterone ($+208\%$, $P < 0.01$) (Bartsch et al. 1999) which, like other glucocorticoids, is an inducer of the hepatic catabolizing enzyme tryptophan 2,3-dioxygenase (TDO) (Feigelson and Greengard 1962). Further experimental results, however, raise doubts whether a depression of circulating tryptophan may indeed limit pineal melatonin production since animals with passage 2 tumors exhibited an enhanced secretion of melatonin in spite of a clear tryptophan depression (-46%, $P < 0.001$). Alternative modes of explanation for the depression of circulating melatonin in animals with tumors of passage 12 consider an involvement of the extra-hepatic isozyme of TDO, namely indolamine 2,3-dioxygenase (IDO), which has been detected in tumor tissue and, in contrast to TDO, is able to metabolize melatonin as well (Yoshida and Hayaishi 1984).

These considerations underline that the cause for the depression of circulating melatonin in tumor-bearing animals probably does not lie within the pineal gland but rather in metabolic processes taking place in, e.g., liver or tumor. This view is supported by quantitative ultrastructural studies on pinealocytes of animals bearing tumors of the 14th passage of the above-described DMBA model system, where no changes among cell organelles involved in the biosynthesis of melatonin were found (Karasek et al. 1994). Leone and Skene (1994), however, obtained evidence that sera from tumor-bearing mice are able to inhibit the in vitro secretion of melatonin. Similar results were obtained with pineal glands of healthy rats perifused with supernatants of cultured sarcoma cells derived from an advanced passage of our DMBA model system (Schmidt 1996, Schmidt et al. 1997) giving evidence for the existence of tumor cell-derived factors that are able to reversibly inhibit the production/secretion of pineal melatonin. One such factor might be TGF-β, which has been reported to be secreted by transformed mesenchymal cells such as sarcomas (Kloen et al. 1997). TGF-β acts as a growth inhibitor for epithelial cells (Goustin et al. 1986) and shows immunosuppressive effects (Antonia et al. 1998), which could thus explain the significant depression of macrophage-derived biopterin observed in animals with passage 12 tumors (-29%, $P < 0.005$) (Bartsch et al. 1995, 1999). Maestroni and Conti (1996), in their studies on human breast cancer tissues, found that the concentration of the pineal hormone is three orders of magnitude higher in neoplastic mammary tissue than in the serum of the patients concerned, which indicates that melatonin may be produced and/or concentrated by these cancer cells. Also in case of estradiol, high concentrations have been reported within human breast tumor tissue, even in the case of postmenopausal women, probably involving very high affinity receptor-mediated uptake mechanisms (Masamura et al. 1997). Since experimental evidence exists that estrogen is formed from its sulfated catabolite estrone sulfate via the action of estrone sulfatase in tumor tissue (Masamura et al. 1996), it is conceivable that also aMT6s may be reconverted to melatonin by malignant tissue and subsequently concentrated within the tumor.

The above-mentioned considerations indicate that independent processes seem to exist acting in parallel to depress circulating melatonin in animals with passage 12 tumors of our model system. These mechanisms may also render explanations for the observation that both exogenous melatonin (Bartsch H et al. 1994a) and physiological pinealectomy by constant light (H. Bartsch et al., unpublished results) are unable to affect the growth of tumors of passage 12 or 14 under in vivo conditions.

8.4 Studies with Chemically Induced Colon Cancers

Anisimov et al. (1999) analyzed the day–night increment of serum melatonin between 10 A.M. and 12 P.M. in rats with colon cancer induced by 1,2-dimethylhydrazine. They found an overall elevation of circulating melatonin in tumor-bearing animals, particularly in the morning samples compared to tumor-free controls. In a preceding study, Anisimov et al. (1997) observed that the administration of melatonin inhibits the development of these chemically induced tumors, perhaps via an anti-initiation effect (Musatov et al. 1999). Therefore, the elevation of endogenous melatonin in animals with colon cancer may be interpreted as an effort of the system to control tumor development, which could be fostered by administration of exogenous melatonin. Since patients with colon cancer also tend to show an elevated production of melatonin (Chap. 7 of this volume) it is conceivable that melatonin administration might favorably influence the course of the disease.

8.5 Conclusions and Considerations

The above-described findings on experimental animals show that the levels of circulating melatonin are modulated by tumors of different histological types and stages of growth. Comparable observations have been made in cancer patients as summarized in the preceding chapters. According to the studies on experimental animals, it appears that the mechanisms involved in the up- or down-regulation of circulating melatonin are highly complex consisting of overlapping synergistic as well as antagonistic processes involving not only the pineal gland and the tumor but also components of the neuroimmunoendocrine network. The measurable levels of circulating melatonin could be viewed as a net effect of multiple and multi-directional vectorial processes governed by the presence of a tumor of defined histopathological type as well as developmental stage. This net effect is transient and is affected by the histological appearance of the tumor, its size, and the formation of metastases, leading to a modulation of circulating melatonin characterized by phases of elevation, depression, or lack of changes compared to tumor-free controls. Endocrine imbalances affecting seven different hormones including insulin, prolactin, growth hormone, thyroxine, as well as gonadal steroids were also reported by Besedovsky et al. (1985) in rats with syngeneic transplants derived from DMBA-induced tumors, and Vaswani et al. (1986) observed a decrease of opioid peptide levels within the central nervous system of animals with DMBA-induced mammary tumors.

It may therefore be assumed that measurements of melatonin and of other hormones as well as of immunological parameters could offer valuable information about the physiological state of the host affected by tumor growth. More detailed knowledge about the significance of these analytical findings could aid in predicting onco-therapeutic results. For the administration of melatonin, it appears that normal or elevated levels of circulating melatonin correlate with a good responsiveness of the tumor whereas low levels of melatonin predict unresponsiveness (Bartsch H et al.

1994a). It requires to be investigated whether this conclusion, mainly derived from investigations in only one experimental model system, namely serial transplants of a DMBA-induced mammary cancer in female rats, is generally applicable or not. If this assumption is correct, it would mean that, e.g., breast or prostate cancer patients with a depressed nocturnal melatonin surge (Bartsch C et al. 1994) could not expect any tumor growth inhibitory effects by exogenous melatonin.

It will be necessary to extend experimental analytical as well as parallel therapeutic studies on the role of melatonin in the control of cancer growth to other in vivo tumor model systems to achieve a more reliable predictability regarding the therapeutic efficacy of melatonin. Furthermore, it should be tested whether circulating melatonin levels in tumor-bearing animals or cancer patients deviating from the normal range of age-matched healthy controls may also possess a general predictive value for cancer patients concerning their prognosis and survival as well as their responsiveness to established and experimental oncotherapies. Such considerations appear to be justified in view of the fact that pronounced circadian endocrine disturbances are observed in cancer patients with low circulating melatonin (Bartsch et al. 1989, 1998) and that the production of melatonin correlates with cancer cell proliferation (Bartsch et al. 1997) thus reflecting important processes both at the level of the tumor and within the neuroimmunoendocrine network of the host. A stimulation of the levels of endogenous melatonin during early phases of tumor growth may be interpreted as an effort of the system to control these cancers (Bartsch H et al. 1994a) either directly or by fostering immunological control. Depressed levels of circulating melatonin during advanced tumor growth most probably are not functionally involved in enhanced primary tumor growth (Bartsch H et al. 1994) but may, however, facilitate metastatic spread and thus would essentially determine survival.

Acknowledgements. We would like to thank the Deutsche Forschungsgemeinschaft for their generous help in supporting these studies for several years (Li 328/3 – 1,3,5).

References

el Abed A, Kerdelhue B, Castanier M, Scholler R (1987) Stimulation of estradiol-17β secretion by 7,12-dimethylbenz(a)anthracene during mammary tumor induction in Sprague-Dawley rats. J Steroid Biochem 26:733–738

Aldhous ME, Arendt J (1988) Radioimmunoassay for 6–sulphatoxymelatonin in urine using an iodinated tracer. Ann Clin Biochem 25:298–303

Anisimov VN, Popovich IG, Zabezhinski MA (1997) Melatonin and colon carcinogenesis: I. Inhibitory effect of melatonin on development of intestinal tumors induced by 1,2-dimethylhydrazine in rats. Carcinogenesis 18:1549–1553

Anisimov VN, Kvetnoy IM, Chumakova NK, Kvetnaya TV, Molotkov AO, Pogudina NA, Popovich IG, Popuchiev VV, Zabezhinski MA, Bartsch H, Bartsch C (1999) Melatonin and colon carcinogenesis. II. Intestinal melatonin-containing cells and serum melatonin level in rats with 1,2-dimethylhydrazine-induced colon tumors. Exp Toxic Pathol 51:47–52

Antonia SJ, Extermann M, Flavell RA (1998) Immunologic nonresponsiveness to tumors.Crit Rev Oncog 9:35–41

Bartsch C (1988) Untersuchungen zur Funktion der Zirbeldrüse und ihrer Beziehung zum endokrinen System im Brust- und Prostatacarcinom: tierexperimentelle und humanbiologische Untersuchungen. Ph.D. Thesis at the Faculty of Chemistry and Pharmacy, University of Tübingen, Tübingen.

Bartsch C, Bartsch H, Fuchs U, Lippert TH, Bellmann O, Gupta D (1989) Stage-dependent depression of melatonin in patients with primary breast cancer. Correlation with prolactin, thyroid stimulating hormone, and steroid receptors. Cancer 64:426–433

Bartsch C, Bartsch H, Lippert TH, Gupta D (1990) Effect of the mammary carcinogen 7,12-dimethylbenz[a]anthracene on pineal melatonin biosynthesis, secretion and peripheral metabolism. Neuroendocrinology 52:538–544

Bartsch C, Bartsch H, Gupta D (1990a) Pineal melatonin synthesis and secretion during induction and growth of mammary cancer in female rats. In: Neuroendocrinology: New Frontiers, Gupta D, Wollmann HA, Ranke MB, eds., Brain Research Promotion, London and Tübingen, pp 326–332

Bartsch C, Bartsch H, Bellmann O, Lippert TH (1991) Depression of serum melatonin in patients with primary breast cancer is not due to an increased peripheral metabolism. Cancer 67:1681–1684

Bartsch C, Bartsch H, Schmidt A, Ilg S, Bichler KH, Flüchter SH (1992) Melatonin and 6-sulfatoxymelatonin circadian rhythms in serum and urine of primary prostate cancer patients: evidence for reduced pineal activity and relevance of urinary determinations. Clin Chim Acta 209:153–167

Bartsch C, Bartsch H, Flüchter SH, Mecke D, Lippert TH (1994) Diminished pineal function coincides with disturbed circadian endocrine rhythmicity in untreated primary cancer patients. Consequence of premature aging or of tumor growth? Ann N Y Acad Sci 719:502–525

Bartsch C, Bartsch H, Buchberger A, Rokos H, Mecke D, Lippert TH (1995) Serial transplants of DMBA-induced mammary tumors in Fischer rats as model system for human breast cancer. IV. Parallel changes of biopterin and melatonin indicate interactions between the pineal gland and cellular immunity in malignancy. Oncology 52:278–283

Bartsch C, Kvetnoy I, Kvetnaia T, Bartsch H, Molotkov A, Franz H, Raikhlin N, Mecke D (1997) Nocturnal urinary 6-sulfatoxymelatonin and proliferating cell nuclear antigen-immunopositive tumor cells show strong positive correlations in patients with gastrointestinal and lung cancer. J Pineal Res 23:90–96

Bartsch C, Bartsch H, Bichler K-H, Flüchter S-H (1998) Prostate cancer and tumor stage-dependent circadian neuroendocrine disturbances. Aging Male 1:188–199

Bartsch C, Bartsch H, Buchberger A, Stieglitz A, Effenberger-Klein A, Kruse-Jarres JD, Besenthal I, Rokos H, Mecke D (1999) Serial transplants of DMBA-induced mammary tumors in Fischer rats as a model system for human breast cancer. VI. The role of different forms of tumor-associated stress for the regulation of pineal melatonin secretion, Oncology 56:169–176

Bartsch C, Szadowska A, Karasek M., Bartsch H, Geppert M, Mecke D (2000) Serial transplants of DMBA-induced mammary tumors in Fischer rats as model system for human breast cancer. V. Myoepithelial-mesenchymal conversion during passaging as possible cause for modulation of pineal-tumor interaction. Exp Toxic Pathol 52:93–101

Bartsch H, Bartsch C, Mecke D, Lippert TH (1993) The relationship between the pineal gland and cancer: seasonal aspects. In: Light and Biological Rhythms in Man, Wetterberg L ed, Pergamon Press, Oxford, pp 337–347

Bartsch H, Bartsch C, Mecke D, Lippert TH (1994) Seasonality of pineal melatonin production in the rat: possible synchronization by the geomagnetic field. Chronobiol Int 11:21–26

Bartsch H, Bartsch C, Mecke D, Lippert TH (1994a) Differential effect of melatonin on early and advanced passages of a DMBA-induced mammary carcinoma in the female rat. In: Advances in Pineal Research, Vol.7, Maestroni GJM, Conti A, Reiter RJ, eds., John Libbey, London, pp 247–252

Besedovsky HO, del Rey A, Schardt M, Sorkin E, Normann S, Baumann J, Girard J (1985) Changes in plasma hormone profiles after tumor transplantation into syngeneic and allogeneic rats. Int J Cancer 36:209–216

Birchmeier C, Birchmeier W, Brand-Saberi B (1996) Epithelial-mesenchymal transitions in cancer progression. Acta Anat (Basel) 156:217–226

Blask DE (1987) Neuroendocrine aspects of neoplastic growth: a review. Neuroendocrinol Lett 9:63–73

Burenin IS, Piskareva NK (1984) Plasma hormone levels in rats with DMBA-induced mammary tumors (in Russian) Eksp Onkol 6:24–27

Chen HJ, Bradley CJ, Meites J (1977) Stimulation of growth of carcinogen-induced mammary cancers in rats by thyrotropin-releasing hormone. Cancer Res 37:64–66

Chiba A, Akema T, Toyoda J (1994) Effects of pinealectomy and melatonin on the timing of the proestrous luteinizing hormone surge in the rat. Neuroendocrinology 59:163–168

Christou M, Savas U, Schroeder S, Shen X, Thompson T, Gould MN, Jefcoate CR (1995) Cytochromes CYP1A1 and CYP1B1 in the rat mammary gland: cell-specific expression and regulation by polycyclic aromatic hydrocarbons and hormones. Mol Cell Endocrinol 115:41–50

DiGiovanni J, Juchau MR (1980) Biotransformation and bioactivation of 7, 12-dimethylbenz[a]anthracene (7, 12-DMBA). Drug Metab Rev 1:61–101

Esquifino A, Agrasal C, Velazquez E, Villanua MA, Cardinali DP (1997) Effect of melatonin on serum cholesterol and phospholipid levels, and on prolactin, thyroid-stimulating hormone and thyroid hormone levels, in hyperprolactinemic rats. Life Sci 61:1051–1058

Evans JJ, Janmohamed S, Forsling ML (1999) Gonadotrophin-releasing hormone and oxytocin secretion from the hypothalamus in vitro during pro-oestrus: the effects of time of day and melatonin. Brain Res Bull 48:93–97

Feigelson P and Greengard O (1962) Immunochemical evidence for increased titers of liver pyrrolase during substrate and hormonal enzyme induction. J Biol Chem 237:3714–3717

Guarino M (1995) Epithelial-to-mesenchymal change of differentiation. From embryogenetic mechanism to pathological patterns. Histol Histopathol 10:171–184

Goodman AD, Hoekstra SJ, Marsh PS (1980) Effects of hypothyroidism on the induction and growth of mammary cancer induced by 7,12-dimethylbenz(a)anthracene in the rat. Cancer Res 40:2336–2342

Goustin AS, Leof EB, Shipley GD, Moses HL (1986) Growth factors and cancer. Cancer Res 46:1015–1029

Hallberg E (1990) Metabolism and toxicity of xenobiotics in the adrenal cortex, with particular reference to 7,12-dimethylbenz(a)anthracene. J Biochem Toxicol 5:71–90

Huggins C, Briziarelli G, Sutton H (1959) Rapid induction of mammary carcinoma in the rat and the influence of hormones on the tumor. J Exp Med 109:25–42

Karasek M, Marek K, Zielinska A, Swietoslawski J, Bartsch H, Bartsch C (1994) Serial transplants of 7,12-dimethylbenz[a]anthracene-induced mammary tumors in Fischer rats as model system for human breast cancer. 3. Quantitative ultrastructural studies of the pinealocytes and plasma melatonin concentrations in rats bearing an advanced passage of the tumor. Biol Signals 3:302–306

Kerdelhue B, Peck EJ Jr (1981) In vitro LHRH release: correlation with the LH surge and alteration by a mammary carcinogen. Peptides 2:219–222

Kerdelhue B, El Abed A (1979) Inhibition of preovulatory gonadotropin secretion and stimulation of prolactin secretion by 7,12-dimethylbenz(a)anthracene in Sprague-Dawley rats. Cancer Res 39:4700–4705

Kothari L, Subramanian A (1992) A possible modulatory influence of melatonin on representative phase I and II drug metabolizing enzymes in 9,10-dimethyl-1,2-benzanthracene induced rat mammary tumorigenesis. Anticancer Drugs 3:623–628

Klein DC (1985) Photoneural regulation of the mammalian pineal gland. In: Photoperiodism, melatonin, and the pineal. Ciba Foundation Symposium 117. Evered D and Clark S eds., Pitman, London, pp 38–51

Kloen P, Gebhardt MC, Perez-Atayde A, Rosenberg AE, Springfield DS, Gold LI, Mankin HJ (1997) Expression of transforming growth factor-beta (TGF-beta) isoforms in osteosarcomas: TGF-beta3 is related to disease progression. Cancer 80:2230–2239

Lapin V (1974) Influence of simultaneous pinealectomy and thymectomy on the growth and formation of metastases of the Yoshida sarcoma in rats. Exp Pathol (Jena) 9:108–112

Lapin V and Frowein A (1981) Effects of growing tumours on pineal melatonin levels in male rats. J Neural Transm 52:123–136

Lee C, Lapin V, Oyasu R., Battifora H (1981) Effect of ovariectomy on serially transplanted rat mammary tumors induced by 7,12-dimethylbenz(a)anthracene. Eur J Cancer Clin Oncol 17:801–808

Leone AM, Skene D (1994) Melatonin concentrations in pineal organ culture are suppressed by sera from tumor-bearing mice. J Pineal Res 17:17–19

Maestroni GJ (1993) The immunoneuroendocrine role of melatonin. J Pineal Res 14:1–10

Maestroni GJ, Conti A (1996) Melatonin in human breast cancer tissue: association with nuclear grade and estrogen receptor status. Lab Invest 75:557–561

Masamura S, Santner SJ, Gimotty P, George J, Santen R (1997) Mechanism for maintenance of high breast tumor estradiol concentrations in the absence of ovarian function: role of very high affinity tissue uptake. Breast Cancer Res Treat 42:215–226

Masamura S, Santner SJ, Santen RJ (1996) Evidence of in situ estrogen synthesis in nitrosomethylurea-induced rat mammary tumors via the enzyme estrone sulfatase. J Steroid Biochem Mol Biol 58:425–429

Meites J (1980) Relation of the neuroendocrine system to the development and growth of experimental mammary tumors. J Neural Transm 48:25–42

Moore CJ, Tricomi WA, Gould MN (1988) Comparison of 7,12-dimethylbenz[a]anthracene metabolism and DNA binding in mammary epithelial cells from three rat strains with differing susceptibilities to mammary carcinogenesis. Carcinogenesis 9:2099–2102

Musatov SA, Anisimov VN, Andre V, Vigreux C, Godard T, Sichel F (1999) Effects of melatonin on N-nitroso-N-methylurea-induced carcinogenesis in rats and mutagenesis in vitro (Ames test and COMET assay). Cancer Lett 138:37–44

Olcese J, Reuss S, Semm P (1988) Geomagnetic field detection in rodents. Life Sci 42:605–613

Pasqualini C, Sarrieau A, Dussaillant M, Corbani M, Bojda-Diolez F, Rostene W, Kerdelhue B (1990) Estrogen-like effects of 7,12-dimethylbenz(a)anthracene on the female rat hypothalamo-pituitary axis. J Steroid Biochem 36:485–491

Praast G, Bartsch C, Bartsch H, Mecke D, Lippert TH (1995) Hepatic hydroxylation of melatonin in the rat is induced by phenobarbital and 7,12-dimethylbenz[a]anthracene–implications for cancer etiology. Experientia 51:349–355

Reiter RJ, Tan DX, Cabrera J, D'Arpa D, Sainz RM, Mayo JC, Ramos S (1999) The oxidant/antioxidant network: role of melatonin. Biol Signals Recept 8:56–63

Schmidt U (1996) Untersuchungen zum Einfluß des Tumorwachstums auf die Melatoninsekretion der Rattenzirbeldrüse. Ph. D.Thesis at the Faculty of Chemistry and Pharmacy, University of Tübingen, Tübingen

Schmidt U, Bartsch C, Bartsch H, Mecke D (1997) Pineal melatonin secretion seems to be reversibly inhibited by a tumor-derived melatonin inhibiting factor (abstract). Cancer Biotherapy 12:429

Shah PN, Mhatre MC, Kothari LS (1984) Effect of melatonin on mammary carcinogenesis in intact and pinealectomized rats in varying photoperiods. Cancer Res 44:3403–3407

Shou M, Gonzalez FJ, Gelboin HV (1996) Stereoselective epoxidation and hydration at the K-region of polycyclic aromatic hydrocarbons by cDNA-expressed cytochromes P450 1A1, 1A2, and epoxide hydrolase. Biochemistry 35:15807–15813

Subramanian A, Kothari L (1991) Suppressive effect by melatonin on different phases of 9,10-dimethyl-1,2-benzanthracene (DMBA)-induced rat mammary gland carcinogenesis. Anticancer Drugs 2:297–303

Tamarkin L, Cohen M, Roselle D, Reichert C, Lippman M, Chabner B. (1981) Melatonin inhibition and pinealectomy enhancement of 7,12-dimethylbenz(a)anthracene-induced mammary tumors in the rat. Cancer Res 41:4432–4436

Tejwani GA, Gudehithlu KP, Hanissian SH, Gienapp IE, Whitacre CC, Malarkey WB (1991) Facilitation of dimethylbenz[a]anthracene-induced rat mammary tumorigenesis by restraint stress: role of beta-endorphin, prolactin and naltrexone. Carcinogenesis 12:637–641

Vaswani KK, Tejwani GA, Abou-Issa HM (1986) Effect of 7,12-dimethylbenz[a]anthracene-induced mammary carcinogenesis on the opioid peptide levels in the rat central nervous system. Cancer Lett 31:115–122

Vorherr H (1987) Endocrinology of breast cancer. Maturitas 9:113–122

Welsch CW (1985) Host factors affecting the growth of carcinogen-induced rat mammary carcinomas: a review and tribute to Charles Brenton Huggins. Cancer Res 45:3415–3443

Yoshida R and Hayaishi O (1984) Overview superoxygenase. Methods Enzymol 106:61–70

Zimmermann RC, McDougle CJ, Schumacher M, Olcese J, Mason JW, Heninger GR, Price LH (1993) Effects of acute tryptophan depletion on nocturnal melatonin secretion in humans. J Clin Endocrinol Metab 76:1160–1164

9 The Pineal Gland, Melatonin, and Neoplastic Growth: Morphological Approach

Michal Karasek

Abstract

The relationship between the pineal gland and neoplastic growth was suggested as early as 1929. However, real progress in this area of research has only been made in the last two decades. Although studies on the link between the pineal gland and tumorigenesis do not always yield consistent results, the majority of reports have pointed to the oncostatic action of the pineal gland. However, very few studies on tumor morphology after pinealectomy or melatonin administration and on pineal morphology in tumor-bearing animals have been performed. In this review, the data (including the author's own) on morphological aspects of the pineal gland and melatonin influence on neoplastic growth as well as on pineal morphology in tumor-bearing animals are presented.

9.1 Introduction

Many studies, both experimental and clinical, indicate a link between the pineal gland and neoplastic disease, although the results are sometimes conflicting (Blask 1984, 1993; Blask and Hill 1988; Reiter 1988; Karasek and Fraschini 1991; Karasek 1994, 1996). There is a bulk of evidence showing an influence of melatonin or pinealectomy on development and/or growth of a variety of experimentally induced malignant tumors. Despite some discrepancies, the majority of reports indicate an oncostatic action of the pineal gland, most probably exerted by its hormone melatonin. It has been demonstrated in various tumor types in vivo as well as in vitro (Blask 1984, 1993; Pawlikowski and Karasek 1990; Karasek 1997).

The influence of the pineal gland on tumorigenesis has recently been extensively studied. However, very few studies on tumor morphology after pinealectomy or melatonin administration have been performed. Studies on pineal morphology in tumor-bearing animals are equally rare. In the present paper, the results of morphological studies on the relationship between the pineal gland and neoplastic growth, including the author's own studies, are presented and discussed.

9.2 Morphological Studies of Tumors Following Melatonin Treatment and Pinealectomy

The majority of studies on the influence of melatonin or pineal removal on development and/or growth of malignant tumors are based on observations of tumor size (diameter), weight or volume, survival time of the animals, and proliferation of neoplastic cells. Methods employing microscopic anatomy are very rare.

The proliferation of neoplastic cells has been studied using various methods. Most commonly, incorporation of [^{3}H]thymidine into DNA or counting the number of cells (in vitro) was used as an index of cell proliferation. However, morphological studies based on mitotic activity rate (number of metaphases) or on incorporation of bromodeoxyuridine (BrDU) into DNA were also performed. The results of the later studies are briefly summarized below.

Lewinski et al. (1993) have demonstrated that melatonin administration decreased the mitotic activity of Guerin epithelioma in rats (both intact and pinealectomized), not affecting the life span of the animals. On the contrary, pinealectomy resulted in small, statistically insignificant enhancement of the mitotic activity in the same tumor. Pinealectomy also increased the mitotic index in mice bearing Ehrlich's tumor (Billittieri and Bindoni 1969).

Inhibition of proliferation of tumor cells (using incorporation of BrDU as an index) has been shown in Kirkman-Robbins hepatoma in Syrian hamsters in vivo (Fig. 9.1 a) but not in vitro (Fig. 9.1 b) (Karasek et al. 1992), in murine transplantable Colon 38 adenocarcinoma, both in vivo (Fig. 9.1 c) and in vitro (Karasek et al. 1998), and in diethylstilbestrol (DES)-induced prolactinoma in vivo (Fig. 9.1 d) (M. Karasek et al., unpublished data). In the last two tumor types, effects similar to those exerted by melatonin have been observed after treatment with the putative melatonin agonist CGP 52608 (Fig. 9.1 c, d) (Karasek et al. 1998; M. Karasek et al., unpublished data). Moreover, in Kirkman-Robbins hepatoma, the proliferation of tumor cells was significantly higher in hamsters kept in long photoperiod than in those maintained in short photoperiod (Karasek et al. 1994 a).

Melatonin was shown to decrease the number of ME-180 human cervical cancer cells (Chen et al. 1995) and M-6 melanoma cell line in vitro (Ying et al. 1993).

Also, the number of MCF-7 human breast cancer cells after melatonin administration to the culture medium was 20–60% lower than in the control group (Hill and Blask 1988; Cos and Blask 1994; Crespo et al. 1994). Additionally, melatonin treatment resulted in a decrease in MCF-7 cell size (both in soma and nuclear areas). Moreover, melatonin significantly altered certain ultrastructural features of MCF-7 cells. Reduced development of microvilli, presence of numerous bundles of tonofilaments, and an increased amount of rough endoplasmic reticulum, Golgi complexes, and secretory granules after melatonin treatment indicate greater differentiation of tumor cells. On the other hand, mitochondrial swelling with disruption of cristae, cytoplasmic vacuolation, increase in the number of autophagic vacuoles, and nuclear chromatin disaggregation suggest an eventual process of cell death (Hill and Blask 1988; Crespo et al. 1994). The changes observed after melatonin were reversed by subsequent addition of estradiol to the culture medium (Crespo et al. 1994).

The morphology of the mammary gland has also been studied in another type of mammary tumor, i.e., the spontaneous mammary tumor in C3H/Jax mice (Subramanian and Kothari 1991). It has been shown that prolonged melatonin treatment alters the architecture of the mammary gland, modulating the degree of develop-

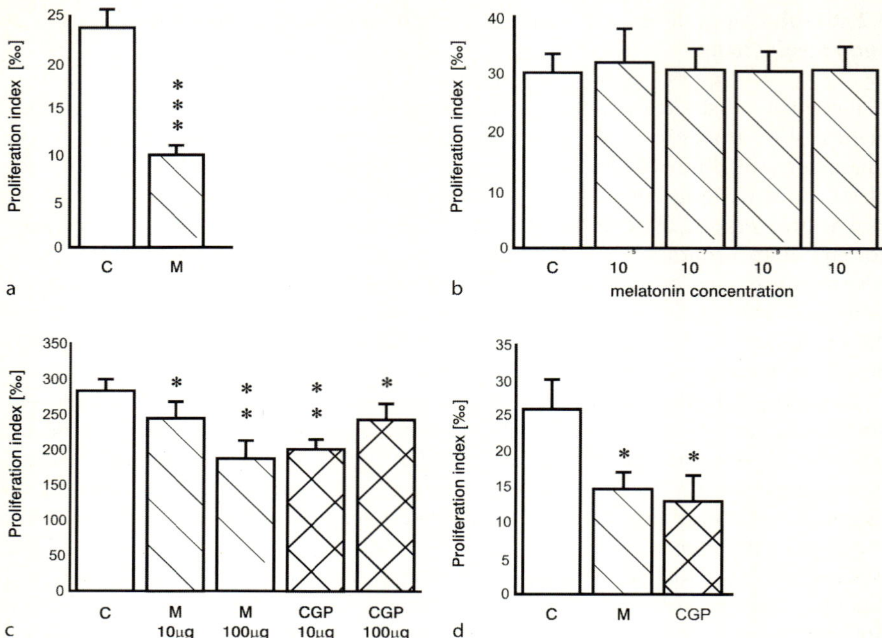

Fig. 9.1a–d. Proliferation of Kirkman-Robbins hepatoma cells in vivo (**a**), and in vitro (**b**), Colon 38 adenocarcinoma cells in vivo (**c**), and diethylstilbestrol-induced prolactinoma cells in vivo (**d**) in control *(C)*, melatonin-treated *(M)*, and CGP 52608-treated *(CGP)* animals. Data are expressed as a number of BrdU-positive nuclei per 1000 cells (proliferation index); $* P < 0.05$, $** P < 0.01$, $*** P < 0.001$. (Based on Karasek et al. 1992; 1998; Karasek et al., unpublished data; modified)

ment of the mammary epithelium, and subsequently reducing spontaneous mammary tumor incidence in these high-risk mice. In whole-mount preparations, mammary glands of intact tumor-bearing mice showed extensive ductal branching with terminal end buds and a few hyperplastic alveolar nodules, whereas melatonin-treated animals showed sparse ductal branching almost without terminal end buds and a complete absence of hyperplastic alveolar nodules.

Pinealectomy resulted in an increased density of terminal end buds and a decrease in the density of alveolar buds in mammary cancer induced by dimethylbenzanthracene (DMBA) (Shah et al. 1984).

It is noteworthy that melatonin was shown also to partially inhibit postnatal development of normal murine mammary gland by reducing the number of epithelial structures representing sites of growth, and increasing the number of structures representing the final stage of ductal growth in virgin animals (Mediavilla et al. 1992).

9.3 Pineal Morphology in Tumor-Bearing Animals

Data on pineal morphology in neoplastic disease are scarce. This is true for both experimentally induced animal tumors and human neoplasia. However, alterations in gross and microscopic anatomy of the pineal gland in malignancy have been reported.

A marked increase in size and number of pinealocytes containing vacuolated cytoplasm and large, vesicular nuclei was observed in rats bearing transplanted melanoma in comparison with intact animals, and these changes were reversed by the administration of melatonin (El-Domeri and Das Gupta 1976). In rats bearing carcinogen-induced fibrosarcoma, no changes in either pineal weight or nuclear diameter were found (Tapp 1980).

In our laboratory, attention has recently been focused on ultrastructural changes in pinealocytes of animals bearing various types of experimentally induced malignant tumors, including DES-induced prolactinoma in the rat (Karasek et al. 1991), transplantable Kirkman-Robbins hepatoma in the Syrian hamster (Karasek et al. 1993), an advanced passage of a DMBA-induced mammary tumor in the rat (Karasek et al. 1994b), and transplantable Colon 38 adenocarcinoma in the mouse (M. Karasek et al., unpublished results). The results of these studies are briefly summarized here.

In pinealocytes of rats bearing DES-induced prolactinoma (Fig. 9.2), ultrastructural features of increased cell activity were observed. It was exemplified by increased relative volumes of mitochondria, lysosomes, Golgi apparatus, granular endoplasmic reticulum, lipid droplets, and vacuoles containing flocculent material as well as by an increased number of dense-core vesicles (Karasek et al. 1991).

On the contrary, pinealocytes of Syrian hamsters bearing Kirkman-Robbins hepatoma (Fig. 9.3) showed ultrastructural patterns of decreased cell activity at night, exemplified by a decrease of the relative volumes of mitochondria, Golgi apparatus, granular endoplasmic reticulum, lipid droplets, and vacuoles with flocculent content as well as a decrease in the number of dense-core vesicles. However, there were no differences in the ultrastructure of pinealocytes between control and tumor-bearing animals during daytime (Karasek et al. 1993).

Ultrastructure of the pinealocytes of rats bearing DMBA-induced mammary tumors of an advanced passage did not show significant differences as compared to control animals, either during the daytime or at night, despite the fact that plasma melatonin concentrations in tumor-bearing animals killed at night were suppressed by 35% when compared to the control animals killed at the same time (Karasek et al. 1994b).

In preliminary studies of pinealocytes of mice bearing Colon 38 adenocarcinoma, a decrease in the nocturnal relative volume of mitochondria and an increase in the relative volume of lipid droplets have been observed in comparison with control tumor-free animals. The relative volumes of Golgi apparatus, granular endoplasmic reticulum, and lysosomes did not differ between tumor-bearing and control animals (M. Karasek et al., unpublished preliminary data).

It should be emphasized that different ultrastructural patterns of the pinealocyte activity were observed in various experimental tumor types. Therefore, it seems that the nature of pinealocyte response to the presence of malignancy, like the influence of the pineal gland on tumor development and/or growth, may in fact be tumor-specific.

9.4 Pineal Morphology in Human Malignancy

Morphological studies of the pineal gland in humans in relation to neoplastic disease are even more rare than animal studies.

It has been shown that pineal glands from patients dying of malignant tumors were greater in size and weight than those of patients succumbing to non-malignant

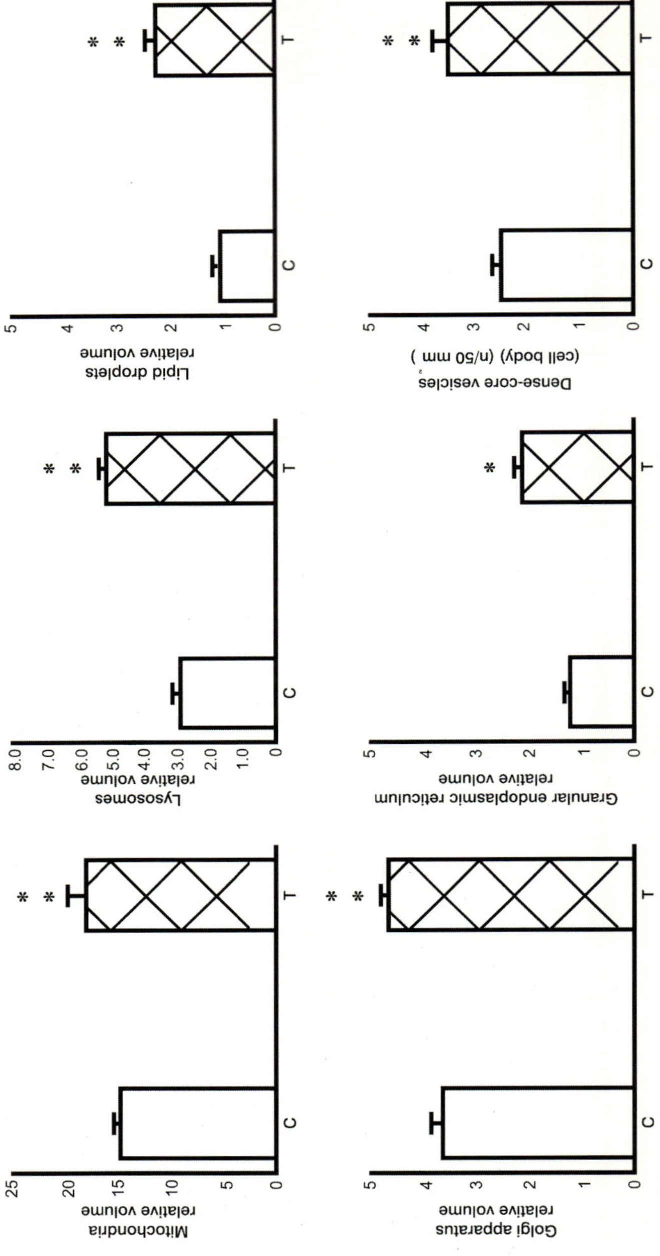

Fig. 9.2. The relative volumes of various cell components, and numerical density of dense-core vesicles in control rats (*C*), and in rats bearing diethylstilbestrol-induced prolactinoma (*T*) (day). Mean ± SD; * *P* < 0.05, ** *P* < 0.01. (Modified from Karasek et al. 1991)

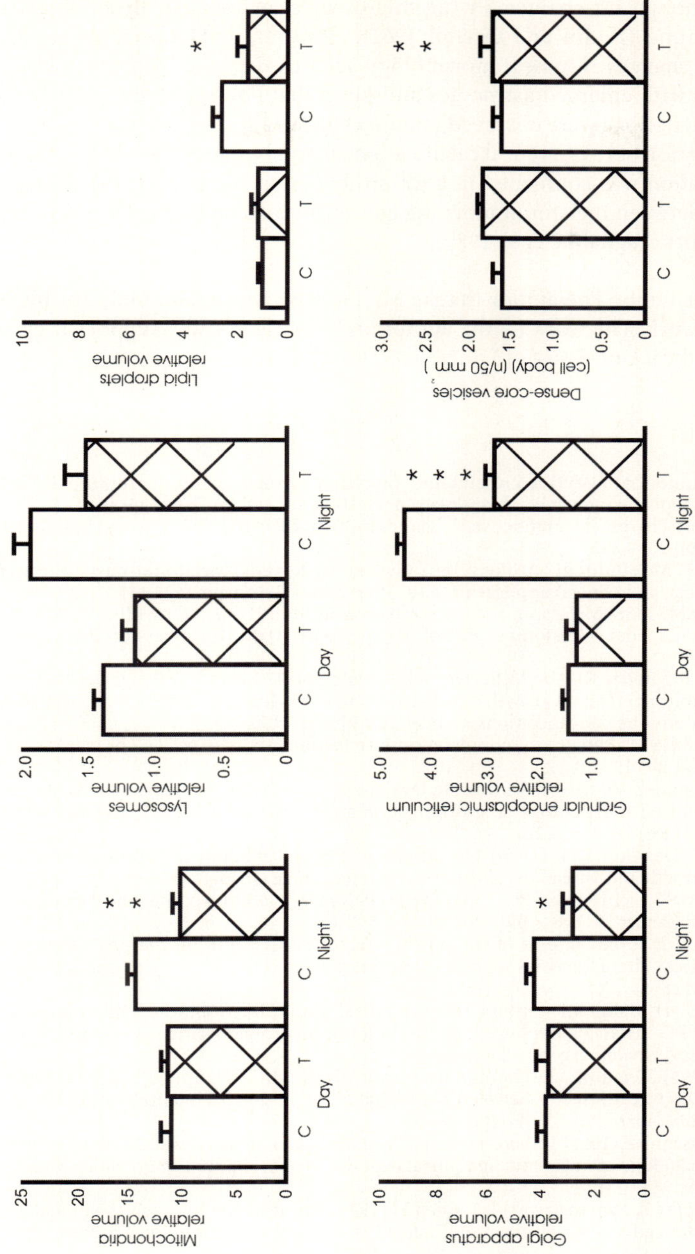

Fig. 9.3. The relative volumes of various cell components, and numerical density of dense-core vesicles in control hamsters (*C*), and in hamsters bearing Kirkman-Robbins hepatoma (*T*) (day and night). Mean ± SD; * $P < 0.05$, ** $P < 0.01$, *** $P < 0.001$. (Modified from Karasek et al. 1993)

diseases, although no changes in the microscopic anatomy of these pineal glands have been found (Rodin and Overall 1967). However, in 11% of patients dying of malignancy, changes in pineal morphology such as deposits of pleomorphic structures, gliosis with enlarged astrocytes and Rosenthal fibers, and the presence of slit-like cystic cavitations were observed (Hadju et al. 1972).

No correlation between pineal calcification and malignancy was found (Tapp 1982). This observation was confirmed in a CT study (Tagliabue et al. 1989). Moreover, no correlation between melatonin morning concentrations and pineal size was found in cancer patients (Tagliabue et al. 1989).

Acknowledgements. The author thanks Mr. Jacek Swietoslawski M.Sc. for his help in the present work. The work by the author and this review was supported by a grant from the Medical University of Lodz, No. 503–129–0.

References

Billittieri A, Bindoni M (1969) Accrescimento e moltiplicazione cellular del tumore di Ehrlich nel topo, dopo asportazione della ghiandola pineale. Boll Soc Ital Biol Sper 45:1647–1650

Blask DE (1984) The pineal: an oncostatic gland. In: Reiter RJ (ed) The pineal gland. Raven Press, New York, pp 253–285

Blask DE (1993) Melatonin in oncology. In: Yu HS, Reiter RJ (eds) Melatonin-biosynthesis, physiological effects, and clinical application. CRC Press, Boca Raton, pp 447–475

Blask DE, Hill SM (1988) Melatonin and cancer: basic and clinical aspects. In: Miles A, Philbrick DRS, Thompson C (eds) Melatonin–Clinical Perspectives. Oxford University Press, New York, pp 128–173

Chen LD, Leal BZ, Reiter RJ, Abe M, Sewerynek E, Melchiori D, Meltz ML, Poeggeler B (1995) Melatonin's inhibitory effect on growth of ME-180 human cervical cancer cells is not related to intracellular glutathione concentrations. Cancer Lett 91:153–159

Cos S, Blask DE (1994) Melatonin modulates growth factor activity in MCF–7 human breast cancer cells. J Pineal Res 17:25–32

Crespo D, Fernandez-Viadero R, Verduga R, Ovejero V, Cos S (1994) Interaction between melatonin and estradiol on morphological and morphometric features of MCF-7 human breast cancer cells. J Pineal Res 16:215–222

El-Domeri AA, Das Gupta TK (1976) The influence of pineal ablation and administration of melatonin on growth and spread of hamster melanoma. J Surg Oncol 8:197–205

Hadju SI, Dorro RS, Liebermann PH, Foote FW Jr (1972) Degeneration of the pineal gland of patients with cancer. Cancer 29:706–709

Hill SM, Blask DE (1988) Effects of the pineal hormone melatonin on the proliferation and the morphological characteristics of human breast cancer cells (MCF-7) in culture. Cancer Res 48:6121–6126

Karasek M (1994) Malignant tumors and the pineal gland. In: Gupta D, Wollmann HA, Fedor-Freybergh PG (eds) Pathophysiology of immune-neuroendocrine communication circuit. Mattes Verlag, Heidelberg, pp 225–244

Karasek M (1996) Melatonin and human neoplastic disease. Int J Thymol 4, suppl 1:75–79

Karasek M (1997) Relationship between the pineal gland and experimentally induced malignant tumors. Front Horm Res 23:99–106

Karasek M, Fraschini F (1991) Is there a role for the pineal gland in neoplastic disease? In: Fraschini G, Reiter RJ (eds) Role of melatonin and pineal peptides in neuroimmunomodulation. Plenum Press, New York, pp 243–251

Karasek M, Marek K, Swietoslawski J, Stepien H (1991) Relationship between the pineal gland and estrogen-induced prolactinoma in the Fisher 344 rat with special reference to the ultrastructure of pinealocytes. Adv Pineal Res 6:27–35

Karasek M, Liberski P, Kunert-Radek J, Bartkowiak J (1992) Influence of melatonin on the proliferation of hepatoma cells in the Syrian hamster: in vivo and in vitro study. J Pineal Res 13:107–110

Karasek M, Marek K, Swietoslawski J, Liberski P, Bartkowiak J (1993) Ultrastructure of the pinealocytes of the Syrian hamster bearing Kirkmann–Robbins hepatoma: a quantitative study. Neuroendocrinol Lett 15:213–219

Karasek M, Liberski P, Bartkowiak J (1994a) Influence of photoperiod on the proliferation of hepatoma cells in the Syrian hamster. Adv Pineal Res 7:243–246

Karasek M, Marek K, Zielinska A, Swietoslawski J, Bartsch H, Bartsch C (1994b) Serial transplants of 7,12-dimethylbenz[a]anthracene-induced mammary tumors in Fischer rats as model system for human breast cancer. 3. Quantitative ultrastructural studies of the pinealocytes and plasma melatonin concentrations in rats bearing an advanced passage of the tumor. Biol Signals 3:302–306

Karasek M, Winczyk K, Kunert-Radek J, Wiesenberg I, Pawlikowski M (1998) Antiproliferative effects of melatonin and CGP 52608 on the murine Colon 38 adenocarcinoma in vitro and in vivo. Neuroendocrinol Lett 19:71–78

Lewinski A, Sewerynek E, Wajs E, Lopaczynski W, Sporny S (1993) Effects of the pineal gland on the growth processes of Guerin epithelioma in male Wistar rats. Cytobios 73:89–94

Mediavilla MD, Dan Martin M, Sanchez-Barcelo EJ (1992) Melatonin inhibits mammary gland development in female mice. J Pineal Res 13:13–19

Pawlikowski M, Karasek M (1990) The pineal gland and experimentally-induced tumors. Adv Pineal Res 4:259–265

Reiter RJ (1988) Pineal gland, cellular proliferation and neoplastic growth: an historical account. In: Gupta D, Attanasio A, Reiter RJ (eds) The pineal gland and cancer. Brain Res Promotion, London, pp 41–64

Rodin AE, Overall J (1967) Statistical relationship of weight of the human pineal gland to age and malignancy. Cancer 20:1203–1214

Shah PN, Mhatre MC, Kothari LS (1984) Effect of melatonin on mammary carcinogenesis in intact and pinealectomized rats in varying photoperiods. Cancer Res 44:3403–3407

Subramanian A, Kothari L (1991) Melatonin, suppressor of spontaneous murine mammary tumors. J Pineal Res 10:136–140

Tagliabue M, Lissoni P, Barni S, Macchi I, Tancini G, Crispino S, Ferri L, Salvini E (1989) A radiologic study by CT scan of pineal size in cancer patients: correlation to melatonin blood levels. Tumori 75:226–228

Tapp E (1980) Pineal gland in rats suffering from malignancy. J Neural Transm 48:131–135

Tapp E (1982) The pineal gland in malignancy. In: Reiter RJ (ed) The pineal gland, Vol III. CRC Press, Boca Raton, pp 171–188

Ying SW, Niles LP, Crocker C (1993) Human malignant melanoma cells express high-affinity receptors for melatonin: antiproliferative effects of melatonin and 6-chloromelatonin. Eur J Pharmacol 246:89–96

Section III
Effects of Melatonin and of Unidentified Pineal Products on Tumor Growth

10 In Vitro Effects of Melatonin on Tumor Cells

Samuel Cos, Emilio J. Sánchez-Barceló

Abstract

The evidence of direct oncostatic actions of melatonin on different neoplasms is reviewed in this article. The most interesting findings concern melatonin's effects on breast cancer cells and come from experiments carried out on the human breast cancer tumor line MCF-7. Melatonin, at physiological doses, inhibits cell proliferation and invasiveness of these cells, and lengthens the duration of the cell cycle, supporting the possible therapeutic value of melatonin in mammary cancer. In different melanoma cell lines where estradiol receptors are present, such as MCF-7, the growth-inhibitory effects of melatonin have also been demonstrated. Other neoplasms on which melatonin oncostatic properties have been studied include: cancer cell lines from female reproductive organs (endometrial and cervical carcinoma cells, trophoblastic tumor of placenta), human neuroblastoma cells, pituitary tumor cells, larynx carcinoma cells, human epidermoid oral carcinoma, bladder carcinoma, rat pheochromocytoma cells, and immune cell neoplasms. In all these cases, the high doses of indolamine required to obtain significant effects, as well as the variability of these effects, make further studies necessary to clarify whether melatonin could be useful in the treatment of these malignancies.

10.1 Introduction

Since Georgiou (1929), who first suggested that the pineal gland might influence the growth and spread of malignant tumors, a great number of studies have found that substances from the pineal gland, particularly its principal hormone melatonin, inhibit carcinogenesis. In 1977, Vera Lapin organized in Vienna the first international workshop focusing on the relationship between the pineal gland and cancer: "The Pineal Gland as a New Approach to the Neuroendocrine Control Mechanism in Cancer" (a survey of the summarized results of this workshop can be found in Lapin and Ebels 1981). This important event served to systematize all the knowledge generated until that moment and can be considered as the starting point of present day studies on the role of the pineal gland in the etiology and pathogenesis of neoplastic diseases, always with the ultimate goal of finding therapeutic tools useful in the treatment of human neoplasias.

Since the pineal gland modulates the activity of several aspects of endocrine physiology, particularly those relating to reproduction (Reiter 1980), the most common models used to study the relationship between pineal function and tumorigenesis are hormone responsive tumors, particularly mammary tumors (Blask 1984; Reiter 1988). The oncostatic role of the pineal gland has been studied in different animal models of tumorigenesis, by using the classical experimental designs of changes of photoperiod, pinealectomy, or administration of melatonin. Contrary to Georgiou's original proposal that the pineal stimulates tumor growth, the most common conclusion accepted at present is that either experimental manipulations which activate the pineal gland or the administration of melatonin reduce the incidence and development of tumors, while pinealectomy usually stimulates cancer growth (see reviews from Blask and Hill 1988, Blask 1993, Sánchez-Barceló et al. 1997, Panzer and Viljoen 1997).

Different mechanisms have been proposed to explain melatonin antitumoral effects. These include: indirect neuroendocrine mechanisms, such as regulation of some pituitary and gonadal hormones that control tumor growth (Blask 1984; Sánchez-Barceló et al. 1988, 1997; Blask et al. 1989); modulation of the immune response to the presence of a malignant neoplasm (Maestroni 1993); actions as an endogenous hydroxyl-radical scavenger (Poeggeler et al. 1993; Tan et al. 1993); and, direct actions at the cellular level (Blask and Hill 1986; Cos et al. 1991).

Due to the complex interplay of hormones and growth factors involved in the regulation of tumor growth as well as the pleiotropic actions of melatonin, it is difficult to draw conclusions about the mechanism of melatonin antitumor actions from in vivo experiments. This fact encouraged numerous researchers to study melatonin anticancer actions by using tumor cell cultures as a model. Evidence of direct melatonin antiproliferative effects has been reported in a variety of neoplastic cell cultures which include basically mammary adenocarcinomas and melanomas, but also other neoplasms.

In this paper, we will review the state of the art of melatonin effects on different cell tumor lines in vitro.

10.2 Effects of Melatonin on Breast Cancer Cells

The direct effects of melatonin on mammary cancer have been studied in vitro, basically by using the MCF-7 human breast cancer cell line as a model. This cell line derives from the pleural effusion of a woman with metastatic carcinoma of the breast (Soule et al. 1973), and contains both estrogen and progesterone receptors (Brooks et al. 1973; Horwitz and McGuire 1978). MCF-7 cells are dependent upon estrogen for optimal growth and it has been shown that the levels of some RNAs and proteins therein are controlled by estrogens (Rochefort 1983). Thus, these cells represent a good model for the study of the molecular mechanisms underlying estrogen action in breast cancer.

10.2.1 Effects of Melatonin on Proliferation of Breast Cancer Cells In Vitro

The effects of melatonin on MCF-7 cell growth were firstly described by Blask and Hill in 1986. They found that melatonin is able to inhibit the in vitro proliferation of MCF-7 cells in an anchorage-dependent monolayer culture system (Blask and Hill 1986; Hill and Blask 1988). Interestingly, the inhibitory effect of melatonin is reversible and the logarithmic growth of MCF-7 cells is restored after melatonin-containing medium is replaced with fresh medium lacking melatonin (Hill and Blask 1988). Melatonin precursors, metabolites, and other pineal methoxyindoles lack the antitumoral properties of melatonin (Blask and Hill 1986) and have greater cytotoxicity than melatonin on MCF-7 cells (Shellard et al. 1989). The findings that MCF-7 cells do not metabolize melatonin (Danforth et al. 1983) and that melatonin could still be detected with a radio-immunological assay hours after it was added to MCF-7 cells (L'Hermite-Baleriaux and de Launoit 1992) confirm the finding of Blask and Hill (1986) that melatonin itself is responsible for the antiproliferative effect. However, these inhibitory effects of melatonin on MCF-7 cell growth have not been confirmed by other investigators (Shellard et al. 1989; Bartsch et al. 1992; L'Hermite-Baleriaux and de Launoit 1992; Panzer et al. 1998). Different sources of MCF-7 cells (Osborne et al. 1987) and different experimental culture conditions could be responsible for these discrepancies, as we will comment later.

The susceptibility of MCF-7 cells to the growth inhibitory effect of melatonin depends on numerous factors, such as:

1. *Concentration of melatonin in the culture medium.* Only melatonin concentrations within the physiological range, and especially 1 nM (corresponding to night peak levels), inhibit MCF-7 cell growth, as measured by either DNA content or hemocytometer cell counts, while either supra- (0.1 µM and 10 µM) or sub-physiological (10 pM and 0.1 pM) concentrations are less or completely ineffective. The dose–response curve of antiproliferative effects of melatonin on MCF-7 cells has a bell-shaped form with a maximum effect coinciding with the 1-nM concentration (Blask and Hill 1986; Hill and Blask 1988).
2. *The patterns of exposure to melatonin.* Since rhythmicity is the most important feature defining the secretory and circulating profiles of melatonin, our group considered it worthwhile studying whether different patterns (rhythmic or not) of exposure to melatonin differently affected the proliferation of human breast cancer cells. In MCF-7 cell cultures, a periodic exposure to melatonin (1 nM/0 nM alternating every 12 h) is as effective as a continuous exposure (1 nM melatonin throughout the experiment) in terms of antiproliferative effects, although the amount of melatonin and the time of cell exposure to the hormone is obviously only half under pulsatile exposure. Interestingly, the highest antiproliferative effect of melatonin is achieved when its concentration in culture media is changed every 12 h from 1 nM to 10 pM thus reproducing the night/day oscillation of serum melatonin in most mammals (Cos and Sánchez-Barceló 1994).
3. *The prevailing culture conditions.* Although only melatonin concentrations within the physiological range inhibit MCF-7 cell growth in monolayer cultures, in an anchorage-independent clonogenic system, melatonin antiproliferative actions exhibit a linear dose–response curve with pharmacological concentrations producing a maximal inhibition (Cos and Blask 1990). The number and size of colonies formed decrease linearly as the dose of melatonin increases. This fact suggests that cellular attachment, with associated changes in cell shape, may influence the response of MCF-7 to melatonin (Cos and Blask 1990). Changes in cell-surface adhesion molecules induced by melatonin on MCF-7 cells, which we will describe later, could also explain the differential response of MCF-7 cells to melatonin depending on the culture conditions. This finding may also explain why melatonin is more effective in vivo at pharmacological doses.

4. *The estrogen receptor status of the cells.* As well as in MCF-7 cells, melatonin inhibits proliferation in other estrogen-responsive cell lines (T47D, ZR-75–1), but has no effects on estrogen-insensitive breast tumor cell lines (BT-20, MDA-MB-231, MDA-MB-364, Hs587t, T47D$_{co}$) (Hill et al. 1992). In addition, among the estrogen-responsive cell lines, the MCF-7 cells are very sensitive to the antiproliferative effects of melatonin, responding to nanomolar and picomolar concentrations, whereas the other cell lines respond only to micromolar concentrations of melatonin. These data suggest a direct relation between the overall cellular responsiveness to estrogen (receptor content) and the cell's responsiveness to melatonin. This relation is further strengthened by the fact that LY-2 cells, a tamoxifen-resistant estrogen-insensitive subclone of MCF-7, is unresponsive to melatonin (Hill et al. 1992). L'Hermite-Baleriaux and de Launoit (1992) described no effect of melatonin at physiological concentrations on estrogen-responsive mammary cancer cell lines (MCF-7, ZR-75–1, and T-47D). These discrepancies may be explained by the different experimental culture conditions and, particularly, by the different cell density on the initial cultures which influence the cell proliferation rate and the effects of melatonin on cell growth (Cos and Sánchez Barceló 1995).

5. *The presence of hormonal and/or growth factors in the culture medium.* The antiproliferative action of melatonin requires the presence of serum in the culture media. In serum-free media, melatonin loses its antimitogenic capabilities unless MCF-7 cells are also simultaneously exposed to some mitogen, such as estradiol, prolactin, or epidermal growth factor (Blask and Hill 1986; Cos and Blask 1994). Melatonin blocks the mitogenic effect of estradiol (Blask and Hill 1986) as well as the stimulatory effects of prolactin (Blask et al. 1993; Lemus-Wilson et al. 1995) and epidermal growth factor (Cos and Blask 1994) on MCF-7 cell proliferation.
Human breast cancer cells (MCF-7) have been shown to produce a variety of polypeptide growth factors which act as autocrine and paracrine growth factors. The addition of melatonin to cultures of MCF-7 cells with conditioned medium from estradiol-treated cells significantly inhibits the growth stimulatory activity of this conditioned medium from which residual estradiol had been removed, suggesting that melatonin inhibits cell proliferation by blocking the action of estrogen-induced autocrine growth stimulatory factors. Furthermore, conditioned medium from melatonin-treated cells (from which melatonin had been removed) significantly inhibits cell proliferation as compared with either conditioned medium from control cells or unconditioned medium. An additional supply of melatonin to these cultures has an even greater inhibitory effect (Cos and Blask 1994). All these results suggest that physiological levels of melatonin have the capacity to act on MCF-7 human breast cancer cells to inhibit their proliferation through actions on the synthesis/secretion and/or the action of autocrine growth stimulatory or inhibitory factors (Cos and Blask 1994).

6. *The cell proliferation rate.* By plating MCF-7 cells at different densities (from 50,000 to 100,000 cells/60-mm diameter dish) it is possible to obtain different rates of cell proliferation. There is a highly significant inverse correlation between the number of plated cells and the population doubling time (PDT) in MCF-7 cells cultured in media with stripped serum (Cos and Sánchez-Barceló 1995; Jakesz et al. 1984). The inhibitory actions of melatonin show a good correlation with the cell proliferation rate of MCF-7 cells. Melatonin significantly decreases the proliferation of fast-growing cells (PDT 33 – 40 h) but is not effective when cells proliferate slowly (PDT

58–75 h). Melatonin also increases PDT in those cultures with a high proliferation rate (and consequently low PDT), but does not modify the PDT of slowly proliferating cells, which is already long (Cos and Sánchez-Barceló 1995).

The above changes in PDT can be explained by the fact that melatonin is able to increase MCF-7 cell cycle duration by 15% as compared with that of control cells, a non-evident action in cells in which the PDT is already long (Cos et al. 1996a). After 5 days of incubation with melatonin (1 nM), the fraction of cells in G_1 phase of the cell cycle increases as compared to the control plates and, simultaneously, a 50% reduction is caused in the proportion of cells in S phase (Cos et al. 1991). Furthermore, melatonin-treated synchronized cells show a decrease of DNA synthesis when compared to untreated cells (Cos et al. 1996b). All these results strongly support a growth inhibition process of melatonin through direct cell-cycle effects. Thus, melatonin increases the duration of the cell cycle by increasing the length of the G_1 phase and delaying the entrance of cells into the S phase. Therefore, melatonin allows MCF-7 tumor cells to achieve greater differentiation as demonstrated by morphological and morphometric studies (Crespo et al. 1994).

As previously indicated, some authors did not find inhibitory actions of melatonin on MCF-7 cell proliferation. The explanation to these discrepancies could be found by the different experimental culture conditions and, particularly, by the different cell density in the initial cultures, which influence the cell proliferation rate and the effects of melatonin on cell growth (Cos and Sánchez-Barceló 1995). Thus, Panzer et al. (1998) failed to observe inhibition of MCF-7 cell proliferation under physiological concentrations of melatonin. However, these authors used an initial cell density of 15×10^4 cells per 25 cm^2 (6000 cells/cm^2), which corresponds to low proliferation rate, and it has been shown that melatonin with this proliferation rate has no effect. Antiproliferative effects of melatonin can only be appreciated from an initial density of 15,000 cells/cm^2 (Cos and Sánchez-Barceló 1995), a 2.5-fold greater density than that used by Panzer et al. (1988). L'Hermite-Baleriaux and de Launoit (1992) also described no effect of melatonin on MCF-7 cell growth, but the initial cell density is not indicated in the paper. Bartsch et al. (1992) only found inhibitory effects of melatonin on MCF-7 cells at concentrations higher than 1 mM. The initial number of cells used in this experiment was 3000 per well, and they do not indicate the growth area of the plate. A low cell proliferation rate could be responsible for the lack of effects of physiological melatonin concentrations in these studies.

10.2.2 Effects of Melatonin on the Metastatic Behavior of Breast Cancer Cells

Since the antitumor actions of melatonin are, at least in part, exerted by interactions with estradiol and this gonadal steroid stimulates the invasive and the metastatic potential of cancer cells (Osborne et al. 1985; Albini et al. 1986), it was reasonable to hypothesize a possible effect of melatonin on the metastatic behavior of breast cancer cells. We dealt with this problem in a series of experiments (Cos and Sánchez-Barceló 1997; Cos et al. 1998) in which we evaluated the antimetastatic properties of melatonin on MCF-7 cells in vitro.

Our results show that, in vitro, melatonin (1 nM) reduces the number of cells which traverse a reconstituted basement membrane (Matrigel) by 44%, whereas both sub-(0.1 pM) and supra-physiological (10 µM) concentrations of melatonin fail to inhibit cell invasion. Pretreatment of MCF-7 cells with 1 nM melatonin increases the response of tumor cells to the anti-invasive effects of this indolamine, and the same concentration of melatonin reduces the invasion of MCF-7 cells by up to 74%. It is well known (Albini et al. 1986) that estradiol (1 µM or 10 nM) induces changes in MCF-7 cells that are characteristic of the malignant phenotype, such as a greater attachment to the basement membrane, a greater ability to migrate toward laminin, a greater proliferation in culture in the presence of the basement membrane matrix and a much greater ability to invade the barriers of a reconstituted basement membrane. In our experiments, melatonin (1 nM) is able to counteract the estradiol-induced increase in the number of cells that invade the Matrigel membrane.

After having found that melatonin reduces the invasiveness of tumor cells, it was necessary to define the role of this indolamine within each step of the metastatic process: (a) the attachment of MCF-7 cells to the basement membrane, (b) the chemotactic response of tumor cells into the target tissue and, (c) the secretion of proteolytic enzymes by tumor cells.

Melatonin reduces the ability of MCF-7 cells to attach to the basement membrane, this effect being greater at 1 nM than at 10 µM or 0.1 pM concentration. Estradiol (1 µM or 10 nM) increases the adhesion of cells to the basement membrane; the simultaneous addition of estradiol and melatonin completely abolishes the effect of the former (Cos and Sánchez-Barceló 1997).

The chemotactic response of MCF-7 cells towards fibronectin is also reduced by melatonin at physiological concentrations (1 nM). The simultaneous addition of estradiol (which stimulates chemotactic migration) and melatonin results in a significantly lower chemotactic response than that of estradiol-treated cells but not of untreated cells. Pretreatment of MCF-7 cells with melatonin (1 nM) for 5 days is more effective in lowering cell migration as compared to unpretreated cells (Cos and Sánchez-Barceló 1997).

Although previous studies indicate that the regulation of MCF-7 cell invasiveness by antiestrogens may be mediated by an increase in collagenase IV activity (Thompson et al. 1988), melatonin does not cause any significant change in type IV collagenolytic activity of MCF-7 cells, which indicates that it is not among the mechanisms involved in its anti-invasive effects.

Tumor cell motility and invasion are adhesion-dependent phenomena related to the presence of cell-surface adhesion molecules for both cell–cell and cell–matrix interactions. A down-regulation or a loss of expression at the cell surface correlates with an increase of the invasiveness of tumor cells as well as with a poor cell differentiation and bad prognosis of the tumor process (Sommers et al. 1994; Millon et al. 1989; Gui et al. 1997). On MCF-7 cells, melatonin increases the expression of E-cadherin (Cos et al. 1998), a calcium-dependent membrane protein responsible for cell–cell contact. The expression of E-cadherin has been inversely correlated with in vitro invasion and tumor cell differentiation (Sommers et al. 1994; Yagasaki et al. 1996). In addition, melatonin also increases the expression of β_1 integrin, a subunit of integrins, receptors that regulate interaction between cells and the extracellular matrix (Cos et al. 1998).

The invasive progression of human breast cancer cells in terms of the acquisition or loss of markers which represent the differentiation status of the cell, traverses a spectrum of stages from the poorly invasive cells (estrogen receptor positive cells) through the loss of cellular adhesion molecules, such as E-cadherin and integrins, and the acquisition of vimentin expression. Cancer cells that are very invasive are poorly differentiated, lose estrogen receptors and E-cadherin expression, have a lower expression of integrins, and express vimentin. Taken together, these results suggest that melatonin shifts MCF-7 human breast cancer cells to a lower invasive status increasing β_1 integrin subunit expression and E-cadherin expression and promoting the differentiation of tumor cells, as has been previously demonstrated by morphological and morphometric studies (Crespo et al. 1994).

The in vitro anti-invasive effect of melatonin is also correlated with an in vivo decrease of the tumorigenicity of MCF-7 cells (Cos et al. 1998).

10.2.3 Influence of Melatonin on Active Cell Death of Mammary Cancer Cells

Most explanations for the antitumor action of melatonin refer to the antiproliferative actions of this hormone. However, tumor growth should be considered as a balance between cell proliferation and cell death. Recent studies have demonstrated that the treatment of DMBA-induced rat mammary cancer with antiestrogens causes some cancer cells to fall into apoptosis with a subsequent regression of the tumor (Watanabe et al. 1995). The treatment of MCF-7 cells with tamoxifen (a nonsteroidal antiestrogen) also induces apoptosis (Welsh 1994). On this basis, we considered it worthwhile studying whether melatonin may play a role in the control of programmed cell death in MCF-7 human breast cancer cells.

Melatonin decreases the viability of MCF-7 cells and increases the percentage of dead cells, as assessed by the trypan blue dye exclusion test. In previous morphological studies (Hill and Blask 1988; Crespo et al. 1994), it has been demonstrated that some melatonin-treated MCF-7 cells display degenerative features such as mitochondrial swelling with disruption of cristae, cytoplasmic vacuolation, nuclear chromatin disaggregation, and cell lysis, all signs suggestive of apoptosis. In situ, the labeling of apoptosis-induced DNA strand breaks by the TUNEL reaction showed that melatonin is able to activate apoptotic cell death in this tumor cell line (Cos et al. 1997).

Regulation of apoptosis involves a large number of genes that can be classified into three categories: (a) suppressors of apoptosis, such as *bcl-2* and *bcl-x$_L$*; (b) genes that act as effectors of apoptosis, such as *bax* and the interleukin-1 converting enzyme genes, and (c) intermediate genes upstream such as Fas/Fas ligand, *p53*, *myc,* and WAF1. The *p53* tumor suppressor gene delays cell cycle progression in response to DNA damage and arrests the cell cycle in G_1 phase before the initiation of DNA synthesis. This function is executed by accumulation of *p53* followed by induction of WAF1 gene. The expression of both *p53* and WAF1 genes in MCF-7 cells is increased by melatonin (Cos et al. 1997). So far, changes induced by melatonin in the expression of some genes suppressors (*bcl-2* and *bcl x$_L$*) or promoters (*Bax*) of apoptosis have not yet been found. These findings support the hypothesis that melatonin treatment increases the number of MCF-7 cells which fall into apoptosis, probably by interacting with *p53*-mediated mechanisms.

10.2.4 Mechanisms of the Oncostatic Action of Melatonin In Vitro

As it has been shown in this review, there is strong evidence for oncostatic actions of melatonin on breast cancer cells in vitro, at least as far as the MCF-7 cell line, the most studied, is concerned. These antitumor actions are manifested by an inhibition of cell proliferation, an increased number of cells falling into apoptosis, and an inhibition of metastatic spread. At this point, the question that we should ask is how the oncostatic action of melatonin is transmitted inside the cell, and which intracellular mechanisms are involved in the antitumor effects.

Although melatonin membrane receptors coupled with G proteins and cyclic nucleotide as second messengers have been described in different animal tissues (Dubocovich 1995), the attempts to characterize melatonin receptors in MCF-7 cell membranes have been unsuccessful. Alternatively, nuclear binding sites specific for melatonin have been considered. Since melatonin is a highly lipophilic hormone (Costa et al. 1995) it can, therefore, easily cross the cytoplasmic and nuclear membranes of the cell. MCF-7 cells endogenously express various nuclear receptors, including some members of the RZR/ROR family, considered to be nuclear receptors for melatonin (Steinhilber et al. 1995). Recent reports of a regulatory effect of melatonin in cell cycle kinetics confirm the nucleus of MCF-7 cells as a possible target for melatonin (Cos et al. 1991, 1996b).

Since only human breast cancer cell lines that express estrogen receptors have been found to be responsive to the antimitogenic effects of melatonin, the current hypothesis is that the oncostatic actions of melatonin are mediated via its effects on the tumor cells estrogen-response pathway (Cos et al. 1991; Hill et al. 1992; Hill and Blask 1988) by behaving as a naturally occurring antiestrogen. The link between the antiproliferative effect of melatonin on MCF-7 cell growth and the estrogen-response pathway is further supported by: (a) the ability of melatonin to block, under different culture systems (monolayer and clonogenic soft agar culture), the mitogenic effect of estradiol (Blask and Hill 1986; Cos et al. 1991); (b) melatonin blockade of the estrogen-rescue of tamoxifen-inhibited cells in clonogenic soft agar and monolayer cultures (Cos et al. 1991); (c) melatonin modulation of estrogen receptor expression in MCF-7 cells (Hill et al. 1992; Molis et al. 1993); and (d) melatonin modulation of estrogen regulated proteins, growth factors, and proto-oncogenes in human breast cancer cells (Cos and Blask 1994; Molis et al. 1995).

Melatonin effects on estrogen receptors are controversial. Although this subject is being discussed in depth in another chapter of this book, we will briefly comment on some aspects of the problem. A first report by Danforth et al. (1983) indicated that in vitro incubation of MCF-7 cells with melatonin (1 nM) increased the number of cytosolic and nuclear estrogen receptors, without changing their affinity. However, later studies by Molis et al. (1993) found that melatonin has a biphasic effect where supraphysiological concentrations (1 µM – 0.1 µM) enhance estrogen receptor expression, while physiological concentrations (10 nM – 10 pM), those which inhibit MCF-7 cell growth, significantly inhibit estrogen receptor expression and decrease estrogen binding activity in a dose-specific and time-dependent manner. Differences in methodologies, such as the duration of exposure to melatonin or the clone of MCF-7 cells used, might explain the discrepancies between these reports. These effects appear not to be mediated by the direct binding of melatonin to the estrogen receptor (Molis et al. 1994). However, although melatonin does not bind to the estrogen receptor, it could modulate estrogen receptor expression through an intermediary factor or by interacting with its own receptors to start the events which induce the down-regulation of the estrogen receptor expression (Molis et al. 1993). In this way, it is interesting to emphasize that from the identified nuclear receptors to melatonin, the subtype RZR/ROR is expressed by MCF-7 cells (Steinhilber et al. 1995). The effects of supraphysiological concentrations of melatonin could be explained as a down-regulation of melatonin receptors, thus suppressing the signal transduction events for estrogen receptor down-regulation.

Molis et al. (1995) proposed that the mechanism of action of melatonin is more complex than mere inhibition of estrogen receptor transcription and suggested that a distinction should be made between early and late events. The former, including signal transduction effects, might account for enhanced expression of TGF-α, TGF-β, and pS2, as well the proto-oncogenes *c-myc* and *c-fos*. The late events include estrogen receptor down-regulation with subsequent modulation of estrogen-related growth factors. They found that melatonin stimulates the expression of the growth inhibitory factor TGF-β, which is normally down-regulated by estrogens, and decreases that of *c-fos*, a proto-oncogene related to cell proliferation, which is up-regulated by steroids. These data could support the hypothesis that melatonin inhibits growth of MCF-7 cells by interacting with the estrogen-response pathway. However, melatonin also increases levels of growth factors which are normally up-regulated by estradiol and have mitogenic actions on MCF-7 cells (TGF-α, c-myc, and pS2).

One of the most interesting aspects of the anticancer activity of melatonin is its ability to inhibit the growth of the tumor cell by blocking the stimulatory effects of other hormones and growth factors. In addition to estrogen, breast cancer cell growth is under the complex regulatory influence of estrogen-inducible polypeptide hormones such as pituitary prolactin as well as autocrine growth factors. Lemus-Wilson et al. (1995) demonstrated that melatonin can block the mitogenic effects of tumor-promoting prolactin or prolactin receptor antibodies on MCF-7 and ZR75-1 cells. Melatonin interrupts the prolactin receptor-mediated growth signal, suggesting that the oncostatic activity of melatonin may also be linked with an antagonism of prolactin action. Similarly, melatonin attenuates epidermal growth factor stimulated MCF-7 cell growth in serum-free medium. The inhibitory effect of melatonin increased as the dose of epidermal growth factor increased. This non-antiestrogenic inhibitory effect of melatonin was reversed by estradiol, but not by the epidermal growth factor itself, suggesting that melatonin requires accessible estrogen receptor sites for its inhibitory activity on the growth-stimulating action of epidermal growth factor (Cos and Blask 1994). Furthermore, melatonin not only blocks estradiol-induced growth factor stimulation of cell proliferation but also increases the levels of inhibitory growth factors (probably TGF-β) in a medium conditioned by MCF-7 cells (Cos and Blask 1994). These effects of melatonin on growth factors may explain the finding that the antiproliferative action of melatonin is directly proportional to the initial density of the cultured cells (Cos and Sánchez-Barceló 1995).

Among the mechanisms by which melatonin causes a decrease in cell proliferation, we can include a direct cell cycle effect. Melatonin action on the cell cycle involves several aspects. On the one hand, melatonin increases the duration of the cell cycle by enlarging the G_1 phase, thereby augmenting the fraction of cells in this phase, delaying the entrance of cells in the S phase (Cos et al. 1991, 1996a), and allowing tumor cells to achieve a greater differentiation (Crespo et al. 1994). On the other hand, melatonin can also decrease the DNA synthesis in cells which are not quiescent and are in the S phase of the cell cycle (Cos et al. 1996b).

Indirect oncostatic actions of melatonin, mediated by its properties as antioxidant, have also been proposed. Thus, Blask and Wilson (1995) showed that the inhibitory effects of physiological concentrations of melatonin on MCF-7 cell proliferation require the presence of reduced glutathione (GSH). Furthermore, some cell lines which are resistant to the antiproliferative effects of melatonin become sensitive by inhibiting the activity of the enzyme glutathione-*S*-transferase thus increasing the levels of GSH.

A possible role of nitric oxide (NO) in the melatonin inhibition of human breast cancer growth has also been suggested by Blask and Wilson (1994). Based on the previous experiment by Maragos et al. (1993) demonstrating the antiproliferative effects of nitric oxide (NO) on the A375 melanoma cells, Blask and Wilson incubated MCF-7 cells for 5 days with 1 nM melatonin either in the presence or in the absence of NMMA (an inhibitor of the enzyme NO synthetase). In the presence of NMMA, melatonin lacked its antiproliferative activity. The authors postulate that a critical threshold level of NO might be required for melatonin to exert its oncostatic actions on MCF-7 cells or, alternatively, that melatonin might be increasing NO production by inducing NO synthetase activity (Blask and Wilson 1994).

Finally, there are other possible mechanisms which could be involved in the oncostatic actions of melatonin on breast cancer cells, although they have not been tested on these tumor models. One of them is the possibility that melatonin antineoplastic actions could be related to its effects on cytoskeletal organization, by acting as a Ca^{2+} calmodulin antagonist (Benitez-King and Anton-Tay 1993; Benitez-King et al. 1990; Huerto-Delgadillo et al. 1994). Another interesting mechanism to be studied is the possible effect of melatonin on the gap junctional intercellular communication. Most, if not all, cancer cells have some dysfunction in gap-junction-mediated intercellular communication, either because of defects in cell adhesion or inability to have functional gap junctional communication. In addition, most, if not all, tumor-promoting chemicals and conditions down-regulate gap junction function, while some anti-tumor-promoting chemicals can up-regulate gap junctional communication (Trosko et al. 1990). If melatonin can increase gap junctional communication in breast cancer cells it may suppress tumor growth by allowing the transfer of other molecules with the ability to suppress tumor growth. Recently, Ubeda et al. (1995) found that physiological concentrations (10 pM – 10 nM) of melatonin increased gap junctional intercellular communication in a cell line of mouse embryo fibroblasts.

10.3 Effect of Melatonin on Melanoma Cells

Melanocytes are cells of neuroendocrine origin characterized by the ability to synthesize melanin from the amino acid L-tyrosine. Uncontrolled proliferation of melanocytes in the skin can lead to the development of melanomas, one of the fastest growing neoplasms in humans. From the pioneering studies of Das Gupta's group showing that pinealectomy favored the development of primary tumors and tumor metastases of pigmented malignant melanoma transplanted to Syrian hamsters (Das Gupta and Terz 1967), and that the effects of pineal ablation were reversed by melatonin (El-Domeiri and Das Gupta 1973), numerous experiments have been performed in clonogenic melanoma cell lines from both animal and human tumors, to assess possible direct actions of melatonin on these tumor cells. In one of the first such studies, millimolar concentrations of melatonin were shown to cause a modest 25% inhibition in the growth of a cloned melanoma cell line (B7) derived from a spontaneously arising tumor in a male golden hamster. Interestingly, micromolar amounts of this indole caused a 37% stimulation of cell growth suggesting that melatonin may have a dose-dependent biphasic effect (Walker et al. 1978). In this experiment, physiological doses of melatonin had no effects. The influence of melatonin doses on the nature of the tumor cell response (stimulatory or inhibitory) has been also described by other authors. Thus, Bartsch and Bartsch (1984), by using human melanoma cells, found that micromolar concentrations of melatonin caused a 60% inhibition of cell proliferation while a tenfold higher dose produced a moderate stimulation of cell growth. Noteborn et al. (1988) isolated a peptidic fraction from a pineal aqueous extract which inhibits the growth of human melanoma cells. The activity of this pineal compound differs from other substances present in the pineal such as melatonin, pteridines, and β-carbolines. Other pineal indoles, especially methoxytryptamine, seem to be more potent than melatonin in inhibiting the incorporation of [^{3}H]thymidine, [^{3}H]uridine, and [^{3}H]leucine by B16 mouse melanoma cells (Sze et al. 1993). In any case, the concentration of pineal indoles used in this experiment was too high (4–500 µM).

Meyskens and Salmon (1981) observed that the sensitivity to melatonin of clonogenic soft agar cultured human melanoma cells from patients' biopsies varied largely. Out of 11 patients studied, no effects were seen in 3 cases; in 6 patients, total cloning efficiency was decreased with varying concentrations of melatonin (0.001 pM–10 µM), and in 2 patients, melatonin stimulated the formation of melanoma colonies. Bartsch et al. (1986) also described significant differences in the melatonin sensitivity of a melanoma cell line depending on the passages tested. The different sensitivity of the human melanoma cells towards melatonin seems to be dependent on the rate of cell growth. Thus, melatonin exhibits a direct growth-inhibiting activity on slow-growing melanoma cells (from the earlier passages) losing its activity when the cells grow faster (in passages of more than 150) (Bartsch et al. 1986). In cells from passages of more than 150, the melatonin concentration to reduce proliferation by 50% in relation to the controls is 1 mM (Bartsch et al. 1987). The differences in response to melatonin observed by Meyskens and Salmon (1981) could be due to the presence or absence of steroid receptors in the tumor cells from the different patients. In the same way, Bartsch et al. (1986) suggested that the loss of sensitivity towards melatonin in melanoma cells could be explained by either loss of steroid receptors or by a selection of cell clones without receptors during passaging. Recently, it has been demonstrated (Hu and Roberts 1997), that melatonin significantly inhibits the growth of human uveal melanoma cell lines in vitro at a concentration comparable to that which inhibits the growth of cutaneous melanoma cells in vitro (0.1 nM–100 nM), and similar also to the melatonin concentration in the aqueous humor of normal eyes. These results indicate that melatonin could be a new treatment for metastatic uveal melanoma, one of the most common intraocular malignant tumors.

Differential and unparalleled effects of melatonin on cell proliferation and melanogenesis have been described on S91 mouse melanoma cell line and AbC1 hamster melanoma cell line (Slominski and Pruski 1993). At low (physiological) concentrations (0.1–10 nM), melatonin inhibited cell proliferation and was without any effect on melanogenesis; however, at high (pharmacological) concentrations (>0.10 µM), melatonin inhibited the induction of melanogenesis but was either without an effect or even stimulated (100 µM) proliferation, suggesting that melatonin can regulate or modify both processes via different mechanisms. Concerning melanogenesis, melatonin, at pharmacological doses, diminishes the number of α-MSH receptors. However, at physiological concentrations, it affects melanogenesis by inhibiting the induction of tyrosinase de novo synthesis (Valverde et al. 1995).

The mechanism of the growth-inhibiting effect of melatonin on melanoma cells is not clear. It may be mediated through melatonin receptors of the melanoma cells (Pickering and Niles 1992; Slominski and Pruski 1993; Helton et al. 1993; Eison and Mullins 1993; Ying et al. 1993). Studies of [125I]melatonin-specific binding to particulate membrane suspensions from Syrian hamster RPMI 1846 melanoma cells revealed a K_d of 0.89 nM and a B_{max} of 6.2 fmol/mg protein, values consistent with expression of a melatonin receptor (Pickering and Niles 1992). Specific 2-[125I]melatonin binding has also been observed in murine B16 melanoma as well as SK Mel 28 and SK Mel 30 human melanoma cells (Helton et al. 1993). Binding studies in the human malignant melanoma (M-6) cell line showed the coexistence of 2-[125I]melatonin binding sites with picomolar and nanomolar affinities (Ying et al. 1993). By using [3H]melatonin as a probe, Slominski and Pruski (1993) detected saturable and specific binding sites in crude membranes and purified nuclei of S91 mouse melanoma cells. However, other authors have not found melatonin binding sites in murine B16 melanoma cells nor in normal human melanocytes cultured in vitro (Mengeaud et al. 1994).

Melatonin could also inhibit melanoma cell growth by interacting with the estrogen receptors which have been described in some human melanoma specimens as well as in some human and rodent melanoma cell lines (Karakousis et al. 1980; Taylor et al. 1984; Gill et al. 1984; Walker et al. 1987).

Zisapel and Bubis (1994) found that melatonin (0.1–10 nM) inhibits growth of murine melanoma M2R cells in a dose-dependent manner but only during the first 2 days after plating, later decreasing. Furthermore, melatonin was ineffective in conditioned media, suggesting its actions occurred subsequent to a decrease in production and/or secretion of autocrine factors. Recently they (Bubis and Zisapel 1995a; Bubis et al. 1996) have demonstrated that melatonin modulates constitutive protein secretion from murine melanoma M2R cells in vitro. The nature of the response depends on cell density: nanomolar concentrations of melatonin inhibit protein release early after plating or at low cell density whereas it facilitates the release later on or at high cell density (Bubis and Zisapel 1995b).

10.4 Effect of Melatonin on Cancer Cells from Female Reproductive Organs

The presence of melatonin receptors on hypothalamic and pituitary areas involved in the control of reproduction (Cardinali et al. 1979), as well as in organs of the female reproductive system such as ovaries (Cohen et al. 1978), encouraged numerous authors to study the possible effects of melatonin on the development of tumors affecting tissues of the reproductive tract. We will review the experiments carried out with cell lines from ovarian, endometrial, cervical, or placental carcinomas.

Leone et al (1988) observed that while melatonin and *N*-acetylserotonin inhibited in vitro growth of SK-OV-3 and JA-1 ovarian carcinoma cell lines only at mM concentrations, other pineal compounds (5-methoxytryptamine) and the metabolite 6-hydroxymelatonin were five times more effective. However, the effects of these methoxyindoles are cytotoxic rather than cytostatic, since, by using soft agar clonogenic cell survival assay, the concentration of these compounds required to reduce survival by 50% of the controls corresponded well with those derived from growth inhibition assays (Leone et al. 1988; Shellard et al. 1989). Kikuchi et al. (1989) did not find a dose-dependent inhibition by melatonin of the proliferation of KF cells derived from a human serous cystadenocarcinoma of the ovary, and even high melatonin doses (200 µM) stimulated rather than inhibited the cell proliferation. Finally, Bartsch et al. (1992) found hardly any effect of melatonin (1 mM – 0.1 pM) on EFO-27 ovarian carcinoma cells; only melatonin concentrations higher than 1 mM were able to reduce the cell number by 50 % in relation to the untreated controls.

Although micromolar concentrations of melatonin inhibit the proliferation of the estrogen responsive human endometrial adenocarcinoma cell line (RL95-2), physiological levels of the indolamine (1 nM) had no effect on cell growth (Hill and Blask 1988). Curiously, a significant correlation between melatonin deficiency and endometrial cancer in woman has been recently demonstrated (Grin and Grünberger 1998). This finding should be the basis of further experiments to investigate a possible role of melatonin in the growth of endometrial carcinomas.

Melatonin, in concentrations up to 1 mM shows no effect on mitosis in cultures of HeLa cells, derived from a human cervical carcinoma (Fitzgerald and Veal 1976; Panzer et al. 1998). However, when HeLa cells were preincubated with 100 µM melatonin prior to the addition of 0.1 µM colchicine, a reduction in the mitotic index was observed in comparison to colchicine alone (Fitzgerald and Veal 1976). Using ME-180 human cervical cancer cells, Chen et al. (1995) observed that melatonin at a concentration of 2 mM inhibited the growth of the cells after 48-h incubation. However, at concentrations of 2 µM or 0.1 mM melatonin had no effect on ME-180 cell proliferation.

The JAR cell line was established directly from a trophoblastic tumor of the placenta. Melatonin and various pineal indoles inhibit [³H]thymidine incorporation by JAR cells in a concentration range from 4 µM to 500 µM (Sze et al. 1993).

10.5 Effect of Melatonin on Other Neoplasms

The bulk of studies on the antitumor effects of melatonin have been focused on mammary cancer and, to a lesser extent, on melanomas. There is also a group of studies focusing on the effects of melatonin on tumors of the female reproductive organs, as we have commented in the previous section. Here, we will review the studies carried out in vitro with cell lines of different neoplasms not included in the above-mentioned studies.

Human Neuroblastoma Cells In Vitro. Cos et al. (1996c) found that 10 pM and 1 nM concentrations of melatonin significantly inhibited the proliferation of the human neuroblastoma cell line SK-N-SH in vitro. Subphysiological (0.1 pM) or supraphysiological (0.1 µM and 10 µM) concentrations of melatonin lacked this effect. After 8 days of exposure to melatonin (1 nM) SK-N-SH cells showed significantly smaller cell and nuclear sizes than control cells. Melatonin-treated cells presented greater differentiation, as evidenced by neurite outgrowth, than control cells (Cos et al. 1996c).

Some authors have been using neuroblastoma cells as a model to study a possible use of melatonin in the treatment of Alzheimer's disease. Melatonin, at pharmacological doses (3 – 4 mM), inhibited in neuroblastoma cells the secretion of the beta amyloid precursor protein (β-APP), a precursor of the amyloid beta-peptide, the component of senile plaques which are one of the major pathological hallmarks of Alzheimer's disease (Song and Lahiri 1997). On this line of research, Papolla et al. (1997) demonstrated that melatonin is remarkably effective in preventing death of cultured neuroblastoma cells as well as oxidative damage and intracellular Ca^{2+} increase induced by a cytotoxic fragment of β-APP. Finally, Lezoualch et al. (1998) showed the protective effects of melatonin and N-acetylserotonin against oxidative stress-induced death of SK-N-MC neuroblastoma cells.

Most of the experiments demonstrating the role of melatonin on the cytoskeleton have been performed on neuroblastoma cells. Thus, it has been reported that physiological concentrations of melatonin induce an increase of microtubules in neuroblastoma N1E-115 cells (Benitez-King et al. 1990; Huerto-Delgadillo et al. 1994; Melendez et al. 1996).

Pituitary Tumor Cells In Vitro. The effects of different concentrations of melatonin on the proliferation of rat pituitary prolactin-secreting tumor cells have been evaluated in vitro by measuring the incorporation of [^{3}H]thymidine by the tumor cells. Melatonin is only effective in a fairly narrow concentration range, 0.1 µM – 10 pM, inhibiting the incorporation of labeled thymidine (Karasek et al. 1988).

Larynx Carcinoma Cells. Melatonin inhibits the proliferation of Hep-2 larynx carcinoma cells, but only in concentrations higher than 0.5 µM (Bartsch et al. 1992).

Human Epidermoid Oral Carcinoma Cells. Melatonin, in concentrations up to 1 mM, shows no effect on mitosis in cultures of KB cells (derived from an epidermoid carcinoma in the mouth). However, when KB cells were preincubated with 100 µM melatonin, the antimitotic effects of 0.1 µM colchicine were significantly higher than in cells which had not been previously exposed to the pineal indolamine (Fitzgerald and Veal 1976).

Bladder Carcinoma Cells. Melatonin exerts significant in vitro growth inhibition against RT112 cells, derived from a human transitional carcinoma of the bladder, but only at relatively high concentrations of approximately 0.5 mM (Shellard et al. 1989).

Rat Pheochromocytoma Cells. Melatonin produces a biphasic response with respect to the rat pheochromocytoma cells (PC12). At low concentrations (1 – 10 nM) melatonin suppresses PC12 growth whereas at higher concentrations (10 µM) it prevents apoptosis in a time- and concentration-dependent way (Roth et al. 1997). In the same cell line, Song and Lahiri (1997) described that melatonin, at mM doses, decreases the expression of β-APP, β-actin, and glyceraldehyde-3-phosphate dehydrogenase. Melatonin was also found to potentiate the nerve growth factor-mediated differentiation in PC12 cells (Song and Lahiri 1997).

Immunological Cell Lines. Melatonin exerts widespread stimulatory effects on both cellular and humoral immunity, which could be a determinant of its antitumoral actions in neoplastic immune cells (see review from Maestroni 1993; Conti and Maestroni 1995). Furthermore, the possible role of melatonin on the proliferation of cell lines from tumors of the immune system has been studied yielding controversial results. Thus, Bartsch et al. (1992) showed melatonin inhibitory effects on K562 erythroleukemia cell line. The concentration of melatonin at which the number of treated cells is 50% of that of the controls was in the micromolar range. Sze et al. (1993) studied the effects of melatonin and other pineal indoles on the proliferation of several tumor cell lines including sarcoma (S180, derived from Swiss Webster Sarcoma 180 ascites) and macrophage-like cell (PU5, adapted to culture from a spontaneous BALB/c mouse lymphoid tumor). In both cases, methoxytryptamine exhibited the highest antiproliferative effect (measured as a reduction of [^{3}H]thymidine incorporation), with melatonin having a weaker effect. In any case, the concentrations of these pineal indoles necessary to obtain significant decreases of [^{3}H]thymidine incorporation were in the micromolar range. Recently, Persengiev and Kyurkchiev (1993) described that melatonin (200 µM) significantly inhibited the incorporation of [^{3}H]thymidine into both normal mouse and human lymphocytes and T lymphoblastoid cell lines. In contrast, melatonin provoked an increase of myeloma cell proliferation in the same study. Panzer et al. (1998) showed no differences in the growth, cell cycle, or morphology between melatonin-exposed (from 0.1 µM to 0.1 nM) and control cells on MG-63 human osteogenic sarcoma cells and TK6 human lymphoblastoid cells.

10.6 Conclusions

In this chapter we have reviewed in detail the direct actions of melatonin on different neoplasms under in vitro conditions. The most interesting data concern the effects of melatonin on breast cancer cells. Although the fact that most studies have been performed on the same tumor cell line (MCF-7 cells) makes the generalization of the results somewhat conflictive, the characteristics of these actions, comprising different aspects of tumor biology such as proliferation, apoptosis and metastasis, as well as the doses (physiological range) at which the effects are accomplished, gives a special value to these findings. Due to the strength of these data, the scarce number of clinical studies focusing on the possible therapeutic value of melatonin on mammary cancer is surprising.

Melanoma cells have in common with MCF-7 cells the presence of estradiol receptors and, as with the breast cancer cells, respond to melatonin. Despite some discrepancies, the inhibitory effects of melatonin at physiological concentrations on the growth of different melanoma cell lines, including human melanoma cells, have been demonstrated, and this could be important for future therapeutic applications.

In relation to the other neoplasms in which the oncostatic properties of melatonin have been investigated, the high doses of indolamine necessary to obtain significant effects, as well as the variability of these effects, make further studies necessary to clarify whether melatonin could be useful in the treatment of these malignancies.

Acknowledgements. This work was supported by grants from the Spanish DGICYT (PM97–0042) and "Marqués de Valdecilla" Foundation (6/98).

References

Albini A, Graf J, Kitten GT, Kleinman HK, Martin GR, Veillette A, Lippman ME (1986) 17β-estradiol regulates and v-Ha-*ras* transfection constitutively enhances MCF-7 breast cancer cell interactions with basement membrane. Proc Natl Acad Sci USA 83:8182–8186

Bartsch C, Bartsch H (1984) The link between the pineal gland and cancer: an interaction involving chronobiological mechanisms. In: Halberg F, Reale L, Tarquini B (eds) Chronobiological approach to social medicine. Istituto Italiano di Medicina Sociale, Roma, pp 105–126

Bartsch H, Bartsch C, Flehmig B (1986) Differential effect of melatonin on slow and fast growing passages of a human melanoma cell line. Neuroendocrinol Lett 8:289–293

Bartsch H, Bartsch C, Noteborn HPJM, Flehmig B, Ebels I, Salemink CA (1987) Growth-inhibiting effect of crude pineal extracts on human melanoma cells in vitro is different from that of known synthetic pineal substances. J Neural Transm 69:299–311

Bartsch H, Bartsch C, Simon WE, Flehmig B, Ebels I, Lippert TH (1992) Antitumor activity of the pineal gland: effect of unidentified substances versus the effect of melatonin. Oncology 49:27–30

Benitez-King G, Antón-Tay F (1993) Calmodulin mediates melatonin cytoskeletal effects. Experientia 49:635–641

Benitez-King G, Huerto-Delgadillo L, Antón-Tay F (1990) Melatonin effects on the cytoskeletal organization of MDCK and neuroblastoma N1E-115 cells. J Pineal Res 9:209–220

Blask DE (1984) The pineal: an oncostatic gland? In: Reiter RJ (ed) The Pineal Gland. Raven Press, New York, pp253–284

Blask DE (1993) Melatonin in Oncology. In: Yu H-S, Reiter RJ (eds) Melatonin: Biosynthesis, Physiological Effects and Clinical Applications. CRC Press, Boca Raton, FL, pp 447–495

Blask DE, Hill SM (1986) Effects of melatonin on cancer: studies on MCF-7 human breast cancer cells in culture. J Neural Transm Suppl 21:433–449

Blask DE, Hill SM (1988) Melatonin and cancer: basic and clinical aspects. In: Miles A, Philbrick DRS, Thompson C (eds) Melatonin Clinical Perspectives. Oxford University Press, New York, pp 128–173

Blask DE, Wilson ST (1994) Melatonin and oncostatic signal transduction: evidence for a novel mechanism involving glutathione and nitric oxide. In: Moller M, Pevet P (eds) Advances in Pineal Research. Vol 8, John Libbey & Co Ltd, London, pp 465–471

Blask DE, Wilson ST (1995) Melatonin action on human breast cancer cells: involvement of glutathione metabolism and the redox environment. In: Fraschini F, Reiter RJ, Stankov B (eds) The Pineal Gland and its Hormones: Fundamentals and Clinical Perspectives. Plenum Press, New York, pp 209–217

Blask DE, Hill SM, Pelletier DB, Anderson JM, Lemus-Wilson A (1989) Melatonin: an anticancer hormone of the pineal gland. In: Reiter RJ, Pang SF (eds) Advances in Pineal Research, Vol 3, John Libbey & Co Ltd, London, pp 259–293

Blask DE, Lemus-Wilson AM, Wilson ST, Cos S (1993) Neurohormonal modulation of cancer growth by pineal melatonin. In: Touitou Y, Arendt J, Pévet P (eds) Melatonin and the Pineal Gland – From Basic Science to Clinical Application. Elsevier Science Publishers BV, Amsterdam-London-New York-Tokyo, pp 303–310

Brooks SC, Locke ER, Soule HD (1973) Estrogen receptor in a human breast tumor cell line (MCF-7) from breast carcinoma. J Biol Chem 248:6251–6256

Bubis M, Zisapel N (1995a) Modulation by melatonin of protein secretion from melanoma cells: is cAMP involved? Mol Cell Endocrinol 112:169–175

Bubis M, Zisapel N (1995b) Facilitation and inhibition of G-protein regulated secretion by melatonin. Neurochem Int 27:177–183

Bubis M, Anis Y, Zisapel N (1996) Enhancement by melatonin of GTP exchange and ADP ribosylation reactions. Mol Cell Endocrinol 123:139–148

Cardinali D, Vacas MI, Boyer EE (1979) Specific binding of melatonin in bovine brain. Endocrinology 105:437–441

Chen LD, Leal BZ, Reiter RJ, Abe M, Sewerynek E, Melchiorri D, Meltz ML, Poeggeler B (1995) Melatonin's inhibitory effect on growth of ME-180 human cervical cancer cells is not related to intracellular glutathione concentrations. Cancer Lett 91:153–159

Cohen M, Roselle D, Chabner B, Schmidt TJ, Lippman M (1978) Evidence for a cytoplasmic melatonin receptor. Nature 274:894–895

Conti A, Maestroni GJM (1995) The clinical neuroimmunotherapeutic role of melatonin in oncology. J Pineal Res 19:103–110

Cos S, Blask DE (1990) Effects of the pineal hormone melatonin on the anchorage-independent growth of human breast cancer cells (MCF-7) in a clonogenic culture system. Cancer Lett 50:115–119

Cos S, Blask DE (1994) Melatonin modulates growth factor activity in MCF-7 human breast cancer cells. J Pineal Res 17:25–32

Cos S, Sánchez-Barceló EJ (1994) Differences between pulsatile or continuous exposure to melatonin on MCF-7 human breast cancer cell proliferation. Cancer Lett 85:105–109

Cos S, Sánchez-Barceló EJ (1995) Melatonin inhibition of MCF-7 human breast cancer cell growth: influence of cell proliferation rate. Cancer Lett 93:207–212

Cos S, Sánchez-Barceló EJ (1997) Melatonin reduces the invasive capacity of MCF-7 human breast cancer cells. In vitro studies. In: Webb SM, Puig-Domingo M, Moller M, Pevet P (eds) Pineal Update. From Molecular Mechanisms to Clinical Implications. PJD Publications Limited, New York, pp377–382

Cos S, Blask DE, Lemus-Wilson A, Hill AB (1991) Effects of melatonin on the cell cycle kinetics and "estrogen-rescue" of MCF-7 human breast cancer cells in culture. J Pineal Res 10:36–42

Cos S, Recio J, Sánchez-Barceló EJ (1996a) Modulation of the length of the cell cycle time of MCF-7 human breast cancer cells by melatonin. Life Sci 58:811–816

Cos S, Fernández F, Sánchez-Barceló EJ (1996b) Melatonin inhibits DNA synthesis in MCF-7 human breast cancer cells in vitro. Life Sci 58:2447–2453

Cos S, Verduga R, Fernández-Viadero C, Megías M, Crespo D (1996c) Effects of melatonin on the proliferation and differentiation of human neuroblastoma cells in culture. Neurosci Lett 216: 113–116

Cos S, Mediavilla MD, Fernández R, Güezmes A, Sánchez-Barceló EJ (1997) Melatonin and MCF-7 human breast cancer cells: a short review. Cancer Biotherapy & Radiopharmaceuticals 12:420

Cos S, Fernández R, Güezmes A, Sánchez-Barceló EJ (1998) Influence of melatonin on invasive and metastatic properties of MCF-7 human breast cancer cells. Cancer Res (in press)

Costa EJX, Harzer Lopes R, Lamy-Freund MT (1995) Permeability of pure lipid bilayers to melatonin. J Pineal Res 19:123–126

Crespo D, Fernández-Viadero C, Verduga R, Ovejero V, Cos S (1994) Interaction between melatonin and estradiol on morphological and morphometric features of MCF-7 human breast cancer cells. J Pineal Res 16:215–222

Danforth DN Jr, Tamarkin L., Lippman ME (1983) Melatonin increases oestrogen receptor binding activity of human breast cancer cells. Nature 305:323–325

Das Gupta TK, Terz J (1967) Influence of pineal gland on the growth and spread of melanoma in the hamster. Cancer Res 27:1306–1311

Dubocovich ML (1995) Melatonin receptors: are there multiple subtypes? Trends Pharmacol Sci 16:50–56

Eison AS, Mullins UL (1993) Melatonin binding sites are functionally coupled to phosphoinositide hydrolysis in Syrian hamster RPMI 1846 melanoma cells. Life Sci 53:393–398

El-Domeiri AAH, Das Gupta TK (1973) Reversal by melatonin of the effect of pinealectomy on tumor growth. Cancer Res 33:2830–2833

Fitzgerald TJ, Veal A (1976) Melatonin antagonizes colchicine-induced mitotic arrest. Experientia 32:372–373

Georgiou E (1929) Über die Natur und die Pathogenese der Krebstumoren. Radikale Heilung des Krebses bei weiblichen Mäusen. Z Krebsforschung 38:562–572

Gill PG, De Young NJ, Thompson A, Keightley DD, Horsfall DJ (1984) The effect of tamoxifen on the growth of human malignant melanoma in vitro. Eur J Cancer Clin Oncol 20:807–815

Grin W, Grünberger W (1998) A significant correlation between melatonin deficiency and endometrial cancer. Gynecol Obstet Invest 45:62–65

Gui GPH, Puddefoot JR, Vinson GP, Wells CA, Carpenter R (1997) Altered cell-matrix contact: a prerequisite for breast cancer metastasis? Br J Cancer 75:623–633

Helton RA, Harrison WA, Kelley K, Kane MA (1993) Melatonin interactions with cultured murine B16 melanoma cells. Melanoma Res 3:403–413

Hill SM, Blask DE (1988) Effects of the pineal hormone melatonin on the proliferation and morphological characteristics of human breast cancer cells (MCF-7) in culture. Cancer Res 48:6121–6126

Hill SM, Spriggs LL, Simon MA, Muraoka H, Blask DE (1992) The growth inhibitory action of melatonin on human breast cancer cells is linked to the estrogen response system. Cancer Lett 64:249–256

Horwitz KB, McGuire WL (1978) Estrogen control of progesterone receptor in human breast cancer: correlations with nuclear processing of estrogen receptors. J Biol Chem 253:2223–2228

Hu DN, Roberts JE (1997) Melatonin inhibits growth of cultured human uveal melanoma cells. Melanoma Res 7:27–31

Huerto-Delgadillo L, Antón-Tay F, Benitez-King G (1994) Effects of melatonin on microtubule assembly depend on hormone concentration: role of melatonin as calmodulin antagonist. J Pineal Res 17:55–62

Jakesz R, Smith CA, Aitken S, Huff K, Schuette W, Schackney S, Lippman M (1984) Influence of cell proliferation and cell phase on expression of estrogen receptor in MCF-7 breast cancer cells. Cancer Res 44:619–625

Karakousis CP, Lopez RE, Bhakoo HS, Rosen F, Moore R, Carlson M (1980) Estrogen and progesterone receptors and tamoxifen in malignant melanoma. Cancer Treat Rep 64:819–827

Karasek M, Kunert-Radek J, Stepien H, Pawlikowski M (1988) Melatonin inhibits the proliferation of estrogen-induced rat pituitary tumor cells in vitro. Neuroendocrinol Lett 10:135–140

Kikuchi Y, Kita T, Miyauchi M, Iwano I, Kato K (1989) Inhibition of human ovarian cancer cell proliferation in vitro by neuroendocrine hormones. Gynecol Oncol 32:60–64

Lapin V, Ebels I (1981) The role of the pineal gland in neuroendocrine control mechanisms of neoplastic growth. J Neural Transm 50:275–282

Lemus-Wilson A, Kelly PA, Blask DE (1995) Melatonin blocks the stimulatory effects of prolactin on human breast cancer cell growth in culture. Br J Cancer 72:1435–1440

Leone AM, Silman RE, Hill BT, Whelan RDH, Shellard SA (1988) Growth inhibitory effects of melatonin and its metabolites against ovarian tumour cell lines in vitro. In: Gupta D, Attanasio A, Reiter RJ (eds) The Pineal Gland and Cancer, Brain Research Promotion, London-Tübingen, pp 273–281

Lezoualch F, Sparapani M, Behl C (1998) N-acetylserotonin (normelatonin) and melatonin protect neurons against oxidative challenges and suppress the activity of the transcription factor NF-kappaB. J Pineal Res 24:168–178

L'Hermite-Baleriaux M, de Launoit Y (1992) Is melatonin really an in vitro inhibitor of human breast cancer cell proliferation. In vitro Cell Dev Biol 28 A:583–584

Maestroni GJM (1993) The immunoneuroendocrine role of melatonin. J Pineal Res 14:1–10

Maragos C, Wang JM, Hrabie JA, Oppenheim JJ, Keefer LK (1993) Nitric oxide/nucleophile complexes inhibit the in vitro proliferation of A375 melanoma cells via nitric oxide release. Cancer Res 53:564–568

Melendez J, Maldonado V, Ortega A (1996) Effect of melatonin on beta-tubulin and MAP2 expression in N1E-115 cells. Neurochem Res 21:653–658

Mengeaud V, Skene D, Pevet P, Ortonne JP (1984) No high affinity melatonin binding sites are detected in murine melanoma cells and in normal human melanocytes cultured *in vitro*. Melanoma Res 4:87–91

Meyskens FL Jr, Salmon SE (1981) Modulation of clonogenic human melanoma cells by follicle-stimulating hormone, melatonin, and nerve growth factor. Br J Cancer 43:111–115

Millon R, Nicora F, Muller D, Eber M, Klein-Soyer C, Abecassis J (1989) Modulation of human breast cancer cell adhesion by estrogens and antiestrogens. Clin Exp Metastasis 7:405–415

Molis TM, Walters MR, Hill SM (1993) Melatonin modulation of estrogen receptor expression in MCF-7 human breast cancer cells. Int J Oncol 3:687–694

Molis TM, Spriggs LL, Hill SM (1994) Modulation of estrogen receptor mRNA expression by melatonin in MCF-7 human breast cancer cells. Mol Endocrinol 8:1681–1690

Molis TM, Spriggs LL, Jupiter Y, Hill SM (1995) Melatonin modulation of estrogen-regulated proteins, growth factors, and proto-oncogenes in human breast cancer. J Pineal Res 18:93–103

Noteborn HPJM, Bartsch H, Bartsch C, Mans DR, Weusten JJ, Flehmig B, Ebels I, Salemink CA (1988) Partial purification of (a) lower molecular weight ovine pineal compound (s) with an inhibiting effect on the growth of human melanoma cells *in vitro*. J Neural Transm 73:135–155

Osborne CK, Hobbs K, Clark GM (1985) Effect of estrogens and antiestrogens on growth of human breast cancer cells in athymic nude mice. Cancer Res 45:584–590

Osborne CK, Hobbs K, Trent JM (1987) Biological differences among MCF-7 human breast cancer cell lines from different laboratories. Breast Cancer Res Treat 9:111–121

Panzer A, Viljoen M (1997) The validity of melatonin as an oncostatic agent. J Pineal Res 22:184–202

Panzer A, Lottering M-L, Bianchi P, Glencross DK, Stark JH, Seegers JC (1998) Melatonin has no effect on the growth, morphology or cell cycle of human breast cancer (MCF-7), cervical cancer (HeLa), osteosarcoma (MG-63) or lymphoblastoid (TK6) cells. Cancer Lett 122:17–23

Pappolla MA, Sos M, Omar RA, Bick RJ, Hickson-Bick DL, Reiter RJ, Efthimiopoulos S, Robakis NK (1997) Melatonin prevents death of neuroblastoma cells exposed to the Alzheimer amyloid peptide. J Neurosci 17:1683–1690

Persengiev S, Kyurkchiev S (1993) Selective effect of melatonin on the proliferation of lymphoid cells. Int J Biochem 25:441–444

Pickering DS, Niles LP (1992) Expression of nanomolar-affinity binding sites for melatonin in syrian hamster RPMI 1846 melanoma cells. Cell Signal 4:201–207

Poeggeler B, Reiter RJ, Tang DX, Chen LD, Manchester LC (1993) Melatonin, hydroxyl radical-mediated oxidative damage and aging: a hypothesis. J Pineal Res 14:151–168

Reiter RJ (1980) The pineal and its hormones in the control of reproduction in mammals. Endocr Rev 1:109–131

Reiter RJ (1988) Pineal gland, cellular proliferation and neoplastic growth: an historical account. In: Gupta D, Attanasio A, Reiter RJ (eds) The Pineal Gland and Cancer. Brain Research Promotion, London-Tübingen, pp 41–64

Rochefort H (1983) In: McKems KW (ed) Regulation of gene expression by hormones. Plenum Press, New York, pp 27–37

Roth JA, Rabin R, Agnello K (1997) Melatonin suppression of PC21 cell growth and death. Brain Res 768:63–70

Sánchez-Barceló EJ, Cos S, Mediavilla MD (1988) Influence of pineal gland function on the initiation and growth of hormone-dependent breast tumors. Possible mechanisms. In: Gupta D, Attanasio A, Reiter RJ (eds) The Pineal Gland and Cancer. Brain Research Promotion, London-Tübingen, pp 221–232

Sánchez-Barceló EJ, Mediavilla MD, Cos S (1997) Effects of melatonin on experimental mammary cancer development. In: Webb SM, Puig-Domingo M, Moller M, Pevet P (eds) Pineal Update. From Molecular Mechanisms to Clinical Implications. PJD Publications Limited, New York, pp 361–368

Shellard SA, Whelan RDH, Hill BT (1989). Growth inhibitory and cytotoxic effects of melatonin and its metabolites on human tumour cell lines in vitro. Br J Cancer 60:288–290

Slominski A, Pruski D (1993) Melatonin inhibits proliferation and melanogenesis in rodent melanoma cells. Exp Cell Res 206:189–194

Sommers CL, Byers SW, Thompson EW, Torri JA, Gelmann EP (1994) Differentiation state and invasiveness of human breast cancer cell lines. Breast Cancer Res Treat 31:325–335

Song W, Lahiri DK (1997) Melatonin alters the metabolism of the β-amyloid precursor protein in the neuroendocrine cell line PC12. J Mol Neurosci 9:75–92

Soule HD, Vazquez J, Long A, Albert S, Brennan MJ (1973) A human cell line from a pleural effusion derived from a breast carcinoma. J Natl Cancer Inst 51:1409–1412

Steinhilber D, Brungs M, Werz O, Wiesenberg I, Danielsson C, Kahlen JP, Nayeri S, Schrader M, Carlberg C (1995) The nuclear receptor for melatonin represses 5-lipoxygenase gene expression in human B lymphocytes. J Biol Chem 270:7037–7040

Sze SF, Ng TB, Liu WK (1993) Antiproliferative effect of pineal indoles on cultured tumor cell lines. J Pineal Res 14:27–33

Tan DX, Chen LD, Poeggeler B, Manchester LC, Reiter RJ (1993) Melatonin: a potent endogenous hydroxyl radical scavenger. Endocrine J 1:57–60

Taylor CM, Blanchard B, Zava DT (1984) Estrogen receptor-mediated and cytotoxic effects of the antiestrogens tamoxifen and 4-hydroxytamoxifen. Cancer Res 44:1409–1414

Thompson EW, Reich R, Shima TB, Albini A, Graf J, Martin GR, Dickson RB, Lippman ME (1988) Differential regulation of growth and invasiveness of MCF-7 breast cancer cells by antiestrogens. Cancer Res 48:6764–6768

Trosko JE, Chang CC, Madhukar BV, Klaunig JE (1990) Chemical, oncogene and growth factor inhibition gap junctional intercellular communication: an integrative hypothesis of carcinogenesis. Pathobiology 58:265–278

Ubeda A, Trillo MA, House DE, Blakeman CF (1995) Melatonin enhances junctional transfer in normal CH3/10T1/2 cells. Cancer Lett 91:241–245

Valverde P, Benedito E, Solano F, Oaknin S, Lozano JA, García-Borrón JC (1995) Melatonin antagonizes alpha-melanocyte-stimulating hormone enhancement of melanogenesis in mouse melanoma cells by blocking the hormone-induced accumulation of the c locus tyrosinase. Eur J Biochem 232:257–263

Walker MJ, Chaduri PK, Beattie CW, Tito WA, Das Gupta TK (1978) Neuroendocrine and endocrine correlates to hamster melanoma growth in vitro. Surg Forum 29:151–152

Walker MJ, Beattie CW, Patel MK, Ronan SM, Das Gupta TK (1987) Estrogen receptor in malignant melanoma. J Clin Oncol 5:1256–1261

Watanabe Y, Sawada N, Isomura H, Satoh H, Hirata K, Mori M (1995) Estrogen-depleted condition induces apoptosis of rat mammary cancer cells after entering the S-phase of the cell cycle. Cell Struct Funct 20:125–132

Welsh JE (1994) Induction of apoptosis in breast cancer cells in response to vitamin D and anti-estrogens. Biochem Cell Biol 72:537–545

Yagasaki R, Noguchi M, Minami M, Earashi M (1996) Clinical significance of E-cadherin and vimentin co-expression in breast cancer. Int J Oncol 9:755–761

Ying SW, Niles LP, Crocker C (1993) Human malignant melanoma cells express high-affinity receptors for melatonin: antiproliferative effects of melatonin and 6-chloromelatonin. Eur J Pharmacol 246:89–96

Zisapel N, Bubis M (1994) Inhibition by melatonin of protein secretion and growth of melanoma cells in culture. In: Maestroni GJM, Conti A, Reiter RJ (eds) Advances in Pineal Research, Vol 7, John Libbey & Co Ltd, London, pp 259–268

11 Melatonin and Colon Carcinogenesis

Vladimir N. Anisimov

Abstract

It was shown earlier that melatonin (MLT) inhibits spontaneous and chemically induced mammary carcinogenesis and transplantable mammary tumor growth in vivo and in vitro. We have studied the effect of MLT on colon carcinogenesis. Three-month-old male LIO rats were exposed to 5 or 15 weekly s.c. injections of 1,2-dimethylhydrazine (DMH, 21 mg/kg). Part of the DMH-treated rats were given MLT in tap water (20 µg/ml) at nighttime during the period of carcinogen treatment. Rats were sacrificed 6 months after the first DMH injection. The exposure to MLT was followed by a decrease of intestinal tumor incidence in the jejunum and ascending colon as compared to controls. MLT inhibits the multiplicity and size of colon tumors as well as increasing tumor differentiation. The levels of immunohisto-chemically detected MLT in epithelium of the intestinal tract of rats exposed to DMH were significantly reduced as compared to intact rats, while in rats treated with DMH + MLT it was within the range of the controls. In the serum of rats exposed to DMH alone, the levels of diene conjugates (DC) and Schiff's bases (SB) were significantly increased as compared with controls. In colon tissue of DMH-treated rats, the levels of DC, SB, and carbonyl derivatives of amino acid NO-synthase activity were significantly increased, while total antioxidative activity was decreased as compared to controls. In rats exposed to DMH + MLT, a normalization of free radical processes in serum and colon has been observed. MLT also inhibits the mutagenic effect of DMH in vivo (chromosome aberration and sperm head anomalies tests) and in vitro (Ames test). MLT also exerts some normalizing influence on glucose and lipid metabolism in rats exposed to DMH and inhibits proliferation and stimulates apoptosis in DMH-induced colon tumors. Thus, our data have shown that MLT affects initiating, promoting, and progression stages of colon carcinogenesis.

11.1 Introduction

The role of the pineal gland in tumor development has been under intensive study during the last years (Anisimov and Reiter 1990; Bartsch et al. 1992; Blask 1993; Panzer and Viljoen 1997). In cancer patients, the morphological signs of pineal function decrease and disturbances in the circadian secretion pattern of the main hormone of the pineal gland, melatonin, have been observed (Anisimov and Reiter 1990; Blask 1993). The growth of a variety of tumors is accelerated in pinealectomized animals, whereas treatment with melatonin inhibits development and growth of some types of tumors, mainly mammary carcinomas, both in vitro and in vivo (Blask 1993). There are few data on the effect of melatonin on tumors of other localizations (Bartsch et al. 1992; Blask 1993). One of the possible targets for melatonin action could be colon carcinogenesis. It is well known that intestines are one of the important sources of melatonin production in mammals (Bubenik 1999; Kvetnoy et al. 1997). Thus, the relationship between melatonin in microenvironment (in bowels) and in circulation, on the one hand, and the development of tumors in bowels, on the other, could be an important aspect of the *pineal gland and cancer* problem.

Colorectal cancer is one of the most common human cancers all over the world (Coleman et al. 1993). It is worthy to note the disturbances in the circadian rhythm of melatonin excretion in colon cancer patients (Kvetnoy and Levin 1987). Intestinal tumors induced in rats by 1,2-dimethylhydrazine (DMH) and its carcinogenic metabolites azoxymethane (AOM) and methylazoxymethanol are at present the most popular models used in experimental oncology to study various aspects of morphology, pathogenesis, prevention, and treatment of colorectal cancer (Pozharisski et al. 1979). It was shown that colon carcinomas induced by DMH are morphologically similar to human colon tumors and, like other malignancies, a colon carcinoma needs at least several stages (events) for development in bowel mucosa (Maskens 1987).

The aim of our study was to evaluate the effect and mechanisms of melatonin action on colon carcinogenesis induced by DMH in rats. DMH is an alkylating agent that is significant for its carcinogenic effect (Pozharisski et al. 1979). There is some evidence for a role of free radicals in DMH-induced carcinogenesis (Salim 1993). Recently, it was found that melatonin is a potent scavenger of free radicals (Reiter et al. 1997), which could be one of the mechanisms of its possible antitumor effect. Some natural and synthetic antioxidants have been shown to act as inhibitors of DMH-induced carcinogenesis (Balanski et al. 1992), but the effect of melatonin on colon carcinogenesis has never been studied. The role of hormones, disturbances in lipid and carbohydrate metabolism, genetic predisposition, oncogene and antioncogene mutations in colon cancer development are intensively studied (Berstein 1995; Zusman 1997). At the same time there are a lot of data on an effect of melatonin on hormonal, carbohydrate, and lipid metabolism as well as on the immune system (Yu and Reiter 1993). Thus, in this paper we will try to describe the modifying effects of melatonin on "early stages" and "late stages" of colon carcinogenesis.

11.2 Effect of Melatonin on Colon Carcinoma Development

11.2.1 Inhibitory Effect of Melatonin on 1,2-Dimethylhydrazine-Induced Colon Carcinogenesis in Rats

In our experiments, female rats were exposed to 15 (experiment 1) or 5 (experiment 2) weekly s.c. injections of DMH, each consisting of 21 mg/kg body weight (Anisimov et al. 1997a). From the day of the first injection, part of the rats from both series of the experiment were given melatonin dissolved in tap water, 20 mg/l, 5 days a week during nighttime (from 1800 to 800 hours). The experiment was finalized 6 months after the first injection of the carcinogen.

In experiment 1, melatonin failed to influence the total incidence of colon tumors. However, the incidence of carcinomas in ascending colon was significantly reduced ($P < 0.01$) (Table 11.1). The multiplicity of total colon tumors per rat as well as the mean number of tumors in ascending and descending colon per rat were also decreased under the influence of melatonin ($P < 0.01$). In the same experiment, melatonin slightly decreased the depth of tumor invasion and increased the number of highly differentiated colon carcinomas induced by DMH. The percentage of small tumors in descending colon among rats exposed to DMH + melatonin was higher than that in rats exposed to DMH alone. Treatment with melatonin was followed also by a decrease in the multiplicity of DMH-induced tumors of the duodenum ($P < 0.05$) and by a decrease in the incidence of jejunum and ileum tumors ($P < 0.05$). In experiment 2, the inhibitory effect of melatonin on DMH-induced colon carcinogenesis was much more expressed than that in experiment 1 (Table 11.1). Thus, the incidence of total colon tumors, ascending and descending, was significantly decreased in the DMH + melatonin-treated group in comparison to rats exposed to DMH alone. Melatonin also reduced the number of tumors per rat in ascending and descending parts of the colon. The number of colon tumors invading only the mucosa was significantly higher in rats treated with DMH + melatonin than in rats exposed only to the carcinogen. The ratio of highly differentiated tumors was increased and the ratio of low-differentiated tumors was decreased in rats exposed to DMH + melatonin as compared with rats treated with DMH alone.

Table 11.1. Intestinal tumor localization, incidence, and multiplicity in female rats exposed to 1,2-dimethylhydrazine (DMH) and melatonin (MLT)

Parameters	Experiment 1		Experiment 2	
	DMH × 15	DMH + MLT	DMH × 5	DMH + MLT
Number of rats	25	23	21	21
Tumor localization:				
Duodenum:				
No. of tumor-bearing rats	8 (32.0%)	5 (21.7%)		
No. of tumors per rat	0.3 ± 0.10	0.2±0.05 *	–	–
Jejunum and ileum:				
No. of tumor-bearing rats	4 (16.0%)	1 (4.3%) *	–	–
No. of tumors per rat	0.2 ± 0.10	0.04 ± 0.05		
Colon (all parts):				
No. of tumor-bearing rats	25 (100%)	23 (100%)	21 (100%)	14 (66.6%) **
No. of tumors per rat	9.9 ± 0.81	6.0 ± 0.65 ***	3.8 ± 0.2	1.5 ± 0.2 ***
Ascending colon:				
No. of tumor-bearing rats	23 (92.0%)	15 (65.2%) **	12 (57.1%)	3 (4.32%) **
No. of tumors per rat	2.2 ± 0.36	0.7 ± 0.11 ***	0.8 ± 0.2	0.1 ± 0.51 **
Descending colon:				
No. of tumor-bearing rats	25 (100%)	23 (100%)	20 (95.2%)	12 (57.1%) **
No. of tumors per rat	6.9 ± 0.66	4.4 ± 0.43 **	2.8 ± 0.4	1.2 ± 0.3 **
Rectum:				
No. of tumor-bearing rats	11 (44.0%)	11 (47.8%)	7 (33.3%)	4 (19.1%)
No. of tumors per rat	0.7 ± 0.20	0.8 ± 0.16	0.4 ± 0.1	0.2 ± 0.05

The difference with experiment-matched group exposed to DMH alone is significant, * $P < 0.05$; ** $P < 0.01$; *** $P < 0.001$.

Thus, these results demonstrate an inhibitory effect of melatonin on intestinal carcinogenesis induced by DMH in rats. This effect was manifested by a decrease of the incidence and multiplicity of bowel tumors, mainly of the colon, by decreasing the invasion rate and dimensions of colon carcinomas and by increasing its differentiation. The inhibitory effect of melatonin was more expressed in the proximal parts of the colon (ascending colon) and was less expressed in the rectum, it was also inversely correlated with the total dose of the carcinogen. This observation suggests that the anticarcinogenic effect of melatonin is realized, at least in part, by its interaction with active metabolites of DMH and/or its adducts with DMH in the bowel lumen. The effect of melatonin on free radical processes involved in the realization of the carcinogenic capacity of DMH must also be taken into consideration. In our experiments, melatonin was given to rats from the day of the first DMH injection onwards, during the whole period of exposure to the carcinogen, and also after the injections of DMH were stopped. Thus, it is reasonable to suggest that the anticarcinogenic effect of melatonin could be realized both on early and late stages of colon carcinogenesis.

11.2.2 Effect of Melatonin on Colon Tumor Growth In Vitro and In Vivo

DLD-1 human colon carcinoma cells are insensitive to high concentrations of melatonin (10^{-4} M) (Blask 1993). Using the CaCo-2 human colon carcinoma cell line, Pentney (1995) failed to observe an antiproliferative effect of melatonin in physiological concentrations. The only effect occurred at suprapharmacological concentrations (10^{-3} M). In our experiments, Balb/c female mice were transplanted with AKATOL mouse colon carcinoma cells and, starting from the next day after grafting, were treated with saline or with melatonin (100 mg/kg, s.c. at 1000–1100 hours for 10 days). There was no effect of melatonin on tumor growth or survival of tumor-bearing animals.

Barni et al. (1995) reported that 9 out of 25 patients treated with melatonin plus interleukin-2 (IL-2) survived for 1 year, while among 25 patients of the control group, only 3 patients survived for 1 year. Analysis of the reported trials with treatment of 112 colorectal cancer patients with melatonin combined with IL-2 has shown zero cases with complete response, 11 cases with partial or minor response, 35 cases of stable disease, and 66 (50%) cases of progressive disease (Panzer and Viljoen 1997). A recent report by Neri et al. (1998) failed to show a significant effect of melatonin treatment in 31 cancer patients, including 9 enteric cancers.

Thus, melatonin has no sufficient inhibitory effect on growth of colon cancer both in vitro and in vivo. The results of treatment of colon cancer patients with melatonin are also not very impressive. Our observations suggest that the inhibitory effect of melatonin on colon carcinogenesis is realized at stages related to initiation, promotion, and progression of tumor development.

11.3 Early Stages of 1,2-Dimethylhydrazine-Induced Colon Carcinogenesis As Targets for the Effect of Melatonin

Stages of DMH metabolism leading to the initiation of carcinogenesis are as follows: formation of "active" metabolites (methylazoxymethanol or methyldiazohydrate) mainly in the liver; binding of the metabolites to glucuronic acid; delivery of conjugates to the intestine via blood flow; liberation of "active" metabolites due to enzymic activity of intestinal flora (β-glucuronidase); formation of carbonium ion (CH_3^+); specific methylation of enterocyte macromolecules (mainly DNA at O^6-position of guanine); and miscoding effect (Pozharisski et al. 1979). These events result in mutations followed by the activation of the oncogene Ki-*ras* and inactivation of the anti-oncogene *p53* (Zusman 1997). There is evidence for an important role of free radical processes in DMH-induced carcinogenesis (Salim 1993). It was also shown that shortly after treatment with various chemical carcinogens, significant disturbances developed in the neuroendocrine and immune systems, in carbohydrate and lipid metabolism. These changes, creating specific macro- and microenvironments, are necessary for survival of cells damaged by carcinogens (initiated) (for review see Anisimov 1987, 1998). To study the mechanism of the inhibitory effect of melatonin on early stages of colon carcinogenesis, some experiments with a single dose of DMH have been conducted.

11.3.1 Effect of Melatonin on the Genotoxic Action of 1,2-Dimethylhydrazine

Because the genotoxic effect is the key event in the initiation of carcinogenesis, the effect of melatonin on the genotoxic action of DMH should be studied for an investigation of the critical point(s) in the mechanism of the inhibitory effect of melatonin on DMH-induced colon carcinogenesis. Melatonin has been found to inhibit X-ray-induced mutagenesis in human lymphocytes in vitro (Vijayalaxmi et al. 1996), to reduce *cis*-platinum-induced genetic damage in the bone marrow of mice (Koratkar et al. 1992), to decrease hepatic DNA adduct formation caused by safrole in rats (Tan et al. 1994), and to protect rat hepatocytes from chromium(VI)-induced DNA single-strand breaks in vitro (Susa et al. 1997). Awara et al. (1998) have shown that pretreatment of human lymphocytes with melatonin reduced the rate of chromosome aberrations and sister chromatid exchanges induced by the mutagen carbamazepine.

We have studied the effect of melatonin on the induction of chromosome aberrations (ChA) and sperm head anomalies (SHA) in mice treated with DMH in vivo (Musatov et al. 1997). Both in vivo tests showed similar results. Melatonin alone failed to increase both ChA and SHA incidence in mice in comparison to those treated with normal saline, but significantly decreased the incidence of DMH-induced ChA by 68.4% and of SHA by 48.2% ($P < 0.01$).

Since melatonin can protect cells directly as an antioxidant and indirectly through receptor-mediated activation of antioxidative enzymes, we applied two different in vitro test systems: the Ames mutagenicity test and the COMET assay for clastogenicity (Musatov et al. 1998). As oxidative mutagens, we used DMH, bleomycin (with S9-mix consisting of Aroclor 1254 induced rat liver S9 fraction), and mitomycin C (without S9-mix), which are believed to generate oxygen radical species. Additionally, we tested eight other intercalating and alkylating agents both direct-acting and requiring metabolic activation. Melatonin inhibited the mutations induced by promutagens DMH (Table 11.2), 7,12-dimethylbenz(a)anthracene (DMBA), benzo(a)pyrene, 2-aminofluorene, and bleomycin in all the strains used. The mutagenicity of the direct-acting mutagens 4-nitroquinoline-*N*-oxide, 2,4,7-trinitro-9-fluorenone, 9-aminoacridine, *N*-nitroso-methylurea, mitomycin C, and sodium azide remained unaffected by melatonin. It is to be noticed that melatonin was effective as an antimutagen only at

Table 11.2. Effect of melatonin on 1,2-dimethylhydrazine (DMH)-induced His+ revertants in the Ames test

DMH, µg/plate	Melatonin, µmol/plate	Net no. of revertants/plate TA 102
500	0	527 ± 12.7[a] (2.3)[b]
	0.25	468 ± 14.1
	0.5	423 ± 43.8
	1	248 ± 39.6
	2	175 ± 21.2

Incubation performed in the presence of rat liver S9 microsomal mix. No mutagenicity detected in the non-toxic range in the preincubation assay used with TA 97, TA 98, and TA 100.
[a] Mean of triplicate plates.
[b] Ratio between induced and spontaneous revertants. The spontaneous revertants for TA 102 + S9 were 136–190.

extremely high doses (0.25–2 µmol/plate). The modifying activity of melatonin was not related to the mechanism of action of the mutagens, i.e., frameshift mutations or base-pair substitutions. As melatonin displayed its protective effect towards promutagens only, we speculate that it can exert its activity on the metabolic activation process, perhaps by inhibiting the cytochrome P-450-dependent monooxygenase enzyme system of S9-mix. However, based on the present data, we cannot suggest any precise mechanism of the interaction between melatonin and metabolizing enzymes. It is known that melatonin administration causes a significant decrease in hepatic microsomal contents of cytochrome b_5 and P-450 in rats (Kothari and Subramanian 1992). This may be related to the suppressive effect of melatonin on safrole-induced hepatic DNA adduct formation in rats (Tan et al. 1994), DMBA-induced mammary carcinogenesis (Kothari and Subramanian 1992), and DMH-induced colon carcinogenesis (Anisimov et al. 1997a).

The S9-mediated effect of melatonin could be the result of the action of a metabolite of melatonin, such as 6-hydroxymelatonin, instead of melatonin itself. 6-Hydroxymelatonin was found to be able to inhibit the thiobarbituric-induced lipoperoxidation in mouse brain homogenates, which was imputed to the production of oxygen free radicals (Pierrefiche et al. 1993). In the Ames test *Salmonella typhimurium* tester strains TA 97, TA 98, TA 100, and TA 102 were used (Musatov et al. 1998). In this test, melatonin itself was neither mutagenic nor toxic towards the four strains at the concentrations tested (0.25–2.0 µM/plate). The mutagenicity of DMH was inhibited by melatonin in a dose-dependent manner. The maximum inhibition observed was 67% for TA 102.

The results of the COMET assay are in accordance with those of the Ames test. Melatonin modulated the clastogenicity of DMBA, benzo(a)pyrene, and cylcophosphamide. Compared to the data obtained in the Ames test, melatonin inhibited the clastogenicity of the chemicals at lower concentrations (0.1–1 nM) (Musatov et al. 1998). Probably, these findings reflect the fact that the protective effect of melatonin is receptor-mediated in this model. Melatonin has been shown to increase the level of glutathione in both the liver and the mammary gland, and to stimulate glutathione peroxidase activity in these organs (Kothari and Subramanian 1992) and in the brain of rats (Barlow-Walden et al. 1995). Glutathione peroxidase, in turn, is responsible for the detoxification of mutagens and their reactive metabolites. Finally, melatonin is structurally related to indole alkaloids, that have been shown to bind directly to DNA. Melatonin was revealed to intercalate with the double-stranded DNA and renature the DNA denatured by Cu^{2+} administration and heating. Hence, the stabilization of the

Watson-Crick helix is probably another mechanism by which melatonin could modulate the mutagenicity of carcinogens (Tan et al. 1994).

11.3.2 Effect of a Single 1,2-Dimethylhydrazine Administration on Free Radical Processes in Rats

In our experiments, rats were exposed to a single s.c. dose of DMH (21 mg/kg) and after 24 h the parameters of free radical processes were studied in serum, liver, and colon mucosa (Arutyunyan et al. 1997). A significant decrease of the levels of active forms of oxygen was observed in serum but not in colon of rats exposed to the carcinogen as compared to controls. However, it did not involve any changes in the concentration of such products of lipid peroxidation as diene conjugates and Schiff's bases. Protein peroxidation was increased by 39.9% in the colon mucosa under influence of the carcinogen. At the same time, the total antioxidative activity was decreased by 27.9% in colon and was similar to the control value in serum. The activity of nitric oxide (NO)-synthase was decreased by 32.8% ($P < 0.05$) and the activity of Cu, Zn-SOD was decreased by 42.1% ($P < 0.001$), whereas the activity of catalase was increased by 19.6% ($P < 0.05$) in colon of DMH-exposed rats in comparison to control values. The serum concentration of ceruloplasmin was decreased by 30.3% ($P < 0.05$) under influence of the carcinogen. Thus, significant changes in free radical processes in colon mucosa, which is the target tissue for the carcinogen, were found at an early stage of DMH-induced carcinogenesis.

11.3.3 Effect of Melatonin on Free Radical Processes in Rats

There are a lot of data at present on an antioxidant potential of melatonin in vitro and in vivo (Reiter et al. 1997). Besides its ability to scavenge reactive oxygen species (ROS), the antioxidant activity of melatonin is augmented by its ability to stimulate enzymes related to the antioxidative defense system. In order to study the anticarcinogenic effect of melatonin, we have administered it to rats through drinking water (20 mg/l) during nighttime (Anisimov et al. 1997a). In a series of experiments, we studied the effect of melatonin given in this way on free radical processes in vivo (Anisimov et al. 1995, 1996, 1997). It was found that in the dose used, melatonin significantly inhibits ROS formation, lipid peroxidation, superoxide dismutase (SOD) activity, and ceruloplasmin levels in serum of rats. At the same time, the total antioxidative activity of serum as well as the levels of nitrites were increased as compared to controls. A slight increase in protein peroxidation was also observed in melatonin-treated rats. Thus, our findings confirm the antioxidative potential of melatonin.

11.3.4 Carbohydrate and Lipid Metabolism in Rats Exposed to a Single Dose of 1,2-Dimethylhydrazine

Many carcinogens influence the metabolism of carbohydrates and lipids at early stages of carcinogenesis (Anisimov 1987). It was suggested that these disturbances contribute to the promotion of initiated cells. In our experiments it was shown that the serum level of cholesterol was increased by 43.5% and the level of glucose and triglycerides was unchanged in rats exposed to a single s.c. administration of DMH (21 mg/kg) as compared to controls (Anisimov et al. 1980). In the same experiment, a pronounced decrease of the tolerance to glucose was observed in DMH-treated rats.

11.3.5 Effect of 1,2-Dimethylhydrazine on the Neuroendocrine System of Rats

It was shown that 3 h after treatment with [^{3}H]DMH, the rate of accumulation of specific radioactivity was significantly higher in the hypothalamus than in the brain stem, cerebral cortex, or liver in rats (Anisimov et al. 1976). DMH produced an anti-gonadotropic effect both in male and female rats and increased the threshold of sensitivity of the hypothalamus to inhibition by estrogens (Pozharisski and Anisimov 1975). The data on a pronounced influence of DMH on the level of biogenic amines in the hypothalamus of rats correspond to the above-stated observation (Anisimov et al. 1976a). It was shown that 3 h after a single injection of DMH, the hypothalamic level of norepinephrine was only 66% of control levels. For dopamine, serotonin, and 5-hydroxyindoleacetic acid these figures were 32, 71, and 57%, respectively. DMH could be supposed to selectively inhibit the activity of aromatic L-amino acid decarboxylase in the hypothalamus. DMH does not influence the level of biogenic amines in the brain stem and hemispheres of rats. It was shown that the exposure to constant light which inhibits melatonin synthesis and secretion also increases the threshold of sensitivity of the hypothalamo-pituitary system to feedback regulation by estrogens (Dilman and Anisimov 1979).

In our recent study, the levels of serotonin, N-acetylserotonin and melatonin were estimated by high pressure liquid chromatography at night (between 0000 and 0100 hours) in pineal glands of male rats 37 h after s.c. injection of a single dose of 21 mg/kg DMH. It was observed that pineal levels of serotonin and 5-acetylserotonin were unchanged while the concentration of melatonin decreased by 43.5% ($P < 0.05$) (V.K. Pozdeev et al., unpublished data).

Thus, melatonin was able to influence several steps of the cancer-initiating effect of DMH. Our experiments failed to elucidate the exact mechanism of the antimutagenic effect of the hormone. Melatonin was shown to be a scavenger of hydroxyl radicals. Likewise, melatonin can stimulate glutathione peroxidase activity in different tissues, thereby reducing the generation of the OH· by neutralizing its precursor H_2O_2. These properties allow melatonin to preserve macromolecules including DNA, protein, and lipid from oxidative damage resulting from ionizing radiation exposure and chemical carcinogen administration. Being both lipophilic and hydrophilic, melatonin may prove to play an important role in the antioxidant defense system in all cells and tissues of the body (Reiter et al. 1997). Some early events induced by DMH treatment in the neuroendocrine system as well as in carbohydrate and lipid metabolism could be factors which create the macro- and microenvironments facilitating survival of carcinogen-damaged target cell(s) (Anisimov 1987). These disturbances could be alleviated, at least in part, by melatonin treatment.

11.4 Effect of Melatonin on Late Stages of 1,2-Dimethylhydrazine-Induced Colon Carcinogenesis

In this section we shall describe different characteristics of the homeostasis in rats with DMH-induced colon tumors and effects of long-term treatment on these parameters.

11.4.1 Free Radical Processes in Rats with Colon Tumors Induced by 1,2-Dimethylhydrazine: Effect of Melatonin

Some parameters of free radical processes have been studied in colon, liver, and serum of rats exposed to DMH alone or to DMH and melatonin 6 months after the start of treatment with the carcinogen (Arutyunyan et al. 1997). In DMH-induced colon carcinoma-bearing rats, the intensity of hydroperoxide-induced chemoluminescence in all tissues studied was practically in the same range as in controls. At the same time, the level of lipid and protein peroxidation was increased in serum and colon mucosa of rats exposed to DMH alone. The activity of NO-synthase in colon mucosa of these rats was significantly increased as well. It was found that long-term treatment with melatonin significantly inhibits lipid and protein peroxidation in rats exposed to DMH (Table 11.3) (Anisimov et al. 1996a). Thus, our findings confirm a high antioxidative potential of melatonin in a model of colon carcinogenesis.

11.4.2 Effect of Melatonin on Proliferative Activity and Apoptosis in Colon Mucosa and Colon Tumors Induced by 1,2-Dimethylhydrazine in Rats

An inhibitory effect of melatonin on the proliferation of mammary epithelium has been confirmed by several studies (Blask 1993). In our experiments, we observed a significant increase in proliferative activity of enterocytes in the colon mucosa in rats exposed to DMH as compared with intact controls: both mitotic index and number of Ki-67 positive cells were approximately twice as high in the carcinogen-treated groups. Treatment with melatonin significantly reduced the mitotic index in colon tumors (Anisimov et al. 2000). The number of Ki-67 positive cells was much higher in colon tumors than in colon mucosa from the same rats. Treatment with melatonin failed to change the relative number of Ki-67 positive epithelial cells in these tissues. The apoptotic index (AI) was similar in the colon mucosa of intact rats and in the colon mucosa of rats treated with DMH and with DMH + melatonin. The value of AI significantly increased in colon adenocarcinomas as compared to the colon mucosa. Treatment with melatonin was followed by an increase of AI in colon tumors.

Our findings are in agreement with observations on the inhibitory effect of melatonin on cell proliferation in the rodent colon (Lewinski et al. 1991), in mammary tumors, and hepatoma cells (Blask 1993). It was also shown that pinealectomy was followed by an increase of crypt cell proliferation in rat bowels (colon including), persisting at least 6 months after operation (Callaghan 1995). The presence of specific binding sites for melatonin in mouse colon has been demonstrated as well (Bubenik 1999). Melatonin enhances cell-to-cell junction contacts (Ubeda et al. 1995). This aspect must be specially studied.

Table 11.3. Free radical processes in colon and serum of rats with colon tumors induced by 1,2-dimethylhydrazine (DMH) and treated or not treated with melatonin

Indexes	Serum			Colon		
	Control	DMH	DMH + melatonin	Control	DMH	DMH + melatonin
ROS, units/mg protein	0.53 ± 0.17	0.78 ± 0.14	0.87 ± 0.17	84.7 ± 8.6	115.1 ± 20.5	79.5 ± 9.0
Diene conjugates, nmol/mg protein	3.63 ± 0.27	4.57 ± 0.20*	3.64 ± 0.26*	26.2 ± 1.5	48.2 ± 3.6*	15.6 ± 2.4*
Schiff bases, units/mg protein	21.0 ± 1.4	28.0 ± 0.8*	20.8 ± 1.0*	289.0 ± 39.5	482.5 ± 21.0*	295.4 ± 45.7*
Protein peroxidation, µmol CO derivatives/mg protein	0.822 ± 0.033	0.940 ± 0.072	0.697 ± 0.082	1.89 ± 0.15	2.74 ± 0.06*	NE
NADPH-diaphorase, inhibited by L-arginine in nmoles of NST formasane/min per mg of protein	Absent	Absent	Absent	13.96 ± 1.32	11.60 ± 1.27	10.73 ± 0.96
NO-synthase NADPH-diaphorase, inhibited by L-arginine in nmoles of NST formasane/min per mg of protein	Absent	Absent	Absent	0.74 ± 0.08	1.41 ± 0.31*	0.51 ± 0.05*
TAA, units/mg protein	1.2 ± 0.04	0.84 ± 0.09	0.95 ± 0.06	1.27 ± 0.04	1.64 ± 0.09*	1.10 ± 0.09*
Ceruloplasmin, mg%	39.9 ± 4.2	37.4 ± 1.6	46.3 ± 2.2	Absent	Absent	Absent
SOD, units/mg protein	2.7 ± 1.39	3.6 ± 0.42	3.3 ± 0.68	30.1 ± 2.0	32.5 ± 4.8	20.6 ± 2.4*
Catalase, mmol/H_2O_2 per min/mg protein	NE	NE	NE	4.34 ± 0.25	4.94 ± 0.30*	3.93 ± 0.24*

The difference with tissue-matched control is significant, * $P < 0.05$.
ROS, reactive oxygen species; TAA, total antioxidative activity; NADPH diaphorase inhibited by L-arginine; SOD, superoxide dismutase; NE, not estimated.

We failed to find a significant difference in AI between normal colon mucosa and the colon mucosa of DMH-treated rats (Anisimov et al. 2000). No significant difference has been observed in bcl-2, bax, and bcl-X protein expression in the non-tumoral mucosa of azoxymethane (AOM)-treated rats and in the normal mucosa of saline-treated rats (Hirose et al. 1997). In our experiment, we observed an increase of AI in colon tumors of rats exposed to DMH alone as compared with AI in the colon mucosa of the same rats as well as of controls. It was suggested that apoptosis is inhibited during the progression of colon tumorigenesis (Bedi et al. 1995). A more detailed study on the model of AOM-induced colon carcinogenesis in rats revealed a decrease of the expression of the apoptotic repressor bcl-2 and an increase of the expression of the apoptotic accelerator protein bax in colon tumors as compared to colon mucosa from tumor-free rats treated with carcinogen and to saline-treated colon mucosa (Hirose et al. 1997). Our data seem to be relevant to this observation and contradictory to findings of Bedi et al. (1995). It is clear that further studies on the dysregulation of apoptosis in colon carcinogenesis are needed.

In our study, melatonin increased the value of AI in DMH-induced colon tumors and failed to change its levels in colon mucosa of the same rats (Anisimov et al. 2000). This effect is in correlation with the anticarcinogenic effect of melatonin (Anisimov et al. 1997a). It was shown that some chemopreventive agents with antioxidative properties increased the rate of apoptosis in AOM-induced colon tumors in rats, whereas the tumor promoter 6-phenylhexyl isothiocyanate blocked the process of apoptosis during colon carcinogenesis (Samaha et al. 1997). There is evidence of the important role of free radicals in the realization of DMH-induced carcinogenesis (Salim 1993). We have shown that melatonin inhibited lipid and protein peroxidation and stimulated antioxidant defense systems in rats during DMH-induced colon carcinogenesis (Anisimov et al. 1996a; Arutyunyan et al. 1997). At the same time, there are reports on the in vivo protection against kainate- and single oxygen-induced apoptosis in brain by melatonin (Cagnoli et al. 1995; Reiter et al. 1997). It was suggested that the anti-oxidant action of melatonin might be operative in protecting target cells from cyto-toxicity and apoptosis. Thus, melatonin inhibits apoptosis in colon tumors and protects the brain against oxygen-induced apoptosis. These data are in agreement with the suggestion that the effect of antioxidants is realized via p53-mediated apoptosis (Tendler et al. 1997).

11.4.3 Pineal Function in Rats with Colon Tumors Induced by 1,2-Dimethylhydrazine

There is a lot of evidence for a decrease of pineal function in cancer patients and that pinealectomy or inhibition of pineal function facilitate tumor processes (Anisimov and Reiter 1990; Bartsch et al. 1992; Blask 1993). There are no data available on pineal function in colon tumor-bearing animals.

11.4.3.1 Electrophysiological Study

In our experiments, the pineal glands of intact rats and rats with DMH-induced colon tumors were studied electrophysiologically (Kovalenko et al. 1998). The intensity of pineal peptide secretion was determined according to extracellularly recorded cellular spike frequency. "Slow", with spikes frequency < 2 Hz, and "fast", with spikes frequency > 4 Hz, types of pineal cell activity were revealed. Intact rats usually demonstrated only slow type of activity (frequency 0.86 ± 0.49 Hz, regularly and irregularly discharging cells). Data obtained indirectly indicate comparatively low pineal peptide secretory activity in intact rats. In colon tumor-bearing animals, pineal cells with fast pattern activity were revealed. This type of activity was absent in intact rats. In tumor-bearing rats slow cells increased spike frequency four to six times due to switching to pattern discharging, fast-type activity also appeared (5.71 ± 0.73 Hz).

The results of our electrophysiological study of intact and colon tumor-bearing animals confirm the existence of several cell types in the pineal gland (Reyes-Vazquez et al. 1986). The ratio of cells with different types of activity were distinguished in colon tumor-bearing and in intact rats (Kovalenko et al. 1998). The main part of cells in intact rats (91%) were presented by "regularly" and "irregularly" active cells, thus explaining the low summary frequency of spikes. In tumor-bearing rats the number of "pattern" and fast cells increased and consequently increased the summary frequency. A threefold decrease of the number of regular cells and simultaneous 550% increase of the number of pattern cells also points to switching of cells to pattern type of activity. This fact, the decreased number of irregular cells and increased number of fast cells, reflects intense pineal secretory processes in rats with tumors in the large intestine. These results correspond to the data on higher serum melatonin levels in rats with DMH-induced colon tumors as compared to intact animals (see Sect. 11.4.3.2; Anisimov et al. 1999).

The pineal secretory activation during tumor processes in the colon may be due to an activation of the sympathetic system, for example, during stress (Lewczuk and Przybylska-Gornowicz 1997). Therefore, we suppose a higher electric activity of pinealo-cytes, which possibly reflects a peptide secretion (Semm 1981), and higher blood melatonin concentration (Anisimov et al. 1999) to be an adaptive reaction against tumor formation, because both melatonin and pineal peptides are known to have anti-neoplastic properties (Anisimov and Reiter 1990; Petrelli 1992; Anisimov et al. 1994).

Thus, the results of our studies have demonstrated an activation of pineal peptide secretion in rats with colon tumors. This activation can be inhibited by electric olfactory stimulation. The similarity of parameters of electric activity in stress-subjected and in tumor-bearing rats allows one to suppose that the tumor process in the colon created stress-like conditions in the organism, and led to an increase of pineal secretory processes even in daytime (Kovalenko et al. 1998).

11.4.3.2 Serum Melatonin in 1,2-Dimethylhydrazine-Induced Colon Tumor-Bearing Rats

To study the relationship between pineal functional activity and serum melatonin levels, serum melatonin was estimated in rats with DMH-induced colon tumors in the daytime and at night (Anisimov et al. 2000). There is a significant (2.7-fold, $P < 0.005$) elevation of the night levels of melatonin as compared to the morning levels in control rats. In rats with DMH-induced colon tumors, the morning level of melatonin was increased ($P < 0.005$) as compared to time-matched controls. However, there was no significant elevation of night levels of melatonin in comparison to the morning levels in colon tumor-bearing animals. Our data have shown that in rats with DMH-induced colon tumors, the morning levels of serum melatonin are increased in comparison to the controls. At the same time there is no clear-cut circadian rhythm of melatonin in colon tumor-bearing animals. These findings coincide with clinical observations which demonstrated that the serum levels of this indole hormone are increased in colorectal cancer patients and that there were disturbances of the circadian rhythms of melatonin excretion in such patients (Kvetnoy and Levin 1987; Bartsch et al. 1997).

11.4.4 Melatonin-Containing Cells in the Intestinal Mucosa of Rats with 1,2-Dimethylhydrazine-Induced Colon Tumors: Effect of Exogenous Melatonin

Immunohistochemical studies revealed that the number of melatonin-containing cells (M-cells) was different in various parts of the gastrointestinal tract and was distributed as stomach > duodenum > colon > ileum. In all tissues of rats with DMH-induced colon tumors, the number of M-cells was decreased in comparison to corresponding controls: by 2.0 times in stomach, 1.8 in duodenum, 1.3 in ileum, and 1.8 in colon (Anisimov et al. 1999). In the ileum and colon of rats treated with DMH + melatonin, the number of M-cells was similar to control levels, whereas in the stomach and duodenum this number was significantly higher than that in rats treated with DMH alone, but less than in corresponding controls. A calculation of the relative content of melatonin in enterochromaffin cells of all parts of the gastrointestinal tract evaluated as optical density of the cells reveals a tendency towards a decrease in rats exposed to DMH alone in comparison to controls, and was similar to normal levels in rats treated with DMH + melatonin. At the same time, a decrease in the number of M-cells was observed in all parts of the gastrointestinal tract of animals with DMH-induced colon tumors and a relative decrease of melatonin content was seen in these cells. It is possible to suggest that increased serum levels of melatonin in DMH-treated rats are a compensatory reaction of the pineal gland to the DMH-induced decrease in local levels of melatonin in the gastrointestinal tract. These data are in accordance with observations on an increased electrophysiological activity of pinealocytes in colon tumor-bearing rats (Kovalenko et al. 1998; Sect. 11.4.3.1). It has been shown that the mammalian gastrointestinal tract contains much more melatonin than the pineal gland and enterochromaffin cells are the main source of melatonin in the organism (Kvetnoy et al. 1997; Bubenik 1999). In spite of data demonstrating an active participation of melatonin in adaptive responses, the normal function of extrapineal melatonin as well as the feedback mechanisms between pineal and gastrointestinal melatonin production are largely unknown.

It is possible to suggest that the decreased number of M-cells and their functional activity in the colon of rats exposed to DMH could play an important role in its carcinogenic effect. Treatment with melatonin prevents the decrease of M-cells in the gastrointestinal tract of rats exposed to DMH that correlated with inhibitory effects of the hormone on colon tumor development.

11.4.5 Disturbances in Carbohydrate and Lipid Metabolism During Carcinogenesis Induced by 1,2-Dimethylhydrazine in Rats

Besides the effect of a single DMH treatment, we have studied also the changes of glucose and lipid levels during carcinogenesis induced by DMH in rats (Anisimov et al. 1980). It was shown that weekly administration of DMH for 1, 4, or 6 months did not cause changes in the basal blood levels of glucose, however, a decrease in the tolerance to glucose loading was observed. A tendency to elevated levels of serum cholesterol and triglycerides was traced in all exposed groups. The levels of insulin in the serum of rats exposed to DMH for 6 months was observed to be higher than in the control group being evaluated both after 18-h starvation and 20 min after i.v. glucose loading (Anisimov et al. 1980). In rats subjected to long-term treatment with DMH (15×15 mg/kg with weekly intervals), triglyceride levels in serum were decreased by 18% in the 16th and 24th week of the experiment and after 48 weeks, when the animals had developed colon tumors, they were increased by 33% (Windle and Bell 1982).

Thus, significant disturbances of carbohydrate and lipid metabolism developed during DMH-induced carcinogenesis. These changes are involved in the mechanism of metabolic immunodepression (Dilman et al. 1977) and are factors promoting tumor growth and progression (see Anisimov 1987, 1998).

11.4.6 Effect of Melatonin on Carbohydrate and Lipid Metabolism in Rats with Colon Tumors Induced by 1,2-Dimethylhydrazine

The levels of blood glucose and serum cholesterol in colon tumor-bearing rats were increased in comparison with intact controls, whereas in rats exposed to long-term treatment with melatonin starting from the day of the first injection of DMH (experiment 2, see Sect. 11.2), blood glucose levels were not significantly changed as compared to controls. Melatonin treatment was followed by some decrease in serum cholesterol and triglyceride levels in colon tumor-bearing rats as compared to animals exposed to the carcinogen alone (Anisimov et al. 2000). Thus, melatonin exerts some normalizing influence on glucose and lipid metabolism in rats exposed to DMH.

11.4.7 Possible Effect of Melatonin on the Immune System in Rats Exposed to 1,2-Dimethylhydrazine

It was shown that the response of peripheral blood lymphocytes of patients with colon cancer to phytohemagglutinin (PHA) in vitro is markedly decreased (Pozharisski et al. 1979). On removal of the tumor, the level of stimulation by PHA was restored. The cytotoxic effect of immune lymphocytes on cell cultures of DMH-induced colon carcinoma in rats has been described (Steele and Sjorgen 1974). An antigenic cross-reaction between colon tumors and fetal intestines, which points to the presence of fetal organospecific antigens in carcinoma, has been observed (Steele and Sjorgen 1974). The authors suggest that the tumor cells produced neoantigens which were the products of reexpressed fetal genes. Okulov and Pozharisski (1975) detected an antigen in perchloric acid extracts of DMH-induced tumors of rats. It appeared to resemble human carcinoembryonal antigen in some physicochemical and biological properties.

In our experiments, some features of significant immunodepression were observed in rats 1 week after exposure to 4 weekly s.c. injections of DMH of 21 mg/kg each (Dilman et al. 1977). At this time point, only few morphologically detected colon tumors could be observed. The level of the mitogenic response of peripheral blood lymphocytes to PHA was only 4% of that in intact controls and to *E. coli* lipopoly-saccharide 18% of the control value. The titer of the antibodies to sheep erythrocytes in DMH-exposed rats was 62.8% less than that in intact controls. A decrease of phago-cytic activity of macrophages (by 43.3%) was observed in rats treated with the carcinogen as compared to intact controls (Dilman et al. 1977).

The morphometric analysis carried out in our experiments has shown that melatonin significantly increased the size of lymphoid infiltrates in both colon mucosa located far from a tumor and in ascending colon tumors in rats exposed to DMH alone. Long-term treatment with melatonin was followed by a decrease in the size of lymphoid infiltrates in the colon mucosa. This parameter was practically similar be-tween colon mucosa in intact rats and colon tumors in rats treated with DMH and melatonin (Anisimov et al. 1999).

It was shown that intestinal tumors induced by DMH very frequently arise in the area of solitary lymphoid follicles (Bandaletova 1990). Factors responsible for this effect might be a high susceptibility to the carcinogen due to the rapid renewal of epithelial cells at this site, and changes in the immunological surveillance in the peri-lymphatic space (Roos 1982). The increased sensitivity of epithelial cells of this zone to DMH was attributed to the direct contact of the cells with lymphocytes (Shimamoto and Vollmer 1987). A decrease in the proliferative activity of enterocytes in T-depen-dent zones of lymphoid patches during DMH-induced colon carcinogenesis was observed by Bandaletova (1990). It was shown that melatonin activated the function of T-cell-mediated immunity (Panzer and Viljoen 1997). Thus, it could be speculated that the reduction of lymphoid infiltration in the colon of melatonin-treated rats may reflect the normalizing influence of the hormone on the immunosurveillance in the target tissue. This effect could be indirect and realized, at least in part, by the stimulat-ing effect of melatonin on apoptosis in colon tumors. Some recent data on the immuno-modulatory role of melatonin in relation to its oncostatic effect has been comprehen-sively reviewed (Yu and Reiter 1993; Panzer and Viljoen 1997).

11.5 Conclusion

An inhibitory effect of melatonin on DMH-induced colon carcinogenesis in rats has been shown in our experiments. This effect was expressed by a decrease of incidence and multiplicity of colon tumors, by a decrease in tumor size and invasiveness, and by an increase in tumor differentiation. All these effects were more expressed when animals were exposed to 5 weekly injections of the carcinogen than in animals expos-ed to 15 injections of DMH (see Sect. 11.2). These observations suggest an influence of melatonin on both early and late stages of DMH-induced colon carcinogenesis. It seems that effects of melatonin on late stages of carcinogenesis have less significance for its anticarcinogenic effect than its effect on early stages. Data on the absence of an inhibitory effect of melatonin on the growth of colon tumors in vivo and in vitro are in accordance with this suggestion.

In Table 11.4 we have summarized some data on early and late events in DMH-induced colon carcinogenesis and effects of melatonin on it. The picture is not yet complete, but it seems clear that practically at any step involved in the process, mela-tonin exerts effects which counteract carcinogenesis. We believe that future research will present new evidence supporting this conclusion.

Table 11.4. Early and late events in 1,2-dimethylhydrazine (DMH)-induced colon carcinogenesis: effect of melatonin

Parameters	Early stages		Late stages	
	DMH	Melatonin	Colon tumor	Melatonin treatment
Level of P-450 and b_5 in liver	↑	↓		
Formation of active metabolites	↑	?		
DNA alkylation at O^6 position	↑	?		
Reactive oxygen species	↑	↓		
Protein peroxidation	↓	?		
Lipid peroxidation	↑	↓		
Mutagenic effect	↑	↓		
DNA repair	↓	?		
Clonal proliferation of stem cells	↑	↓	↑	↓
Apoptosis	↓	↓	↓	↑
Cell-to-cell communication	↓	↑	↓	↑?
Growth factors	↑	?	↑	?
Insulin	↑	↓	↑	?
Glucose tolerance	↓	↑	↓	↑
Serum cholesterol	↑	↓	↑	↓
T-cell immunity	↓	↑	↓	↑
Pineal melatonin level	↓	↑	↑	?
Serum melatonin level	?	↑	↑	↑
Melatonin in colon mucosa	↓	↑	↓	↑
Hypothalamic threshold to steroid feedback regulation	↑	↓	↑	↓?

↓, decreases; ↑, increases; ?, no or scarce data.

Acknowledgements. The author is very thankful to Drs. M. A. Zabezhinski, I. G. Popovich, A. V. Arutyunyan, C. Bartsch, H. Bartsch, R. I. Kovalenko, I. M. Kvetnoy, S. A. Musatov, T. I. Oparina, D. A. Sibarov, F. Sichel, I. Zusman for fruitful help in this work and for valuable discussion and comments. This work was supported in part by grant 6/96 from The Russian Ministry of Science State Program "National Priorities in Medicine and Health" and grant 99 from the Russian Foundation for Basic Research.

References

Anisimov VN (1987) Carcinogenesis and Aging. Vols. 1 & 2. CRC Press, Boca Raton

Anisimov VN (1998) In: Balducci L, Lyman GH, Ershler BW (eds) Comprehensive Geriatric Oncology. Harwood Acad Publ, Amsterdam, pp 157–178

Anisimov VN, Reiter RJ (1990) Pineal function during cancer and ageing. Vopr Oncol 36:259–268

Anisimov VN, Azarova MA, Dmitrievskaya AY, Likhachev A, Petrov AS (1976) Disssemination of H3-dialkylhydrazines in the neuroendocrine system and their antigonadotropic effect in rats. Bull Exp Biol Med 82:1473–1475

Anisimov VN, Pozdeev VK, Dmitrievskaya AY, Gracheva GM, Il'in AP (1976a) Effect of the enterotropic carcinogen 1,2-dimethylhydrazine on the level of biogenic amines in the hypothalamus of rats. Bull Exp Biol Med 82:1359–1361

Anisimov VN, Pozharisski KM, Dilman VM (1980) Effect of phenformin on blastomogenic effect of 1,2-dimethylhydrazine. Vopr Onkol 26:54–58

Anisimov VN, Khavinson VKh, Morozov VG (1994) Twenty years of study on effects of pineal peptide preparation: epithalamin in experimental gerontology and oncology. Ann NY Acad Sci 719:483–493

Anisimov VN, Khavinson VKh, Prokopenko VM (1995) Melatonin and epithalamin inhibit free radical oxidation in rats. Proc Russian Acad Sci 343:557–559

Anisimov VN, Arutyunian AV, Khavinson VKh (1996) Melatonin and epithalamin inhibit the process of lipid peroxidation in rats. Proc Russian Acad Sci 348:265–267

Anisimov VN, Zabezhinski MA, Popovich IG et al. (1996a) Pineal hormone melatonin inhibits 1,2-dimethylhydrazine-induced colon carcinogenesis in rats: the role of free radical mechanisms. Int J Oncol 9(Suppl):834

Anisimov VN, Arutyunian AV, Khavinson VKh (1997) Effect of melatonin and epithalamin on the activity of the antioxidant defense system in rats. Proc Russian Acad Sci 352:831–833

Anisimov VN, Popovich IG, Zabezhinski MA(1997a) Melatonin and colon carcinogenesis: I. Inhibitory effects of melatonin on development of intestinal tumors induced by 1,2-dimethylhydrazine in rats. Carcinogenesis 18:1549–1553

Anisimov VN, Kvetnoy IM, Chumakova NK, Kvetnaia TV, Molotkov AO, Pogudina NA, Popovich VV, Popuchiev VV, Zabezhinski MA, Bartsch H, Bartsch C (1999) Melatonin and colon carcinogenesis. II. Intestinal melatonin-containing cells and serum melatonin level in rats with 1,2-dimethylhydrazine-induced colon tumors. Exp Toxicol Pathol 51:47–52

Anisimov VN, Popovich IG, Shtylik AV, Zabezhinski MA, Ben-Huh H, Gurevich P, Berman V, Tendler Y, Zusman I (2000) Melatonin and colon carcinogenesis. III. Effect of melatonin on proliferative activity and apoptosis in colon mucosa and colon tumors induced by 1,2-dimethylhydrazine in rats. Exp Toxicol Pathol 52:71–76

Arutyunyan AV, Prokopenko VM, Burmistrov SO, Opatina TI, Frolova EV, Zabezhinskii MA, Popovich IG, Anisimov VN (1997) Free radical processes in serum, liver and colon during carcinogenesis induced by 1,2-dimethylhydrazine in rats. Vopr Onkol 43:618–622

Awara WM, El-Gohary M, El-Nabi SH, Fadel WA (1998) In vivo and in vitro evaluation of the mutagenic potential of carbamazepine: does melatonin have anti-mutagenic activity? Toxicology 125:45–52

Balanski R, Blagoeva P, Mircheva Z, Pozharisski K, de Flora S (1992) Effect of metabolic inhibitors, methylxanthines, antioxidants, alkali metals, and corn oil on 1,2-dimethylhydrzine carcinogenicity in rats. Anticancer Res 12:933–940

Bandaletova TY (1990) Concerning primary development of 1,2-dimethyl-hydrazine-induced cancer in lymphoid patches of the large bowel. Vopr Onkol 36:1456–1460

Barlow-Walden LR, Reiter RJ, Abe M, Pablos M, Menendez-Pelaez A, Chen LD, Poeggeler B (1995) Melatonin stimulates brain glutathione peroxidase activity. Neurochem Int 26:497–502

Barni S, Lissoni P, Cazzaniga M, Ardizzoia A, Mergalli S, Fossati V, Fumagalli L, Brivio F, Tancini G (1995) A randomized study of low-dose subcutaneous interleukin-2 plus melatonin versus supportive care alone in metastatic colon cancer patients progressing under 5-fluorouracil and folates. Oncology 52:243–245

Bartsch C, Bartsch H, Lippert TH (1992) The pineal gland and cancer: facts, hypotheses and perspectives. The Cancer J 5:194–199

Bartsch C, Kvetnoy I, Kvetnaya T, Bartsch H, Molotkov A, Franz H, Raikhlin N, Mecke D (1997) Nocturnal urinary 6-sulfatoxymelatonin and proliferating cell nuclear antigen-immunopositive tumor cells show strong positive correlations in patients with gastrointestinal and lung cancer. J Pineal Res 23:90–96

Bedi A, Pasricha PJ, Akhtar AJ, Barber JP, Bedi GC, Giardiello FM, Zehnbauer BA, Hamilton SR, Jones RJ (1995) Inhibition of apoptosis during development of colorectal cancer. Cancer Res 55:1811–1816

Berstein LM (1995) The endocrinology of large bowel cancer. Endocrine-Related Cancer 2:171–185

Blask DE (1993) Melatonin in oncology. In: Yu HS and Reiter RJ (eds) Melatonin Biosynthesis, Physiological Effects, and Clinical Applications. CRC Press, Boca Raton, pp 447–475

Bubenik GA (1999) Localization and physiological significance of gastrointestial melatonin. In: Watson RR (ed) Melatonin in Health Promotion. CRC Books, Boca Raton, pp 21–39

Cagnoli CM, Atabay C, Kharlamova E, Manev H (1995) Melatonin protects neurons from singlet oxygen-induced apoptosis. J Pineal Res 18:222–226

Callaghan BD (1995) The long-term effect of pinealectomy on the crypts of the rat gastrointestinal tract. J Pineal Res 18:191–196

Coleman MP, Esteve J, Damiecki P, Arslan A, Renard H (1993) Trends in Cancer Incidence and Mortality. IARC Sci Publ 121:1–806

Dilman VM, Anisimov VN (1979) Hypothalamic mechanisms of ageing and of specific age pathology – I. Sensitivity threshold of hypothalamo-pituitary complex to homeostatic stimuli in the
reproductive system. Exp Gerontol 14:161–174

Dilman VM, Sofronov BN, Anisimov VN et al. (1977) Withdrawal of 1,2-dimethylhydrazine-induc-
ed immunodepression by phenformin in rats. Vopr Onkol 23:50–54

Hirose Y, Yoshimi N, Suzui M, Kawabata K, Tanaka T, Mori H (1997) Expression of bcl-2, bax, and
bcl-X_L proteins in azoxymethane -induced rat colonic adenocarcinomas. Mol Carcinogenesis
19:25–30

Koratkar R, Vasudha A, Ramesh G et al. (1992) Effect of melatonin on cis-platinum induced genetic
damage to the bone marrow cells of mice. Med Sci Res 20:179–180

Kothari L, Subramanian A (1992) A possible modulatory influence of melatonin on representative
phase I and II drug metabolizing enzymes in 9,10-dimethyl-1,2-benzanthracene induced rat
mammary tumorigenesis. Anti-Cancer Drugs 3:623–628

Kovalenko RI, Novikova IA, Sibarov DA et al. (1998) Adaptive morphofunctional changes of pineal
cells during tumor process in large intestine. Pathophysiology 5(Suppl 1):17

Kvetnoy IM, Levin IM (1987) Diurnal melatonin excretion in stomach and colon cancer patients.
Vopr Onkol 33:29–32

Kvetnoy I, Sandvik AK, Waldum HL (1997) The diffuse neuroendocrine system and extrapineal
melatonin. J Mol Endocr 18:1–3

Lewczuk B, Przybylska-Gornowicz B (1997) Effects of sympatholytic and sympathicomimetic
drugs on pineal ultrastructure in the domestic pig. J Pineal Res 23:198–208

Lewinski A, Rybicka I, Wajs E, Szkudlinski M, Pawlikowski M (1991) Influence of pineal indolamines on the mitotic activity of gastric and colonic mucosa epithelial cells in the rat: interaction
with omeprazole. J Pineal Res 10:104–108

Maskens AP (1987) Multistep carcinogenesis in the colorectal mucosa. In: Maskens AP, Ebbesen P,
Burny A (eds) Concepts and Theories in Carcinogenesis. Excerpta Medica, Amsterdam,
pp 221–242

Musatov SA, Anisimov VN, Andre V, Vigreux C, Godard T, Gauduchon P, Sichel F (1998) Modulatory
effects of melatonin on genotoxic response of reference mutagens in the Ames test and the
COMET assay. Mutat Res 417:75–84

Musatov SA, Rosenfeld SV, Togo EF, Mikheev VS, Anisimov VN (1997) The influence of melatonin
on mutagenicity and antitumor action of cytostatic drugs in mice. Vopr Onkol 43:623–627

Neri B, De Leonardis V, Gemelli MT, di Loro F, Mottola A, Ponchietti R, Raugei A, Cini G (1998) Melatonin as biological response modifier in cancer patients. Anticancer Res 18:1329–1332

Okulov VB, Pozharisski KM (1975) Carcinoembryonic antigen in experimental intestinal tumours.
Bull Exp Biol Med 80:78–81

Panzer A, Viljoen M (1997) The validity of melatonin as an oncostatic agent. J Pineal Res 22:184–202

Pentney P (1995) An investigation of melatonin in the gastrointestinal tract. PhD Thesis, University
of Guelph, pp 1–161

Petrelli C, Moretti P, Petrelli F, Bramucci M (1992) A low molecular weight calf pineal peptide with
an inhibiting effect on the growth of L1210 and HL60 cells. Cell Biol Int Rep 16:967–974

Pierrefiche G, Topall G, Courboin G, Henriet I, Laborit H (1993) Antioxidant activity of melatonin
in mice. Res Commun Chem Pathol Pharmacol 80:211–223

Pozharisski KM, Anisimov VN (1975) On the role of endocrine system in development of experimental colon tumors in rats. Pathol Physiol 1:47–50

Pozharisski KM, Likhachev AJ, Klimashevski VF, Shaposhnikov JD (1979) Experimental intestinal
cancer research with special reference to human pathology. Adv Cancer Res 30:165–237

Reiter RJ, Carneiro RC, Oh C-S (1997) Melatonin in relation to cellular antioxidative defense mechanisms. Horm Metab Res 29:363–372

Reyes-Vazquez C, Prieto-Gomez B, Aldes LD, Dafny N (1986) Rat pineal exhibits two electrophysiological patterns of response to microiontophoretic norepinephrine application. J Pineal
Res 3:213–222

Roos JS (1992) Experimental large intestinal adenocarcinoma induced by hydrazine and human
colorectal cancer: a comparative study. In: Malt RA, Williamson RCN (eds) Colon Carcinogenesis.
Proc. of 31st Falk Symp, Titisee Lancaster, pp 187–210

Salim AS (1993) The permissive role of oxygen-derived free radicals in the development of colonic
cancer in the rat. A new theory for carcinogenesis. Int J Cancer 53:1031–1035

Samaha HS, Kelloff GJ, Steele V, Rao CV, Reddy BS (1997) Modulation of apoptosis by sulindac,
curcumin, phenylethyl-3-methylcaffeate, and 6-phenylhexyl isothiocyanate: apoptotic index as
a biomarker in colon cancer chemoprevention and promotion. Cancer Res 57:1301–1305

Semm P, Schneider T, Vollrath L (1981) Morphological and electrophysiological evidence for
habenular influence on the guinea-pig pineal gland. J Neural Transm 50:247–266

Shimamoto F, Vollmer E (1987) Changes in intestinal mucosa above lymph follicles during carcinogenesis in rats. A light and electron microscopic study. J Cancer Res Clin Oncol 113:41–50

Steele G, Sjogren HO (1974) Cross-reacting tumor-associated antigen(s) among chemically induced rat colon carcinomas. Cancer Res 34:1801–1807
Susa N, Ueno S, Furukawa Y, Ueda J, Sugiyama M (1997) Potent protective effect of melatonin on chromium(VI)-induced DNA single-strand breaks, cytotoxicity, and lipid peroxidation in primary cultures of rat hepatocytes. Toxicol Appl Pharmacol 144:377–384
Tan D, Reiter RJ, Chen L, Poeggeler B, Manchester LC, Barlow-Walden LR (1994) Both physiological and pharmacological levels of melatonin reduce DNA adduct formation induced by the carcinogen safrole. Carcinogenesis 15:215–218
Tendler Y, Ben-Hur H, Gurevich P, Berman V, Sandler B, Nyska A, Zion H, Zinder O, Zusman I (1997) Role of apoptosis, proliferating cell nuclear antigen and p53 protein in the immune response of the rat colon cells to cancer and vaccination with the anti-p53 polyclonal antibodies. Anticancer Res 17:4653–4658
Ubeda A, Trillo MA, House DE, Blackman CF (1995) Melatonin enhances junctional transfer in normal C3H/10T1/2 cells. Cancer Lett 91:241–245
Vijayalaxmi, Reiter RJ, Leal BZ, Meltz ML (1996) Effect of melatonin on mitotic and proliferation indices, and sister chromatid exchange in human blood lymphocytes. Mutat Res 351:187–192
Windle R, Bell PRF (1982) Lipid clearance in a colonic tumour model in rats. Br J Cancer 46:515
Yu HS, Reiter RJ (eds) (1993) Melatonin Biosynthesis, Physiological Effects, and Clinical Applications. CRC Press, Boca Raton
Zusman I (1997) The role of p53 tumor-associated protein in colon cancer detection and prevention (review). Int J Oncol 10:1241–1249

12 Role of Extrapineal Melatonin and Related APUD Series Peptides in Malignancy

Igor M. Kvetnoy, Tatiana V. Kvetnaia, Vadim V. Yuzhakov

Abstract

During the last decade, attention has centered on melatonin (MT), one of the hormones of the diffuse neuroendocrine system (DNES) which for many years was considered only as a hormone of the pineal gland. Currently, MT has been identified not only in the pineal gland but also in extrapineal tissues such as the retina, harderian gland, gut mucosa, cerebellum, airway epithelium, liver, kidney, adrenals, thymus, thyroid, pancreas, ovary, testis, carotid body, placenta, and endometrium as well as in nonendocrine cells like mast cells, natural killer cells, eosinophilic leukocytes, platelets, and endothelial cells. The above list of the cells storing MT indicates that MT has a unique position among the hormones of the DNES, being found in practically all organ systems. Functionally, MT-producing cells are certain to be part and parcel of the DNES as a universal system of response and control of protection to the organism. Taking into account the large number of MT-producing cells in many organs, the wide spectrum of its biological activities and, especially, its main property as a universal regulator of biological rhythms, it should be possible to consider extrapineal MT as a key paracrine signal molecule for the local coordination of intercellular relationships. Analysis of the experimental results described here shows the direct participation and active role of extrapineal MT and related APUD hormones in both the pathogenesis of tumor growth and the modification of antitumor therapy.

12.1 Introduction

Among many fundamental problems in modern oncobiology, the analysis of systems that control and integrate adaptation processes associated with tumor growth is of great interest. One of these systems is the diffuse neuroendocrine system (DNES). Current morphofunctional views on DNES are based on the APUD concept proposed and developed in detail by the English histochemist and pathologist A. G. E. Pearse (1968).

In 1968–1969, Pearse suggested that a specialized, highly organized cell system exists in organisms, whose main feature is the capability of component cells to produce peptide hormones and biogenic amines. The concept was based on the extensive series of experiments to distinguish endocrine cells in different organs by a detailed cytochemical and ultrastructural identification of endocrine cell-generated products. Different types of cells, widely dispersed throughout the organism, have a common ability to take up and decarboxylate monoamine precursors to produce biogenic amines. This ability accounts for the term "APUD," an acronym for amine precursor uptake and decarboxylation, used by Pearse to designate these special cell series (1969). The existence of both biogenic amines and active peptides, i.e., regulatory peptides, in neurons of the central and peripheral nervous system and in APUD cells located in different organs gave support to that concept (Polak and Bloom 1986; Sundler et al. 1989; Hakanson et al. 1990).

To date, the APUD series includes over 60 types of cells localized in gut, pancreas, urogenital tract, airway epithelium, pineal gland, adenohypophysis and hypothalamus, thyroid gland, adrenals, carotid body, skin, sympathetic ganglia, thymus, placenta, and other organs (Kvetnoy 1987; Raikhlin and Kvetnoy 1994). Data on the identification and localization of monoamines and regulatory peptides in both neural and endocrine cells of different organs suggest that these elements are incorporated into a common but diffuse regulating system, namely the DNES. Located in practically all organs and producing vitally important, biologically active substances, DNES cells fulfill the role of tissue regulators of homeostasis, controlling a multitude of physiological processes via neuroendocrine, endocrine, and paracrine mechanisms of messenger molecule effects on target cells (Larsson 1980; Kvetnoy 1987).

In recent years, special attention has largely focused on a particular DNES hormone, melatonin (MT). Featuring a wide activity spectrum, this hormone plays a key role in the control of biological rhythms (Reiter 1991), thus essentially affecting the nervous, endocrine, and immune systems as well as the organism as a whole.

In the late 1950s, the group around the American dermatologist Aaron Lerner of Yale University first identified MT as the pineal substance responsible for bleaching of frog skin. Melatonin was found to be the 5-methoxy-N-acetylated derivative of serotonin (ST) or 5-hydroxytryptamine (Lerner et al. 1958). Two years later, Axelrod and Weissbach (1960) identified the key enzymes of MT synthesis as N-acetyltransferase (NAT) and hydroxyindole-O-methyltransferase (HIOMT).

The identification of MT stimulated researchers' interest in the physiology of the pineal gland, and a wide spectrum of biological activities of pineal MT was shown. The most important of these are the following: control of biological rhythms, antigonadotrophic effects, stimulation of immune processes, scavenging of free radicals, and cytostatic and antiproliferative effects in vitro and in vivo (Banerjee and Margulis 1973; Reiter 1980; Maestroni and Conti 1990; Reiter et al. 1993; Arendt 1994; Blask et al. 1994; Reiter and Robinson 1995).

As soon as highly sensitive techniques of analysis and identification became available, MT and its precursors as well as the catalytic enzymes NAT and HIOMT began to be found in extrapineal tissues, primarily those anatomically connected with the visual system – retina and harderian gland (Bubenik et al. 1976, 1978). Progress concerning extrapineal sources of MT was based on developing a technique of obtaining highly specific antibodies to indolealkylamines (Grota and Brown 1974).

Using biological and radioimmunological methods as well as thin-layer chromatography, important data were obtained at the same time which indicated that after removal of the pineal gland, MT could still be identified in blood plasma and urine of laboratory animals (Ozaki and Lynch 1976; Kennaway et al. 1977), indicating significant extrapineal synthesis of MT.

12.2 Extrapineal Melatonin: Cellular Localization, Role, and Significance Within the Diffuse Neuroendocrine System

Taking into account that gut enterochromaffin (EC) cells are the main ST depot in the organism (Erspamer and Asero 1952; Barter and Pearse 1955; Vialli 1966), we were the first to suggest MT production (Raikhlin et al. 1975) and later to identify it in these cells (Raikhlin and Kvetnoy 1976). The presence of MT and its precursors in EC cells in human appendix, stomach, and duodenum as well as in stomach and duodenum of experimental animals (dogs, rabbits, rats, and mice) has been shown immunohisto-chemically (Raikhlin et al. 1976; Kvetnoy and Yuzhakov 1993). These results were sub-sequently confirmed by Bubenik (1980), who detected MT in practically all parts of the rat gastrointestinal tract. The localization of the key MT-synthesizing enzyme HIOMT in gut confirmed the occurrence of MT synthesis rather than passive accumulation (Quay and Ma 1976).

Quantitative analysis showed that the total number of EC cells throughout the gut is significantly larger than the possible number of MT-producing cells of the pineal gland (Kvetnoy 1987). Recently, Huether (1993) showed that avian and mammalian gastrointestinal tracts contain at least 400 times more MT than the pineal gland. These data and the fact that EC cells account for 95% of ST (Verhofstad et al. 1983), the principal precursor of MT, allow us to consider gut EC cells as the main source of MT in humans and animals.

To date, the functional morphology of EC cells has been studied in sufficient detail. Enterochromaffin cells can serve as a classic example of APUD cells in which biogenic amines (ST and MT) and peptide hormones (substance P, motilin and enkephalins) coexist (Solcia et al. 1989). In our studies, we also observed colocalization of MT, endothelin, and calretinin in gastric parietal cells, MT and histamine in mast cells, MT, somatostatin, and beta-endorphins in the natural killer cells, and MT and prosta-glandin F_2 in thymic reticuloepithelial cells (Kvetnoy and Yuzhakov 1993, 1994).

Evidence that MT is related not only to the pineal gland and appears very early in evolution was supported by the discovery of MT production in a primitive metazoan (Mechawar and Anctil 1997) and unicellular organisms (Tilden et al. 1997).

It has become increasingly evident that the pineal gland is not the only site of MT production. The MT content in the organism and its concentration in blood is account-ed for not only by the pineal gland but also by extrapineal sources of synthesis, changes in the extracellular fluid volume, hormone binding with blood proteins, and meta-bolism and excretion rates depending on different external and internal regulatory factors. Actually, the cells synthesizing MT and ST exist not only in gut but also in airway epithelium, liver, thyroid, adrenals, paraganglia, gallbladder, kidneys, ovary, endometrium, placenta, skin, and the inner ear (Kvetnoy et al. 1997).

In recent years, MT was found also in nonendocrine cells, i.e., mast cells, natural killer cells, eosinophilic leukocytes, pancreatic acinar cells, reticuloepithelial cells in thymus, some endothelial cells (Kvetnoy and Yuzhakov 1995), and platelets (Champier et al. 1997). MT and its key enzymes (NAT and HIOMT) are measurable in hamster skin culture (Slominski et al. 1996) and in rat testis, where they were localized pre-dominantly in interstitial cells (Tijmes et al. 1996).

The above list of cells storing MT (Table 12.1) indicates that there are considerable prospects for the future search of potential MT-producing cells. Thus, MT has a unique position among the hormones of the DNES, being found in practically all organ

Table 12.1. Extrapineal localization of MT

MT immunopositive endocrine cells	MT immunopositive nonendocrine cells
Gut (EC cells)	Photoreceptors (retina)
Airway epithelium	Cells of ciliary body
Liver	Cells of harderian gland
Kidneys	Mast cells
Adrenals	Natural killer (NK) cells
Thyroid	Eosinophilic leukocytes
Paraganglia	Platelets
Gallbladder	Acinar cells (pancreas)
Ovary	Reticuloepithelial cells (thymus)
Endometrium	Endothelial cells (blood vessels)
Placenta	Interstitial cells (testis)
Inner ear	Purkinje's cells (cerebellum)
Skin	Melatoninergic neurons (brain)

systems. Functionally, MT-producing cells are certain to be part and parcel of the DNES as a universal system of response, control, and organism protection.

In spite of data showing the active participation of MT in adaptive response as well as in pathophysiology, the functional significance of extrapineal MT-producing cells remains practically unknown. As a hypothesis, variations in the blood MT level controlled through changes in lighting at the central location of these cells, namely the pineal gland, affect the peripheral MT-producing cells dispersed throughout different organs by "catching" to a certain extent the cell secretion rhythm. However, pinealectomy results in no demonstrable effect on the endogenous MT content in gut, and no daily variations of MT were found in different parts of digestive tract (Bubenik 1980).

During the last 15 years, our team has studied the functional morphology and behavior of extrapineal MT-producing cells as well as other main APUD cells in different pathologies and environmental conditions (e.g., ionizing and nonionizing radiation, tumor growth, cytostatic therapy and radiotherapy for malignant tumors, autoimmune and gastrointestinal diseases, and pharmacological and toxicological influence). The data obtained testify to an active participation of extrapineal MT as well as other APUD hormones in the pathogenesis of various diseases (Kvetnoy and Yuzhakov 1994; Raikhlin and Kvetnoy 1994; Kvetnoy et al. 1994; Raikhlin et al. 1994; Kvetnoy et al. 1995).

Hence, taking into account the large number of MT-producing cells in many organs, the wide spectrum of MT's biological activities and especially its main property as a universal regulator of biological rhythms, it seems possible to suppose that extrapineal MT plays a key role as a paracrine signal molecule for the local coordination of intercellular relationships.

The additional confirmation of this point of view is that many neighboring cells in different organs where MT was found also have MT membrane receptors (Vanecek and Vollrath 1990; Krause and Dubocovich 1991; Pang et al. 1993; Barrett et al. 1994; Reppert et al. 1994; Dubocovich and Masana 1998). It seems that this property is the main physiological distinction between extrapineal and pineal MT, the latter always acting as a typical hormone, reaching widely spread target cells through the bloodstream. In both cases, some nonendocrine cells such as mast cells and eosinophilic leukocytes may take up MT from the blood or intercellular space for transport to sites where it acts.

12.3 Extrapineal MT and APUD Series Peptides: Possible Participation in Endogenous Mechanisms of Tumor Growth

Analysis of the biological characteristics of many physiologically active substances produced by DNES cells (MT, ST, gastrin, insulin, glucagon, somatostatin, etc.) suggests an important role of these cells and hormones in tumor growth (Raikhlin and Kvetnoy 1994). Thus, the study of the role and significance of DNES and, in particular, extrapineal MT-producing cells in the pathogenesis of tumor development lends a new perspective to the interpretation of endogenous mechanisms of tumor-induced responses in several organ and tissue systems as well as in the organism as a whole.

We have documented the hyperplasia of EC cells producing MT, pancreatic B and D cells (producing insulin and somatostatin, respectively) as well as adrenal NEP cells (producing norepinephrine) in initial stages, and a decrease in these cells' number in late stages of carcinogenesis in humans (Figs. 12.1, 12.2). Our studies also showed similar behavior of these cells in Lewis lung carcinoma in mice. In nonmetastasizing tumors, hypoplasia and decreased functional activity occur for gut ECL cells (histamine), G cells (gastrin) of the stomach, A cells (glucagon) of the pancreas, and adrenal EP cells (epinephrine). These cells increased significantly in number during advanced stages of cancer (Figs. 12.3, 12.4).

The data about intensification of the binding of exogenous ^{3}H-MT as well as ^{125}I-MT in many vital organs of tumor animals (gut, airway epithelium, liver, kidneys, adrenals, and pancreas of male Wistar rats with sarcoma-45) compared to the intact group (Petrova et al. 1994) can be considered additional confirmation of the activation of extrapineal MT-producing cells in initial stages of tumor development. Biologically, the organism may attempt to normalize proliferative processes through MT's cytostatic properties.

These data show that participation of APUD cells is important in the endogenous mechanisms of tumor growth, because a number of hormones synthesized by APUD

Fig. 12.1. EC cells of appendix in patients with stomach cancer with (mts+) and without (mts-) metastases (83 and 74 patients, respectively). The control group included 49 patients with nononcological pathology (myocardial infarction and cerebral injury). The samples for immunohistochemical studies were taken during autopsy not later than 4 h after death. The group medians ± errors of the medians are given. The differences between both tumor groups and controls are significant ($p < 0.05$)

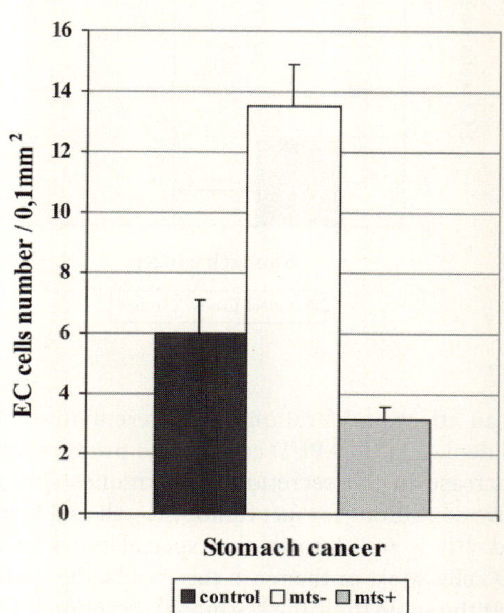

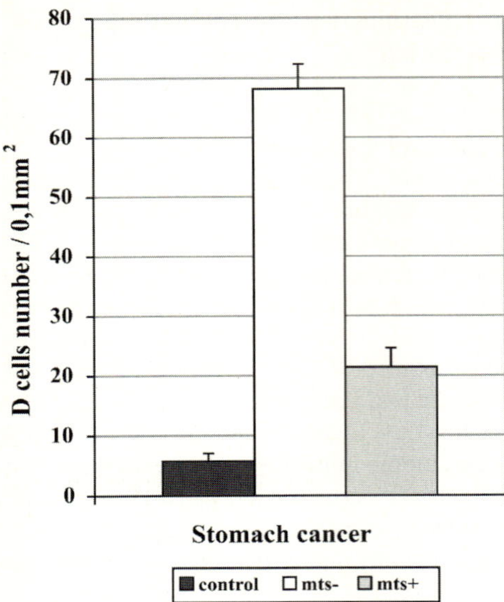

Fig. 12.2. Pancreatic D cells in patients with stomach cancer with (mts+) and without (mts-) metastases (83 and 74 patients, respectively). The control group included 49 patients with nononcological pathology (myocardial infarction and cerebral injury). The samples for immunohistochemical studies were taken during autopsy not later than 4 hours after death. The group medians ± errors of the medians are given. The differences between both tumor groups and the controls are significant ($p < 0.05$)

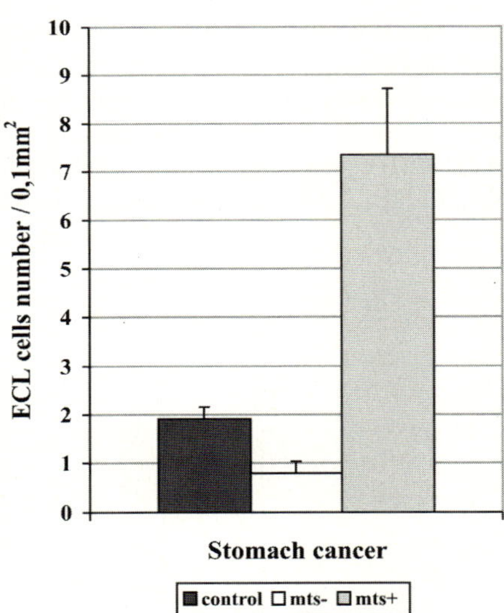

Fig. 12.3. Gastric ECL cells in patients with stomach cancer with (mts+) and without (mts-) metastases (83 and 74 patients, respectively). The control group included 49 patients with non-oncological pathology (myocardial infarction and cerebral injury). The samples for immunohistochemical studies were taken during autopsy not later than 4 hours after death. The group medians ± errors of the medians are given. The differences between both tumor groups and controls are significant ($p < 0.05$)

cells can affect proliferation and differentiation of tumor cells. Generally, the "functional depletion" of APUD cells, which produce antiproliferative hormones, as well as the increase in the secretion of hormones able to stimulate cell proliferation can improve conditions for fast tumor growth and formation of metastases.

Today it is well-known that special types of tumors (apudomas) develop from APUD cells. Most of them are carcinoids, the typical neoplasms originating from EC cells. At the same time, the hormonal secretion by nonendocrine carcinomas has great

Fig. 12.4. Gastric G cells in patients with stomach cancer with (mts+) and without (mts-) metastases (83 and 74 patients respectively). The control group included 49 patients with non-oncological pathology (myocardial infarction and cerebral injury). The samples for immunohistochemical studies were taken during autopsy not later than 4 hours after death. The group medians ± errors of the medians are given. The differences between both tumor groups and controls are significant ($p < 0.05$)

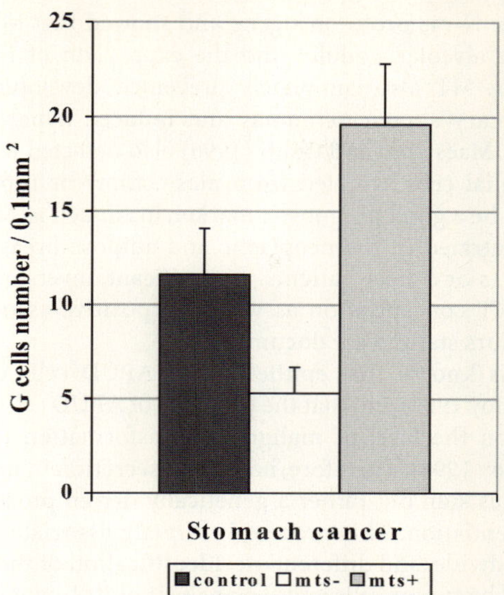

theoretical and practical significance. Abnormalities in protein synthesis, hybridization of malignant nonendocrine cells with normal endocrine cells, and expression of the corresponding genes resulting from malignant growth have been considered as possible explanations for this phenomenon (Bosman 1984). However, the presence of endocrine cells is demonstrable in different nonendocrine epithelial malignant tumors (Bosman 1984; Maluf and Koerner 1994; Bonkhoff et al. 1995).

We studied 182 human carcinomas of different histological types and localizations (cancer of larynx, lung, breast, stomach, colon, and rectum). We found that 60 tumors (30% of all nonendocrine carcinomas studied) contained endocrine cells and, among them, 37 neoplasms (60% of hormone-producing nonendocrine tumors) contained MT immunopositive cells (Raikhlin and Kvetnoy 1994). The identification of endocrine cells in metastases of the carcinomas studied confirms the malignant nature of these cells. In our investigations, we documented strong correlations between the histological type of the tumors and the biological properties of the hormones produced by them, i.e., MT, ST and somatostatin (known as antiproliferative substances). These substances were more often found within highly differentiated tumors (adenocarcinomas and squamous cell carcinomas with keratinization), whereas catecholamines, histamine, STH, insulin, and gastrin (substances known to stimulate cell growth) were usually observed within poorly differentiated tumors (solid carcinomas, squamous cell carcinomas without keratinization). These facts allow us to suppose that the production of MT and related APUD peptides in situ in nonendocrine carcinomas plays a key role in autocrine mechanisms of tumor homeostasis, promoting or preventing tumor progression and the formation of metastases.

An illustration of this postulate is provided by our data about increases in the number of MT immunopositive cells in human breast adenocarcinoma without metastases (Raikhlin and Kvetnoy 1994). In confirmation, Mediavilla et al. (1997) studied the oncostatic effect of MT on the mammary gland in transgenic mice carrying the N-ras proto-oncogene and showed that MT reduced the incidence of hyperplastic alveolar nodules and the expression of N-ras protein in focal hyperplastic lesions. MT also completely prevented development of epithelial cell atypia and mammary adenocarcinomas and reduced hyperplasia of the mammary lymphoid tissue. Maestroni and Conti (1996) also detected MT production in situ by mammary epithelial cells sampled from mastectomy or tumorectomy and supposed that MT might be a good prognostic marker, inasmuch as MT was 3 orders of magnitude more concentrated in the neoplastic and adipose breast tissue than in sera from healthy subjects or cancer patients. A significant, inverse correlation between nuclear atypia and MT concentration as well as a positive association between MT and estrogen receptors status were documented.

It is known that epithelial and APUD cells develop from common stem cells (Kvetnoy 1987) and that the presence of APUD cells in non-endocrine tumors depends only on the level of malignant transformation (Kvetnoy et al. 1994; Raikhlin and Kvetnoy 1994). Therefore, hormonal secretion in nonendocrine tumors is not an autonomous sign but rather a genetically driven process determined by cell genesis and differentiation. This process is directly associated with the potential of the cells to grow, divide, and differentiate. Identification of the chemical type of hormone(s) produced by tumor cells and the analysis of its biological features may be very important for tumor prognosis.

Together with Drs. C. and H. Bartsch (University of Tübingen, Germany), we recently studied the excretion of 6-sulphatoxymelatonin (aMT6s) in urine, expression of proliferating cell nuclear antigen (PCNA), and numbers of apoptotic (APT) and MT immunopositive cells in 26 primary human gut and lung tumors (cancer of colon, rectum, stomach, and lung) without metastases (Kvetnoy et al. 1996; Bartsch et al. 1997). The results showed strong positive correlations between the expression of PCNA in tumors and aMT6 s excretion in urine as well as between MT immunopositive cells and APT cells in tumors. Strong negative correlations were observed between MT immunoreactivity and proliferative activity of tumor cells. All results were independent of the age of the patients and histological type and localization of the tumor.

It is now well-established that the proliferative activity of tumor cells plays a key role in neoplastic growth, invasiveness, and metastatic formation (Sagar et al. 1993). Therefore, the measurement of proliferative capacity was proposed as an effective means of assessing the malignant potential of various carcinomas. From this standpoint, PCNA is one of the most suitable markers. It is an immunohistochemical marker of proliferative activity and its determination is possible only in tissue specimens from tumors obtained during surgery.

Our findings of a strong correlation between the proliferative activity of tumors and aMT6s excretion in urine, obtained by a combination of immunohistochemical and radioimmunological methods, offers a new noninvasive method for determining the degree of tumor proliferation at different stages of malignant disease in daily clinical practice.

In spite of many studies of the inhibiting effect of MT on tumor growth, the mechanism of its role in regulating tumor cells' proliferative activity remains unclear. In general, the effects of MT are considered only as a function of pineal MT and do not take into account the extrapineal synthesis of MT. Considering the wide distribution of MT in different tissues and its possible influence on target cells via paracrine mechanisms, the views on active participation of extrapineal MT in the endogenous regulation of tumor growth deserve attention.

There are some ideas on the mechanisms underlying these phenomena. Lemus-Wilson et al. (1995) suggested that the oncostatic activity of MT on MCF-7 cells is linked to an antagonism of prolactin's action, as indicated by the interruption by MT of prolactin-mediated growth of human breast carcinoma. Cos et al. (1996) examined whether physiological doses of MT (1 nM) modified DNA synthesis in MCF-7 human breast cancer cells. Exponentially growing MCF-7 cells were incubated for 24 h with thymidine (2 mM) to block mitosis and synchronize the cell division cycle. A flow cytometry study indicated that 82.3 % of the cells were in phase G_1. Synchronized cells were pulsed for 1 h with ^{3}H-deoxythymidine (^{3}H-dThy) or ^{3}H-dThy plus MT at 0 h, 3 h, 6 h, 9 h, 12 h, 15 h, or 24 h after the release of mitotic arrest. Exposing these synchronized MCF-7 cells to MT for only 1 h significantly inhibited ^{3}H-dThy incorporation when DNA precursor incorporation was the highest and the number of cells in the S phase was maximal. The authors concluded that MT antiproliferative effects on MCF-7 cells could be mediated at least in part by a reduction of DNA synthesis.

It is known that gap junctions mediate the communication between adjacent cells and are closely related to cellular growth. In this connection, the data showing that MT is able to inhibit the proliferation of rat hepatocytes and induce Cx32 gap junction protein expression and gap junctional intercellular communication (Kojima et al. 1997) appear to be of interest. The tubulin polymerization process may be one of the intracellular targets of MT action on tumor cells. Melendez et al. (1996) showed that physiological concentrations of MT induce an increase of microtubules in neuroblastoma NIE-115 cells and that this effect was due to an increase in the polymerization state of tubulin. Manev et al. (1996) recently reported that MT protected neuronal cultures from excitotoxicity mediated via kainate-sensitive glutamate receptors and from oxidative stress-induced apoptosis.

Taking into account the direct connection between the levels of MT in blood and aMT6s in urine, the determination of the latter appears to represent a reliable marker of the degree of MT synthesis in the organism. It is possible to assume that a high urinary excretion of aMT6s is an evidence of an increase of MT secretion by pinealocytes and extrapineal sources, which in turn leads to a decreased binding of MT in the tumor. Due to a deficiency of MT in the tumor, the proliferative activity of tumor cells may increase and the metastatic potential may become stronger. On the other hand, a decrease in urinary aMT6s excretion parallels reduced secretion of MT from cellular sources into the blood, and MT binding to the tumor increases under such conditions and may suppress tumor cell proliferation via paracrine mechanisms.

Therefore, the quantity of MT produced by APUD tumor cells may have great significance for prognosis, since, at least in some studies, a dose-dependent inhibitory effect of MT on tumor growth was established (Carossino et al. 1996; Hu and Roberts 1997).

12.4 Extrapineal Melatonin and Tumor Radiosensitivity: New Approaches for Modification of Antitumor Therapy

The study of ways to modify antitumor therapy is one of the central topics in current oncology. This is accounted for by the fact that the efficiency of radiation therapy and cytostatic substances is low, due to both the lack of selectivity of their effect and their high toxicity in relation to intact organs. Today, the evaluation of therapeutic significance of biological modifiers for the treatment of malignant tumors appears to be a promising trend in radiobiology and oncology. The biological modification methods of radiation therapy and chemotherapy are more significant than other approaches because of the existence of endogenous biologically active substances synthesized in the living body that show radiomodifying properties and exert stabilizing effects on metabolic disorders developing during the oncological process.

The application of MT as a potential response modifier is of great interest, especially because, besides having immunomodulating activity and stimulating cytotoxic function of natural killer lymphocytes and interferon production, MT has a wide tissue distribution and spectrum of biological activity, regulates cell differentiation and division, and in some cases inhibits the growth of experimental tumors. Being a direct metabolite of ST, which has radioprotective features, MT is able to act in a similar way after exposure to radiation. Analysis of the biological features of MT indicates that it possesses radiomodifying properties which, in combination with its regulatory and proliferative effects, render MT a potentially useful drug in radiobiology and oncoradiology.

To study such a phenomenon in experimental conditions, Ehrlich ascites carcinoma (EAC) cells growing intraperitoneally were used. They were maintained by weekly passages of 0.2 ml of ascitic fluid given intraperitoneally and passaged in male mice $F_1 \times CBA \times C_{57}BL_6$. Intact and irradiated animals (gamma-rays ^{60}Co, dose 6 Gy) with 7 days' EAC were used.

Melatonin was dissolved in 10% dimethylsulfoxide, and 0.2 ml, containing 0.025 mg, was injected intraperitoneally 15 min, 30 min, 1 h, 2 h, and 4 h before sampling the tumor ascitic fluid. The cloning of tumor cells was carried out in vivo in diffusion chambers placed in the abdominal cavity of syngenic mice recipients. Morphological analysis was performed using staining with H & E, silver nitrate according to Grimelius, the Sevki reaction, and immunohistochemical techniques with antibodies to MT (1:150) (CIDtech Res, Mississauga, Canada). The following values served as criteria for the morphofunctional evaluation:

1. The number of Sevki-positive cells in EAC to evaluate the autocrine secretory function of the tumor
2. The number of MT-immunopositive eosinophilic leukocytes (EL) in the tumor
3. The index of mitotic activity of EAC cells
4. The state of the argyrophilic nucleolar organizer (AgNOR) in tumor cells, which was divided into three different types:
    a) Type one, as usually found in highly differentiated cells (e. g., normal cells and tumor cells in nonmalignant neoplasms)
    b) Type two, characteristic of slow growing malignant cells
    c) Type three, as found in highly malignant cells with low differentiation at high growth rate

In addition, we also evaluated the survival of EAC cells.

The identification of Sevki-positive cells in EAC is of great interest. We first identified these cells within the EAC cellular population. Sevki-positive cells within EAC, stained by violet color, indicate the presence of norepinephrine (NEP). They can be considered as tumor cells with autocrine NEP secretion. We also obtained MT in the majority of EL infiltrating EAC. Since NEP has specific radioprotective features and

MT can be involved in biological processes mediated by the activation or inhibition of NEP synthesis, Sevki-positive NEP cells in EAC could be taken as those morphological structures which respond distinctly to radiation and MT action.

A short time after injection (15–30 min), MT alone and in combination with radiation reduces the number of NEP cells (and in turn the NEP content in EAC), which leads to an increase in tumor radiosensitivity. The number of MT-containing EL in EAC was also reduced. However, after a certain time the NEP quantity again increased, thus reducing tumor radiosensitivity. It should be noted that this occurred at about a 1 hr interval.

Taking into account the biochemistry of MT metabolism in living organisms, it seemed possible that MT was metabolized 60–120 min after injection to 6-hydroxy-melatonin. Thus, the observed effects could be attributed to both the influence of MT alone 15–30 min after injection, and of its metabolite – 6-hydroxymelatonin – 60 min or more after injection. MT also optimized the effect of ionizing radiation, causing increased survival of EAC cells (Fig. 12.5).

Another interesting observation derived from analysis of the mitotic activity of EAC cells and the state of AgNOR, each of which corresponded strictly to cell proliferation and differentiation, was that ionizing radiation reduces the mitotic activity of EAC cells, showing both the second and the third types of AgNOR, which are characteristic of highly malignant cells with rapid division and low differentiation. The mitotic activity of cells was reduced by MT. Cells with an AgNOR of the first type (quiet, highly differentiated) were identified for up to 1 hr after injection. After relapsing to the initial level of mitotic activity, the cells with second and third degree AgNOR appeared. In this case, the combination of MT and radiation sharply reduced the mitotic activity of EAC cells for up to 2 h of exposure. At that time, the AgNOR type changed again from the

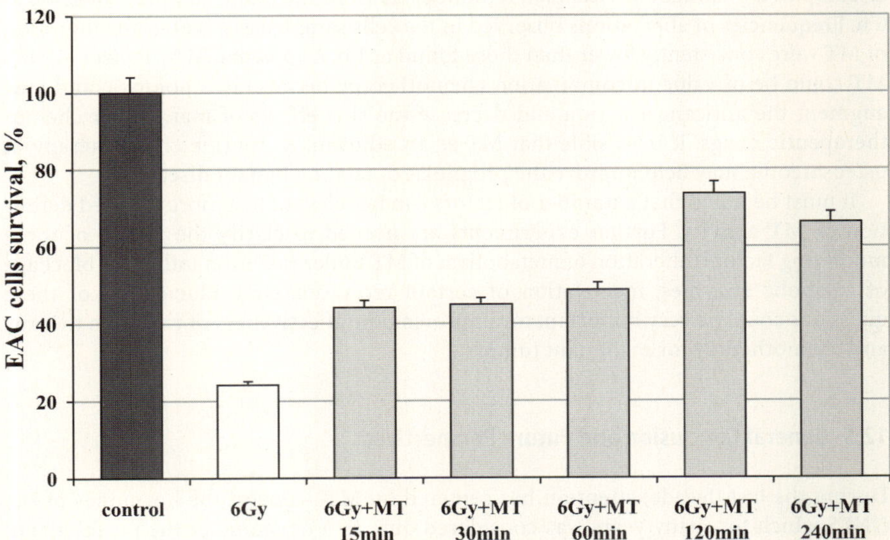

Fig. 12.5. Survival of EAC cells under influence of MT and radiation. The parameter is expressed as percent of the mean number of cell colonies per 10^2 tumor cells inoculated. The differences between all treated groups and control group are significant ($p < 0.05$)

first to second type. The third type of AgNOR was not found in these short-term studies. At 30–60 min, only the first type of AgNOR was found. After 120 min with maximal mitotic activity, only the second type of AgNOR was revealed, thus confirming the ability of EAC cells to proliferate for up to 240 min.

Our data agree with the results of Wright et al. (1996) which showed that by sampling at 8-hr intervals for 24 h after MT treatment (15 µg/day for 12 days) of premetamorphic *Rana pipiens* tadpoles, thyroid follicle cell proliferation was not depressed. In a second experiment conducted under the same conditions but sampling at 3-hr intervals for 24 h, MT significantly lowered follicle cell proliferation.

Ronco and Halberg (1996) also showed that MT can be an oncostatic or oncostimulating hormone, depending on the time of administration. They postulated that MT circadian rhythm is altered in cancer patients, possibly as a consequence of certain therapies.

Melatonin may be useful not only as a factor directly acting on tumor proliferation but also for improving the general status of cancer patients. Lissoni et al. (1996a) suggested that MT was effective in the treatment of neoplastic cachexia by decreasing blood concentrations of tumor necrosis factor (TNF). Neoplastic cachexia shows metabolic characteristics different from other common causes of malnutrition and is mainly due to abnormal TNF secretion, the levels of which are often high in patients with advanced neoplasia. Lissoni et al. (1996b) postulated that MT may enhance tamoxifen antitumor efficacy and be useful in the palliative treatment of metastatic neoplasms and in untreatable metastatic solid tumors.

Vijayalaxmi et al. (1996) reported about the protective effect of MT from genetic damage. Lymphocytes in blood samples collected at 1 and 2 h after a single MT oral ingestion (300 mg) and exposed in vitro to 150 cGy of ^{137}Cs gamma radiation exhibited a significant decrease in the incidence of chromosome aberrations and micronuclei, as compared to similarly irradiated lymphocytes from the blood samples collected at 0 h. Frequencies of aberrations observed in the cells sampled at 2 h after the ingestion of MT were consistently lower than those found at 1 h. As postulated by Panzer (1997), MT could be of value in combination chemotherapy because it is nontoxic and may augment the anticancer action and decrease the side effects of many other chemotherapeutic drugs. It is feasible that MT as an adjuvant to routine chemotherapy in osteosarcoma may help improve the prognosis of fatal malignant disease.

It must be noted that a number of factors can be relevant to a more detailed definition of MT activity. Further experiments are needed to clarify the nature of these modifying factors (alteration of metabolism of MT under radiation influence, blockage of catabolic enzymes, inactivation of certain receptors, etc.). Elucidation of these questions could be very important for improving the effectiveness of radiation therapy and chemotherapy for malignant tumors.

12.5 General Conclusion and Future Perspectives

During the last decade, attention has centered on MT – one of the hormones of the DNES which for many years was considered only as a hormone of the pineal gland. As soon as highly sensitive antibodies to indolealkylamines became available, MT was identified not only in the pineal gland, but also in extrapineal tissues such as the retina, harderian gland, gut mucosa, cerebellum, airway epithelium, liver, kidney, adrenals, thymus, thyroid, pancreas, ovary, testis, carotid body, placenta, and endometrium as well as in non-neuroendocrine cells like mast cells, natural killer cells, eosinophilic leukocytes, platelets, and endothelial cells. The above list of the cells storing MT indicates that MT has a unique position among the hormones of the DNES, being found in practically all organ systems. Functionally, MT-producing cells are certain to be part and parcel of the DNES as a universal system of response, control, and protection of the organism.

Taking into account the large number of MT-producing cells, the wide spectrum of biological activities of MT, and especially its main property as a universal regulator of biological rhythms, it should be possible to consider extrapineal melatonin as a key paracrine signal molecule for the local coordination of intercellular communication. This warrants future detailed studies of DNES in oncology.

Analysis of the experimental results described herein suggests a direct participation and active role of extrapineal MT and related APUD hormones in both the pathogenesis of tumor growth and modification of antitumor therapy. Taking into account the fact that MT and some hormones synthesized and produced by APUD cells have a radiomodifying effect and are able to activate or suppress cell proliferation, we consider it important to carry out detailed studies on the development of tumor radiosensitivity after changing the functional activity of MT-producing cells and other APUD cells located within different carcinomas.

For practical reasons, the pharmacological influence on the hormonal functions of APUD cells located inside and outside the tumors as well as on the activity of non-endocrine cells involved in hormonal transport and tumor growth promotion may appear helpful for modification and optimization of chemo- and radiation therapies used for malignant diseases.

In conclusion, the available data on the structural and functional organization of the DNES, the role and significance of pineal and extrapineal MT, other biogenic amines, and peptide hormones in homeostasis regulation suggest an important role for APUD cells in the endogenous mechanisms of tumor growth and progression. It could be very important for modern oncology and general pathology to pursue these studies further.

Acknowledgements. The authors are grateful to Prof. Gregory M. Brown (Clarke Institute of Psychiatry, Toronto) for permanent support by providing CIDtech ultra-specific rabbit melatonin antiserum for immunocytochemical studies. We thank our colleagues Drs. Christian and Hella Bartsch (University of Tübingen, Germany) for long-term fruitful cooperation and friendly discussion. The study of hormonal functions of nonendocrine cells was partly supported by a grant of the Russian Foundation of Fundamental Research (RFFR 98–04–48763).

References

Arendt J (1994) Human responses to light and melatonin. In: Moller M, Pevet P (eds) Advances in pineal research: 8. John Libbey and Company, London, pp 439–441

Axelrod J, Weissbach H (1960) Enzymatic O-methylation of N-acetyl-serotonin to melatonin. Science 131:1312–1318

Banerjee S, Margulis L (1973) Mitotic arrest by melatonin. Exp Cell Res 78:314–318

Barrett P, Fraser S, Macdonald A, Helliweill RJA, Morgan PJ (1994) Strategies for cloning and characterizing the melatonin and other G-protein coupled receptors expressed in the pars tuberalis. In: Moller M, Pevet P (eds) Advances in pineal research: 8. John Libbey and Company, London, pp 309–319

Barter R, Pearse AGE (1955) Mammalian enterochromaffin cells as the source of serotonin (5-hydroxytryptamine). J Pathol Bacteriol 69:25–31

Bartsch C, Kvetnoy IM, Kvetnaia TV, Bartsch, H, Molotkov AO, Franz H, Raikhlin NT, Mecke D (1997) Nocturnal urinary 6-sulphatoxymelatonin and proliferative cell nuclear antigen-immunopositive tumor cells show strong positive correlations in patients with gastrointestinal and lung cancer. J Pineal Res 23:90–96

Blask DE, Wilson ST, Lemus-Wilson AM (1994) The oncostatic and oncomodulatory role of the pineal gland and melatonin. In: Maestroni GJM, Conti A, Reiter RJ (eds) Advances in pineal research: 7. John Libbey and Company, London, pp 235–242

Bonkhoff H, Stein U, Remberger K (1995) Endocrine-paracrine cell types in the prostate and prostatic adenocarcinoma are postmitotic cells. Hum Pathol 26:167–170

Bosman F (1984) Neuroendocrine cells in non-endocrine tumours. In: Falkmer S, Hakanson R, Sundler F (eds) Evolution and tumour pathology of the neuroendocrine system. Elsevier Science Publishers B. V., Amsterdam, pp 519–543

Bubenik GA (1980) Localization of melatonin in the digestive tract of the rat. Effect of maturation, diurnal variation, melatonin treatment and pinealectomy. Horm Res 6:313–323

Bubenik GA, Brown GM, Grota LJ (1976) Immunohistochemical localization of melatonin in the rat harderian gland. J Histochem Cytochem 24:1173–1177

Bubenik GA, Purtill R, Brown GM (1978) Melatonin in the retina and harderian gland. Ontogeneity, diurnal variations and melatonin treatment. Exp Eye Res 27:323–333

Carossino AM, Lombardi A, Matucci-Cerinic M, Pignone A, Cagnoni M (1996) Effect of melatonin on normal and sclerodermic skin fibroblast proliferation. Clin Exp Rheum 14:493–498

Champier J, Claustrat B, Besancon R, Eymin C, Killer C, Jouvet A, Chamba G, Fevremontange M (1997) Evidence for tryptophan hydroxylase and hydroxyindole-O-methyltransferase mRNAs in human blood platelets. Life Sci 60:2191–2197

Cos S, Fernandez F, Sanchez-Barcelo EJ (1996) Melatonin inhibits DNA synthesis in MCF-7 human breast cancer cells in vitro. Life Sci 58:2447–2453

Dubocovich ML, Masana MI (1998) The efficacy of melatonin receptor analogues is dependent on the level of human melatonin receptor subtype expression. In: Touitou Y (ed) Biological clocks. Mechanisms and applications. Elsevier Science BV, Amsterdam, pp 289–294

Erspamer V, Asero B (1952) Identification of enteramine, the specific hormone of the entero-chromaffine cell system, as 5-hydroxytryptamine. Nature 169:800–801

Grota LJ, Brown GM (1974) Antibodies to indolealkylamines: serotonin and melatonin. Canad J Biochem 52:196–203

Hakanson R, Bottcher G, Ekblad E, Grunditz T, Sundler F (1990) Functional implications of messenger coexpression in neurons and endocrine cells. In: Schwartz TW, Hilsted LM, Rehfeld JF (eds) Neuro-peptides and their receptors. Alfred Benzon Symposium 29. Munksgaard, Copenhagen, pp 211–232

Hu DN, Roberts JE (1997) Melatonin inhibits growth of cultured human uveal melanoma cells. Melanoma Res 7:27–31

Huether G (1993) The contribution of extrapineal sites of melatonin synthesis to circulating mela-tonin levels in higher vertebrates. Experientia 49:1–6

Kennaway DJ, Frith RG, Phillipou G, Matthews CD, Seamark RF (1977) A specific radioimmuno-assay for melatonin in biological tissue and fluids and its validation by gas chromatography-mass spectrometry. Endocrinology 110:2186–2194

Kojima T, Mochizuki C, Mitaka T, Mochizuki Y (1997) Effects of melatonin on proliferation, oxida-tive stress and Cx32 gap junction protein expression in primary cultures of adult rat hepato-cytes. Cell Struct and Funct 22:347–356

Krause DN, Dubocovich ML (1991) Melatonin receptors. Ann Rev Pharmacol Toxicol 31:549–568

Kvetnoy IM (1987) The APUD system (structure – function organization and biological significance in normal and pathological states). Adv Physiol Sci 18:84–102

Kvetnoy IM, Yuzhakov VV (1993) Extrapineal melatonin: advances in microscopical identification of hormones in endocrine and non-endocrine cells. Microscopy and Analysis 21:27–29

Kvetnoy IM, Yuzhakov VV (1994) Extrapineal melatonin: non-traditional localization and possible significance for oncology. In: Maestroni GJM, Conti A, Reiter RJ (eds) Advances in pineal re-search: 7. John Libbey and Company, London, pp 199–212

Kvetnoy IM, Yuzhakov VV (1995) Hormones in non-endocrine cells. Proc Royal Microscop Soc 30, 2:123–124

Kvetnoy IM, Kvetnaia TV, Konopljannikov AG, Tsyb AF, Yuzhakov VV (1994) Melatonin and tumor growth: from experiments to clinical application. In: Kvetnoy IM, Reiter RJ (eds) Melatonin: general biological and oncoradiological aspects. MRRC Press, Obninsk, pp 17–23

Kvetnoy IM, Bartsch C, Bartsch H, Kvetnaia TV, Medvedev VN, Raikhlin NT (1996) Melatonin and proliferative activity of tumours: successful combination of immunohistochemical and radio-immunological methods for definition of prognosis in humans. Acta Histochem Cytochem 29:304–305

Kvetnoy IM, Sandvik AK, Waldum HL (1997) The diffuse neuroendocrine system and extrapineal melatonin. J Mol. Endocrinol 18:1–3
Kvetnoy IM, Yuzhakov VV, Tsyb AF (1995) Diffuse neuroendocrine system: effects of non-lethal doses of ionizing radiation (clinical and experimental aspects). Pathol Res Pract 191:705–706
Larsson LI (1980) On the possible existence of multiple endocrine, paracrine and neurocrine messengers in secretory system. Invest Cell Pathol 3:73–85
Lemus-Wilson A, Kelly PA, Blask DE (1995) Melatonin blocks the stimulatory effects of prolactin on human breast cancer cell growth in culture. Brit J Cancer 72:1435–1440
Lerner A, Case J, Takahashi J (1958) Isolation of melatonin, the pineal gland factor that lightens melanocytes. J Amer Chem Soc 81:6084–6086
Lissoni P, Paolorossi F, Tancini G, Barni S, Ardizzoia A, Brivio F, Zubelewicz, Chatikhine V (1996a) Is there a role for melatonin in the treatment of neoplastic cachexia? Eur J Cancer 32 A: 1340–1343
Lissoni P, Paolorossi F, Tancini G, Ardizzoia A, Barni S, Brivio F, Maestroni GJM, Chilelli M (1996b) A phase II study of tamoxifen plus melatonin in metastatic solid tumour patients. Brit J Cancer 74:1466–1468
Maestroni GJM, Conti A (1990) The melatonin-immune system-opiod network. In: Reiter RJ, Lukaszyk A (eds) Advances in pineal research: 4. John Libbey and Company, London, pp 233–241
Maestroni GJM, Conti A (1996) Melatonin in human breast cancer tissue: association with nuclear grade and estrogen receptor status. Lab Invest 75:557–561
Maluf HM, Koerner FC (1994) Carcinomas of the breast with endocrine differentiation: a review. Virch Arch 425:449–457
Manev H, Uz T, Kharlamov A, Cagnoli CM, Franceschini D, Giusti P (1996) In vivo protection against kainate-induced apoptosis by the pineal hormone melatonin: effect of exogenous melatonin and circadian rhythm. Restorative Neurology and Neuroscience 9:251–256
Mechawar N, Anctil M (1997) Melatonin in a primitive metazoan: seasonal changes of levels and immunohistochemical visualization in neurons. J Comp Neurol 387:243–254
Mediavilla MD, Guezmez A, Ramos S, Kothari L, Garijo F, Barcelo EJS (1997) Effects of melatonin on mammary gland lesions in transgenic mice overexpressing N-ras proto-oncogene. J Pineal Res 22:86–94
Melendez J, Maldonado V, Ortega A (1996) Effect of melatonin on beta-tubulin and MAP2 expression in NIE-115 cells. Neurochem Res 21:653–658
Ozaki Y, Lynch H (1976) Presence of melatonin in plasma and urine of pinealectomised rats. Endocrinology 99:641–644
Pang SF, Dubocovich ML, Brown GM (1993) Melatonin receptors in peripheral tissues: a new area of melatonin research. Biol Signals 2:177–180
Panzer A (1997) Melatonin in osteosarcoma: an effective drug? Med Hypoth 48:523–525
Pearse AGE (1968) Common cytochemical and ultrastructural characteristics of cells producing polypeptide hormones (the APUD series) and their relevance to thyroid and ultimobranchial C-cells and calcitonin. Proc Royal Soc B 170:71–80
Pearse AGE (1969) The cytochemistry and ultrastructure of polypeptide hormone-producing cells of the APUD series and the embryologic, physiologic and pathologic implications of the concept. J Histochem Cytochem 17:303–313
Petrova GA, Dedenkov AN, Kvetnoy IM, Yuzhakov VV, Konopljannikov AG, Kvetnaia TV, Garbuzov PI, Levin IM (1994) Experimental and clinical study of diagnostic usefulness of melatonin determination in cancer patients. In: Kvetnoy IM, Reiter RJ (eds) Melatonin: general biological and oncoradiological aspects. MRRC Press, Obninsk, pp 72–74
Polak JM, Bloom SR (1986) Immunocytochemistry of the diffuse neuroendocrine system. In: Polak JM, Van Noorden S (eds) Immunocytochemistry: modern methods and applications. John Wright and Sons, Bristol, pp 328–348
Quay WB, Ma YH (1976) Demonstration of gastrointestinal hydroxyindole-O-methyltransferase. IRCS Med Sci 4:563
Raikhlin NT, Kvetnoy IM (1976) Melatonin and enterochromaffine cells. Acta Histochem 55:19–25
Raikhlin NT, Kvetnoy IM (1994) The APUD system (diffuse endocrine system) in normal and pathological states. Physiol Gen Biol Rev 8, part 4:1–44
Raikhlin NT, Kvetnoy IM, Tolkachev VN (1975) Melatonin may be synthesised in enterochromaffine cells. Nature 255:344–345
Raikhlin NT, Kvetnoy IM, Kadagidze ZG, Sokolov AV (1976) Immunomorphological studies on synthesis of melatonin in enterochromaffine cells. Acta Histochem Cytochem 11:75–77
Raikhlin NT, Kvetnoy IM, Blinova SA, Solomatina TM, Yakovleva ND, Yuzhakov VV, Denisov IN (1994) APUD system and melatonin: structure-function relationship in the norm and pathology. In: Kvetnoy IM, Reiter RJ (eds) Melatonin: general biological and oncoradiological aspects. MRRC Press, Obninsk, pp 9–16

Reiter RJ (1980) The pineal and its hormones in the control of reproduction in mammals. Endocr Rev 1:109–131

Reiter RJ (1991) Melatonin: that ubiquitously acting pineal hormone. News Physiol Sci 6:223–228

Reiter RJ, Robinson J (1995) Melatonin. Bantam Books, New York

Reiter RJ, Poeggeler B, Tan DX, Chen LD, Manchester LC, Guerrero JM (1993) Antioxidant capacity of melatonin: a novel action not requiring a receptor. Neuroendocrinol Lett 15:103–116

Reiter RJ, Tan DX, Poeggeler B, Chen LD, Menendez-Pelaez A (1994) Melatonin, free radicals and cancer initiation. In: Maestroni GJM, Conti A, Reiter RJ (eds) Advances in pineal research: 7. John Libbey and Company, London, pp 211–228

Reppert SM, Weaver DR, Ebisawa T (1994) Cloning and characterization of a mammalian melatonin receptor that mediates reproductive and circadian responses. Neuron 13:1177–1185

Ronco AL, Halberg F (1996) The pineal gland and cancer. Anticancer Res 16:2033–2039

Sagar SM, Klassen GA, Barclay KD, Aldrich JE (1993) Tumour blood flow: measurement and manipulation for therapeutic gain. Cancer Treat Rev 19:299–349

Solcia E, Usellini L, Buffa R, Rindi G, Villani L, Aguzzi A, Silini E (1989) Endocrine cells producing regulatory peptides. In: Polak JM (ed) Regulatory Peptides. Birkhauser Verlag, Basel, pp 220–246

Slominski A, Baker J, Rosano TG, Guisti LW, Ermak G, Grande M, Gaudet SJ (1996) Metabolism of serotonin to N-acetylserotonin, melatonin, and 5-methoxytryptamine in hamster skin culture. J Biol Chem 271:12281–12286

Sundler F, Bottcher G, Ekblad E, Hakanson R (1989) The neuroendocrine system of the gut. Acta Oncol 283:303–314

Tilden AR, Becker MA, Amma LL, Arciniega J, Mcgaw AK (1997) Melatonin production in an aerobic photosynthetic bacterium: an evolutionarily early association with darkness. J Pineal Res 22:102–106

Tijmes M, Pedraza R, Valladares L (1996) Melatonin in the rat testis: evidence for local synthesis. Steroids 61:65–68

Vanecek J, Vollrath L (1990) Localization and characterization of melatonin receptors. In: Reiter RJ, Lukaszyk A (eds) Advances in pineal research: 4. John Libbey and Company, London, pp 147–154

Verhofstad AAJ, Steinbusch HWM, Joosten HWJ, Penke B, Varga J, Goldstein M (1983) Immunocytochemical localization of noradrenaline, adrenaline and serotonin. In: Polak JM, Van Noorden S (eds) Immunocytochemitry. Practical Applications in Biology and Pathology. Wright PSG, Bristol, pp 143–168

Vialli M (1966) Histology of the enterochromaffin cells. In: Erspamer V (ed) 5-Hydroxytryptamine and related indolealkylamines: handbook of experimental pharmacology 19. Springer, Berlin, pp 1–65

Vijayalaxmi, Reiter RJ, Herman TS, Meltz ML (1996) Melatonin and radioprotection from genetic damage: in vivo/in vitro studies with human volunteers. Mutat Res 371:221–228

Wright ML, Pikula A, Cykowski LJ, Kuliga K (1996) Effect of melatonin on the anuran thyroid gland: follicle cell proliferation, morphometry, and subsequent thyroid hormone secretion in vitro after melatonin treatment in vivo. Gen Comp Endocrinol 103:189–191

13 A Survey of the Evidence That Melatonin and Unidentified Pineal Substances Affect Neoplastic Growth

Ietskina Ebels, *Bryant Benson*

Abstract

A survey has been made of the evidence that the pineal hormone melatonin and unidentified substances affect neoplastic growth. Although a handful of scientists provided early contributions in this area, the potential role of the pineal gland in the development and growth of tumors was not acknowledged until the 1970s with the pioneering work of Vera Lapin. She and her associates concluded from experimental and clinical data that the pineal gland was involved in the development and growth of tumors. Full recognition of this potential relationship was evident with the first pineal-cancer meeting that took place in Vienna in 1977. Since that time, the majority of experimental work has focused on the relationship between the pineal indolamine melatonin and tumor growth. According to several reviews of the literature written in the interim, melatonin administration has proven to be an effective modulator of tumor growth in a variety of animal models and in human malignancy. Current work focuses on pineal effects in relation to malignant growth, stress, and the immune system. Additionally, suppressive effects of melatonin on tumor cell growth have been demonstrated in vitro. A potentially significant advancement was made by Georges Maestroni and coworkers in their suggestion that the pineal gland participates in regulating immunity through circadian melatonin production. There is growing related evidence, derived in large portion from the work of Christian and Hella Bartsch and their colleagues, that the depression of rhythmic melatonin secretion favors malignant growth by a generalized loss of synchronization of endocrine and immune functions. Although most work has focused on the relationship between melatonin and cancer, unidentified pineal compounds may play a yet undefined role in relation to neoplastic growth. Compounds extracted and purified from rat, ovine, and bovine pineal glands have been shown to inhibit cancer cell growth in a variety of in vitro systems. We have carried out related experiments that demonstrate the presence of a substance in extracts of bovine pineal glands which inhibits the growth of MCF-7 human breast cancer cells. Similar to the results of other studies, the inhibitory substance appears to be a small peptide that can be purified by aqueous ethanol extraction, molecular sieve ultrafiltration and chromatography, and high-performance liquid chromatography. Determination of the chemical structure(s) of these unidentifed antigrowth pineal factors could provide synthetic analogs for further testing in animal models of cancer. In conclusion, the possible relationship between the pineal gland and tumor growth suggested more than 60 years ago has become widely accepted. A significant regulatory role for the pineal hormone melatonin is suggested by its measurements in both cancer patients and experimental animal models. Although attention has focused primarily on the relationship between melatonin and malignant growth, it is also clear that other, unidentified compounds that inhibit cancer cell growth can be extracted and purified from pineal glands, but their chemical structures and modes of action are yet to be elucidated.

13.1 Introduction

A possible relationship between the pineal gland and tumor growth was suggested more than 60 years ago by Engel (1934, 1935). At that time, a small aggregate of clinical observations and experimental studies indicated that the pineal gland influenced the growth of experimental tumors and neoplastic growth in humans. Experimental work in this area was extended by Bergmann and Engel (1950), Hofstätter (1950), and Engel and Bergmann (1952). Shortly thereafter, in 1954, Kitay and Altschule published their landmark book *The Pineal Gland*, which charted potential courses for future pineal gland research. Their comprehensive summary suggested that the pineal is involved in gonadal function, light-pigmentation responses, and possibly in brain/behavioral modulation. Two decades passed, however, before a potential role in cancer was widely acknowledged.

In 1972, Hajdu et al. noted the degeneration of the pineal glands of patients with cancer. Gross and microscopic changes of the pineal glands from 275 cancer patients were studied. Multiple slit-like cystic cavities, gliosis, and large numbers of Rosenthal fibers were found in the enlarged pineals of 31 patients, mostly children and adolescents. Shortly thereafter, Lapin (1974) studied the influence of simultaneous pinealectomy and thymectomy on the growth of tumors and the formation of metastases of the Yoshida sarcoma in rats. Besides growth and metastases, the period of survival following inoculation of Yoshida sarcoma was studied in adult rats subjected to pinealectomy, thymectomy, or simultaneous pinealectomy and thymectomy within 24 hours after birth. The period of survival was significantly reduced in pinealectomized rats as compared to control animals, and predominantly pancreas metastases were observed in pinealectomized animals, whereas metastases occurred mainly in the livers of thymectomized rats; in simultaneously thymectomized and pinealectomized animals, metastases appeared coincidentally in the pancreas and the liver. Lapin reviewed the relation of the pineal gland to malignancy in 1976 and concluded from experimental as well as clinical data that the pineal gland and its compounds may be involved in the development and growth of neoplasia.

In the following year, the first pineal-cancer workshop was held in Vienna. This historical meeting culminated with the conclusion that the neuroendocrine function of the pineal gland should be considered as a factor in relation to oncogenic processes (Lapin and Ebels 1981). About that time, Tapp (1980a) studied the human pineal gland in malignancy and observed that patients dying from carcinoma of the breast and from melanoma have much larger pineals than those dying from sarcoma. In experiments with rats suffering from 9,10-dimethyl-1,2-benzanthracene-induced fibrosarcoma, the weights of the pineal glands did not differ between treated and control animals. However, the pinealocytes and their nuclei in the pineal glands of tumor-bearing animals were larger and contained more lipid than those in the pineal glands of control animals (Tapp 1980b). A number of years later, Logan et al. (1988) showed that the growth and metastatic spread of renal tumors induced in Syrian hamsters by diethylstilbestrol were attenuated by light deprivation and that pinealectomy appeared to counter this effect (see also Logan and Benson 1990).

13.2 The Role of the Pineal Hormone Melatonin

Pineal gland indole synthesis and metabolism have been studied extensively since the isolation of melatonin from bovine pineal glands by Lerner et al. in 1958. Synthetic analogs of this indoleamine have been available for extensive experimentation and the indoleamine has been measured in tissues and body fluids by sensitive methods such as spectrofluorometry, radioimmunoassay, and mass spectrometry. The status of pineal gland function has also been studied through measurements of the activities of the enzymes required for melatonin biosynthesis, e. g., hydroxyindole-O-methyltransferase or serotonin-N-acetyltransferase.

Blask (1984) made an extensive study of the literature concerning the effects of the pineal gland and its secretagogues on the proliferation of normal and neoplastic cells and tumor growth and development in animal models and human malignancy. This author concluded that the bulk of evidence supported a pineally mediated neuro-endocrine modulation of tumor growth as exemplified by hormone-dependent mammary carcinoma. Accordingly, the pineal gland would transmit its oncostatic message to tumorous tissues through a complex, interacting network of hormonal, immune, and metabolic signals. Since that time, experimental work on the relationship of the pineal gland to cancer has focused on the effects of the pineal hormone melatonin in relation to malignant growth, stress, and the immune system in vivo and in vitro, and only a small number of investigations have been carried out on the effects of unidentified pineal substances on the growth of malignant cells in culture.

13.2.1 The Effects of Melatonin in Relation to Malignant Growth

A number of important papers on melatonin and the growth of malignant cells in experimental animals and humans have appeared. A cytostatic effect of melatonin on human melanoma cells cultured in vitro was observed by Lapin and Vetterlein (cited as an unpublished observation in Lapin and Ebels 1981). In the same year, Lapin and Frowein (1981) observed a negative correlation between pineal melatonin content and the growth of transplanted Yoshida sarcoma. In a survey published that year, Lapin and Ebels (1981) concluded from their and others' experiments that, when available, melatonin may contribute to the restriction of tumor growth.

Stanberry et al. (1983) showed that pinealectomy and melatonin influence the growth of melanoma in young adult male golden hamsters. These authors concluded from their experiments that the photoperiod under which the hamsters are kept dictates the growth rate of the melanoma tumors as well as the effect of pinealectomy on tumor behavior. Shah et al. (1984) studied the effect of melatonin on mammary carcinogenesis in intact and pinealectomized rats in varying photoperiods and concluded that the pineal must be present to obtain an impressive antitumor effect. These authors concluded that the probable site of melatonin action may be the pineal gland itself and the hypothalamus.

A relationship was observed between the pineal gland and neoplastic growth not only in experimental animals but also in humans. In 1981, Bartsch et al. observed a correlation between melatonin levels and the growth of breast cancer. It was discovered that the 24-hour urinary excretion of radioimmunoassayable melatonin in these patients was lower than in controls, whereas serum cortisol levels were higher. These studies suggest that pineal melatonin secretion may be modified both in quantity as well as in secretory rhythmicity. In another study by the same group, the circadian rhythms of serum melatonin, prolactin, and growth hormone were compared in patients with benign and malignant tumors of the prostate and in nontumor-bearing controls. The results revealed that rhythmicity of melatonin secretion was absent in elderly men with malignant tumors (Bartsch et al. 1983). These studies were extended and the authors concluded from their experiments that low melatonin levels at night were accompanied by endocrine changes, particularly in the rhythmicity of prolactin and growth hormone secretion, which changed from circadian to ultradian rhythms (Bartsch and Bartsch 1984). They state that the observed changes in prolactin levels point to disturbances in hypothalamic dopaminergic neuron function. For details, see Bartsch and Bartsch (1988a).

Blask and Leadem summarized existing literature on neuroendocrine aspects of neoplastic growth in 1987. These investigators suggested that several sites exist along the neuroendocrine-tumor axis where either endogenous neuroendocrine processes or exogenous neuroendocrine manipulations might influence the course of hormonally responsive tumor growth. One class of the tumors discussed were prolactinomas. It was observed that stimulation of the pineal by light deprivation caused inhibition of estrogen-induced prolactinoma growth in rats (Leadem 1986); the effect was also obtained by the administration of melatonin (Leadem and Burns 1987).

Hill and Blask (1988) first studied the effects of melatonin on the proliferation and morphological characteristics of MCF-7 human breast cancer cells in culture. The results of their experiments support the hypothesis that melatonin at physiological concentrations exerts a direct but reversible anti-proliferative effect on MCF-7 cell growth in culture. This antiproliferative effect is associated with striking changes in the ultrastructural features of these cells suggestive of a sublethal cellular injury (Shellard et al. 1989). Initially, others could not confirm these experiments (Bartsch et al. 1992b). However, there is evidence that MCF-7 cell stocks are not all identical, may respond differently to various hormones, and express varying levels of receptors to estrogen and melatonin (Klotz et al. 1995). Cos et al. (1997) have shown that inhibition of MCF-7 cell growth is best at physiological concentrations and that the effect depends on prevailing culture conditions, the presence of hormonal factors in the culture medium, and the pattern of exposure to melatonin. The inhibition of MCF-7 cell growth by melatonin was also confirmed by Liburdy (1997).

Feuer and Kerenyi (1989) studied the role of the pineal gland in the development of malignant melanoma. These authors observed that levels of radioimmunoassayable melatonin in blood increased 4- to 5-fold in patients during the active phase of malignant melanoma. Post mortem histological examinations of isolated pineal glands showed that several glands contained a pigment called neuromelanin. These authors concluded that their data is suggestive of the possibility of pineal action on carcinogenesis related to endocrine control. In particular, the endocrine environment plays a role in the development and growth of melanoma, and steroid hormone receptors are involved in these processes. The data confirm that increased pineal activity exists in human malignant melanoma and, according to these authors, plasma melatonin levels may serve as a tumor marker.

Bartsch et al. (1989) extended previous investigations by determining the circadian rhythms of serum melatonin, prolactin, and thyroid stimulating hormone with respect to steroid receptor concentrations of the tumors in different clinical groups of breast cancer patients. From their experiments and the literature, the authors concluded that the observed changes of serum melatonin in primary breast cancer patients affect the endocrine and perhaps also the immune systems, favoring malignant growth by a general dysregulation which includes changes in temporal organization of hormonal secretion. However, pineal-tumor interaction appears to be a dynamic process with phases of inhibition and restimulation. These authors believe that this may be a vital part of the host-tumor interaction and may play a role in the course of malignant diseases including human breast cancer. Furthermore, it was observed that the reduction of serum melatonin in patients with primary breast cancer is not due to increased peripheral metabolism, i.e., a degradation to 6-sulfatoxymelatonin in the liver, but that it may be caused by reduced pineal gland activity (Bartsch et al. 1991).

In experiments studying the effects of the mammary carcinogen 7,12-dimethyl-benzanthracene (DMBA) on pineal melatonin biosynthesis, secretion, and peripheral metabolism, a depression of nocturnal plasma melatonin in DMBA-treated animals was seen to be due to an enhanced and modified metabolism to 6-sulfatoxymelatonin and other compounds in the liver (Bartsch et al. 1990a). According to these investigators, the depression of circulating melatonin constitutes an important part of the endocrine disturbance due to the carcinogen, rendering mammary tissue more susceptible to tumor initiation. In this regard, melatonin may act as an endogenous protective hormone against chemical carcinogens, suggesting an important role of the central neuroendocrine mechanism in malignancy.

The main objective of another study was to investigate pineal melatonin biosynthesis and secretion during induction and growth of mammary cancer in female rats. A steep increase of pineal melatonin content and plasma concentrations was observed during the growth of tumors < 10 cm^3, indicating a response of pineal melatonin production and secretion to the growing tumor mass. However, a decline in both pineal and plasma melatonin was observed in animals with total tumor volume > 10 cm^3. According to Bartsch et al. (1990b), a possible biphasic relationship between melatonin and tumor volume needs further investigation.

Bartsch et al. (1990c) summarized the seasonal variations of endogenous defense mechanisms against cancer. From their own experiments and the results of others, they formulated the hypothesis that endogenous defense mechanisms against cancer, such as known and yet unknown substances in the pineal and human urine as well as immune factors which undergo circannual rhythms even under relatively constant environmental conditions, may be at least a partial cause for seasonal variations in the occurrence of cancer (see also Bartsch et al. 1993). In their article entitled "Diminished pineal function coincides with disturbed circadian endocrine rhythmicity in untreated primary cancer patients with prostate and breast cancer," Bartsch et al. (1994) postulate that a depression of pineal function in cancer patients could be an essential point within a vicious cycle supporting the growth and spread of malignant tumors, leading ultimately to the destruction of its host. Based on their own experimental work and the data obtained by other investigators, these authors propose that a substitutional therapy with melatonin may help prevent if not restore central endocrine imbalances as described in cancer patients with depressed melatonin. Another possible therapeutic effect of melatonin might involve the stimulation of pineal antitumor activity, ostensibly present in low molecular weight substances in pineal glands. Such unidentified substances effectively inhibit the growth of a wide range of experimental tumor cells in vitro, some of which are unresponsive to melatonin. For details, see Bartsch et al. (1992a, 1994).

Bartsch et al. (1997a, b) recently extended the experiments on nocturnal 6-sulfatoxymelatonin in breast cancer patients and patients with gastrointestinal and lung cancer. As to breast cancer, the recent results confirm previous findings of decreased pineal melatonin secretion as well as an inverse relationship with tumor size, excluding a possible distortion due to age. The results obtained in patients with gastrointestinal and lung cancer provide support for the concept of pineal gland involvement in malignancy and suggest that 6-sulfatoxymelatonin measurements may be considered as a noninvasive tool for estimating tumor cell proliferation. Negative correlations found between urinary 6-sulfatoxymelatonin and melatonin in tumor cells could be interpreted as an effort of the pineal gland to secrete melatonin in compensation for the decrease in number of melatonin-immune positive cells within tumor tissue, where it may possess important regulatory functions. An interesting observation is that phenobarbital and 7,12-dimethylbenzanthracene cause a depletion of circulating melatonin as a result of the microsomal monooxygenases that catalyze the 6-hydroxylation of melatonin (Praast et al. 1995).

More recently, it was shown that melatonin treatment of male Syrian hamsters maintained in long photoperiods suppressed the growth and metastasis of DES-induced renal tumors (Benson 1997). According to Schmidt et al. (1997), it appears that pineal melatonin secretion may be reversibly inhibited by a tumor-derived melatonin inhibiting factor. If this be the case, the relationship between melatonin and malignant growth becomes even more complicated.

13.2.2 The Effects of Melatonin in Relation to the Immune System and Stress

Maestroni and coworkers have studied melatonin, stress, and the immune system (1989, 1993, 1996, 1997). These authors concluded from their experiments that the pineal gland participates in regulating immune reactivity through the circadian production of melatonin. They propose that melatonin possesses immune-enhancing and antistress properties. This proposed relationship between melatonin and the immune system is the subject of another chapter in this volume and will not be discussed in detail here. It is thought that the melatonin-endogenous-opioid system connection may be a main efferent arm, as opioid peptides mimic the immune-enhancing and antistress effect of melatonin. Believing that their findings constitute a significant step towards a better understanding of the physiological functions of the pineal gland, these authors recommend the extensive investigation of melatonin in all clinical situations where physiologic up-regulation of the immune reactivity is desirable.

13.3 Effects of Yet Unknown Pineal Substances on Malignant Cells in Culture

Although most attention has been focused on the relationship between melatonin and cancer, it cannot be denied that other, unidentified pineal compounds may also play a role in relation to neoplastic growth (see Bindoni et al. 1976, Dilman et al. 1979, Anisimov 1988). Unfortunately, these compounds have not been isolated in pure form, are not easily measurable, and are unavailable in synthetic form for experimental research. In the following section, experiments will be described and discussed which point to the fact that low molecular weight substances which differ from melatonin are present in pineal extracts that affect neoplastic growth (Ebels et al. 1988).

13.3.1 Extracts of Ovine Pineal Glands

Lapin and Ebels (1976) studied the effects of low molecular weight sheep pineal fractions and melatonin on different tumors in rats and mice. These authors observed that certain of these fractions demonstrated effects on some of the tumor models employed. However, the experimental models required an excessive amount of partially purified fractions per animal and consequently were not suitable for use in a guiding test for the isolation of the active pineal compound(s).

Following the original finding of a cytostatic effect of melatonin on human melanoma cells cultured in vitro (unpublished results by Lapin and Vetterlein, cited in Lapin and Ebels 1981), Bartsch et al. (1986) studied the growth of human melanoma cells more extensively and observed a differential effect of melatonin on slow- and fast-growing passages of a human melanoma cell line. In these studies, it was seen that early passages were inhibited by micromolar concentrations of melatonin, whereas late passages were affected only by millimolar concentrations. These results indicate that the melanoma cells lost their sensitivity to melatonin during passaging. This knowledge was used to test yet unknown pineal substances. In other experiments in which the effects of crude ovine pineal extracts were tested on human melanoma cells in culture, it was observed that the growth-inhibiting activity was different from that of known synthetic indolamines including melatonin, serotonin, and others, and also of β-carbolines, pteridines, and the synthetic peptide arginine vasotocin. These identified pineal substances and analogues showed an inhibitory effect only at non-physiologically high concentrations; however, crude pineal fractions were more active than the synthetic substances tested. It was concluded that these melatonin-free pineal fractions contained a compound which may have a tumor-inhibiting potency comparable to that of methotrexate, but with a different mechanism of action (Bartsch et al. 1987 a).

The active ovine pineal fraction was further purified by Noteborn et al. (1988). A combination of paper chromatography, ion exchange, and reverse-phase high-performance liquid chromatography was employed coupled with a post column assay. This permitted identification of an ovine pineal factor (MW < 500) which inhibited the growth of human melanoma cells in vitro. The substance was partially purified 1000-fold compared to the IC-100 value of the starting material. The growth inhibition of human melanoma cells in culture was complete at a dose of 0.1 µg of the purified pineal factor per ml of culture medium. In another study it was observed that the activity was not destroyed by treatment with proteolytic enzymes.

Continuing these studies, Noteborn et al. (1989) investigated the activity of an ovine pineal polypeptide extract that suppresses the growth of human melanoma cells in culture. Again, a combination of gel filtration and reverse-phase high-performance liquid chromatography was coupled to postcolumn antigrowth assay. In this case, as before, inhibition of the growth of a melatonin-insensitive human melanoma cell strain was used to guide the purification of the antigrowth activity identified in aqueous extracts with a M_r between 1000 and 10,000 Da. The study yielded preliminary identification of a pineal factor (M_r between 2000 and 6000 Da) that inhibited the growth of human melanoma cells. Evidence for the peptide nature of the antitumor factor was also presented.

The results of these combined studies showed that ovine pineal glands contain substances in addition to melatonin that inhibit the growth of human melanoma cells in culture. It is not known whether these antitumorigenic compounds act as hormonal inhibitors of cellular proliferation by modifying the binding sites of growth factors or antimitotic agents that interfere with mitosis and the cytoskeleton. Moreover, it is also unclear whether the investigated pineal antitumorigenic compound is the same factor that was observed to suppress the hypophysiotropic activity of the hypothalamus (see Ebels et al. 1975, Ebels 1980).

13.3.2 Extracts of Rat Pineals

In 1988, Dwyer et al. reported on a computer-enhanced comparative study of brain region polypeptides and proteins separated by two-dimensional gel electrophoresis. Acetic acid protein extractions of pineal gland, retina, hypothalamus, and cerebral cortex from groups of Sprague-Dawley rats were compared. A reproducible, quantitative strategy for identifying tissue-specific proteins was established. It was observed that 17 newly identified acidic proteins were unique to the pineal gland. Several others were common to the retina but not to the other regions examined. Further study of these and other regionally specific proteins may be of interest under conditions in which biological mechanisms involving the pineal gland are altered and especially in situations involving neoplastic growth.

In other studies, the antigrowth activity of ethanolic extracts of rat pineal glands was examined on the K562 line of human erythroleukemia which were unresponsive to melatonin (Bartsch et al. 1990c). The results of these experiments indicated that circannual changes existed in extractable inhibiting activity. Highest amounts of growth-inhibiting activity were detected in summer and least in winter when stimulatory activity was sometimes observed. In another study, the antitumor activity in rat pineal glands and urine were tested on three transplantable murine tumors. It was found that an unidentified antitumor activity in the rat pineal gland shows a circannual rhythmicity similar to that of the tumor-inhibiting activity present in human urine. The relationship between the pineal gland and cancer and their seasonal aspects were further discussed by Bartsch et al. (1992c) and Bartsch and Bartsch (1993, 1997).

13.3.3 Studies on an Unidentified Bovine Pineal Substance Which Inhibits MCF-7 Cell Growth in Vitro

We have carried out experiments over several years that demonstrate the presence of a substance(s) in extracts of bovine pineal glands that inhibits the growth of MCF-7 human breast cancer cells in vitro. In most regards, the results corroborate the presence of antitumor activity seen in ovine and rat pineals by Bartsch et al. (1987a, b, 1990d, 1992a), Bartsch and Bartsch (1988b, 1988), Ebels et al. (1988) and Noteborn et al. (1988, 1989).

13.3.3.1 Test for MCF-7 Cell Growth-Inhibiting Activity

A stock of low passage, estrogen receptor-positive MCF-7 cells was provided for these studies by Dr. David E. Blask, presently in the Research Institute of The Mary Imogene Bassett Hospital in Cooperstown, New York. After thawing, the cells were maintained in low glucose, pyruvate-containing Dulbecco's modified eagles medium (DMEM) and 5.0% fetal bovine serum. The MCF-7 cells were maintained at 37°C in an incubator under a constant flow of 95% O_2 and 5% CO_2. For plating, the medium was removed and the cells covered with a sterile phosphobuffered saline solution (PBS) containing ethylenediamine tetra-acetic acid (EDTA). The PBS/EDTA solution was neutralized by the addition of fresh medium and the cells transferred to 50 ml conical tubes and concentrated by low speed centrifugation, with the medium removed, fresh medium added, and the cell concentrations determined. Medium containing 10% fetal bovine serum was added in quantities sufficient to bring the concentration of cells to desired density. After dilution, 2×10^5 cells/3.0 ml of medium were placed routinely in each well of a Falcon 3814 plate with six 35 mm wells in the morning between 9.00 and 11.00 hours. The cells were allowed to attach to the bottom of the wells before an additional 1.0 ml of medium containing the dissolved test or control medium was added in the late afternoon between 16.00 and 20.00 hours. The pH of medium containing test substances was determined routinely and adjusted when necessary before addition to the cells in culture.

Twelve wells were established for each control or test substance. Six of these were harvested at 2 days and the number of cells determined; another six were harvested at 4 days after plating. The culture medium was also collected after concentration of the cells for counting and the numbers of unattached or suspended cells determined by hemocytometer counts. Trypan blue exclusion was utilized to determine cell viability and the number of dead or viable cells recorded. A standard growth curve for MCF-7 cells in culture is shown in Fig. 13.1.

13.3.3.2 Bovine Pineal Extracts

Small batches of fresh frozen bovine pineal glands (15–20 g) were first homogenized in water and lyophilized. The dry pineal powder was extracted with a variety of aqueous or lipid solvents at different pH values and temperatures. After centrifugation, the supernatant portion of each extract was lyophilized and a weighed amount of the residue added to monolayers of MCF-7 human breast cancer cells in culture as described above.

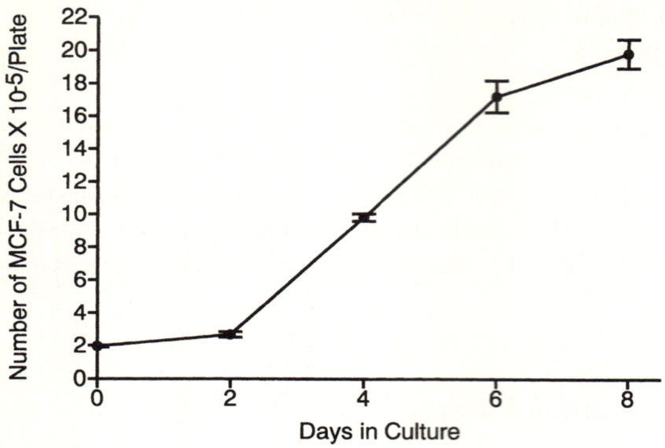

Fig. 13.1. A standard MCF-7 cell growth curve is shown. Each point represents the mean value ± SEM for six wells

More than 2 dozen different extraction methods were explored. It was learned that the extraction methods yielding residues with maximum inhibition of MCF-7 cell growth were those that incorporated solutions with high concentrations of ethanol (>75%) coupled with extractions carried out over several hours at 37 °C. The extraction and partial purification scheme is shown in Fig. 13.2. After centrifugation, recovery of the first lyophilizate (residue 1) was ~ 15 mg/g wet weight pineal tissue. The crude extract (extract ②) significantly inhibited growth of MCF-7 cells after 2 and 2 days of culture in doses ranges between 10^{-4} to 10^{-6} g/l incubation medium. However, a dose response relationship was not observed with these crude extracts, with the lowest concentrations often showing the greatest effects (Fig. 13.3). In these studies, significant differences between control and treatment groups was determined by ANOVA with post hoc comparisons made with the Newman-Keuls test.

In a series of experiments, ethanolic extracts of pineal glands were diluted to 50% (v/v) with either deionized H_2O or 1% formic acid and subjected to serial ultrafiltration

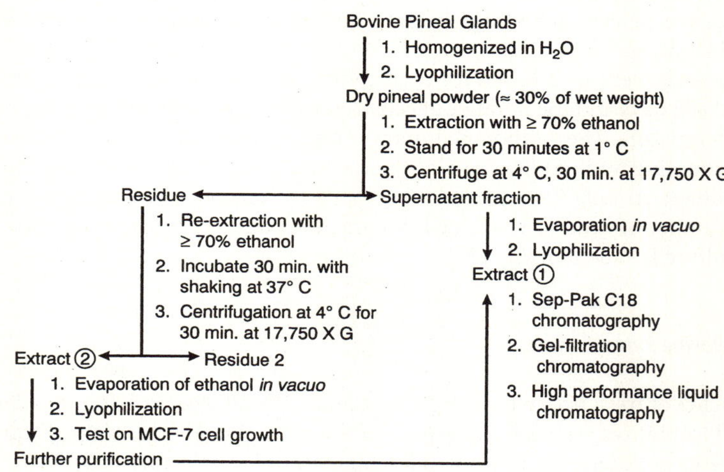

Fig. 13.2. Extraction and purification scheme for bovine pineal gland-derived anti-MCF-7 growth factors

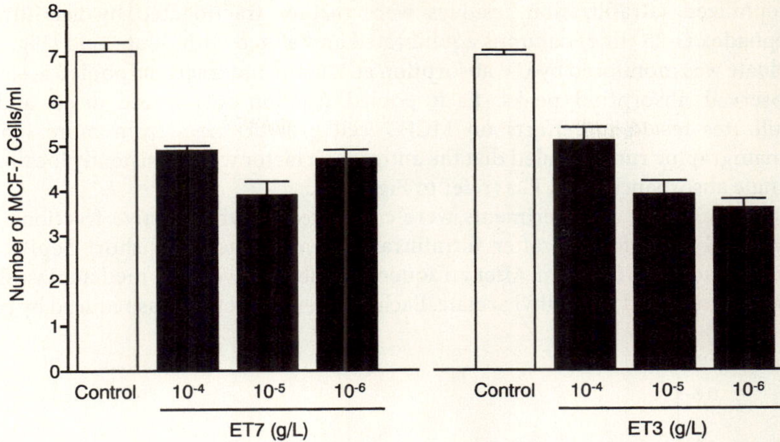

Fig. 13.3. Effects of two crude extracts (extracts type 2 in Fig. 13.2) on MCF-7 cell growth. Cells were harvested after 4 days in culture. Attenuation of growth at the doses indicated was without apparent cytotoxic effects as determined by vital dye staining and electron microscopy

through Amicon ultrafiltration membranes XM100, YM30, PM10, YM1, and UM05. A portion of each ultrafiltration membrane retentate or residue (R) and filtrate (F) was saved, the solvents removed by rotary evaporation in vacuo, water added, the residue freeze dried, and the lyophilizates tested for effects on MCF-7 cell growth. It was discovered that the growth-inhibiting factor(s) was localized primarily to the ultrafiltration residue YM2-R, suggesting a M_r in the approximate range of 1000 to 10,000 (Fig. 13.4). This finding agrees well with those of Noteborn et al. (1988, 1989), who found both melanoma and MCF-7 cell growth-inhibiting activity in partially purified, low molecular weight fractions ($M_r \sim 2000-6000$) from peptide extracts of ovine pineal gland extracts. Some growth-inhibiting activity was seen in the UM05-R ($M_r > 500-1000$). This was especially true with ultrafiltrates carried out with 1% formic acid.

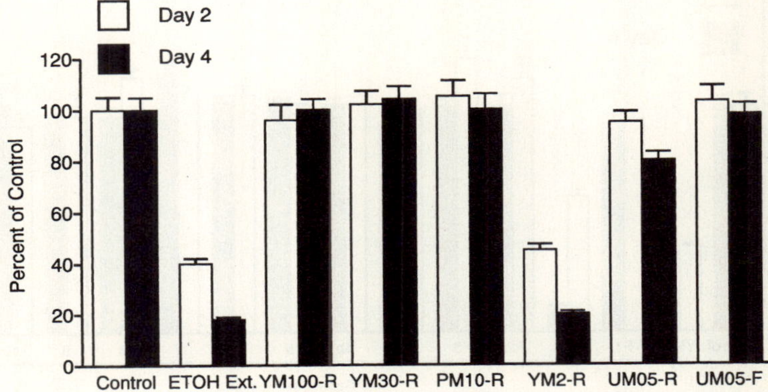

Fig. 13.4. Effects of ultrafiltrate residues on MCF-7 cell growth. Growth-inhibiting factor(s) was localized primarily to the YM2-R, which had an approximate M_r between 1000 and 10,000

Lyophilized ultrafiltration residues were further fractionated by gel filtration on Sephadex G-25 (fine) columns equilibrated and eluted with dilute acid (Fig. 13.5). The eluate was monitored by UV absorption at 280 nm and fractions pooled according to observed absorption peaks. Each pooled fraction was freeze dried and the lyophilizates tested for effects on MCF-7 cell growth. Tests from more than 20 chromatographic runs revealed that the antigrowth factor was consistently localized to the single absorbance peak, F2a (refer to Fig. 13.6).

A further series of experiments were conducted in which active fractions from Sephadex G-25, before or after ultrafiltration, were placed on short SepPac C18 columns in aqueous medium. After an aqueous wash, followed by a methanol wash, the columns were eluted with ethyl acetate. Each recovered fraction was reduced by rotary

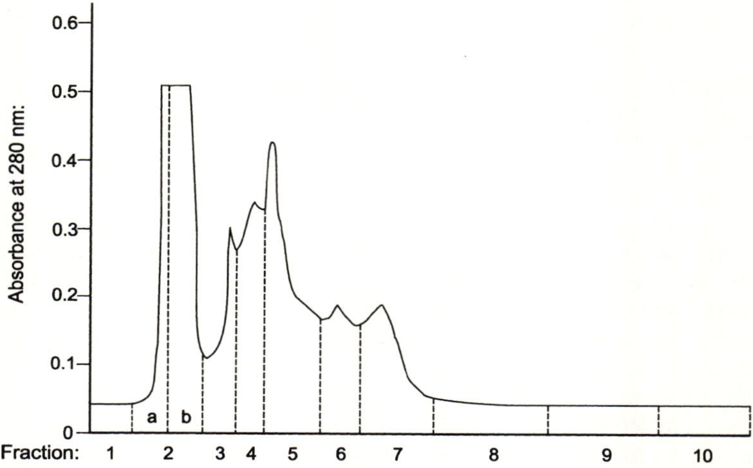

Fig. 13.5. Sephadex 25 chromatography of a sample of YM2-R

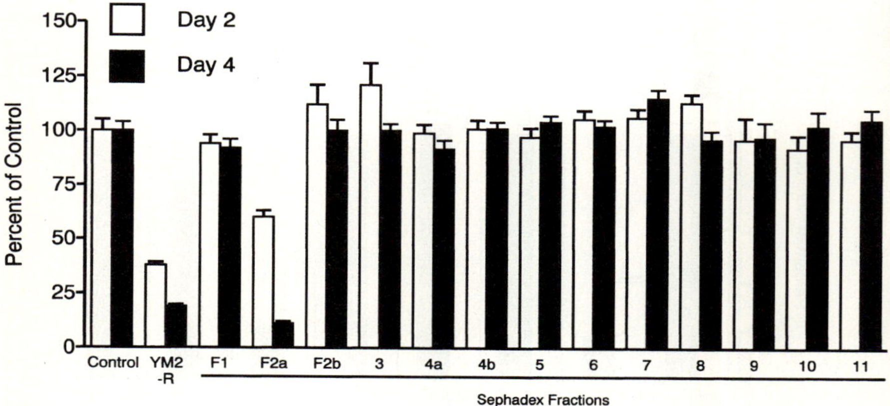

Fig. 13.6. Effects of Sephadex fractions on MCF-7 cell growth. The antigrowth factor(s) was consistently localized to the absorbance peak F2a

evaporation in vacuo before water was added and the fractions lyophilized and tested. It was seen that neither the aqueous nor methanol washes contained material that significantly affected the growth of MCF-7 cell growth in vitro. On the other hand, the ethyl acetate eluate contained antigrowth material (data not shown). This was a consistent result from several experiments in which it was estimated that more than 75 % of inactive material could be removed in this manner. On the other hand, there was considerable loss of active material in the process and, consequently, this procedure was excluded as a possible step for larger scale preparations. However, the chromatographic behavior of the substance with antigrowth activity suggests that the structure of the active component(s) in the partially purified material was that of a peptide.

The growth-inhibiting substance was further purified by high-performance reverse-phase liquid chromatography. A Spectra-Physics model SP8800–01 was utilized equipped with ternary gradient pumps and semipreparative heads. Solvents included: solvent A: 0.01 M trifluoroacetic acid (TFA) in acetonitrile (ACN); solvent B: 0.01 M TFA; solvent C: 0.01 M TFA in H_2O/ACN (1:1, v/v) with programmed mobile phase component changes incorporated that increase the concentrations of ACN over a period of 40 minutes at a flow rate of 2.0 ml/min. Column eluents were monitored at 210 nm and fractions collected and pooled according to absorbance peaks. The pooled fractions were concentrated by rotary evaporation, lyophilized, and stored at – 80 °C until testing for growth inhibition in the MCF-7 cell test. Fraction 2, as illustrated in Fig. 13.7, contained growth-inhibiting activity.

A number of untreated MCF-7 cells or others treated with several partially purified bovine pineal fractions in vitro were fixed and examined by electron microscopy. Compared to either untreated controls or cells treated with fractions that did not

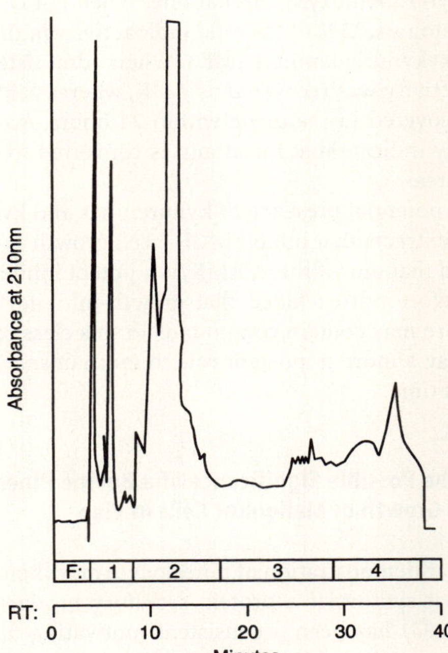

Fig. 13.7. High-performance reverse-phase liquid chromatography of a Sephadex fraction F2a. Anti-MCF-7 cell growth was localized to the impure fraction 2

inhibit growth, the cells treated with fractions that inhibited growth showed characteristic morphological changes which included cytoplasmic withdrawal or "rounding up" and an apparent reduction in cell size. The observed morphological changes, as well as viability tests with trypan blue throughout these experiments, suggest that the growth-inhibiting effect may be antimitotic rather than cytotoxic.

In summary, methods were established for the extraction of a bovine pineal substance(s) which attenuates the growth of MCF-7 human breast cancer cells in vitro. Recovery of active material was greatest when solvents containing >75% ethanol were incorporated and, after extraction, the growth-inhibiting factor(s) was partially purified by Sephadex G-25 gel filtration chromatography, serial ultrafiltration, and high-performance chromatography. The chemical structure of the bovine pineal growth-inhibiting factor was not determined but appears to be a small molecular weight peptide(s).

13.4 Possible Presence of Kynurenines and Kynurenamines in Pineal Extracts

During the study of bovine pineal extracts, partially purified fractions that inhibited the growth of MCF-7 human breast cancer cells in vitro were subjected to scanning spectrometry. It was learned that many of the partially purified inhibitory fractions demonstrated absorption maxima in the range exhibited by kynurenines and kynurenamines. Little attention has been paid to the work of several Japanese investigators who studied particularly the formation of these compounds from indoles in the pineal gland. Hirata et al. (1974) reported on the formation of new metabolites of melatonin in vitro and in vivo and found that the major metabolite in the rat brain was N-acetyl-5-methoxykynurenamine. When [14 C]-melatonin was injected intracisternally into rats, 35% of the total radioactivity in the urine was recovered as N-acetyl-5-methoxykynurenamine (AMK). When administered intravenously, 15% of the total radioactivity was recovered as AMK, whereas 65% of the radioactivity administered was recovered in the urine within 24 hours. According to these authors, the results strongly indicate that melatonin is converted to AMK in the pineal gland and other brain areas.

The potential presence of kynurenines and kynurenamines in the purified bovine pineal extracts that inhibit MCF-7 cell growth is of interest, as Kelly et al. (1984) discovered that one of these, AMK, is a potent inhibitor of prostaglandin biosynthesis. It is therefore not excluded that growth-inhibitory pineal substances of unidentified structure may contain compounds in this class of chemical structures and that these may play a more important role in mechanisms of cell growth than is known at the present time.

13.5 The Possible Significance of a Bovine Pineal Gland-Derived Decapeptide for the Growth of Malignant Cells in Vivo

The clear demonstration of pinealocyte morphological characteristics consistent with cells that synthesize a protein secretory product (Pévet et al. 1981, Ariens Kappers 1981, 1983) has been a consistent motivating factor in the search for unique antigonadotropic peptides synthesized and secreted by the pineal gland. The history of this endeavor is long and arduous, as a number of individuals and groups have worked on this problem over many years. In general, their results have yielded more questions than answers. A long list of neuropeptides of known structure have been identified in the mammalian pineal gland with various experimental techniques (Benson 1999a), yet for the most part their individual functions remain enigmatic.

Since the early 1970s we have observed antigonadotropic effects of partially purified peptide extracts of rat, human, bovine, and ovine glands when tested in rodents. Surveys of the early work have been presented previously (Benson 1977, Ebels and Benson 1978, Benson and Ebels 1978). Based on the purification methods derived from these early studies, a large scale extraction was completed with the aim of determining the structure of peptides with neurohypophysial hormone-like and/or antigonadotropic activity. The structure of an antigonadotropic decapeptide (AGD) without neurohypophysial hormone-like activity was revealed and the presence of the putative antigonadotropic decapeptide arginine vasotocin excluded (Benson et al. 1990, Benson and Ebels 1994).

Purified AGD and a synthetic analog prepared by solid state methods reduced circulating levels of prolactin and LH after intra-atrial or lateral ventricle injection in male rats (Benson and Ebels 1994, Benson and Machen 1994, Benson et al. 1996). Lateral ventricle AGD infusion was recently seen to inhibit pulsatile LH secretion in sheep, especially in an apparent steroid-independent manner (Lee 1998). Other recent experiments suggest that AGD attenuates stress-related increases in prolactin secretion potentially via stimulatory effects on hypothalamic dopamine turnover (Benson 1999b).

In preliminary experiments, a variety of AGD concentrations did not inhibit the growth of MCF-7 breast cancer cells in vitro. However, as this decapeptide demonstrates an influence on pituitary and blood levels of prolactin and gonadotropins in vivo, the possibility exists that AGD could influence certain malignant cells which are either influenced by or dependent on prolactin and/or gonadotropins for their growth. Since its structure is known and analogs are available for such studies, a novel approach would be to study the influence of the pineally derived decapeptide on tumor growth in vivo.

13.6 Summary

A substantial amount of long-standing evidence that the pineal gland affects neoplastic growth has culminated in the search for a pineal secretory product capable of suppressing malignant growth. Measurements of the hormone melatonin in cancer patients and experimental animal models of cancer suggest a significant role for this indoleamine. When administered in vivo, melatonin has been shown to suppress the development of certain types of neoplasia, and the growth of a variety of types of cancer cells is inhibited by the addition of melatonin in vitro.

Although most attention has focused primarily on the relationship between melatonin and malignant growth, it is clear that other, unidentified pineal compounds which inhibit cancer cell growth in vitro can be extracted and purified from several species of mammalian pineals. The physiological significance of these substances is yet to be determined.

Novel evidence obtained in our laboratory suggests that a small molecular weight substance, extracted and purified from bovine glands, inhibits the growth of MCF-7 human breast cancer cells in vitro. Similar to findings in other species, the bovine pineal gland antigrowth substance(s) appears to have the characteristics of a small peptide. Future determination of the structure of this antigrowth factor will permit the synthesis of analogs in quantities sufficient for testing in animal models in vivo as well as studies on mechanisms of action.

References

Anisimov VN (1988) Pineal gland, aging and carcinogenesis. In: Gupta D, Attanasio A, Reiter RJ (eds) The Pineal Gland and Cancer. Brain Research Promotion, London, pp 107–118

Ariëns Kappers J (1981) Evolution of pineal concepts. In: Oksche A, Pévet P (eds) The Pineal Organ. Photobiology, Biochemistry and Endocrinology. Elsevier/Amsterdam, pp 3–23

Ariëns Kappers J (1983) Comparative gross and fine morphology of the mammalian pineal gland. In: Axelrod J, Fraschini F, Velo GP (eds) The Pineal Gland and its Endocrine Role. Plenum, New York, pp 37–59

Bartsch H, Bartsch C (1981) Effect of melatonin on experimental tumors under different photoperiods and times of administration. J Neural Transm 52:269–279

Bartsch C, Bartsch H (1984) The link between the pineal gland and cancer: An interaction involving chronobiological mechanisms. In: Halberg F, Reale L, Tarquini B (eds) Chronobiological approach to social medicine. Instituto Italiano di Medicina Sociale, Rome, pp 105–126

Bartsch C, Bartsch H (1988a) Melatonin in human cancer patients. In: Gupta D, Attanasio A, Reiter RJ (eds) The pineal gland and cancer. Brain Research Promotion, London, pp 361–368

Bartsch H, Bartsch C (1988b) Unidentified pineal substances with antitumor activity. In: Gupta D, Attanasio A, Reiter RJ (eds) The pineal gland and cancer. Brain Research Promotion, London, pp 369–376

Bartsch H, Bartsch C (1993) Antitumor activity in pineal glands and in urine display similar circannual rhythmicity. In: Gutenbrunner C, Hildebrandt G, Moog R (eds) Chronobiology and Chronomedicine. Peter Lang GmbH, Frankfurt am Main, pp 390–395

Bartsch C, Bartsch H (1997) Pineal gland and cancer: an old concept with potential for the future. Neuroendocrinology Lett 18:63–72

Bartsch C, Bartsch H, Jain AK, Laumas KB, Wetterberg L (1981) Urinary melatonin levels in human breast cancer patients. J Neural Transm 52:281–294

Bartsch C, Bartsch H, Flüchter S-H, Harzmann A, Attanasio A, Bichler K-H, Gupta D (1983) Circadian rhythms of serum melatonin, prolactin and growth hormone in patients with benign and malignant tumors of the prostate and in non-tumor controls. Neuroendocrinology Lett 5: 377–386

Bartsch H, Bartsch C, Flehmig B (1986) Differential effect of melatonin on slow and fast growing passages of a human melanoma cell line. Neuroendocrinology Lett 8:289–293

Bartsch H, Bartsch C, Noteborn HPJM, Flehmig B, Ebels I, Salemink CA (1987a) Growth-inhibiting effect of crude pineal extracts on human melanoma cells in vitro is different from that of known synthetic pineal substances. J Neural Transm 69:299–311

Bartsch H, Bartsch C, Flehmig B (1987b) Pineal anti-tumor activity (PATA) of rats under different physiological conditions. In: Trentini GP, De Gaetani C, Pévet P (eds) Fundamentals and Clinics in Pineal Research. Raven Press, New York, pp 381–384

Bartsch C, Bartsch H, Fuchs U, Lippert TH, Bellmann O, Gupta D (1989) Stage-dependent depression of melatonin in patients with primary breast cancer. Cancer 64:426–433

Bartsch C, Bartsch H, Lippert TH, Gupta D (1990a) Effect of the mammary carcinogen 7,12-dimethylbenz[a]anthracene on pineal melatonin biosynthesis, secretion and peripheral metabolism. Neuroendocrinology 52:538–544

Bartsch C, Bartsch H, Gupta D (1990b) Pineal melatonin synthesis and secretion during induction and growth of mammary cancer in rats. In: Gupta D, Wollmann HA, Ranke MB (eds) Neuroendocrinology: New Frontiers. Brain Research Promotion, Tübingen, pp 326–332

Bartsch H, Bartsch C, Gupta D (1990c) Seasonal variations of endogenous defence mechanisms against cancer. In: Gupta D, Wollmann HA, Ranke MB (eds) Neuroendocrinology: New Frontiers. Brain Research Promotion, Tübingen, pp 333–339

Bartsch H, Bartsch C, Gupta D (1990d) Tumor inhibiting activity in the rat pineal gland displays a circannual rhythm. J Pineal Res 9:171–178

Bartsch C, Bartsch H, Bellmann O, Lippert TH (1991) Depression of serum melatonin in patients with primary breast cancer is not due to an increased peripheral metabolism. Cancer 67:1681–1684

Bartsch C, Bartsch H, Schmidt A, Ilg S, Bichler K-H, Fluchter S-H (1992a) Melatonin and 6-sulfat-oxymelatonin circadian rhythms in serum and urine of primary prostate cancer patients: evidence for reduced pineal activity and relevance of urinary determination. Clin Chim Acta 209:153–167

Bartsch H, Bartsch C, Simon WE, Flehmig B, Ebels I, Lippert TH (1992b) Antitumor activity of the pineal gland. Effect of unidentified substances versus effect of melatonin. Oncology 49: 27–30

Bartsch C, Bartsch H, Lippert TH (1992c) The pineal gland and cancer: facts, hypotheses and perspectives. The Cancer Journal 5:194–199

Bartsch H, Bartsch C, Mecke D, Lippert TH (1993) The relationship between the pineal gland and cancer: seasonal aspects. In: Wetterberg L (ed) Light and biological rhythms in man. Pergamon Press Ltd., London, pp 337–347

Bartsch C, Bartsch H, Flüchter S-H, Mecke D, Lippert TH (1994) Diminished pineal function coincides with disturbed circadian endocrine rhythmicity in untreated primary cancer patients. Ann N Y Acad Sci 719:502–525

Bartsch C, Bartsch H, Karenovics A, Franz H, Peiker G, Mecke D (1997a) Nocturnal urinary 6-sulfat-oxymelatonin excretion is decreased in primary breast cancer patients compared to age-matched controls and shows negative correlation with tumor size. J Pineal Res 23:53–58

Bartsch C, Kvetnoy I, Kvetnaia T, Bartsch H, Molotkov A, Franz H, Raikhlin N, Mecke D (1997b) Nocturnal urinary 6-sulfatoxymelatonin and proliferating cell nuclear antigen immuno-positive tumor cells show strong positive correlations in patients with gastrointestinal and lung cancer. J Pineal Res 23:90–96

Benson B (1977) Current status of pineal peptides. Neuroendocrinology 24:241–258

Benson B (1997) Melatonin retards the growth of diethylstilbestrol-induced renal tumors in male Syrian hamsters. Cancer Biother Radiopharm 12:415

Benson B (1999a) Pineal gland neuropeptides. In Watson RR (ed) Melatonin in Health Promotion. CRC Press, Boca Raton, 137–149

Benson B (1999b) Pineal gland-derived antigonadotropic decapeptide. In Watson RR (ed) Melatonin in Health Promotion. CRC Press, Boca Raton,151–160

Benson B, Ebels I (1978) Pineal peptides. J Neural Transm Suppl 13:157–173

Benson B, Ebels I (1981) Other pineal peptides and related substances – physiological implications for reproductive biology. In: Reiter RJ (ed) The Pineal Gland. Reproductive Effects Vol II, pp 165–187

Benson B, Ebels I (1994) Structure of a pineal-derived antigonadotropic decapeptide. Life Sci 54:PL437–443

Benson B, Machen N (1994) Infusion of a pineal gland-derived antigonadotropic decapeptide into the lateral ventricle depresses prolactin levels in male rats. Life Sci 55:363–368

Benson B, Matthews MJ, Rodin AE (1972) Studies on a non-melatonin pineal antigonadotropin. Acta Endocrinol 69:257–266

Benson B, Ebels I, Hruby VJ (1990) Isolation and structure elucidation of bovine arginine vasopressin: arginine vasotocin not identified. Int J Pep Prot Res 36:109–121

Benson B, Machen N, Dunn AM, Wise ME (1996) Chronic infusion of a pineal gland-derived antigonadotropic decapeptide alters pulsatile secretion of LH in rats. Life Sci 58:1083–1090

Bergmann W, Engel P (1950) Über den Einfluss von Zirbelextrakten auf Tumoren bei weissen Mäusen und bei Menschen. Wien Klin Wschr 62:79–82

Bindoni M, Jutisz M, Ribot G (1976) Characterisation and partial purification of a substance in the pineal gland which inhibits cell multiplication in vitro. Biochim Biophys Acta 437:577–588

Blask DE (1984) The pineal gland: an oncostatic gland? In: Reiter RJ (ed) The Pineal Gland, Raven Press, New York, pp 253–284

Blask DE, Leadem CA (1987) Neuroendocrine aspects of neoplastic growth: a review. Neuro-endocrinol Lett 9:63–73

Cos S, Mediavilla MD, Fernandez R, Güezmes A, Sanchez-Barcelo EJ (1997) Melatonin and MCF-7 human breast cancer cells: a short review. Cancer Biother Radiopharm 12:420

Dilman VM, Anisimov VH, Ostroumova MN, Morosov VG, Khavinson VKh, Azanova MA (1979) Study of the antitumor effect of polypeptide pineal extract. Oncology 36:274–280

Dwyer VG, Benson B, Kwan KH, Humphreys RC, Ko WJ, Lin NH, Sammons DW (1988) A computer-enhanced comparative study of brain region polypeptides and proteins by two dimensional gel electrophoresis. J Pharmac & Biomed Anal 6:793–779

Ebels I (1980) A survey of the location, isolation and identification of indoles, pteridines and some unknown active substances in sheep pineals. The possible significance of pteridines for the neuroendocrine control of neoplastic growth. J Neural Transm 49:87–105

Ebels I, Benson B (1978) A survey of the evidence that unidentified pineal substances affect the reproductive system in mammals. Progr Reprod Biology 4:51–89

Ebels I, Citharel A, Moszkowska A (1975) Separation of pineal extracts by gelfiltration. III. Sheep pineal factors acting either on the hypothalamus or on the anterior hypophysis of mice and rats in in vitro experiments. J Neural Transm 36:281–302

Ebels I, Noteborn HPJM, Bartsch H, Bartsch C (1988) Effects of low molecular weight pineal compounds on neoplastic growth. In: Gupta D, Attanasio A, Reiter RJ (eds) The pineal gland and cancer. Brain Research Promotion, London/Tübingen, pp 261–272

Engel P (1934) Über den Einfluss von Hypophysenvorderlappen-Hormonen und Epiphysen-Hormonen auf das Wachstum von Impftumoren. Z Krebsforschung 41:281–291

Engel P (1935) Wachstumsbeeinflussende Hormone und Tumorwachstum. Z Krebsforschung 41:488–496

Engel P, Bergmann W (1952) Die Physiologische Funktion der Zirbeldrüse und ihre therapeutische Anwendung. Z Vitamin Hormon Fermentforschung IV (6):564–594

Feuer GM, Kerenyi NA (1989) Role of the pineal gland in the development of malignant melanoma. Neurochem Int 14:265–273

Hajdu SJ, Porro RS, Liebermann PH, Foote Jr FW (1972) Degeneration of the pineal gland of patients with cancer. Cancer 29:706–709

Hill SM, Blask DE (1988) Effects of the pineal gland hormone melatonin on the proliferation and morphological characteristics of human breast cancer cells (MCF-7) in culture. Cancer Res. 48:6121–6126

Hirata F, Hayaishi O (1974) In vitro and in vivo formation of two new metabolites of melatonin. J Biol Chem 249:1311–1313

Hofstätter R (1950) Beitrag zur therapeutischen Verwendung von Zirbelextrakten. Wien Klin Wschr 62:1–4

Kelly RW, Amato F, Seamark RF (1984) N-acetyl-5-methoxy kynurenamine, a brain metabolite of melatonin, is a potent inhibitor of prostaglandin biosynthesis. Biochem Biophys Res Commun 121:372–379

Kitay JI, Altschule MD (1954) The Pineal Gland, a Review of the Physiological Literature, Harvard University Press, Cambridge

Klotz DM, Castles CG, Fuqua SAW, Spriggs LL, Hill SM (1995) Differential expression of wild-type and variant ER mRNAs by stocks of MCF-7 breast cancer cells may account for differences in estrogen-responsiveness. Biochem Biophys Res Comm 210:609–615

Lapin V (1974) Influence of simultaneous thymectomy and pinealectomy on the growth and formation of metastases of Yoshida sarcoma in rats. Exp Pathol Jena 9:108–112

Lapin V (1976) Pineal gland and malignancy. Österr Z Onkol 3:51–60

Lapin V, Ebels I (1976) Effects of some low molecular weight sheep pineal fractions and melatonin on different tumors in rats and mice. Oncology 33:110–113

Lapin V, Ebels I (1981) The role of the pineal gland in neuroendocrine control mechanisms of neoplastic growth. J Neural Transm 50:275–282

Lapin V, Frowein A (1981) Effects of growing tumors on pineal melatonin level in rats. J Neural Transm 52:123–136

Leadem CA (1986) Pineal-mediated inhibition of estrogen-induced pituitary tumor growth in Fisher 344 rats. Abstracts 68th Ann Mtg Endocr Soc, p 46

Leadem CA, Burns DM (1987) Pineal-induced inhibition of prolactinoma growth in F344 rats: effects of blinding and melatonin treatment. Abstracts Ann Mtg Endocr Soc, p 344

Lee, H-J (1998) Regulation of hypothalamic GnRH secretion by a novel decapeptide (antigonadotropic decapeptide). Master's thesis, University of Arizona

Lerner AB, Case JD, Takahashi Y, Lee TH (1958) Isolation of melatonin, the pineal factor that lightens melanocytes. J Am Chem Soc 80:2587

Liburdy RP (1977) In vitro effects of ELF magnetic fields on melatonin function in human breast cancer cells. Cancer Biother Radiopharm 12:426

Logan JL, Benson B (1990) Light deprivation retards the growth of the diethylstilbestrol-induced renal tumor in hamsters. Growth Dev Aging 54:30–43

Logan JL, Leadem CA, Benson B (1988) Blinding retards the growth and metastatic spread of the DES-induced renal tumor in male Syrian hamsters. In Gupta D, Attanasio A, Reiter RJ (eds) The Pineal Gland and Cancer. Brain Research Promotion, London/Tübingen, pp 307–319

Maestroni GJM (1993) Mini-review. The immunoneuroendocrine role of melatonin. J Pineal Res 14:1–10

Maestroni GJM (1997) Melatonin affects immune-hematopoietic system via novel T-helper cells opioid cytokines acting on kappa-opioid receptor in bone marrow stroma. Cancer Biother Radiopharm 12:428

Maestroni GJM, Conti A (1996) Melatonin in human breast cancer tissue: association with nuclear grade and estrogen receptor status. Lab Invest 75:557–561

Maestroni GJM, Conti A, Pierpaoli W (1989) Melatonin, stress and the immune system. In: Reiter RJ (ed) Pineal Research Reviews 7, Alan Liss, New York, pp 203–226

Matthews MJ, Benson B (1973) Inactivation of pineal antigonadotropin by proteolytic enzymes. J Endocr 56:339–340

Matthews MJ, Benson B, Rodin AE (1971) Antigonadotropic activity in a melatonin-free extract of human pineal glands. Life Sci 10:1375–1379

Noteborn HPJM, Bartsch H, Bartsch C, Mans DRA, Weusten JJAM, Flehmig B, Ebels I, Salemink CA (1988) Partial purification of (a) low molecular weight ovine pineal compound(s) with an inhibiting effect on the growth of human melanoma cells in vitro. J Neural Transm 73:135–155

Noteborn HPJM, Weusten JJAM, Bartsch H, Bartsch C, Flehmig B, Ebels I, Salemink CA (1989) Partial purification of a polypeptide extract derived from ovine pineal that suppresses the growth of human melanoma cells in vitro. J Pineal Res 6:385–396

Pévet P, Buijs RM, Dogterom J, Vivien-Roels B, Holder F C, Guermé J M, Reinharz A, Swaab DF, Ebels I, Neacsu C (1981) Peptides in the mammalian pineal gland. In: Matthews CD, Seamark RF (eds) Pineal Function, Elsevier/North-Holland Biomedical Press, pp 173–184

Praast G, Bartsch C, Bartsch H, Mecke D, Lippert TH (1995) Hepatic hydoxylation of melatonin in the rat is induced by phenobarbital and 7,12-dimethylbenz[a]anthracene– implications for cancer etiology. Experientia 51:349–355

Schmidt U, Bartsch C, Bartsch H, Mecke D (1997) Pineal melatonin secretion seems to be reversibly inhibited by a tumor-derived melatonin-inhibiting factor. Cancer Biother Radiopharm 12:429

Shah P, Mhatre M, Kothari I (1984) Effects of melatonin on mammary carcinogenesis in intact and pinealectomised rats in varying photoperiods. Cancer Res 44:3403–3407

Shellard SA, Whelan RDH, Hill BT (1989) Growth inhibitory and cytotoxic effects of melatonin and its metabolites on human tumor cell lines in vitro. Br J Cancer 60:288–290

Stanberry LR, Das Gupta TK, Beattle CW (1983) Photoperiod control of melanoma growth in hamsters; influence of pinealectomy and melatonin. Endocrinology 113:469–475

Tapp E (1980a) The human pineal gland in malignancy. J Neural Transm 48:119–129

Tapp E (1980b) Pineal gland in rats suffering from malignancy. J Neural Transm 48:131–135

14 Experimental Studies of the Pineal Gland Preparation Epithalamin

Vladimir Kh. Khavinson, V. G. Morozov, Vladimir N. Anisimov

Abstract

Twenty-five years of study have shown a wide spectrum of high biological activity of the pineal peptide preparation epithalamin. Long-term exposure to epithalamin was followed by an increase in the mean and maximum life spans and slower rates of aging of rats, mice, and *D. melanogaster*. Epithalamin increases pineal synthesis of serotonin, N-acetylserotonin, and melatonin and night pineal secretion of melatonin in adult and old rats. The pineal preparation decreases the luteinizing hormone and prolactin levels in adult male rats as well as the threshold of the hypothalamopituitary complex to feedback inhibition by estrogens in old female rats; it slows down age-related cessation of estrous function in rats and induces the recurrence of estrous cycles and fertility in old, persistently estrous rats. Epithalamin increases the levels of tri-iodothyronine and decreases thyroxine in serum of adult rats. It further decreases the levels of corticosterone in the serum of mice and increases the susceptibility of the hypothalamo-pituitary complex to the homeostatic inhibition of adrenocorticotropic function by glucocorticoids in old rats. Serum insulin and triglyceride levels in rabbits are decreased by epithalamin and the tolerance to glucose and diuresis are increased. With respect to immune function, it was found that T and B cell-mediated immunity in adult and old mice as well as the titer of thymic serum factor and the titer of thymosin-like compounds in old mice are stimulated by the pineal peptide preparation in the same way as the colony-forming activity of splenocytes in pinealectomized rats. Epithalamin inhibits spontaneous and induced carcinogenesis and is a potent antioxidant, decreasing lipid peroxidation and stimulating the activity of CuZn superoxide dismutase. The obtained results demonstrate a high efficiency of epithalamin therapy for prophylaxis of age-related pathology, including cancer, showing a new physiological way to slow down pathological processes and to extend human life spans.

14.1 Introduction

During the last decade, a number of reports appeared on the role of the pineal gland in aging and cancer development (Anisimov 1988; Armstrong and Redman 1991; Pierpaoli 1991; Trentini et al. 1991). Thus, a modulating effect of the pineal gland on the neuroendocrine and the immune system was shown to change during aging (Reiter 1981, 1988; Maestroni et al. 1987, 1989) and tumor development (Blask 1990; Anisimov and Reiter 1990; Bartsch et al. 1992a, b). Pinealectomized rats showed a reduced life span (Malm et al. 1959) and increased tumor development (Aubert et al. 1980; Anisimov and Reiter 1990). The administration of the pineal hormone melatonin to old mice or the grafting of pineal glands from young donors into thymus of old mice prolonged the life span of the latter (Pierpaoli et al. 1991). The antitumor effect of melatonin has been highlighted (Blask 1990; Bartsch et al. 1992a, b). Most investigators invoked melatonin as a primary mediator of the endocrine capabilities of the pineal gland. However, some of the effects of the pineal gland obviously might have resulted from pineal peptide secretion (Blask et al. 1983). Some crude peptide extracts or purified peptides isolated from pineal glands were shown to have antigonadotropic, metabolic, and antitumor activity (Lapin and Ebels 1976; Benson 1977; Bartsch and Bartsch 1988; Bartsch et al. 1992a, b; Noteborn et al. 1988). In 1955, Parhon reported on the effect of pineal extracts in prolonging the life span of old rats. No details were given either on the method of pineal extraction or on the experiment itself. Twenty years ago, we published the first data describing the results of administrating low molecular weight pineal peptide extracts, the commercial drug form of which was later named epithalamin. We observed recovery of the estrus cycle in old female rats with persistent estrous syndrome as well as the lowering of threshold sensitivity of the hypothalamo-pituitary complex to the feedback inhibition by estrogens in old animals (Anisimov et al. 1973a). Since that time, the effect of epithalamin on neuroendocrine and the immune system functions has been continuously studied by us and others demonstrating its high biological activity. Long-term treatment with the preparation prolonged the life spans of animals, slowed down aging of the reproductive system, improved parameters of immunity, and inhibited the development of spontaneous, chemically, and radiation-induced or transplanted tumors (Anisimov 1987; Anisimov et al. 1973b, 1982a, b, c, 1987, 1988, 1989, 1990, 1992; Morozov and Khavinson 1974; Anisimov and Reiter 1990). This chapter summarizes the results of 20 years of studies on the biological effects of epithalamin, mainly in experimental and clinical gerontology.

14.2 Epithalamin – a Low Molecular Weight Peptide Preparation

An officinal preparation – epithalamin – contains a complex of biologically active peptides isolated from cattle pineal glands. (Morozov and Khavinson 1971, 1974). The native tissues were kept in acetone at $-4\,°C$ for 48 hours. The acetone was poured off, the tissues homogenized, and extraction performed in 3% acetic acid (1:6 v/v) in the presence of $ZnCl_2$ for 72 hours. Upon final extraction and centrifugation for 20 minutes at $400 \times g$, acetone was added to the supernatant at $-4\,°C$ (1:8 v/v). Following precipitate formation, the acetone was poured off. The precipitate was treated with acetone and ether in the filter until white powder was formed. The powder was dissolved, sterilized, and lyophilized. Studying the prepared substances by ion exchange chromatography, gel filtration, and electrophoresis on cellulose, it was found that the above substances were complexes of polypeptides. The ratio of three main fractions in the complete preparation was 74:16:10 (in percent), with molecular weights of 250, 11000, and 1200, respectively, and isoelectric points of 3.0, 5.0, and 10.5, respectively. The biological assay of the activity of epithalamin was designed according to Ostroumova et al. (1976) using a test with human chorionic gonadotropin (HCG) as follows.

Infantile (7–9 g weight) female mice were treated during 3 consecutive days with saline, HCG (0.5 IU for 3 days), or HCG plus epithalamin (50, 100, or 500 µg per mouse per day). The suppressive effect of epithalamin on HCG-induced uterine weight gain was calculated and evaluated as significant when greater than 25–30% as compared to mice treated with HCG plus saline. Epithalamin was filtered on Sephadex G-50 and the low and high molecular weight fractions were studied in the test with HCG (Prokopenko et al. 1989). It was shown that the antigonadotropic effect of the low molecular weight fraction was higher than that of the high molecular weight fraction. The officinal drug form of the crude pineal peptide preparation (epithalamin) was permitted by the Pharmaceutical Committee of the USSR Ministry of Health for medical use and the results of the experiments with this preparation are reported here.

14.3 Effect of Epithalamin on the Life Span of Mice, Rats, and *Drosophila melanogaster*

The effect of epithalamin on the life span of C3H/Sn and SHR female mice was studied in two sets of experiments (Table 14.1). Chronic treatment of C3H/Sn female mice with epithalamin in a single dose of 0.5 mg per mouse begun at the age of 3.5 months significantly prolonged their mean life span (by 40%) and increased their maximum life span by 3.5 months (Anisimov et al. 1982a, c). The survival curves of the mice exposed to epithalamin were significantly shifted to the right. The calculation of parameters of the Gompertz equation showed that the aging rate of mice treated with epithalamin was significantly decreased. Long-term epithalamin administration to SHR mice in a single daily dose of 0.1 mg per mouse slightly increased their mean life span but failed to change their maximum life span (Anisimov et al. 1987). The exposure to epithalamin was followed by a significant increase in life span of tumor-free mice of both strains (41% and 26%, respectively). A comparative effect of long-term epithalamin treatment on the survival of SHR female mice starting at ages of 3.5 and 12 months was estimated (Anisimov et al. 1989). Epithalamin was shown to increase mean life span by 14% and 17% in young and middle-aged groups, respectively, as compared to controls treated with saline. Epithalamin treatment shifted survival curves of both age groups distinctly to the right, as compared to corresponding controls. Calculation of the aging rate in young and middle-aged groups treated with epithalamin revealed a slowing down of aging in the young group in comparison with the middle-aged group. Administration of epithalamin was followed by an increased life span of tumor-free young and middle-aged mice by 20% and 36%, respectively.

Epithalamin was administered also to female outbred rats beginning at the age of 3.5 months for 20 months in daily doses of 0.1 or 0.5 mg per animal (Dilman et al. 1979a). The mean life span of rats exposed to epithalamin in these doses increased by 25% and 45%, respectively; however, the maximum life span increased by 2 months only in rats exposed to the highest dose of the preparation (Table 14.1). Analysis of survival curves of rats treated with epithalamin suggested a reduced rate of aging under the influence of pineal peptides. It is worth noting that the age-related dynamics of body weight in controls and treated rats was similar.

Table 14.1. The effect of epithalamin on the life spans of female mice and rats

Age at start (months)	Animal group	Number of animals	Life span (days)	
			Mean	Maximum
C3H/Sn mice				
3.5	Control	21	487 ± 29	776
	Epithalamin	22	640 ± 33* (± 40%)	885
SHR mice				
3.5	Control	31	564 ± 22	843
	Epithalamin	32	627 ± 21* (± 14%)	827
12	Control	41	576 ± 18	795
	Epithalamin	33	612 ± 14 (± 17%)	768
Outbred LIO rats				
3.5	Control	75	618 ± 15	1054
	Epithalamin (0.1 mg)	39	749 ± 20* (± 25%)	936
	Epithalamin (0.5 mg)	33	852 ± 34* (± 45%)	1112
15	Control	44	695 ± 18	888
	Epithalamin	47	738 ± 22 (± 18%)	972

In another experiment, female rats were injected chronically with epithalamin (a single dose of 0.5 mg per rat) starting at the age of 15 months (Anisimov and Khavinson 1991; Anisimov et al. 1992). It was shown that this treatment increased their life span insignificantly by 6.2% when calculated from birth and by 18% when calculated from the onset of the experiment. At the same time, 23% of the epithalamin-treated rats lived longer than the longest living control rat. The maximum life span in the epithalamin-treated group was 3 months longer than in controls. Calculation of the parameters of the Gompertz equation showed that epithalamin treatment slowed down the aging rate.

Thus, we observed the capacity of epithalamin to increase the mean and sometimes the maximum life spans of mice and rats and to slow the aging of the treated population. These effects were less pronounced when epithalamin treatment was started after the ages of 1 year in mice and 15 months in rats.

The antioxidative activity of melatonin and epithalamin was compared in a series of experiments (Anisimov et al 1995, 1996). Administration of melatonin with tap water (20 µg/ml) to rats for 5 days during night hours from 06:00 P.M. to 08:00 A.M. or of epithalamin subcutaneously in the morning between 10:00 A.M. and 11:00 A.M. (0.5 mg per rat) produced a considerable inhibiting effect on the intensity of peroxide chemoluminescence. The impact of melatonin and epithalamin on lipid peroxide oxidation was estimated with respect to the content of diene conjugates and Schiff bases in serum, while protein peroxide oxidation was estimated by the contents of carbonyl derivatives of amino acids in serum. Melatonin was observed to decrease the levels of diene conjugates 3.5-fold and those of Schiff bases by 30.4% and to increase slightly the contents of CO derivatives in serum, while epithalamin considerably increased (by 35%, $p < 0.01$) the general antioxidative activity (AOA) of serum. While the activity of CuZn superoxide dismutase (SOD) and the serum level of ceruloplasmin (CP) decreased under the influence of melatonin by 37% and 16%, respectively, the same indices increased by 20% and 6% after epithalamin treatment. SOD and CP levels were significantly higher in rats exposed to epithalamin as compared to those treated by melatonin (by 75% and 27%, respectively). We studied the capacities of melatonin and epithalamin with experiments in vitro to suppress the luminol-dependent chemoluminescence stimulated by different forms of oxygen (hypochlorite, superoxide radical, and hydrogen peroxide) as well as by mixtures of various forms of oxygen generated by neutrophils.

In case of hypochlorite, the greatest suppression was observed due to melatonin (up to 30%). Epithalamin was less efficient than melatonin (up to 17%), but more active than glutathione (up to 12%). In the presence of hydrogen peroxide, melatonin suppressed chemoluminescence up to 45%, epithalamin up to 33%, and glutathione up to 20%. The range of capacities to suppress the formation of the superoxide radical within the xanthine oxidase reaction was as follows: melatonin (up to 30%), epithalamin (up to 22%), and glutathione (up to 16.5%). Chemoluminescence induced by activated neutrophils was also reduced up to 20% by melatonin, epithalamin, or glutathione. However, the effect of these substances was less pronounced in this case than in the experiments with separate, distinct forms of oxygen.

The results of these experiments manifested a higher activity of melatonin than of epithalamin and glutathione. These differences could probably be explained by the structural peculiarities of these compounds, the antioxidative properties of melatonin being related to the indole ring. Melatonin's antioxidative properties were stronger, as compared to epithalamin under in vitro conditions, while in vivo the effect of the latter was much more pronounced. This could be explained by the ability of epithalamin to stimulate synthesis and secretion of melatonin in animals (Anisimov et al. 1994) and to affect enzyme systems of antioxidative defense.

There is every reason to suppose that the properties of melatonin and epithalamin to increase animal life span, stimulate immune response, and inhibit the development of neoplasia may be mediated to some extent by their inhibitory influence on free radical processes in the organism, which play a key role in the molecular mechanisms of aging and age-related pathologies such as arteriosclerosis, immunodepression, and cancer.

The effect of epithalamin on the life span of *D. melanogaster* was also studied (Anisimov et al 1997). Epithalamin was given during the larval stage of fly life. The exposure to epithalamin failed to change significantly any of the life span parameters in male flies. However, in females, it significantly increased the mean life span (by 17%, $p < 0.02$), medium (by 26%) and maximum longevity (by 14%), and decreased both population mortality rate (2.12 times, $p < 0.01$) and mortality rate doubling time (by 32%), as compared to controls. The survival curve was shifted to the right and the slope of the Gompertz plot was decreased only in female flies exposed to epithalamin.

We also studied the effect of epithalamin on lipid peroxidation and antioxidative enzyme activity in *Drosophila melanogaster*. The tissue levels of conjugated hydroperoxides (CHP) and ketodienes (KD) were significantly higher in control males than in females (40% and 49%, respectively), which inversely correlated with the mean life span in flies. Most significantly, the contents of CHP and KD were decreased in female flies exposed to epithalamin (2.3 and 3.4 times, respectively, $p < 0.001$), as compared to controls. The catalase activity was twice as high in control males than in control females, whereas the activity of SOD was the same in both sexes. The exposure to epithalamin was followed by a significant elevation of catalase activity in males (by 20%) and in females (by 7%) and by an increase of the SOD activity in males (by 41%).

Thus, epithalamin treatment significantly increased the mean and maximum life span of female *D. melanogaster* and reduced the aging rate of the population more than twofold. These effects are in good agreement with the strong inhibitory effect of epithalamin on lipid peroxidation and its stimulating effect on the activity of antioxidative enzymes.

14.4 The Influence of Epithalamin on the Function of the Nervous, Endocrine, and Immune Systems of Young and Old Rats

It was shown that the sensitivity threshold of the hypothalamo-pituitary complex to feedback suppression by estrogens increased with age in female rats (Dilman and Anisimov 1979). This mechanism was suggested to be crucial for the age-related cessation of reproductive function in rats. Subcutaneous injections of epithalamin to old rats increased the sensitivity of the hypothalamo-pituitary complex to estrogen suppression and restored cyclic estrous activity in rats with persistent estrus (Anisimov et al. 1973a; Anisimov and Dilman 1975). Injection of epithalamin into the third brain ventricle was followed by a similar effect. Pinealectomy or exposure to constant light were followed by an elevated threshold of the hypothalamic feedback sensitivity to inhibition by estrogen (Dilman and Anisimov 1979).

The study of the estrous function of female rats showed that 38% of the control animals aged 16–18 months had persistent estrous, anestrous, or repeated pseudopregnancies, whereas only 7% of rats treated with epithalamin from the age of 3.5 months onwards showed estrous cycle disturbances. Of 16 old rats which remained sterile after mating with adult males, four females became pregnant and gave birth to 5–9 fetuses per litter after a 2-week course of epithalamin administration (Dilman et al. 1979a).

Daily cytological studies of vaginal smears in rats treated with epithalamin from the age of 15 months onwards revealed a slowing down of the age-related switching-off of estrous function as compared to controls. While at the age of 18 months the number of rats with disturbances of estrous function (persistent estrus and anestrus) was the same in both groups, at the age of 27 months these disturbances were observed in 50% of controls and 30% of females treated with the pineal preparation ($p < 0.05$) (Anisimov and Khavinson, 1991). Thus, we observed a slowing of the age-related reduction in estrous function in female rats exposed to epithalamin. Epithalamin was also capable of restoring estrous function, ovulation, and fertility in rats with persistent estrous disturbance. These effects were suggested to depend on the capacity of the preparation to prevent an age-related increase in the threshold sensitivity of the hypothalamo-pituitary complex to feedback regulation by estrogen.

Epithalamin also restored regular estrous cycles in young and old rats with persistent estrous syndrome induced by the exposure to constant illumination (Gadzhieva and Blinova 1980). A significant decrease in the number of follicular cysts in rats treated with epithalamin in comparison to untreated controls and a partial luteinization of the follicular cyst wall and the appearance of corpora lutea in ovaries were observed under the influence of epithalamin.

A single administration of epithalamin in the morning was followed by decreased levels of serum luteinizing hormone (LH) and failed to influence serum testosterone levels in 4–5-month-old male rats, but increased the levels of both hormones in 16-month-old animals 30 min after injection. Daily epithalamin injections in the morning (between 10.00 A.M. and 11.00 A.M.) for 5 consecutive days failed to influence serum LH levels in young and old male rats and were followed by a significant decrease of serum testosterone levels in young rats at noon and 5:00 P.M. No influence on serum testosterone was registered in older males at any time of day (Anisimov et al. 1994). Epithalamin was tested for its effect on pineal androgen receptors and showed a trend to reduce cytosolic receptors but had no effect on the nuclear 5-alpha-dihydrotestosterone receptor in rat pineal gland (Gupta et al. 1993).

It was shown that the susceptibility of the hypothalamo-pituitary complex to feedback inhibition by testosterone is elevated in aged male rats (Riegle and Miller 1978; Meites 1988). Our data support this observation. Thus, if epithalamin decreases the threshold sensitivity of the hypothalamo-pituitary complex to the inhibitory effect of testosterone, it cannot be expected to decrease LH levels in response to elevated testosterone levels, but rather to lead to increased LH levels. This phenomenon was in fact observed in our experiment.

Administration of epithalamin to adult rats during 5 consecutive days was followed by an increased activity of succinate, alpha-ketoglutarate, and pyruvate dehydrogenases in rat brain and by a 39% increase of learning capacity in labyrinth and other behavioral tests (Belozertsev et al. 1987). Arushanian et al. (1990) observed an increase in the amplitude and shifts in the acrophase of circadian locomotor activity towards late hours in rats exposed to long-term administration of epithalamin. Epithalamin treatment was also followed by changes in the time course of forced swimming and by a decrease in the rhythmic index of depression. Intravenous administration of epithalamin revealed a sedative effect in dogs (Slepushkin and Pashinski 1982).

Ten-day-long administration of epithalamin to adult male rats was followed by significant activation of neurosecretory elements in the hypothalamic nucleus paraventricularis, lower activation in the nucleus supraopticus, and an increased content of neurosecretory substances in the neurohypophysis (Khavinson et al. 1977). Ultramicroscopic studies showed that a single morning injection of epithalamin into adult rats induced signs of activated pinealocyte function (Bondarenko et al. 1992).

A 1- or 2-week-long administration of epithalamin failed to change the levels of dopamine, norepinephrine, serotonin, or 5-hydroxyindole acetic acid (HIAA) in the hypothalamus of adult male rats (Ostroumova and Anisimov, unpublished data). In other experiments of ours, the effect of epithalamin on serotonin metabolism in the pineal gland of young and old male rats was studied (Anisimov et al. 1990, 1991; Bondarenko and Anisimov 1990, 1992). It appeared that a single injection of epithalamin in the morning or a 5-day course of epithalamin treatment to 4–5-month-old rats was followed by an increase in night levels of serotonin, N-acetylserotonin, and melatonin in the pineal gland. In 18–20-month-old rats, similar treatment caused only a tendency to increase pineal melatonin level. At the same time, increased serum melatonin levels in young adult and old rats treated with epithalamin were observed. In young as well as in old rats, the exposure to epithalamin failed to influence direct O-methylation reaction of serotonin into 5-methoxytryptamine and the oxidative deamination with subsequent O-methylation and 5-HIAA and 5-methoxyindole acetic acid. These data suggest the existence of an ultrashort loop of regulation between pineal peptides and indoles. Pineal peptides may influence the metabolic reactions of tryptophan into serotonin and its subsequent transformation into melatonin, which declines with aging. It is worth noting that epithalamin given at 6:00 P.M. entailed a decrease in night levels of serotonin, N-acetylserotonin, and melatonin in the rat pineal gland.

The effect of epithalamin on the biosynthesis of melatonin was studied in rat pineal glands in vitro. Pineals were collected from rats killed in the light period (9:00 A.M.) and incubated for 3 hours with either isoproterenol (1 nM), epithalamin (25 ng/ml) or isoproterenol and epithalamin in equal concentrations. Additional pineals were incubated in the absence of these drugs. Epithalamin as well as isoproterenol significantly increased melatonin levels in the rat pineal gland, whereas epithalamin did not potentiate the effect of isoproterenol (R. J. Reiter, personal communication).

We also observed a stimulating effect of a 5-day course of epithalamin treatment on serum tri-iodothyronine (T3) levels and an inhibitory effect of this treatment on serum thyroxine (T4) in 4–5-month-old male rats. These data suggest that the pineal preparation influences the metabolism of T3 into T4. Direct measurement of the activity of 5′-deiodinase in the thyroid gland of rats treated with epithalamin may help to explain these results. In 16-month-old male rats, the course of epithalamin treatment was followed by a decrease in T3 and T4 levels in serum (Anisimov et al. 1994).

In male CBA/Ca mice, we observed an age-related elevation of the serum levels of corticosterone and corticoliberin (CRF). In 2-month- and 8-month-old mice, the CRF levels were 1.62 ± 0.45 pg/ml and 12.62 ± 4.4 pg/ml, respectively, ($p < 0.05$), whereas the levels of corticosterone were 12.38 ± 3.44 µg/ml and 80.3 ± 18.9 µg/ml, respectively, ($p < 0.05$). A 5-day course of epithalamin given to younger mice was followed by a 3.2-fold decrease in serum corticosterone in comparison to controls (Labunets 1992; Anisimov et al. 1994).

Treatment of rabbits with epithalamin was followed by significantly increased glucose tolerance, but the tolerance to insulin loading was unaffected (Ostroumova and Vasilyeva 1976). The authors also observed that 3-week-long exposure to epithalamin led to decreased levels of both serum insulin and triglycerides. Slepushkin and Pashinski (1982) showed a significant influence of epithalamin on electrolyte and water metabolism: exposure of dogs or rats to the preparation was followed by an increase in urine excretion, accompanied by hyposodium-, hyperpotassium-, hypercalcium-, and hypermagnesiumuria.

With respect to the action of epithalamin on the immune system it was found that the pineal preparation increased the number of antibody-forming cells generated in the spleen as well as the level of serum hemagglutinins in response to immunization with sheep red blood cells (Belokrylov et al. 1976). Epithalamin treatment increased survival of CC57Br/Mv mice infected with 1.5×10^9 bacterium *Salmonella typhimurium*, survival of skin allografts in mice and rats, stimulated the reaction of hypersensitivity of delayed type in guinea pigs, and stimulated the phagocytic activity of blood neutrophils (Slepushkin et al. 1990). Epithalamin was also shown to restore the level of proliferation of granulocytes and macrophages (CFU-GM) in pinealectomized rats up to the levels of intact animals (Gupta et al 1993).

The treatment with epithalamin of 2-month-old and 2-year-old mice with an intact thymus led to increases in the levels of the thymic serum factor (TSF), the compounds with thymosin-like activity (CTLA), and thymic index and to an increase of spleen cellularity in old animals (Labunets 1992; Anisimov et al 1994). Epithalamin also induced an increase of TSF and CTLA titers in serum of old mice bearing spontaneous hepatomas discovered at autopsy, but to a smaller extent than in old tumor-free animals. Melatonin administration to young mice was also followed by increased levels of TSF, CTLA, and relative weight and cellularity of the thymus. A stimulating effect of melatonin on CTLA titer and thymic index and its cellularity, but not on TSF titer, was observed in old mice. The CTLA titer of splenocyte supernatants taken from 2-month-old mice and cultivated in vitro with normal saline was 2.0 + 0.30, but only 1.0 + 0.32 in the culture of splenocytes from 2-year-old mice. Incubation of splenocytes from young mice with melatonin (5 pg/ml and 20 pg/ml) or epithalamin (0.1 mg/ml and 1.0 mg/ml) was followed by CTLA titer increases of up to 6.7 ± 1.06, 6.3 ± 0.42, 6.5 ± 0.59, and 4.8 ± 0.71, respectively ($p < 0.05$). After in vitro incubation of splenocytes of old mice with melatonin (25 pg/ml), levels of CTLA were significantly higher than in controls. The titers of CTLA in the supernatant of old splenocytes incubated with epithalamin in doses of 0.1 mg/ml and 1.0 mg/ml were 1.7 ± 0.35 and 6.2 ± 1.47, respectively, demonstrating that epithalamin induced an increase of TSF and CTLA serum titers which was accompanied by morphological features of thymus and spleen function stimulation.

Grinevich et al. (1992) observed an activation of the proliferation and maturation of thymic cortical thymocytes, hyperplasia, and medullar differentiation of epithelial cells in the thymus and epithelioid cells in the spleen of AKR mice under long-term epithalamin treatment in comparison to controls. In female C3H/Sn mice exposed to long-term treatment with epithalamin starting at the age of 3.5 months, we observed a delay in the age-related decrease in the phytohemagglutinin-induced blast transformation reaction of T-lymphocytes, in comparison to controls (Anisimov et al. 1982 c).

14.5 The Effect of Epithalamin on Spontaneous Tumor Development in Rats and Mice

In our first experiment, female outbred rats were injected subcutaneously with 0.2 ml of saline or 0.1 or 0.5 mg of epithalamin 5 times per week for 20 months, starting at the age of 3.5 months. All animals were autopsied and the tumors discovered were studied histologically. It was shown that any dose of epithalamin failed to change either overall tumor incidence or occurrence of malignancies in female rats, while tumor latency was significantly increased (Dilman et al. 1979 b).

In another experiment, female rats were exposed to long-term subcutaneous injections with saline or 0.5 mg epithalamin starting at the age of 15 months. In this case, the overall tumor incidence in the epithalamin-treated group was 1.6 times lower in comparison to the control group ($p < 0.004$), while the incidence of malignant tumors was 2.7 times lower ($p < 0.04$) (Anisimov and Khavinson 1991).

Female C3H/Sn mice aged 3.5 months were treated subcutaneously with 0.2 ml of saline (control group) or with 0.5 mg of epithalamin dissolved in 0.2 ml of saline for 5 consecutive days once a month up to their natural death. It was shown that the overall tumor incidence in epithalamin-treated C3H/Sn females decreased 2.1 times in comparison with saline-treated control animals. The antitumor effect of epithalamin was most pronounced in relation to mammary adenocarcinomas. Their incidence decreased 2.9 times; the multiplicity of mammary adenocarcinomas (number of tumors per mouse) also decreased significantly (Anisimov et al. 1982 a, c).

Outbred female Swiss-derived SHR mice were treated subcutaneously with saline or 0.1 mg of epithalamin dissolved in saline for 5 consecutive days every month beginning from the ages of 3.5 or 12 months. Administration of epithalamin to young mice was followed by a significant reduction in spontaneous tumor incidence, first of all due to a 2.6-fold decrease in the incidence of mammary adenocarcinomas

($p < 0.025$). The exposure to epithalamin caused a shift to the right of the curves of tumor-free survival in young mice whereas, in the middle-aged group, the effect was similar but less marked (Anisimov et al. 1989).

The incidence of spontaneous thymic lymphomas constitutes 39% among 12-month-old intact AKR mice and only 14% among mice given long-term epithalamin treatment. Treated mice revealed increased TSF serum levels in comparison with controls, accompanied by morphological features giving evidence of activation of thymus and spleen cellular elements (Grinevich et al. 1992).

Thus, during the last 2 decades, a wide spectrum of biological activities of the low molecular weight pineal preparation epithalamin was observed. Administration of epithalamin to old animals led to normalization of some parameters of the endocrine, reproductive, and immune systems and some other physiological functions, and to the inhibition of tumor growth and reduced formation of metastases. Long-term treatment with epithalamin was followed by an increased life span in mice and rats, a slowing of age-related reductions in reproductive and immune functions, and inhibition of spontaneous as well as chemically or irradiation-induced carcinogenesis. The mechanisms of the biological effects of epithalamin consist mainly of stimulation of the night peak of melatonin in the pineal gland and serum, lowering the hypothalamic threshold sensitivity to homeostatic feedback stimuli, and modulation of some T and B cell-mediated immune functions. Melatonin was shown to be a very potent and efficient scavenger of endogenous radicals (Poeggeler et al. 1993). This mechanism could play a critical role in the geroprotective and antitumor effects of melatonin and epithalamin.

The documented experimental data support the expediency of epithalamin applications in clinical practice for prevention and treatment of age-related pathologies, including cancer. The results of clinical trials of epithalamin were reported elsewhere (Karpov et al. 1985; Slepushkin et al. 1990; Khavinson and Morozov 1992; Morozov and Khavinson 1996).

References

Anisimov VN (1987) Pineal gland preparations. In: Anisimov VN (ed) Carcinogenesis and Aging. Vol. 2. CRC Press Inc, Boca Raton, pp 69–72

Anisimov VN (1988) Pineal gland, aging and carcinogenesis. In: Gupta D, Attanasio A, Reiter RJ (eds) The Pineal Gland and Cancer. Brain Research Promotion, London and Tübingen, pp 107–118

Anisimov VN, Dilman VM (1975) Increase of hypothalamic sensitivity to inhibiting effect of estrogens induced by administration of L-DOPA, diphenylhydantoin, epithalamin and phenformin in old rats. Bull ExpBiol Med 80:96–98

Anisimov VN, Khavinson VK (1991) Effect of polypeptide pineal preparation on life span and spontaneous tumor incidence in old female rats. Dokl Akad Nauk USSR 319:250–254

Anisimov VN, Reiter RJ (1990) The function of pineal gland in cancer and aging Vopr Onkol 36:259–268

Anisimov VN, Khavinson VKh, Morozov VG, Dilman VM (1973a) The lowering of the threshold of sensitivity of hypothalamo-pituitary system to estrogen action under influence of pineal extract in old female rats. Dokl.Acad.Nauk USSR 213:483–485

Anisimov VN, Morozov VG, Khavinson VKh, Dilman VM (1973b) Comparison of anti-tumor activity of pineal and hypothalamus extracts, melatonin and sygetin in mice with transplanted mammary carcinoma. Vopr Onkol 19:99–101

Anisimov VN, Khavinson VKh, Morozov VG (1982a) Carcinogenesis and aging IV. Effect of low-molecular factors of thymus, pineal gland and anterior hypothalamus on immunity, tumor incidence and life span of C3H/Sn mice. Mech Ageing Dev 19:245–258

Anisimov VN, Miretski GI, Morozov VG, Khavinson VKh (1982b) Effect of polypeptide factors of thymus and epiphysis on radiation carcinogenesis. Bull Exp Biol Med 94:26–29

Anisimov VN, Morozov VG, Khavinson VKh (1982c) Increase of life span and decrease of tumor incidence in C3H/Sn mice under influence of polypeptide factors of thymus and pineal gland. Dokl Akad Sci USSR 263:742–745

Anisimov VN, Morozov VG, Khavinson VKh (1987) Effect of polypeptide factors of thymus, bone marrow, epiphysis and blood vessels on life span and tumor development in mice. Dokl Akad Nauk USSR 293:1000–1004

Anisimov VN, Loktionov AS, Morozov VG, Khavinson VKh (1988) Increase of life span and decrease of tumor incidence in mice treated with polypeptide factors of thymus and pineal gland in various age. Dokl Akad Nauk USSR 302:473–476

Anisimov VN, Loktionov AS, Khavinson VKh, Morozov VG (1989) Effect of low-molecular-weight factors of thymus and pineal gland on life span and spontaneous tumour development in female mice of different age. Mech Ageing Dev 49:245–257

Anisimov VN, Bondarenko LA, Khavinson VKh (1990) The pineal peptides: interaction with indoles and the role in aging and cancer. In: Gupta D, Wollmann HA, Ranke MB (eds) Neuroendocrinology: New Frontiers. Brain Research Promotion, London and Tübingen, pp 317–325

Anisimov VN, Bondarenko LA, Khavinson VKh (1991) Effect of pineal peptide preparation (epithalamin) on life span and pineal and serum melatonin in rats. In: Abstracts of the 4th Int. Congress of International Association of Biomedical Gerontology, Ancona, June 20–24 1991, p 238

Anisimov VN, Bondarenko LA, Khavinson VKh (1992) Effect of pineal peptide preparation (epithalamin) on life span and pineal and serum melatonin level in old rats. Ann NY Acad Sci 673:53–57

Anisimov VN, Khavinson VKh, Morozov VG (1994) Twenty years of study on effects of pineal peptide preparation: Epithalamin in experimental gerontology and oncology. Ann NY Acad Sci 719:483–493

Anisimov VN, Prokopenko VM, Khavinson VKh (1995) Melatonin and epithalamin inhibit process of free radical oxidation in rats. Dokl Akad Nauk 343:557–559

Anisimov VN, Arutyunian AV, Khavinson VKh (1996)) Melatonin and epithalamin inhibit process of lipid peroxidation in rats. Dokl Akad Nauk 348:265–267

Anisimov VN, Mylnikov SV, Oparina TI, Khavinson VKh (1997) Effect of melatonin and pineal peptide preparation epithalamin on life span and free radical oxidation in Drosophila melanogaster. Mech Ageing Dev 91:81–91

Armstrong SM, Redman JR (1991) Melatonin: a chronobiotic with anti-aging properties? Med Hypotheses 34:300–309

Arushanian EB, Baturin VA, Ovanesov KB (1990) Chronic administration of pineal peptides change the circadian locomotor activity and time-course of forced swimming in rats. J Pineal Res 9:271–277

Aubert C, Janiaud P, Lecalvez J (1980) Effect of pinealectomy and melatonin on mammary tumor growth in Sprague-Dawley rats under different conditions of ligthing. J Neural Transm 47: 121–130

Bartsch H, Bartsch C (1988) Unidentified pineal substances with anti-tumor activity. In: Gupta D, Attanasio A, Reiter RJ (eds) The Pineal Gland and Cancer. Brain Res Promotion, London and Tübingen, pp 369–376

Bartsch C, Bartsch H, Lippert TH (1992a) The pineal gland and cancer: facts, hypotheses and perspectives. The Cancer J 5:194–199

Bartsch H, Bartsch C, Simon WE, Flehmig B, Ebels I, Lippert TH (1992b) Antitumor activity of the pineal gland: effect of unidentified substances versus the effect of melatonin. Oncology 49:27–30

Belokrylov GA, Morozov VG, Khavinson VKh, Sofronov BN (1976) Effect of low-molecular extracts from heterologous thymus, epiphysis and hypothalamus on the immune response of mice. Bull Exp Biol Med 81:202–204

Belozertsev YA, Tertyshnik ID, Spinko AA, Grudinin AN (1987) Restoration by pineal cytomedins of learning and homeostatic disturbances. In: Abstracts of Scientific Conference "The Role of Peptide Bioregulators (Cytomedins) in Regulation of Homeostasis". Yakovlev GM (ed) Military Medical Academy Publ, Leningrad, pp 13–14

Benson B (1977) Current status of pineal peptides. Neuroendocrinology 24:241–248

Blask DE (1990) The emerging role of the pineal gland and melatonin in oncogenesis. In: Wilson BW, Stevens RG, Anderson LE (eds) Extremely Low Frequency Electromagnetic Fields: the Question of Cancer. Battelle Press, Columbus, pp 319–335

Blask DE, Vaughan MK, Reiter RJ (1983) Pineal peptides and reproduction. In: Relkin R (ed) The Pineal Gland. Elsevier, Amsterdam, pp 201–224

Bondarenko LA, Anisimov VN (1990) Effect of polypeptide pineal preparation epithalamin on serotonin metabolism in the pineal gland of rats. Bull Exp Biol Med 110:150–151

Bondarenko LA, Anisimov VN (1992) Age-related peculiarities of effect of epithalamin on metabolism of serotonin in pineal gland of rats. Bull Exp Biol Med 113:194–195

Bondarenko LA, Lukashova OP, Rom-Boguslavkaya ES (1992) Chronopharmacological aspects of the effect of pineal bioregulator epithalamin on morpho-functional activity of pineal gland. In: Yakovlev GM (ed) Abstracts of the Symposium "Peptide Bioregulators – Cytomedins" Medical Military Academy Publ, St.Petersburg, pp 34–35

Dilman VM, Anisimov VN (1979) Hypothalamic mechanisms of ageing and of specific age pathology – I. Sensitivity threshold of hypothalamo-pituitary complex to homeostatic stimuli in the reproductive system. Exp Gerontol 14:161–174

Dilman VM, Anisimov VN, Ostroumova MN, Khavinson VKh, Morozov VG (1979a) Increase in lifespan of rats following polypeptide pineal extract treatment. Exp Pathol 17:539–545

Dilman VM, Anisimov VN, Ostroumova MN, Khavinson VKh, Morozov VG, Azarova MA (1979b) Study of the anti-tumor effect of polypeptide pineal extract. Oncology 36:274–280

Gadzhieva TS, Blinova TS (1980) Effect of epithalamin on gonadotropic function of pituitary and on ovaries. Akush Gynekol 9:15–18

Grinevich YuA, Bendyug GD, Bobrov LI, Labunets IF, Kireeva ES (1992) Effect of polypeptide factors of thymus and pineal gland on spontaneous development of lymphomas in AKR mice. In: Yakovlev GM (ed) Abstracts of the Symposium "Peptide Bioregulators – Cytomedines". Medical Military Academy Publ, St.Petersburg, pp 47–48

Gupta D, Haldar C, Coeleveld M, Roth J (1993) Ontogeny, circadian rhythm pattern, and hormonal modulation of 5-alpha-dihydrotestosterone receptors in the rat pineal. Neuroendocrinology 57:45–53

Karpov RS, Slepushkin VD, Mordovin VF, Khavinson VKh, Morozov VG, Grischenko VI (1985) The Use of Pineal Preparations in Clinical Practice. Tomsk Univ Publ, Tomsk

Khavinson VKh, Morozov VG (1992) The Preparations of Epiphysis and Thymus in Gerontology. Institute of Bioregulation and Gerontology Publ, St. Petersburg

Khavinson V.Kh, Morozov VG, Grintzevich II, Knorre ZD (1977) Influence of polypeptide extracted from the anterior hypothalamus and the epiphysis on the cells of the hypothalamo-hypophyseal neurosecretory system. Arch Anat Histol Embryol 73:100–104

Labunets IF (1992) Effect of epithalamin on immune system of young and aged animals. In: Yakovlev GM (ed) Abstracts of the Symposium "Peptide Bioregulators – Cytomedins", Medical Military Academy Publ, St. Petersburg, pp 85–86

Lapin V, Ebels I (1976) Effect of some low molecular weight sheep pineal fractions and melatonin on different tumours in rats and mice. Oncology 33:11–13

Maestroni GJM, Conti A, Pierpaoli W (1987) The pineal gland and the circadian, opiatergic, immunoregulatory role of melatonin. Ann NY Acad Sci 496:67–77

Maestroni GJM, Conti A, Pierpaoli W (1989) Melatonin, stress and the immune system. Pineal Res Rev 7:203–226

Malm OJ, Skaug OE, Lingjaerde P (1959) The effect of pinealectomy on bodily growth, survival rate and 32P uptake in the rat. Acta Endocrinol 30:22–28

Meites J (1988) Neuroendocrine basis of aging in the rat. In: Everitt AV, Walton JR (eds) Regulation of Neuroendocrine Aging (Interdisciplinary Topics in Gerontology, Vol. 24), Karger, Basel, pp 37–50

Morozov VG, Khavinson VKh (1971) The influence of extracts from hypothalamus and pineal gland on some host functions. Proc of the Conference of the Medical Military Academy, Leningrad, pp 127–128

Morozov VG, Khavinson VKh (1974) Effect of pineal extract on the development of experimental tumors and leukemias. Exp Khirurgia 1:34–38

Morozov VG, Khavinson VKh (1996) Peptide Bioregulators (25-years Experience of Experimental and Clinical Study). Nauka, St.Petersburg.

Noteborn HPJM, Bartsch H, Bartsch C, Mans DRA, Weusten JJAM, Flehmig B, Ebels I, Salemnik CA (1988) Partial purification of a low molecular weight ovine pineal compound(s) with an inhibiting effect on the growth of human melanoma cells in vitro. J Neural Transm 73:135–155

Ostroumova MN, Vasilyeva IA (1976) Effect of pineal extract on the regulation of fat-carbohydrate metabolism. Probl Endocrinol 22:66–69

Ostroumova MN, Anisimov VN, Syromyatnikov AA, Dilman VM (1976) Decrease of antigonadotropic factor excretion in patients with cancer of cervix uteri and prostate. Antitumor effect of pineal extract in the experiment. Vopr Onkol 22:39–43

Parhon CI (1955) Biologia Virstelor – cercetari clinici si experimentali. Acad R.P.R, Bucharest

Pierpaoli W (1991) The pineal gland: a circadian or seasonal aging clock? Aging 3:99–101

Pierpaoli W, Dall'Ara A, Pedrini E, Regelson W (1991) The pineal control of aging: the effects of melatonin and pineal grafting on the survival of older mice. Ann NY Acad Sci 621:291–313

Poeggeler B, Reiter RJ, Tan D-X, Chen L-D, Manchester LC (1993) Melatonin, hydroxyl radical-mediated oxidative damage, and aging: a hypothesis. J Pineal Res 14:151–168

Prokopenko NP, Tigranyan RA, Muratovam GL, Plokhoy VI, Khavinson VKh (1989) Properties of the acid extract fraction from epiphysis. Vopr Med Khimii 4:79–81

Reiter RJ (1981) The mammalian pineal gland: structure and function. Amer J Anat 162:287–313

Reiter RJ (1988) Neuroendocrinology of melatonin. In: Miles A, Philbrick DRS, Tompson C (eds) Melatonin – Clinical Perspectives. Oxford University Press, Oxford, pp 1–42

Riegle GD, Miller AE (1978) Aging effects on the hypothalamic-hypophyseal-gonadal control system in the rat. In: Schneider EL (ed) The Aging Reproductive System (Aging, Vol. 4), Raven Press, New York, pp 159–192

Slepushkin VD, Pashinski VG (1982) The Epiphysis and Adaptation of Organism. Tomsk University Publ, Tomsk

Slepushkin VD, Anisimov VN, Khavinson VKh, Morozov VG, Vasiliev NV, Kosykh VA (1990) The Pineal Gland, Immunity and Cancer (Theoretical and Clinical Aspects), Tomsk University Publ Tomsk

Trentini GP, De Gaetani C, Criscuolo M (1991) Pineal gland and aging. Aging 3:103–116

Section IV
Mechanisms of Action
of Melatonin on Tumor Cells

A. Actions Via the Endocrine System

15 An Overview of the Neuroendocrine Regulation of Experimental Tumor Growth by Melatonin and Its Analogues and the Therapeutic Use of Melatonin in Oncology

David E. Blask

Abstract

Scientific support for melatonin as a physiological, neuroendocrine regulator of a wide spectrum of carcinogenic processes and as a therapeutic anticancer agent in experimental and clinical settings continues to increase. In the experimental setting, melatonin and some of its analogues have efficacy in inhibiting the growth of a wide variety of solid neoplasms, particularly those related to hormone-responsive tumors. This is due primarily to a cytostatic action of melatonin although, under certain experimental conditions, melatonin may have cytotoxic effects as well. Based on its chronobiologic properties, melatonin is most efficacious in vivo when administered near the end of the light phase prior to the rise in endogenous melatonin secreted by the pineal gland. Numerous in vitro studies and some in vivo investigations have provided valuable insights into the potential mechanisms underlying melatonin's action as an oncostatic hormone. These mechanisms include melatonin modulation of hormone receptor expression and regulation, signal transduction events, transcriptional control, genomic interactions, redox processes, and metabolic activities. Melatonin has also been tested in a number of clinical trials for its therapeutic efficacy as either a single agent or, more commonly, in combination with established chemo- or radiotherapeutic approaches.

15.1 Introduction

The emergence of melatonin as an important neuroendocrine regulator of neoplastic growth has helped usher in a new era of pineal gland research into basic mechanisms of melatonin action at the cellular, molecular, and organismal levels. Today, oncology is one of the few areas of clinical medicine in which melatonin is being used therapeutically, albeit to a limited extent in clinical trials (Blask 1993). In fact, melatonin found its first clinical application as a treatment for cancer little more than a decade after its discovery (Starr 1970). Interestingly, Hamilton (1969) reported the first effects of melatonin on experimental cancer in animals a year earlier. Prior to that time, pineal extracts and grafts from farm animals were being used in Europe to treat human cancer (Lapin 1976). However, it is not clear whether therapeutic responses were due to melatonin or other pineal-derived oncostatic substance(s) (Bartsch et al. 1987).

Over the past 2 decades, we have witnessed an explosion in melatonin research, particularly as it relates to the binding of melatonin to its cognate receptor and the intracellular signal transduction pathways that may mediate many of this indole's diverse actions (Krause and Dubocovich 1990). As part of this exponential increase in melatonin research, there has been increased interest in the use of melatonin as a therapeutic modality in the treatment of cancer (Gonzalez et al. 1991; Lissoni et al. 1994a). Parallel to this interest in and use of melatonin as a cancer treatment, there has been a burgeoning number of basic in vivo and in vitro science studies aimed at determining the mechanisms by which melatonin modulates neoplastic growth (Blask 1993). There are probably few areas of melatonin research in which the link between basic and clinical studies is closer.

The present chapter will review, at times critically, the effects of melatonin and its analogues on the growth of various types of neoplasms in order to provide the reader with a frame of reference for understanding the rationale for the use of melatonin in clinical trials for the treatment of cancer. A substantial part of the discussion will also consider data on melatonin receptor binding and potential signal transduction mechanisms in neoplastic tissues that may mediate the action of melatonin on carcinogenesis. Many of the more current studies summarized in the present overview or inadvertently overlooked by the author are covered in more detail in other chapters prepared for this book.

15.2 Effects of Melatonin and Its Analogues on Experimental Cancer Growth

Several types of experimental neoplasms have been studied with respect to the ability of melatonin and its analogues to influence cancer growth both in vivo and in vitro. These neoplasms include tumors of the breast, skin, prostate, uterus, cervix, pituitary, liver, ovary, bladder, neural tissue, colon, connective tissues and lung (Lapin 1976; Kerenyi 1979; Tapp 1982; Blask 1984; Blask and Hill 1988; Blask 1993b). The in vivo investigations have almost exclusively used animal models of human tumor growth, while the in vitro work has focused primarily on human cancer cells themselves, although a number of transformed and non-transformed animal cell lines have been employed as well. The nature of the neoplastic growth response to indolamines depends to a large extent on a variety of experimental factors including dose, timing and duration of treatment, route of administration, photoperiod, and culture conditions (Blask 1993b). In general, however, melatonin and related molecules are inhibitory to tumor growth in vivo and in vitro and thus have been referred to by the author as oncostatic agents (Blask 1984, 1993; Blask and Hill 1988). Nevertheless, there are instances when melatonin has either no effect on or actually stimulates tumorigenesis depending on the time of day it is administered suggesting that it can function as a "chrono-oncostatic or oncomodulatory" neurohormone (Bartsch and Bartsch 1981; Blask 1993a, 1994; Blask et al. 1994).

15.2.1 Melatonin and Breast Cancer

15.2.1.1 Animal Studies

Numerous studies in vivo have demonstrated a marked and consistent inhibitory effect of melatonin at doses ranging from 25 µg to 1 mg, injected on a daily basis for 7 to 335 days on transplantable (Anisimov et al. 1973; Karamali et al. 1978), spontaneous (Wrba et al. 1986; Subramanian and Kothari 1991a), or carcinogen-induced (Aubert et al. 1980; Tamarkin et al. 1981; Shah et al. 1984; Blask et al. 1986, 1991a) mammary cancer in either female mice or rats. Clearly, the 7,12-dimethylbenzanthracene (DMBA) and N-nitroso N-methylurea (NMU) models of chemically-induced mammary carcinogenesis have provided the most complete picture of the oncostatic effects of melatonin in vivo (Blask 1993b, 1984; Blask and Hill 1988). While the earliest study of melatonin's action on DMBA-induced mammary tumors showed a stimulatory effect of morning injections of this indole (Hamilton 1969), subsequent investigations in this model system have demonstrated an inhibitory effect of daily, late afternoon injections of melatonin (Tamarkin et al. 1981; Shah et al. 1984; Blask et al. 1986), suggesting that these tumors exhibit a diurnal rhythm of sensitivity to melatonin (Wrba et al. 1986). We have observed a similar diurnal rhythm of sensitivity of NMU-induced mammary tumors to melatonin as well (Blask et al. 1990; 1992). The oral administration of melatonin via drinking water is also quite effective in suppressing DMBA-induced mammary tumor growth (Kothari 1987; Subramanian and Kothari 1991b).

Some studies have shown that variations in environmental factors such as nutritional intake or photoperiodic exposure significantly influence the tumorigenic response to melatonin in rats with DMBA-induced mammary cancers. For example, in rats placed on a modest underfeeding regimen (40% reduction in food intake per day), melatonin is much more effective in suppressing tumorigenesis than in animals fed ad libitum (Blask 1984; Blask et al. 1986). Additionally, the anticarcinogenic effects of melatonin appear to be somewhat greater in rats maintained on a constant lighting regimen as compared with their counterparts maintained on a short photoperiod of 10 hours of light (L) and 14 hours of darkness (D) (Shah et al. 1984; Kothari 1987).

Since melatonin administration in most of the earlier studies in the DMBA model encompassed both the initiation and promotion phases of carcinogenesis, it was impossible to determine whether melatonin acted on tumor initiation, promotion, or both (Blask 1993b). In the initial study of this issue, biweekly, subcutaneous, late afternoon injections of melatonin (200 µg) over a 2-week period caused a regression in the number and size of established DMBA mammary tumors (Aubert et al. 1980). Studies by Subramanian and Kothari (1991b) in the DMBA model examined the influence of photoperiod and pinealectomy on the ability of melatonin to inhibit tumor initiation or promotion. In the initiation experiments, vehicle or melatonin (200 µg/day) was administered via the drinking water for 1 week prior to and 1 week following DMBA treatment (10 mg at 55 days of age) to female rats maintained on either constant light or short photoperiod (10L:14D). The promotion phase study was identical to the initiation study, except that melatonin treatment was delayed until 1 week following DMBA administration and continued until the end of the experiment 27 weeks later.

In the initiation experiments, melatonin treatment resulted in a 55% to 65% decrease in tumor incidence in intact rats on constant light or short photoperiods. However, in pinealectomized animals, melatonin was ineffective in suppressing tumor development in animals on either photoperiod. In the promotion study, melatonin effectively suppressed tumorigenesis by 55% to 77%, irrespective of photoperiod or pinealectomy. No effects of melatonin treatment, photoperiod, pinealectomy, or combinations of these treatments were observed on tumor number; however, melatonin treatment did delay the development of tumors in each of these groups. These results indicate that, in this model system, an intact pineal gland and presumably its endogenous melatonin rhythm are required for exogenous melatonin's oncostatic effects on tumor initiation but not promotion (Subramanian and Kothari 1991b). Interestingly, DMBA itself has been shown to suppress circulating levels of melatonin (Bartsch et al. 1990). Although we previously reported a suppressive effect of melatonin on tumor promotion in this model system (Blask et al. 1986), we have not observed melatonin inhibition of tumor initiation (unpublished results).

The initiation versus promotion issue has also been examined in the NMU mammary tumor model, in which these tumors are more growth-responsive to estrogen than DMBA-induced tumors, which are more dependent upon prolactin (Welsch 1985). Contrary to the findings of Kothari and her coworkers, we failed to observe an inhibitory effect of daily, afternoon melatonin injections (500 µg/day) restricted to the initiation phase of NMU-induced mammary carcinogenesis (Blask et al. 1991a). However, melatonin was effective in inhibiting tumorigenesis when its administration was restricted to the promotion phase of carcinogenesis. The antipromotion effect of melatonin in this model system was further confirmed by the ability of melatonin (20–500 µg/day) to completely block estradiol-induced regrowth of NMU tumors that had regressed in response to ovariectomy (Blask et al. 1991a).

Additionally, pinealectomy alters the responsiveness of NMU tumors so that, at certain doses of melatonin, they are less sensitive to its oncostatic effects (Blask et al. 1992). Furthermore, pinealectomy causes these tumors to become responsive to the oncostatic effects of morning injections of melatonin which are ineffective in intact animals. However, late afternoon melatonin is still more effective than late morning melatonin in suppressing tumorigenesis in pinealectomized animals (Blask et al. 1990; Blask et al. 1991b).

The reasons why Kothari and her group were able to inhibit the initiation phase of DMBA-induced carcinogenesis while we failed to affect tumor initiation in either the DMBA or NMU models could be explained in a number of ways. For example, in the DMBA system, we administer the carcinogen via the intravenous route and melatonin by subcutaneous injection, whereas Kothari and associates typically administer DMBA via the intragastric route and melatonin orally via the drinking water (Kothari 1987; Subramanian and Kothari 1991b). Therefore, following the oral administration of DMBA and melatonin, these compounds would be subjected to first-pass metabolism by the liver subsequent to their absorption from the gastrointestinal tract (Cardinali 1981). Thus, melatonin might inhibit tumor initiation by limiting the absorption of DMBA or by decreasing its activation and increasing its detoxification by phase I (cytochromes b5 and P450) and II (glutathione-S-transferase) xenobiotic metabolizing enzymes in the liver (Kothari and Subramanian 1992).

Another means by which melatonin has been proposed to prevent tumor initiation is via the inhibition of adduct formation between carcinogens, such as safrole, and DNA (Reiter et al. 1994; Tan et al. 1994). This could also be true in the case of DMBA, since the ultimate carcinogenic epoxide derivative of DMBA forms DNA adducts (DiGiovanni and Juchau 1980) in mammary epithelial cells.

However, if such a mechanism were operating, we would expect inhibition of tumor initiation in both our DMBA and NMU studies as well, since NMU also forms DNA adducts (Kumar et al. 1990). Perhaps in our studies, any effects of melatonin on liver metabolism of DMBA were minimized by the parenteral route of administration of both substances. Unlike DMBA, which is metabolically activated in the liver and mammary gland to its ultimate carcinogenic form, 3,4-dihydrodiol-1,2-epoxide (DiGiovanni and Juchau 1980), NMU directly alkylates DNA and activates H-ras and K-ras oncogenes (Kumar et al. 1990). Since NMU does not require metabolic activation to cause malignant transformation, melatonin might not be expected under these circumstances to have an impact on the ability of this nitrosoamine to initiate mammary carcinogenesis.

Melatonin treatment of female rats or mice results in a decrease in the number of terminal end and alveolar buds as well as in the amount of mammary gland DNA synthesis to create a more "differentiated" ductal system which might then render the mammary epithelium refractory to the effects of a carcinogen such as DMBA (Shah et al. 1984; Kothari 1988; Subramanian and Kothari 1991a). Studies in vitro suggest that such an effect of pharmacological doses of melatonin could act directly on mammary tissue to maintain a differentiated state (Sanchez-Barcelo et al. 1990). Recent findings by Crespo et al. (1994) also indicate that physiological concentrations of melatonin may induce a partially differentiated phenotype in MCF-7 breast cancer cells.

Other evidence suggests that melatonin suppression of circulating prolactin, estradiol, and insulin-like growth factor levels may be responsible for the inhibition of murine mammary tumorigenesis (Tamarkin et al. 1981; Shah et al. 1984; Subramanian and Kothari 1991a; Scaglione et al. 1992), since these hormones are endogenous promoters of mammary cancer development and growth (Freiss et al. 1993). However, we have never observed a melatonin-induced decrease in prolactin or estradiol levels in rats with either DMBA- or NMU-induced tumors (Blask et al. 1986, 1991a). Rather, to us it appears more likely that melatonin may be acting directly at the level of breast cancer cells to inhibit the mitogenic actions of estradiol, prolactin, and other growth factors (Hill and Spriggs 1992; Cos and Blask 1994; Lemus-Wilson et al. 1995).

15.2.1.2 Cell Culture Studies

The demonstration that melatonin inhibits the growth of monolayer cultures of estrogen receptor-positive (ER+) human breast cancer cells (MCF-7) supplemented in vitro with 10% fetal bovine serum (FBS) established for the first time that physiologically relevant levels of this hormone have direct oncostatic effects on breast cancer cells, while higher or lower concentrations of this indole exert little or no influence (Hill and Blask 1988). These findings formed the basis of a model system of human breast cancer growth for probing the cellular and molecular mechanisms of melatonin's oncostatic action.

Melatonin inhibits the proliferation not only of monolayer cultures of MCF-7 cells but also of cells cultured in an anchorage-independent system (Cos and Blask 1990). However, the bell-shaped dose-response curve characteristic of melatonin action on monolayer cultures changes to a linear dose-response relationship in soft agar culture. Additionally, as the amount of FBS is reduced in the medium, MCF-7 cells lose their sensitivity to melatonin until they are totally refractory in serum-free, chemically defined medium (Blask and Hill 1986a; Hill 1986; Hill and Blask 1986). In fact, we have found that the MCF-7 cell response to melatonin can dramatically vary from robust to little or no response with different batches of FBS obtained from different or even the same commercial sources; we find the same to be true for tamoxifen as well (unpublished observations). In addition to serum factors, different growth rates of MCF-7 cells, as determined by different initial plating densities, have a profound impact on the responsiveness of these cells to melatonin. For example, melatonin is more effective in inhibiting the growth of fast-growing (short doubling time) MCF-7 cells (high initial plating density) than slower growing cells (long doubling time) initially plated at low density (Cos and Sanchez-Barcelo 1995). These findings emphasize the importance of culture conditions, particularly the growth rates as determined by serum components and plating density, in setting the level of responsiveness of MCF-7 cells to melatonin inhibition.

In response to continuous exposure to melatonin, MCF-7 cells also exhibit characteristic ultrastructural changes which include reduced numbers of surface microvilli, cytoplasmic and ribosomal shedding, disruption of mitochondrial cristae, vesiculation of smooth endoplasmic reticulum, and increased numbers of autophagic vacuoles (Hill and Blask 1988). This morphological profile suggests that these cells undergo transition into a more differentiated state in the presence of melatonin (Crespo et al. 1994). Melatonin exerts its differentiating antiproliferative effect by delaying the progression of cells from the G_0/G_1 to the S phase of the cell cycle (Cos et al. 1991). Other studies using partially synchronized MCF-7 cells suggest that they are most susceptible to inhibition by melatonin during the S phase (Cos et al. 1996). Furthermore, the application of melatonin to MCF-7 cells in 12-hour pulses rather than continuously is also an effective regimen for inhibiting cell growth (Cos and Sanchez-Barcelo 1994). In fact, this circadian-like presentation of physiological levels of melatonin to MCF-7 cells is more effective in inhibiting cell growth than continuous exposure of the cells to this indole.

The fact that melatonin inhibits other ER+ human breast cancer cell lines and has little or no effect on ER– human breast cancer cells indicates that the estrogen response pathway is crucial for melatonin's oncostatic action (Hill et al. 1992). Melatonin's ability to block estradiol-stimulated MCF-7 cell growth as well as the estradiol-induced rescue of tamoxifen-inhibited cells further attests to the importance of the estrogen response system in melatonin's oncostatic action in vitro (Cos et al. 1991; Hill et al. 1992). Interestingly, estradiol is also capable of rescuing MCF-7 cells from melatonin-induced growth arrest (Cos et al. 1991) and essentially reverses the differentiating effects melatonin has on the morphology of these cells (Crespo et al. 1994).

The hormonal regulation of breast cancer growth involves not only estrogen but encompasses a complex array of other estrogen-inducible endocrine, autocrine, and paracrine substances including prolactin, epidermal growth factor (transforming growth factor α), insulin-like growth factor I (IGF-1) and cathepsin D (Freiss et al. 1993). In serum-free, chemically-defined medium, melatonin inhibits the release of several proteins synthesized by MCF-7 cells, including an estradiol-inducible 52 kD glycoprotein corresponding to the autocrine growth stimulator cathepsin-D (Hill 1986; Blask and Hill 1988). Additionally, melatonin increases the levels of inhibitory growth factor activity (i.e., transforming growth factor β), in medium conditioned by MCF-7 cells (Cos and Blask 1994) and the expression of transforming growth factor β

mRNA (Molis et al. 1995). Furthermore, melatonin inhibits the capacity of estradiol-induced growth factor activity as well as prolactin (Lemus-Wilson et al. 1995) and epidermal growth factor to stimulate MCF-7 cell growth (Cos and Blask 1994). Taken together, these results indicate that melatonin exerts its oncostatic effects in vitro by modulating the secretion and/or action of estradiol-inducible endocrine, paracrine, or autocrine growth factors.

Another aspect of melatonin's action on human breast cancer cell growth deserves a comment, since it has important implications for the treatment of breast cancer. Although melatonin appears to have important interactions with endogenous factors germane to the regulation of breast cancer growth, melatonin also modulates the oncostatic activity of exogenous factors such as cytostatic and cytotoxic chemo-therapeutic agents in vitro (Blask et al. 1993). For example, the preincubation of MCF-7 cells with a physiological concentration of melatonin followed by exposure of the cells to the cytostatic antiestrogen, tamoxifen, increases the potency of the subsequent growth inhibitory effects of this agent (Wilson and Blask 1992). This suggests that melatonin, through unknown mechanisms, increases the sensitivity of MCF-7 cells to the inhibitory effects of the most important compound used in the hormonal treat-ment of ER+ breast cancer (Jordan and Murphy 1991). Moreover, when melatonin and tamoxifen are used in combination in cells exposed to the growth-stimulating effects of estradiol, cell growth is inhibited to a greater extent than that achieved with either agent alone (Cos et al. 1991).

Paradoxically, in the absence of exogenously added estradiol, the combination of melatonin and tamoxifen is no more effective than either agent alone (Hill et al. 1992). However, 1 nM melatonin in conjunction with a suboptimal growth-inhibiting dose of a chemotherapeutic regimen consisting of cyclophosphamide, methotrexate, and 5-fluorouracil (CMF) to treat MCF-7 cells exhibiting decreased sensitivity to melatonin is much more effective than either melatonin or CMF alone in suppressing cell pro-liferation (Blask et al. 1993). Interestingly, it was recently reported that 5-fluorouracil alone attenuates the oncostatic action of physiological melatonin on MCF-7 cell growth (Furuya et al. 1994).

Melatonin's direct inhibitory effect on MCF-7 cell growth in culture has been repro-duced by a number of other independent laboratories (Shellard et al. 1989; de Launoit et al. 1990; L'Hermite-Baleriaux et al. 1990; Bartsch et al. 1992; Liburdy et al. 1993; Cos and Sanchez-Barcelo 1994, 1995; Crespo et al. 1994; Furuya et al. 1994; Blackman et al. 1996; Papazisis et al. 1998). While the majority of these investigators report an inhibi-tory effect of physiological melatonin, the remaining authors detect a cytotoxic or cytostatic effect of melatonin only at pharmacological levels (Shellard et al. 1989; Bartsch et al. 1992; Papazisis et al. 1998). Nevertheless, one of these same groups (L'Hermite-Baleriaux and de Launoit 1992) has questioned the direct oncostatic effects of physiological concentrations of melatonin on MCF-7 cell growth. In spite of their two earlier reports (de Launoit et al. 1990; L'Hermite-Baleriaux et al. 1990), this group and more recently Panzer et al. (1998) failed to observe an inhibitory effect of mela-tonin on the growth of MCF-7 cells as well as other ER+ human breast cancer cell lines (T47D and ZR75-1) over all concentrations tested from 10^{-13} M to 10^{-6} M in RPMI 1640 medium supplemented with either charcoal-stripped FBS (CS-FBS) or 10% FBS in the absence or presence of 0.3 nM estradiol (L'Hermite-Baleriaux and de Launoit 1992). However, the culture conditions they used differed in several respects from those under which we typically see inhibition with melatonin on MCF-7 cell growth (Hill and

Blask 1988). Not only is our initial plating density 6 times higher than theirs, but the inhibitory effect of melatonin begins to disappear at lower densities as well (Hill 1986; Cos and Sanchez-Barcelo 1995). Furthermore, they incubated their cells with melatonin for 12 days, whereas we incubate our cells for only 5 to 7 days, since MCF-7 cells escape from melatonin inhibition after this time period (Blask, unpublished results). In our experiments, we typically use Dulbecco's modified Eagle's medium (DMEM), whereas this group used RPMI 1640 medium.

It is not surprising that they also failed to observe melatonin inhibition of MCF-7 cell growth in either 1% FBS or 10% CS-FBS, since we originally reported that melatonin loses its oncostatic activity in either CS-FBS or as the percentage of FBS in the medium decreases, indicating the serum dependence of this response (Hill 1986; Hill and Blask 1986; Blask and Hill 1986a). Another confounding factor is that these workers used MCF-7 cells which were apparently hyper-responsive to estradiol, since they achieved a 20-fold increase in MCF-7 cell growth with estradiol in CS-FBS and a twofold increase in the presence of 10% FBS in which cell growth is already maximally stimulated. This suggests that they may have inadvertently tested a clone which was already biased towards growth stimulation by endogenous estrogens in the FBS rather than growth inhibition by melatonin. In this regard, it has been demonstrated that MCF-7 cells from different laboratories vary widely in their basal and estradiol-stimulated growth rates, responses to inhibition by the antiestrogen tamoxifen, and their ER and progesterone receptor content (PgR) (Osborne et al. 1987). There is no reason to believe that similar variations could not also exist among MCF-7 cells from different laboratories with respect to their sensitivity to melatonin.

One additional factor that these investigators did not take into account is the intensity of the magnetic fields in their incubator. For example, magnetic fields present in cell culture incubators can block the ability of MCF-7 cells to respond to the inhibitory effects of melatonin in vitro (Blask et al. 1993a, b; Liburdy et al. 1993). Therefore, it is possible that magnetic fields in their incubator may have been of sufficient intensity to prevent the oncostatic effects of melatonin.

15.2.1.3 Melatonin Analogs

Neither 6-hydroxymelatonin, serotonin, N-acetylserotonin, 5-methoxytryptophol, nor 5-methoxytryptamine exhibit any oncostatic activity in the MCF-7 monolayer culture system (Blask and Hill 1986a). However, the halogenated melatonin analogue, 6-chloromelatonin (1 nM), is just as potent as native melatonin in inhibiting MCF-7 cell growth (Blask and Hill 1986b). Using a soft agar clonogenic growth inhibition assay, Shellard et al. (1989) evaluated the effectiveness of melatonin, N-acetylserotonin, 5-methoxytryptamine, 6-hydroxymelatonin, tryptamine hydrochloride, and 5,6-dihyroxytryptamine on MCF-7 cell growth over a 3-day period. They found that it required anywhere from 23 µg/ml to 400 µg/ml of these compounds to elicit a 50% inhibition of cell growth as compared with controls (IG_{50}). Interestingly, in their test system, melatonin ($IG_{50} = 400$ µg/ml) and 6-hydroxymelatonin ($IG_{50} = 305$ µg/ml) were the least potent in inhibiting cell growth, whereas 5,6-dihydroxytryptamine was the most potent ($IG_{50} = 23$ µg/ml). On the other hand, tryptamine hydrochloride, 5-methoxytryptamine and N-acetylserotonin were all 2 to 4 times more potent than melatonin in inhibiting MCF-7 cell growth.

15.2.2 Melatonin and Melanoma

15.2.2.1 Animal Studies

The inhibitory effects of melatonin on melanoma growth were initially examined in Syrian hamsters with transplantable hormone-responsive hamster melanomas (MM1) (El-Domieri et al. 1973; Ghosh et al. 1973; El-Domieri et al. 1976). While either 100 µg or 4 mg of melatonin per day for 2 to 3 weeks had no effect on melanoma growth in intact animals, melatonin inhibited pinealectomy-induced growth and metastases of these tumors. Subsequent studies by these same investigators (Stanberry et al. 1983) showed that either morning or late afternoon injections of melatonin (5–50 µg/day) for 5 weeks had a stimulatory effect on the growth of MM1 melanomas in long-day exposed hamsters. Interestingly, constant release Silastic implants of melatonin (14–18 µg/day) stimulated tumor growth in long photoperiods, while they were inhibitory in animals exposed to short days. In athymic nude mice, melatonin (5 µg/g BW) supplied in the drinking water suppressed transplantable B16 mouse melanoma growth following almost 6 weeks of treatment (Narita and Kudo 1985).

15.2.2.2 Cell Culture Studies

One of the earliest studies on the effects of melatonin on cancer cell growth in vitro demonstrated that millimolar concentrations of melatonin inhibited hamster melanoma cell (B7) growth, while micromolar levels actually stimulated growth (Walker et al. 1978). More recently using a different hamster melanoma cell line, Slominski and Pruski (1993) found that the growth of AbC1 hamster amelanotic melanoma cells was completely blocked after only 3 days of exposure to as little as 100 pM of melatonin. In fact, they observed a bell-shaped dose-response curve of cell growth to the inhibitory action of melatonin similar to what we reported for MCF-7 human breast cancer cells (Hill and Blask 1988). They also reported that physiological levels of melatonin (100 pM) inhibited the growth of the mouse melanoma cell line S91 by 50% after only 24 hours of incubation (Slominski and Pruski 1993). In both the hamster AbC1 and mouse S91 melanoma cell lines, 1.0 nM of melatonin was also quite effective in suppressing cell growth. These findings were corroborated in murine melanoma cells (M2R), which were shown to be inhibited by melatonin (10 nM) over a 5-day culture period (Zisapel and Bubis 1994). Additional dose-response studies revealed that the inhibitory effects of melatonin on M2R cell growth increased linearly with the dose (10^{-13} M to 10^{-3} M), with maximum growth suppression being achieved with micro- and millimolar concentrations; however, these oncostatic effects of melatonin are apparently not cell cycle-specific. Although melatonin also inhibited the constitutive secretion of proteins from these cells in a dose-response manner, melatonin failed to affect the growth enhancing action of conditioned medium on melanoma cell growth. These results suggest that melatonin's antiproliferative effects may involve an inhibition of the secretion but not the mitogenic action of autocrine growth regulatory factors of this cell line (Zisapel and Bubis 1994).

The clonogenic growth of human melanoma cells derived from biopsy specimens is inhibited by melatonin in a linear dose-response manner (10^{-15} M to 10^{-5} M) in the majority of tumors, while colony formation is either stimulated or unaffected by melatonin in the remaining cases (Meyskens and Salmon 1981). Human melanoma cells in monolayer culture also exhibit interesting biphasic responses to different doses of melatonin, depending on whether the cells are slow-growing early passage or fast-growing late passage cells. Millimolar concentrations of melatonin stimulated the growth of slow-growing human melanoma cells, while micromolar levels inhibit growth; however, doses in the physiological range are without effect. In contrast to slow-growing cells, millimolar concentrations of melatonin inhibited fast-growing human melanoma cells while lower concentrations had no effect (Bartsch et al. 1986). An interesting study by Hu and Roberts (1997) demonstrated that melatonin, at concentrations comparable to the concentration of melatonin in the aqueous humor of human eyes (100 pM to 100 nM), inhibited the proliferation of human uveal melanoma cells in vitro.

15.2.2.3 Melatonin Analogues

A variety of indolamines have been compared with melatonin at pharmacological doses for their ability to inhibit the uptake of $[^3H]$-thymidine into B16 mouse melanoma cells following 2 days of incubation (Sze et al. 1993). Melatonin itself weakly inhibited thymidine uptake at all concentrations tested, while serotonin and 5-methoxytryptophol demonstrated modest inhibition of thymidine uptake in a dose-response manner. At all but the lowest dose, 5-methoxytryptamine exhibited almost 100% inhibition of thymidine uptake by these cells. Both 5-methoxytryptamine and 5-methoxytryptophol, at millimolar concentrations, were half as potent as melatonin in suppressing human melanoma cell proliferation after 6 days of culture, while neither serotonin nor N-acetylserotonin had any effect (Bartsch et al. 1987). In another human malignant melanoma cell line (M6), 6-chloromelatonin was much more potent than melatonin in suppressing cell growth as evidenced by an IC_{50} of 5 nM for 6-chloromelatonin and an IC_{50} of 30 nM for melatonin. Maximal inhibition of cell proliferation (65% to 85%) by these compounds was achieved at approximately 10 µM (Ying et al. 1993). In S91 hamster melanoma cells, a pharmacological concentration of 5-methoxytryptamine (0.1 µM) was effective in totally blocking cell growth after 3 days of culture, whereas neither physiological nor subphysiological levels of this compound had any effect. In the case of N-acetylserotonin, neither physiological nor subphysiological concentrations had any effect on cell growth, whereas pharmacological levels actually induced a fourfold stimulation of growth over controls (Slominski and Pruski 1993).

15.2.3 Melatonin and Sarcoma

15.2.3.1 Animal Studies

Daily melatonin injections (50 µg/day) following the induction of fibrosarcomas in DBF1 mice with the chemical carcinogen methylcholanthrene (MCA) result in a suppression of tumorigenesis during the early stages (Lapin and Ebels 1976). In the first investigation to examine a possible circadian stage-dependent response of tumor growth to melatonin, Bartsch and Bartsch (1981) demonstrated that transplantable MCA-induced fibrosarcomas in Swiss mice exhibit a diurnal rhythm of sensitivity to melatonin (100 µg/day) injected at different times of the day in animals maintained on photoperiods of either 13L:11D or 12L:12D. Whereas morning intraperitoneal injections of melatonin stimulated tumorigenesis, afternoon injections were oncostatic, while subcutaneous injections were ineffective at any time.

These results were corroborated and extended in BALB/c mice bearing transplantable Meth-A sarcomas maintained on 12L:12D and treated with either melatonin (1 mg/kg BW) or interleukin-2 (IL-2) (4800 U) alone or in combination (Sanchez de la Pena et al. 1992). Either melatonin or IL-2 administered alone 18 hours after lights on stimulated tumor growth while both compounds inhibited growth when given during the end of the light phase (10 hours after lights on). Interestingly, when given in combination with IL-2, melatonin severely blunted the oncogenic effect of IL-2 given during the dark phase, while it potentiated the oncostatic effect of this cytokine given just before lights off.

Taken together with the results we obtained with NMU-induced mammary cancers (Blask et al. 1992), these findings suggest that melatonin can be either oncogenic or oncostatic in vivo, depending on the circadian stage during which it is administered (Blask 1993a, 1994).

When BALB/c mice are transplanted with leukemia cells (LSTRA), melatonin (100 µg/day) injections over a 2-week period are leukemostatic (Buswell 1973). However, when melatonin (4 mg/kg BW) is injected in the afternoon into C57BL/6 mice with leukemia induced by the intrathymic injection of radiation leukemia virus (RadLV), the leukemogenic process is accelerated, as evidenced by the decreased survival time of these animals. This leukemogenic effect of melatonin was blocked by naltrexone, indicating that it is mediated via the melatonin-induced release of endogenous opioids (Conti and Maestroni 1994).

15.2.3.2 Cell Culture Studies

The growth of P388 human leukemia cells cultured in DMEM supplemented with 10% FBS for 6 days was found to be unresponsive to melatonin (10^{-6} M or 10^{-4} M) in vitro (Leone 1991).

15.2.3.3 Melatonin Analogues

5-Methoxytryptophol was found to be weakly inhibitory to P388 human leukemia cells at 10^{-4} M, while 6-hydroxymelatonin was inactive at any concentration tested. On the other hand, 6-methoxyharmalan inhibited cell growth by 80% at 10^{-4} M and 54% at 10^{-6} M (Leone 1991). In the mouse sarcoma cell line (S180), 5-methoxytryptamine was the most efficacious of any melatonin analogue tested in causing almost complete inhibition of [^{3}H]-thymidine uptake at concentrations ranging from 20 µM to 500 µM (Sze et al. 1993). At the highest dose, melatonin, 5-methoxytryptophol, 5-methoxyindoleacetic acid, and 5-hydroxytryptophol were all effective inhibitors of [^{3}H]-thymidine uptake, whereas neither serotonin nor 5-hydroxyindoleacetic acid had an effect at any concentration tested.

15.2.4 Melatonin and Other Tumors

15.2.4.1 Animal Studies

Studies on the effects of melatonin on transplantable prostate adenocarcinoma (Dunning R3327) in male Copenhagen-Fischer rats that are either intact or olfactoribulbectomized have revealed that daily, late afternoon injections of melatonin at doses of 25–200 µg/day decrease the growth of androgen-sensitive and androgen-insensitive tumors. However, not all androgen-sensitive variants of this tumor respond to melatonin, and in some cases melatonin stimulates the growth of some androgen-insensitive variants (Toma et al. 1987; Buzzell 1988; Buzzell et al. 1988; Philo and Berkowitz 1988).

Pituitary tumors (prolactinomas) induced with diethylstilbestrol (DES) in either intact or olfactoribulbectomized female Fischer 344 rats are growth-inhibited by either daily afternoon melatonin injections (50 µg/day) or melatonin (1 mg) continuously provided by beeswax implants (Leadem and Burns 1987). In another transplantable pituitary tumor model (mT/F4), tumors demonstrated a diurnal rhythm of sensitivity to daily melatonin injections (50 µg/day), with morning injections having no effect while afternoon injections inhibited tumor growth (Chaterjee and Banerji 1989).

Conflicting results have been obtained with respect to melatonin's influence on the growth of a transplantable Lewis lung carcinoma in mice. For example, in one study daily melatonin injections (3.5 mg/kg BW) actually stimulated tumor growth, as evidenced by a decrease in survival time (Lapin and Ebels 1976) while, in another study, melatonin (1.25 mg/kg BW) administered in the drinking water decreased the size of the primary tumor, metastases to the lungs, and the number of animals with tumors (Perissin et al. 1994). In this tumor model, perhaps the route of melatonin administration was critical in determining the outcome of experiments.

Recently, Anisimov et al. (1997) demonstrated that melatonin administered in the drinking water (20 mg/L) inhibited colon carcinogenesis induced by 1,2-dimethyl-hydrazine (DMH) in female rats as manifested by a decrease in the incidence, multiplicity, invasiveness, and size of colon adenocarcinomas and by an increase in their degree of differentiation. These investigators postulated that melatonin suppressed DMH-induced tumors by virtue of its antioxidant properties. The growth of transplantable Colon 38 colon adenocarcinoma (derived from DMH colon adenocarcinoma) in mice was also inhibited by daily late afternoon injections of melatonin (10–100 µg/day) as evidenced by decreased labeling of tumor cells with bromodeoxyuridine (BrDU) (Karasek et al. 1998). With respect to metastatic colon adenocarcinoma, twice daily injections (A.M. and P.M.) of melatonin (250 µg or 50 mg/kg BW), either alone or in combination with IL-2 (20,000 U/injection), have been shown to reduce the development and growth of colon adenocarcinoma cells (CL26) that metastasize to the lungs of female BALB/c mice (Maestroni and Conti 1993). These results indicate not only that melatonin and IL-2 alone are equally effective in decreasing lung colonies but that their individual oncostatic effects are additive when they are administered in combination.

In rats, transplantable Guerin uterine tumors have been shown to respond to daily, afternoon melatonin injections (25 µg) with a decrease in mitotic activity (Lewinski et al. 1990). Similarly, the growth of a transplantable Kirkman-Robbins hamster hepatoma is inhibited by melatonin under the same experimental conditions (Karasek et al. 1992).

The effects of melatonin on the growth of Yoshida tumors (an undifferentiated tumor) in rats depended on whether the animals were pinealectomized or intact

(Lapin and Ebels 1976). Melatonin (6 mg/kg or 9 mg/kg BW) injected in the afternoon daily for 3 to 7 weeks had no effect on tumor growth in intact animals, whereas it was oncostatic in pinealectomized rats, suggesting that the presence or absence of the endogenous melatonin rhythm somehow modulates the responsiveness of these tumors to exogenously administered melatonin.

In an interesting circadian study, melatonin injections (100 µg/day) had differing effects on the growth of Ehrlich's solid tumors (an undifferentiated tumor) in Swiss mice, depending on photoperiod and timing of the injection during the day (Bartsch and Bartsch 1981). In animals maintained on a long photoperiod (13L:11D), mela-tonin treatment in the morning stimulated tumor growth, while later afternoon melatonin was oncostatic. However, when this indole was given at mid-afternoon, no effect on tumor growth was observed. In mice maintained on a short photoperiod (8L:16D), melatonin was oncostatic irrespective of whether it was administered during the morning or late afternoon hours, indicating that short photoperiod exposure, perhaps as a result of a longer duration of nocturnal melatonin secretion, obliterated the diurnal variation in tumor sensitivity to melatonin present in long day animals.

15.2.4.2 Cell Culture Studies

In spite of its inhibitory action on breast cancer, melanoma, and sarcoma cells, mela-tonin has no effect on the growth in vitro of a variety of neoplastic cell types including HeLa human cervical carcinoma (Fitzgerald and Veal 1976), KB human epidermoid carcinoma cells (Bindoni et al. 1976), DLD-1 human colon carcinoma cells (Leone 1991), and Kirkman-Robbins hamster hepatoma cells (Karasek et al. 1992). However, Karasek et al. (1998) did observe a substantial inhibitory effect of melatonin (10^{-9} M to 10^{-5} M) on [^{3}H]-thymidine incorporation into DNA of Colon 38 cancer cells in vitro. Pharmacological doses of melatonin (300–500 µg/ml) have been reported to elicit a 50% inhibition of the growth of SK-OV-3 and JA-1 human ovarian carcinoma cells and RT112 human transitional bladder carcinoma cells in culture (Shellard et al. 1989).

Interestingly, DES-induced rat pituitary tumor cell proliferation is inhibited by physiological concentrations of melatonin, whereas sub- or supraphysiological con-centrations have little or no effect (Karasek et al. 1988), similar to what we reported for MCF-7 cells (Blask and Hill 1986a; Hill and Blask 1988). Even more surprising is the biphasic response to melatonin exhibited by NIE-115 murine neuroblastoma cells (Benitez-King et al. 1994): during the first 3 days of logarithmic growth, melatonin (10^{13}-10^{-5} M) stimulates cell growth in a bell-shaped dose-response manner, with physiological levels (10^{-9} M) of melatonin being the most effective; however, during the last 3 days of the plateau phase, these concentrations of melatonin reduce cell number in a bell-shaped dose-response manner, again with physiological levels of melatonin being the most effective.

15.2.4.3 Melatonin Analogues

In the first published study of the effects of melatonin analogues on tumor growth in vivo, Lapin and Frowein (1981) reported that N-acetylserotonin (9 mg/kg BW) inject-ed daily in the late afternoon for 3 weeks caused a 50% reduction in the volume of

Yoshida tumors. However, serotonin (15 mg/kg BW) administered only during the morning hours was the most effective inhibitor of tumor growth in both intact and pinealectomized animals. More recently, it was demonstrated that the melatonin analogue CGP 52608, a member of a novel class of thiazolidinediones, inhibited Colon 38 colon adenocarcinoma both in vivo (10 µg and 100 µg/day) and in vitro (10^{-9} M to 10^{-7} M) (Karasek et al. 1998). This compound presumably specifically activates the nuclear receptor RZR/RORα (Wiesenberg et al. 1995) suggesting that CGP 52608, and perhaps even melatonin may inhibit the proliferation of Colon 38 adenocarcinoma via a nuclear receptor-mediated mechanism (Karasek et al. 1998).

In vitro experiments using a clonogenic colony-forming growth inhibition assay demonstrated that N-acetylserotonin was the least potent inhibitor of either JA human ovarian carcinoma cells (IG_{50} = 371 µg/ml) or RT112 human transitional bladder carcinoma cells (IG_{50} = 525 µg/ml), whereas 5,6-dihydroxytryptamine was the most potent oncostatic agent with IG_{50}s of 18 and 4 µg/ml, respectively (Shellard et al. 1989). Melatonin analogues 5-methoxytryptamine, 6-hydroxymelatonin, and tryptamine hydrochloride had intermediate potencies for inhibition of JA cell growth, with IG_{50}s of approximately 100 µg/ml. These same compounds were even more effective in inhibiting RT112 cell growth and exhibited IG_{50}s in the range of 24–39 µg/ml. Like melatonin, none of the analogues tested had an effect on the growth of DLD-1 human colon carcinoma cells in vitro (Leone 1991). Cell growth inhibition and survival assays of SK-OV-3 human ovarian carcinoma demonstrated basically the same rank order of potencies for these melatonin analogues as well; however, as in the case of JA cells, melatonin was more potent than N-acetylserotonin. Neither the 6-sulfatoxy nor 6-glucuronide conjugates of melatonin exerted any modulatory effect on the growth of these cell lines (Shellard et al., 1989). In the JAr human choriocarcinoma cell line, melatonin and a variety of its analogues all exerted a dose-dependent inhibition of [^{3}H]-thymidine uptake, with 5-hydroxytryptophol being most effective at the lower concentrations (Sze et al. 1993).

15.3 Melatonin Binding in Neoplastic Tissues

15.3.1 Mouse Melanoma

Melatonin binding sites were first reported by Stankov et al. (1989) in membrane preparations of transplantable B16 mouse melanoma using 2-[^{125}I]iodomelatonin (2-IMel) as the labeled ligand. Binding was described as being stable, saturable, specific, and of high affinity (Kd = 200 pM) and low capacity (Bmax = 7.5 fmol/mg protein). Similarly, Helton and Kane (1991) demonstrated saturable high affinity (Kd = 216 pM) and low capacity (53.2 fmol/106 cells) 2-IMel binding at 4 °C in cultured B16 mouse I melanoma cells. By characterizing the subcellular distribution of 2-IMel in these cells, these same investigators (Helton and Kane 1992) found that 96% of 2-Imel localized to the soluble compartment, while 3.5% and 0.5% of 2IMel was found in nonnuclear membranes and nuclear fraction, respectively. Additional experiments revealed that 2-IMel was neither bound to a high molecular weight moiety nor taken up by bulk diffusion. Melatonin analogues more closely related to melatonin induced the greatest inhibition of 2-IMel uptake, suggesting a specific uptake system for melatonin in these cells.

15.3.2 Hamster Melanoma

Studies by Pickering and Niles (1992) characterized 2-IMel binding in an epithelial-like melanotic melanoma cell line (RPMI 1846) originally derived from a spontaneous malignant melanoma from an aged Syrian hamster; this cell line apparently retains its ability to produce melanin in vitro. Therefore, when crude membrane preparations rather than whole cell homogenates were used in the binding assays, specific 2-IMel binding increased from 20 % to approximately 50 %, presumably due to the elimination of melanin. Scatchard analysis of saturable 2-IMel membrane binding at 0 °C revealed the relative lower affinity (as compared with mouse melanoma) of a single class of binding sites (Kd = 0.89 nM) with a Bmax of 6.2 fmol/mg protein. The fact that neither guanosine triphosphate (GTP) nor Na^+ altered 2-IMel binding in these preparations verified the low-affinity nature of these binding sites. Furthermore, these sites are not coupled to adenylate cyclase and do not affect cAMP levels. However, another study suggests that 2-IMel binding sites are linked to the hydrolysis of phosphoinositides in these cells (Eison and Mullins 1993).

15.3.3 Human Melanoma

The first preliminary report of 2-IMel binding to intact human melanoma cells was by Helton and Kane (1991), who detected specific binding in two human melanoma cell lines, SK Mel 28 and 30. Ying et al. (1993) examined the binding of 2-IMel to crude membranes of the human malignant melanoma cell line. Interestingly, the membranes of these cells exhibited two affinity states, one of high affinity (Kd = 290 pM) and the other of low affinity (Kd = 6.53 nM). Competition studies with melatonin analogues revealed the rank order of potency for inhibition of 2-IMel binding indicative of a melatoninergic process. Guanosine triphosphate reduced the high affinity binding sites to a low affinity state, indicating that they are linked to a G protein typical of high affinity binding sites in brain tissue.

15.3.4 Murine Mammary Cancer

In preliminary studies (Blask et al. 1990; Burns et al. 1990), we originally described 2-IMel binding sites in mammary cancers generated by the chemical carcinogen NMU in adult female Sprague Dawley rats. Crude membrane fractions of NMU mammary tumors demonstrated highly specific 2-IMel binding that increased linearly with increasing tissue concentrations. Nonlinear regression analysis of the saturation isotherm revealed a single class of binding sites with relatively low affinity (Kd = 9.6 nM) and moderate capacity (Bmax = 67.8 fmol/mg protein). In order to assess the selectivity of these binding sites for various melatonin-related compounds, incubations of NMU tumor membranes were performed in the presence of a single concentration of 2-IMel along with various concentrations of indolamines including unlabeled 2-IMel, mela-tonin, serotonin, and N-acetylserotonin. The rank order of potency for competition with 2-IMel was indicative of a melatoninergic process: unlabeled 2-IMel > melatonin > N-acetylserotonin > serotonin.

Although we detected 2-IMel binding sites in these tumors, we still don't know whether these binding sites mediate the oncostatic signal provided by either exogenous or endogenous melatonin in rats with NMU-induced tumors (Blask et al. 1991). However, if these binding sites are in fact involved in the signal transduction mechanisms mediating melatonin's direct oncostatic effects on the growth of such tumors in vivo, then the low-affinity state of these binding sites may explain why pharmacological levels of exogenous melatonin (in the high microgram range) are required to inhibit the growth of carcinogen-induced mammary tumors in rats (Stankov et al. 1991).

As alluded previously, the administration of melatonin either orally or by injection has been shown to inhibit the development of spontaneous adenocarcinomas in female mice (Wrba et al. 1986; Subramanian and Kothari 1991b). Furthermore, melatonin suppresses the normal development of mammary parenchyma in mice as well (Sanchez-Barcelo et al. 1990). While melatonin binding sites have not been described in mouse mammary tumors, 2-IMel binding sites were recently characterized in crude membrane preparations of mammary glands from normal female BALB/c mice (Recio et al. 1994). Saturation studies were performed on mammary gland membranes collected at different times during an alternating 14L:10D cycle. Scatchard analysis of the saturation isotherms revealed a diurnal rhythm in binding affinity and receptor number such that the highest affinity binding (Kd = 1.33 nM) and lowest receptor number (Bmax = 55 fmol/mg protein) occurred during light phase at 1800 hours (4 hours prior to lights off), while the lowest affinity binding (Kd = 3.05 nM) and highest receptor number (Bmax = 267 fmol/mg protein) occurred during the dark at 5:00 A.M. (3 hours prior to lights on). The question remains as to whether the diurnal differences in binding affinity are actually due to different affinity states. The Kd (1.28 nM) derived from the association:dissociation rate constants was consistent with high affinity binding.

A large number of indolaminergic and nonindolaminergic compounds were surveyed in 2-IMel competition experiments. Interestingly, compounds such as hydroxyindoles and dopaminergic agents as well as 6-hydroxymelatonin bound with higher affinity (in the nanomolar range) to 2-IMel binding sites than did either melatonin or 6-chloromelatonin, which bound in the micromolar range. These results certainly argue against these sites being authentic melatonin receptors. Nevertheless, as with NMU mammary tumors in rats, the low affinity of these binding sites might also explain why large doses of melatonin are required to inhibit mammary carcinogenesis in mice (Recio et al. 1994).

15.3.5 Human Breast Cancer

There have been only a few reports on the presence of melatonin binding sites in human breast cancer tissue or cells. Kerenyi et al. (1988) claimed that out of 500 breast cancer specimens removed at mastectomy, 50% were positive for melatonin receptors while the other 50% were negative. The melatonin receptor-positive tumors were restricted to postmenopausal women, while melatonin receptor-negative cancers were limited to women in the perimenopausal years (40–45 years of age). There was no indication as to the steroid receptor status of these tumors. Since the nighttime surge of melatonin decreases with age, particularly after age 60 (Waldhauser and Waldhauser 1988), it is tempting to speculate that the expression of melatonin binding in human breast cancer tissue is related to this phenomenon. Virtually all benign breast neoplasms were positive for melatonin receptors. These investigators stated that melatonin receptors were detected using a technique, unfamiliar to this author, of immunoperoxidase staining following melatonin saturation. Since no other details of this method were given, it is difficult to evaluate whether melatonin receptors were actually evaluated in this study.

Stankov and associates (1991) examined 2-IMel binding in the membranes of tumors from eight breast cancer patients. Two of the breast cancer specimens that were ER+ were also positive for 2-IMel binding, while two ER– tumors were negative for 2-IMel binding. These results are interesting in light of the fact that the nocturnal elevation of melatonin in patients with ER+ breast cancer is blunted whereas, in healthy age-matched controls or in patients with ER– breast cancer, melatonin levels at night were normal (Tamarkin et al. 1982). Perhaps there is some important relationship between the expression of 2-IMel binding in breast cancer in ER+ versus ER– tumors and altered secretory patterns of melatonin in breast cancer patients. Half of the remaining tumors, in which steroid receptor status was not assessed, were positive for 2-IMel binding while the other half were negative. The ER+ human breast cancer cell line MCF-7 was weakly positive for 2-IMel binding in both membrane and nuclear fractions (Blask et al. 1990; Stankov et al. 1991).

15.3.6 Human Benign Prostatic Hyperplasia

Stankov and his associates (1991) found the most consistent 2-IMel binding in prostatic tissue obtained from patients with benign prostatic hypertrophy. However, when saturation studies were performed on crude membrane preparations of this tissue, saturation was not achieved when concentrations of 2-IMel as high as 250 pM were used. Cold saturation experiments with unlabeled 2-IMel revealed a low affinity binding site (Kd = 300 nM) and a Bmax of approximately 900 fmol/mg protein. In competition experiments, only unlabeled 2-IMel competed effectively for 2-IMel binding, whereas melatonin and 6-chloromelatonin demonstrated low affinity in the micromolar range. Other compounds such as serotonin and N-acetylserotonin failed to compete for binding. Men with benign prostatic hypertrophy have a completely normal melatonin rhythm, whereas those with clinically evident prostate cancer have no detectable nocturnal melatonin rhythm (Bartsch et al. 1985).

In contrast, Laudon et al. (1996) and Gilad et al. (1996) recently found specific binding sites for 2-IMel in human benign prostatic tissue and cultured epithelial cells. The binding affinity (Kd = 120 pM) was inhibited by GTP analogues, which suggested coupling of these binding sites to a G protein. Incubation of cells with physiological concentrations of melatonin resulted in a transient inhibition of DNA synthesis, protein synthesis, and cell viability. These investigators also reported that protein kinase C (PKC) desensitizes melatonin receptor binding in these cells, suggesting that the transient inhibitory effect of melatonin on cell proliferation may be the result of melatonin activation of endogenous PKC (Gilad et al. 1997).

15.3.7 Other Immortalized Cell Lines

A number of microglial and monocytic immortalized cell lines have been surveyed for their ability to bind 2-IMel (Stankov et al. 1991). The monocytic cell lines M2/1 and MT2C11 demonstrated 2-IMel binding, whereas cell line MS1C14 was negative. Of the three microglial cell lines N3, N11, and N13, only crude membranes from N13 bound 2-IMel. Cold saturation experiments revealed a single class of binding sites with a Kd = 8.4 nM and a Bmax = 80 fmol/mg protein. Only unlabeled 2-IMel competed for 2-IMel binding in the nanomolar range, whereas melatonin and 6-chloromelatonin competed in the micromolar range; serotonin, N-acetylserotonin, and catecholamines were ineffective.

15.4 Potential Melatonin Signal Transduction Mechanisms in Neoplastic Cells

15.4.1 Cyclic AMP and G Proteins

In human melanoma cells, although melatonin failed to affect basal cAMP levels, it did significantly inhibit forskolin-stimulated cAMP formation at concentrations that also inhibited cell growth in culture (Ying et al. 1993). Since 2-IMel binding in these cells is apparently linked to a G protein, melatonin's inhibitory effect on the proliferation of the M-6 human melanoma cell line may be mediated through a suppression of cAMP production via inhibition of a stimulatory G protein.

Although there is no evidence for cAMP mediation of melatonin's oncostatic effect on mammary cancer, this indolamine and related methoxyindoles suppress cAMP and stimulate cGMP in normal murine mammary glands. Moreover, this effect is maximal near the end of the light phase of an alternating light:dark cycle (Cardinali et al. 1992). Whether melatonin's ability to suppress the development of normal mammary gland tissue (Sanchez-Barcelo 1990) as well as spontaneous mammary cancer in female mice (Subramanian and Kothari 1991b) involves cAMP-related and/or cGMP-related signal transduction mechanisms is unknown.

15.4.2 Phosphoinositide Metabolism

The generation of phosphoinositides as second messenger molecules has an impact on numerous aspects of cellular functions, including proliferation (Stoclet et al. 1987). Recent evidence indicates that, in RPMI hamster melanoma cells, the hydrolysis of phosphoinositides is functionally linked to lower affinity (nanomolar range) 2-IMel binding (Eison and Mullins 1993). Melatonin and several melatonin analogues stimulated phosphoinositide hydrolysis in dose-dependent and saturable manner. The rank order of potency for these compounds was as follows: *N*-acetylserotonin > unlabeled 2-IMel > 6-chloromelatonin > melatonin. The EC_{50}s of these compounds for the induction of phosphoinositide hydrolysis ranged from 290 nM to 450 nM, which is consistent with the lower affinity of 2-IMel binding sites in these cells. However, major questions remain as to whether melatonin actually affects the proliferation of this cell line and, if so, whether melatonin binding sites mediate such proliferative responses via phosphoinositide metabolism.

15.4.3 Nonmelatonin Receptor Expression

Evidence for melatonin regulation of heterologous receptor systems in neoplastic cells continues to emerge. Direct effects of melatonin on estrogen binding in human breast cancer cells (MCF-7) were first documented by Danforth et al. (1983). They reported that melatonin at physiological concentrations augmented both cytoplasmic and nuclear estrogen binding activity within 40 minutes of exposure, an increase that lasted for 5 hours. Overall, the increase in receptor number in melatonin-treated cells was around 80% over controls, while the binding affinity of the receptor for estradiol was unchanged.

More recently, Molis and associates (1995) performed extensive studies on melatonin's ability to modulate ER expression in MCF-7 cells. Contrary to the work of Danforth and colleagues (1983), these investigators found that a melatonin-induced, progressive reduction in estrogen binding capacity occurred from 6 to 48 hours of continuous incubation of MCF-7 cells in steroid-depleted medium, with the greatest decrease (90%) occurring after 24 hours of incubation; binding affinity was unaffected. Neither melatonin nor one of its analogues, 2-bromomelatonin, directly competed with estradiol for binding to the ER. Furthermore, melatonin was unable to inhibit the transcriptional regulatory activity of completely activated ER in a yeast expression assay, providing additional evidence that it does not bind to the ER. Melatonin also caused dose- and time-dependent reductions in of immunoreactive ER levels, with maximal suppression occurring with 1 nM melatonin following 12 hours of incubation. Surprisingly, pharmacological concentrations of melatonin caused a 50% increase in immunoreactive ER levels following 48-hour incubation. The failure of melatonin to bind directly to the ER indicated that its ability to regulate ER expression is exerted via indirect mechanisms, perhaps through the induction of a new gene product. Even though mammary tumor ER expression is unaltered by melatonin in rats with NMU-induced tumors (Blask et al. 1991a), immunocytochemical evidence indicates that melatonin inhibits the increase in epidermal growth factor receptor and c-erb2 oncoprotein expression that occurs during DMBA-induced mammary tumorigenesis in female rats (Scaglione et al. 1992).

15.4.4 Transcriptional Regulation

Using an RNase protection assay, Molis and coworkers (1994) determined whether melatonin's suppression of ER expression is exerted at the transcriptional level. As in the ER expression experiments, physiological concentrations of melatonin were effective in substantially reducing steady-state ER mRNA levels in MCF-7 cells to 30% of control levels as early as 1 hour and as late as 48 hours after starting incubation. In fact, the same bell-shaped dose-response curve demonstrated for melatonin's inhibition of cell proliferation (Hill and Blask 1988) also characterizes its action on ER mRNA expression in MCF-7 cells. Nuclear run-on experiments revealed that melatonin's inhibitory action on ER mRNA expression is mediated at the transcriptional level. Apparently, down-regulation of the ER by melatonin does not involve a direct interaction with the ER, since melatonin failed to alter the transcriptional regulatory capacity of the fully activated wild type ER as well as a constitutively active ER variant with a deletion in the hormone-binding domain. Furthermore, melatonin had no influence on the stability of the ER transcript (Molis et al. 1994).

Melatonin has also been reported to increase the expression of c-*fos* (Hull et al. 1993; Molis et al. 1995) and c-*myc* (Molis et al. 1995), immediate-early genes important in estradiol-induced proliferation of MCF-7 cells in culture (Wilding et al. 1988). In fact, the proto-oncogene product of c-*fos* is a transcription factor that is a key target of signal transduction in processes such as cell proliferation (van der Burg et al. 1990). Molis and associates (1995) took this even further and demonstrated that physiological melatonin modulates steady-state mRNA expression of a variety of other estrogen-regulated proteins, including the progesterone receptor and pS2 as well as transforming growth factors α and β. These latter findings with transforming growth factors receive further credence from the evidence of melatonin's modulatory action on growth factor activity reported by Cos and Blask (1994).

The results of Molis et al. (1995) indicate that melatonin causes an early but transient increase in the expression of an estrogen-inducible, immediate-early gene important in cell proliferation. Whether the melatonin-induced modulation of ER and c-*fos* mRNA expression in MCF-7 breast cancer cells described above can be ascribed unequivocally to either alterations in transcriptional mechanisms per se or changes in mRNA stability, or both, is impossible to ascertain at this point, since experiments specifically designed to address these issues were not performed.

15.4.5 Genomic Interactions

Recent immunohistochemical studies indicate that melatonin appears to accumulate preferentially in the nuclear compartments of cells of many types of organs and tissues (Menendez-Pelaez et al. 1993). Within the nucleus, melatonin appears to be confined to the chromatin (Menendez-Pelaez and Reiter 1993). Additionally, high affinity (Kd = 180 pM) and low capacity (9.19 fmol/mg protein) 2-IMel binding sites have been identified in the purified nuclei of rat livers (Acuna-Castroviejo et al. 1993). Because of its intrinsic and highly potent antioxidant activity and propensity for nuclear localization, it has been proposed that melatonin may confer DNA molecules with on-site protection against DNA damage resulting from the generation of free radicals and reactive oxygen intermediates (Reiter et al. 1994). Furthermore, Reiter and colleagues (1994) have postulated that melatonin's oncostatic actions may be in part due to its ability to act directly at the level of nuclear DNA to inhibit certain types of carcinogen-induced cancer initiation via a reduction in the formation of adducts between the carcinogen and DNA. This may be the case for safrole-induced carcinogenesis, since pharmacological and physiologically relevant levels of melatonin prevent DNA adduct formation, perhaps via stabilization of DNA (Tan et al. 1994).

15.4.6 Calcium/Calmodulin (Ca^{2+}/CaM) and the Cytoskeleton

Calcium/CaM plays an important role not only in the proliferation of normal cells but in the growth of neoplastic cells, presumably via effects on intracellular processes including cell cycle progression and cytoskeletal integrity (Rasmussen and Means 1987). It is known that agents that increase CaM levels and disrupt cytoskeletal integrity stimulate cell proliferation, whereas the opposite effects are manifested with CaM antagonists (Stoclet et al. 1987). Therefore, the ease of melatonin's entry into the intracellular compartment coupled with its high affinity for CaM (Benitez-King and Anton-Tay 1993) make this a logical intracellular signal transduction pathway through which melatonin may influence cancer cell growth. In the neoplastic, murine neuroblastoma cell line NIE-115, in which melatonin is first stimulatory and then inhibitory to cell proliferation in vitro, changes in the intracellular content of CaM occur in response to physiological concentrations of melatonin (Benitez-King et al. 1994). During logarithmic phase growth, when melatonin is stimulatory, CaM levels are increased whereas during the stationary phase, when melatonin inhibits cell growth, CaM levels are substantially diminished. Melatonin-induced changes in the intracellular distribution of CaM may also be involved in transmitting growth-modulatory information in these cells as well (Benitez-King and Anton-Tay 1993).

With respect to cytoskeletal integrity, the cytoplasmic microtubule complex is thought to act as a signal transduction mechanism by which depolymerization of microtubules stimulates DNA synthesis and thus cell proliferation. Moreover, Ca^{2+}/CaM promotes microtubule disassembly, presumably via the phosphorylation of tubulin and/or microtubule-associated proteins (MAPs) via CaM-dependent protein kinase (Rasmussen and Means 1987; Stoclet et al. 1987). At a physiological concentration, melatonin may inhibit the growth of neoplastic cells such as NIE-115 by its ability to bind to CaM and block MAPs/CaM and tubulin/CaM complex formation and thus also inhibit cytoskeletal disruption (Benitez-King and Anton-Tay 1993).

15.4.7 Redox Mechanisms: Glutathione and Nitric Oxide

15.4.7.1 Glutathione

Augmented intracellular concentrations of reactive oxygen species and free radicals can create a pro-oxidant state within cells that can lead not only to cancer initiation but to promotion as well (Ceruti 1985). Since melatonin is an important antitumor-promoting hormone both in vivo and in vitro (Blask 1993), some of our recent efforts have focused on the potential role played by the redox state of the intracellular environment in governing melatonin's inhibitory mode of action on breast cancer at the cellular level (Blask et al. 1994b; Blask et al. 1997). As the most ubiquitous non-protein thiol produced by mammalian cells, glutathione (gamma-glutamylcysteinyl-glycine, or GSH) is the most potent and crucial antioxidant molecule in the intracellular compartment by providing it with a reducing environment. Glutathione is synthesized via the glutamyl cycle from its constituent amino acids in two consecutive enzymatic steps involving the rate-limiting enzyme gamma-glutamylcysteine synthetase and GSH synthetase (Meister 1991). We have taken advantage of a highly specific inhibitor of GSH synthesis, L-buthionine sulfoximine (BSO), which blocks the rate-limiting step of GSH synthesis and depletes cells of their GSH content (Meister 1991), to determine whether GSH is involved in the mechanism of melatonin's direct inhibitory effect on MCF-7 human breast cancer cell growth in culture.

The simultaneous and continuous incubation of MCF-7 cells with melatonin (1 nM) and BSO (1–10 µM) completely blocks the inhibitory effect of melatonin on cell growth, indicating that the depletion of GSH somehow renders these cells unable to respond to melatonin (Blask et al. 1994b; Blask et al. 1997). Similar results were reported with respect to the cytotoxic effects of the diterpene drug taxol in this same cell line (Leibmann et al. 1993). That the depletion of GSH rather than an interaction with BSO with melatonin is responsible for the elimination of melatonin's oncostatic effect is substantiated by the ability of GSH added back to the cultures to restore melatonin's ability to inhibit cell proliferation in the presence of BSO. In the face of intracellular and extracellular GSH manipulations, the same types of proliferative responses to melatonin are obtained in other ER+ human breast cancer cell lines such as ZR75-1 (Blask et al. 1997). However, apparently not all cancer cells rely on GSH for melatonin inhibition of growth, since L-BSO failed to block the inhibitory effect of high doses of melatonin (2 µM or 1.0 mM) on the proliferation of ME-180 human cervical cancer cells in vitro (Chen et al. 1995). Thus, the oncostatic action of melatonin in ER+ breast cancer cells appears to depend on the presence of an adequate, perhaps threshold level of intracellular GSH. In fact, melatonin itself substantially increases GSH levels in MCF-7 cells, which effect is completely blocked with BSO (Blask et al. 1994b, 1997). Under conditions in which melatonin inhibits Ca^{2+}-stimulated MCF-7 cell growth and markedly elevates intracellular GSH levels, BSO also completely blocks these effects, suggesting an important interaction between melatonin, Ca^{2+}, and GSH (Blask 1997). Surprisingly, ER– human breast cancer cells, which ordinarily do not respond to the oncostatic effects of melatonin (Hill et al. 1992), become responsive to melatonin when GSH-S-transferase (GST), another important enzyme in GSH metabolism, is inactivated by ethacrynic acid (Blask et al. 1997).

Other investigators have reported that melatonin injections raise not only GSH levels but levels of GST in the mammary gland and liver tissues in female rats treated with the mammary carcinogen DMBA. It has been proposed that melatonin's suppression of the initiation step of DMBA-induced mammary carcinogenesis may be brought about in part by an increased detoxification of DMBA by elevated GSH and GST levels. Additionally, a melatonin-induced inhibition of phase I enzymes, cytochromes b5 and P450, may also lead to decreased activation of DMBA as well (Kothari and Subramanian 1992).

15.4.7.2 Nitric Oxide

Nitric oxide (NO) has emerged as an extremely important intracellular signal transduction molecule (Moncada and Higgs 1993). Not only does it function in a variety of physiological processes, but it also acts either as a free radical or antioxidant species (Stamler et al. 1992). This fascinating molecule has been shown to exert cytostatic antiproliferative effects on melanoma (Meragos et al. 1993) and breast cancer cells (Lepoivre et al. 1989), presumably via an inhibition of ribonucleotide reductase, the rate-limiting step in DNA synthesis (Bitonti et al. 1994). The redox state of the intracellular environment can determine specific redox forms of NO which, in turn, may target specific cellular processes such as DNA synthesis and elicit a particular biological response such as cytostasis (Stamler et al. 1992; Girard and Potier 1993). Since melatonin alters the redox state of breast cancer cells by elevating GSH levels (Blask et al. 1994b, 1997), we wanted to determine whether NO is involved in the mechanisms mediating the oncostatic effects of melatonin on MCF-7 cells in vitro.

We approached this question by blocking the synthesis of NO and determining what effects this manipulation had on the antiproliferative action of melatonin in MCF-7 cells. The synthesis of NO can be inhibited by N-monomethyl-L-arginine (NMMA), a potent inhibitor of both inducible and constitutive forms of NO synthase (Moncada and Higgs 1993). In the presence of NMMA, melatonin fails to inhibit MCF-7 cell growth in culture. However, when increasing concentrations of sodium nitroprusside (SNP), which spontaneously releases NO into solution (Moncada and Higgs 1993), are added to cell cultures containing melatonin plus NMMA, melatonin's inhibition of cell growth is restored in a dose-dependent manner (Blask and Wilson 1994; Blask 1997). These results suggest that NO is involved in mediating melatonin's direct oncostatic action in human breast cancer cells, perhaps through an interaction with GSH. Thus, the entrance of melatonin into MCF-7 cells to elevate GSH levels may create the appropriate microenvironment for the formation of a redox form of NO that inhibits ribonucleotide reductase and thus DNA synthesis. Therefore, melatonin in the intracellular compartment may act as part of a mechanism to couple GSH and NO, perhaps via Ca^{2+}/CaM, which regulates the constitutive form of NO synthase (Stamler et al. 1992) to inhibit cell proliferation.

15.4.8 Tumor Linoleic Acid Uptake and Metabolism

The increased dietary intake of linoleic acid (LA), an n-6 essential polyunsaturated fatty acid, results in the growth progression of both murine and human transplantable tumors, particularly of rat hepatoma 7288CTC (Sauer et al. 1997). The growth of tissue-isolated rat hepatoma 7288CTC in vivo is stimulated by the uptake of LA and its metabolism to 13-hydroxyoctadecadienoic acid (13-HODE) (Sauer et al. 1997, unpublished results), an important mitogenic signaling molecule that amplifies EGF-responsive mitogenesis (Glasgow and Eling 1994). Since no data are available on the mechanism(s) by which melatonin inhibits tumor growth in vivo, we used the tissue-isolated hepatoma 7288CTC. Arteriovenous measurements across tissue-isolated rat hepatoma 7288CTC at four-hour intervals over a 24-hour period revealed maximal LA uptake and metabolism to 13-HODE during the light phase, when plasma melatonin levels were lowest, and minimal uptake and metabolism during the mid-dark phase, when melatonin levels were at their peak. The elimination of the nocturnal decrease of tumor LA metabolism in pinealectomized rats indicated that it was a melatonin-driven circadian rhythm (Blask et al. 1997a, 1998). Reverse transcriptase-polymerase chain reaction (RT-PCR) analysis showed that hepatoma 7288CTC expresses mRNA transcripts for melatonin receptors (mt), while northern blot analysis showed that fatty acid transport protein (FATP) was overexpressed in this tumor. Melatonin receptor and FATP mRNA expression of tumor-bearing rats were not affected by treatment with daily afternoon melatonin injections (200 µg/day) or exposure to constant light (unpublished results). However, perfusion of tissue-isolated tumors in situ with melatonin (1 nM) directly, rapidly, and reversibly inhibited the uptake of plasma fatty acids, including LA and its metabolism to 13-HODE (Blask et al. 1997a, 1998). The inhibitory effects of melatonin on tumor fatty acid uptake and 13-HODE release were completely reversible by perfusion of tumors in situ with pertussis toxin (PTX), forskolin, or 8-bromo-cAMP (unpublished results). Pinealectomy or constant light exposure stimulated tumor growth, while daily late afternoon melatonin injections or melatonin feeding (200 µg/day) inhibited tumor growth as well as LA uptake, storage, and metabolism to 13-HODE (Blask et al. 1997a; 1998). These results support a novel mechanism of melatonin-induced tumor growth inhibition in vivo involving a G protein-coupled melatonin receptor-mediated suppression of cAMP levels that compromises the activity of a fatty acid transport protein such as FATP. Melatonin's inhibition of these signal transduction events culminates in a suppression of LA uptake and metabolism to the mitogenic signaling molecule 13-HODE, resulting in an inhibition of cancer growth.

15.5 Melatonin in the Chemoendocrine Therapy of Human Malignancies

The basic science studies reviewed above provide a solid biological rationale for exploring melatonin's use as a "chrono-oncostatic or chronomodulatory" agent in cancer therapy. As reviewed extensively elsewhere (Blask and Hill 1988; Blask 1993), a majority of studies have demonstrated that the nocturnal secretion of melatonin is altered in patients with a variety of cancers, further suggesting that this neurohormone plays an important role in neoplastic growth. As a result, over the past several years we have witnessed a slow but steady increase in the number of clinical trials utilizing melatonin in the treatment of a variety of human malignancies (Blask 1993). A few of these clinical trials will be reviewed here in some detail to give the reader a taste for what has been accomplished and what needs to be done in the future in this important area of clinical application.

Although the original reports of Starr (1970) and DiBella et al. (1979) claimed that melatonin could be effectively used to treat a diverse number of human malignancies, Lissoni and his associates in Monza, Italy, have pioneered a more systematic approach to examining the clinical utility of melatonin in cancer therapy (Lissoni et al. 1987, 1989, 1991). In their initial clinical studies with melatonin, these oncologists treated patients with carcinoma of the breast, lung, stomach, colon liver, pancreas, and cervix as well as individuals with soft tissue sarcoma and nonseminomatous testicular cancer; all patients had metastatic disease in distant organs. Melatonin (20 mg/day) was administered intramuscularly during a 2-month induction phase followed by oral melatonin (18 mg/day) for an additional 4-month maintenance phase. Virtually half of the patients treated with melatonin experienced a stabilization in their disease, whereas a partial regression was observed in one patient with pancreatic cancer. The quality of life and immune status of the individuals with stable disease treated with melatonin was increased. The survival time was increased in two of six patients treated with melatonin for myelodysplastic syndrome secondary to treatment with chemotherapy for Hodgkin's or non-Hodgkin's lymphoma or breast cancer (Viviani et al. 1990).

In a phase II trial by this group, the efficacy of melatonin in combination with IL-2 was evaluated as a first-line therapy for metastatic nonsmall cell lung cancer (NSCLC), which is notoriously resistant to chemotherapy and has a poor prognosis (Lissoni et al. 1992). Interleukin-2 therapy alone has little effect on the course of advanced NSCLC. Therefore, 20 patients (14 males and 6 females 38–70 years of age) began receiving melatonin (10 mg/day) orally at 8 P.M. every evening for 1 week prior to the onset of IL-2 treatment (3×10^6 IU/m^2 twice a day) and then continued with one cycle of therapy for an additional 4 weeks. In responder patients or those with stable disease, a second cycle was administered following a rest period of 3 weeks, during which melatonin alone was given. A partial response (at least 50% tumor regression) was observed in four of the patients, while 10 additional patients had stable disease (no tumor regression or increase greater than 25%) in response to this neuro-immunotherapy, which was well-tolerated.

In the largest clinical trial involving melatonin in cancer therapy conducted to date, Lissoni and colleagues (1994b) soon followed-up their original investigation of the efficacy of combination IL-2/melatonin therapy by studying 80 consecutive patients with locally advanced or metastatic solid tumors including NSCLC, colorectal cancer, hepatocarcinoma, gastric carcinoma, pancreatic adenocarcinoma, and breast cancer. The patients were randomized to receive either IL-2 (3×10^6 IU) alone or IL-2 plus oral melatonin (40 mg/day) at 8 P.M. every day for 4 weeks; patients receiving combination therapy began receiving melatonin 1 week prior to the beginning of IL-2 administration. Of the group of 39 patients receiving IL-2 alone, 11 experienced stabilization of their disease and one had a partial response, while the remaining 27 patients progressed. However, in the 41 patients receiving IL-2 plus melatonin, three had a complete response (complete disappearance of disease for at least 1 month) and eight had a partial response, while 12 experienced stabilization of their disease. Therefore, the percent of patients with stable disease (28%) was the same in both treatment groups; however, the percent of patients with complete or partial responses was 3% in the IL-2 group and 26% in the IL-2 plus melatonin group. Even more interesting was that 46% of the patients receiving IL-2 plus melatonin were alive after 1 year, whereas only 15% were alive then in the IL-2 group. The results of this study emphasize the ability of melatonin to enhance the action of IL-2 to increase tumor regression, prolong progression-free survival, and improve overall survival in patients with a variety of malignancies with or without metastases. Moreover, it is also important to recall that melatonin enhanced the oncostatic effects of IL-2 therapy in mice with solid tumors as well (Sanchez de la Pena et al. 1992).

Yet another example of the ability of melatonin to enhance the efficacy of a more established anticancer therapy again comes from Lissoni and his colleagues (1995), who examined, in a small phase II trial, the ability of melatonin to modulate the breast cancer response in women with metastatic disease who either failed at the outset to respond to tamoxifen or progressed following initial disease stabilization with the drug. The rationale for this study was based on in vitro findings from our laboratory showing that melatonin enhances the sensitivity of MCF-7 cells to the oncostatic effects of tamoxifen (Cos et al. 1990; Wilson et al. 1992). Tamoxifen-resistant patients that had been treated with tamoxifen for 1 to 3 months were taken off of their medication for 1 month. Following this washout period, the patients took melatonin p.o. (20 mg) every evening at 8 P.M. for 1 week. Subsequent to this induction phase, the patients again began taking tamoxifen p.o. (20 mg) every day at 12 noon while continuing their melatonin in the evening. All patients receiving the combination of melatonin with tamoxifen were followed-up for at least 1 year with regular CT scans and routine laboratory tests.

While no complete responses were observed in any of the patients, a partial response (at least 50% regression of lesion size) was observed in four of the 14 patients (28.5%), irrespective of ER status. Eight other patients experienced stabilization of their disease, while the remaining two progressed. Interestingly, ten of the 14 patients (71%) survived longer than 1 year following the onset of treatment. Both serum PRL and IGF-1 levels significantly decreased with combination melatonin and tamoxifen therapy as compared with pretreatment levels. However, it is difficult to discern whether these were treatment effects due to melatonin and/or tamoxifen or whether they would have occurred in spite of treatment, since no placebo controls were included. Nevertheless, this encouraging preliminary study shows not only that melatonin may augment the therapeutic efficacy of tamoxifen in women with metastatic breast cancer, as we originally demonstrated in vitro with MCF-7 cells (Cos et al. 1990; Wilson et al. 1992), but that melatonin may actually be able to reverse tamoxifen resistance.

The Lissoni group combined melatonin with other chemotherapeutic agents including etoposide, cisplatin, cisplatin plus etoposide, or interferon. Their results with these strategies generally suggest that melatonin acts synergistically with these agents to enhance their efficacy by promoting the stabilization of disease and inducing objective tumor regression in a few cases. Moreover, the presence of melatonin in these combinatorial approaches appeared to prolong both progression-free and overall 1-year survival and to reduce the toxicity of chemotherapy while improving the quality of life and performance status of these patients. Interestingly, these workers also reported that melatonin in combination with radical or adjuvant radiation therapy for untreatable glioblastoma multiforme substantially increased the 1-year survival rate compared with radiation therapy alone, with a concomitant decrease in radiation-induced toxicity (Bubenik et al. 1998).

The therapeutic effect of melatonin as a single agent was again examined by Lissoni and coworkers (1994a) in 50 consecutive patients with brain metastases originating from primary cancers of the lung, breast, colon, kidney, and uterine cervix as well as from melanoma or sarcoma patients who had progressed following radiation therapy and chemotherapy. The patients were randomized to be treated with supportive care alone or supportive care plus melatonin. Melatonin (20 mg/day) was given orally at 8 P.M. until the progression of brain metastases. The mean period until that progression was significantly higher in patients receiving melatonin (5.9 months) than in those receiving supportive care alone (2.7 months). The survival at 1 year was also significantly higher in patients treated with melatonin (37%) versus those receiving only supportive care (12%). Additionally, mean survival time was significantly higher in patients receiving melatonin (9.2 months) versus those not receiving this indole (5.5 months). There was also a significant improvement in quality of life for those patients on melatonin as compared with those receiving supportive care alone. Furthermore, melatonin significantly reduced the frequency of steroid-related complications (i.e., infections) as well. Thus, melatonin alone may be effective in reducing brain metastases, prolonging survival, and improving the quality of life of patients with metastatic disease due to a variety of solid tumors.

The first clinical trial of melatonin in the treatment of cancer in the United States was by Robinson and his group (Gonzalez et al. 1991) from the University of Colorado in Denver. They administered melatonin orally to 42 patients with a diagnosis of melanoma. Patients received melatonin in doses ranging from 5 mg/m^2/day to 700 mg/m^2/day in four divided doses. After a median follow-up of 5 weeks, six patients had partial responses (at least a 50% decrease in the size of all lesions), while six had stable disease (no lesion increase or decrease). Sites of response to melatonin included the brain, subcutaneous tissue, and lung. The authors concluded that melatonin had a modest antitumor effect, minimal toxicity, and can be administered safely to humans. They also suggested that the oncostatic potential of melatonin should be investigated further.

15.6 Conclusions

There is no question that melatonin plays an important role as a unique endogenous and exogenous neurohormonal regulator of neoplastic growth. Although this author originally endorsed melatonin as primarily a "naturally occurring chrono-oncostatic neurohormone," based on the numerous reports of its ability to inhibit tumor growth in vivo and in vitro (Blask 1993b, 1984; Blask and Hill 1988), circadian-oriented in vivo studies have revealed that melatonin may also be oncogenic or have no effect at all, depending on the time of day it is administered (Bartsch and Bartsch 1981; Wrba et al. 1986; Blask et al. 1992; Sanchez de la Pena et al. 1992). Therefore, it would be more accurate to refer to melatonin as a "chrono-oncomodulatory neurohormone" to account for these circadian stage-dependent actions.

Remarkable strides have recently been made in elucidating and understanding the cellular mechanisms of action in neoplastic cells, particularly mammary cancer, neuroblastoma, and melanoma cells. Although high and low affinity melatonin binding sites exist in a number of neoplastic tissues, there is no proof as yet that these binding sites are coupled to appropriate signal transduction mechanisms important in regulating cell growth. As we learn more about the signal transduction mechanisms with which melatonin interacts, one of the challenges of future research will be to link these pathways to the complex process of cell proliferation. In this regard, the antioxidant properties of melatonin itself as well as its ability to boost intracellular levels of other antioxidant molecules in cancer cells appear to be especially promising. However, it remains to be determined whether the oncostatic effects of melatonin that occur at the cellular level in vitro actually translate to the in vivo situation. One such link is beginning to emerge from research on the ability of melatonin to inhibit LA and 13-HODE-dependent tumor growth in vivo, apparently via melatonin receptor-mediated inhibition of cAMP. As we learn more about melatonin mechanism(s) of action in cancer cells, it may become possible to design superpotent melatonin analogues that specifically target these various pathways either to inhibit cell growth or to modulate the neoplastic cell response to other chemotherapeutic agents.

An encouraging sign is the gradual increase in the number and quality of clinical trials using melatonin as a single oncostatic agent or as an oncomodulatory hormone in combination with other chemotherapeutic drugs or biological response modifiers. Unfortunately, these studies still suffer from the lack of a double blind, placebo-controlled design. The need for more and better studies notwithstanding, this author predicts that, as melatonin undergoes a natural evolution toward acceptance in the mainstream of basic and clinical cancer research, this unique molecule will receive much more serious attention in the next few years as a neuroendocrine-based approach for the prevention and treatment of human malignancy.

References

Acuna-Castroviejo D, Pablos MI, Menendez-Pelaez, A, Reiter RJ (1993) Melatonin receptors in purified cell nuclei of liver. Res Commun Chem Pathol Pharmacol 82:253–256

Anisimov VN, Morozov VG, Khavinson VK, Dilman VM (1973) Correlations of anti-tumour activity of pineal and hypothalamic extract, melatonin and sygethin in mouse transplantable mammary tumours. Vopr Onkol 19:99–101

Anisimov VN, Popovich IG, Zabezhinski MA (1997) Melatonin and colon carcinogenesis: I. inhibitory effect of melatonin on development of intestinal tumors induced by 1,2-dimethylhydrazine in rats. Carcinogenesis 18:1549–1553

Aubert C, Janiaud P, Lecalvez J (1981) Effect of pinealectomy and melatonin on mammary tumour growth in Sprague Dawley rats under different conditions of lighting. J Neural Transm 47:121–130

Bartsch H, Bartsch C (1981) Effect of melatonin on experimental tumours under different photoperiods and times of administration. J Neural Transm 52:269–279

Bartsch C, Bartsch H, Fluchter SH, Attanasio A, Gupta, D (1985) Evidence for modulation of melatonin secretion in men with benign and malignant tumors of the prostate: relationship with pituitary hormones. J Pineal Res 2:121–132

Bartsch H, Bartsch C, Flehmig B (1986) Differential effect of melatonin on slow and fast growing passages of a human melanoma cell line. Neuroendorinol Lett 8:289–293

Bartsch H, Bartsch, H, Noteborn, HPJM, Flehmig B, Ebels I, Salemink CA (1987) Growth-inhibiting effect of crude pineal extracts on human melanoma cells in vitro is different from that of known synthetic pineal substances. J Neural Transm 69:299–311

Bartsch C, Bartsch H, Lippert TH, Gupta, D (1990) Effect of the mammary carcinogen 7,12-dimethylbenz(a)anthracene on pineal melatonin biosynthesis, secretion and peripheral metabolism. Neuroendocrinology 52:538–544

Benitez-King G, Anton-Tay F (1993) Calmodulin mediates melatonin cytoskeletal effects. Experientia 49:635–641

Benitez-King G, Huerto-Delgadillo L, Samano-Coronel L, Anton-Tay F (1994) Melatonin effects on cell growth and calmodulin synthesis in MDCK and NlE-115 cells. In: Maestroni GJM, Conti A, Reiter RJ (eds) Advances in pineal research. Vol. 7. John Libbey, London, pp 57–61

Bindoni M, Jutisz M, Ribot G (1976) Characterization and partial purification of a substance in the pineal gland which inhibits cell multiplication in vitro. Biochim Biophys Acta 437:577–588

Bitonti AJ, Dumont JA, Bush TL, Cashmain EA, Cross-Doersen DE, Wright PS, Matthews DP, McCarthy JR, Kaplan, DA (1994) Regression of human breast tumor xenografts in response to (E)-2′-deoxy-2′-(fluoromethylene)cytidine, an inhibitor of ribonucleoside-diphosphate reductase. Cancer Res 54:1485–1490

Blask, DE (1984) The pineal: an oncostatic gland? In: Reiter RJ (ed) The pineal gland. Raven Press, New York, pp 253–284

Blask DE (1990) The emerging role of the pineal gland and melatonin in oncogenesis. In: Wilson BW, Stevens RG, Anderson LE (eds) Extremely low frequency electromagnetic fields: the question of cancer. Battelle Press, Columbus, pp 319–335

Blask, DE (1993a) Integration of neuroendocrine-circadian, pineal, adrenocortical and immune functions in homeokinesis: implications for host–cancer interactions. Neuroendocrinol Lett 15:117–133

Blask, DE (1993b) Melatonin in oncology. In: Yu HS, Reiter RJ (eds) Melatonin biosynthesis, physiological effects, and clinical applications. CRC Press, Boca Raton, pp 447–475

Blask, DE (1994) Neuroendocrine aspects of circadian pharmacodynamics. In: Hrushesky WJM (ed) Circadian cancer therapy. CRC Press, Boca Raton, pp 39–54

Blask, DE (1997) Systemic, cellular, and molecular aspects of melatonin action on experimental breast carcinogenesis. In: Stevens RG, Wilson, BW, Anderson LE (eds) The melatonin hypothesis – breast cancer and the use of electric power. Battelle Press, Columbus, pp 189–230

Blask DE, Hill SM (1986a) Effects of melatonin on cancer: studies on MCF-7 human breast cancer cells in culture. J Neural Transm (Suppl) 21:433–449

Blask DE, Hill, SM (1986b) Inhibition of human breast and endometrial cancer cell growth in culture by the pineal hormone melatonin and its analogue 6-chloromelatonin. Biol. Reprod. 34 (Suppl1) Abs 254, p 176

Blask DE, Hill SM (1988) Melatonin and cancer: basic and clinical perspectives. In: Miles A, Philbrick DRS, Thompson C (eds) Melatonin – clinical perspectives. Oxford Univ Press, New York, pp 128–173

Blask DE, Wilson ST (1994) Melatonin and oncostatic signal transduction: evidence for a novel mechanism involving glutathione and nitric oxide. In: Moller M, Pevet P (eds) Advances in pineal research. Vol 8. John Libbey, London, pp 465–471

Blask DE, Hill SM, Orstead KM, Massa JS (1986) Inhibitory effects of the pineal hormone melatonin and underfeeding during the promotional phase of 7,12dimethylbenzanthracene (DMBA)-induced mammary tumorigenesis. J Neural Transm 6:125–138

Blask DE, Hill SM, Pelletier DB (1988) Oncostatic signaling by the pineal gland and melatonin in the control of breast cancer. In: Gupta D, Attanasio A, Reiter RJ (eds) The pineal gland and cancer. Brain Research Promotion, London, pp 221–232

Blask DE, Hill SM, Pelletier DB, Anderson J.M, Lemus-Wilson A (1989) Melatonin: an anticancer hormone of the pineal gland. In: Reiter RJ, Pang SF (eds) Advances in pineal research. Vol 3. John Libbey, London, pp 259–263

Blask DE, Cos S, Hill SM, Burns DM, Lemus-Wilson S, Pelletier DB, Liaw L, Hill A (1990) Breast cancer: a target site for the oncostatic actions of pineal hormones. In: Reiter RJ, Lukaszyk A (eds) Advances in pineal research. Vol 4. John Libbey, London, pp 267–274

Blask DE, Pelletier DB, Hill SM, Lemus-Wilson A, Grosso DS, Wilson ST, Wise ME (1991a) Pineal melatonin inhibition of tumor promotion in the N-nitroso-N-methylurea model of mammary carcinogenesis: potential involvement of antiestrogenic mechanisms in vivo. J Cancer Res Clin Oncol 117:526–532

Blask DE, Cos S, Hill SM, Burns DM, Lemus-Wilson A, Grosso DS (1991b) Melatonin action on oncogenesis. In: Fraschini F, Reiter RJ (eds) Role of melatonin and pineal peptides in neuroimmunomodulation. Plenum Press, New York, p 233

Blask DE, Wilson ST, Cos S, Lemus-Wilson AM, Liaw L (1992) Pineal and circadian influence on the inhibitory growth response of experimental breast cancer to melatonin. Endocr Soc Abst 1025, p 308

Blask DE, Lemus-Wilson, AM, Wilson ST (1993) Neurohormonal modulation of cancer growth by pineal melatonin. In: Touitou Y, Arendt J, Pevet P (eds) Melatonin and the pineal gland – from basic science to clinical application. Elsevier, Amsterdam, pp 303–310

Blask DE, Wilson ST, Saffer JD, Wilson MA, Andreson LE, Wilson BW (1993a) Culture conditions influence the effects of weak magnetic fields on the growth-response of MCF-7 human breast cancer cells to melatonin in vitro. Ann Rev Res Biol Effects of Electric Magnetic Fields. Abst P–45, p 6

Blask DE, Wilson ST, Valenta K (1993b) Alterations in the growth response of human breast cancer cells to the inhibitory action of a physiological concentration of melatonin in incubators with different ambient 60 Hz magnetic fields. Ann Rev Res Biol Effects of Electric Magnetic Fields. Abst A–28, p 96

Blask DE, Wilson ST, Lemus-Wilson AM (1994) The oncostatic and oncomodulatory role of the pineal gland and melatonin. In: Maestroni GJM, Conti A, Reiter RJ (eds) Advances in pineal research. Vol 7. John Libbey, London, pp 235–241

Blask DE, Wilson ST, Zalatan F (1997) Physiological melatonin inhibition of human breast cancer cell growth in vitro: evidence for a glutathione-mediated pathway. Cancer Res 57:1909–1914

Blask DE, Sauer LA, Dauchy RT (1997a) Melatonin regulation of tumor growth and the role of fatty acid uptake and metabolism. Neuroendocrinol Lett 18:59–62

Blask DE, Sauer LA, Dauchy RT, Holowachuk, EW, Ruhoff MS (1998) Circadian rhythm in experimental tumor carcinogenesis and progression and the role of melatonin. In: Touitou Y (ed) Biologial clocks – mechanisms and applications. Elsevier, Amsterdam, pp 469–474

Bubenik GA, Blask, DE, Brown, GM, Maestroni GJM, Pang SF, Reiter RJ, Viswanathan M, Zisapel N (1998) Prospects of the clinical utilization of melatonin. Biol Signals Recept 7:195–219

Burns DM, Cos Corral S, Blask D (1990) Demonstration and partial characterization of 2-iodomelatonin binding sites in experimental mammary cancer. Endocr Soc Abst 151, p 62

Buswell RS (1973) The pineal gland and neoplasia. Lancet 2:34

Buzzell GR (1988) Studies on the effects of the pineal hormone melatonin on an androgen-insensitive rat protatic adenocarcinoma, the Dunning R3327 HIF tumor. J Neural Transm 72:131–140

Buzzell GR, Amerogen HM, Toma JG (1988) Melatonin and the growth of the Dunning R3327 rat prostatic adenocarcinoma. In: Gupta D, Attanasio A, Reiter RJ (eds) The pineal gland and cancer. Brain Research Promotion, London, pp 295–306

Cardinali DP (1981) Melatonin: a mammalian pineal hormone. Endocr Rev 2:327–346

Cardinali DP, Rey B, Mediavilla MD, Sanchez-Barcelo EJ (1992) Diurnal changes in cyclic nucleotide response to pineal indoles in murine mammary glands. J Pineal Res 13:111–116

Ceruti PA (1985) Prooxidant states and tumor promotion. Science 227:375–381

Chaterjee S, Banerji TK (1989) Effects of melatonin on thegrowth of MtT/F4 anterior pituitary tumor: evidence for inhibition of tumor growth dependent upon the time of administration. J Pineal Res 7:381–391

Chen LD, Leal BZ, Reiter RJ, Abe M, Sewerynek E, Melchiorri D, Meltz ML, Poeggeler B (1995) Melatonin's inhibitory effect on growth of ME-180 human cervical cancer cells is not related to intracellular glutathione concentrations. Cancer Lett 91:153–159

Conti A, Maestroni GJM (1994) Melatonin-induced immunoopioids: role in lymphoproliferative and autoimmune diseases. In: Maestroni GJM, Conti A, Reiter RJ (eds) Advances in pineal research. Vol 7. John Libbey, London, pp 83–100

Cos S, Blask DE (1990) Effects of melatonin on anchorage-independent growth of human breast cancer cells (MCF-7) in a clonogenic culture system. Cancer Lett 50:115–119

Cos S, Blask DE (1994) Melatonin modulates growth factor activity in MCF-7 human breast cancer cells. J Pineal Res 17:25–32

Cos S, Sanchez-Barchelo EJ (1994) Differences between pulsatile or continuous exposure to melatonin on MCF-7 human breast cancer cell proliferation. Cancer Lett 85:105–110

Cos S, Sanchez-Barcelo EJ (1995) Melatonin inhibition of MCF-7 human breast cancer cells growth: influence of cell proliferation rate. Cancer Lett 93:207–212

Cos S, Blask DE, Lemus-Wilson A, Hill AB (1991) Effects of melatonin on the cell cycle kinetics and estrogenrescue of MCF-7 human breast cancer cells in culture. J Pineal Res 10:36–42

Cos S, Fernandez F, Sanchez-Barcelo EJ (1996) Melatonin inhibits DNA synthesis in MCF-7 human breast cancer cells in vitro. Life Sci 58:2447–2453

Crespo D, Fernandez-Viadero C, Verduga R, Ovegero V, Cos S (1994) Interaction between melatonin and estradiol on morphological and morphometric features of MCF-7 human breast cancer cells. J Pineal Res 16:215–222

Danforth DN, Tamarkin L, Lippman M (1983) Melatonin increases oestrogen receptor hormone binding activity of human breast cancer cells. Nature 305:323–325

de Launoit Y, Pasteels JL, L'Hermite M, L'Hermite-Baleriaux M (1990) In vitro effect of melatonin on cell cycle kinetics of mammary cancer cell lines. Endocr Soc Abst 140, p 5

DiBella L, Scalera G, Rossi MT (1979) Perspectives in pineal function. In: Kappers JA, Pevet P (eds) The Pineal gland of vertebrates including man. Progress in Brain Research. Vol 52. Elsevier, Amsterdam, pp 475–478

DiGiovanni J, Juchau MR (1980) Biotransformation and bioactivation of 7,12-dimethylbenz(a)-anthracene (7,12-DMBA). Drug Metab Rev 11:61–101

Eison AS, Mullins UL (1993) Melatonin binding sites are functionally coupled to phosphoinositide hydrolysis in Syrian hamster RPMI 1846 melanoma cells. Life Sci 53:393–398

El-Domieri AAH, Das Gupta, TK (1973) Reversal by melatonin of the effect of pinealectomy on tumor growth. Cancer Res 33:2830–2833

El-Domieri AAH, Das Gupta TK (1976) The influence of pineal ablation and administration of melatonin on growth and spread of hamster melanoma. J Surg Oncol 8:197–205

Fitzgerald TJ, Veal A (1976) Melatonin antagonizes colchicine-induced mitotic arrest. Experientia 32:372–373

Freiss G, Pretois C, Vignon F (1993) Control of breast cancer cell growth by steroids and growth factors: interactions and mechanisms. Breast Cancer Res Treat 27:57–68

Furuya Y, Yamaoto K, Kohno N, Ku YS, Saitoh Y (1994) 5-fluorouracil attenuates an oncostatic effect of melatonin on estrogen-sensitive human breast cancer (MCF-7). Cancer Lett 81:95–98

Ghosh BC, El-Domieri AAH, Das Gupta TK (1973) Effect of melatonin on hamster melanoma. Surg Forum 24:121–122

Gilad E, Laudon M, Matzkin H, Pick E, Sofer M, Braf Z, Zisapel N (1996) Functional melatonin receptors in human prostate epithelial cells. Endocrinology 137:1412–1417

Gilad E, Matzkin H, Zisapel N (1997) Inactivation of melatonin receptors by protein kinase C in human prostate epithelial cells. Endocrinology 138:255–261

Girard P, Potier P (1993) NO, thiols and disulfides. Fed Eur Biochem Soc 320:7–8

Glasgow WC, Eling TE (1994) Structure-activity relationship for potentiation of EGF-dependent mitogenesis by oxygenated metabolites of linoleic acid. Arch Biochem Biophys 311:286–292

Gonzalez R, Sanchez A, Ferguson J, Balmer C, Daniel C, Cohn A, Robinson WA (1991) Melatonin therapy of advanced human malignant melanoma. Melanoma Res 1:237–243

Hamilton T (1969) Influence of environmental light and melatonin upon mammary tumour induction. Br J Surg 56:764–766

Helton RA, Kane M.A (1991) Demonstration of a putative melatonin receptor on cultured melanoma cells. Proc Am Assoc Cancer Res 32:Abst 4, p 1

Helton RA, Kane MA (1992) Properties of a specific melatonin uptake system in B16 murine melanoma cells. Proc Am Assoc Cancer Res 33:Abst 114, p 19

Hill SM (1986) Antiproliferative effect of the pineal hormone melatonin on human breast cancer cells in vitro. Doctoral Dissertaion, Univeristy of Arizona, Tucson, pp 1–206

Hill SM, Blask DE (1986) Melatonin inhibition of MCF-7 breast cancer cell proliferation: influence of serum factors, prolactin and estradiol. Endocr Soc Abst 863, p 246

Hill SM, Blask DE (1988) Effects of the pineal hormone melatonin on the proliferation and morphological characteristics of human breast cancer cells (MCF-7) in culture. Cancer Res 48:6121–6126

Hill SM, Spriggs LL, Simon MA, Muraoka H, Blask DE (1992) The growth inhibitory action of melatonin on human breast cancer-cells is linked to the estrogen response system. Cancer Lett 64:249–256

Hu DN, Roberts JE (1997) Melatonin inhibits growth of cultured human uveal melanoma cells. Melanoma Res 7:27–31

Hull KM, Maher, TJ, Jorgenson KL (1993) Melatonin increases cfos mRNA in MCF-7 human breast cancer cells. Soc Neurosci 19:773.5

Jordan VC, Murphy CS (1991) Endocrine pharmacology of antiestrogens as antitumor agents. Endocr Rev 11:578–610

Karasek M, Kunert-Radek J, Stepien H, Pawlikowski M (1988) Melatonin inhibits the proliferation of estrogen-induced rat pituitary tumor cells in vitro. Neuroendocrinol Lett 10:135–140

Karasek M, Liberski P, Kunert-Radek J, Bartkowiak J (1992) Influence of melatonin on the proliferation of hepatoma cells in the Syrian hamster: in vivo and in vitro study. J Pineal Res 13:107–110

Karasek M, Winczyk K, Kunert-Radek J, Wiesenberg I, Pawlikowski M (1998) Antiproliferative effects of melatonin and CGP 52608 on the murine Colon 38 adenocarcinoma in vitro and in vivo. Neuroendocrinol Lett 19:71–78

Karmali RA, Horrobin DF, Ghayur T (1978) Role of the pineal gland in the aetiology and treatment of breast cancer. Lancet 2:1002

Kerenyi NA (1979) Tumors and the pineal gland. In: Kellen JA, Hilf R (eds) Influences of hormones in tumor development. Vol I. CRC Press, Boca Raton, pp 155–165

Kerenyi NA, Feurer G, Pandula E (1988) The presence of melatonin receptors in carcinoma of the breast. Am J Clin Pathol 89: Abst 25, pp 434–435

Kothari L (1988) Effect of melatonin on the mammary gland morphology, DNA synthesis, hormone profiles and incidence of mammary cancer in rats. In: Gupta D, Attanasio A, Reiter RJ (eds) The pineal gland and cancer. Brain Research Promotion, London, pp 210–219

Kothari L (1987) Influence of chronic melatonin on 9,10 dimethyl-1,2-benzanthracene (DMBA)-induced mammary tumorigenesis. Oncology 44:64–66

Kothari L, Subramanian A (1992) A possible modulatory influence of melatonin on representative phase I and phase II drug metabolizing enzymes in 9,10-dimethyl-1,2-benzanthracene induced rat mammary tumorigenesis. Anti-Cancer Drugs 3:623

Krause DN, Dubocovich, ML (1990) Regulatory sites in the melatonin system of mammals. Trends Neurosci 13:464–470

Kumar R, Sukumar S, Barbacid M (1990) Activation of ras oncogenes preceding the onset of neoplasia. Science 248:1101

Lapin V (1976) Pineal gland and malignancy. Oster Z Onkol 3:51–60

Lapin V, Ebels I (1976) Effects of some low molecular weight sheep pineal fractions and melatonin on different tumors in rats and mice. Oncol 33:110–113

Lapin V, Frowein A (1981) Effects of growing tumors on pineal melatonin levels in male rats. J Neural Transm 52:123–126

Laudon M, Gilad E, Matzkin H, Braf Z, Zisapel N (1996) Putative melatonin receptors in benign human prostate tissue. J Clin Endocrinol Metab 81:1336–1342

Leadem CA, Burns DM (1987) Pineal-induced inhibition of prolactinoma growth in F344 rats: effects of blinding and melatonin treatment. Endocr Soc Abst 344, p 107

Lemus-Wilson A, Kelly PA, Blask DE (1995) Melatonin blocks the stimulatory effects of prolactin on human breast cancer cell growth in culture. Br J Cancer 72:1435–1440

Leone AM (1991) The effects of melatonin and melatonin analogues on the P388, DLD-1 and MCF-7 tumour cell lines. In: Fraschini F, Reiter RJ (eds) Role of melatonin and pineal peptides in neuroimmunomodulation. Plenum Press., New York, pp 241–242

Lepoivre M, Boudbid H, Petit, JF (1989) Antiproliferative activity of gamma-interferon combined with lipopolysaccharide on murine adenocarcinoma: dependence on an L-arginine metabolism with production of nitrite and citrulline. Cancer Res 49:1970–1976

Lewinski A, Sewerynek E, Lopaczynsk, W, Wajs E (1990) Influence of the pineal gland on the growth and the pathomorphological picture of Guerin's tumor in male Wistar rats. Eur Pineal Study Grp, p 163

L'Hermite-Baleriaux M, de Launoit Y (1992) Is melatonin really an in vitro inhibitor of human breast cancer cell proliferation? In Vitro Cell Dev Biol 28 A:583–584

L'Hermite-Baleriaux M, L'Hermite M, Pasteels JL, de Launoit Y (1990) Effect of melatonin on the proliferation of human mammary cancer cells in culture. Eur Pineal Study Grp, p 161

Liburdy RP, Sloma TR, Sokolic R, Yaswen P (1993) ELF magnetic fields, breast cancer, and melatonin: 60 Hz fields block melatonin's oncostatic action on ER+ breast cancer cell proliferation. J Pineal Res 14:89–97

Liebmann JE, Hahn SM, Cook JA, Lipschultz C, Mitchell JB, Kaufman DC (1993) Glutathione depletion by L-buthionine sulfoximine antagonizes taxol cytotoxicity. Cancer Res 53:2066–2070

Lippman ME (1988) Steroid hormone receptors and mechanisms of growth regulation of human breast cancer. In: Lippman ME, Lichter AS, Danforth DN (eds) Diagnosis and management of breast cancer. Saunders, Philadelphia, pp 326–347

Lissoni P, Barni S, Tancini G, Crispino S, Paolorossi, F, Lucini V, Mariani M, Cattaneo G, Esposti D, Esposti G, Fraschini F (1987) Clinical study of melatonin in untreatable advanced cancer patients. Tumori 73:475–480

Lissoni P, Barni S, Crispino S, Tancini G, Fraschini F (1989) Endocrine and immune effects of melatonin therapy in metastatic cancer patients. Eur J Clin Oncol 25:789–795

Lissoni P, Barni S, Cattaneo G, Tancini G, Esposti G, Esposti D, Fraschini F (1991) Clinical results with the pineal hormone melatonin in advanced cancer resistant to standard antitumour therapies. Oncology 48:448–450

Lissoni P, Tisi E, Barni S, Ardizzoia A, Rovelli F, Rescaldan R, Ballabio D, Benenti C, Angeli M, Tancini G, Conti A, Maestroni GJM (1992) Biological and clinical results of a neuroimmunotherapy with interleukin-2 and the pineal hormone melatonin as a first line treatment in advanced non-small cell lung cancer. Br J Cancer 66:155–158

Lissoni P, Barni S, Ardizzoia A, Tancini G, Conti A, Maestroni G (1994a) A randomized study with the pineal hormone melatonin versus supportive care alone in patients with brain metastases due to solid neoplasms. Cancer 73:699–701

Lissoni P, Barni S, Tancini G, Ardizzoia A, Ricci G, Aldeghi R, Brivio F, Tisi E, Rovelli F, Rescaldani R, Quadro G, Maestroni G (1994b) A randomised study with subcutaneous low-dose interleukin. 2 alone vs interleukin 2 plus the pineal neurohormone melatonin in advanced solid neoplasms other than renal cancer and melanoma. Br J Cancer 69:196

Lissoni P, Barni S, Meregalli, S, Fossati V, Cazzinga M, Esposti D, Tancini G (1995) Modulation of cancer endocrine therapy by melatonin: a phase II study of tamoxifen plus melatonin in metastatic breast cancer patients progressing under tamoxifen alone. Br J Cancer 71:854–856

Maestroni GJM, Conti A (1993) Melatonin in relation with the immune system. In: Yu,HS, Reiter RJ (eds) Melatonin biosynthesis, physiological effects, and clinical application. CRC Press, Boca Raton, pp 289–311

Maragos CM, Wang JM, Hrabie JA, Oppenheim JJ, Keefer LK (1993) Nitric oxide/nucleophile complexes inhibit the in vitro proliferation of A375 melanoma cells via nitric oxide release. Cancer Res 53:564–568

Meister A (1991) Glutathione deficiency produced by inhibition of its synthesis, and its reversal; applications in research and therapy. Pharmacol Ther 51:1551–1944

Menendez-Pelaez A, Reiter RJ (1993) Distribution of melatonin in mammalian tissues: the relative importance of nuclear versus cytosolic localization. J Pineal Res 15:59–69

Menendez-Pelaez A, Poeggeler B, Reiter RJ, Barlow-Walden L, Pablos MI, Tan DX (1993) Nuclear localization of melatonin in different mammalian tissues: immunocytochemical and radioimmunoassay evidence. J Cell Biochem 53:373–382

Meyskens FI, Salmon S (1981) Modulation of clonogenic human melanoma cells by follicle-stimulating hormone, melatonin and nerve growth factor. Br J Cancer 43:111–115

Molis T, Cockerham Y, Hill SM (1992) Regulation of ER mRNA expression by melatonin in MCF-7 human breast cancer cells. Endocr Soc Abst 1030, p 309

Molis T, Walters, MR, Hill SM (1993) Melatonin modulation of estrogen receptor expression in MCF-7 human breast cancer cells. Int J Oncol 3:687–694

Moncada S, Higgs A (1993) The L-arginine–nitric oxide pathway. New Engl J Med 329:2002–2012

Narita T, Kudo H (1985) Effect of melatonin on B16 melanoma growth in athymic mice. Cancer Res 45:4175–4177

Osborne CK, Hobbs K, Trent JM (1987) Biological differences among MCF-7 human breast cancer cell lines from different laboratories. Breast Cancer Res Treat 9:111–121

Panzer A, Lottering M-L, Bianci P, Glencross DK, Stark JH, Seegers JC (1998) Melatonin has no effect on the growth, morphology or cell cycle of human breast cancer (MCF-7), cervical cancer (HeLa), osteosarcoma (MG-63) or lymphoblastoid (TK6) cells. Cancer Lett 122:17–23

Perissin L, Zorzet S, Rapozzi V, Giraldi T (1994) Stress, melatonin and tumour metastasis in mice bearing Lewis lung carcinoma. In: Maestroni GJM, Conti A, Reiter RJ (eds) Advances in pineal research. Vol 7. John Libbey, London, pp 253–257

Philo R, Berkowitz AS (1988) Inhibition of Dunning tumor growth by melatonin. J Urology 139:1099–1102

Pickering DS, Niles LP (1992) Expression of nanmolar affinity binding sites for melatonin in Syrian hamster RPMI 1846 melanoma cells. Cell Signal 4:201–207

Papazisis KT, Kouretas D, Geromichalos GD, Sivridis E, Tsekreli OK, Dimitriadis KA, Kortsaris AH (1998) Effects of melatonin on proliferation of cancer cell lines. J Pineal Res 25:211–218

Rasmussen CD, Means AR (1987) Calmodulin is involved in regulation of cell proliferation. EMBO J 6:3961–3968

Recio J, Mediavilla MD, Cardinali DP, Sanchez-Barcelo EJ (1994) Pharmacological profile and diurnal rhythmicity of 2-[^{125}I]-iodomelatonin binding sites in murine mammary tissue. J Pineal Res 16:10–17

Reiter RJ, Tan DX, Poeggler B, Chen LD, Menendez-Pelaez A (1994) Melatonin, free radicals and cancer initiation. In: Maestroni GJM, Conti A, Reiter RJ (eds) Advances in pineal research. Vol 7. John Libbey, London, pp 211

Sanchez-Barcelo EJ, Mediavilla MD, Tucker HA (1990) Influence of melatonin on mammary gland growth: in vivo and in vitro studies. Proc Soc Exp Biol Med 194:103–107

Sanchez de la Pena S, Hrushesky WJM, Vyzula R, Lobo S, Wood P (1992) Circadian dependent interleukin-2 tumor growth interaction is favorably modulated by melatonin. Int Soc Chronobiol p IV–16

Sauer LA, Dauchy RT, Blask DE (1997) Dietary linoleic acid intake controls the arterial blood plasma concentration and the rates of growth and linoleic acid uptake and metabolism in hepatoma 7288CTC in Buffalo rats. J Nutr 127:1412–1421

Scaglione F, Dermartini, G, Lucini V, Capsoni S, Esposti D (1992) Melatonin and DMBA induced tumors: action on growth factors. Endocr Soc Abst 1028, p 308

Shah PN, Mhatre MC, Kothari LS (1984) Effect of melatonin on mammary carcinogenesis in intact and pinealectomized rats in varying photoperiods. Cancer Res 44:3403–3407

Shellard SA, Whelan RDH, Hill BT (1989) Growth inhibitory and cytotoxic effects of melatonin and its metabolites on human tumour cell lines in vitro. Br J Cancer 60:288–290

Slominski A, Pruski D (1993) Melatonin inhibits proliferation and melanogenesis in rodent melanoma cells. Exp Cell Res 206:189–194

Stamler JS, Singel DJ, Loscalzo J (1992) Biochemistry of nitric oxide and its redox-activated forms. Science 258:1898

Stanberry LR, Das Gupta TK, Beattie CW (1983) Photoperiodic control of melanoma growth in hamster: influence of pinealectomy and melatonin. Endocrinology 113:469–475

Stankov B, Scaglione F, Lucini V, Fraschini F (1989) Specific binding in vitro and in vivo effects of melatonin on melanoma B16 and its metastatic growth. In: Rubinstein E, Adams D (eds) Recent advances in chemotherapy, Proc. 16th International Congress of Chemotherapy. Lewin-Epstein, Jerusalem, p 1

Stankov B, Lucini V, Scaglione F, Cozzi B, Righi M, Canti G, Demartini G, Fraschini F (1991) 2[^{125}I]iodo-melatonin binding in normal and. neoplastic tissues. In: Fraschini F, Reiter RJ (eds) Role of melatonin and pineal peptides in neuroimmunomodulation. Plenum Press, New York, pp 117–125

Starr KW (1970) Growth and new growth: environmental carcinogens in the process of human ontogeny. Prog Clin Cancer 4:1–29

Stoclet JC, Gerard D, Kilhoffer MC, Lugnier C, Miller R, Schaeffer P (1987) Calmodulin and its role in intracellular calcium regulation. Prog Neurobiol 29:321

Subramanian A, Kothari L (1991a) Melatonin, a suppressor of spontaneous murine mammary tumors. J Pineal Res 10:136–140

Subramanian A, Kothari L (1991b) Suppressive effect by melatonin on different phases of 9,10-dimethyl-1,2-benzanthracene (DMBA)-induced rat mammary gland carcinogenesis. Anti-Cancer Drugs 2:297–303

Sze SF, Ng TB, Liu WK (1993) Antiproliferative effect of pineal indoles on cultured tumor cell lines. J Pineal Res 14:27–33

Tamarkin L, Cohen M, Roselle D, Reichert C, Lippman M, Chabner B (1981) Melatonin inhibition and pinealectomy enhancement of 7,12-dimethylbenz(a)anthracene-induced mammary tumors in the rat. Cancer Res 41:4432–4436

Tamarkin L, Danforth D, Lichter A, De Moss E, Cohen M, Chabner B, Lippman M (1982) Decreased nocturnal plasma melatonin peak in patients with estrogen receptor positive breast cancer. Science 216:1003–1005

Tan DX, Reiter, RJ, Chen LD, Poeggeler B, Manchester LC, Barlow-Walden L (1994) Both physiological and pharmacological levels of melatonin reduce DNA adduct formation induced by the carcinogen safrole. Carcinogenesis 15:215–218

Tapp E (1982) The pineal gland in malignancy. In: Reiter RJ (ed) The pineal gland. Vol III. CRC Press, Boca Raton, pp 171

Toma JG, Amerongen HM, Hennes SC, O'Brien MG, McBlain WA, Buzzell GR (1987) Effects of olfactory bulbectomy, melatonin, and/or pinealectomy on three sublines of the Dunning R3327 rat prostatic adenocarcinoma. J Pineal Res 4:321–338

van der Burg B, de Groot RP, Isbrucker L, Kruijer W, de Laat SW (1990) Stimulation of TPA-responsive element activity by a cooperative action of insulin and estrogen in human breast cancer cells. Mol Endocrinol 4:1720–1726

Viviani S, Negretti E, Orazil A, Sozzi G, Santoro A, Lissoni P, Esposti G, Fraschini F (1990) Preliminary studies on melatonin in the treatment of myelodysplastic syndromes following cancer chemotherapy. J Pineal Res 8:347

Waldhauser F, Waldhauser M (1988) Melatonin and aging. In: Miles A, Philbrick DRS, Thompson C (eds) Melatonin – clinical perspectives. Oxford Press, London, pp 174–189

Walker MJ, Chauduri P.K, Beattie CW, Tito WA, Das Gupta, TK (1978) Neuroendocrine and endocrine correlates to hamster melanoma growth in vitro. Surg Forum 29:151–152

Welsch CW (1985) Host factors affecting the growth of carcinogen-induced rat mammary carcinomas: a review and tribute to Charles Benton Huggins. Cancer Res 45:3415–3443

Wiesenberg I, Missbach, Kahlen J-P, Schraeder M, Carlberg C (1995) Transcriptional activation of the nuclear receptor RZR by the pineal hormone melatonin and identification of CGP 52608 as a synthetic ligand. Nucleic Acid Res 23:327–333

Wilding G, Lippman ME, Gelmann EP (1988) Effects of steroid hormones and peptide growth factors on protooncogene c-fos expression in human breast cancer cells. Cancer Res 48:802–805

Wilson ST, Blask DE, Lemus-Wilson AM (1992) Melatonin augments the sensitivity of MCF-7 human breast cancer cells to tamoxifen in vitro. J Clin Endocrinol Metab 75:669–670
Wrba H, Halberg F, Dutter A (1986) Melatonin circadian stage-dependently delays breast tumor development in mice injected daily for several months. Chronobiologia 13:123–126
Ying SW, Niles, LP, Crocker C (1993) Human malignant melanoma cells express high-affinity receptors for melatonin: antiproliferative effects of melatonin and 6-chloromelatonin. Eur J Pharmacol 246:89–96
Zisapel N, Bubis M (1994) Inhibition by melatonin of protein secretion and growth of melanoma cells in culture. In: Maestroni GJM, Conti A, Reiter RJ (eds) Advances in pineal research. Vol 7. John Libbey, London, pp 259–267

16 Modulation of the Estrogen Response Pathway in Human Breast Cancer Cells by Melatonin

Steven M. Hill, Todd Kiefer, Stephenie Teplitzky, Louaine L. Spriggs, Prahlad Ram

Abstract

The pineal hormone melatonin has been shown to exert an inhibitory influence only on estrogen receptor (ER)-positive human breast tumor cell lines and to modulate ER expression, suggesting that melatonin's actions may be linked to the cell's estrogen response pathway. Two distinct melatonin receptors have recently been identified, a membrane-bound G protein-coupled receptor (Mel1a) and a nuclear orphan receptor (RZR/RORα) belonging to the retinoic acid receptor class. Melatonin's antitumorigenic effects in MCF-7 cells have been shown to be mediated via the Mel1a receptor and appear to be serum-dependent. We investigated whether melatonin, alone or in conjunction with serum factors such as EGF and insulin, can modulate ER activity by affecting ER phosphorylation and transactivation. We found that melatonin alone decreased ER phosphorylation but did not transactivate the ER. However, a combined treatment with melatonin followed by either EGF or insulin stimulated mitogen-activated protein kinase (MAPK) activity and induced ER phosphorylation, leading to transactivation of the human ER. Taken together, these findings suggest the possibility of cross talk between the melatonin receptor pathway and the tyrosine kinase receptor pathways.

16.1 Introduction

As an extension of the visual system, the pineal gland translates photoperiodic messages into a neuroendocrine hormone, melatonin, which is emitted as a nighttime signal to other organs and tissues (Reiter 1988). Melatonin (N-acetyl-5-methoxytryptamine), a lipophilic indolamine, exhibits daily rhythms within the pineal gland and serum in a variety of mammalian species, including humans (Reiter 1991). Under appropriate photoperiodic conditions, it functions as a hormonal transducer, signaling the neuroendocrine system to modulate physiological events critical to the regulation of endocrine functions, particularly those controlling seasonal reproduction (Reiter 1980, 1987).

Melatonin, the major secretory product of the pineal gland, is unique in that not only is it a central regulator of chronobiology, but it also exhibits antineoplastic activity in numerous systems. Melatonin's oncostatic actions have been observed both in vitro and in vivo in a broad spectrum of tumor types, such as melanoma, ovarian, prostate, pituitary, uterine, and breast. However, the mechanism(s) through which melatonin acts to mediate these effects has not been clearly elucidated.

We are just now beginning to unravel the molecular and signal transduction pathways that mediate the actions of melatonin at the cellular level. For example, in cells of the pars tuberalis (PT) of the pituitary gland, melatonin's signal is apparently transmitted by the interaction of melatonin with a membrane-bound receptor coupled to an inhibitory guanine nucleotide (G)-binding protein (Morgan et al. 1989). Although there is presently a great deal of information about melatonin's actions on the neuroendocrine substrates controlling reproductive physiology, little is known regarding the signaling and downstream gene modulatory events initiated or controlled by melatonin that produce its oncostatic effects.

16.2 Melatonin and Breast Cancer

Via its hormone, melatonin, the pineal gland has been shown to exert an inhibitory influence on the development and growth of carcinogen-induced hormone-responsive rat mammary tumors (Welsch and Nagasawa 1977; Tamarkin et al. 1982; Chang et al. 1985). In the 7,12-dimethylbenzathracene (DMBA)-induced rat mammary tumor model, stimulation of pineal melatonin secretion by blinding and olfactory bulbectomy significantly decreases both the incidence and volume of tumors and induces the regression of established mammary tumors. Blinding and bulbectomy also markedly reduce the levels of estradiol and prolactin, both of which are critical regulators of mammary tumor growth (Blask 1984; Chang et al. 1985). In addition, chronic daily late afternoon injections of melatonin (250 μg to 1 mg/day) have been shown to suppress both the development and growth of tumors in the N-nitroso-N-methylurea (NMU)-induced rat mammary tumor model (Blask et al. 1991).

In contrast to animal studies, only a marginal amount of work has been conducted with respect to the role of the pineal and melatonin in human breast cancer. Bartsch et al. (1981) found that postmenopausal Indian women with advanced breast cancer have diminished urinary levels of melatonin as compared to healthy controls. Tamarkin et al. (1982) have shown that the normal nocturnal rise in plasma melatonin is significantly reduced in women with estrogen receptor (ER)-positive breast tumors as compared to patients with ER-negative breast tumors and age-matched controls. This group also showed an inverse correlation between ER levels and peak nighttime melatonin values. We demonstrated that physiological concentrations of melatonin corresponding to the peak nighttime plasma levels seen in humans are able to inhibit the growth (50%–75% vs. controls) of estrogen-responsive MCF-7 human breast cancer cells in vitro (Hill et al. 1988). Melatonin has been shown to block the estrogen rescue of MCF-7 cells inhibited by the antiestrogen, tamoxifen (Sutherland et al. 1983; Cos 1991). The antiproliferative effects of melatonin, like that of tamoxifen, are cell cycle-specific, inducing a G0/G1-S transition delay in MCF-7 human breast cancer cells (Cos et al. 1991). Recently, it was shown that only ER-positive human breast tumor cell lines are responsive to the antiproliferative effects of melatonin (Hill et al. 1992). Taken together, these data suggest that melatonin's actions may be linked to the cell's estrogen response pathway.

16.3 Melatonin Receptors

Melatonin is highly lipophilic (Lerner et al. 1959) and, like the steroid hormones, is capable of diffusing through both the cell and nuclear membranes and entering intracellular compartments, including the nucleus (Menendez-Pelaez 1993). The iodinated form of melatonin, 2-[^{125}I]-iodomelatonin (I-MEL), has been used for ligand-binding analyses and, because of its high specific activity, ensures the detection of low abundance receptors with a high affinity for melatonin. I-MEL retains its biological activity (Weaver et al. 1988) and has been used extensively to determine both tissue and subcellular localization of melatonin binding. Using this approach, high affinity binding sites for melatonin have been identified in the hypothalamus of the rat brain (Vanecek et al. 1987; Morgan et al. 1994), the liver, skin and breast, and in certain neoplasms, including mammary tumors. Two different melatonin binding sites with significantly different binding affinities have been reported – a G protein-coupled high affinity binding site with a K_d in the picomolar range, and a low affinity binding site with a K_d in the nanomolar range (Morgan et al. 1989; Rivkees et al. 1989; Laitinen 1990; Dubocovich 1991).

Although most melatonin binding occurs at the cell membrane, the lipophilic nature of melatonin suggests that melatonin receptors could also be located in intracellular compartments. Within the last few years, two distinct melatonin receptors have been identified. The first to be characterized is a membrane-bound G protein-linked receptor (Reppert et al. 1994), and the second is a nuclear orphan receptor belonging to the class of retinoic acid receptors which are members of the steroid/thyroid hormone receptor superfamily (Carlberg 1994).

16.3.1 Membrane-Associated, G Protein-Linked Melatonin Receptors

The cDNA for the melatonin membrane receptor, Mel1a, was isolated from dermal melanophores of *Xenopus laevis* (Ebisawa et al. 1994) and exhibits a high affinity for I-MEL, with a K_d of approximately 0.06 nM (Reppert et al. 1994). The cloned Mel1a receptor inhibits adenylate cyclase through a pertussis toxin-sensitive mechanism (Reppert et al. 1994) and thus appears to be synonymous with the high affinity binding sites observed in earlier ligand-binding studies (Morgan et al. 1990; Stankov 1992; White 1987). Transfection and expression of this receptor into COS-7 cells shows an I-MEL binding affinity of 24 pM with a Bmax of about 210 fmol/mg of protein for the human receptor (Reppert et al. 1994). Structural analysis of this Mel1a receptor indicates that it belongs to the G protein-linked superfamily of surface receptors which utilizes heterotrimeric G proteins to transmit signals between the receptor and effector enzymes (Birnbaumer 1990). A number of the receptors in this superfamily share common features, including an extracellular ligand binding domain located at the amino terminus, a transmembrane domain containing seven hydrophobic segments, and a G protein-coupled intracellular domain at the carboxy terminus (Pearson 1990).

The Mel1a cDNA encodes a 426 amino acid protein receptor in *Xenopus*, a 366 amino acid protein in sheep, and, in humans, a receptor of approximately 350 amino acids with a predicted molecular mass of 39 kD. The human Mel1a melatonin membrane receptor shows 65% sequence identity to the *Xenopus* receptor and 85% identity to the sheep receptor (Reppert et al. 1994). Other melatonin receptors have recently been identified, including Mel1b, found only in the central nervous system, and Mel1c, expressed only in birds. Even though the three melatonin receptors share considerable sequence identity, their genes reside on separate chromosomes. All of these receptor proteins contain a consensus site for N-linked glycosylation, a common feature of G protein-associated receptors. The Mel1a receptor also exhibits a high degree of homology with the mu opioid and somatostatin type 2 receptors (25% identity) (Ebisawa et al. 1994), and its carboxy terminus contains consensus sites for protein kinase C phosphorylation, which may play a critical role in receptor regulation.

In situ hybridization studies using the cloned Mel1a receptor cDNA demonstrated a specific distribution of this receptor within the suprachiasmatic and paraventricular nuclei of the thalamus (Vanecek 1987; Williams et al. 1989; Morgan et al. 1989; Bitman and Weaver 1990) and in the pars tuberalis of the hypothalamus (Reppert et al. 1994). Northern blot analysis of RNA isolated from the pars tuberalis identified a major transcript of >9.5 kb and a minor message of 4 kb (Reppert 1994). Melatonin binding sites have also been found on the cell membranes of mammary tumor cells (Burns et al. 1986). MCF-7 human breast cancer cells were shown to express Mel1a mRNA by RT-PCR and Southern blot analyses, and Mel1a receptor protein by immunoblot analysis (Ram et al. 1997) (Fig. 16.1).

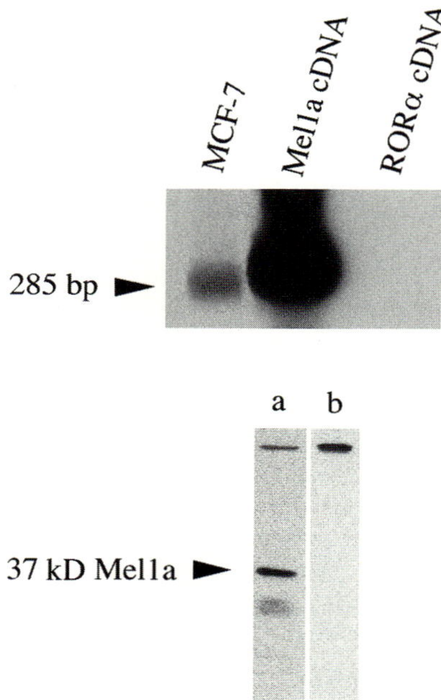

Fig. 16.1. Expression of Mel1a in MCF-7 cells. *Above:* RT-PCR/Southern blot analysis of Mel1a mRNA in MCF-7 human breast cancer cells. Five micrograms of total RNA was used for the RT-PCR. The Mel1a receptor cDNA was used as a positive control and the cDNA for RORα was used as a negative control. RT-PCR/Southern analysis was performed 3 times with different samples of RNA. *Below:* Immunoblot analysis of Mel1a melatonin receptor in MCF-7 cells. Fifty micrograms of MCF-7 whole cell extract was electrophoretically separated by size on a 12% SDS-polyacrylamide gel and electroblotted onto HyBond-C membranes. The Mel1a receptor was detected using a polyclonal antibody against the receptor and visualized using the ECL system. *Lane a,* antiMel1a antibody. *Lane b,* antiMel1a antibody preincubated with 0.5 mg/ml Mel1a peptide

16.3.2 Nuclear Melatonin Receptors

Nuclear binding sites for melatonin have been found in various tissues (Stankov et al. 1987; Burns et al. 1990; Acuna-Castroviejo et al. 1993, 1994). Becker-André et al. (1994) reported that I-MEL can specifically bind to and transactivate a class of nuclear orphan receptors, the retinoid Z receptors α and β (RZRα and β), and the retinoic orphan receptor α (RORα), for which no ligand had previously been identified (Wiesenberg et al. 1995). The RZRs and RORs belong to the steroid/thyroid hormone receptor super-family and show considerable sequence identity with the retinoic acid receptors RAR and RXR (Becker-André et al. 1993; Giguere et al. 1994). The RZR/RORα receptors have a lower binding affinity for I-MEL than the Mel1a receptor, with a K_d of approximately 1–10 nM. Like other steroid hormone receptors, the RZR/RORαs exhibit three key functional domains – an activation function domain (AF-1) located at their N terminus, an internal DNA binding domain (DBD), and a C terminus hormone binding domain (HBD). There are also various subdomains located within the HBD and DBD that are critical for the functioning of the receptor.

There are two species of RZR receptors, the RZRα and RZRβ (Becker-André et al. 1993) and four isoforms of the RORα receptors, RORα1, RORα2, RORα3, and RORα4 (Giguere et al. 1994). RZRα is, in fact, a splice variant of the ROR family and, as such, can be considered a member of the RORα family and the same as RORα4. The tissue distribution of RZRβ is considerably different from that of RZRα and RORα. RZRβ has only been identified in central nervous tissue and does not appear to be expressed elsewhere in the body (Becker-André et al. 1994). On the other hand, RZR/RORα are expressed in a number of tissues including the brain, ovary, testes, kidney, spleen, prostate, thymus, liver, lung, heart, and, most abundantly, in peripheral blood leukocytes (Becker-André et al. 1993; Forman et al. 1994). In addition, both Carlberg et al. (1995) and our laboratory (Ram et al. 1997) have identified RORα transcripts in the MCF-7 human breast cancer cell line.

Most activated steroid hormone receptors recognize and bind cis-acting domains in the upstream region of responsive genes called hormone response elements (HREs) as either hetero- or homodimers (inverted repeat palindromic sequences). However, the RZR/RORαs bind DNA as monomers or, possibly, homodimers and do not form heterodimers with the RXR, RAR, thyroid receptor (TR), or vitamin D receptor (VDR) (Carlberg 1994; Giguere 1994).

Transcriptional activation of RZR/RORα by melatonin occurs in the low nanomolar range (Becker-André et al. 1994; Wiesenberg et al. 1995; Steinhilber et al. 1995). With the exception of RORα1, all the isoforms of RZR and RORα bind and are transcriptionally activated by melatonin. In human B lymphocytes, RZRα/RORα are biologically responsive to melatonin treatment, modulating the transcription of the 5-lipoxygenase gene (Steinhilber et al. 1995). Although no selective gene activation has been noted for the various receptor isoforms, differential DNA-binding sites (RORE) have been described (Giguere et al. 1994), suggesting that each receptor isoform may recognize a specific RORE and thus may activate different sets of genes.

Whether the RZR/RORα receptors function as mediators of melatonin's action is not yet resolved. Tini et al. (1995) were not able to repeat melatonin's binding to the RORα receptors, and our laboratory has been unable to demonstrate melatonin binding or activation of the RORα receptors. However, we are able to block the growth inhibitory effects of melatonin on human breast cancer cells using a melatonin membrane receptor antagonist, CBPT, provided by David Sugden (Kings College, UK). An interesting side note to this story is that Kane et al. (1998) have recently reported in abstract form that RORαs can be transactivated by Ca^{++} and calmodulin-dependent protein kinase IV and that, once activated, these receptors utilize the cAMP-associated coactivator molecules P300/CBP to support gene transcription.

16.4 Melatonin and the Estrogen Response Pathway

The mechanism(s) by which estrogen promotes the growth of breast cancer is not completely understood; however, considerable evidence indicates that polypeptide growth factors and even oncogenes mediate estrogen's mitogenic effects (Dickson et al. 1987b). Human breast cancer cells in culture synthesize and secrete a variety of growth factors, such as transforming growth factors alpha (TGF-α) and beta (TGF-β), insulin-like growth factors (IGFs), and cathepsin-D, a lysosomal protease. Estrogen modulates the expression and, sometimes, secretion of these growth factors. It also induces expression of the progesterone receptor (PgR) (Horwitz et al. 1978), the immediate early genes, c-myc (Dubik et al. 1988), c-fos (Weisz et al. 1988), and the protein product of pS2 mRNA (Brown et al. 1984).

As a member of the steroid/thyroid hormone receptor superfamily, ER represents a class of *trans*-acting regulatory proteins which bind to the genome and activate specific sets of responsive genes involved in regulating cell proliferation, differentiation, and physiological function (Dickson et al. 1987). The expression of ER is regulated both in vivo and in vitro by a number of factors including cell cycle, cell density, growth factors, and hormones (Katzenellenbogen 1980). Insulin and progesterone can alter ER expression to various degrees (Murphy and Sutherland 1985; Shafie et al. 1977) and thus affect the growth of breast cancer cells. The presence or absence of the ER is used to identify human breast tumors which are hormone-responsive (Edwards et al. 1978; McGuire et al. 1975) and will successfully respond to endocrine therapy. Approximately 60%–70% of all primary breast tumors are ER-positive and, of these, approximately 60% respond to endocrine therapy directed toward blockade of estrogenic stimulation (Edwards et al. 1978). Breast tumor cell lines, mammary cancer in animal models, and breast tumors maintained in organ culture are for the most part estrogen-responsive (Daniel and Prichard 1963; Lippman et al. 1976). For example, the MCF-7 human breast cancer cell line is responsive to estrogen but is not totally dependent on estrogen for growth in culture. However, when MCF-7 cells are implanted into ovariectomized nude mice, cell growth and solid tumor formation are dependent on concomitant estrogen administration.

16.4.1 Melatonin Modulation of ER Expression

There appears to be a correlation between melatonin and ER expression. Melatonin's suppression of estrogen-induced growth of MCF-7 human breast cancer cells in vitro is associated with an increase in estrogen-binding activity (Danforth et al. 1983a). In rats, activation of the pineal by blinding and olfactory bulbectomy results in a significant decrease in estrogen-binding activity in DMBA-induced mammary tumors (Sanchez-Barcelo et al. 1988). Melatonin-induced suppression of estrogen-binding activity occurs in other tissues as well, including the hamster brain (Lawson et al. 1992; Danforth et al. 1983b). These data suggest that, via melatonin, the pineal gland inhibits the growth of estrogen-responsive breast cancer through modulation of ER expression and, thus, the cell's estrogen response pathway. In MCF-7 breast cancer cells, nanomolar concentrations of melatonin decrease transcription of the ER gene, resulting in suppression of both ER mRNA and protein levels in a time-dependent manner (Molis et al. 1993, 1994, 1995). Melatonin does not, however, compete with labeled estrogen for binding to the ER (Molis et al. 1993). Thus, it does not appear that the down-regulation of ER by melatonin is mediated through melatonin's binding to the hormone-binding domain of the ER.

16.4.2 Melatonin Modulation of Estrogen-Regulated Genes

The estrogen-induced proliferation of breast tumor cells may be mediated by induction of a number of growth stimulatory factors (Dickson et al. 1987) and inhibition of specific growth inhibitory factors. In MCF-7 cells, estrogen-induced growth is accompanied by increased expression of several mitogenic factors, including PgR (Eckert and Katzenellenbogen 1982), TGF-α (Bates et al. 1986, 1988), and the lysosomal protease, cathepsin D (Cavailles et al. 1988). Conversely, estrogen down-regulates the expression of its own receptor (Saceda et al. 1988) and decreases the synthesis and expression of the potent growth inhibitory factor, TGF-β.

Since melatonin can down-regulate ER expression, its inhibition of breast cancer cell proliferation may occur via suppression of the tumor cell's estrogen response pathway. One mechanism could be melatonin's modulation of expression of downstream estrogen-regulated gene products, including critical growth factors such as TGF-α and TGF-β, c-myc and c-fos, PgR, and pS2. Melatonin treatment of MCF-7 cells induces a biphasic response in TGF-α, with a transient increase in mRNA after 3 h and 6 h followed by a transient but significant decrease at 12 h and a return to control values by 24 h. Thus, melatonin does not appear to produce a significant long-term effect on TGF-α mRNA expression. Of particular interest is that the expression of TGF-β, a potent growth inhibitor of breast tumor cells which is suppressed by estrogen, is rapidly induced by melatonin in MCF-7 cells (Molis et al. 1995). A key mechanism by which melatonin inhibits MCF-7 breast cancer growth is clearly via the induction of this growth inhibitory factor.

Melatonin treatment induces a significant long-term increase in the steady-state c-myc mRNA level, beginning as early as 1 h after addition and continuing through 48 h. Conversely, the long-term expression (6–48 h) of c-fos steady-state mRNA is significantly decreased (Molis et al. 1995). Considering that both c-myc and c-fos are regulators of cellular proliferation and induced by estrogen, the induction of c-myc by the growth inhibitor, melatonin, is somewhat confusing. However, Armstrong et al. (1994) have shown that c-myc expression is significantly increased in response to the growth inhibitory effects of epidermal growth factor (EGF) in the ER-negative MDA-MB-468 breast tumor cell line. Thus, the induction of c-myc expression may correlate with growth inhibition.

Also of interest is that PgR mRNA levels, which are up-regulated by estrogen, are suppressed by melatonin treatment (Molis et al. 1994) in contrast to the pS2 transcript, another estrogen-induced mRNA, which is stimulated after 3 h of melatonin treatment.

The rapid modulation of most of these transcripts (3 h to 6 h) occurs prior to melatonin's down-regulation of the ER, indicating that melatonin may affect the expression of these gene products directly and that melatonin's growth inhibitory effects may not be mediated exclusively via the estrogen response pathway. It is probable that later effects beginning at 12 h and 24 h are mediated via melatonin's regulation of ER levels and subsequent downstream effects on the cell's estrogen response pathway. Studies by Cos et al. (1994) support this concept because the growth inhibitory effects of melatonin on MCF-7 cells in cultured media supplemented with epidermal growth factor (EGF) are reversed by estrogen but not by EGF.

16.4.3 Melatonin Modulation of ER Transactivation

The ER, as a phosphoprotein member of the family of steroid/thyroid receptors, functions as a ligand-activated transcription factor. In the presence of estrogen, ER phosphorylation increases approximately three- to fourfold, an event that is important for activation of the AF-1 domain and, thus, gene transcription (Le Goff et al. 1994; Ali et al. 1993). Recently, it was demonstrated that estrogen-independent phosphorylation and activation of the ER is mediated via the mitogen-activated protein kinase (MAPK) pathway (Kato et al. 1995). This observation raises the possibility that the ER can be activated by a number of growth factors through the MAPK or other signal transduction pathways.

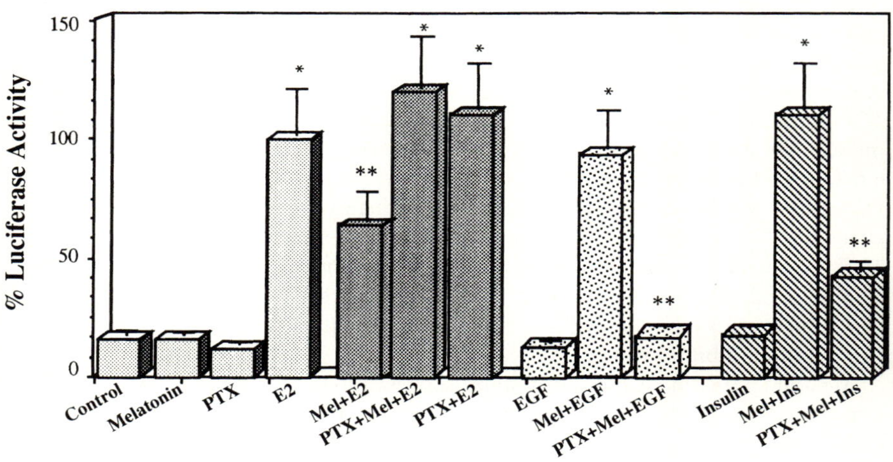

Fig. 16.2. ER transactivation assay in breast cancer cells treated with melatonin and/or estradiol, EGF, insulin, and pertussis toxin. MCF-7 cells were transiently transfected with an ERE-luciferase reporter construct containing three EREs and treated with either 0.001 % ethanol, 1 nM melatonin, 1 nM E2, 100 ng/ml EGF, 1 nM insulin, or 100 ng/ml pertussis toxin or pretreated for 5 min with melatonin followed by E2, EGF, or insulin, or PTX for 2 h followed by melatonin, E2, melatonin + E2, melatonin + EGF, or melatonin + insulin. Cells were cultured for 18 h following the final treatment to allow expression of the reporter gene. Luciferase activity in cells treated with 1 nM E2 was set at 100. The Student's t-test was used to determine the statistical difference in luciferase expression for each treatment versus vehicle-treated control (*) or versus vehicle and E2 treated cells (**). $N = 9, p < 0.001$

Melatonin can bind and activate the Mel1a melatonin receptor, a $G_{\alpha i}$ protein-coupled receptor (Ebisawa et al. 1994; Reppert et al. 1994) and can modulate ER expression. We investigated whether melatonin modulates ER activity by affecting phosphorylation and transactivation via its G protein-coupled Mel1a receptor, either by itself or in conjunction with factors that transduce their signals via tyrosine kinase receptors and the MAPK pathway. To examine the transcriptional activity of the ER in response to various growth factors and hormones, MCF-7 human breast tumor cells were transiently transfected with an ERE-luciferase reporter construct and treated with 1 nM estradiol (E_2). The cells showed significant ER transactivation, as demonstrated by expression of the luciferase reporter gene (Fig. 16.3). Melatonin alone (1 nM) did not transactivate the ER, and melatonin added simultaneously with E_2 did not alter estrogen's transactivation of the ER. However, estrogen's transactivation of the ER was blunted in MCF-7 cells pretreated with melatonin for 1 hour prior to the addition of E_2. These effects appear to be mediated via the Mel1a receptor, since pertussis toxin, a G_i protein inhibitor, blocks the decrease in ER transactivation seen from a 1 h pretreatment with melatonin (Fig. 16.2).

Since the presence of serum is necessary for the full growth-modulatory effects of melatonin (Blask and Hill 1986), the potential interaction between melatonin and serum factors such as EGF and insulin in modulating ER transactivation was examined using an ERE-luciferase reporter assay. There is no transactivation of the ER in MCF-7 cells cultured in estrogen-deficient medium by either EGF (100 ng/ml) or insulin (Fig. 16.2).

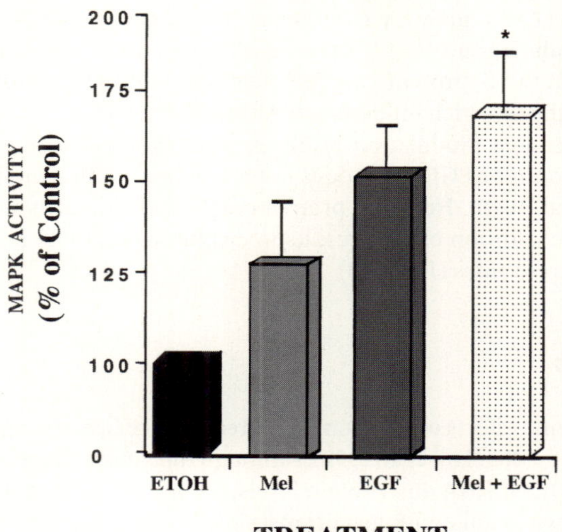

Fig. 16.3. MAPK activity in MCF-7 cells treated with melatonin and/or EGF. MCF-7 cells grown in serum-free media for 24 h were treated for 5 min with 0.001 % ethanol, 1 nM melatonin, 100 ng/ml EGF, or pretreated for 5 min with melatonin followed by EGF for 5 min. Cells were lysed and mixed with protein G sepharose, and MAPK was immunoprecipitated. The immunoprecipitated MAPK protein G sepharose complex was incubated with a 20 mg/sample of myelin basic protein, 2 mCi/sample of [^{32}P]-γ-ATP, 40 ml kinase buffer/sample, and 50 mM ATP/sample for 10 min. MAPK activity in the samples was quantified by scintillation counting and the data are expressed as percent of control. The Student's t-test was used to compare MAPK activity for each treatment group vs. controls. * $P < 0.05$ vs. controls

However, pretreatment of cells with melatonin (1 nM) for 5 min, followed by either EGF or insulin (1 nM), induces a significant ($p < 0.001$) five- to sixfold increase in luciferase activity. Melatonin pretreatment for longer times (1 h and 24 h), followed by the addition of EGF, also results in a significant transactivation of the ER in the absence of E2, which can be blocked by pertussis toxin, an inhibitor of G_i protein-coupled receptors (Ram et al. 1998). However, unlike the melatonin/EGF paradigm, only pretreatment with melatonin for 24 h (but not 1 h) followed by insulin transactivates the ER.

Pretreatment with a wide range of melatonin concentrations, from 0.1 mM to 1 pM, followed by EGF can effectively transactivate the ER, with the highest level of ER transactivation (131 % of estradiol controls) occurring with 0.1 mM melatonin (Ram et al. 1998). As the concentration of melatonin decreases from 0.1 mM to 1 pM, there is a corresponding diminution of ER transactivation (from 131 % to 58 % of estradiol control). In addition, incubation of MCF-7 cells with the antiestrogen, ICI 164,384 blocks ER transactivation by 1 nM melatonin pretreatment followed by EGF.

Changes in the phosphorylation status of the ER can also affect its transactivation. The basal level of ER phosphorylation (in the absence of ligand) increases approximately threefold upon stimulation with E2 (Aronica and Katzenellenbogen 1993). Melatonin treatment alone induces a 50 % decrease in basal ER phosphorylation levels ($p < 0.03$). Insulin (1 nM) and EGF (100 ng/ml) do not significantly increase ER phosphorylation individually, but when the cells are pretreated with melatonin for 5 min followed by either EGF or insulin, ER phosphorylation increases significantly, approximately threefold ($p < 0.001$) (Ram et al. 1998).

Ligand-independent transactivation of the ER in COS-1 cells can be induced via the activation of the MAPK pathway (Kato et al. 1995). Given that $G_{\alpha i}$ protein-coupled receptors can modulate the MAPK cascade (Pace et al. 1995), it is logical to suspect cross talk between the G_i protein-coupled melatonin Mel1a receptor pathway and the EGF signaling pathway which influences MAPK activity. Such cross talk could result in the ligand-independent modulation of ER transactivation. Treatment of MCF-7 cells with either melatonin or EGF alone does not result in a significant increase in MAPK activity (Ram et al. 1998). However, pretreatment of MCF-7 cells with melatonin for 5 min prior to the addition of EGF results in a significant (75 %) induction of MAPK activity over control values (Fig. 16.3).

16.5 Conclusions

The antitumorigenic effects of melatonin on breast cancer are strongly correlated with the expression of the ER (Hill et al. 1992). Although melatonin decreases the expression of the ER at both the mRNA and protein levels, it does not bind to the ER (Molis et al. 1993, 1994) to down-regulate ER gene transcription (Secada et al. 1994). The fact that melatonin's effects in MCF-7 cells are mediated via the Mel1a receptor, a G protein-coupled receptor which can interact with steroid hormone signaling pathways, raises the question of a potential interaction between the melatonin signaling pathway and the estrogen response pathway. Pretreatment of ER-positive MCF-7 cells with melatonin can blunt E2 transactivation of the ER (Fig. 16.2) in a temporally specific manner. These effects of melatonin are not directly associated with the down-regulation of ER protein levels, which are not significantly decreased until 6 h post treatment with melatonin (Molis et al. 1995).

Previous studies have shown that the melatonin-induced inhibition of MCF-7 cell growth is a serum-dependent phenomenon (Cos et al. 1991). The ER can be trans-activated in a ligand-independent manner by serum factors such as EGF and insulin only in cells primed by melatonin (Fig. 16.2). Neither melatonin nor EGF alone can transactivate the ER. However, melatonin (0.1 mM – 1.0 pM) in combination with EGF can affect ER transactivation in a dose-responsive manner, which can be blocked by pretreatment with the antiestrogen ICI 164,384 (Ram et al. 1998). The exact mechanism(s) by which melatonin enhances the sensitivity of the ER to the effects of EGF or insulin is not known. MCF-7 cells express relatively low levels of EGF receptor (EGF-R) and are not typically responsive to the mitogenic influence of EGF (Hafner et al. 1996). Thus, it is possible that melatonin alters the sensitivity of the ER to EGF by increasing the number of EGF-R or by amplifying the EGF signaling pathway distal to the EGF-R.

The effect of melatonin pretreatment on ER transactivation is quite rapid (5 min) and correlates with those seen with G protein-coupled receptors. These observations suggest that the E_2-independent transactivation of the ER may depend on several events and that a single factor or event may not be sufficient for the transactivation of this receptor in MCF-7 cells. These data also suggest that, although the EGF-R is expressed at very low levels and has been suggested to have no major mitogenic functions in MCF-7 cells, the EGF-R pathway, in conjunction with other signaling pathways, is capable of influencing MCF-7 cells.

Although insulin alone does not transactivate the ER in MCF-7 cells, ER trans-activation does occur in response to insulin following pretreatment with melatonin in a biphasic manner (Ram et al. 1998). The data indicate that, in cells pretreated with melatonin followed by either E_2 or insulin but not EGF, there is an initial rapid receptor-mediated effect that directly modulates ER activity and a more protracted secondary response possibly reflecting the expression of gene products that regulate ER transcriptional activity. The differential temporal response seen with melatonin and insulin or EGF suggests that different cell signaling pathways or different aspects of related pathways are utilized by insulin and EGF.

Melatonin treatment of MCF-7 cells apparently results in certain modifications in the intracellular milieu of signal transduction molecules and transcription factors, one of which is the ER. One possible mechanism by which melatonin can affect ER function is in altering its phosphorylation state and allowing other growth factors and hormones to phosphorylate and activate the ER via their own signal transduction pathways. This concept is supported by studies from McNulty et al. (1994) showing that melatonin inhibits the forskolin-induced phosphorylation of the cyclic AMP response element binding protein (CREB) in ovine pars tuberalis cells. We examined the ability of melatonin alone and in combination with EGF or insulin to alter the phosphorylation state of the ER in MCF-7 cells. The fact that melatonin can decrease the basal levels of ER phosphorylation indicates that melatonin may activate phosphatases or possibly inhibit kinases resulting in this change in ER phosphorylation. Melatonin also diminishes cAMP activity (Vanecek, 1995), which, in turn, could lead to a decrease in ER phosphorylation. Upon binding E2, the ER undergoes a transient dephosphorylation prior to its hyperphosphorylation, which leads to the transcriptionally active state (Arnold et al. 1995). It is possible that the dephosphorylation event induced by melatonin represses full-blown estrogen-induced ER transactivation but, under E2-deficient conditions, primes the ER such that specific residues become available for phosphorylation by EGF or insulin. This concept is supported by the fact that neither EGF nor insulin alone significantly alters ER phosphorylation, but pretreatment with melatonin followed by either EGF or insulin significantly enhances it. Additional studies are currently underway to define critical residues on the ER which are (1) dephosphorylated in response to melatonin and (2) phosphorylated in response to melatonin followed by EGF. Such studies are essential to determining the mechanisms by which melatonin modulates ER function.

Via its tyrosine kinase receptor and the MAPK pathway, EGF phosphorylates and transactivates the ER. Our data show that EGF stimulation of MAPK activity is greatly increased when the cells are treated with melatonin prior to the addition of EGF. Cells treated with melatonin alone demonstrate no significant increase in MAPK activity, suggesting that the increase in MAPK activity observed in response to the sequential treatment with melatonin and EGF results from the interaction between the two receptor pathways.

A variety of second messenger molecules, such as cAMP, intracellular calcium, protein kinase C (PKC), and MAPK are implicated in modulating melatonin's intracellular signals (Vanecek 1995; Hazelrigg et al. 1996; McNulty et al. 1994). The recent identification and cloning of the Mel1a receptor has provided an important tool with which to begin elucidating the molecular mechanisms of melatonin's actions. The Mel1a receptor mRNA and protein are expressed in melatonin-responsive MCF-7 cells. Melatonin's effects on ER transactivation appear to be mediated via this G protein-coupled receptor, since pertussis toxin, an inhibitor of G_i-coupled receptors, blocks the transactivation of the ER by the melatonin/EGF paradigm, and cross talk has recently been demonstrated between a G protein-coupled receptor and the EGF receptor signaling pathway (Luttrell et al. 1997).

In summary, the ER in MCF-7 human breast tumor cells can be transactivated and the estrogen response pathway activated in the absence of E2 by the interaction and possible cross talk between the G protein-coupled Mel1a receptor pathway and the EGF or insulin tyrosine kinase receptor pathways. Our data show that the combined treatment of melatonin followed by either EGF or insulin can stimulate MAPK activity, induce ER phosphorylation, and lead to the transactivation of the human ER. The exact role that this ER transactivation plays relative to melatonin's ability to inhibit MCF-7 cell growth is not defined as yet. However, it is possible that, even though melatonin and EGF do transactivate the ER, they do so in a manner that renders the receptor less responsive to E_2 and less efficient in regulating the transcription of estrogen-responsive genes essential for MCF-7 cell proliferation. Site-specific phosphorylation of the ER in response to growth factors, antiestrogens, and estrogenic ligands induce various levels of transcriptional activation and inhibition (Le Goff et al. 1994). Melatonin/EGF treatment may produce a pattern of ER phosphorylation different from that generated by estradiol. Thus, a focus of future studies will be to examine and define the residues phosphorylated on the ER in response to melatonin and EGF. Moreover, it has been shown that, under certain conditions and in certain cell- and promoter-specific contexts, antiestrogens such as tamoxifen can transactivate the ER and differentially modulate the expression of ERE-reporter constructs (Tzukerman et al. 1994). It should be noted, however, that in our studies only the perfect vitellogenin ERE sequence was used (Klein-Hitpass et al. 1986) and that different levels of activity may be observed using other nonperfect EREs. Finally, it is our hypothesis that the transactivation of the ER by melatonin in combination with EGF or insulin makes the ER refractory to estradiol stimulation and thus protects breast tumor cells from the mitogenic effects of estradiol.

References

Acuna-Castroviejo D, Pablos MI, Menendez-Pelaez A, Reiter RJ (1993) Melatonin receptors in purified cell nuclei of liver. Res Commun Chem Pathol Pharmacol 82:253–256

Ali S, Metzger D, Bornet J-M, Chambon P (1993) Modulation of transcriptional activation by ligand-dependent phosphorylation of the human oestrogen receptor A/B region. EMBO J 12:1153–1160

Armstrong DK, Issacs JT, Ottaviano YL, Davidson NE (1992) Programmed cell death during regression of the MCF-7 human breast cancer following estrogen ablation. Cancer Res 51:3418–3424

Arnold SF, Obourn JD, Yudt MR, Carter TH, Notides AC (1995) In vivo and in vitro phosphorylation of the human estrogen receptor. J Steroid Biochem Mol Biol 52:159–171

Aronica SM, Katzenellenbogen BS (1993) Stimulation of estrogen receptor-mediated transcription and alteration in the phosphorylation state of the rat uterine estrogen receptor by estrogen, cyclic adenosine monophosphate, and insulin-like growth factor. Mol Endocrinol 7:743–752

Bartsch C, Bartsch H, Jain AK, Laumas KR, Wetterberg L (1981) Urinary melatonin levels in human breast cancer patients. J Neural Transm 52:269–279

Bates S, Davidson NE, Valverius EM, Freter CE, Dickson RB, Tam JP, Kudlow JE, Solomon DS, Lippman ME (1988) Expression of transforming growth factor alpha and its mRNA in human breast cancer: its regulation by estrogen and its possible functional significance. Mol Endocrinol 2:543–555

Becker-André M, Wiesenberg I, Schaeren-Wiemers M, Saurat JH, Carlberg C (1994) Pineal gland hormone melatonin binds and activates an orphan of the nuclear receptor superfamilly. J Biol Chem 269:28531–28536

Birnbaumer LG (1990) G-proteins in signal transduction. Ann Rev Pharmacol Toxicol 30:675–705

Bittman EL, Weaver DL (1990) The distribution of melatonin binding sites in neuroendocrine tissue of the ewe. Biol Reprod 43:986–993

Blask DE (1984) The pineal: An oncostatic gland? In: Reiter RJ (ed) The Pineal Gland. Raven Press, New York, pp 253–284

Blask DE, Hill SM (1986) Effects of melatonin on cancer: studies on MCF-7 human breast cancer cells in culture. J Neural Transm 29:406–412

Blask DE, Hill SM, Orstead KM, Massa JS (1986) Inhibitory effects of the pineal hormone melatonin and underfeeding during the promotional phase of 7,12-dimethylbenzanthracene (DMBA)-induced mammary tumorigenesis. J Neural Transm 67:125–138

Blask DE, Pelletier DB, Hill SM, Lemus-Wilson A, Grosso DS, Wilson ST, Wise ME (1991) Pineal melatonin inhibition of tumor promotion in the N-nitroso-N-methylurea model of mammary carcinognesis: potential involvement of antiestrogenic mechanisms in vivo. J Cancer Res Clin Oncol 117:526–532

Brown AM, Jeltsch JM, Roberts M, Chambon P (1984) Activation of pS2 gene transcription is a primary response to estrogen in the human breast cancer cell line MCF-7. Proc Natl Acad Sc. U.S.A. 81:6344–6388

Burns DM, Cos CS, Blask D (1990) Demonstration and partial characterization of 2-iodomelatonin binding sites in experimental mammary carcinoma. Program of the 72nd Annual Meeting of the Endocrine Society, San Antonio, TX, p 62 (abstract)

Cavailles V, Augereau P, Garcia M, Rochefort H (1988) Estrogens and growth factors induce the mRNA of the 52K-pro-cathepsin-D secreted by breast cancer cells. Nucleic Acids Res 16:1903–1919

Carlberg C, Hooft van Huijsduijnen R, Staple JK, De Lamarter JF, Becker-Andre M (1994) RZRs, a new family of retinoid-related orphan receptors that function as both monomers and homodimers. Mol Endocrinol 8:757–770

Carlson LL, Weaver DR, Reppert SM (1989) Melatonin signal transduction in hamster brain: inhibition of adenyl cyclase by pertusssis toxin-sensitive G protein. Endocrinology 125:2670–2676

Chang N, Spaulding TS, Tseng MT (1985) Inhibitory effects of superior cervical ganglionectomy on dimethylbenzanthracene-induced mammary tumors in the rat. J Pineal Res 2:331–340

Cos S, Blask DE, Lemus-Wilson A, Hill AB (1991) Effects of melatonin on the cell cycle kinetics and "estrogen-rescue" of MCF-7 human breast cancer cells in culture. J Pineal Res 10:36–42

Danforth DN, Tamarkin L, Lippman ME (1983a) Melatonin increases oestrogen receptor binding activity of human breast cancer cells. Nature 305:323–325

Danforth DN, Tamarkin L, Do R, Lippman ME (1983b) Melatonin induced increase in cytoplasmic estrogen receptor activity in hamster uteri. Endocrinology 113:81–85

Daniel PH, Prichard M L (1963) The response of experimentally induced mammary tumors in rats to ovariectomy. Br J Cancer 17:687–693

Dickson RB, Lippman ME (1987) Estrogenic regulation of growth and polypeptide growth factor secretion in human breast carcinoma. Endocr Rev 8:29–43

Dubocovich ML (1991) Pharmacological characterization of melatonin binding sites. Advances in Pineal Res 5:167–173

Ebisawa T, Karne S, Lerner MR, Reppert SM (1994) Expression cloning of a high-affinity melatonin receptor from Xenopus dermal melanophores. Proc Natl Acad Sci USA 91:6133–6137

Eckert RL, Katzenellenbogen BS (1982) Effects of estrogens and antiestrogens on estrogen receptor dynamics and the induction of progesterone receptor in MCF-7 human breast cancer cells. Cancer Res 42:139–144

Edwards DP, Chamness GC, McGuire WL (1979) Estrogen and progesterone receptor proteins in breast cancer. Biochem Biophys Acta 5460:457–486

Forman BM, Chen J, Blumberg B, Kliewer SA, Henshaw R, Ong ES, Evans RM (1994) Cross-talk among RORα1 and the Rev-erb family of orphan nuclear receptors. Mol Endocrinol 8: 1253–1261

Godson C, Reppert SM (1997) The Mel1a melatonin receptor is coupled to parallel signal transduction pathways. Endocrinology 138:397–404

Giguere V, Tini M, Flock G, Ong E, Evans RM, Otulakoski G (1994) Isoform-specific amino-terminal domains indicate DNA-binding properties of RORα, a novel family of orphan hormone nuclear receptors. Genes Dev 8:538–553

Hafner F, Holler E, von Angerer E (1996) Effect of growth factors on estrogen receptor mediated gene expression. J Steroid Biochem Mol Biol 58:385–393

Hazelrigg DG, Thompson M, Hastings MH, Morgan PJ (1996) Regulation of mitogen-activated protein kinase in the pars tuberalis of ovine pituitary: interactions between melatonin, insulin-like growth factor-1, and forskolin. Endocrinology 137:210–218

Hill SM, Blask DE (1988) Effects of the pineal hormone melatonin on the proliferation and the morphological characteristics of human breast cancer cells (MCF-7) in culture. Cancer Res 48:6121–6126

Hill SM, Spriggs LL, Simon MA, Muraoka H, Blask DE (1992) The growth inhibitory action of melatonin on human breast cancer cells is linked to the estrogen response system. Cancer Letters 64:249–256

Horwitz KB, McGuire WL (1978) Estrogen control of progesterone receptor in human breast cancer: correlations with nuclear processing of estrogen receptors. J Biol Chem 253:2223–2228

Kane CD, Means AR (1998) Ligand independent activation of orphan receptor mediated transcription by Ca++/calmodulin-dependent protein kinase IV. 80th Ann Mtg Endocrine Society, New Orleans, LA, Abstract # OR14–1, p 71

Kato S, Endoh H, Masuhiro Y, Kitamoto T, Uchiyama S, Sasaki H, Masushige S, Gotch Y, Nishide E, Chambon P (1995) Activation of the estrogen receptor through phosphorylation by the mitogen activated protein kinase. Science 270:1491–1494

Katzenellenbogen BS (1980) Dynamics of steroid hormone receptor action. Ann Rev Physiol 42:17–35

Klein-Hitpass L, Schorpp M, Wagner U, Ryffel GU (1986) An estrogen-responsive element derived from the 5′-flanking region of the Xenopus vitellogenin A2 gene functions in transfected human cells. Cell 46:1053–1061

Laitinen JT, Saavedra JM (1990) Characterization of melatonin receptors in the rat suprachiasmatic nuclei: modulation of affinity with cations and guanine nucleotides. Endocrinology 126: 2210–2215

Lawson NO, Wee B, Blask DE, Castles C, Spriggs LL, Hill SM (1992) Modulation of hypothalamic estrogen receptors by melatonin in female LSH/SsLak golden hamsters. J Biol Reprod 47:1082–1090

Le Goff P, Montano MM., Shodin DJ, Katzenellenbogen BS (1994) Phosphorylation of the human estrogen receptor. J Biol Chem 269:4458–4466

Lerner AB, Case JD, Heinzelman RV (1959) Structure of melatonin. J Am Chem Soc 81:6084–6085

Lippman ME, Bolan G, Huff K (1976) The effects of estrogen and antiestrogen on hormone-responsive human breast cancer in long-term tissue culture. Cancer Res 36:4595–4601

Luttrell LM, Della Rocca GJ, van Biesen T, Lattrell DK, Lefkowitz RJ (1997) G$\beta\gamma$ subunits mediate Src-dependent phosphorylation of the epidermal growth factor receptor. A scaffold for G protein-coupled receptor-mediated Ras activity. J Biol Chem 272:4637–4644

McNulty S, Morgan PJ, Thompson M, Davidson G, Lawson W, Hastings MH (1994) Phospholipases and melatonin signal transduction in the ovine pars tuberalis. Mol Cell Endocrinol 99:73–79

McNulty S, Ross AW, Barrett P, Hastings MH, Morgan PJ (1994) Melatonin regulates the phosphorylation of CREB in ovine pars tuberalis. J Neuroendocrinol 6:523–532

Menendez-Pelaez A, Reiter RJ (1993) Distribution of melatonin in mammalian tissues: the relative importance of nuclear versus cytoplasmic localization. J Pineal Res 15:59–69

Molis TM, Spriggs LL, Hill SM (1994) Modulation of estrogen receptor mRNA expression by melatonin in MCF-7 human breast cancer cells. Mol Endocrinol 8:1681–1690

Molis TM, Spriggs LL, Jupiter Y, Hill SM (1995) Melatonin modulation of estrogen-regulated proteins, growth factors, and proto-oncogenes in human breast cancer. J Pineal Res 18:93–103

Molis TM, Walters MR, Hill SM (1993) Melatonin modulation of estrogen receptor expression in MCF-7 human breast cancer cells. Int J Oncol 3:687–694

Morgan PJ, Lawson W, Davidson G, Howell HE (1989) Guanine nucleotides regulate the affinity of melatonin receptors in the ovine pars tuberalis. Neuroendocrinology 50:359–362

Morgan PJ, Perry B, Edward HH, Helliwell R, Sugden D (1994) Melatonin receptors, localization, molecular pharmacology and physiological significance. Neurochem Int 24:101–146

Murphy LC, Sutherland RL (1985) Differential effects of tamoxifen and analogs with non-basic side chains on cell proliferation in vitro. Endocrinology 116:1071–1078

Pace AM, Faure M, Bourne HR (1995) Gi1-mediated activation of the MAP kinase cascade. Mol Biol Cell 6:1685–1695

Pearson WR (1990) Rapid ans sensitive sequence comparison with FASTP and FASTA. Methods Enzymol 183:63–98

Ram PR, Hill SM (1995) Melatonin's inhibition of breast cancer cell proliferation is mediated through the RORα receptor pathway. Seventh Int. Congress on Hormones and Cancer, Quebec, Canada, p 62

Ram PT, Kiefer T, Silverman M, Song Y, Brown GM, Hill SM (1998) Estrogen receptor transactivation in MCF-7 breast cancer cells by melatonin and growth factors. Mol Cell Endocrinol 141:53–64

Reiter RJ (1980) The pineal and its hormones in the control of reproduction in mammals. Endocr Rev 1:109–131

Reiter RJ (1987) The melatonin message: duration versus coincidence hypothesis. Life Science 40:2119–2124

Reiter RJ (1988) Neuroendocrinology of melatonin. In: Miles A, Philbrick DR, Thompson C (eds) Melatonin, Clinical Perspectives. Oxford University Press, New York, pp 1–42

Reiter RJ (1991) Melatonin: That ubiquitously acting pineal hormone. Psychological Science 6:223–228

Reppert SM, Weaver DR, Ebisawa T (1994) Cloning and characterization of a melatonin receptor that mediates reproductive and circadian responses. Neuron 13:1177–1185

Rivkees SA, Carlson LL, Reppert SM (1989) Guanine nucleotide binding protein regulation of melatonin receptors in lizard brain. Proc Natl Acad Sci USA 86:11543–11546

Rothenberg PL, Kahn CR (1988) Insulin inhibits pertussis toxin-catalyzed ADP-ribosylation of G-proteins. Evidence for a novel interaction between insulin receptors and G-proteins. J Biol Chem 263:15546–15552

Saceda M, Lippman ME, Chambon P, Lindsey RL, Ponglikitmongkol M, Puente M, Martin MB (1988) Regulation of estrogen receptor in MCF-7 cells by estradiol. Mol Endocrinol 2:1157–1162

Sanchez-Barcelo EJ, Corral SC, Mediavilla MD (1988) Influence of pineal gland function on the initiation and growth of hormone-dependent breast tumors. Possible mechanisms. In: Gupta D, Attanasio A, Reiter RJ (eds) The Pineal Gland and Cancer. Brain Res. Promotion, Tübingen, Germany, pp 221–232

Shafie SM, Gibson SL, Hilf R (1977) Effects of insulin and estrogen on hormone binding in the R3230 AC mammary adenocarcinoma. Cancer Res 37:4641–4649

Song Y, Chan CWY, Brown GM, Pang SF, Silverman M (1997) Studies of the renal actions of melatonin: evidence that the effects are mediated by 37 kD receptors of the Mel1a subtype localized primarily to the basolateral membrane of the proximal tubule. FASEB J 11:93–100

Stankov B, Lucini V, Scaglione F (1991) 2-[125]iodomelatonin binding in normal and neoplastic tissues: In: Fraschini F, Reiter RJ (eds) Role of Melatonin and Pineals Peptides in Neuroimmunomodulation. Plenum, New York, pp 117–125

Steinhilber D, Brungs M, Werz O, Wiesenberg I, Danielsson C, Kahlen JP, Nayeri S, Schrader M, Carlberg C (1995) The nuclear receptor for melatonin represses 5-lipoxygenase gene expression in human B lymphocytes. J Biol Chem 270:7037–7040

Sutherland RL, Green MD, Hall RE, Reddle RR, Taylor IW (1983) Tamoxifen induces accumulation of MCF-7 human mammary carcinoma cells in G_0/G_1 phase of the cell cycle. Eur J Cancer Clin Oncol 19:615–621

Tamarkin L, Danforth D, Lichter A, DeMoss E, Cohen M, Chabner B, Lippman ME (1982) Decreased nocturnal plasma melatonin peak in patients with estrogen receptor positive breast cancer. Science 216:1003–1005

Tapp E (1982) The pineal gland in malignancy. In: Reiter RJ (ed) The Pineal Gland, Vol. 3, CRC Press, Boca Raton, pp 171–188

Thompson EW, Katz D, Shima TB, Wakeling AE, Lippman ME, Dickson RB (1989) ICI 164,384, a pure antagonist of estrogen-stimulated MCF-7 cell proliferation and invasiveness. Cancer Res 49:6929–6934

Tini M, Fraser A, Giguere V (1995) Functional interactions between retinoic acid receptors, related orphan receptor (RORα) and the retinoic acid receptors in the regulation of the μF-crytalline promoter. J Biol Chem 270:20156–20162

Tzukerman MT, Esty A, Santiso-Mere D, Danielian P, Parker MG, Stein RB, Pike JW, McDonnell DP (1994) Human estrogen receptor transcriptional capacity is determined by both cellular and promoter context and mediated by two functionally distinct intramolecular regions. Mol Endocrinol 9:21–30

Vanecek J, Pavlik A, Illnerova H (1987) Hypothalamic receptor sites revealed by autoradiography. Brain Res 435:359–363

Vanecek J (1995) Melatonin inhibits increase of intracellular calcium and cyclic AMP in neonatal rat pituitary via independent pathways. Mol Cell Endocrinol 107:149–153

Weaver DR, Namboordiri MAA, Reppert SM (1988) Iodinated melatonin mimics melatonin action and reveals discrete binding sites in fetal brain. FEBS Lett 228:123–127

Weisz A, Breciani F (1988) Estrogen induces expression of c-fos and c-myc protooncogenes in rat uterus. Mol Endocrinol 2:816–824

Welsch CW, Nagasawa H (1977) Prolactin and murine mammary tumorigenesis: a review. Cancer Res 37:951–963

Wiesenberg I, Missbach M, Kahlen JP, Schrader M, Carlberg C (1995) Transcriptional activation of the nuclear receptor RZRα by the pineal gland hormone melatonin and identification of CGP 52608 as a synthetic ligand. Nucleic Acid Res 23:327–333

Williams LM, Morgan PJ, Hastings MH, Lawson W, Davidson G, Howell HE (1989) Melatonin receptor sites in the Syrian hamster brain and pituitary: localization and characterization using [125I]iodinated melatonin. J Neuroendocrinol 1:315–329

17 Benign and Tumor Prostate Cells as Melatonin Target Sites

Nava Zisapel

Abstract

Melatonin, secreted nocturnally by the pineal gland into the circulation, affects behavioral and physiological adaptation of the organism to the light and dark cycles. To date, the physiological study of melatonin's role has focused primarily on the pituitary-gonadal axis, and there is little information on the direct effects of melatonin on peripheral organs. The prostate gland is apparently a peripheral target for melatonin. Recent data indicate that human benign and tumor prostate epithelial cells express functional melatonin receptors. At physiological concentrations, melatonin is able to suppress DNA synthesis transiently and inhibit cyclic guanosine monophosphate (cGMP) in the benign prostate epithelial cells. These effects are mediated via heterotrimeric guanosine triphosphate (GTP)-binding proteins and involve TGFβ1. The transient nature of melatonin responses in the prostate cells reflects inactivation of the receptors via a protein kinase C (PKC)-mediated pathway. Recent data demonstrate differential interaction of melatonin with androgen receptor-negative and -positive prostate cells. In the androgen-sensitive LNCaP cells, melatonin attenuates cell growth. As in the benign epithelial cells, this attenuation is transient, probably due to inactivation of the receptors via a PKC-mediated pathway. The androgen insensitive PC3 cells express low affinity melatonin binding sites. Unlike the benign cells, melatonin receptors in the PC3 cells are coupled to pertussis toxin (PTX)-sensitive G proteins to induce cell density-dependent changes in cGMP, cyclic adenosine monophosphate (cAMP) and cell growth. The data imply that androgen receptors play a crucial role in the melatonin-mediated regulation of prostate cell growth. The possibility that melatonin may impinge on androgen-dependent pathways in the prostate is extremely intriguing, since it may lead to the development of new strategies to manipulate benign and tumor prostate growth.

17.1 Effects of Melatonin on the Prostate Gland In Vivo

The nocturnal production of melatonin in the pineal gland and its release into the circulation convey a darkness signal to the organism. Melatonin is a low molecular weight, uncharged, lipophilic molecule. Thus, it is permeable to biological membranes and is present in all biological fluids and, essentially, in every cell. Certain parts of the brain, especially the hypothalamus, have been implicated as the sites of melatonin's antigonadal and neuroendocrine activities (Glass and Lynch 1981).

High and low affinity membrane-associated melatonin binding sites have been found in the mammalian brain, retina, pars tuberalis of the pituitary, and the newborn rat pituitary (Reiter 1991). High affinity receptors of the Mel-1 family have been cloned from a number of species and shown to belong to the superfamily of guanosine triphosphate (GTP)-binding protein-coupled receptors (Reppert et al. 1996). The existence of peripheral melatonin receptors has been given little scientific attention, in spite of the fact that the melatonin signal is widely distributed within the organism via the blood circulation and is essentially available in and to every cell.

There is some evidence for direct effects of melatonin on the prostate, at least in rodents. Oral administration of melatonin delayed pubertal development of male and female rats with the most pronounced effects on prostate development (Laudon et al. 1988). Administration of high doses of melatonin (> 8 mg per day) to intact rats (Yamada 1992), castrated rats treated with testosterone (Debeljuk et al. 1970), and hypophysectomized rats treated with human chorionic gonadotropin (Debeljuk et al. 1971) decreased prostate weight in these animals. Orally administered melatonin effectively prevented the testosterone-mediated regrowth of the prostate in adult castrated rats (Gilad et al. 1996). Melatonin inhibited the growth of R3327H Dunning prostatic adenocarcinoma in rats (Philo and Berkowitz 1988). Melatonin enhanced growth of a transplantable androgen-insensitive prostatic adenocarcinoma in rats (Buzzell 1988). This view is further substantiated by recent findings on specific interactions of melatonin with human benign and tumor prostate cells.

17.2 Melatonin Action in Prostate Epithelial Cells In Vitro

17.2.1 Benign Prostate Hypertrophy Cells

Benign prostatic hyperplasia (BPH) is a disease of aging men characterized by benign enlargement of the prostate which eventually causes urinary obstruction. BPH frequently requires surgery to relieve the urinary retention, nocturia, and micturition. In vitro autoradiography revealed specific binding sites for the physiologically active ^{125}I-labeled melatonin (I-mel) in human benign prostate tissue (Laudon et al. 1996). Equilibrium binding and competition experiments revealed reversible, saturable, and specific binding of I-mel (apparent half saturation at 120 pM) to sites primarily associated with the microsomal fraction of the prostate epithelial cells. The I-mel binding to the microsomal fraction was specifically inhibited by GTP analogs, suggesting that the melatonin receptors in human prostatic epithelial cells were coupled to GTP-binding proteins (Gilad et al. 1997b).

High affinity melatonin receptors have been cloned and shown to be coupled to pertussis toxin (PTX)-sensitive GTP-binding proteins, presumably of the Gi/Go types (Reppert et al. 1996). The involvement of cholera toxin (CTX)-sensitive GTP-binding proteins, presumably of the Gs type, has also been demonstrated in melatonin responses (Bubis and Zisapel 1995a; Gilad et al. 1997b).

Microsomes may be the site of receptor processing, targeting, and recycling (Steinsapir and Muldoon 1991). Some G protein-coupled receptors undergo recycling through intracellular endosomes as part of a ligand-induced desensitization mechanism. It is thus possible that a microsomal melatonin binding in the prostate cell represents recycled G protein-coupled receptors. Binding of hormones to receptors coupled to heteromeric G proteins enhances the exchange of guanine nucleotides and GTP hydrolase activity of the coupled G proteins, whereas GTP analogs inhibit high affinity agonist binding to the receptors (Pfeufer 1977; Birnbaumer et al. 1990). Melatonin (1 nM) enhanced binding of the nonhydrolyzable GTP analog guanosine 5′-O-(3-^{35}S-thiotriphosphate (GTPγ^{35}S) to the microsomal fraction by about 40 % (Zisapel et al. 1998). The nonhydrolyzable guanine nucleotide analogs GDPβS and GTPγ S (100 nM each) inhibited I-mel and basal GTPγ^{35}S binding by about 50 % and prevented the enhancement by melatonin of GTPγ^{35}S binding to the microsomal fraction (Zisapel et al. 1998). Pretreatment of the microsomal fraction with PTX and cholera toxins (CTX), which mediate adenosine diphosphate (ADP) ribosylation of the GTP-binding proteins of the Gi/Go and Gs types, respectively, suppressed I-mel binding. In addition, CTX ablated the melatonin-mediated enhancement of GTPγ^{35}S binding, whereas PTX did not (Zisapel et al. 1998). These data are compatible with the hypothesis that melatonin binding sites in the microsomal fraction of the prostate epithelial cells are coupled to CTX-sensitive G proteins.

Cultured epithelial cells from the human benign prostate display high affinity melatonin binding (Gilad et al. 1996). The whole cell sites resembled the microsomal sites in apparent affinity, specificity, and sensitivity to GTP analogs (Gilad et al. 1996). In the cultured benign prostate epithelial cells, melatonin, at physiological concentrations, suppressed ^{3}H-thymidine incorporation and cyclic guanosine monophosphate (cGMP) levels. The effects of melatonin on ^{3}H-thymidine incorporation were transient, suggesting inactivation of the receptors (Gilad et al. 1996). Treatment of benign prostate epithelial cells in culture with the protein kinase C (PKC) activator, phorbol-12-myristate-13-acetate (TPA), markedly reduced the apparent affinity of I-mel (Gilad et al. 1997a). In addition, TPA ablated the cell responses to melatonin, namely, the suppression of ^{3}H-thymidine incorporation and cGMP levels. Pretreatment of the cells with the PKC inhibitor bisindolylmaleimide (GF-109203) prevented the TPA effects on ^{125}I-melatonin binding and responses. In addition, GF-109203 slowed down the inactivation of the melatonin-mediated inhibition of ^{3}H-thymidine incorporation (Gilad et al. 1997a). These data show that melatonin receptors are desensitized by PKC. The data also imply that the transient response to melatonin may be the outcome of a direct or indirect melatonin-mediated activation of endogenous PKC.

Our previous studies on murine melanoma cells indicated that melatonin modulates the constitutive secretion of proteins from cells (Bubis and Zisapel 1995b; Bubis et al. 1996). Accordingly, some of the responses to melatonin in the prostate cells might be the outcome of its influence on trafficking and secretion of proteins in general and, particularly, growth factors. Our studies were focused on transforming growth factor-β1 (TGFβ1). TGFβ1 immunoreactivity has been exclusively identified in the secretory epithelial cells with total lack of immunoreactivity among the stroma elements (Kyprianou et al. 1996; Story et al. 1996; Perry et al. 1997). TGFβ1 appears to be an inhibitor of normal epithelial cell proliferation (Steiner 1995) and may induce apoptotic prostate epithelial cell death (Sutkowski et al.1992).

Incubation of the prostate epithelial cells with TGFβ1 resulted in a time- and dose-dependent inhibition of ^{3}H-thymidine incorporation into the cells (Rimler et al. 1999). Melatonin (10–500 pM) inhibited ^{3}H-thymidine incorporation, and its effects were attenuated at higher (1–10 nM) concentrations. In the presence of submaximal doses of TGFβ1, the inhibitory effect of melatonin was maintained over the entire concentration range tested (10 pM to 10 nM). The inhibition of ^{3}H-thymidine incorporation by TGFβ1 was more pronounced in the absence of dihydrotestosterone (DHT) than in its presence, and melatonin had no further effect. Melatonin enhanced the release of proteins from the cells, among them proteins recognized by specific TGFβ1 antisera (Rimler et al. 1999). The TGFβ1-neutralizing antisera prevented the inhibitory action of melatonin on ^{3}H-thymidine incorporation into the cells (Rimler et al. 1999). These data indicate a role for TGFβ1 in the melatonin-mediated attenuation of benign prostate epithelial cell growth.

17.2.2 Prostate Cancer Cells

Prostate cancer (PC) is a major problem in the aging human male and a leading cause of male death in the Western world. Since androgens stimulate the growth of hormone-dependent PC, complete removal of androgens by means of functional castration (e. g., use of potent LHRH agonists or estrogens) and of antiandrogens has become the most prominent treatment for PC. Despite hormonal therapy, PC progresses from an androgen-dependent to an androgen-independent phase.

Nocturnal melatonin production is diminished in a number of human malignancies (breast, liver, kidney, upper respiratory tract, and skin) including prostate carcinoma (Touitou et al. 1985; Bartsch et al. 1992, 1997). Administration of melatonin has recently been shown to reverse the clinical resistance of metastatic prostate cancer patients to the elimination of testicular androgens (that is, chemical castration using luteinizing hormone-releasing hormone analogs) (Lissoni et al. 1997).

Some evidence suggests that melatonin differentially affects hormone-responsive and -nonresponsive prostate tumors: The growth of the androgen-sensitive R3327H Dunning prostatic adenocarcinoma is attenuated by melatonin (Philo and Berkowitz 1988). whereas melatonin enhanced growth of a transplantable androgen-insensitive prostatic adenocarcinoma in rats (Buzzell 1988). We have therefore explored melatonin responses in two prostate tumor cell lines, LNCaP, which is androgen-sensitive, and PC3, which is insensitive to androgens.

In the androgen-sensitive LNCaP cells, melatonin inhibited ^{3}H-thymidine incorporation at physiological concentrations (Lupowitz and Zisapel 1999). This response decayed within 24 hr. The inactivation of the response slowed down in the presence of the PKC inhibitor GF-109203X. Suppression of DNA content was observed in cells treated for 2 days with melatonin (0.1 nM). In addition, melatonin slightly decreased cell viability (Lupowitz and Zisapel 1999). Thus, the interaction of melatonin with the LNCaP cells, as in the benign epithelial cells, leads to attenuation of cell growth. As in the benign cells, this attenuation is transient, probably due to inactivation of the receptors via a PKC-mediated pathway.

PC3 cells bound I-mel with low affinity (Kd ca. 0.9 nM), regardless of whether the cultures were subconfluent (low density) or confluent (high density) (Gilad et al. 1999). Melatonin enhanced cGMP and ^{3}H-thymidine incorporation at low cell density but attenuated them at high cell density. In addition, melatonin inhibited cAMP at low density but augmented it at high cell density (Gilad et al. 1999). These effects were associated with an increase in cell count at low but not high density cultures. PTX treatment suppressed ^{125}I-melatonin binding and ablated all the effects of melatonin on ^{3}H-thymidine incorporation, cAMP, and cGMP at both cell densities (Gilad et al. 1999). CTX treatment failed to block the effects of melatonin on ^{3}H-thymidine incorporation but prevented the modulation by melatonin of cAMP at low and cGMP at high cell density. The cGMP analog 8-Br-cGMP inhibited melatonin's effects on ^{3}H-thymidine incorporation at both cell densities. H89, a protein kinase A inhibitor, prevented melatonin's effects on ^{3}H-thymidine incorporation at low but not high cell density (Gilad et al. 1999). These results indicated that, unlike the benign cells, the melatonin receptors in the PC3 cells are coupled to PTX-sensitive G proteins to induce cell density-dependent changes in cGMP, cAMP, and cell growth. They also imply a differential interaction of melatonin with androgen receptor-negative and -positive prostate cells.

17.3 Melatonin Action in Prostate Epithelial Cells: Towards a Unifying Hypothesis

Androgens are known to regulate epithelial cell function, (i. e., production of secretory proteins like prostate specific antigen and human glandular kallikrein), but it is still controversial whether they directly stimulate epithelial cell growth. Androgen ablation leads to an increase in TGFβ1 and TGF receptors. Accordingly, the net effect of androgen withdrawal, which leads to prostate involution, might be due at least in part to the increased apoptotic death induced by TGFβ1 (Sutkowski et al.1992; Kim et al. 1996). While the role of TGFβ1 in melatonin-mediated inhibition of LNCaP cell growth has yet to be demonstrated, it may be useful to present at this stage a unifying, testable hypothesis to explain the relationship between melatonin, androgens, TGFβ1, and the regulation of prostate epithelial cell growth and survival. Androgens may inhibit TGFβ1 expression and, besides, may have a direct stimulatory effect on cell growth. Melatonin may act to suppress androgen receptor level or function and, besides, to enhance TGFβ1 secretion from the cells. The inhibition by melatonin of androgen receptor activity will result in attenuated cell growth. Under the functional absence of androgen (due to reduced androgen receptor activity), an increase in TGFβ1 secretion may lead to apoptotic death of the cells.

Prostate cells may serve as useful models in which the cross talk between melatonin growth factor and hormone systems can be thoroughly explored. In addition, the possibility that melatonin may impinge on androgen-dependent pathways in the prostate may lead to the development of new strategies to regulate benign and tumor prostate growth.

References

Bartsch C, Bartsch H, Schmidt A, Ilg S, Bichler KH, Flüchter SH (1992) Melatonin and 6-sulfatoxyme-latonin circadian rhythms in serum and urine of primary prostate cancer patients: evidence for reduced pineal activity and relevance of urinary determinations. Clin Chim Acta 209:153–167

Bartsch C, Bartsch H, Karenovics A, Franz H, Peiker G, Mecke D (1997) Nocturnal urinary 6-sulfat-oxymelatonin excretion is decreased in primary breast cancer patients compared to age-matched controls and shows negative correlation with tumor-size. J Pineal Res 23:53–58

Birnbaumer L, Abramowitz J, Brown AM (1990) Receptor-effector coupling by G proteins. Biochim Biophys Acta 1031:163–224

Bubis M, Zisapel N (1995a) Facilitation and inhibition of G-protein regulated protein secretion by melatonin. Neurochem Int 27:177–183

Bubis M, Zisapel N (1995b) Modulation by melatonin of protein secretion from melanoma cells: is cAMP involved? Mol Cell Endocrinol 112:169–175

Bubis M, Anis Y, Zisapel N (1996) Enhancement by melatonin of GTP exchange and ADP ribosyla-tion reactions. Mol Cell Endocrinol 123:139–148

Buzzell GR (1988) Studies on the effects of the pineal hormone melatonin on an androgen-insensitive rat prostatic adrenocarcinoma, the Dunning R 3327 HIF tumor. J Neural Transm 72:131–140

Debeljuk L, Feder VM, Paulucci OA (1970) Effects of melatonin on changes induced by castration and testosterone in sexual structures of male rats. Endocrinology 87:1358–1360

Debeljuk L, Vilchez JA, Schnitman MA, Paulucci OA Feder VM (1971) Further evidence for a peripheral action of melatonin. Endocrinology 89:1117–1119

Gilad E, Laudon M, Matzkin H, Pick E, Sofer M, Braf Z, Zisapel N(1996) Functional melatonin receptors in human prostate epithelial cells. Endocrinology 137:1412–1417

Gilad E, Matzkin H, Zisapel N (1997a) Inactivation of melatonin receptors by protein kinase C in human prostate epithelial cells. Endocrinology 138:4255–4261

Gilad E, Pick E, Matzkin H, Zisapel N (1997b) Melatonin receptors in benign prostate epithelial cells: evidence for the involvement of cholera and pertussis toxins-sensitive G proteins in their signal transduction pathways. Prostate 159:1069–1073

Gilad, E, Matzkin H Zisapel N (1998) Evidence for local action of melatonin on rat prostate. J Urol 159:1069–1073

Gilad E, Laufer M, Matzkin H, Zisapel N (1999) Melatonin receptors in PC3 human prostate tumor cells. J Pineal Res 26:211–220

Glass JD, Lynch JR (1981) Melatonin: identification of antigonadal action in mouse brain Science 214:821–823

Kim IY, Kim JH, Zelner DJ, Ahn HJ, Sensibar JA, Lee C (1996) Transforming growth factor $\beta1$ is mediator of androgen-regulated growth arrest in an androgen-responsive prostatic cancer cell line, LNCaP. Endocrinology 137:991–999

Kyprianou N, Huacheng T, Jacobs SC (1996) Apoptotic versus proliferative activities in human benign prostate hyperplasia. Human Pathol 27:668–675

Laudon M, Yaron Z, Zisapel N (1988) N-3,5-dinitrophenyl)-5-methoxytryptamine, a novel melatonin antagonist: effects on sexual maturation of the male and female rat and on oestrous cycles of the female rat. J Endocrinol 116:43–53

Laudon M, Gilad E, Matzkin H, Braf Z, Zisapel N (1996) Putative melatonin receptors in benign human prostate tissue. J Clin Endocrinol Metab 81:1336–1342

Lissoni P, Cazzaniga M, Tancini G, Scardino E, Musci R, Barni S, Maffezzini M, Meroni T, Rocco F, Conti A, Maestroni G (1997) Reversal of clinical resistance to LHRH analogue in metastatic prostate cancer by the pineal hormone melatonin: efficacy of LHRH analogue plus melatonin in patients progressing on LHRH analogue alone. Eur Urol 31:178–181

Lupowitz Z, Zisapel N (1999) Hormonal interactions in human prostate tumor LNCaP cells. J Steroid Biochem Mol Biol 68:83–88

Perry KT, Anthony CT, Steiner MS (1997) Immunohistochemical localization of TGF$\beta1$, TGF$\beta2$ and TGF$\beta3$ in normal and malignant human prostate. Prostate 33:133–140

Pfeufer T (1977) GTP binding proteins in membranes and the control of adenylate cyclase activity. J Biol Chem 25:7224–7234

Philo R, Berkowitz AS (1988) Inhibition of Dunning tumor growth by melatonin. J Urol 139:1099–1102

Reiter RJ (1991) Pineal melatonin: Cell biology of Its synthesis and of Its physiological interactions. Endocr Rev 12:151–180

Reppert S, Weaver D, Godson C (1996) Melatonin receptors step into the light: cloning and classification of subtypes. Trends Pharmacol Sci 17:100–102

Rimler A, Matzkin H, Zisapel N (1999) Cross talk between melatonin and TGF$\beta1$ in human benign prostate epithelial cells. Prostate 40:211–217

Steiner MS (1995) Review of peptide growth factors in benign prostatic hyperplasia and urological malignancy. J Urol 153:1085–1096

Steinsapir J, Muldoon GM (1991) Role of microsomal receptors in steroid hormone action. Steroids 56:66–71

Story M, Hopp K, Molter M (1996) Expression of transforming growth factor beta-1 (TGF$\beta1$)-$\beta2$, and -$\beta3$ by cultured human prostate cells. J Cell Physiol 169:97–107

Sutkowski DM, Fong CJ, Sensibar JA, Rademaker AW, Sherwood ER, Kozlowski (1992) Interaction of epidermal growth factor and transforming growth factor beta in human prostatic epithelial cells in culture. Prostate 21:133–139

Touitou Y, Fevre-Montange M, Proust J, Klinger E, Nakache J (1985) Age- and sex-associated modification of plasma melatonin concentrations in man. Relationship to pathology, malignant or not, and autopsy findings. Acta Endocrinol 108:135–144

Yamada K (1992) Effects of melatonin on reproductive and accessory reproductive organs in male rats. Chem Pharm Bull 40:1066–1068

Zisapel N, Matzkin H, Gilad E (1988) Melatonin receptors in human prostate epithelial cells. In: Touitou Y (ed) Biological clocks – mechanisms and applications. Excerpta Medica International Congress series 1152, Elsevier, Amsterdam pp 295–299

B. Actions Via the Immune System

18 Neuroimmunomodulation Via the Autonomic Nervous System

Konrad Schauenstein, Ingo Rinner, Peter Felsner, Albert Wölfler, Helga S. Haas, J. Ross Stevenson, Werner Korsatko, Peter M. Liebmann

Abstract

The existence of an intense dialogue between the brain and the immune system is well-established, and the role of the autonomic nervous system as the translator for both systems is becoming increasingly evident. The development and functions of lymphocytes are influenced by transmitters of the autonomic nervous system. On the other hand, the "feedback" of the activated immune system to the brain is also modulated by autonomic mechanisms. Finally, besides expressing adrenergic and cholinergic receptors, lymphocytes were recently shown to synthesize and release catecholamines and acetylcholine, pointing to a possible role of these mediators in the intrinsic regulation of the immune system. In this review, we will discuss data to these points as obtained in experimental studies in the rat model.

18.1 Introduction

The extrinsic regulation of the immune system through neuroendocrine signals is well-established, as is the fact that the immune system in turn informs the brain about contacts with antigens via "immunotransmitters," i.e., cytokines and/or hormones with central effects (Cotman et al. 1987). These data that have accumulated during the past 20 years contributed to the vision of the immune system as "the sixth sense" (Blalock 1994). While there is certainly more work needed to define the physiology of this concept in all details, strong evidence has been obtained that the immune-neuro-endocrine dialogue is of relevance for the homeostasis of the immune response, as defects in the activation of the hypothalamopituitary-adrenal (HPA) axis by immune signals were found to be associated with and/or to predispose to spontaneously occurring (Schauenstein et al. 1987) and experimentally induced autoimmune diseases in animal models (Sternberg et al. 1989; Mason et al. 1990), and there is evidence that the same is also true in humans (Berczi et al. 1993). A large body of recent literature data strongly suggests that this dialogue involves not only the hypothalamus but several other brain areas, notably the structures of the "limbic system" (for review, see Haas and Schauenstein 1997).

One of the main goals of our studies is to define the relevance of the autonomic nervous system within this bi-directional relationship between brain and immune system (Schauenstein et al 1994). In the following, we will review experimental data of our studies in the rat model which were performed to examine if and how in vivo disturbances in the adrenergic/cholinergic balance may affect the immune-neuro-endocrine dialogue and to understand better the functional relationships between the immune and the adrenergic and cholinergic systems.

18.2 Adrenergic In Vivo Effects on Immune Functions

Beginning in the early 1970s, a vast literature on morphological and functional inter-relationship between the sympathetic nervous system and the immune system has accumulated. Lymphoid organs are strongly innervated by adrenergic fibers, the terminals of which come in close "synapsis-like" apposition with lymphoid cells (Felten et al. 1987). Furthermore, lymphocytes express β-adrenergic receptors in dependence of lineage and activation stage (Khan et al. 1986) and the presence of α-adrenergic receptors by radioligand binding has been documented for human lymphocytes (Titinchi and Clark 1984) and splenic lymphocytes in the rat (Fernandez-Lopez and Pazos, 1994). Concerning adrenergic effects on immune functions, the data are numerous and conflicting. Both enhancement and suppression of lymphocyte functions have been ascribed to either α- or β-adrenergic mechanisms, which may be due to the complexity of both systems involved but also to differences in the experimental approach, particularly between in vivo and in vitro studies.

In our own studies, we sought to define in vivo effects of chronically (20 h) increased levels of peripheral catecholamines such as epinephrine (E) and norepinephrine (NE) on lymphocyte functions in the rat model. The experimental approach consisted in the s. c. implantation of retard tablets that continuously release defined amounts of adrenergic agonists or antagonists into the circulation (Korsatko et al 1982). This technology avoids handling stress effects due to repeated injections that may interfere with effects of the adrenergic treatment. The data obtained with this system (Felsner et al. 1992) are summarized as follows: (1) a 20-h treatment with catecholamines had no (E) or only a marginal (NE) suppressive effect on the ex vivo proliferative response of peripheral blood lymphocytes (PBL) to concanavalin A, (2) combining either catecholamine with propranolol or other β-blockers resulted in profound suppression of this T cell responsiveness, whereas concomitant α-blockade had no effect and β- or α-blockade per se had no measurable effect, and (3) in contrast to PBL, spleen cells were consistently resistant to chronic (20 h) α-adrenergic treatment. In the following, we have shown that this α-mediated suppression of lymphocyte activation does not represent a general stress symptom due to metabolic waste, nor is it due to secondarily elicited glucocorticoids. Furthermore, we could confirm this effect with the synthetic α_2-agonist clonidine, which, again only in combination with β-blockade, led to significant suppression of T cell proliferation (Felsner et al. 1995), whereas selective α_1- or β-agonists had no effect. Moreover, we observed an analogous suppressive effect on the proliferative T cell response after a single i.p. injection of the indirect sympatho-mimetic drug tyramine in combination with β-blockade. Thus, this phenomenon is not an artifact inherent to treatment with exogenous catecholamines but is likewise observed under enhanced release of endogenous noradrenaline (NA) from peripheral terminals. Interestingly, the effect of tyramine was again restricted to PBL, leaving the mitogenic reactivity of splenic T lymphocytes unaffected, which seemed puzzling, in view of the mentioned intense sympathetic innervation of the spleen (Felsner et al. 1995). These data were in agreement with earlier studies of Besedovsky et al. (1979) and contradicted, at least in the rat model, the significance of a β-adrenergic immuno-suppression in vivo.

Strong evidence was obtained in this experimental model that the pineal hormone melatonin plays an important and physiological role to protect the reactivity of lymphocytes from α-adrenergic inhibition (Liebmann et al. 1996). Experiments are presently underway to examine the mechanism of action of melatonin (for review see Liebmann et al. 2000) and if or to what extent this concept is valid in humans.

More recently, we started investigations into mechanisms of α_2-mediated in vivo effects on rat lymphocyte functions. As a first approach, we studied the time-dependence of adrenergic effects on the differential white blood cell count and on lymphocyte responsiveness in a range from 2 to 40 h of α-adrenergic treatment. The yet unpublished results of these experiments can be summarized as follows: (1) the s.c. implantation of a retard tablet containing 5 mg NE induces a long-lasting (>40 h) seven- to tenfold increase in NE plasma levels, (2) little or no changes in white blood cell counts were noted in the early phase, i.e., 2 and 4 h after application of NE and propranolol whereas, between 6 and 20 h of treatment, a significant lymphopenia and pronounced granulocytosis became evident, (3) after 40 h, the catecholamine effects on leukocyte distribution disappeared while NE levels were still elevated, (4) a pronounced inhibition of the reactivity of CD3$^+$ cells in the peripheral blood was seen after 2 and 4 h of treatment, in which phase splenic T lymphocytes were likewise suppressed by the α-adrenergic in vivo treatment, and (5) in the later phase, i.e., at 12 and 20 h of treatment, the effects on the T cell reactivity per se were no longer detectable, and the previously observed suppression of the concanavalin A response of PBL was obviously caused primarily by a quantitative decrease in the peripheral blood T cell number.

To conclude this section, our data in the rat model so far suggest that, via α_2-receptor activation, a chronic increase in circulating plasma catecholamines leads to a relatively short-lived suppression of peripheral lymphocyte functions, which is due both to shifts in leukocyte distribution and negative influences on the cellular responsiveness. These effects exhibit differences in their kinetics. In vitro studies are currently being performed to define adrenergic effects on T cell signaling and the mRNA and protein expression of lymphokines.

18.3 Cholinergic Immunomodulation

Until recently, much less was known about cholinergic-immune system interrelationships. The thymus has been reported to be innervated by cholinergic fibers (Bullock 1988), which was questioned by other authors (Nance et al. 1987). Concerning the spleen, there is no evidence of cholinergic innervation, although low activities of both acetylcholine (ACh) and its synthetic enzyme choline acetyltransferase (ChAT) have been detected in the spleens of several species (Felten and Felten 1991). Human and rodent lymphocytes from thymus, spleen, and peripheral blood were reported to express muscarinic and nicotinic cholinergic receptors (Maslinski 1989, Genaro et al. 1993), as well as a cellular acetylcholine esterase (AChE) (Szelenyi et al. 1987).

Recent results from our group obtained with a semiquantitative RT-PCR approach (Schauenstein et al., in press) suggest that rat lymphocytes derived from different lymphoid organs express mRNA for muscarinic m3 and nicotinic receptors (brain type). It was interesting to note that, for both receptors, the relative expression was highest in the thymus. This may be seen in context with reports that cholinergic signals are involved in normal thymocyte development. Vagotomy significantly decreases the export of immature thymus cells from the thymus, indicating that the cholinergic system regulates the trafficking of thymus cells (Antonica 1994). Singh and Fatani (1988) reported that thymic lymphopoiesis is associated with the presence of cholinergic nerves, and we have shown that cholinergic stimulation affected thymocyte apoptosis in murine fetal organ cultures via a nicotinergic effect on cortical thymic epithelial cells (Rinner et al. 1994, 1999).

Cholinergic agents were reported to influence immune functions in vivo and in vitro (Iliano et al. 1973; Strom et al. 1974; Rossi et al. 1989), although the physiological source of the specific ligand in immune organs, i.e., ACh, remained obscure. In our own studies, to investigate chronic cholinergic stimulation in the rat model using a similar technology as described for the adrenergic treatment, we obtained evidence that the cholinergic system is intrinsically involved in the immunoneuroendocrine dialogue (Rinner and Schauenstein 1991). This was concluded from the following data: (1) altering the endogenous cholinergic tonus by treatment with the AChE inhibitor physostigmine or the muscarinic antagonist atropine influenced lymphocyte functions of different compartments in different ways, suggesting an enhancing cholinergic effect on lymphocytes from thymus and spleen but not in the peripheral blood, (2) chronic physostigmine treatment prior to immunization with antigen was found to abrogate activation of the HPA axis due to immunization, and (3) 3 days after immunization with antigen, a transient increase in affinity and decreased numbers of muscarinic receptors in the gyrus hippocampalis was observed.

More recently, we obtained evidence that lymphocytes not only react to but also produce the neurotransmitter acetylcholine. Using a radioenzymatic method (Fonnum 1975), we were able to detect the synthesis of acetylcholine in homogenates of isolated rat lymphocytes from thymus, spleen, and peripheral blood, as well as in several murine and human lymphoid cell lines (Rinner and Schauenstein 1993). The activity of ChAT found in rat thymus lymphocytes was quantitatively very similar to what has been reported elsewhere with murine thymus tissue homogenates (Badamchian et al. 1992), and we concluded that this activity may have been entirely due to lymphocyte-derived ChAT rather than to cholinergic innervation. In very recent studies using a set of four different primers, we detected the ChAT mRNA message in thymic and peripheral rat lymphocytes by means of reverse transcriptase polymerase chain reaction techniques and showed that the synthesis and release of acetylcholine are increased in polyclonally activated cells (Rinner et al. 1998).

18.4 Conclusions

Our experimental results as reviewed above confirm that both the adrenergic and cholinergic autonomic nervous systems take part in the immune-neuroendocrine network. The integration of the immune system with the parasympathetic nervous system may be even closer and more direct than the one with the adrenergic system. Cholinergic in vivo stimulation was found to interfere directly with the activation of the HPA axis due to immunization, which may facilitate unwanted immune responses, but no such effect was found under adrenergic treatment (Felsner et al., unpublished observations). Acetylcholine was found to influence survival and differentiation of thymic lymphocytes, whereas catecholamines had no effect (Rinner et al., unpublished observations) and, finally, immune cells were found not only to express functional cholinergic receptors but also to be equipped to synthesize the specific ligand to these surface molecules, i.e., ACh, which may be instrumental to communication with the autonomic nervous system and contribute to autocrine immunoregulatory mechanisms. Even though catecholamines, too, have been detected in rat and human lymphocytes (Badamchian et al. 1992: Josefsson et al. 1996), our own attempts to detect the mRNA expression of synthesizing enzymes, i.e., tyrosine hydroxylase and dopamine β-hydroxylase, were unsuccessful (Rinner et al., unpublished results).

Studies are underway to investigate the mechanisms by which immune cells release acetylcholine and to define possible physiological role(s) of non-neuronal, lymphocyte-derived neurotransmitters in the regulation of the immune system and its dialogue with the peripheral and central nervous system.

Acknowledgements. The experiments reviewed herein were supported by the Austrian Science Foundation (Projects 7509, 7038, 9925, 11130, and 12679) and by the "Jubiläumsfonds der Österreichischen Nationalbank" (Projects 3556 and 4349).

References

Antonica A, Magni F, Mearini L, Paolocci N (1994) Vagal control of lymphocyte decrease from rat thymus. J Autonom Nerv Syst 48:187–197

Badamchian M, Damavandy H, Radojcic T, Bulloch K (1992) Choline O-acetyltransferase (ChAT) and muscarinic receptor in the Balb/C mouse thymus. Abstract, Satellite meeting of the 8th International Congress of Immunology "Advances in Psychoneuroimmunology", Budapest

Berczi I, Baragar FD, Chalmers IM, Keystone EC, Nagy E, Warrington RJ(1993) Hormones in self tolerance and autoimmunity: a role in pathogensis of rheumatoid arthritis? Autoimmunity 16:45–56

Besedovsky HO, Del Rey A, Sorkin E, Da Prada M, Keller HH (1979) Immunoregulation mediated by the sympathetic nervous system. Cell Immunol 48:346–355

Blalock JE (1994) The immune system – our sixth sense. The Immunologist 2:8–15

Bulloch K (1988) A comparative study of the autonomous nervous system innervation of the thymus in the mouse and chicken. Int J Neurosci 40:129–140

Cotman CW, Brinton RE, Galaburda A, MC Ewen B, Schneider D (1987) The neuro-immune-endocrine connection. Raven Press, New York

Felsner P, Hofer D, Rinner I, Mangge H, Gruber M, Korsatko W, Schauenstein K (1992) Continuous in vivo treatment with catecholamines suppresses in vitro reactivity of rat peripheral blood T-lymphocytes via a-meditated mechanisms. J Neuroimmunol 37:47–57

Felsner P, Hofer D, Rinner I, Porta S, Korsatko W, Schauenstein K (1995) Adrenergic suppression of peripheral blood T cell reactivity in the rat is due to activation of peripheral α_2-receptors. J Neuroimmunol 57:27–34

Felten SY, Felten DL (1991) Innervation of lymphoid tissue. In: Ader R, Felten DL, Cohen N (eds) Psychoneuroimmunology, 2nd edn. Academic Press, New York, pp 27–69

Felten DL, Felten SY, Bellinger DL, Carlson SL, Ackermann KD, Madden KS, Olschowski JA, Livnat S (1987) Noradrenergic sympathetic neural interactions with the immune system: structure and function. Immunol Rev 100:225–260

Fernandez-Lopez A, Pazos A (1994) Identification of alpha2-adrenoceptors in rat lymph nodes and spleen: an autoradiographic study. Eur J Pharmacol 252:333–336

Fonnum FA (1975) Rapid radiochemical method for the determination of choline acetyltransferase. J Neurochem 24:407–409

Genaro AM, Cremaschi GA, Borda ES (1993) Muscarinic cholinergic receptors on murine lymphocyte subpopulations. Selective interactions with second messenger response system upon pharmacological stimulation. Immunopharmacology 26:21–29

Haas HS, Schauenstein K (1997) Neuroimmunomodulation via limbic structures – the neuroanatomy of psychoimmunology. Progr Neurobiol 52:195–222

Iliano G, Tell GPE, Segal MI, Cuatrecasas P (1973) Guanosine 3″5′-cyclic monophophate and the action of insulin and acetylcholine. Proc Natl Acad Sci USA 70:2443–2447

Josefsson E, Bergquist J, Ekman R, Tarkowski A (1996) Catecholamines are synthesized by mouse lymphocytes and regulate function of these cells by induction of apoptosis. Immunology 88: 140–146

Khan MM, Sansoni P, Silverman ED, Engleman EG, Melmon KL (1986) Beta-adrenergic receptors on human suppressor, helper and cytolytic lymphocytes. Biochem Pharm 7:1137–1142

Korsatko W. Porta S, Sadjak A, Supanz S (1982) Implantation von Adrenalin-Retardtabletten zur Langzeituntersuchung in Ratten. Pharmazie 37:565–568

Liebmann P, Hofer D, Felsner P, Wölfler A, Schauenstein K (1996) Beta-blockade enhances adrenergic immunosuppression in rats via inhibition of melatonin release. J Neuroimmunol 67:137–142

Liebmann PM, Wölfler A, Schauenstein K (2000) Melatonin and immune functions. This volume

Maslinski W (1989) Cholinergic receptors on lymphocytes. Brain Behav Immun 3:1–14

Mason D, MacPhee I, Antoni F (1990) The role of the neuroendocrine system in determining genetic susceptibility to experimental allergic encephalomyelitis in the rat. Immunology 70:1–5

Nance DM, Hopkins DA, Bieger D (1987) Re-investigation of the innervation of the thymus gland in mice and rats. Brain Behav Immun 1:134–147

Rinner I, Schauenstein K (1991) The parasympathetic nervous system takes part in the immuno-neuroendocrine dialogue. J Neuroimmunol 34:165–172

Rinner I, Schauenstein K (1993) Detection of choline-acetytransferase activity in lymphocytes. J Neurosci Res 35:188–191

Rinner I, Kukulansky T, Felsner P, Skreiner E, Globerson A, Kasai M, Hirokawa K, Korsatko W, Schauenstein K (1994) Cholinergic stimulation modulates apoptosis and differentiation of murine thymocytes via a nicotinic effect on thymic epithelium. Biochem Biophys Res Comm 203:1057–1062

Rinner I, Kawashima K, Schauenstein K (1998) Rat lymphocytes produce and secrete acetylcholine in dependence of differentation and activation. J Neuroimmunol 81:31–37

Rinner I, Globerson A, Kawashima K, Korsatko W, Schauenstein K (1999) A possible role for acetyl-choline in the dialogue between thymocytes and thymic stroma. Neuroimmunomodulation 6:51–55

Rossi A, Tria MA, Baschieri S, Doria G, Frasca D (1989) Cholinergic agonists selectively induce pro-liferative responses in the mature subpopulation of murine thymocytes. J Neurosci Res 24: 369–373

Schauenstein K, Faessler R, Dietrich H, Schwarz S, Kroemer G, Wick G (1987) Disturbed immune-endocrine communication in autoimmune disease. Lack of corticosterone response to immune signals in obese strain chickens with spontaneous autoimmune thyroiditis. J Immunol 139: 1830–1833

Schauenstein K, Rinner I, Felsner P, Hofer D, Mangge H, Skreiner E, Liebmann P, Globerson A (1994) The role of the adrenergic/cholinergic balance in the immune-neuroendocrine circuit. In: Berci I, Szeleny J (eds) Advances in Psychoneuroimmunology. Plenum Press, New York, pp 349–356

Schauenstein K, Felsner P. Rinner I, Liebmann PM, Stevenson JR, Westermann J, Haas HS, Cohen RL, Chambers DA. In vivo immunomodulation by peripheral adrenergic and cholinergic agonists/antagonists in rat and mouse models. Ann N Y Acad Sci, in press

Singh U, Fatani J (1988) Thymic lymphopoesis and cholinergic innervation. Thymus 11:3–13

Sternberg EM, Hill JM, Chrousos GP, Kamilaris T, Listwak SJ, Gold PW, Wilder RL (1989) Inflamma-tory mediator-induced hypothalamic-pituitary-adrenal axis activation is defective in strepto-coccal cell wall arthritis-susceptible Lewis rats. Proc Natl Acad Sci USA 86:4771–4775

Strom TB, Sytkowski AT, Carpenter CB, Merill JB (1974) Cholinergic augmentation of lymphocyte mediated cytotoxicity. A study of the cholinergic receptor of cytotoxic T lymphocytes. Proc Natl Acad Sci USA 71:1330–1333

Szelenyi J, Palldi-Haris P, Hollan S (1987) Changes in the cholinergic system due to mitogenic stimulation. Immunol Lett 16:49–54

Titinchi S, Clark B (1984) Alpha 2-adrenoceptors in human lymphocytes: Direct characterization by (^{3}H) yohimbine binding. Biochem Biophys Res Commun 121:1–7

19 Melatonin and Immune Functions

Peter M. Liebmann, Albert Wölfler, Konrad Schauenstein

Abstract

Evidence has accumulated suggesting that melatonin, the major endocrine product of the pineal gland – as a well-preserved molecule through evolution – is involved in the feedback between neuroendocrine and immune functions. At present, we are beginning to understand the mechanisms of action by which melatonin affects cellular functions. In this article, we present a critical review of the numerous reports on the influence of melatonin on immune functions, focussing especially on possible underlying molecular pathways.

19.1 Evidence of an Immunoregulatory Role of Melatonin (MEL)

The first evidence of functional relationships between the pineal gland and the immune system were reported by Csaba and coworkers (1965, 1968, 1975), and Jankovic et al. (1970), who found a disorganization of thymic structures after surgical pinealectomy in newborn rats. Later, Maestroni and Pierpaoli (1981) kept mice under constant light for 3–4 generations to suppress the production of melatonin (MEL). The third and fourth generation of these mice grew poorly and showed an impaired ability to mount antibody responses to T cell-dependent antigens accompanied by cellular depletion of the thymic cortex and atrophy of the white splenic pulp. Further studies with inbred mice showed that evening administration of the beta-blocking agent propranolol or daily injections of p-chlorophenylalanine, both of which suppress the synthesis of MEL, resulted in a depression of the primary antibody response to sheep red blood cells (SRBC) and a reduced autologous mixed lymphocyte reaction (Maestroni et al. 1986). Evening administration of MEL reversed these effects. In untreated animals, MEL augmented the number of splenic plaque-forming cells (PFC) and enhanced the primary and secondary antibody response to SRBC (Maestroni et al. 1987). However, more recent findings partially contradict these results: adult mice and hamsters receiving chronic subcutaneous MEL injections in the evening showed enhanced cell-mediated but not humoral immune functions (Champney et al. 1997, 1998; Demas and Nelson 1998). It is emphasized that these effects were seen only when MEL was administered in the evening.

In mice, MEL was found to antagonize the depression of antibody production induced by acute restraint stress or treatments with corticosteroids or cyclophosphamide

(Maestroni et al. 1986, 1988). Concomitant administration of the specific opioid antagonist naltrexone abolished these antistress effects of MEL (Maestroni et al. 1987), whereas opioids such as beta-endorphin and dynorphin 1–13 mimicked them (Maestroni and Conti 1989). In following studies, Maestroni and Conti (1990, 1991) showed that physiological concentrations of MEL stimulated CD4$^+$ lymphocytes in vitro to release mediators which competed with the specific binding of ^{3}H-naloxone to mouse brain and thymus membranes. These so far unidentified lymphocyte-derived opioid receptor ligands had stress-protective effects comparable to those of MEL on thymus cellularity and primary antibody response in vivo and cross-reacted with antibodies specific for beta-endorphin and met-enkephalin but not with those directed against leu-enkephalin or dynorphin. These data are partly in contradiction to results of Ovadia et al. (1989), who were unable to show specific opioid binding to cultured rat thymocytes, and also to results of Wajs et al. (1995), who demonstrated that, whereas MEL stimulated the gene expression of pro-opiomelanocortin (POMC), the precursor protein of beta-endorphin and other opioids, in rat lymph nodes and bone marrow, an effect on spleen and thymus was not detectable.

Maestroni et al. (1994a, 1994b) reported that MEL may rescue bone marrow cells from apoptosis induced in vivo or in vitro by the chemotherapeutic drugs etoposide or cyclophosphamide. Their results indicate that this effect is mediated via MEL receptors on bone marrow T helper cells by stimulating the endogenous production of a both IL-4- and dynorphin-like cytokine, which may then stimulate adherent stromal cells via κ-opioid receptors to produce granulocyte/macrophage colony-stimulating factor (GMCSF) (Maestroni 1998). Rapozzi et al. (1998), however, at least partially related the protective effects of MEL against bone marrow and lymphocytic toxicity of adriamycin, another chemotherapeutic drug, to its antioxidant properties.

Ben-Nathan et al. (1995) showed that MEL reduced viremia and postponed the onset of disease and death in mice after infection with Semliki Forest virus. These results confirmed findings of Maestroni et al. (1988) showing that MEL blocked the restraint stress-induced lethality after infection with sublethal doses of encephalomyocarditis virus in inbred mice. Furthermore, MEL antagonized the reduction of thymus weight induced by corticosterone in mice immunized with *Vaccinia* virus.

The pineal gland is an endogenous "Zeitgeber" by transducing light/dark signals into chemical signals such as MEL (Axelrod 1974). However, it is important to note that the susceptibility to MEL effects also varies during the day and even during the seasons of the year: Giordano and Palermo (1991) found that MEL could enhance antibody-dependent cellular cytotoxicity (ADCC) in mice, but only during the summer (Giordano et al. 1993); it had no effect during the rest of the year. This finding appears confirmed by data from Vermeulen et al. (1993) showing that neonatal pinealectomy reduces ADCC.

19.2 Protective Effects of MEL on α_2-Adrenergic Immunosuppression in Rats

Results from our group suggest that MEL plays a significant and physiologically relevant role in the protection of immune cells against adrenergic stress. In a rat model, Felsner et al. (1992) showed that continuous in vivo treatment with adrenaline or noradrenaline (NA) had a strong suppressive effect on the ex vivo T and B cell mitogen response of peripheral blood lymphocytes (PBL), provided that a beta-blocker was concomitantly administered. Using receptor-selective adrenergic agonists, Felsner et al. (1995) further identified the receptor involved as belonging to the α_2 subtype. Interestingly, however, the immunosuppressive activity of the specific α_2 agonist clonidine turned out to be likewise dependent on the concomitant treatment with beta-blockers. These results suggested that, in this model, catecholamines exert their in vivo immunosuppressive effect via α_2-receptors and that the β-adrenergic blockade additionally antagonizes a putative endogenous immunoprotective mediator. One candidate for such an immunoprotective substance was MEL, which is released from the pineal gland mainly via β-adrenergic stimulation (Klein et al. 1979; McLeod et al. 1995). This hypothesis was tested by investigating the effect of oral or intraperitoneal treatment with MEL on the immunosuppressive effect of NA or clonidine in combination with propranolol. Secondly, the effect of clonidine in combination with a "functional pinealectomy", i. e., constant light throughout 24 hours, on the in vitro responsiveness of PBL was examined. It turned out that both functional pinealectomy and beta-blockade strongly enhance the suppressive effect of NA or clonidine on one hand, and MEL treatment or a regular light/dark cycle likewise counteract this effect on the other (Liebmann et al. 1996, 1997). These results suggest that the endogenous MEL tone persistently protects PBL from α_2-adrenergic suppression and that beta-blockers enhance the suppressive adrenergic effect on lymphocytes via inhibition of MEL release.

At the same time, MEL had a slight but consistent inhibitory effect on PBL mitogen stimulation from otherwise untreated rats. Both regimens to inhibit MEL release, i. e., beta-blockade or constant light, led to a slight increase of mitogen responses, and MEL counteracted this effect as well. This result is in accordance with Konakchieva et al. (1995), who reported that MEL and also 5-methoxytryptophol inhibited the concanavalin A (ConA)-induced ^{3}H-thymidine incorporation in PBL and tonsillar lymphocytes, and with Vijayalaxmi et al. (1996), who demonstrated that supraphysiological MEL concentrations led to a concentration-dependent decrease in mitotic index and alteration in proliferation kinetics in human PBL. This also relates to findings, according to which MEL in high concentrations inhibited mitogen induced interferon-gamma (IFN-γ) production by lymphocytes (Arzt et al. 1988, DiStefano et al. 1994). IFN-γ in turn enhanced the lymphocytic production of serotonin and MEL, suggesting an immunoregulatory circuit (Finocchiaro et al. 1988).

19.3 CD4$^+$ Lymphocytes and Monocytes/Macrophages as Targets of MEL Activity

Based on their findings, Maestroni (1995) and Lissoni et al. (1993) postulated that two of the main target cells for MEL activity on immune functions in humans may be CD4$^+$ lymphocytes and monocytes/macrophages. This suggestion is corroborated by several findings: in aged mice or mice immunodepressed by cyclophosphamide treatment, chronic administration of MEL enhanced the antibody response via induction of T helper cell activity, augmented mitogenic T cell responses, induced higher levels of IFN-γ and IL-2 in splenocytes (Caroleo et al. 1992, 1994; Sze et al. 1993), and enhanced the antigen presentation by splenic macrophages to T cells by increasing the expression of MHC class II molecules and the production of IL-1 and TNF-α (Tumor Necrosis Factor) (Pioli et al. 1993). Furthermore, MEL was shown to boost the activity of IL-2 in host antitumor responses in humans (Lissoni et al. 1992a, b).

According to results of Garcia-Maurino et al. (1997), MEL somewhat influences the activity of the Th_1 subtype of CD4[+] lymphocytes, since MEL increased IL-2 and IFN-γ but not IL-4 production, which is mostly produced by Th_2 cells. Interestingly, the balance of Th_1- and Th_2-cytokine production of human PBL after mitogenic stimulation shows a circadian rhythmicity, whereby IFN-γ production peaks during the early morning and thus correlates positively with MEL serum concentrations (Petrovsky et al. 1997). In contrast, MEL enhanced secretion of IL-4 in antigen-specific activated mouse T lymphocytes but down-regulated that of IL-2 and IFN-γ. In lymphocytes of unprimed mice, MEL had no effect on cytokine production (Shaji et al. 1998).

In human monocytes, MEL was shown to have a synergistic effect with lipopolysaccharide (LPS) on the cytotoxic activity, which resulted in induction of IL-1, IL-6, and TNF-α production as well as formation of reactive oxygen species (Morrey et al. 1994). This effect was only found in freshly isolated monocytes binding radioactively labeled MEL, but not in monocytes cultured for several days, which had lost their MEL binding sites (Barjavel et al. 1998). However, these results are conflicting to findings of two other groups: Williams et al. (1998) could not find any effect of MEL on cytotoxic activity as well as on TNF-α or IL-6 production of LPS-activated rat alveolar macrophages and RAW 264.7 macrophages. In cultured murine macrophages, MEL exhibited anti-inflammatory effects by inhibiting the activity of the transcription factor NFκB, thus decreasing the production of proinflammatory substances like nitric oxide and 6-keto-prostaglandin F1α, which is the major stable breakdown product of prostacyclin (Gilad et al. 1998).

19.4 Possible Mechanisms of Action of MEL on Immune Cells

Although several in vivo and in vitro effects of MEL have been reported for decades, the respective mechanisms of action are still controversial.

In 1994 a high affinity MEL membrane receptor was cloned for the first time by expression cloning from *Xenopus laevis* dermal melanophores (Ebisawa et al. 1994). Subsequently, a high affinity membrane receptor (MEL-1a) which showed in its primary structure 60% homology with the frog receptor was cloned from the sheep and human, by using a PCR approach based on the *Xenopus* sequence (Reppert et al. 1994). One year later, a second human MEL receptor, designated MEL-1b, was cloned (Reppert et al. 1995a). A third MEL receptor (1c) was found in invertebrates but so far has not been detected in mammalian tissues (Reppert et al. 1995b). These MEL receptor subtypes are coupled to an inhibitory guanosine nucleotide binding protein (G_i protein). Activation of these receptors results in inhibition of intracellular cyclic adenosine monophosphate (cAMP) formation (Reppert and Weaver 1995). Besides cAMP, these receptors also may regulate intracellular concentrations of other second messengers such as cyclic guanosine monophosphate, diacylglycerol, and inositol triphosphate (Vanecek 1998).

In the year of the first MEL membrane receptor cloning, another mechanism of MEL action was found: Becker-André et al. (1994) showed that MEL binds and activates the nuclear retinoid Z receptors-alpha and -beta (RZR-α, RZR-β). Therefore, MEL-induced transcriptional control is proposed to mediate MEL functions in the brain, where RZR-β is expressed, but also in peripheral tissues, where RZR-α was found (Steinhilber et al. 1995). However, although mRNA for RZR-α was detected in ovine anterior pituitary, no binding of 2-^{125}I-MEL to nuclear extracts could be found (Hazlerigg et al. 1996). It was also stated recently that the lab where the binding of MEL to RZR was found could not reproduce this in later experiments (Becker-André et al. 1997).

Benitez-King et al. (1993) described a specific, high affinity binding of ^{3}H-MEL to calmodulin (CaM), an essential protein for intracellular calcium signaling. Associated influences on cytoskeletal changes in a CaM-antagonistic manner were observed in various cell lines (Benitez-King and Anton-Tay 1993). Furthermore, it was demonstrated that MEL, by virtue of binding to CaM, inhibited the CaM-dependent phosphodiesterase and CaM-dependent kinase II (Benitez-King et al. 1991, 1996). Interestingly, the inhibition of these enzymes with MEL was much more effective than that seen with several well-characterized pharmacological CaM antagonists such as trifluoperazine (TFP) or W7. Referring to these results, Pozo et al. (1994) found a calcium-dependent inhibition of nitric oxide synthase by MEL in rat cerebellum. Therefore, MEL was considered to exert its physiological effect in part by inhibiting intracellular CaM activity (Pozo et al. 1997a; Anton-Tay et al. 1993).

Besides binding to specific receptors and to CaM, in recent years MEL has been reported to be one of the most potent endogenous antioxidants (Reiter 1998). In biochemical assays, MEL was reported to scavenge the highly toxic hydroxyl and peroxyl radicals in a more efficient manner than equimolar concentrations of glutathione and α-tocopherol, respectively (Hardeland et al. 1993; Pieri et al. 1994). In vivo, MEL was shown to protect tissues against toxic effects induced by a variety of agents and processes which may exert their effects at least in part via reactive oxygen species (ROS) formation: MEL was able to overcome kainate-induced neurotoxicity in rat brain (Melchiorri et al. 1995) and cerebellum (Giusti et al. 1996) and to counteract safrole-induced DNA damage in rat hepatocytes (Tan et al. 1994). Furthermore, MEL may influence distinct cellular functions via this mechanism of action (Reiter et al. 1996, Gilad et al. 1998), since ROS are known to be important elements of intracellular signaling cascades (Sen and Packer 1996, Suzuki et al. 1997, Lander 1997).

In addition to possible direct effects of MEL on immune cells, indirect immuno-modulating effects of MEL via influencing the synthesis and release of other hormones and modulating their effects are most likely to exist.

In the following paragraphs, the available data about mechanism of MEL actions in immune cells are shortly reviewed and their possible implications on lymphocyte functions are critically discussed.

19.4.1 MEL Membrane Receptors on Immune Cells

Although Maestroni and Conti (1993), who described direct effects of MEL on lymphocytes in vitro (see above), could not detect any specific 2-^{125}I-MEL binding to activated or nonactivated mouse splenocytes, there is sufficient evidence of specific MEL binding sites on immune cells in higher vertebrates to assume direct effects of MEL on the immune system (Calvo et al 1995). In humans, Lopez-Gonzalez et al. (1992a, b, c) characterized high affinity MEL binding sites in membranes of peripheral blood lymphocytes (PBL). Further investigations revealed differences between subsets of human PBL: binding of 2-^{125}I-MEL was tenfold higher in peripheral blood T cells than in B lymphocytes. Among T cells, binding of MEL was mostly found in CD4$^+$ cells rather than in CD8$^+$ cells (Gonzalez-Haba et al. 1995). According to Konakchieva et al.

(1995), the specific 2-^{125}I-MEL binding to human PBL and tonsillar lymphocytes significantly increased after 72-h stimulation with ConA, suggesting a functional role of these binding sites. Recently, binding sites for MEL have been demonstrated only in freshly isolated monocytes but not after several days of culture (Barjavel et al. 1998).

The dissociation constant (K_d) values of these receptors are in the 1 nM range, which is 5–50 times higher than those found in rodent or human brain areas (Vanecek 1988; Reppert et al. 1989, 1996; Laitinen et al. 1990) or those found in lymphatic tissues of birds (Yu et al. 1991; Pang et al. 1992). Since this K_d is far higher than maximum serum MEL concentrations at night in humans and rodents (200 pM), the relevance of these receptors to bind circulating MEL is questionable. However, considering the ability of human lymphocytes to produce MEL (Finocchiaro et al. 1988, 1991, 1995), higher concentrations of this hormone can be anticipated in the lymphocytic microenvironment.

Recently, mRNA expression of the MEL-1a membrane receptor was detected in T as well as B lymphocytes of rat thymus and spleen (Pozo et al. 1997b), suggesting that the same receptor found in distinct brain regions is also responsible for specific 2-^{125}I-MEL binding in lymphocytes. Interestingly, mRNA of the MEL-1b receptor was not found in these cells. Chen et al. (1998) demonstrated that mRNA of a MEL receptor is expressed in activated mast cells. Unfortunately, in their publication, this group did not specify the subtype of the receptor found.

Concerning second messenger systems involved in signal transduction of a putative lymphocytic MEL receptor, guanosine nucleotides were shown to decrease the specific binding of 2-^{125}I-MEL to crude lymphocyte membranes, suggesting the coupling of these binding sites to a G protein (Lopez-Gonzalez et al. 1992a). Recently, Garcia-Perganeda et al. (1997) showed that MEL can reduce cAMP formation in peripheral blood lymphocytes via inhibition of the adenyl cyclase but also can stimulate phospholipase C in these cells via a pertussis toxin-sensitive signaling pathway. Together with findings of Rafii El-Idrissi et al. (1995), who reported a decrease of forskolin-induced cAMP production after MEL treatment in rat splenocytes, these results argue in favor of an involvement of a G_i protein-coupled signaling pathway of lymphocytic MEL membrane receptors.

In contrast, Lopez-Gonzalez et al. (1992a, b, c) demonstrated a potentiating effect of MEL on cAMP formation after stimulation with Vasoactive Intestinal Peptide (VIP) in human lymphocytes, although MEL alone did not affect intracellular cAMP concentration in these cells. This effect could be explained via an inhibitory effect of MEL on cellular phosphodiesterase activity.

Taken together in spite of somewhat conflicting data, it seems likely that effects of MEL on lymphoid cells may be transduced by the same G protein-coupled membrane receptors which were found in distinct brain regions (Reppert and Weaver 1995). However, a clear connection between activation of these binding sites on lymphocytes and MEL effects on lymphocytic functions remains to be demonstrated.

19.4.2 MEL as a Ligand for the Nuclear Receptors RZR-α and RZR-β

Due to its lipophilic nature, MEL is also able to pass the cell membrane and bind to intracellular sites. A new perspective for direct effects of MEL on cells of the immune system was opened up when MEL was shown to be the natural ligand of the nuclear retinoid orphan receptors RZR-α and RZR-β (Becker-André et al. 1994). Concerning immune functions, RZR-α has been previously demonstrated to bind to a response element in the promoter of the 5-lipoxygenase gene (Wiesenberg et al. 1995). In B lymphocytes, which express RZR-α, MEL down-regulated the expression of this key enzyme in the biosynthesis of leukotrienes (Steinhilber et al. 1995). In human Th$_1$-lymphocytes, MEL-induced stimulation of IL-2 and IFN-γ production was mimicked by CGP 52608, which is a synthetic ligand of RZR-α. Although no binding site for RZR-α in the promoter region of these cytokine genes is known, the authors postulated that MEL may regulate transcription of these genes via binding to the nuclear receptor RZR-α (Garcia-Maurino et al. 1997). Rafii-El-Idrissi et al. (1998) demonstrated a specific, high affinity binding of 2-^{125}I-MEL to cell nuclei of rat thymus and spleen which was displaced by CGP 52608, suggesting the nuclear receptor RZR to be responsible for this MEL binding.

The K$_d$ value for this nuclear receptor is in the high pico- to nanomolar range, which again may put in question the relevance of these receptors for endocrine MEL but which may suggest that MEL acts also as an intracellular modulator of lymphocyte functions.

19.4.3 MEL as a Calmodulin Antagonist

Besides acting via specific receptors, MEL has been reported to influence the activity of calmodulin, which plays an important role in signal transduction pathways in immune cells. Benitez-King et al. (1993) described a specific, high affinity binding of ^{3}H-MEL (K$_d$ 188 γM) to calmodulin (CaM). On the cellular level, cytoskeletal changes in a CaM-antagonistic manner were observed in various nonlymphoid cell lines after MEL treatment (Benitez-King and Anton-Tay 1993; Huerto-Delgadillo et al 1994). In biochemical assays, it was demonstrated that MEL, by virtue of its binding to CaM, inhibited the CaM-dependent phosphodiesterase and CaM-dependent kinase II (Benitez-King et al. 1991, 1996) more effectively than pharmacological CaM antagonists such as TFP or W7. In human lymphocytes, Lopez-Gonzalez et al. (1992a, b), as mentioned above, reported that MEL can potentiate cAMP production after VIP stimulation. They suggested that this effect of MEL can be explained by its influence on intracellular phosphodiesterase activity, which is CaM-dependent.

Activation of CaM by Ca^{2+} plays a crucial role in various lymphocyte functions. It is involved in T lymphocyte activation after antigenic stimulation, provided that a costimulatory signal is also delivered to the cell (Weiss and Littman 1994). In absence of this costimulatory signal, elevation of intracellular Ca^{2+} concentration and subsequent CaM activation leads to T lymphocyte anergy, which is characterized by a block of IL-2 production, even after complete restimulation (Schwartz 1990; Jenkins and Miller 1992; Nghiem et al. 1994). Thus, substances which interact with CaM have an important impact on lymphocyte activity.

We therefore investigated the influence of MEL on CaM-dependent IL-2 production and proliferation of activated T lymphocytes (Wölfler et al. 1998). It was found that, in contrast to the pharmacologically well-characterized CaM antagonists TFP and W7, MEL neither inhibited the IL-2 production of activated leukemic Jurkat T cells nor decreased the mitogen response of PBL: while TFP significantly reduced IL-2 production of activated Jurkat cells already after 4 hours, MEL did not affect it at any time of incubation. Coincubation with MEL and TFP revealed that MEL does not compete with TFP for the binding to CaM, which has been suggested previously.

Our results were confirmed by other groups: NMR spectroscopic and biochemical studies by Ouyang and Vogel (1998) demonstrated that MEL binds to CaM only in extremely high concentrations (K_d 2–20 mM). Specific ^{125}I-MEL binding to nuclei and cell membranes of human peripheral blood lymphocytes was not affected by addition of various anti-CaM antibodies, which induce conformational changes in the molecule and thus inhibit interaction with its respective enzymes and pharmacological antagonists (Garcia-Maurino et al. 1998). Taken together, these results contradict the concept of MEL being a general antagonist of CaM.

19.4.4 MEL as Antioxidant

Considering the influence of reactive oxygen species (ROS) on intracellular signal transduction in lymphocytes (Flescher et al. 1994; Buttke and Sandstrom 1994; Goldstone et al. 1995; Nakamura et al. 1997), a further possible mechanism of MEL action affecting immune functions is suggested by its reported potent ROS-scavenging effects (Melchiorri et al. 1995; Reiter 1995; Reiter et al. 1997). A significant, concentration-dependent decrease in the frequency of radiation-induced chromosome damage was shown in human PBL which were pretreated with MEL, whereby MEL was found to be more potent than dimethylsulfoxide, a known free radical scavenger (Vijayalaxmi et al. 1995a, b, 1996). Recently, micromolar concentrations of MEL in murine macrophages were demonstrated to inhibit activation of NFκB (Gilad et al. 1998), which is tightly regulated by the cellular redox state (Sen and Packer 1996), also pointing to an antioxidant activity of MEL.

Using Jurkat cells, a human leukemic T cell line, we also investigated antioxidant features of MEL and its relevance and significance in lymphocytic function. It turned out that, in contrast to glutathione, MEL could not protect against the impairment of mitogen-induced calcium mobilization as well as IL-2 production in Jurkat cells exposed to oxidative stress (Wölfler et al. submitted). In resting Jurkat cells, micro- to millimolar concentrations of MEL even exhibited pro-oxidant activity, as was shown by an increased intracellular oxidation of ROS-sensitive fluorescent dyes. This pro-oxidant activity led to a dose-dependent promotion of fas- and ROS-induced cell death, which could be antagonized by addition of trolox, a water-soluble analogue of α-tocopherol with well-described antioxidant features. These results suggest that MEL can modulate the intracellular redox status but not necessarily in an antioxidant manner.

Prompted by these unexpected results, the effect of MEL on ROS formation in cultured human PBL was investigated (Wölfler et al. submitted). In contrast to what was found in Jurkat cells, MEL exhibited antioxidant properties in human PBL: micro- to millimolar concentrations of MEL counteracted t-butylated hydroperoxide (t-BHP)- and diamide-induced ROS-formation, although the scavenging capacity of MEL in comparison to trolox and N-acetylserotonin (NAS), its immediate precursor, was rather weak, since NAS turned out to be about 3 to 10 times more effective than MEL.

The functional basis of the opposite effect of MEL on the intracellular redox status in Jurkat cells and PBL has to be investigated in further studies. Recent reports on variations in the redox status and the susceptibility to oxidant-induced changes in cell functions among different cell types and cell lines (Baeuerle and Henkel 1994; Sen and Packer 1996; Ginn-Pease and Whisler 1998) as well as on the low oxidation potential of MEL (Barsacchi et al. 1998) may contribute to an explanation of these conflicting results.

19.5 Conclusion

In this article, we wanted to review the evidence that MEL is part of the extrinsic regulation of the immune system. The available data suggest multiple and highly complex effects that are still difficult to reconcile with the existence of one distinct physiological role of MEL in immunoregulation. Nevertheless, MEL appears to contribute to the homeostasis of the immune system, which becomes evident under conditions that endanger nonspecific and specific defense mechanisms. It can be anticipated that MEL acts on immune cells at several levels and in direct and indirect ways. In view of the variety of possible direct effects on intracellular signaling, the existence of different types of MEL receptors and the fact that lymphocytes themselves may produce MEL in response to certain stimuli, MEL, in addition to its endocrine role, can be expected to have a physiological impact as a paracrine or even intracellular modulator within the immune system. Much has still to be learned about the specific and relevant mechanisms by which MEL can interact with the immune system before we can begin to understand the numerous phenomena described in vitro and in vivo and before we can envisage possibilities to make therapeutical use of this hormone in immunopathological conditions.

References

Anton-Tay F, Huerto-Delgadillo L, Ortega-Corona B, Benitez-King G (1993) Melatonin antagonism to calmodulin may modulate multiple cellular functions. In: Touitou Y, Arendt J, Pevet P (eds) Melatonin and the pineal gland: From science to clinical applications. Elsevier, Amsterdam, pp 41–46

Arzt ES, Fernandez-Castelo S, Finocchiaro LM, Criscuolo ME, Diaz A, Finkielman S, Nahmod VE (1988) Immunomodulation by indoleamines: serotonin and melatonin action on DNA and interferon-gamma synthesis by human peripheral blood mononuclear cells. J Clin Immunol 8:513–520

Axelrod J (1974) The pineal gland: a neurochemical transducer. Science 184:1341–1348

Baeuerle PA, Henkel T (1994) Function and activation of NF-κB in the immune system. Annu Rev Immunol 12:141–179

Barjavel MJ, Mamdouh Z, Raghbate N, Bakouche O (1998) Differential expression of the melatonin receptor in human monocytes. J Immunol 160:1191–1197

Barsacchi R, Kusmic C, Damiani E, Carloni P, Greci L, Donato L (1998) Vitamin E consumption induced by oxidative stress in red blood cells is enhanced by melatonin and reduced by N-acetyl-serotonin. Free Radical Biol Med 24:1187–1192

Becker-Andre M, Wiesenberg I, Schaeren-Wiemers N, Andre E, Missbach M, Saurat JH, Carlberg C (1994) Pineal gland hormone melatonin binds and activates an orphan of the nuclear receptor superfamily. J Biol Chem 269:28531–28534

Becker-Andre M, Wiesenberg I, Schaeren-Wiemers N, Andre E, Missbach M, Saurat JH, Carlberg C (1997) Pineal gland hormone melatonin binds and activates an orphan of the nuclear receptor superfamily. J Biol Chem 272:16707

Benitez-King G, Antón-Tay F (1993) Calmodulin mediates melatonin cytoskeletal effects. Experientia 49:635–641

Benitez-King G, Huerto-Delgadillo L, Anton-Tay F (1991) Melatonin modifies calmodulin cell levels in MDCK and N1E-115 cell lines and inhibits phosphodiesterase activity in vitro. Brain Res 557:289–292

Benitez-King G, Huerto-Delgadillo L, Antón-Tay F (1993) Binding of ^{3}H-melatonin to calmodulin Life Sci 53:201–207

Benitez-King G, Rios A, Martinez A, Anton-Tay F (1996) In vitro inhibition of Ca2+/calmodulin-dependent kinase II activity by melatonin. Biochem Biophys Acta 1290:191–196

Ben-Nathan D, Maestroni GJ, Lustig S, Conti A (1995) Protective effects of melatonin in mice infected with encephalitis viruses. Arch Virol 140:223–230

Buttke TM, Sandstrom PA (1994) Oxidative stress as a mediator of apoptosis. Immunol Today 15:7–10

Calvo JR, Rafii-el-Idrissi M, Pozo D, Guerrero JM (1995) Immunomodulatory role of melatonin: specific binding sites in human and rodent lymphoid cells. J Pineal Res 18:119–126

Caroleo MC, Doria G, Nistico G (1994) Melatonin restores immunodepression in aged and cyclophosphamide-treated mice. Ann N Y Acad Sci 719:343–352

Caroleo MC, Frasca D, Nistico G, Doria G (1992) Melatonin as immunomodulator in immunodeficient mice. Immunopharmacology 23:81–89

Champney TH, Prado J, Youngblood T, Appel K, McMurray DN (1997) Immune responsiveness of splenocytes after chronic daily melatonin administration in male Syrien hamsters. Immunol Lett 58:95–100

Champney TH, Allen GC, Zannelli M, Beausang LA (1998) Time-dependent effects of melatonin on immune measurements in male Syrian hamsters. J Pineal Res 25:142–146

Chen H, Centola M, Altschul SF, Metzger H (1998) Characterisation of gene expression in resting and activated mast cells. J Exp Med 188:1657–1668

Csaba G, Barath P (1975) Morphological changes of thymus and the thyroid gland after postnatal extirpation of pineal body. Endocrinol Exp 9:59–67

Csaba G, Bodoky M, Toro I (1965) Hormonal relationships of mastocytogenesis in lymphatic organs. II. Effect of epiphysectomy on the genesis of mast cells. Acta Anat Basel 61:289–296

Csaba G, Dunay C, Fischer J, Bodoky M (1968) Hormonal relationships of mastocytogenesis in lymphatic organs. 3. Effect of the pineal body-thyroid-thymus system on mast cell production. Acta Anat Basel 71:565–580

Demas GE, Nelson RJ (1998) Exogenous melatonin enhances cell-mediated, but not humoral, immune function in adult male deer mice (Peromyscus maniculatus). J Biol Rhythms 13:245–252

DiStefano A, Paulesu L (1994) Inhibitory effect of melatonin on production of IFN gamma or TNF alpha in peripheral blood mononuclear cells of some blood donors. J Pineal Res 17:164–169

Ebisawa T, Karne S, Lerner MR, Reppert SM (1994) Expression cloning of a high-affinity melatonin receptor from Xenopus dermal melanophores. Proc Natl Acad Sci USA 91:6133–6137

Felsner P, Hofer D, Rinner I, Mangge H, Gruber M, Korsatko W, Schauenstein K (1992) Continuous in vivo treatment with catecholamines suppresses in vitro reactivity of rat peripheral blood T-lymphocytes via α-mediated mechanisms. J Neuroimmunol 37:47–57

Felsner P, Hofer D, Rinner I, Porta S, Korsatko W, Schauenstein K (1995) Adrenergic suppression of peripheral blood T cell reactivity in the rat is due to activation of peripheral α_2-receptors. J Neuroimmunol 57:27–34

Finocchiaro LME, Arzt ES, Fernandez-Castelo S, Criscuolo M, Finkielman S, Nahmod VE (1988) Serotonin and melatonin synthesis in peripheral blood mononuclear cells: stimulation by interferon-γ as part of an immunomodulatory pathway. J Interferon Res 8:705–716

Finocchiaro LM, Nahmod VE, Launay JM (1991) Melatonin biosynthesis and metabolism in peripheral blood mononuclear leucocytes. Biochem J 280:727–731

Finocchiaro LM, Polack E, Nahmod VE, Glikin GC (1995) Sensitivity of human peripheral blood mononuclear leukocytes to visible light. Life Sci 57:1097–1110

Flescher E, Ledbetter JA, Schieven GL, Vela-Roch N, Fossum D, Dang H, Ogawa N, Talal N (1994) Longitudinal exposure of human T lymphocytes to weak oxidative stress suppresses transmembrane and nuclear signal transduction. J Immunol 153:4880–4889

Garcia-Mauriño S, Gonzalez-Haba MG, Calvo JR, Rafii-El-Idrissi M, Sanchez-Margalet V, Goberna R, Guerrero JM (1997) Melatonin enhances IL-2, IL-6, and IFN-γ production by human circulating CD4+ cells. J Immunol 159:574–581

Garcia-Mauriño S, Gonzalez-Haba MG, Calvo JR, Goberna R, Guerrero JM (1998) Involvement of nuclear binding sites for melatonin in the regulation of IL-2 and IL-6 production by human blood mononuclear cells. J Neuroimmunol 92:76–84

Garcia-Perganeda A, Pozo D, Guerrero JM, Calvo JR (1997) Signal transduction for melatonin in human lymphocytes: involvement of a pertussis toxin-sensitive G protein. J Immunol 159:3774–3781

Gilad E, Wong HR, Zingarelli B, Virág L, O'Connor M, Salzman AL, Szabo C (1998) Melatonin inhibits expression of the inducible isoform of nitric oxide synthase in murine macrophages: role of inhibition of NFκB activation. FASEB J 12:685–693

Ginn-Pease ME, Whisler RL (1998) Redox signals and NF-κB activation in T cells. Free Radic Biol Med 25:346–361

Giordano M, Palermo MS (1991) Melatonin-induced enhancement of antibody-dependent cellular cytotoxicity. J Pineal Res 10:117–121

Giordano M, Vermeulen M, Palermo MS (1993) Seasonal variations in antibody-dependent cellular cytotoxicity regulation by melatonin. FASEB J 7:1052–1054

Giusti P, Lipartiti M, Franceschini D, Schiavo N, Floreani M, Manev H (1996) Neuroprotection by melatonin from kianate-induced excitotoxicity in rats. FASEB J 10:891–896

Goldstone SD, Fragonas JC, Jeitner TM, Hunt NH (1995) Transcription factors as targets for oxidative signalling during lymphocyte activation. Biochim Biophys Acta 1263:114–122

Gonzalez-Haba MG, Garcia-Maurino S, Calvo JR, Goberna R, Guerrero JM (1995) High-affinity binding of melatonin by human circulating T lymphocytes (CD4+). FASEB J 9:1331–1335

Hardeland R, Reiter RJ, Poeggeler B, Tan DX (1993) The significance of the metabolism of the neurohormone melatonin: antioxidative protection and formation of bioactive substances. Neurosci Biobehav Rev 17:347–357

Hazlerigg DG, Barrett P Hastings MH, Morgan PJ (1996) Are nuclear receptors involved in pituitary responsiveness to melatonin? Mol Cell Endocrinol 123:53–59

Huerto-Delgadillo L, Anton-Tay F, Benitez-King G (1994) Effects of melatonin on microtubule assembly depend on hormone concentration: role of melatonin as a calmodulin antagonist. J Pineal Res 17:55–62

Jankovic D, Isakovic K, Petrovic S. (1970) Effect of pinealectomy on immune reactions in the rat. Immunology 18:1–6

Jenkins MK, Miller RA (1992) Memory and anergy: challenges to traditional models of T lymphocyte differentiation. FASEB J 6:2428–2433

Klein DC, Berg GR, Weller J (1979) Melatonin synthesis: adenosine 3′,5′-monophosphate and norepinephrine stimulate N-acetyltransferase. Science 168:979–980

Konakchieva R, Kyurkchiev S, Kehayov I, Taushanova P, Kanchev L (1995) Selective effect of methoxyindoles on the lymphocyte proliferation and melatonin binding to activated human lymphoid cells. J Neuroimmunol 63:125–132

Laitinen JT, Flügge G, Saavedra JM (1990) Characterization of melatonin receptors in the rat area postrema: modulation of affinity with cations and guanine nucleotides. Neuroendocrinology 51:619–624

Lander HM (1997) An essential role for free radicals and derived species in signal transduction. FASEB J 11:118–124

Liebmann PM, Hofer D, Felsner P, Wölfler A, Schauenstein K (1996) Beta-blockade enhances adrenergic immunosuppression in rats via inhibition of melatonin release. J Neuroimmunol 67:137–142

Liebmann PM, Wölfler A, Felsner P, Hofer D, Schauenstein K (1997) Melatonin and the immune system. Int Arch Allergy Immunol 112:203–211

Lissoni P, Barni S, Ardizzoia A, Brivio F, Tancini G, Conti A, Maestroni GJ (1992a) Immunological effects of a single evening subcutaneous injection of low-dose interleukin-2 in association with the pineal hormone melatonin in advanced cancer patients. J Biol Regul Homeost Agents 6:132–136

Lissoni P, Tisi E, Barni S, Ardizzoia A, Rovelli F, Rescaldani R, Ballabio D, Benenti C, Angeli M, Tancini G (1992b) Biological and clinical results of a neuroimmunotherapy with interleukin-2 and the pineal hormone melatonin as a first line treatment in advanced non-small cell lung cancer. Br J Cancer 66:155–158

Lissoni P, Barni S, Tancini G, Rovelli F, Ardizzoia A, Conti A, Maestroni GJ (1993) A study of the mechanisms involved in the immunostimulatory action of the pineal hormone in cancer patients. Oncology 50:399–402

Lopez-Gonzalez MA, Calvo JR, Osuna C, Guerrero JM (1992a) Interaction of melatonin with human lymphocytes: evidence for binding sites coupled to potentiation of cyclic AMP stimulated by vasoactive intestinal peptide and activation of cyclic GMP. J Pineal Res 12:97–104

Lopez-Gonzalez MA, Calvo JR, Osuna C, Rubio A, Guerrero JM (1992b) Melatonin potentiates cyclic AMP production stimulated by vasoactive intestinal peptide in human lymphocytes. Neurosci Lett 136:150–152

Lopez-Gonzalez MA, Calvo JR, Osuna C, Rubio A, Guerrero JM (1992c) Synergistic action of melatonin and vasoactive intestinal peptide in stimulating cyclic AMP production in human lymphocytes. J Pineal Res 12:174–180

Maestroni GJ (1995) T-helper-2 lymphocytes as a peripheral target of melatonin. J Pineal Res 18:84–89

Maestroni GJ (1998) Kappa-Opioid receptors in marrow stroma mediate the hematopoietic effects of melatonin-induced opioid cytokines. Ann N Y Acad Sci 840:411–419

Maestroni GJ, Conti A (1989) β-endorphin and dynorphin mimic the circadian immunoenhancing and anti-stress effects of melatonin. Int J Immunopharmacol 11:333–340

Maestroni GJ, Conti A (1990) The pineal neurohormone melatonin stimulates activated CD4+, Thy-1+ cells to release opioid agonist(s) with immunoenhancing and anti-stress properties. J Neuroimmunol 28:167–176

Maestroni GJ, Conti A (1991) Anti-stress role of the melatonin-immuno-opioid network: evidence for a physiological mechanism involving T cell-derived, immunoreactive β-endorphin and MET-enkephalin binding to thymic opioid receptors. Int J Neurosci 61:289–298

Maestroni GJ, Conti A (1993) Melatonin in relation to the immune system. In: Yu H-S, Reiter RJ (eds) Melatonin Biosynthesis, Physiological Effects, and Clinical Applications. CRC Press, Boca Raton, pp 289–309

Maestroni GJ, Pierpaoli W (1981) Pharmacologic control of the hormonally mediated immune response. In: Ader R (ed): Psychoneuroimmunology. Academic Press, pp 405–413

Maestroni GJ, Conti A, Pierpaoli W (1986) Role of the pineal gland in immunity. Circadian synthesis and release of melatonin modulates the antibody response and antagonizes the immuno-suppressive effect of corticosterone. J Neuroimmunol 13:19–30

Maestroni GJ, Conti A, Pierpaoli W (1987) Role of the pineal gland in immunity. II. melatonin enhances the antibody response via an opiatergic mechanism. Clin Exp Immunol 68:384–391

Maestroni GJ, Conti A, Pierpaoli W (1988) Role of the pineal gland in immunity. III. melatonin antagonizes the immunsuppressive effect of acute stress via an opiatergic mechanism. Immunology 63:465–469

Maestroni GJ, Covacci V, Conti A (1994a) Hematopoietic rescue via T-cell-dependent, endogenous granulocyte-macrophage colony-stimulating factor induced by the pineal neurohormone melatonin in tumor-bearing mice. Cancer Res 54:2429–2432

Maestroni GJ, Conti A, Lissoni P (1994b) Colony-stimulating activity and hematopoietic rescue from cancer chemotherapy compounds are induced by melatonin via endogenous interleukin 4. Cancer Res 54:4740–4743

McLeod SD, Cairncross KD (1995) Preliminary evidence of a synergistic α1- and β1-adrenoceptor regulation of rat pineal hydroxyindole-O-methyltransferase. Gen Comp Endocrinol 97:283–288

Melchiorri D, Reiter RJ, Sewerynek E, Chen LD, Nistico G (1995) Melatonin reduces kainate-induced lipid peroxidation in homogenates of different brain regions. FASEB J 9:1205–1210

Morrey KM, McLachlan JA, Serkin CD, Bakouche O (1994) Activation of human monocytes by the pineal hormone melatonin. J Immunol 153:2671–2680

Nakamura H, Nakamura K, Yodoi J (1997) Redox regulation of cellular activation. Annu Rev Immunol 15:351–369

Nghiem P, Ollick T, Gardner P, Schulman H (1994) Interleukin-2 transcriptional block by multi-functional Ca^{2+}/calmodulin kinase. Nature 371:347–350

Ouyang H, Vogel HJ (1998) Melatonin and serotonin interactions with calmodulin: NMR, spectroscopic and biochemical studies. Biochim Biophys Acta 1383:37–47

Ovadia H, Nitsan P, Abramsky O (1989) Characterization of opiate binding sites on mebranes of rat lymphocytes. J Neuroimmunol 21:93–102

Pang CS, Pang SF (1992) High affinity specific binding of 2-[^{125}I]iodomelatonin by spleen membrane preparations of chicken. J Pineal Res 12:167–173

Petrovsky N, Harrison LC (1997) Diurnal rhythmicity of human cytokine production: a dynamic disequilibrium in T helper cell type 1/T helper cell type 2 balance? J Immunol 158:5163–5168

Pieri C, Marra M, Moroni F, Recchioni R, Marcheselli F (1994) Melatonin: a peroxyl radical scavenger more effective than vitamin E. Life Sci 55:271–276

Pioli C, Caroleo MC, Nistico G, Doria G (1993) Melatonin increases antigen presentation and amplifies specific and non specific signals for T-cell proliferation. Int J Immunopharmacol 15: 463–468

Pozo D, Reiter RJ, Calvo JR, Guerrero JM (1994) Physiological concentrations of melatonin inhibit nitric oxide synthase in rat cerebellum. Life Sci 55:455–460

Pozo D, Reiter RJ, Calvo JR, Guerrero JM (1997a) Inhibition of cerebellar nitric oxide synthase and cyclic GMP production by melatonin via complex formation with calmodulin. J Cell Biochem 65:430–442

Pozo D, Delgado M, Fernandez-Santos JM, Calvo JR, Gomariz RO, Martin-Lacave I, Ortiz GG, Guerrero JM (1997b) Expression of the Mel1a-melatonin receptor mRNA in T and B subsets of lymphocytes from rat thymus and spleen. FASEB J 11:466–473

Rafii-El-Idrissi M, Calvo JR, Pozo D, Harmouch A, Guerrero JM (1995) Specific binding of 2-[^{125}I]iodomelatonin by rat splenocytes: characterization and its role on regulation of cyclic AMP production. J Neuroimmunol 57:171–178

Rafii-El-Idrissi M, Calvo JR, Harmouch A, Garcia-Maurino S, Guerrero JM (1998) Specific binding of melatonin by purified cell nuclei from spleen and thymus of the rat. J Neuroimmunol 86: 190–197

Rapozzi V, Zorzet S, Comelli M, Mavelli I, Perissin L, Giraldi T (1998) Melatonin decreases bone marrow and lymphatic toxicity of adriamycin in mice bearing TLX5 lymphoma. Life Sci 63:1701–1713

Reiter RJ (1995) Oxidative processes and antioxidative defense mechanisms in the aging brain. FASEB J 9:526–533

Reiter RJ (1998) Oxidative damage in the central nervous system: protection by melatonin. Prog Neurobiol 56:359–384

Reiter RJ, Oh CS, Fujimori O (1996) Melatonin: its intracellular and genomic actions. Trends Endocrin Metab 7:22–27

Reiter RJ, Tang L, Garcia JJ, Munoz-Hoyos A (1997) Pharmacological actions of melatonin in oxygen radical pathophysiology. Life Sci 60:2255–2271

Reppert SM, Weaver DR (1995) Melatonin madness. Cell 83:1059–1062

Reppert SM, Weaver DR, Rivkees SAH, Stopa EG (1989) Putative melatonin receptors in a human biological clock. Science 242:78–81

Reppert SM, Weaver DR, Ebisawa T (1994) Cloning and characterization of a mammalian melatonin receptor that mediates reproductive and circadian responses. Neuron 13:1177–1185

Reppert SM, Godson C, Mahle CD, Weaver DR, Slaugenhaupt SA, Gusella JF (1995a) Molecular characterisation of a second melatonin receptor expressed in human retina and brain: the Mel1b melatonin receptor. Proc Natl Acad Sci U S A 92:8734–8738

Reppert SM, Weaver DR, Cassone VM, Godson C, Kolakowski LF Jr (1995b) Melatonin receptors are for the birds: molecular analysis of two receptor subtypes differentially expressed in chick brain. Neuron 15:1003–1015

Reppert SM, Weaver DR, Godson C (1996) Melatonin receptors step into the light: cloning and classification of subtypes. Trend Pharmacol Sci 17:100–102

Schwartz RH (1990) A cell culture model for T lymphocyte clonal anergy. Science 248:1349–1356

Sen CK, Packer L (1996) Antioxidant and redox regulation of gene transcription. FASEB J 10: 709–720

Shaji AV, Kulkarni SK, Agrewala JN (1998) Regulation of secretion of IL-4 and IgG1 isotype by melatonin-stimulated ovalbumin-specific T cells. Clin Exp Immunol 111:181–185

Steinhilber D, Brungs M, Werz O, Wiesenberg I, Danielsson C, Kahlen JP, Nayeri S, Schrader M, Carlberg C (1995) The nuclear receptor for melatonin represses 5-lipoxygenase gene expression in human B lymphocytes. J Biol Chem 270:7037–7040

Suzuki YJ, Forman HJ, Sevanian A (1997) Oxidants as stimulators of signal transduction. Free Radic Biol Med 22:269–285

Sze SF, Liu WK, Ng TB (1993) Stimulation of murine splenocytes by melatonin and methoxytryptamine. J Neural Transm Gen Sect 94:115–126

Tan DX, Reiter RJ, Chen LD, Poeggeler B, Manchester LC, Barlow-Walden LR (1994) Both physiological and pharmacological levels of melatonin reduce DNA adduct formation induced by the carcinogen safrole. Carcinogenesis 15:215–218

Vanecek J (1988) Melatonin binding sites. J Neurochem 51:1436–1440

Vanecek J (1998) Cellular mechanisms of melatonin action. Physiol Rev 78:687–721

Vermeulen M, Palermo M, Giordano M (1993) Neonatal pinealectomy impairs murine antibody-dependent cellular cytotoxicity. J Neuroimmunol 43:97–101

Vijayalaxmi, Reiter RJ, Meltz ML (1995a) Melatonin protects human blood lymphocytes from radiation-induced chromosome damage. Mutat Res 346:23–31

Vijayalaxmi, Reiter RJ, Sewerynek E, Poeggeler B, Leal BZ, Meltz ML (1995b) Marked reduction of radiation-induced micronuclei in human blood lymphocytes pretreated with melatonin. Radiat Res 143:102–106

Vijayalaxmi, Reiter RJ, Leal BZ, Meltz ML (1996) Effect of melatonin on mitotic and proliferation indices, and sister chromatid exchange in human blood lymphocytes. Mutat Res 351:187–192

Wajs E, Kutoh E, Gupta D (1995) Melatonin affects proopiomelanocortin gene expression in the immune organs of the rat. Eur J Endocrinol 133:754–760

Weiss A, Littman DR (1994) Signal transduction by lymphocytes antigen receptors. Cell 76: 263–274

Wiesenberg I, Missbach M, Kahlen JP, Schräder M, Carlberg C (1995) Transcriptional activation of the nuclear receptor RZR alpha by the pineal gland hormone melatonin and identification of CGP 52608 as a synthetic ligand. Nucleic Acids Res 23:327–333

Williams JG, Bernstein S, Prager M (1998) Effect of melatonin on activated macrophage TNF, IL-6, and reactive oxygen intermediates. Shock 9:406–411

Wölfler A, Schauenstein K, Liebmann PM (1998) Lack of calmodulin antagonism of melatonin in T lymphocyte activation. Life Sci 63:835–842

Wölfler A, Abuja PM, Schauenstein K, Liebmann PM (1999) N-acetylserotonin is a better extra- and intracellular antioxidant than melatonin. FEBS Lett 449:206–210

Wölfler A, Caluba HC, Dohr G, Schauenstein K, Liebmann PM: Prooxidant activity of melatonin promotes fas- and ROS-induced cell death in Jurkat cells. Submitted for publication

Yu ZH, Yuan H, Lu Y, Pang SF (1991) [125]Iodomelatonin binding sites in spleens of birds and mammals. Neurosci Lett 125:175–178

20 Melatonin and the Immune System
Therapeutic Potential in Cancer, Viral Diseases, and Immunodeficiency States

Georges J. M. Maestroni

Abstract

Neuroimmunomodulation refers to the continuous interaction between the nervous and immune systems. It is now well recognized that a main actor of this connection is the pineal hormone melatonin. T helper (Th) cells bear G-protein-coupled melatonin cell membrane receptors and, perhaps, melatonin nuclear receptors. Activation of melatonin receptors enhances the release of T helper cell type 1 (Th1) cytokines, such as interferon-γ (IFN-γ) and interleukin-2 (IL-2), as well as of novel opioid cytokines which crossreact immunologically with both interleukin-4 (IL-4) and dynorphin. Melatonin has been reported to also enhance the production of interleukin-1 (IL-1), interleukin-6 (IL-6), and interleukin-12 (IL-12) in human monocytes. These mediators may counteract stress-induced immunodepression and other secondary immunodeficiencies, protect mice against lethal viral and bacterial diseases, synergize with interleukin-2 (IL-2) in cancer patients, and influence hematopoiesis. In cancer patients, melatonin seems to be required for the effectiveness of low-dose IL-2 in neoplasias that are generally resistant to IL-2 alone. Hematopoiesis is apparently influenced by the action of the melatonin-induced-opioids (MIO) on κ-opioid receptors present on stromal bone marrow macrophages. Most interestingly, IFN-γ and colony-stimulating factors (CSFs) may modulate the production of melatonin in the pineal gland. A hypothetical pineal-immune-hematopoietic network is therefore taking shape. From the immunopharmacological and ethical point of view, clinical studies on the effect of melatonin in combination with IL-2 or other cytokines in viral disease including human immunodeficiency virus-infected patients and cancer patients are needed. In conclusion, melatonin seems to play a crucial role in the homeostatic interactions between the brain and the immune-hematopoietic system and deserves to be further studied to identify its therapeutic indications and its adverse effects.

20.1 Neuroimmune Interactions with the Environment

Maintenance of health depends to a significant extent on the ability of the exposed host to respond appropriately and, eventually, to adapt to environmental stressors. It is now well established that inappropriate or maladaptative responses to such stressors weaken the body's resistance to other stimuli from the environment such as pathogenic organisms or cancer cells. It is fair to consider the social environment as part of the general environment which has an impact on the body via redundant and reciprocal interactions between the body and the brain. These interactions are linked by the nervous, endocrine, and immune systems and utilize a large array of chemical messengers including hormones, cytokines, and neurotransmitters (Spector et al. 1996). There is abundant evidence that there are functional afferent nerve endings in the tissue of the immune and hematopoietic systems arising from both the sympathetic and parasympathetic systems (Ader et al. 1991, 1995). It is also clear that many neurotransmitters, neuroendocrine factors, and hormones can drastically change immune functions and that, on the contrary, cytokines derived from immunocompetent cells can profoundly affect the central nervous system (Ader et al. 1991, 1995). As a consequence, any environmental stimulus to the nervous system will affect the immune system and *vice versa* essentially via the endocrine system. On this conceptual basis, it should not be surprising that the day/night photoperiod, which constitutes a basic environmental cue for any organism, can also influence the immune-hematopoietic system. As for many other adaptive responses, a major mediator of this influence seems to be the pineal gland which transduces the light/dark rhythm into the circadian synthesis and release of melatonin (Yu and Reiter 1993).

20.2 Melatonin and the Immune-Hematopoietic System

20.2.1 Functional Effects

Early studies reported controversial results about a possible link between the pineal gland and the immune system. Some papers claimed that the absence of the pineal gland stimulated the proliferation of immunocompetent cells (Bindoni and Cambria 1968; Csaba et al. 1965; Rella and Lapin 1978) and others just the opposite (Barath and Csaba 1974; Jankovic et al. 1970; Vaughan and Reiter 1971). However, most studies agreed that pinealectomy is associated with a precocious involution and histological disorganization of the thymus (Barath and Csaba 1974; Csaba and Barath 1975; Csaba et al. 1965; Vaughan and Reiter 1971). The mechanism of this effect was postulated to depend on increased steroid gonadal hormones. By various pharmacological interventions aimed at inhibiting melatonin synthesis, we provided the first evidence of a possible involvement of endogenous melatonin on humoral and T cell immune reactions, as well as on spleen and thymus cellularity in mice (Maestroni et al. 1986; Maestroni and Pierpaoli 1981). In another report, we showed that pinealectomy inhibits leukemogenesis in a radiation leukemia virus murine model and that melatonin has a promoting effect on the disease (Conti et al. 1992). A number of other authors have then further extended this evidence. Pinealectomized mice were reported to have depressed humoral immunity (Becker et al. 1988). In another report, inhibition of endogenous melatonin in hamster produced a decrease of spleen weight and reduced T cell blastogenesis. Melatonin administration counteracted this effect (Champney and McMurray 1991). IL-2 production and antibody-dependent cellular cytotoxicity were inhibited in pinealectomized mice and exogenous melatonin restored these important functions (Del Gobbo et al. 1989; Palermo et al. 1994).

Endogenous melatonin has been also reported to influence the concentration of bone marrow granulocyte/macrophage colony-forming unity (GM-CFU) (Haldar et al. 1992). An interesting finding which might be associated with and explained by the immunoenhancing action of endogenous melatonin is the widely documented oncostatic role of the pineal gland and of melatonin (Blask 1993). From the pharmacological point of view, melatonin can augment the immune response and correct immunodeficiency states which may follow acute stress, viral diseases, or drug treatment (Maestroni 1993; Maestroni and Conti 1989; 1990, 1993; Maestroni et al. 1986, 1987, 1988, 1994a). This finding has been confirmed and extended in mice and in humans to a variety of immune parameters (Caroleo et al. 1992; Champney and McMurray 1991; Del Gobbo et al. 1989; Giordano and Palermo 1991; Lissoni et al. 1992; Morrey et al. 1994; Palermo et al. 1994; Pioli et al. 1993). Similarly, a very significant biological effect of melatonin is the protection of mice against encephalitis viruses and lethal bacterial infections (Ben-Nathan et al. 1995, 1996, Bonilla et al. 1997).

In general, the immunoenhancing action of melatonin seems to be restricted to T-dependent antigens and to be most pronounced in immunodepressed situations. For example, melatonin may completely counteract thymus involution and the immunological depression induced by stress events or glucocorticoid treatment (Maestroni and Conti 1990). Melatonin is active only when injected in the afternoon or in the evening, i. e., with a schedule consonant with its physiological rhythm (Maestroni et al. 1986, 1987). In addition, melatonin is most active on antigen- or cytokine-activated immunocompetent cells (Maestroni et al. 1986, 1987). Consistent with these requirements, a recent report shows that melatonin may also restore depressed immunological functions after soft-tissue trauma and hemorrhagic shock (Wichmann et al. 1996). Besides acquired immunity, natural immune parameters also seem to be influenced by melatonin. (Angeli et al. 1988; Del Gobbo et al. 1989; Lewinsky et al. 1989). In a tumor model of established lung metastases, we found that melatonin could synergize with the anticancer effect of IL-2 (Maestroni and Conti 1993). More recently, we reported that melatonin may rescue hematopoiesis in mice transplanted with Lewis Lung Carcinoma (LLC) and treated with cancer chemotherapeutic compounds (Maestroni et al. 1994b, c). However, when the mice were tumor-free, melatonin augmented the chemotherapy-induced myelotoxicity. This indicates that melatonin does not have only beneficial or therapeutic effects. As a matter of fact, melatonin has been reported to exaggerate collagen-induced arthritis (Hansson et al. 1992) and to promote T cell leukemia (Conti et al. 1992).

20.2.2 Cytokines Which Mediate the Effect of Melatonin

IL-2, interferon-γ (IFN-γ), and opioid peptides released by activated Th cells seem to mediate, at least in part, the immunopharmacological action of melatonin (Caroleo et al. 1992; Colombo et al. 1992; Del Gobbo et al. 1989; Garcia-Mauriño et al. 1997; Hofbauer and Heufelder 1996; Maestroni and Conti 1990; Pioli et al. 1993). Melatonin also activates human monocytes and stimulates IL-1, IL-6, and IL-12 production (Garcia-Mauriño et al. 1997; Lissoni et al. 1997a; Morrey et al. 1994). However, T lymphocytes seem to be the main target of melatonin in mice (Hofbauer and Heufelder 1996; Maestroni and Conti 1990) and humans in whom physiological concentrations of melatonin stimulate IL-2 and IFN-γ production (Garcia-Mauriño et al. 1997; Guerrero et al. 1996).

We have reported that the immunoenhancing and antistress effect of melatonin is neutralized by the opioid antagonist naltrexone (Maestroni and Conti 1990; Conti 1991). Known opioid peptides could mimic the effects of melatonin, with the κ-agonist dynorphin being the most potent agent (Maestroni and Conti 1989). The hematopoietic protection involved the release of granulocyte/macrophage colony-stimulating factor (GM-CSF) from bone marrow stroma upon stimulation by a Th cell factor induced by melatonin (Maestroni et al. 1994b). This factor was immunologically and biologically indistinguishable from IL-4 (Maestroni et al. 1994c). Nevertheless, further investigations revealed that this Th cell factor comprised of two cytokines of 15 and 67 kDa with the common opioid sequence (Tyr-Gly-Gly-Phe) at their amino terminal and a carboxy-terminal extension which resembles both IL-4 and dynorphin (Maestroni et al. 1996). Both agents activated lymph node Th cells, and bone marrow Th cells released these opioid cytokines, which were named MIOs (Maestroni et al. 1996). Due to their size and unusual immunological characterization, the MIOs might represent novel opioid cytokines. The lower molecular weight MIO (MIO-15) seems to mediate both the antistress and hematopoietic effects of melatonin (Maestroni et al. 1996).

Our data may reflect a physiological requirement for sustained melatonin regulation of hematopoiesis. Most recently, we performed experiments comparing the ability of melatonin to protect hematopoiesis in LLC-bearing mice and in tumor-free normal mice treated with the cytotoxic drug cyclophosphamide. This experiment was prompted by the fact that melatonin added in GM-CFU cultures could directly enhance the number of GM-CFU but only in presence of suboptimal concentration of colony-stimulating factors (CSF), i. e., in presence of activated bone marrow adherent cells (Maestroni et al. 1994b, c). In addition, LLC is known to produce CSF and exert myelopoietic activity in vivo (Young et al. 1988). Melatonin did not exert any hematopoietic protection in tumor-free mice but rather increased the bone marrow toxicity of cyclophosphamide (Maestroni and Conti 1996). However, both in tumor-free and LLC-bearing mice, the effect of melatonin was neutralized by naltrexone which suggested the involvement of MIOs. Most recently, the MIOs were found to be acting on a single opioid binding site present in adherent bone marrow cells. Opioid agonists could mimic the colony-stimulating activity of melatonin in murine adherent bone marrow cells with an order of potency (dynorphin A > ICI 199,441 = U50488H > ICI204,449 = DPDPE = DAMGO) which suggested the presence of a type 1 κ-opioid receptor. Consistently, the specific κ-opioid receptor antagonists norbinaltorphimine neutralized the in vitro effect of melatonin and, most relevant, inhibited regeneration of hematopoiesis in mice treated with carboplatin. Like the MIOs, dynorphin A could exert colony-stimulating activity on adherent bone marrow cells but not on non-adherent cells and only in the presence of GM-CFS. The effect of dynorphin A was abolished by overnight incubation of adherent cells with antisense oligodeoxynucleotide to κ-opioid receptor or by addition of anti-IL-1 monoclonal antibody. The adherent bone marrow cells which express κ-opioid receptors were identified by a specific anti-κ-opioid receptor mAb and found to be macrophages (G. Maestroni, unpublished results).

20.2.3 Melatonin Receptors

The recent cloning of a family of G-protein-coupled receptors for melatonin has been characterized by three melatonin receptors subtypes, Mel1a, Mel1b, and Mel1c with a

K_d ranging from 20 to 160 pM (Reppert and Weaver 1995). Regarding melatonin receptors in immunocompetent cells, high-affinity binding sites for melatonin have been described in the membrane homogenates of thymus, bursa of Fabricius, and spleen of a number of birds and mammals (Poon et al. 1994). We have described a high-affinity binding site in bone marrow Th cells (Maestroni 1995). Another study showed that melatonin binds to human lymphoid cells modulating their proliferative response. Consistent with our findings, T cell activation significantly increased melatonin binding (Konakchieva et al. 1995). Melatonin binding sites and melatonin receptor mRNA was mostly found in human Th cells, but also in CD8[+] T and B cells (Garcia-Mauriño et al. 1997; Garcia-Perganeda et al. 1997; Guerrero et al. 1996). Besides membrane receptors, nuclear receptors for melatonin have been described in human myeloid cells. Melatonin seems to be the natural ligand for nuclear orphan receptors RZR/ROR. It appears that melatonin downregulates the expression of the RZR/ROR responding gene which is 5-lipoxygenase, a key enzyme in allergic and inflammatory disease (Carlberg and Wiesenberg 1995).

20.2.4 Clinical Trials

We showed that melatonin may synergize with IL-2 in controlling tumor growth (Maestroni and Conti 1993). On this basis, Dr. Paolo Lissoni and co-workers in Italy have conducted an impressive series of clinical studies in cancer patients. Over 200 advanced solid tumor patients in whom the standard anticancer therapies were not tolerated or not effective were treated with IL-2 and melatonin. The results obtained show that this neuroimmunotherapeutic strategy may amplify the antitumoral activity of low-dose IL-2, induce objective tumor regression, prolong progression-free time and overall survival and, moreover, the treatment was very well tolerated. It should be stressed that melatonin seems to be required for the effectiveness of low-dose IL-2 in neoplasias that are generally resistant to IL-2 alone (reviewed in Conti and Maestroni 1995). Similar findings were obtained in a smaller study in which melatonin was combined with IFN-γ in metastatic renal cell carcinoma (Neri et al. 1994). In addition, melatonin in combination with low-dose IL-2 was able to neutralize the surgery-induced lymphocytopenia in cancer patients (Lissoni et al. 1995). On the contrary, a recent double blind study investigating the myeloprotective effect of melatonin given in combination with carboplatin and etoposide to lung cancer patients shows that melatonin did not influence the chemotherapy-induced hematopoietic toxicity (Ghielmini, submitted for publication).

20.2.5 Mechanism of Action

Melatonin binds to specific melatonin receptors in Th cells and/or monocytes stimulating the production of IFN-γ, IL-2, MIO, IL-1, IL-6, and IL-12 which in turn upregulate the immune response. Second messengers are not completely understood but include G protein and inhibition of cAMP (Guerrero et al. 1996). The immunotherapeutic effect of melatonin against encephalitis viruses or bacterial infections (Ben-Nathan et al. 1995, 1996; Bonilla et al. 1997) might be explained by the increased production of IL-1, IL-12, IFN-γ, and/or IL-2 as well as by an increased myelopoiesis due to the hematopoietic action of the MIO. A mechanism involving Th type 1 cytokines might also account for the capacity of melatonin to restore immunodeficiency states secondary to aging (Caroleo et al. 1992; Pioli et al. 1993), trauma-hemorrhage (Wichmann et al. 1996), or to synergize with IL-2 in cancer patients (Conti and Maestroni 1995; Lissoni et al. 1995). In this regard, it is noteworthy to recall that melatonin is most active on antigen- or cytokine-activated cells (Konakchieva et al. 1995; Maestroni et al. 1986, 1987). Consistently, *IL-2 treatment in patients results in activation of the whole immune system and creates the most suitable biological background for melatonin*. The finding that melatonin can stimulate IL-12 production from human monocytes only if incubated in the presence of IL-2 further supports this concept (Lissoni et al. 1997). In fact, the plasma concentration of IL-12 was increased in patients who showed a partial response to the melatonin/IL-2 combination (Lissoni et al. 1997b). This finding seems important because IL-12, which is mainly produced by monocyte/macrophages, plays a relevant role in cytokine therapy (Banks et al. 1995).

The ability of melatonin to counteract the thymus involution and immunodepression caused by stress or corticosteroid treatment seems to be mediated by the MIOs. We do not yet know whether MIO action is exerted on peripheral immunocompetent cells and in the thymus or whether the hematopoietic effects of these novel opioid cytokines are also involved. On the contrary, it seems clear that the hematopoietic effects of melatonin depend on a complex series of events which involve the effects of the MIOs on κ-opioid receptors expressed on bone marrow macrophages (G. Maestroni, unpublished results). This seems of considerable relevance for understanding melatonin effects and, in general, the physiology of hematopoiesis. MIOs might belong to a new family of endogenous κ-opioid agonists which, in the case of hematopoietic protection, seem to synergize with GM-CSF on stromal cells kappa-receptors (Maestroni et al. 1996). This would explain why in LLC-bearing mice, MIOs rescue hematopoiesis against the toxic action of cancer chemotherapy. LLC is in fact known to release GM-CSF (Young et al. 1988). Activated Th cells may also produce GM-CSF. This might account for the therapeutic and positive hematopoietic effects of melatonin when administered together with IL-2 in cancer patients (Conti and Maestroni 1995; Lissoni et al. 1995). The cytokines involved in the immune-hematopoietic action of melatonin may exert an influence on the production of melatonin by the pineal gland. The pineal gland is, in fact, located outside the blood–brain barrier and some reports show that IFN-γ may directly affect the synthesis of melatonin in the pineal gland (Withyachumnarnkul et al. 1990).

20.3 Conclusion

Melatonin may be considered as an important endogenous neuroimmunomodulator and as a potential immunotherapeutic agent. A hypothetical pineal gland-immune system physiological network might, therefore, take shape (Fig. 20.1). The proper functioning of such a network might be crucial in the adaptive response of the organism to environmental demands and, thus, in the maintenance of health. However, we are still far from a complete understanding of the mechanism underlying the immunological and hematopoietic action of melatonin. For example, it is not clear whether melatonin acts on Th1 or Th2 cells or on both. In addition, it is not known whether melatonin may induce cytokine gene expression or whether its action is posttranslational only. These

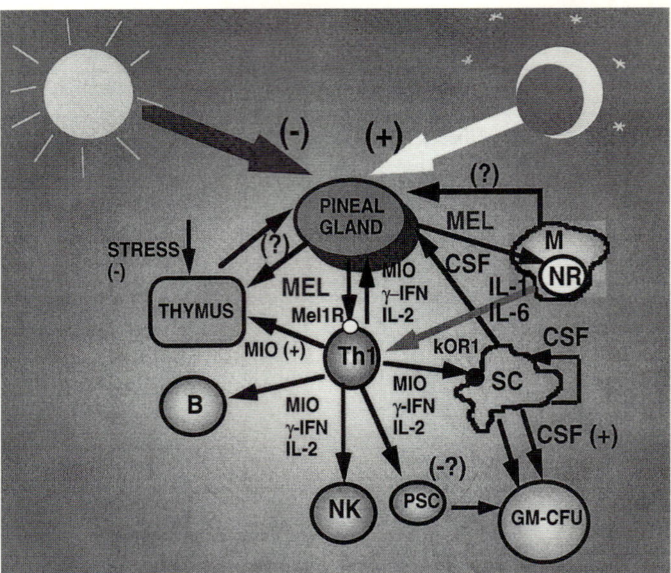

Fig. 20.1. The melatonin–immune network. Melatonin, whose synthesis is regulated by the photo-period, activates its specific membrane receptor (*Mel1R*) in T helper type 1 cells (*Th1*) or nuclear receptors (*NR*) in macrophages (*M*). This stimulates the secretion of cytokines such as IFN-γ, IL-2, MIO, and IL-1, IL-6 respectively. In turn, these cytokines may upregulate immune effectors which may counteract secondary immunodeficiencies and protect against viral and bacterial diseases and cancer. Besides the immune response, MIOs may also exert hematopoietic effects. MIO bind to kappa 1 opioid receptors (*kOR1*) on bone marrow stromal cells (*SC*). Depending on the affinity state of the kOR1 receptor, the effect may be an increased production of CSF which rescue GM-CFU from myelotoxic drugs or, by a still obscure mechanism, an increased toxicity. IFN-γ, IL-2, MIO, IL-1, IL-6, and CSF might modulate melatonin production in the pineal gland. *PSC*, pluripotent stem cells; *NK*, natural killer cells, *B*, B cells

seem to be important questions, as the Th1/Th2 balance and the resulting cytokine production are crucial for a successful immune response and may be relevant in immune-based pathologies (Del Prete et al. 1994). The stimulatory effect of melatonin on IL-2 and IFN-γ and the lack of influence on IL-4 suggest the involvement of Th1 cells. Perhaps the same Th cell type may also produce MIOs which are radically different from the enkephalin-containing molecules reported to be produced by Th2 cells (Hiddinga et al. 1994). On the other hand, the dramatic protection exerted by melatonin in experimental models of viral encephalitis and lethal bacterial infections as well as its capacity to restore depressed immune functions is consonant with a Th1 cell involvement as well as with the action of melatonin on monocyte/macrophage cytokines (Del Prete et al. 1994). Physiologically, it seems possible to distinguish two different roles. The first one occurs in acute conditions during a viral or bacterial infection which produces a substantial activation of the immune system. In that condition, endogenous and/or exogenous melatonin may optimize the immune response by sustaining Th cell and macrophage functions and production of cytokines, part of which (MIO, IL-6, IL-1) may also affect hematopoiesis. A second, more general role may be exerted at the hematopoietic-immune level by a chronic circadian resetting of the immunological machinery to maintain the immune homeostasis. This is suggested by the observation that in healthy mice, i.e., in absence of any infection and immunological activation, only Th cells which sit in the bone marrow are sensitive to melatonin (Maestroni et al. 1996; Maestroni 1993). Products of this melatonin–bone marrow Th cell interaction are the MIOs which may affect hematopoiesis and thymocyte proliferation (Maestroni et al. 1996; Maestroni 1993).

Both the acute and chronic mechanisms might be exploited in the use of melatonin as an immunotherapeutic agent to correct secondary immunodeficiency or fight viral diseases. As we already stated in preceding reviews (Maestroni 1993; Maestroni and Conti 1996), *we would like to stress the need for a large double blind study in human immunodeficiency virus (HIV)-positive patients.* In the presence of normal Th cell counts, the apparent ability of melatonin to sustain Th cell functions and IL-2 and IFN-γ production might result in delayed development or occurrence of AIDS. Reduction of plasma viremia was associated with an increased IL-2 mRNA expression in lymph nodes of HIV-infected patients (Sei et al. 1996). IL-2 is the most potent cytokine capable of inducing the $CD8^+$ T-cell-mediated inhibition of HIV replication which seems to override the ability of IL-2 to stimulate HIV expression (Kinter et al. 1995). If effective, melatonin administration would be a relatively cheap and safe prevention of this devastating disease. Alternatively, melatonin might be associated with low-dose IL-2 which seems to be beneficial in HIV-associated malignancies (Bernstein et al. 1995) or with HIV protease inhibitors. The use of melatonin in combination with IL-2 in cancer neuroimmunotherapy might prolong survival and improve patients' quality of life (Conti and Maestroni 1995). *These encouraging results obtained by Lissoni and co-workers deserve, therefore, to be expanded and challenged in other studies. In addition, melatonin might be useful in enhancing the immune response against tumor antigens, a promising therapeutic strategy for cancer vaccines.* With respect to the use of melatonin in combination with cancer chemotherapeutic drugs, the results obtained so far are disappointing. Melatonin administered alone seems to worsen the bone marrow toxicity of common cancer chemotherapeutic regimens or, at best, be ineffective. This fact calls for further studies to understand the role of melatonin in hematopoiesis and indicates that in certain conditions melatonin may have adverse effects.

Acknowledgements. The work performed in Locarno has been supported by Swiss Nationalfonds grants no. 3.267.0.85; 31.25350.88; 31.36128.92, 31.45532.95 and by the Helmut Horten Foundation.

References

Ader R, Cohen N, Felten D (1995) Psychoneuroimmunology: interactions between the nervous system and the immune system. Lancet 345(8942):99–103

Ader R, Felten DL, Cohen N (eds) (1991) Psychoneuroimmunology II. (ed) Academic Press, San Diego

Angeli A, Gatti G, Sartori ML, Del Ponte D, Cerignola R (1988) Effect of exogenous melatonin on human natural killer (NK) cell activity. An approach to the immunomodulatory role of the pineal gland. In: Gupta D, Attanasio A, Reiter RJ (eds) The Pineal Gland and Cancer. Müller and Bass, Tübingen, pp 145–157

Banks RE, Patel PM, Selby PJ (1995) Interleukin-12: a new player in cytokine therapy. Br J Cancer 71:655–659

Barath P, Csaba G (1974) Histological changes in the lung, thymus and adrenal one and half year after pinealectomy. Acta Biol Acad Sci Hung 25:123–125

Becker J, Veit G, Handgretinger R, Attanasio G, Bruchelt G, Treuner I, Niethammer D, Gupta D (1988) Circadian variations in the immunomodulatory role of the pineal gland. Neuroendocrinol Lett 10:65–80

Ben-Nathan D, Maestroni GJM, Conti A (1995) Protective effect of melatonin in mice infected with encephalitis viruses. Arch Virol 140:223–230

Ben-Nathan D, Maestroni GJM, Conti A (1996) Protective effect of melatonin in viral and bacterial infections. Abstract, 2nd Locarno Meeting on Neuroendocrinoimmunology Therapeutic potential of the Pineal Hormone Melatonin, May 5–8, Locarno, Switzerland

Bernstein ZP, Porter MM, Gould M, Lipman B, Bluman EM, Stewart CC, Hewitt RG, Fyfe G, Poiesz B, Caligiuri MA (1995) Prolonged administration of low-dose interleukin-2 in human immuno-deficiency virus-associated malignancy results in selective expansion on innate immune effectors without significant clinical toxicity. Blood 86:3287–3294

Bindoni M, Cambria A (1968) Effects of pinealectomy on the in vivo and the in vitro biosynthesis of nucleic acids and on the mitotic rate in some organs of the rat. Arch Sci Biol (Bologna) 52: 271–283

Blask DE (1993) Melatonin in oncology. In: Yu H-S, Reiter RJ (eds) Melatonin. Biosynthesis, Physiological Effects, and Clinical Applications. CRC Press, Boca Raton, pp 447–477

Bonilla E, Valerofuenmajor N, Pons H, Chacin-Bonilla L (1997) Melatonin protects mice infected with Venezuelan equine encephalomyelitis virus. Cell Mol Life Sci 53:430–434

Carlberg C, Wiesenberg I (1995) The orphan receptor family RZR/ROR, melatonin and 5-lipoxygenase: an unexpected relationship. J Pineal Res 18:171–178

Caroleo MC, Frasca D, Nistico G, Doria G (1992) Melatonin as immunomodulator in immunodeficient mice. Immunopharmacology 23:81–89

Champney TH, McMurray DN (1991) Spleen morphology and lymphoproliferative activity in short photoperiod exposed hamsters. In: Fraschini F, Reiter RJ (eds) Role of melatonin and pineal peptides in neuroimmunomodulation. Plenum Publ Co, London, pp 219–225

Colombo LL, Chen G-J, Lopez MC, Watson RR (1992) Melatonin induced increase in gamma-interferon production by murine splenocytes. Immunol Lett 33:123–126

Conti A, Maestroni GJM (1995) The clinical neuroimmunotherapeutic role of melatonin in oncology. J Pineal Res 19:103–110

Conti A, Haran-Ghera N, Maestroni GJM (1992) Role of pineal melatonin and melatonin-induced-immuno-opioids in murine leukemogenesis. Med Oncol Tumor Pharmacother 9:87–92

Csaba G, Barath P (1975) Morphological changes of thymus and the thyroid gland after postnatal extirpation of pineal body. Endocrinol exp 9:59–67

Csaba G, Bodocky M, Fischer J, Acs T (1965) The effect of pinealectomy and thymectomy on the immune capacity of the rat. Experientia 22:168–175

Del Gobbo V, Libri V, Villani N, Caliò R, Nstico G (1989) Pinealectomy inhibits interleukin-2 production and natural killer activity in mice. Int J Immunopharmacol 11:567–577

Del Prete G, Maggi E, Romagnani S (1994) Human Th1 and Th2 cells: Functional properties, mechanisms of regulation and role in disease. Lab Invest 70:299–307

Garcia-Mauriño S, Gonzales-Haba MG, Calvo JR, Rafii-El-Idrissi M, Sanchez-Margalet V, Goberna R, Guerrero JM (1997) Melatonin enhances IL-2, Il-6, and IFN-γ production by human circulating CD4+ cells. J Immunology 159:574–581

Garcia-Perganeda A, Pozo D, Guerrero JM, Calvo JR (1997) Signal transduction for melatonin in human lymphocytes. Involvement of a pertussis toxin-sensitive G protein. J Immunol 159:3774–3781

Giordano M, Palermo MS (1991) Melatonin-induced enhancement of antibody dependent cellular cytotoxicity. J Pineal Res 10:117–121

Guerrero J-M, Garcia-Maurino S, Gil-Haba M, Rafii-El-Idrissi M, Pozo D, Garcia-Perganeda A, Calvo JR (1996) Mechanisms of action on the human immune system: Membrane receptors versus nuclear receptors. In: Maestroni GJM, Conti A, Reiter RJ (eds) Therapeutic potential of the Pineal Hormone Melatonin. Karger, Basel, pp 43–52

Haldar C, Haussler D, Gupta D (1992) Response of CFU-GM(colony forming units for granulocytes and macrophages) from intact and pinealectomized rat bone marrow to macrophage colony stimulating factor (rGM-CSF) and human recombinant erythropoietin (rEPO). Prog Brain Res 91:323–325

Hansson I, Holmdahl R, Mattsson R (1992) The pineal hormone melatonin exaggerates development of collagen-induced arthritis in mice. J Neuroimmunol 39:23–31

Hiddinga HJ, Isaak DD, Lewis RV (1994) Enkephalin-containing peptides processed from proenkephalin significantly enhance the antibody-forming cell responses to antigens. J Immunol 152: 3748–3758

Hofbauer LC, Heufelder AE (1996) Endocrinology meets immunology: T lymphocytes as novel targets for melatonin. Eur J Endocrinol 134:424–425

Jankovic BD, Isakovic S, Petrovic S (1970) Effect of pinealectomy on immune reactions in the rat. Immunology 8:1–6

Kinter AL, Bende SM, Hardy EC, Jackson R, Fauci AS (1995) Interleukin 2 induces CD8+ T cell-mediated suppression of human immunodeficiency virus replication in CD4+ T cells and this effect overrides its ability to stimulate virus expression. Proc Natl Acad Sci USA 92:19985–10989

Konakchieva R, Kyurkchiev S, Kehayov I, Taushanova P, Kanchev L (1995) Selective effect of methoxyndoles on the lymphocyte proliferation and melatonin binding to activated human lymphoid cells. J Neuroimmunol 63:125–132

Lewinsky A, Zelazowsky P, Sewerynek E, Zerek-Melen G, Szudlinsky M (1989) Melatonin-induced immunosuppression of humal lymphocyte natural killer activity in vitro. J Pineal Res 7:153–164

Lissoni P, Barni S, Ardizzoia A, Brivio F, Tancini G, Conti A, Maestroni GJM (1992) Immunological effects of a single evening subcutaneous injection of low-dose interleukin-2 in association with the pineal hormone melatonin in advanced cancer patients. J Biol Reg Homeos Agents 6:132–136

Lissoni P, Brivio F, Brivio O, Fumagalli L, Gramazio F, Rossi M, Emanuelli G, Alderi G, Lavorato F (1995) Immune effects of preoperative immunotherapy with high-dose subcutaneous interleukin-2 versus neuroimmunotherapy with low-dose interleukin-2 plus the neurohormone melatonin in gastrointestinal tract tumor patients. J Biol Regul Homeost Agents 9:31–33

Lissoni P, Pittalis S, Barni S, Rovelli F, Fumagalli L, Maestroni GJM (1997a) Regulation of interleukin-2/interleukin-12 interactions by the pineal gland. Int J Thymology 5:443–447

Lissoni P, Rovelli F, Brivio F, Pittalis S, Maestroni GJM, Fumagalli L (1997b) In vivo study of the modulatory effect of the pineal hormone melatonin on interleukin-12 secretion in response to interleukin-2. Int J Thymology 5:482–486

Maestroni GJM (1993) The immunoneuroendocrine role of melatonin. J Pineal Res 14:1–10

Maestroni GJM (1995) T helper 2 lymphocytes as a peripheral target of melatonin. J Pineal Res 18:84–89

Maestroni GJM, Conti A (1989) Beta-endorphin and dynorphin mimic the circadian immuno-enhancing and anti-stress effects of melatonin. Int J Immunopharmacol 11:333–340

Maestroni GJM, Conti A (1990) The pineal neurohormone melatonin stimulates activated CD4+, Thy-1+ cells to release opioid agonist(s) with immunoenhancing and anti-stress properties. J Neuroimmunol 28:167–176

Maestroni GJM, Conti A (1991) Anti-stress role of the melatonin-immuno-opioid network. Evidence for a physiological mechanism involving T cell-derived, immunoreactive? β-endorphin and MET-enkephalin binding to thymic opioid receptors. Int J Neurosci 61:289–298

Maestroni GJM, Conti A (1993) Melatonin in relation to the immune system. In: Yu H-S, Reiter RJ (eds) Melatonin. Biosynthesis, Physiological effects and Clinical Applications. CRC Press, Boca Raton, pp 290–306

Maestroni GJM, Conti A (1996) Melatonin and the immune-hematopoietic system. Therapeutic and adverse pharmacological correlates. Neuroimmunomodulation 3:325–332

Maestroni GJM, Pierpaoli W (1981) Pharmacological control of the hormonally modulated immune response. In: Ader R (eds) Psychoneuroimmunology. Academic Press, New York, pp 405–425

Maestroni GJM, Conti A, Pierpaoli W (1986) Role of the pineal gland in immunity. Circadian synthesis and release of melatonin modulates the antibody response and antagonize the imuno-suppressive effect of corticosterone. J Neuroimmunol 13:19–30

Maestroni GJM, Conti A, Pierpaoli W (1987) Role of the pineal gland in immunity: II. Melatonin enhances the antibody response via an opiatergic mechanism. Clin Exp Immunol 68:384–391

Maestroni GJM, Conti A, Pierpaoli W (1988) Role of the pineal gland in immunity. III. Melatonin antagonizes the immunosuppressive effect of acute stress via an opiatergic mechanism. Immunology 63:465–469

Maestroni GJM, Conti A, Reiter RJ (eds) (1994a) Advances in Pineal Research 7. Reiter RJ (ed). John Libbey & Co., London

Maestroni GJM, Covacci V, Conti A (1994b) Hematopoietic rescue via T-cell-dependent, endogenous GM-CSF by the pineal neurohormone melatonin in tumor bearing mice. Cancer Res 54:2429–2432

Maestroni GJM, Conti A, Lissoni P (1994c) Colony-stimulating activity and hematopoietic rescue from cancer chemothereapy compounds are induced by melatonin via endogenous interleukin 4. Cancer Res 54:4740–4743

Maestroni G, Hertens E, Galli P, Conti A, Pedrinis E (1996) Melatonin-induced T-helper cell hematopoietic cytokines resembling both interleukin-4 and dynorphin. J Pineal Res 21:131–139

Morrey MK, McLachlan JA, Serkin CD, Bakouche O (1994) Activation of human monocytes by the pineal neurohormone melatonin. J Immunol 153:2671–2680

Neri B, Fiorelli C, Moroni F, Nicita G, Paoletti MC, Ponchietti R, Raugei A, Santoni G, Trippitelli A, Grechi G (1994) Modulation of human lymphobalstoid interferon activity by melatonin in metastatic renal cell carcinoma. A phase II study. Cancer 73:3015–3019

Palermo MS, Vermeulen M, Giordano M (1994) Modulation of antibody-dependent cellular cytotoxicity by pineal gland. In: Maestroni GJM, Conti A, Reiter RJ (eds) Advances in Pineal Research. John Libbey, London, pp 143–147

Pioli C, Caroleo MC, Nistico G, Doria G (1993) Melatonin increases antigen presentation and amplifies specific and non specific signals for T-cell proliferation. Int J Immunopharmacol 15:463–469

Poon AM, Liu ZM, Pang CS, Brown GM, Pang SF (1994) Evidence for a direct action of melatonin on the immune system. Biol Signals 3:107–117

Rella W, Lapin V (1978) Immunocompetence of pinealectomized and simultaneously pineal-ectomized and thymectomized rats. Oncology 33:3–6

Reppert SM, Weaver DR (1995) Melatonin madness. Cell 83:1059–1062

Sei S, Akiyoshi H, Bernard J, Venzon DJ, Fox CH, Schwartzentruber DJ, Anderson BD, Kopp JB, Mueller BU, Pizzo PA (1996) Dynamics of virus versus host intercation in children with human immunodeficiency virus type 1 infection. J Infect Dis 173:1485–1490

Spector NH, Dolina S, Cornelissen G, Halberg F, Markovic BM, Jankovic BD (1996) Neuroimmuno-modulation: neuroimmune interactions with the environment. In: Blatteis CM (eds) Handbook of Physiology: The Environment. Amer Physiol Society, Bethesda, pp 1437–1549

Vaughan MK, Reiter RJ (1971) Transient hypertrophy of the ventral prostate and coagulating glands and accelerated thymic involution following pinealectomy in the mouse. Texas Rep Biol Med 29:579–586

Wichmann M, Zellweger R, DeMaso A, Ayala A, Chaudry I (1996) Melatonin administration attenuates depressed immune functions trauma-hemorrhage. J Surg Res 63:256–262

Withyachumnarnkul B, Nonaka KO, Santana C, Attia AM, Reiter RJ (1990) Interferon-gamma modulates melatonin production in rat pineal gland in organ culture. J Interferon Res 10:403–411

Young MRI, Young ME, Kim K (1988) Regulation of tumor-induced myelopoiesis and the associat-ed immune suppressor cells in mice bearing metastatic Lewis lung carcinoma by prostaglandin E2. Cancer Res 48:6826–6831

Yu H-S, Reiter RJ (eds) (1993) Melatonin. Biosynthesis, Physiological Effects, and Clinical Applica-tions. CRC Press, Boca Raton

21 Melatonin Rhythms in Mice: Role in Autoimmune and Lymphoproliferative Diseases

Ario Conti, Georges J. M. Maestroni

Abstract

Production of melatonin (MLT) in the pineal gland (PG) of inbred mice such as C57Bl/6J and AKR strains is still a matter of debate. In previous studies, other authors and we showed that these strains of inbred mice have a clear-cut circadian rhythm of serum MLT and urinary 6-hydroxy-MLT-sulfate. In contrast, other groups claimed these mice are unable to synthesize MLT. These studies were based on RIA measurements and/or estimates of N-acetyltransferase (NAT) and hydroxy-indole-O-methyltransferase (HIOMT) activities. In a recent study, we validated the presence of MLT in the PG of C57Bl/6, BALB/c, and AKR mice by HPLC determinations. We found a short-term MLT peak in the middle of the dark period with a pattern which mirrors that found previously in the serum. The possibility remains, although it seems unlikely, that the pineal MLT rhythm measured here represents MLT produced elsewhere which is then subsequently taken up by the PG. In the last 15 years we demonstrated that the pineal gland and melatonin (MLT) play an immunoregulatory role both in mice and in humans. In particular we found that MLT: (a) counteracts immunosuppression and thymus atrophy induced by stress or corticosteroid treatment; (b) protects mice injected with encephalitogenic viruses; (c) synergizes with interleukin-2 (IL-2) in cancer immunotherapy; and (d) rescues hematopoiesis from cancer chemotherapy toxicity. Regarding the mechanism of action, MLT seems to act directly on CD4$^+$ lymphocytes which release MLT-induced opioid peptides (MIO) with immunoenhancing properties along with other cytokines. These findings prompted us to investigate the role of the pineal gland and MLT in lymphoproliferative and autoimmune diseases. In the first model, C57Bl6 mice were injected with the leukemogenic virus A-RadLV which induces a murine leukemia type T. Mice were surgically pinealectomized and/or treated with MLT, MLT plus naltrexone, naltrexone alone, and saline. MLT accelerated leukemogenesis whereas surgical pinealectomy delayed it. Moreover, the action of MLT was blocked by naltrexone indicating the involvement of MIO in the development of lymphomas. In the second study, we investigated the role of the PG and MLT in the immunopathogenesis of autoimmune diabetes mellitus type I, using female non-obese diabetic (NOD) mice as an experimental model. Mice were pinealectomized or treated chronically with MLT (injected subcutaneously or administered via drinking water). We found that neonatal pinealectomy accelerates the development of disease in female NOD mice while exogenous MLT protects animals. This in spite of the fact that MLT increased the production of insulin autoantibodies (IAA). We conclude that the PG and MLT influence the development of autoimmune diabetes although the mechanism of action needs further investigations.

21.1 Introduction

Before 1982, no data concerning melatonin rhythmicity in serum or pineal gland of mice were available. In 1983, Franz Halberg's group was the first to describe a circadian melatonin rhythm in mice (Brown et al. 1983), further corroborated in subsequent experiments (Brown et al. 1986).

In our own investigations of melatonin levels in mice, using a commercially available radioimmunoassay, we found that three strains of female mice (C57Bl/6, BALB/c, AKR) displayed a nocturnal peak of serum melatonin (Maestroni et al. 1986, 1987), confirming the results of Brown et al. (1983, 1986). Although the pattern of serum melatonin was characterized by a short-term peak, our result showed that inbred mice, as other rodents, have a daily rhythm of serum melatonin (Reiter 1991; Yu and Reiter 1993). Moreover, in the C57Bl6 strain, 6-hydroxymelatoninsulphate, the melatonin metabolite in urine, showed a similar rhythm (Conti et al. 1992). These results were in contrast to other studies which reported that inbred mice such as C57Bl6/J, AKR/J, BALB/c, and NZB/BLNJ have no detectable melatonin in their pineal gland (Ebihara et al. 1986; Vollrath et al. 1988; Goto Oshima et al. 1989).

Between 1992 and 1996, in order to better clarify and to settle this dispute, our group performed a study to validate the presence of melatonin in such inbred mice. We not only confirmed the presence of melatonin in the pineal gland of these mice, but also demonstrated that in the pineal gland melatonin shows a circadian rhythm which is comparable to the endogenous rhythm found in the serum of the same strain of mice (Conti and Maestroni 1996). Over the past several years, according to the fact that mice have an endogenous production of melatonin, it has been recognized that this indolamine affects immune mechanisms (Maestroni et al. 1986, 1987, 1994b, c; Maestroni and Conti 1991; Ben-Nathan et al. 1995) via the synthesis and/or release of opioid peptides, i.e., MIO (Maestroni et al. 1995). Table 21.1 summarizes some of the most important effects of melatonin on the immune system.

On these bases, two groups investigated whether the pineal gland and melatonin may play a role in the development of autoimmunity (Conti and Maestroni 1993, 1996; Hansson et al. 1990, 1992, 1993) and lymphoproliferative diseases (Conti et al. 1992).

Table 21.1. Melatonin effects on the immune system

1. Melatonin counteracts the immunosuppressive effect induced by acute stress and/or corticosteroid treatment on antibody production and thymus cellularity (Maestroni et al. 1986, 1987)
2. Melatonin protects normal or stressed mice from lethal infections with encephalitogenic viruses (Maestroni and Conti 1991; Ben-Nathan et al. 1995)
3. Melatonin rescues the blood-forming system from the toxic effect of cancer chemotherapeutic compounds (Maestroni et al. 1994b, c)
4. Melatonin synergizes with IL-2 in cancer patients and reverses age-associated immune defects (Conti and Maestroni 1995)

Table 21.2. Experimental models

1. Autoimmune diabetes
   Non-obese diabetic mouse (NOD)
   Develop a spontaneous disease
   Disease similar to the insulin-dependent diabetes mellitus type 1 in humans
2. Rheumatoid arthritis
   Collagen-induced arthritis (CIA) in BBA/1 mice
   Disease similar to rheumatoid arthritis in humans
3. Lymphoproliferative diseases
   C57Bl/6 mice inoculated intrathymically with A-RadLV
   Disease similar to the development of human T leukemia

Although in most autoimmune diseases the main effectors appear to be T helper cells, many other underlying conditions such as immune deficiency, hormonal factors, and environmental agents have to be considered. It is important to note that only a combination of these various factors can trigger the disease (Conti and Maestroni 1994). In addition, autoimmune diseases seem to be associated with a host of psychoneuro-endocrinological features, as reported in various studies (Jankovìc et al. 1986; Fabris et al. 1992; Maestroni et al. 1994a). It thus seems evident that a correct approach to the study of autoimmunity should also integrate neuroendocrinoimmunological information (Conti and Maestroni 1993, 1994). Experimental models are listed in Table 21.2.

21.2 Melatonin Endogenous Rhythm in Mice

C57Bl/6NCrlBR, BALB/cAnNCrlBR, AKR, NOD, and BALB/c female mice, aged 9–10 weeks, were maintained under a light–dark cycle of 12 h light and 12 h darkness (0600 hours light on, 1800 hours light off) at $22 \pm 1\,°C$. A colony of NOD mice was established in our animal facilities (Conti and Maestroni 1996): diabetic mice were selected by testing for glycosuria with test strip and colorimetric assay. BALB/c mice were used as control animals. Great care was taken to avoid environmental stress before and during the course of the experiments (noise, smells, cage crowding, etc.). Tap water and fodder were given ad libitum.

The detailed procedure for biological group selection, pineal gland collection, melatonin extraction, and HPLC quantification is described elsewhere (Conti and Maestroni 1996).

Figure 21.1 shows the circadian rhythm of melatonin in the pineal gland of C57Bl/6 and AKR female mice. In C57Bl/6 mice the peak of melatonin is clear and short-term in duration. The highest concentration of melatonin was 344 ± 144 pg/gland, which represents the maximum concentration. The mean concentration of melatonin in the other samples was 17.80 ± 6.26 pg/gland. AKR mice exhibited a nocturnal peak similar to that found in C57Bl/6 mice: the peak appeared at 0030 hours and not at 0045 hours with an amplitude similar to C57Bl/6. The acrophase concentration was 380.0 ± 125 pg/gland, while the basal mean level corresponded to 29.13 ± 13.91 pg/gland. A similar peak has been detected in BALB/c and NOD mice. BALB/c mice present a small and short peak at 0030 hours with 21.21 ± 5.6 pg/gland and a basal value of 12.12 ± 1.43 pg/gland, while the highest concentration of melatonin in the pineal gland of NOD mice was 46.7 ± 7.7 pg/gland and occurred 8.5 h after darkness onset. The mean concentration of the other sampling points was 9.3 ± 2.9 pg/gland.

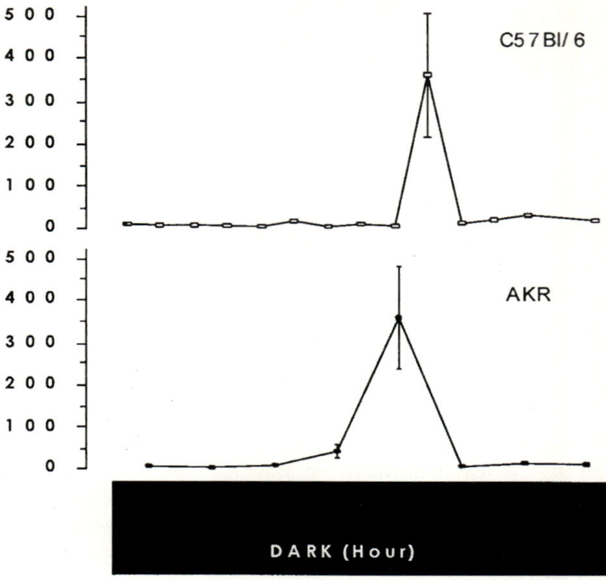

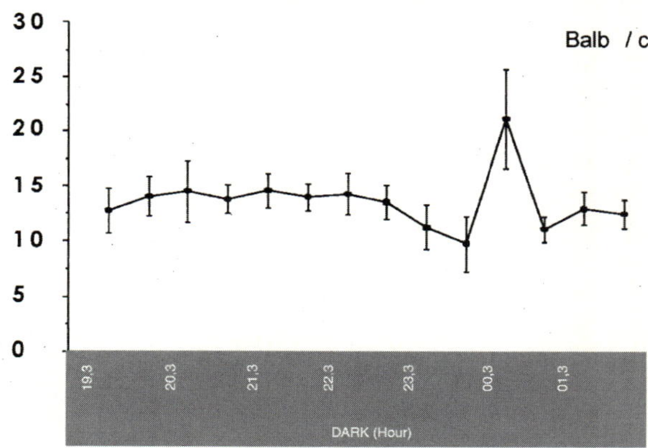

Fig. 21.1. Melatonin endogenous rhythms in inbred mice

21.3 The Role of Pineal Gland and Melatonin in Autoimmune and Lymphoproliferative Diseases

21.3.1 Autoimmune Diabetes

The NOD strain develops insulin-dependent diabetes at a very high rate and represents the main model used for such a study. Relevant to this study is the fact that in NOD mice the T cell repertoire seems to play a major role in mediating the effector phase of the disease. Diabetes develops in B-lymphocyte-depleted NOD mice but does not develop in NOD nu/nu mice. Both splenic CD4[+] and CD8[+] T lymphocytes are required for transfer of the disease. The purpose of the experiments reported here was to assess the effect of surgical pinealectomy and exogenous melatonin treatment on the development of diabetes mellitus type 1 in female NOD mice (Conti and Maestroni 1996). Experimental details and procedures have been published elsewhere (Conti and Maestroni 1996). Briefly, the following different experimental groups of mice have been considered:

1. Neonatal mice ($n = 25$) pinealectomized as described above. After pinealectomy animals were left undisturbed until death.
2. Mice treated with subcutaneous injections of melatonin ($n = 30$) administered chronically at dose of 4 mg/kg b.w. (0.5 ml s.c. at 1630 hours for 5 days a week). The treatment started when the animals were 4 weeks old and ended during the 38th week.
3. Control mice injected subcutaneously with PBS ($n = 25$) in the same schedule and route of administration as group 2 mice.
4. Mice given melatonin in the drinking water at night ($n = 17$). Melatonin was dissolved in 100 % ethanol, diluted in water to a final concentration of 10 µg/ml. Every day, for 5 days a week, the mice drank water containing melatonin from 1630 until 0700 hours.
5. Untreated control mice ($n = 29$).

Figure 21.2 shows, as a survival curve, the effect of newborn surgical pinealectomy, subcutaneous and oral melatonin administration on incidence of diabetes. Figure 21.2 also shows that newborn pinealectomized female NOD mice ($n = 25$) began dying 19 weeks after the surgical intervention. The incidence of disease increased progressively and after 38 weeks, 92 % of all pinealectomized mice were dead. Control mice ($n = 29$) began to die at 18 weeks of age and the slope of the survival curve was flatter. In fact, after 50 weeks of age only 65.5 % of mice were dead. The difference in survival between pinealectomized and unoperated mice was significant, $P < 0.0035$ (Log rank test). Chronic subcutaneous melatonin treatment of female NOD mice ($n = 30$) for

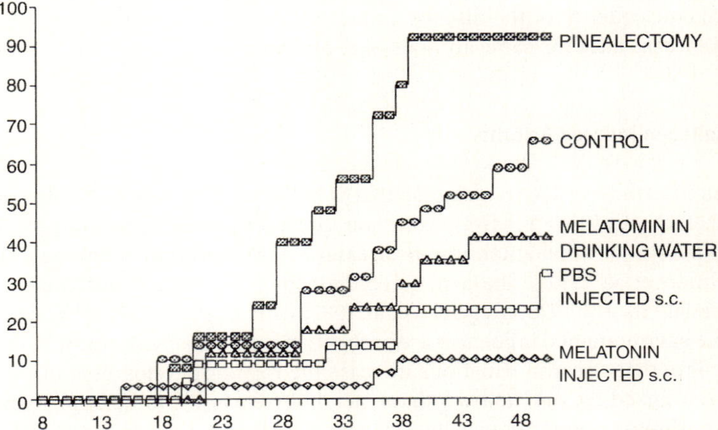

Fig. 21.2. Effect of pinealectomy and melatonin on survival of NOD mice

33 weeks protected the animals against the development of the disease when compared with control mice ($n = 29$) (Log rank test; $P < 0.0001$). However, PBS administration ($n = 22$) showed also a highly significant effect when compared with the untreated group ($P < 0.0023$). When the same data were analyzed by the Chi-square test at the end of the experiment, the difference between melatonin and PBS treatments was significant ($P < 0.0006$). In fact, mice started dying 11 weeks after the beginning of the treatment, i. e., when animals were 15 weeks old, while in the PBS group the first animal died after 16 weeks (i. e., 20 weeks old). After 50 weeks, in the melatonin treated group only 10 % of mice were dead, while in the PBS group the percentage of dead animals was 31.8 %. This experiment shows that melatonin seems to protect NOD mice against the development of autoimmune diabetes.

The effect of a chronic treatment with melatonin administered via drinking water ($n = 17$) compared with the control group ($n = 29$) is not significant when analyzed using the Log rank test ($P = 0.1526$) but it was significant with the chi-square test. Thus, at week 50, 41.2 % of melatonin-treated mice were dead while in the control group the percentage was 65.5 %, the difference between the two groups being significant ($P < 0.0019$).

During the study, the glucose response to subcutaneous or oral melatonin treatment and pinealectomy has been evaluated. Chronic treatment with melatonin and/or PBS from week 4 through week 37 influenced serum glucose levels. In fact, the baseline glucose level was lower than 10 mmol/l not only during treatment period but also afterwards. At the age of 55 weeks, both melatonin- and PBS-treated mice had serum glucose values lower than 10 mmol/l. On the contrary, pinealectomy potentiated the level of glycosuria and this effect reflected the higher incidence of diabetes. Concerning oral melatonin treatment, 58.8 % of mice were still alive after 43 weeks while 32 % of them were glycosuria-positive. In the untreated control group, only 51.7 % of mice were living and 20 % were glucose positive after 43 weeks. At the age of 50 weeks, the percentage of living controls was only 34.5 % and 8 % of them were glucose-positive, while in the melatonin group 58.8 % were living and 32 % were glycosuria-positive. It thus seems evident that the effect of oral administration of melatonin on glucose level was less dramatic than that of subcutaneous administration of melatonin or PBS. However, oral administration of melatonin seems to protect NOD female mice against the lethal consequence of the disease. In fact, although 32 % of mice were diabetic at week 43, 8 weeks later the same animals were still alive.

21.3.2 Collagen-Induced Arthritis

Rheumatoid arthritis (RA) is a disease that is believed to be caused by aberrant auto-immune reactions. It is a relatively common disease affecting approximately 1–2 % of the population. Like other autoimmune diseases, it is more common among women that men (Silman et al. 1992). The genetic component is significant but conclusions are controversial (Stastny 1978, Gregersen et al. 1987; Calin et al. 1989; Wordsworth and Bell 1991). The environmental influences are probably strong, but obviously difficult to study. Many believe that stressful situations increase the risk for development of the disease. Certain sex-linked factors, however, have clearly been shown to be important. During pregnancy, disease symptoms usually decline in severity whereas after partus, symptoms commonly worsen (Hazes et al. 1990; Spector et al. 1990; Silman et al. 1992). Most epidemiologic studies on the influence of contraceptive pills indicate that they have a protective effect, suggesting that estrogen mediates inhibition of disease development.

Between 1990 and 1993, several experiments were performed in order to evaluate effects of the pineal gland and melatonin in animal models (Trentham et al. 1977; Courtenay et al. 1980) of rheumatoid arthritis, i.e., collagen-induced arthritis (CIA). CIA is a chronic, self-perpetuating, inflammatory disease that primarily involves peripheral joints of the limbs. The synovial membrane of the affected joint is characterized by increased MHC class II expression and the infiltration of $CD4^+$ and $CD8^+$ T cells. The arthritic joint space often contains a large number of polymorphonuclear cells. Not all rat or mouse strains are susceptible to CIA. The development of this disease has been shown to be under the control of MHC class II. In mice, only those with the $H-2^q$ (and $H-2^r$) haplotype develop CIA.

There are a number of inbred strains carrying the $H-2^q$ haplotype, like DBA/1, NFR/N, B10Q, and C3HQ. Because DBA/1 is the most susceptible of the strains, it was the one used most frequently in the study of Hansson and colleagues. The immune system is clearly involved in the development of CIA. The importance of both B cells and T cells is reflected in the fact that treatment with either anti-T-cell antibodies (Ranges et al. 1985) or anti-B-cell antibodies blocks arthritis development (Helfgott et al. 1984). CIA exhibits many similarities to RA: for example autoreactive B and T cells are involved in both diseases, both are chronic and restricted to specific MHC class II alleles and produce antibodies to CII.

On this basis and considering the above mentioned immunoregulatory properties of the pineal gland and melatonin, one of the first hypotheses was that the development of psychosomatic and autoimmune disease like RA could be due to disturbed release of melatonin. The first investigation was performed in order to evaluate the effect of constant darkness, which represents a physiological stimulation of melatonin, or constant light, which on the contrary represents a physiological suppression of melatonin synthesis, on the course of CIA in DBA/1 mice. Mice kept in constant darkness develop more severe arthritis than those kept on constant light or in a normal dark/light rhythm (12 h light/12 h dark). Levels of anti-type II collagen antibodies were higher in mice kept in darkness and the spleens of these animals were enlarged. Since castration of female DBA/1 mice enhances the severity of CIA and since melatonin is known to exert an effect on gonadal function, a second series of experiments was carried out using ovariectomized mice. In this case, the same difference in arthritis severity between darkness- and light-exposed mice was observed. Hansson and colleagues concluded that exacerbation of arthritis in darkness is due to darkness-induced changes in levels of critical neurohormonal compound(s), which affect the autoimmune response via gonad-independent mechanisms. Moreover, the exaggerated severity and chronicity of arthritis may be due to higher levels of melatonin in these animals (Hansson et al. 1990). In order to clarify this hypothesis, a third series of experiments has been done. DBA/1 mice kept in constant darkness were analyzed for serum melatonin levels. An increase in background levels in comparison to mice kept in a normal dark/light cycle or in constant light was recorded. Then, different groups of mice (kept in constant light in order to minimize endogenous melatonin levels) were immunized with rat CII to induce arthritis and injected with melatonin. Melatonin injections were performed daily (1 mg/kg b.w.) in the afternoon (at 1600 hours) for 10 days at two different periods: day 1–10 after collagen injection, or at the onset of the disease (day 30–39). Mice injected with melatonin at day 1–10 developed a more severe arthritis, while those injected at onset did not differ significantly from the corresponding controls. This work supports the hypothesis that the pineal gland can exaggerate the development of CIA via a high release of melatonin which could allow an enhancement of T cell priming (Hansson et al. 1992).

The importance of melatonin and the pineal gland in the development of CIA has been confirmed in a fourth series of experiments. Two strains of mice, DBA/1 and NFR/N, were subjected to surgical pinealectomy. Melatonin levels in sera were reduced by approximately 70% by pinealectomy compared with the corresponding sham-operated controls. After 3–4 weeks of rest, the mice were immunized with rat type II collagen to induce autoimmune arthritis and the animals were kept in constant darkness during the experiments. In comparison with the controls, all groups of pinealectomized mice showed reduced severity of the arthritis by means of a slower onset of the disease, a less severe course of the disease as reduced clinical scores, and reduced serum levels of anti-collagen II antibodies. These effects were not significant in all experiments, but the trends were always the same (Hansson et al. 1993).

In conclusion, the work done by Hansson et al. strengthens the hypothesis that high physiological levels of melatonin stimulate the immune system and cause exacerbation of autoimmune collagen II arthritis, while inhibition of melatonin release has a beneficial effect (Hansson et al. 1993).

21.3.3 Induced T Cell Leukemia

The relationship between the pineal gland, melatonin, and melatonin-immuno-induced-opioids (MIO) with the response of C57Bl/6 mice to A-RadLV-induced T cell lymphomas was investigated. Mice were injected at day 0 with A-RadLV and from day 10 they were treated chronically with melatonin 4 mg/kg b.w., naltrexone 1 mg/kg, or phosphate-buffered saline, throughout the experiment. In another protocol, groups of mice were: (a) surgically pinealectomized at day 14, (b) functionally pinealectomized (24 h light) from day 20, and (c) sham pinealectomized.

At day 0, each group was inoculated intrathymically with A-RadLV. The results show that melatonin accelerated ($P < 0.005$) leukemogenesis, whereas the surgical pinealectomy and the functional pinealectomy delayed it ($P < 0.005$ and $P < 0.01$). Moreover, the action of melatonin was blocked by naltrexone ($P < 0.005$), indicating the involvement of MIOs in the development of the lymphomas.

As experimental model, young 5–8 week-old C57Bl/6 mice were inoculated intrathymically with A-RadLV. Mice are highly sensitive to the injection of this leukemogenic virus which produces an impairment of humoral and cellular response to SRBC via a suppressive effect on T helper cells. Leukemic cell infiltrates are found in thymus, mesenteric and axillary lymph nodes, spleen, liver, and kidney. The majority of the A-RadLV-induced leukemias consist of the proliferation of T lymphocytes bearing high levels of surface H-2 antigens. Cellular targets for A-RadLV were shown to be corticosterone and radiation resistant, CD4[+] and/or CD8[+] thymocytes. The aim of this study was in fact to define the role of the pineal gland, melatonin, and MIO in leukemogenesis using the experimental model represented by C57Bl/6 mice inoculated intrathymically with A-RadLV (Conti et al. 1992).

The virus isolate was prepared from thymic lymphomas induced in adult C57Bl/6 mice by A-RadLV as described previously (Haran-Ghera et al. 1977). The same intrathymic route of virus administration (20 µl/injection) was used in all experiments. Mice were injected with A-RadLV 14 days after surgical pinealectomy and 20 days after functional pinealectomy at the age of 66 days. Surgical pinealectomy was performed in adult mice as described (Conti et al. 1992). Sham operation was carried out exactly in the same way but without removal of the pineal gland. The mortality rate of this delicate intervention was rather high. In fact, 15% of the mice died spontaneously within 24 h while another 30% had to be eliminated later because of paralysis symptoms occurring presumably after thrombosis of the local sinuses. However, the surviving mice recovered completely within 10 days after surgery and were never used before day 20. Functional pinealectomy was performed keeping animals under constant light.

The different groups of animals were: (a) melatonin, 4 mg/kg b.w., 0.5 ml s.c. at 1630 hours, (b) naltrexone, 1 mg/kg, 0.5 ml s.c. at 1600 hours. All drugs were injected for 5 days per week throughout the experiment. Surgically and functionally pinealectomized mice were injected with A-RadLV and kept undisturbed till their death. Control groups consisted of either sham pinealectomized mice injected with A-RadLV or normal mice injected with A-RadLV and treated chronically with melatonin.

The development of the disease was checked weekly by palpation of animals and the survival rate was evaluated. To summarize results published elsewhere (Conti et al. 1992) and reported in Fig. 21.3, melatonin treated mice ($n = 26$) began to die 49 days after virus inoculation and all mice died after 76 days ($P < 0.005$). Control, PBS-injected mice ($n = 48$) showed a survival curve displaced to the right: they started to die at day 59 and at day 100 all animals were dead. It seems clear that exogenous melatonin accelerated the leukemogenic process and consequently animals died earlier. Moreover, surgically pinealectomized ($n = 17$) mice started to die at the same time as control PBS-treated mice, but the slope of their survival curve was significantly lower than that of melatonin- or control PBS-injected animals.

In fact, surgically pinealectomized mice started to die at day 51, but the last mouse died after 140 days ($P < 0.005$ surgical pinealectomized mice versus control PBS- and melatonin-treated mice). For more details concerning sham operated, control nontreated and functionally pinealectomized mice, please refer to the original paper (Conti at al. 1992). Concerning the mechanism of action of melatonin on leukemo-

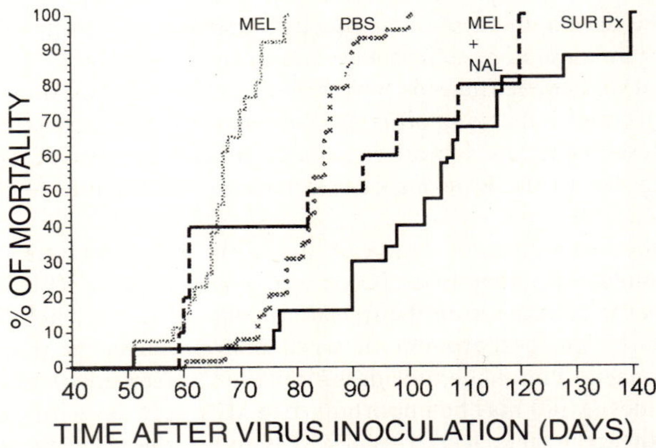

Fig. 21.3. Role of pineal gland and melatonin in lymphoproliferative diseases

genesis, the endogenous opioid system seems to be involved in this case as well. In fact, naltrexone treatment apparently counteracted the accelerating action of melatonin, and shifted the curve to the right ($P < 0.005$, melatonin and naltrexone versus melatonin and PBS).

21.4 Conclusions

This paper reviewed some findings concerning the endogenous melatonin rhythm in inbred mice and summarized some results on the role of the pineal gland and the effect of exogenous melatonin in autoimmune diabetes, rheumatoid arthritis, and mouse T cell leukemia.

A recent study confirmed that some inbred mice, i.e., C57Bl/6, BALB/c, AKR, and NOD strains, have melatonin in their pineal gland which is secreted in a circadian fashion. The peak is short-term and high at the middle of the dark period (Conti and Maestroni 1996). Moreover, these patterns of pineal melatonin content closely reproduce the circadian rhythm of serum melatonin (Maestroni et al.1986, 1987). The existence of an endogenous rhythm of melatonin was also suggested by the determination of 6-hydroxymelatoninsulphate (the main metabolite of melatonin) in urine from surgically pinealectomized and sham pinealectomized C57Bl/6 female mice. In fact, pinealectomy decreased the limit of detection of the concentration of urinary 6-hydroxymelatoninsulphate (Conti et al. 1992).

Based on the well known immunoregulatory role of the pineal gland and melatonin, their role in the development of spontaneous autoimmune type I diabetes in mice was investigated (Conti and Maestroni 1997). We showed that the pineal gland and melatonin influence the development of diabetes mellitus type I: pinealectomy accelerates the development of the disease while melatonin administration postpones it. In fact, as shown in Fig. 21.2, chronic treatment of mice injected subcutaneously with melatonin seems to protect animals against the development of diabetes. However, a similar protection was also observed in mice injected with saline, although the difference with melatonin at 50 weeks of age was significant (Chi-square, $P < 0.0006$). With regard to the astonishing effect of PBS treatment, it should be noted that repeated injections for 37 weeks represent a chronic stress. According to previous reports, stressors seem to modulate the development of spontaneous autoimmune diabetes by influencing immune and/or inflammatory components (Conti and Maestroni 1997). In particular, it seems evident that corticosteroids have beneficial effects on autoimmune diabetes not only in NOD mice but also in other experimental models such as streptozotocin-induced diabetes. On the contrary, adrenalectomy accelerated the onset of the disease, which could be due to the lowering of circulating corticosteroid levels (Conti and Maestroni 1997).

Regarding the mechanism of action of melatonin on autoimmune diabetes, it is possible to formulate two hypotheses (Conti and Maestroni 1997) (Fig. 21.4). The first one is based on the consideration that melatonin acts on CD4$^+$ T helper cells, which, together with macrophages, promote beta cell destruction in the pancreas of NOD mice. In this scenario, it might be considered that the involvement of melatonin-induced opioid peptides (MIO) may be important. Two MIO peptides with 14.4 and 67 kDa molecular weight are in fact released by CD4$^+$ T lymphocytes upon melatonin stimulation. On the other hand, melatonin could involve more classical cytokines. A growing

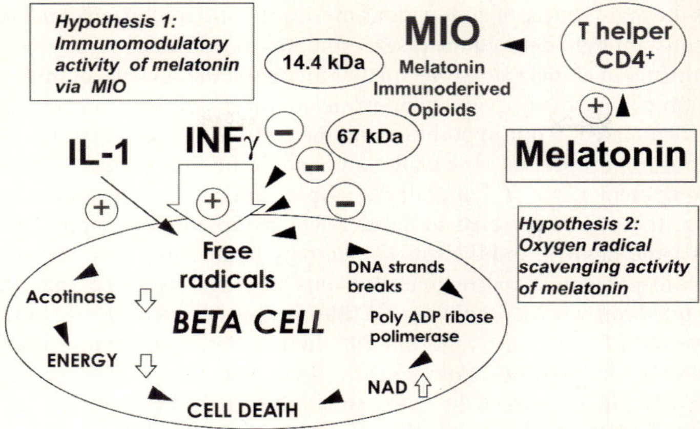

Fig. 21.4. Possible mechanisms of action of melatonin on autoimmune diabetes. For comment see text

body of evidence suggests that cytokines likes TNF-α, IL-1, IL-2, and IFN-γ, are involved in the pathogenesis of autoimmune diabetes (Conti and Maestroni 1997).

Another interesting possibility which might be considered as a second hypothesis relates to the oxygen radical scavenging property of melatonin. In the case of autoimmune diabetes, oxygen free radicals, which are released after lymphocyte infiltration of the pancreas, are known to be the mediators of pancreatic B cell damage. A large number of papers have been published in the last few years showing a protective effect of antioxidants on diabetes. It is indeed the case for lipoic acid which partially suppresses inflammation, vitamin E which delays onset of the disease, or superoxide dismutase-polyethylene glycol and catalase polyethylene glycol which decrease insulitis. On the other hand, a combination of antioxidant and corticosteroid prevents autoimmune diabetes in NOD mice (Conti and Maestroni 1997).

The role of melatonin and the pineal gland in autoimmune diseases seems to be more complex. In fact, in the model of collagen-induced arthritis, melatonin promotes the development of the disease, while pinealectomy protects animals against the development of rheumatoid arthritis.

In conclusion, the effect of melatonin in autoimmune diseases, i.e., diabetes mellitus and rheumatoid arthritis, deserves to be further studied with a focus on its mechanisms of action and, eventually, on its therapeutic potentials.

Beside their effects on autoimmune diseases, melatonin and MIO played an important and negative role in modulating the leukemogenic process. In the model we used, C57Bl/6 control mice injected intrathymically with the highly leukemogenic virus A-RadLV and left undisturbed without any further treatment showed a survival curve almost identical with that described in other reports (Conti et al. 1992). In fact, within 50–60 days after the virus inoculation, a lymphoma appeared in the thymus followed by rapid development of leukemia with metastatic spread to the liver, spleen, and lymph nodes. Surprisingly, chronic treatment of mice with melatonin accelerated the development of the disease and this effect was apparently abolished by the specific opioid antagonist naltrexone. On the other hand, removal of the pineal gland, which represents the main source of endogenous melatonin, and inhibition of melatonin synthesis by constant light delayed the disease and increased the survival time. Moreover, when administered alone, naltrexone further increased the survival of mice suggesting an inhibition of the effect of endogenous melatonin. These results contradict the expectation formulated in our hypothesis and the consensus that melatonin has general tumor-inhibiting properties. One explanation could be that targets of melatonin are immunocompetent $CD4^+$ T lymphocytes. Upon antigen activation and melatonin stimulation, these cells secreted molecules which presented an opioid-like activity both in vivo and in vitro, and their in vivo activity on immune functions was blocked by the opioid antagonist naltrexone. Previous findings indicated that 10–20 days following infection with A-RadLV in C57Bl/6 mice, the infected cells were corticosterone-resistant, $Thy-1^+$, $CD4^+$, and/or $CD8^+$ thymocytes. Furthermore, specific opioid binding sites in membranes from murine thymuses has been demonstrated. The promotion of leukemogenesis by melatonin might thus be explained by an overproduction of MIO which would affect proliferation and differentiation events in the thymus. In conclusion, there is little doubt left that melatonin can promote the development of leukemia by MIO. The detailed mechanisms of the MIO-promoting action in leukemia need to be further studied in the future.

As a more general conclusion, results reviewed here demonstrated that melatonin should not be considered a completely innocuous molecule and could be dangerous for the organism in some circumstances.

Doctors (i.e., general practitioners, consultants, family doctors), chemists, or druggists have to be very careful in prescribing this hormone. On the other hand, scientists have to consider that further studies, both at basic and clinical levels, are needed to better understand and to clarify the actions of the "sophisticated molecule" melatonin.

References

Ben-Nathan D, Maestroni GJM, Lustig S, Conti A (1995) Protective effects of melatonin in mice infected with encephalitis viruses. Archives of Virology 140:223–230

Brown GM, Grota LJ, Sanchez de la Pena S, Halberg F, Halberg E (1983) Circadian melatonin rhythm: part of stimulatory feed-sideward in pineal modulation of adrenal response to ACTH 1–17? Abstracts of papers, Minn Acad Sci 51st Annual Spring Meeting, University of Minnesota-Duluth, April 29–30, p 12

Brown GM, Sanchez de la Pena S, Marques N, Grota LJ, Ungar F, Halberg F (1986) More on the circadian melatonin rhythm in pineals from domesticated B6D2F1 mice. Chronobiologiy 13:335–338

Calin A, Elswood J, Klouda PT (1989) Destructive arthritis, rheumatoid factor and HLA-DR4. Arthr Rheum 32:1221–1225

Conti A, Maestroni GJM (1993) Do the pineal gland and melatonin play a role in autoimmunity? J Immunology and Immunopharmacology 13:42–45

Conti A, Maestroni GJM (1994) Melatonin-induced-immuno-opioids: role in lymphoproliferative and autoimmune diseases. In: Maestroni GJM, Conti A, Reiter RJ (eds) Advances in pineal research: vol 7. John Libbey and Company, London, pp 83–100

Conti A, Maestroni GJM (1995) The clinical neuroimmunotherapeutic role of melatonin in oncology. J Pineal Res 19:103–110

Conti A, Maestroni GJM (1996) HPLC validation of melatonin circadian rhythm in the pineal gland of inbred mice. J Pineal Res 20:138–144

Conti A, Maestroni GJM (1996) Role of the pineal gland and melatonin in the development of autoimmune diabetes in non obese diabetic (NOD) mice. J Pineal Res 20:164–172

Conti A, Haran-Ghera N, Maestroni GJM (1992) Role of pineal melatonin and melatonin-induced-immuno-opioids in murine leukemogenesis. Med Oncol Tumor Pharmacoth 9:87–92

Courtenay JS, Dallman MJ, Dayan AD (1980) Immunization against heterologous type II collagen induced arthritis in mice. Nature 283:666–668

Ebihara S, Marks T, Hudson DJ, Menaker M (1986) Genetic control of melatonin synthesis in the pineal gland of the mouse. Science 231:491–493

Fabris N, Jankovic BD, Markovic BM, Spector NH (1992) Ontogenetic and phylogenetic mechanisms of neuroimmunomodulation. New York, The New York Academy of Sciences

Goto M, Oshima I, Tomita T, Ebihara S (1989) Melatonin content of the pineal gland in different mouse strains. J Pineal Res 7:195–204

Gregersen PK, Silver J, Winchester RJ (1987) The shared epitope hypothesis. An approach to understanding the molecular genetics of susceptibility to rheumatoid arthritis. Arthr Rheum 30:1205–1213

Hansson I, Holmdahl R, Mattson R (1990) Constant darkness enhances auto-immunity to type II collagen and exaggerates development of collagen-induced arthritis in DBA/1 mice. J Neuroimmunol 27:79–84

Hansson I, Holmdahl R, Mattson R (1992) The pineal hormone melatonin exag-gerates development of collagen-induced arthritis in mice. J Neuroimmunol 39:23–31

Hansson I, Holmdahl R, Mattson R (1993) Pinealectomy ameliorates collagen-induced arthritis in mice. Clin Exp Immunol 1993:432–436

Haran-Ghera N, Ben-Yakov M, Peled A (1977) Immunologic characteristic in relation to high and low leukemogenic activity of radiation leukemia virus. J Immunol 118:600–606

Hazes JMW, Dijkmans BAC, Vandernbroucke JP, Devries RRP, Cats A (1990) Pregnancy and the risk of the developing rheumatoid arthritis. Arthr Rheum 33:1770–1775

Helfgott SM, Bazin H, Dessein A, Trentham DE (1984) Suppressive effects of anti-μ serum on the development of collagen arthritis in rats. Clin Immunol Immu-nopathol 31:403–411

Jankovìc BD, Markovic MM, Spector NM (1986) Neuroimmune interactions: proceedings of the second international workshop on neuroimmunomodulation. New York, Annals of the New York Academy of Sciences

Maestroni GJM, Conti A (1991) Anti-stress role of the melatonin-immuno-opioid network. Evidence for a physiological mechanism involving T cell-derived, immunoreactive β-endorphin and metenkephalin binding to thymic opioid receptors. Int J Neurosci 61:289–298

Maestroni GJM, Conti A, Pierpaoli W (1986) Role of the pineal gland in immunity. Circadian synthesis and release of melatonin modulates the antibody response and antagonizes the immuno-suppressive effect of corticosterone. J Neuroimmunol 13:19–30

Maestroni GJM, Conti A, Pierpaoli W (1987) Role of the pineal gland in immunity: II. Melatonin enhances the antibody response via an opiatergic mechanism. Clin Exp Immunol 68:384–391

Maestroni GJM, Conti A, Reiter RJ (1994a) Advances in pineal research: volume 7. John Libbey, London, Paris, Rome

Maestroni GJM, Covacci V, Conti A (1994b) Haematopoietic rescue via T-cell-dependent, endogenous granulocyte-macrophage colony-stimulating factor induced by the pineal neurohormone melatonin in tumor bearing mice. Cancer Research 54:2429–2432

Maestroni GJM, Conti A, Lissoni P (1994c) Colony-stimulating activity and haema-topoietic rescue from cancer chemotherapy compounds are induced by melatonin via endogenous interleukin – 4. Cancer Research 54:4740–4743

Maestroni GJM, Flamigni L, Hertens H, Conti A (1995) Biochemical and functional characterization of melatonin-induced-opioids in spleen and bone marrow T-helper cells. Neuroendocrinol Lett 17:145–152

Ranges GE, Sriram S, Cooper SM (1985) Prevention of type II collagen-induced arthritis by in vivo treatment with anti-L3T4. J Exp Med 162:1105–1110

Reiter RJ (1991) Pineal melatonin: cell biology of its synthesis and of its physiological interactions. Endocrine Reviews 12:151–180

Silman A, Kay A, Brennan P (1992) Timing of pregnancy in relation to onset of rheumatoid arthritis. Arthr Rheum 35:152–155

Spector TD, Roman E, Silman AJ (1990) The pill, parity and rheumatoid arthritis. Arthr Rheum 33:782–789

Stastny P (1978) Association of the B-cell alloantigen DRw4 with rheumatoid arthritis. New Engl J Med 298:869–871

Trentham DE, Townes AS, Kang AH (1977) Autoimmunity to type II collagen: an experimental model of arthritis. J Exp Med 146:857–868

Vollrath L, Huesgen A, Manz B, Pollow K (1988) Day/night serotonin levels in the pineal gland of male BALB/c mice with melatonin deficiency. Acta Endocrinol 117:93–98

Wordsworth P, Bell J (1991) Polygenic susceptibility in rheumatoid arthritis. Ann Rheum Dis 50:343–346

Yu H-S, Reiter RJ (1993) Melatonin. Biosynthesis, physiological effects and clinical applications. CRC Press, Boca Raton

22 Mechanisms Involved in the Immunomodulatory Effects of Melatonin on the Human Immune System

Juan M. Guerrero, Sofia Garcia-Mauriño, David Pozo, Antonio Garcia-Pergañeda, Antonio Carrillo-Vico, Patrocinio Molinero, Carmen Osuna, Juan R. Calvo

Abstract

In this chapter, we review the evidence accumulated during the last 10 years indicating the relationship between the pineal gland and its hormone melatonin and the immune system. A number of experiments have shown that melatonin has immunoenhancing properties that can be observed studying different immune functions in vivo as well as the production of cytokines and other immune messengers in vitro. Moreover, we report evidence for the presence of both typical membrane receptors (Mel$_{1a}$) for melatonin as well as nuclear receptors in immunocompetent cells. Data presented suggest that the nuclear receptor is the primary mechanism of melatonin to participate in the regulation of cytokines by lymphocytes.

22.1 Introduction

The long-held view that homeostatic mechanisms are integrated by the nervous and endocrine systems has recently been expanded by anatomical, physiological, as well as pharmacological evidence that these systems interact with the immune system. Immune responses alter neural and endocrine function, and in turn, neural and endocrine activity modifies immunological function (Reichlin 1993; Gaillard 1994). The communication between neuroendocrine and immune systems includes the use of common signal and recognition molecules. Typical cytokines such as interleukin-1 (IL-1), interleukin-2 (IL-2), interferon (IFN), and tumor necrosis factor (TNF) are produced by neural cells. In an analogous manner, neuroendocrine hormone expression has been reported by cells of the immune system (Blalock 1989, 1994). Furthermore, receptors for common signal molecules are expressed by both neuroendocrine and immune cells (Carr and Blalock 1991; Carr 1992).

These inquiries have not only heightened in neuroimmunology, which is the study of immune reactions involving the central nervous system, but have also led to the coining of several new terms. Thus, "neuroimmunomodulation" refers to the influence of the nervous system on the immune response, "psychoneuroimmunology" to the study of the effects of the psychological status on immune function, and "neuroendocrinimmunology" to the study of the neuroendocrine influences on the function of immunocompetent cells and the way these cells can, in turn, influence neural function and endocrine activity.

22.2 Effects of Melatonin on the Immune System

Among the neuroendocrine factors that affect the immune system, the pineal hormone melatonin appears to be an important one (Guerrero and Reiter 1992). Melatonin has been shown to enhance the natural and acquired immunity (Poon et al. 1994), to activate bone marrow and lymph node cells (Wajs et al. 1995), to activate natural killer (NK) cell activity (del Gobbo et al. 1989) and antibody (Ab)-dependent cellular cytotoxicity (Giordano and Palermo 1991), to enhance the Ab responses in vivo (Maestroni 1995; Poon et al. 1994; Pioli et al. 1993), to restore the impaired Th cell activity in immunodepressed mice (Maestroni et al. 1988; Maestroni and Conti 1990), to increase T cell proliferation (Pioli et al. 1993; Konakchieva et al. 1995), and to activate monocytes (Morrey et al. 1994; Lissoni et al. 1993a). Melatonin also enhances antigen (Ag) presentation by splenic macrophages to T cells concomitant with an increase of the expression of MHC class II molecules (Pioli et al. 1993). Indeed, melatonin in vivo increases the number of Th2 lymphocytes (Lissoni et al. 1994a), and a combination of melatonin and IL-2, compared with IL-2 alone, increases the number of T lymphocytes, NK cells, and eosinophils (Barni et al. 1992; Lissoni et al. 1993b, 1994b).

Finally, in vivo and in vitro studies have shown that melatonin induces the release of lymphokines (Fig. 22.1). Thus, melatonin activates the production of IL-2, IFN-γ, and the melatonin-induced opioid system (MIOS) by T lymphocytes (Maestroni and Conti 1990; Garcia-Maurino et al. 1997). Melatonin also activates the production of IL-1, IL-6, TNF, reactive oxygen intermediates (ROI), and nitric oxide (Morrey et al., 1994; Garcia-Maurino et al. 1997, 1999; Barjavel et al. 1998). In B lymphocytes, melatonin represses the lipoxygenase gene expression (Steinhilber et al. 1995).

Fig. 22.1. Hypothetical scheme of the action of melatonin on human lymphocytes. T lymphocytes (CD3⁺, CD4⁺, Th1), B lymphocytes (CD19⁺), and monocytes (CD14⁺, CD4⁺) are included. *MEL*, melatonin; *Th*, T helper cell; *BL*, B lymphocytes; *M*, monocytes; *NO*, nitric oxide; *ROI*, reactive oxygen intermediates; *MIOS*, melatonin-induced opioid system

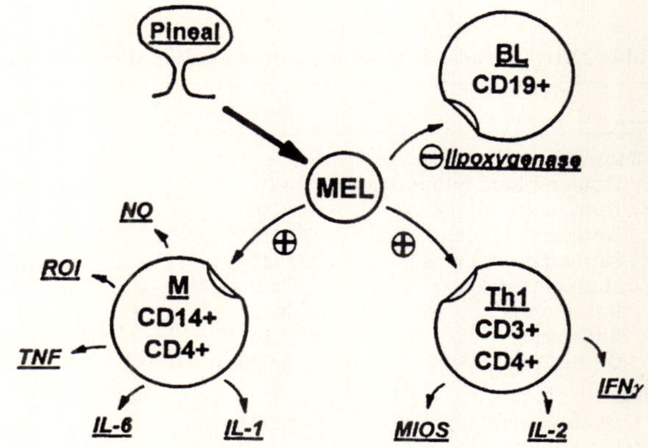

22.3 Mechanisms of Action of Melatonin: The Membrane Receptor

Our understanding about the mechanism of the immunological action of melatonin is still far from complete. Four different mechanisms of action have been suggested for melatonin in different tissues: (a) binding to membrane receptors (Morgan et al. 1989); (b) binding to nuclear receptors (Wiesenberg et al. 1995); (c) melatonin–calmodulin interactions (Benitez-King et al. 1993; Pozo et al. 1997a); and (d) antioxidant activity (Reiter 1995). Thus, there are seemingly receptor and non-receptor mediated actions of melatonin. At least two of them, membrane and nuclear receptors, seem to play a role in the regulation of immune functions.

The direct effect of melatonin on lymphocytes is supported by the existence of specific melatonin binding sites in these cells. Thus, Lopez-Gonzalez et al. (1992), using [^{125}I]melatonin as ligand, have characterized the binding of melatonin to human blood lymphocytes. Binding of [^{125}I]melatonin to lymphocytes fulfills all criteria for binding to a receptor site. The binding was dependent on time and temperature, stable, saturable, as well as specific and reversible. Interpretation of the stoichiometric data by the Scatchard analysis (1949) showed a high-affinity binding site with a K_d in the range of 0.2–1.0 nM. The possible physiological relevance of this receptor deserves some comments. Thus, the affinity of the high-affinity binding site described in lymphocytes is similar to the serum melatonin concentrations at night in adult rats or humans (200 pM), possibly allowing melatonin binding sites on the blood lymphocytes to recognize circulating melatonin. As shown in Table 22.1, binding sites for melatonin have been identified in immunocompetent cells from mammals (including humans) and birds. The cells studied included blood mononuclear cells, T lymphocytes, monocytes, neutrophils, thymocytes, as well as platelets.

Most of the binding studies have been performed using complete cells. For that reason, the actual location of the melatonin receptor has not been identified in immune cells. However, some experiments have shown the presence of specific binding sites for melatonin at membrane level in human lymphocytes (Garcia-Perganeda et al.

Table 22.1. Characteristics of melatonin binding sites in lymphoid cells

	K_d (nM)	Reference
Mammals		
Human blood lymphocytes	1.01	Lopez-Gonzalez et al. 1992
Human granulocytes	2100	Lopez-Gonzalez et al. 1993a
Human CD4$^+$ cells	0.27	Gonzalez-Haba et al. 1995
Human monocytes	0.27	Barjavel et al. 1998
Human platelets	4.00	Vacas et al. 1992
Rat thymus	0.47	Lopez-Gonzalez et al. 1993b
Rat spleen	0.34	Rafii-El-Idrissi et al. 1995
Guinea pig spleen	0.049	Poon and Pang 1992
Birds		
Duck spleen	0.073	Yu and Pang 1991
Duck thymus	0.040	Liu and Pang 1992
Chicken spleen	0.041	Pang and Pang 1992
Chicken bursa of Fabricius	0.043	Liu and Pang 1993
Pigeon bursa of Fabricius	0.073	Liu and Pang 1993
Quail bursa of Fabricius	0.035	Liu and Pang 1993

K_d, dissociation constant.

1997), rat thymus (Lopez-Gonzalez et al. 1993b), and spleen (Rafii-El-Idrissi et al. 1996). The characteristics of the membrane receptor seem similar to those of the Mel_{1a}, which is a member of the guanine nucleotide binding protein-coupled receptor family (Ebisawa et al. 1994). Studies in vertebrates indicate the existence of high-affinity melatonin receptors negatively coupled to adenylyl cyclase by a pertussis toxin-sensitive G protein (Reppert et al. 1996). A high-affinity melatonin receptor from *Xenopus* dermal melanophores has been cloned (Ebisawa et al. 1994), and subsequently from several mammals including humans (Reppert et al. 1994; Slaugenhaupt et al. 1995). The human Mel_{1a} melatonin receptor is coupled to both inhibition of adenylyl cyclase and potentiation of phospholipase activation (Godson and Reppert 1997).

Studies carried out in human lymphocytes by our group strongly suggest the presence of a pertussis toxin-sensitive melatonin signal transduction pathway in human lymphocyte membranes (Garcia-Perganeda et al. 1997). High-affinity melatonin receptors were identified by binding assays. Scatchard analysis revealed the presence of high-affinity receptors with a K_d of 0.45 nM. Binding of melatonin was reduced markedly by GTP and its nonhydrolyzable analogues Gpp(NH)p and GTP-γ-S. Melatonin also inhibited significantly the forskolin-stimulated cyclic AMP production by intact human lymphocytes in a dose-dependent manner and was able to stimulate diacylglycerol production by membranes. Pertussis toxin treatment inhibited melatonin binding to membranes and blocked the ability of melatonin to both inhibit cyclic AMP production and to stimulate diacylglycerol production. Finally, pertussis toxin ADP-ribosylation and Western blot experiments demonstrated the protein expression of $\alpha_{i1/2}$, $\alpha_{i3/o}$, and $\beta\gamma$ complexes of G proteins. The results strongly suggest a pertussis toxin-sensitive melatonin signal transduction pathway in human lymphocytes that involves the inhibition of adenylyl cyclase and the stimulation of phospholipase C.

Moreover, expression of Mel_{1a} melatonin receptor has been shown in both thymus and spleen from the rat (Pozo et al. 1997b). We analyzed the expression of membrane melatonin receptors in rat thymus and spleen by RT-PCR, cDNA cloning, and sequence analysis. Results showed that the melatonin receptor (Mel_{1a}) mRNA is expressed in both thymus and spleen. Moreover, the melatonin receptor mRNA was expressed in all the lymphocyte subpopulations (CD4[+], CD8[+], doubled positive, doubled negative, and B cells) studied from the rat thymus. Southern blot analysis with the melatonin receptor probe and sequence data also showed the identity of the DNA fragments in the thymus, spleen, and lymphocyte subpopulations studied. The melatonin receptor fragments amplified from rat brain, thymus, and spleen share identical nucleotide sequences with the rat Mel_{1a} melatonin receptor subtype. No signal was obtained with primers used for the amplification of the rat Mel_{1b} melatonin receptor subtype in both thymus and spleen. Finally, the melatonin receptor mRNA transcript distribution throughout the rat thymus was examined. Using digoxigenin-labeled cRNA probe to the specific melatonin receptor mRNA, examination of the whole thymus revealed a clear hybridization signal in the cortex and medulla.

22.4 Nuclear Receptors for Melatonin in the Immune System

The notion that melatonin can interact with other subcellular structures in addition to membranes is a very new one. Studies on the subcellular distribution of melatonin by means of radioimmunoassay after cell fractionating and using immunochemistry techniques have shown an accumulation of melatonin in the nuclear fraction of cells from different mammalian organs (Menendez-Pelaez et al. 1993). After addition of either proteinase K or trichloroacetic acid to DNase-treated purified cell nuclear homogenates, the specific binding of melatonin to purified cell nuclei of rat liver disappeared, suggesting a melatonin binding to nuclear proteins (Menendez-Pelaez and Reiter 1993). Therefore, melatonin may exhibit direct effects at the nuclear level (Acuna-Castroviejo et al. 1994).

New data related to the melatonin-nucleus interactions arose from findings from the nuclear receptor field. Nuclear receptors regulate the transcription of complex networks of genes and thereby control diverse aspects of growth, development, and homeostasis. The nuclear receptor superfamily includes receptors for steroid hormones, thyroid hormones, vitamin D, and retinoids, as well as a large number of related proteins for which regulatory ligands have not been identified (nuclear orphan receptors). Nuclear receptors are ligand-regulated transcription factors that control the transcription of target genes in response to extracellular signals (Glass 1994). Three subtypes of the RZR/ROR receptor(α, β, and γ) and four splicing variants of RZR/RORα (RZRα, and ROR α1, α2, and α3) form a subfamily within the superfamily of nuclear hormone receptors. It has been reported that melatonin binds to the orphan receptor RZR/RORα (Becker-Andre et al. 1994; Carlberg and Weisenberg 1995; Steinhilber et al. 1995), that there is a high expression of RZRα mRNA in human peripheral blood mononuclear cells (PBMC) (Becker-Andre et al. 1993), and there is a repression of the 5-lipoxygenase gene expression in human B lymphocytes by binding of melatonin to the RZRα nuclear receptor (Becker-Andre et al. 1994). Melatonin may also bind to and activate the RZRβ receptor, a brain specific receptor whose expression in the rat brain is nearly coincident with binding sites for melatonin (Becker-Andre et al. 1994). In contrast to RZRβ, RZR/RORα is expressed in many tissues and cells outside the brain, and melatonin binds to and activates RZR/RORα with binding specificities in the low nanomolar range (Wiesenberg et al. 1995).

We have recently shown that melatonin interacts with purified cell nuclei from rat spleen and thymus (Rafii-El-Idrissi et al. 1998). Binding of melatonin by cell nuclei fulfills all criteria for binding to a receptor site, such as dependence on time and temperature as well as reversibility, saturability, high affinity, and specificity. The K_d values obtained (68 and 102 pM for rat spleen and thymus purified cell nuclei, respectively) suggest that they may recognize the physiological concentration of melatonin in the tissues. Moreover binding of melatonin by the nuclei from rat spleen and thymus is displaced by CGP 52608, a specific ligand of the putative nuclear melatonin receptor RZR/RORα (Wiesenberg et al., 1995). Similar results were obtained with human lymphocytes (Garcia-Maurino et al. 1998). Results strongly suggest that, in addition to membrane receptor-mediated mechanisms, nuclear receptors may be involved in the regulation of immune system by melatonin.

22.5 The Physiological Role of Membrane Receptors Versus Nuclear Receptors

While trying to find out a physiological role for the membrane receptor in the immune system, the specific membrane receptor ligand S 20098 was used. S 20098 is a potent melatonin agonist at the membrane receptor, as shown by binding studies on ovine pars tuberalis membrane homogenates, and with a biological activity very close to that of melatonin when measured on the basis of inhibition of forskolin-stimulated cyclic AMP production (Depreux et al. 1994). This compound is currently under clinical development and has been used for the re-entrainment of sleep–wake cycles (Redman et al. 1993). When incubated with human lymphocytes, S 20098 binds to membranes and does not bind to cell nuclei, but produces no stimulatory effect on IL-2 or IL-6 production (Garcia-Maurino et al. 1997, 1998). The IC_{50} value for S 20098 binding to PBMCs membranes (1.1 μM) is about 1000 times higher than binding data in pars tuberalis membranes (0.76 pM) (Depreux et al. 1994) and at least 100 times higher than for melatonin in our system (5.1 nM). This low-affinity binding to lymphocyte membranes could account for the lack of effects on cytokine production in our experiments.

Different results were obtained when lymphocytes are incubated with agonists of the nuclear receptor for melatonin. Thus, there are several lines of evidence for a nuclear mechanism primarily involved in the enhancement of cytokine production in human lymphocytes. First, CGP 52608 and other analogues were shown to selectively

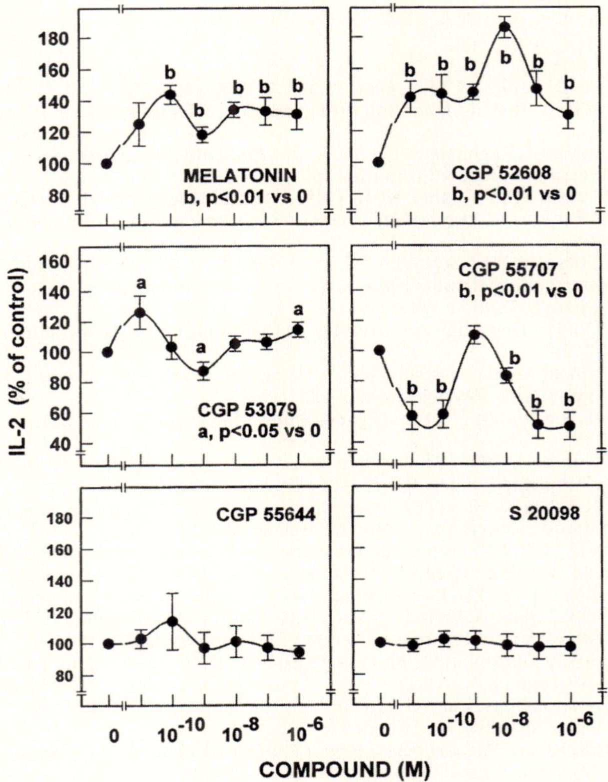

Fig. 22.2. Effect of melatonin, CGP compounds, and S 20098 on IL-2 production by peripheral blood mononuclear cells (PBMC) activated with irradiated PBMCs. Cells were incubated for 72 h in the presence of irradiated PBMCs (0.125 × 10⁶ cell/ml) and the indicated concentrations of each compound. Data are mean ± SEM of eight unrelated donors measured in triplicate. (From Garcia-Maurino et al. 1998)

bind to human lymphocyte nuclei but not to membranes. The IC_{50} values were in the low nanomolar range and similar to that of melatonin. Second, in a *Drosophila* model system designed to demonstrate RZR/RORα receptor transactivation (Wiesenberg et al. 1995; Missbach et al. 1996), melatonin, CGP 52608, and CGP 53079 caused receptor activation in low nanomolar concentrations. These same compounds also increased in nanomolar concentrations IL-2 (Fig. 22.2) and IL-6 production in human lympho-cytes. In contrast, CGP 55644 which was also able to bind to nuclei, did not activate RZR/RORα (Missbach et al. 1996) and had no effect on cytokine production in human lymphocytes. Instead it antagonized the effects of CGP 52608 on IL-2 and IL-6 pro-duction. Third, the specific melatonin membrane receptor ligand S 20098 did not bind to lymphocyte nuclei and did not stimulate cytokine production.

The results reported confirm the primary involvement of nuclear mechanisms in melatonin effects on cytokine production in human lymphocytes. Since the presence of membrane receptors for melatonin is also well characterized, further experiments are required to find out the relevance of these receptors, alone or in combination with nuclear receptors, in terms of their regulation of cytokine production.

References

Acuna-Castroviejo D, Reiter RJ, Menendez-Pelaez A, Pablos MI, Burgos A (1994) Characteriza-tion of high-affinity melatonin binding sites in purified cell nuclei of liver. J Pineal Res 16:100–112

Barjavel MJ, Mamdouh Z, Raghbate N, Bakouche O (1998) Differential expression of the melatonin receptor in human monocytes. J Immunol 160:1191–1197

Becker-Andre M, Andre E, De Lamarter JF (1993) Identification of nuclear receptor mRNAs by RT-PCR amplification of conserved zinc-finger motif sequences. Biochem Biophys Res Commun 194:1371–1379

Becker-Andre M, Wiesenberg I, Schaeren-Wiemers N, Andre E, Missbach M, Saurat JH, Carlberg C (1994) Pineal hormone melatonin binds and activates an orphan of the nuclear receptor super-family. J Biol Chem 269:28531–28534

Benitez-King G, Huerto-Delgadillo L, Anton-Tay F (1993) Binding of ^{3}H-melatonin to calmodulin. Life Sci 53:201–207

Blalock JE (1989) A molecular basis for bidirectional communication between the immune and neuroendocrine systems. Physiol Rev 69:1–32

Blalock JE (1994) The syntax of immune-neuroendocrine communication. Immunol Today 15:504–511

Carlberg C, Wiesenberg I (1995) The orphan family RZR/ROR, melatonin and lipoxygenase: An unexpected relationship. J Pineal Res 18:171–178

Carr DJJ (1992) Neuroendocrine peptide receptors on cells of the immune system. In: Blalock JE (ed) Neuroimmunoendocrinology. Karger, Basel, pp 84–105

Carr DJJ, Blalock JE (1991) Neuropeptide hormones and receptors common to the immune and neuroendocrine systems: Bidirectional pathway of intersystem communication. In: Arder R, Felten DL, Cohen N (eds) Psychoneuroimmunology. Academic Press, San Diego, pp 573–588

Del Gobbo V, Libri V, Villani N, Calio R, Nistico G (1989) Pinealectomy inhibits interleukin-2 pro-duction and natural killer activity in mice. Int J Immuopharmacol 11:567–573

Depreux P, Lesieur D, Mansour HA, Morgan P, Howell HE, Renard P, Caignard DH, Pfeiffer B, Delagrange P, Guardiola B, Yous S, Demarque A, Adam G, Andrieux J (1994) Synthesis and struc-ture-activity relationships of novel naphthalenic and bioisosteric related amidic derivatives as melatonin receptor ligands. J Med Chem 37:3231–3239

Ebisawa T, Karne S, Lerner MR, Reppert S (1994) Expression cloning of a high-affinity melatonin receptor from *Xenopus* dermal melanophores. Proc Natl Acad Sci USA 91:6133–6137

Gaillard RC (1994) Neuroendocrine-immune system interactions. The immune-hypothalamo-pituitary-adrenal axis. Trends Endocrinol Metab 5:303–309

Garcia-Mauriño S, Gonzalez-Haba MG, Calvo JR, Rafii-El-Idrissi M, Sanchez-Margalet V, Goberna R, Guerrero JM (1997) Melatonin enhances IL-2, IL-6 and IFN-γ production by human circulating CD4+ cells: A possible nuclear-receptor mediated mechanism involving Th1 lymphocytes and monocytes. J Immunol 159:574–581

Garcia-Mauriño S, Gonzalez-Haba MG, Calvo JR, Goberna R, Guerrero JM (1998) Involvement of nuclear binding sites for melatonin in the regulation of IL-2 and IL-6 production by human blood mononuclear cells. J Neuroimmunol 92:76–84

Garcia-Perganeda A, Pozo D, Guerrero JM, Calvo JR (1997) Signal transduction for melatonin in human lymphocytes: Involvement of a pertussis-toxin sensitive G protein. J Immunol 159: 3774–3781

Giordano M, Palermo MS (1991) Melatonin-induced enhancement of antibody-dependent cellular cytotoxicity. J Pineal Res 10:117–121

Glass CK (1994) Differential recognition of target genes by nuclear receptor monomers, dimers, and heterodimers. Endocrine Rev 15:391–407

Godson C, Reppert SM (1997) The Mel$_{1a}$ melatonin receptor is coupled to parallel signal transduction pathways. Endocrinology 138:397–404

Gonzalez-Haba MG, Garcia-Maurino S, Calvo JR, Goberna R, Guerrero JM (1995) High-affinity binding of melatonin by human circulating T lymphocytes (CD4+). FASEB J 9:1331–1335

Guerrero JM, Reiter RJ (1992) A brief survey of pineal gland-immune system interrelationships. Endocr Res 18:91–113

Konakchieva R, Kyurkchiev S, Kehayov I, Taushanova P, Kanchev L (1995) Selective effect of metoxyindoles on the lymphocyte proliferation and melatonin binding to activated human lymphoid cells. J Neuroimmunol 63:125–132

Lissoni P, Pittalis S, Bivrio F, Tisi E, Rovelli F, Ardizzoia A, Barni S, Tancini G, Giudici G, Biondi A (1993 a) In vitro modulatory effects of interleukin-3 on macrophage activation induced by interleukin-2. Cancer 71:2076–2081

Lissoni P, Ardizzoia A, Tisi E, Rossini F, Barni S, Tancini G, Conti A, Maestroni GJM (1993 b) Amplification of eosinophilia by melatonin during the immunotherapy of cancer with interleukin-2. J Biol Regul Homeostatic Agents 7:34–36

Lissoni P, Barni S, Tancini G, Bivrio F, Tisi E, Zubelewicz B, Braczkowski R (1994 a) Role of the pineal gland in the control of macrophage functions and its possible implications in cancer: a study of interactions between tumor necrosis factor-alpha and the pineal hormone melatonin. J Biol Regul Homeostatic Agents 8:126–129

Lissoni P, Barni S, Cazzaniga A, Ardizzoia A, Rovelli F, Bivrio F, Tancini G (1994 b) Efficacy of the concomitant administration of the pineal hormone melatonin in cancer immunotherapy with low-dose IL-2 in patients with advanced solid tumors who had progressed on IL-2 alone. Oncology 51:344–347

Liu ZM, Pang SF (1992) Binding of [^{125}I]labeled iodomelatonin in the duck thymus. Biol Signals 1:271–278

Liu ZM, Pang SF (1993) [^{125}I]iodomelatonin binding sites in the bursa of Fabricius of birds: binding characteristics, subcellular distribution, diurnal variations and age studies. J Endocrinol 138:51–57

Lopez-Gonzalez MA, Calvo JR, Osuna C, Guerrero JM (1992) Interaction of melatonin with human lymphocytes: Evidence for binding sites coupled to potentiation of cyclic AMP stimulation by vasoactive intestinal peptide and activation of cyclic GMP production. J Pineal Res 12:97–104

Lopez-Gonzalez MA, Calvo JR, Segura JJ, Guerrero JM (1993 a) Characterization of melatonin binding sites in human peripheral blood neutrophils. Biotech Ther 4:253–262

Lopez-Gonzalez MA, Martin-Cacao A, Calvo JR, Reiter RJ, Osuna C, Guerrero JM (1993 b) Specific binding of 2-[^{125}I]melatonin by partially purified membranes of rat thymus. J Neuroimmunol 45:121–126

Maestroni GJM (1995) T-helper-2 lymphocytes as a peripheral target of melatonin. J Pineal Res 18:84–89

Maestroni GJM, Conti A (1990) The pineal neurohormone melatonin stimulates activated CD4+, Thy-1 cells to release opioid agonist(s) with immunoenhancing and anti-stress properties. J Neuroimmunol 28:167–176

Maestroni GJM, Conti A, Pierpaoli W (1988) Role of the pineal gland in immunity. III. Melatonin antagonizes the immunosuppressive effect of acute stress via an opiatergic mechanism. Immunology 63:465–469

Maestroni GJM, Covacci V, Conti A (1994) Hematopoietic rescue via T-cell-dependent, endogenous granulocyte-macrophage colony-stimulating factor induced by the pineal neurohormone melatonin in tumor-bearing mice. Cancer Res 54:2429–2432

Menendez-Pelaez A, Reiter RJ (1993) Distribution of melatonin in mammalian tissues: the relative importance of nuclear versus cytosolic localization. J Pineal Res 15:59–69

Menendez-Pelaez A, Poeggeler B, Reiter RJ, Barlow-Walden L, Pablos MI, Tan D (1993) Nuclear localization of melatonin in different mammalian tissues: immunocytochemical and radioimmunoassay evidence. J Cell Biochem 53:373–382

Missbach M, Jagher B, Sigg I, Nayeri S, Carlberg C, Wiesenberg I (1996) Thiazolidinediones, specific ligands of the nuclear receptor retinoid Z receptor/retinoid acid receptor-related orphan receptor α with potent antiarthritic activity. J Biol Chem 23:13515–13522

Morgan PJ, Williams LM, Davidson G, Lawson W, Howell E (1989) Melatonin receptors on ovine pars tuberalis: Characterization and autoradiographical localization. J Neuroendocrinol 1:1–14

Morrey K M, McLachan JA, Serkin CD, Bakouche O (1994) Activation of human monocytes by the pineal hormone melatonin. J Immunol 153:2671–2680

Pang CS, Pang SF (1992) High affinity specific binding of 2-[^{125}I]iodomelatonin by spleen membrane preparations of chicken. J Pineal Res 12:167–173

Pioli C, Caroleo MC, Nistico G, Doria G (1993) Melatonin increases antigen presentation and amplifies specific and non specific signals for T-cell proliferation. Int J Immunopharmacol 15: 463–468

Poon AMS, Pang SF (1992) 2[^{125}I]iodomelatonin binding sites in spleens of guinea pigs. Life Sci 50:1719–1726

Poon AM, Liu ZM, Pang CS, Brown GM, Pang SF (1994) Evidence for a direct action of melatonin on the immune system. Biol Signals 3:107–117

Pozo D, Reiter RJ, Calvo JR, Guerrero JM (1997a) Inhibition of cerebellar nitric oxide synthase and cyclic GMP production by melatonin via complex formation with calmodulin. J Cell Biochem 65:430–442

Pozo D, Delgado M, Fernandez-Santos JM, Calvo JR, Gomariz RP, Martin- Lacave I, Ortiz GG, Guerrero JM (1997b) Expression of the Mel$_{1a}$ melatonin receptor mRNA in T and B subsets of lymphocytes from rat thymus and spleen. FASEB J 11:466–473

Rafii-El-Idrissi M, Calvo JR, Pozo D, Harmouch A, Guerrero JM (1995) Specific binding of 2-[^{125}I]iodomelatonin by rat splenocytes: Characterization and its role on regulation of cyclic AMP production. J Neuroimmunol 57:171–178

Rafii-El-Idrissi M, Calvo JR, Giordano M, Guerrero JM (1996) Specific binding of 2-[^{125}I]iodo-melatonin by rat spleen crude membranes: Day-night variations and effect of pinealectomy and continuous light exposure. J Pineal Res 20:33–38

Rafii-El-Idrissi M, Calvo JR, Harmouch A, Garcia-Maurino S, Guerrero JM (1998) Specific binding of melatonin by purified cell nuclei from spleen and thymus of the rat. J Neuroimmunol 86: 190–197

Redman JM, Brown M, Armstrong SM, Guardiola-Lemaitre B (1993) Successful use of S 20098 and melatonin in an animal model of delayed sleep phase syndrome (DSPS). Pharmacol Biochem Behav 46:45–49

Reichlin S (1993) Neuroendocrine-immune interactions. N Engl J Med 329:1246–1253

Reiter RJ (1995) Oxidative processes and antioxidative defense mechanisms in the aging brain. FASEB J 9:526–533

Reppert SM, Weaver DR, Godson C (1996) Melatonin receptors step into the light. Trends Pharmacol Sci 17:100–102

Scatchard G (1949) The attractions of proteins for small molecules and ions. Ann NY Acad Sci 51: 660–672

Slaugenhaupt SA, Roca AL, Liebert CB, Altherr MR, Gusella JF, Reppert SM (1995) Mapping of the gene for the Mel 1a-melatonin receptor to human chromosome 4 (MTNRIA) and mouse chromosome 8 (Mtnr 1a). Genomics 27:355–357

Steinhilber D, Brungs M, Werz O, Wiesenberg I, Danielsson C, Kalen JP, Nayeri S, Schräder M, Carlberg C (1995) The nuclear receptor for melatonin represses 5-lipoxygenase gene expression in human B lymphocytes. J Biol Chem 270:7073–7040

Vacas MI, Del Zar MM, Martinuzzo M, Cardinali DP (1992) Binding sites for [^{3}H]-melatonin in human platelets. J Pineal Res 13:60–65

Wajs E, Kutoh E, Gupta D (1995) Melatonin affects proopiomelanocortin gene expression in the immune organs in the rat. Eur J Endocrinol 133:754–760

Wiesenberg, I, Missbach M, Kahlen JP, Schräder M, Carlberg C (1995) Transcriptional activation of the nuclear receptor RZRα by the pineal gland hormone melatonin and identification of CGP 52608 as a synthetic ligand. Nucleic Acids Res 23:327–333

Yu ZH, Pang SF (1991) [^{125}I]iodomelatonin binding sites in spleens of birds and mammals. Neurosci Lett 125:175–178

C. Actions Via Neural Pathways

23 The Role of the Pineal Gland in Neural Control of Cell Proliferation in Healthy and Malignant Tissue

Brian D. Callaghan

Abstract

Pinealectomy is associated with hyperproliferation of the healthy cells of the spleen, intestinal crypts, liver and adenohypophysis. In the intestinal crypts the effect persists for as long as 6 months, is not influenced by the intestinal contents, and is mediated almost entirely by the autonomic nervous system. Thus, the pineal gland has a suppressant effect on crypt cell proliferation, which is mediated by the hypothalamus and the autonomic nervous innervation of the gut, although it is possible that local changes in mucosal melatonin levels are also involved. Since luminal nutrients have prime importance in the control of crypt cell proliferation, it is suggested that the pineal gland has a role in modulating this mechanism of control to prevent excessive proliferation. Hyperproliferation of the crypts is known to be a precursor of malignant change. It is suggested therefore that the pineal gland may play a role in the prevention of malignant transformation. An identical hyperproliferative effect on the crypts (also mediated by the autonomic nervous system) is associated with various limbic lesions, so that it is possible that the limbic system is involved in the effects of the pineal on the crypts. The pineal may possibly modify the growth of normal tissues also by its effects as a modulator of neuro-endocrine activity.

In support of the role of the pineal gland in the induction or promotion of malignancy, it has been found that melatonin (its principal secretion) has an oncostatic effect on breast, lung, and gastrointestinal tumors, which may be mediated by the immune system, or by its antioxidant effect. Melatonin can thus be considered as a physiological anticancer substance. There is no evidence that the pineal gland influences the growth of malignant tissues by a direct neural effect. Nevertheless, the effect of hypothalamic lesions on inoculated ascitic or solid tumors suggests that it is possible that the pineal gland may influence tumor growth via neural pathways, since the hypothalamus is involved in its effect on normal tissues. Furthermore, changes in the local level of amines have been shown to influence gastrointestinal tumor growth. The possibility that the effects of the pineal gland on malignant growth could be mediated by neural pathways, as well as via melatonin, certainly requires further investigation.

23.1 Introduction

Bindoni (1971), Bindoni and Cambria (1968), Bindoni and Raffaele (1968, 1971) and Giuffrida et al. (1969) found that the cells of the spleen, intestinal crypts, liver, and adenohypophysis from pinealectomized rats had a higher mitotic index than the corresponding cells of sham-operated rats. They suggested that these effects were probably due to a direct effect on the tissues, rather than being mediated by the pituitary gland, because although testicular atrophy was found to follow removal of the pituitary gland, this atrophy was found to be less pronounced in animals in which both the pineal and pituitary gland had been removed. An analogous effect was demonstrated in follicular spleen cells. Using a stathmokinetic technique, Callaghan and Firth (1978) found that there was a great increase in the mitotic rate in the crypts of rat small intestine 2 weeks after pinealectomy. Subsequent investigations (Callaghan 1989a, 1991, 1995a, 1997a) have shown that this increase in mitotic rate is present in the crypts of both small and large intestine, from 1 week to 6 months after pinealectomy. The pineal gland thus appears to have a suppressive effect on mitotic rate, which may possibly be mediated by its neural connections with the autonomic nervous system, via the hypothalamus, or via its principal secretion, melatonin.

23.2 The Role of the Autonomic Nervous System in the Control of Normal and Neoplastic Crypt Cell Proliferation in the Gut

The autonomic nervous system, in particular the hypothalamus, has been found to be involved in the control of intestinal crypt cell proliferation. Jutisz et al. (1974) extracted and concentrated a principle from the sheep hypothalamus which was capable of inhibiting the in vitro multiplication of some cell strains stabilized in a continuous line. They concluded that this substance was not a peptide but could not further identify it. Bindoni et al. (1973) found that radiofrequency lesions of the tuberoinfundibular region of the rat hypothalamus were associated with a very significant increase in small bowel crypt cell proliferation rate but lesions outside this region were not, and there was a correlation between the volume of tissue destroyed and the increase in proliferation rate. It was concluded that these changes were not due to functional changes in the endocrine system because the rats with the greatest increase in proliferation rate were found to have regressive changes in the pituitary gland of the same type as those in rats without any change in proliferation rate. Furthermore, Sharp et al. (1980) found that hypophysectomy reduced mitotic activity in the crypts in rats, and concluded that an intact pituitary gland was required for crypt cell replication leading to intestinal growth. It therefore appears that this hyperproliferative effect was due to the hypothalamic lesion alone and was not mediated via an effect on the pituitary gland. It is perhaps of some interest to note that Pearl et al. (1966) also found that stimulation of the anterior hypothalamus produced marked hyperplasia of the mucous neck, parietal and chief cells of the stomach, which did not occur if the vagi were sectioned before stimulation.

Besides the effect on normal cell proliferation, Bindoni et al. (1986) found a significant increase in cell multiplication in inoculated ascitic and solid tumours in both DBA/2 and C57Bl/6 mice, as well as in Wistar rats after radiofrequency lesions in the median hypothalamus (ventromedial and dorsomedial nuclei; part of the arcuate nucleus). Although they had previously demonstrated a similar proliferative effect following lesions of this area of the hypothalamus on Yoshida ascites tumor in the rat (Bindoni et al.1980), Ehrlich's tumor and L1210 ascites tumor in the mouse (Bindoni et al. 1981), they did not find any similar effect when the anterior or posterior hypothalamus or the cerebral hemispheres were lesioned. It is of interest that they noted a slight decrease in the secretory activity of the adenohypophysis in animals with hypothalamic lesions in the areas mentioned. This might be expected since this part of the hypothalamus is involved in the secretion of several neurohormones governing the function of the hypophysis. However, because a fall in hypophyseal function is generally associated with a fall in cell multiplication rate in normal and neoplastic tissues (Yeh and Moog 1975; Kirschbaum 1957; cited in Bindoni et al. 1986), they suggested that the increase in cell proliferation after hypothalamic lesions was more likely due to suppression of an inhibitory mechanism located in the hypothalamus, which is independent of the hypophysis. However, they concluded that a humoral mechanism was probably involved to some extent in the effect, since it was noted even in cells suspended in ascitic fluid.

Furthermore, Musso et al. (1975), Lachat and Goncalves (1978), Tutton and Helme (1974) and Callaghan (1988, 1991) found that sympathectomy of the normal small bowel (both generalized and local sympathectomy) was associated with a marked fall in the crypt cell mitotic rate. Musso et al. (1975) and Lachat and Goncalves (1978) also noted an initial fall in small intestinal crypt cell proliferation rate after truncal abdominal vagotomy, with a later rise in most cases, but Liavag and Vaage (1972), Alpers and Kinzie (1973) and Callaghan (1979b) found an accelerated crypt cell proliferation rate initially after truncal abdominal vagotomy. Tutton and Barkla (1980) found that the mitotic rate in the colonic crypts was lowered by chemical sympathectomy and was elevated by treatment with metaraminol, a drug with properties similar to those of noradrenaline. However, chemical sympathectomy, α-adrenergic stimulation and α-adrenergic blockade all failed to influence the mitotic rates in dimethylhydrazine-induced colonic carcinomas. They concluded that there was an absence of autonomic control of colonic tumour cells. Further support for the notion that regulation of crypt cell proliferation in the gut is an important function of the sympathetic nervous system was supplied by the work of Tutton and Barkla (1989), who found that inhibition of re-uptake of noradrenaline into the adrenergic nerves of the gut by desipramine accelerated crypt cell proliferation in intact but not in chemically sympathectomized rats.

However, tumor cell proliferation was found to be strongly inhibited by amine uptake inhibitors (Tutton and Barkla 1987 a, b), and it was suggested that some neoplastic cells can, and need to, take up the amine before being stimulated by it. Tutton and Barkla (1976) showed that inhibition of monoamine oxidase did not significantly influence cell proliferation in nonmalignant colon but accelerated cell division in colonic tumors. Tutton (1974) showed that partial depletion of intestinal serotonin content inhibited cell proliferation in the small intestine but not in the colon (Tutton and Helme 1974) so that changes in local levels of serotonin or noradrenaline could possibly be involved in the effect of pinealectomy on the small bowel crypts but it would be necessary to postulate increased release of serotonin or noradrenaline locally following pinealectomy, and any possible mechanism for this remains obscure.

However, although crypt cell proliferation is influenced by changes in the activity of the sympathetic and parasympathetic nervous system supplying the gut, it has been found (Rowinski et al. 1977) that the duration of the cell cycle in the intestinal crypts of mouse ileum isolated from the luminal contents and from its normal innervation or vascularization by implantation beneath the kidney capsule of syngeneic mice was unchanged. This suggests that the normal process of control of crypt cell proliferation does not depend on continuous input from the autonomic nervous system but may be influenced by it.

A distinction should be made between the enteric division of the autonomic nervous system and the remainder of that system. Cooke (1986) has described the enteric nervous division of the autonomic nervous system as an independent integrative system that differs in structure and function from the parasympathetic and sympathetic divisions of the autonomic nervous system and forms a nonganglioned mucosal plexus surrounding the crypts and subjacent to the villus cells. It is believed that the function of the enteric nervous system is continuously modulated by input from the extrinsic parasympathetic and sympathetic nerves. It is known, however, that sympathetic nerve influences on the mucosa are quite complex and it is difficult to completely separate direct neural effects on the crypt cells from those on other factors in the mucosa. Furthermore, Cooke (1986) thought that direct synaptic connections between preganglionic vagal fibres to the small intestine and the enteric motor neurons was unlikely. The role of the enteric plexuses in control of crypt cell proliferation is further complicated by the findings of See et al. (1990) that the myenteric, rather than the mucosal, plexus is able to influence the rate of crypt cell proliferation in the rat jejunum, independently of the extrinsic nerve supply. Thus, although crypt cell proliferation rate can be influenced by changes in the autonomic activity in the gut wall, the means by which this is achieved is not completely clear.

Furthermore, whereas the effects of vagotomy and sympathectomy on crypt cell proliferation were dissimilar, it is interesting to note that when both these forms of autonomic denervation were combined with pinealectomy (Callaghan 1991), the effect in diminishing the hyperproliferative effect of pinealectomy was statistically similar, both qualitatively and quantitatively. Thus arises the question of how separate the vagal and sympathetic pathways to the gut are. In this regard, Brooks (1983) noted that the vagus was an afferent-efferent cable whose actions are extremely complex, and which contains both efferent parasympathetic nerves and sympathetic efferents that liberate adrenergic transmitters. Moreover, since Tutton (1975) found that there were probably functional interactions between the cholinergic and adrenergic postganglionic neurons innervating the intestine, it is not possible to postulate effectively separate vagal and sympathetic neural pathways between the pineal gland and the small intestinal mucosa.

23.3 Efferent Neural Connections Between the Pineal Gland and the Autonomic Nervous System

Korf and Wagner (1980) found that fibres originating in the region of the paraventricular nucleus of the hypothalamus coursed to the distal portion of the guinea pig pineal gland. Furthermore, paraventricular efferents have been found to innervate the autonomic preganglionic nuclei of the lateral horn of the spinal cord (Reuss 1996), thus establishing a possible route for the influence of the pineal gland to be transmitted to the innervation of the bowel via the autonomic nervous system. It is also believed that the direct connection between the paraventricular nuclei and the sympathetic pre-ganglionic neurons provides the central noradrenergic pathway for the integration of neuroendocrine and autonomic mechanisms (Sawchenko and Swanson 1981; cited in Reuss 1996), so that it is possible that both autonomic neural and neuroendocrine mechanisms are involved in the action of the pineal gland on the intestinal crypts.

Furthermore, Smith and Ariens-Kappers (1975) and Ariens-Kappers et al. (1974) found that the pineal gland was involved in the regulation of the function of the parvocellular and magnocellular hypothalamic nuclei, and Dafny (1977) noted that ventromedial hypothalamus stimulation elicited electrophysiological responses in the rat pineal gland. Thus, it was considered possible that the pineal gland could influence crypt cell proliferation via the autonomic nervous system. This was confirmed (Callaghan 1991) by demonstrating that interruption of either the vagal or sympathetic nerve supply to the rat small intestine significantly decreased the usual hyperproliferative effect on the crypt cells following pinealectomy. These results suggested that the suppressant effect of the pineal gland on crypt cell proliferation was probably mediated, at least to a considerable extent, by (1) direct connections between the pineal and the hypothalamus, (2) connections between the hypothalamus and the vagal and sympathetic innervation of the small intestine, and (3) the influence of the autonomic innervation on the activity of the enteric plexuses in relationship to the crypts. However, these results do not exclude other possible pathways for the pineal effect on the crypts, e. g. via changes in the systemic secretion of melatonin, by changes in the level of hypothalamic melatonin (pinealectomy reduces hypothalamic melatonin significantly, Vakkuri et al. 1985) or endocrine effects (e. g. via the pituitary), or combinations of these effects with a neurally mediated effect, e. g. melatonin acting as a local neurotransmitter at the gut wall.

23.4 The Enteric Nervous System and Its Possible Role in Neural Control of Crypt Cell Proliferation in the Normal Gut. Is It Suitably Located?

Neural control of crypt cell proliferation presupposes the presence of nerve fibres located in such a way that they can act directly on the crypts or related blood vessels in close proximity to the crypts, so that they may act via changes in blood flow locally or via secretion into blood vessels.

Although early investigators, such as Bizzozero (1893, cited in Tutton and Barkla 1982), disputed the presence of such suitably located nerve fibres, Dogiel (1895, cited in Tutton and Barkla 1982) and Cajal (1911, cited in Tutton and Barkla 1982) found a rich nerve supply to the mucous membrane of the intestine. More recently, Dahlstrom et al. (1984) demonstrated cells with neuron-like properties immediately beneath the basal lamina of the rat ileal mucosa. Palay and Karlin (1959) found many unmyelinated nerves, associated with blood vessels, in the lamina propria of the rat jejunum. Furthermore, Lane and Rhodin (1964) found that many bundles of axons arising in the submucosal plexuses passed into the villi. Similar fibres were noted to be related to arterioles and venules in the mouse colonic mucous membrane by Silva et al. (1968), the adrenergic innervation being sparse, and these fibres did not enter into or between the cells of the epithelium. Similarly, both noradrenergic and cholinergic nerve fibres have been found immediately subjacent to the basement membrane of the mucosa throughout the intestine of (1) the cat and monkey (Jacobowitz 1965, cited in Tutton and Barkla 1982); (2) the guinea pig (Gabella and Costa 1968, cited in Tutton and Barkla 1982); (3) humans (Baumgarten 1967, cited in Tutton and Barkla 1982); and (4) the rabbit and rat (Costa and Gabella 1971, cited in Tutton and Barkla 1982). The function of these plexuses cannot be directly related to the presence of neural control of crypt cell proliferation but demonstration of their presence is necessary before neural factors can be considered a possibility.

23.5 Neural Control of Normal Cell Proliferation in Organs Other Than the Gastrointestinal System

The concept of neural control of cell proliferation has also been supported by experimental findings in tissues other than the gut. For example, Byron (1975) found that stimulation of proliferation of haemopoietic stem cells of bone marrow (which receive an autonomic innervation) can be produced by a cholinergic mechanism, and Muir et al. (1975) showed that increased proliferation of the acinar cells of the rat parotid could be produced by stimulation of the sympathetic nerve supply of the gland.

23.6 The Precise Role of the Pineal Gland in the Normal Mechanism of Control of Proliferation in the Gastrointestinal Tract

Although its precise role in the control of crypt cell proliferation in the gastrointestinal tract has not yet been defined, it appears quite possible that the pineal gland has a role in modulating the local control mechanism of crypt cell proliferation. It is unlikely that it has a primary role in the control of normal crypt cell proliferation because of the demonstrated prime importance of local luminal factors in this multifactorial control mechanism (Goodlad 1989; Thomas 1995). Curiously, however, although Rijke et al. (1977) and Callaghan (1978) found that the crypt cell proliferation rate was lower in bypassed jejunum which was not in contact with nutrients (as expected), when pinealectomy and jejunal bypass were performed, the expected hypoproliferative effect in the bypassed loop did not occur. In fact, the mitotic rate in the bypassed jejunal loop was increased to a similar level to that seen in other parts of the small bowel after pinealectomy (Callaghan 1989a). The hyperproliferative effects of pinealectomy had completely overridden the hypoproliferative effects of absence of luminal nutrients. These findings can be reconciled if it is postulated that the pineal gland is not part of the normal mechanism of crypt cell proliferation control in the small intestine but has a role in modulating the normal mechanism of control. This may be part of the known, more generalized modulatory effect of the pineal gland on other functions of the central, nervous and endocrine system (Ariens-Kappers 1981).

However, when pinealectomy was combined with defunctioning of a loop of colon by colostomy for 6 months, it was found that the expected hypoproliferative effect of defunctioning the loop (Delvaux et al.1983) was largely, but not completely, overridden by the hyperproliferative effects of pinealectomy, as it had been in the small bowel (Callaghan 1997a). Since the tissue concentration of melatonin is higher in the colon (Bubenik et al.1992), the higher local production of melatonin in the colon (which might be secreted into the lumen) may explain the difference in the effects of pinealectomy in the small bowel and colon. Furthermore, Rijke and Gart (1979) have suggested that there may possibly be a basic difference in the mechanism of crypt cell proliferation control between the large and small bowel. Thus, in keeping with this, it is also possible that the proposed role of the pineal gland in crypt cell proliferation control may be different in the small bowel and the colon.

Thus, it appears that the action of the pineal gland on the crypts is principally via the autonomic nervous system, and is little related to the presence or absence of luminal factors which are involved in the normal control of crypt cell proliferation. In further support of the lack of dependence of the pineal effect on the crypts on luminal factors, it has also been found that, although pinealectomy was associated with hyperproliferation of the hepatocytes of the liver (Bindoni and Raffaele 1971), complete obstruction of the bile duct in the rat was not associated with any diminution of the hyperproliferative effects of pinealectomy (Callaghan 1997b). The issue is further complicated, however, by the finding that the effects of the luminal contents on crypt cell proliferation, in particular the jejunotrophic effects of caecally infused short-chained fatty acids, may possibly be mediated by the autonomic nervous system (Frankel et al. 1992). However, while it has been shown previously that food intake is not increased after pinealectomy (Callaghan 1988; Takahashi et al. 1976), Bubenik and Pang (1994) found that food intake was elevated in mice implanted with melatonin-containing pellets. Therefore, changes in the level of luminal nutrients cannot be totally excluded but it appears that this factor is probably of minimal importance in the effect of pinealectomy on the crypts.

Finally, it is interesting to note that while the effects of pinealectomy on intestinal crypts appear to be mediated, at least to some extent, by the autonomic nervous system, pineal denervation by bilateral excision of the superior cervical ganglia (Callaghan 1979a) did not result in any significant change in crypt cell proliferation rate in the rat small bowel. The principal innervation of the rat pineal gland is composed of sympathetic nerve fibres arising in the superior cervical ganglion (Bowers et al.1984) and melatonin metabolism in the mammalian pineal gland is under the clear influence of sympathetic fibres originating in the superior cervical ganglion (Reuss and Oxenkrug 1989). However, peptidergic (Mikkelson 1989) and histaminergic fibres (Mikkelson et al. 1992) have been described in the proximal region of the gland, and these types of non-sympathetic fibres remain after superior cervical sympathectomy (Moller 1985). The function of these fibres is uncertain.

23.7 The Role of the Pineal Gland in Induction or Promotion of Malignancy: How Important is It?

In the gastrointestinal tract, any mechanism (such as the proposed control of crypt cell proliferation by the pineal gland) which plays a role in the control of crypt cell proliferation may be involved in the production of gastrointestinal malignancy because disturbances of normal crypt cell proliferation may play an important role in colonic carcinogenesis (Deschner and Lipkin 1975), and it is thought that increased crypt cell proliferation is probably a precursor of colonic carcinogenesis (Williamson 1979). Furthermore, Terpstra et al. (1987) found that in patients with localized colonic neoplastic disease an enlargement of the proliferative compartment of the crypts was found throughout the colon. Hyperproliferation has been identified by Preston-Martin et al. (1990, cited in Nugent 1995) as an early manifestation that can lead to the onset of neoplasia, and Lipkin et al. (1984, cited in Nugent 1995) have found that patients at high risk of colon cancer have high cell proliferation rates throughout the colon and rectum. Furthermore, in dimethylhydrazine-treated rats, the development of malignancy is preceded by an increased labelling index of colonic cells, an upward extension of the proliferating zone and an upward shift of the major area of DNA synthesis of epithelial cells (Okada et al. 1996). This is consistent with the observation that increased crypt cell proliferation is associated with an increase in the number of mutant crypt cells exhibiting DNA damage and thus an increased tendency to malignant change (Nugent 1995).

Furthermore, it has been shown that mutations can predispose to the development of many cancers. For example, mutations in the *p53* tumor suppressor gene have been found in many cancers (Pierceall et al. 1991) and it is thought (Tournaletti and Pfeiffer 1994) that slow DNA repair of the mutations in the *p53* suppressor gene may strongly contribute to the development of skin cancer and possibly other cancers (Gao et al. 1994; Service 1994).

Koratkar et al. (1992) have in fact demonstrated that melatonin can prevent genetic damage to the bone marrow cells in mice induced by cisplatin, indicating that melatonin can either suppress mutations or enhance DNA repair, or both. Thus, it is quite possible that melatonin may have a direct oncostatic effect by preventing DNA damage.

The demonstration that the hyperproliferative effects of pinealectomy are present after 6 months, which is a considerable duration in the lifetime of a rat (Callaghan 1995a), also makes it appear more feasible that the pineal gland could be involved in the production of bowel malignancy, since this is known to take a considerable time to develop in most mammals. However, it is interesting to note that in the rat, the proliferation rate after pinealectomy was higher in the small bowel (where mucosal tumors are rare in humans) and lower in the colon (where mucosal tumors are more common in humans, Callaghan 1995a). It is also possible that the suggested modulatory effect of the pineal gland on local mechanisms of control of crypt cell proliferation is greater in the small bowel than in the colon, so that the release of this controlling effect by pinealectomy is more marked in the small bowel, resulting in a greater hyperproliferative response.

Thus, at present, the role of the pineal and its secretions in the development of malignancy is not clear. Earlier work by Georgiou (1929, cited in Tamarkin et al. 1985) suggested that pinealectomy inhibited tumorigenesis and he concluded that the pineal actually stimulated cancer. Lapin (1976) showed that pinealectomy alone, and particularly if combined with thymectomy, accelerated the growth of experimental tumors. Tapp (1980) found increased metabolic activity in the pineal gland of tumor-bearing animals. However, the bulk of more recent experimental evidence suggests that a product of the pineal, melatonin, (the serum levels of which are reduced after pinealectomy) has an oncostatic effect on the growth of a wide range of experimental tumors, and this topic has been well reviewed by Garcia-Patterson et al. (1996).

For instance, a relationship has been reported between pinealectomy and an increased incidence of induced breast cancer in rats, which can be reversed by administration of melatonin (Tamarkin et al. 1981, cited in Garcia-Patterson et al.1996). A similar inhibitory effect has been demonstrated in human breast tumor cell lines in vitro by Blask et al. (1992, cited in Garcia-Patterson et al. 1996). Bartsch and Bartsch (1981) found that melatonin administration suppressed the growth of fibrosarcoma and the growth of Ehrlich's solid tumors. Melatonin administration also suppressed growth in melanoma in athymic mice (Narita and Kudo 1985). Bartsch et al. (1997a) found decreased pineal melatonin secretion in breast cancer patients, as well as an inverse relationship between melatonin levels and tumor size, and suggested that breast cancer itself may in fact lead to an impaired production of pineal melatonin. In fact, it has been suggested (Bartsch et al. 1992; Blask 1993; Blask et al. 1994) that melatonin regulates mammary tumor growth by direct modulation of mitotic activity, as well as inhibition of serum estrogen and prolactin levels and modulation of the immune system. Similarly, depression of the circadian amplitude of serum melatonin in patients with prostate cancer has been found (Bartsch et al. 1994). Although Bartsch et al. (1997b) found a significant positive correlation between the proliferation rate and levels of melatonin production in gastrointestinal and lung cancers, Lissoni et al. (1990) found a negative correlation between cell proliferation in breast cancer and serum melatonin, which Bartsch et al. (1997b) thought could be explained by different phases of the diurnal secretion of melatonin being used in these two studies. They also point out that it is difficult to correlate melatonin levels with tumor growth because of variation in normal melatonin levels with age, sex, or individual variation, and that it is not known if the pineal gland increases its production of melatonin to compensate for lack of intra-tumoral melatonin or whether the two processes are coincidental.

If the effects of pinealectomy in promoting neoplasia are due to a fall in the level of melatonin, which is assumed to be oncostatic, then pineal inhibition in vivo by exposure to constant light should have the same effect. However, Aubert et al. (1980) found that constant light exposure at the time of administration of dimethyl-benz(a)anthracene did not affect the incidence of mammary tumors in the rat, while Hamilton (1969) reported increased tumour incidence under these conditions. However, it is possible that exposure to constant light affects other central nervous system structures besides its effect on the pineal gland and the secretion of melatonin.

However, despite the reported direct cytostatic effects of melatonin by inhibiting the secretion of tumor growth factors (Cos and Blask 1994) and, in particular, the secretion of somatomedins (Smythe et al. 1974), it is thought that these effects of melatonin on the growth of experimental tumors are probably mostly indirect (Karasek et al. 1992) through the stimulation of the immune system by melatonin (Maestroni et al. 1988; Lissoni et al. 1994; Conti and Maestroni 1995). For instance, it is known that pinealectomy, which decreases melatonin levels, induces immunodepression that has been found to be counteracted by melatonin administration in several species (Maestroni 1993; 1995; Poon and Pang 1996).

In the case of malignancy of the gastrointestinal tract, the plasma melatonin concentration has been found to be lower at night in patients with colorectal carcinoma (Khoory and Stemme 1988) and the levels of melatonin were noted to be increased in the preliminary stage of rectal carcinoma (Relkin 1976). Relkin has postulated that in the early phase of hyperplastic growth, the pineal gland reacts with enhanced secretion of an antitumour factor (possibly melatonin) and when the tumor then runs out of control the pineal gland becomes exhausted and its secretion declines. This is one explanation for the falling levels of melatonin observed in carcinoma. In support of the positive effect of melatonin in preventing the development of colonic tumors, Anisimov et al. (1997) demonstrated the inhibitory effect of melatonin on intestinal carcinogenesis induced by 1,2-dimethylhydrazine in rats. They considered that the antioxidative properties of melatonin may be an important factor in its anticarcinogenic action. In the colon it is possible that the effect of pineal secretions is directly opposed to the tendency to increased crypt cell proliferation due to diet-dependent increases in soluble colonic bile acid concentration and luminal lytic activity (Lapre and van der Meer 1992). Since diet is known to be important in the production of colonic carcinoma (Chen et al. 1978), the pineal gland may act in this way to prevent excessive swings of hyperproliferation in the crypts due to the effects of dietary factors (e.g. carcinogens in food). It is known that free radicals are commonly present in the gut lumen and cause extensive damage to the mucosa (Itoh and Guth 1985; Clark et al. 1988). It is also known that melatonin is one of the most effective scavengers of free radicals, and thus capable of counteracting the effect of various carcinogens (Reiter et al. 1993, 1995; Tan et al. 1993; Pieri et al. 1994). However, this may be an oversimplification of the relative roles of dietary fibre and the pineal gland, since increased levels of dietary fibre, which are thought to be associated with stimulation of crypt cell mitosis, have been found to be protective against the development of experimental colonic carcinomas (Chen et al. 1978), which are known to be associated with increased crypt cell proliferation. In an extensive review of the role of melatonin as an oncostatic agent, Panzer and Viljoen (1997) concluded that melatonin could be considered a physiological anticancer substance, based on the available evidence derived from its antiproliferative, antioxidative and immunostimulatory mechanisms of action, from its abnormal levels in cancer patients, and from clinical trials in which it had been administered.

On the other hand, Chang et al. (1985) showed that denervation of the pineal gland by superior cervical ganglionectomy inhibited the growth of dimethylbenz(a)anthracene-induced mammary tumors in the rat. However, it is not clear whether this effect was due to changes in the level of melatonin, since levels of pineal indoles were not measured in this case.

23.8 Further Consideration of the Role of Melatonin in the Control of Normal Crypt Cell Proliferation in the Gut

Gastrointestinal mucosal cells are known to produce melatonin (Raikhlin and Kvetnoy 1976; Bubenik et al. 1977) and the gastrointestinal tract is known to contain at least 400 times more melatonin than the pineal gland (Huether 1993). However, pineal melatonin circulates throughout the body (Reiter 1991), while most of the gastrointestinal melatonin is known to remain in situ (Bubenik 1986; Huether 1993), and it is possible that this latter variety of melatonin is involved in the control of crypt cell proliferation. However, observations on the effects of melatonin on the crypts of the normal small bowel and colon have not been uniform. For example, Wajs and Lewinski (1988; cited in Lewinski et al. 1991) found that melatonin had an inhibitory effect on crypt cell proliferation in the rat jejunum and colon, whereas Zerek-Melen et al. (1987) found that low doses of melatonin (1 – 10 μg) inhibited proliferation but high doses (100 μg) stimulated it. Pentney and Bubenik (1995) also observed goblet cell proliferation in the colon of melatonin-treated mice. Certainly, it is known that there are specific binding sites for melatonin in mouse colon (Bubenik et al. 1993) and melatonin has been shown to be mainly located in the mucosa in rat colon (Holloway et al. 1980; Bubenik 1980). It seems possible therefore that changes in local levels of melatonin in the mucosa, in response to changes in the level of activity of the autonomic innervation of the mucosa, may be involved in the proposed neural mechanism of control of crypt cell proliferation by the pineal gland. Unfortunately, there is no evidence that the nerve endings in the mucosa contain melatonin, but it is possible that the pineal may alter the levels of melatonin in intestinal fluid, perhaps via the autonomic nervous system, and this in turn may affect the intestinal cells.

Thus, it is possible that changes in local melatonin levels in the mucosa associated with local changes in activity of the nerves supplying the mucosa may both be involved in regulation of crypt cell proliferation. In the rat, melatonin has been mostly localized in the mucosa (Bubenik 1980) and Lee and Pang (1993) also found melatonin almost exclusively in the mucosa in the duck. However, it is also known that the effects of melatonin on peristalsis (Bubenik 1986; Harlow and Weekly 1986) are due to its effect on the muscularis, and not the mucosa. This could possibly be explained by the melatonin being produced by the mucosa and taken up by the intestinal nerves, which in turn exert their effect on the muscularis. The concept of melatonin directly affecting the function of the nerves to the muscularis is supported by the findings of Benouali et al. (1993), who found that pinealectomy modified the ileo-caeco-colic electromyogram and that this effect could be abolished by administration of exogenous melatonin.

Cho et al. (1989) found that melatonin administration increased blood flow into the gastric mucosa, so that it appears possible that if melatonin levels are decreased after pinealectomy, this may result in decreased blood flow into the intestines possibly affecting the crypt cell proliferation rate. However, local vascular changes in the mucosa do not seem to be a likely explanation for the effects of pinealectomy on the crypts because Hanson et al. (1977) found that there was no proven relationship between blood perfusion rate, pressure and resistance in the mucosa and intestinal proliferation rate. Certainly, there is no evidence that the pineal gland can directly influence melatonin levels in the rat (Bubenik 1980) or pigeon (Vakkuri et al.1985) intestinal mucosa. Another problem with the possible role of melatonin secretion at the nerve endings approximating the mucosa in mediating the neurally transmitted effects of the pineal gland is the great diversity in distribution of melatonin binding sites in the gastrointestinal tracts of various species (Poon et al. 1997). For instance, Poon et al. (1997) found that 2-[I^{125}]iodomelatonin binding sites (or putative melatonin receptors) could be detected in humans in the mucosa of the colon, caecum, appendix, and on their blood vessels, but not in the ileum.

Furthermore, in other mammals, they demonstrated melatonin receptors only in the rabbit rectum, mouse colon, mouse rectum and guinea pig ileum. In birds, the sites of melatonin receptors varied with the species, and varied in location in the gut wall from species to species. However, although it is generally thought that melatonin only acts on specific receptors in cell membranes (as in the gut wall), the interaction of melatonin with nuclear receptor sites and with intracellular proteins, such as calmodulin or tubulin-associated proteins, as well as its direct antioxidant effects (Cardinali et al. 1997) may be involved in its action in the gut wall and in other tissues. The actual concentration of melatonin that reaches a particular binding site on the cell membrane is also uncertain, and can differ considerably from the circulating hormone concentration, e.g. melatonin can be concentrated several times in the hypothalamus (Cardinali et al. 1997).

23.9 The Possible Role of the Pineal as a Modulator of Neuroendocrine Activity in Controlling Tumor Growth (Rather Than a Direct Effect Via the Autonomic Nervous System or Changes in the Level of Melatonin Secretion)

The pineal gland is generally considered to inhibit the pituitary gland, principally acting on the hypothalamic-anterior pituitary axis and dependent endocrine organs, and the hypothalamic-posterior pituitary axis (Booth 1987). Therefore, it is conceivable that it could modify tumour growth via its effects on the pituitary gland. Williams et al. (1995) found that not only were there melatonin binding sites in the suprachiasmatic nucleus of the hypothalamus but also in the pars tuberalis of the pituitary, and Reiter (1991) found that melatonin may directly affect hypothalamic hormone release. It is thus possible that the pineal acts on the pituitary gland directly, possibly via the action of melatonin. In general, the pineal gland is also believed to exert a suppressive action on hormonal processes of the hypothalamo-hypophyseal axis (Cardinali 1983; Steger et al. 1984). Blask et al. (1976) showed that pituitary hormones are subject to modulation by the pineal gland, and it is also known that these hormones can influence the growth of some tumors in animals and humans (McGuire et al. 1975; Meites 1980). The work of Lapin and Ebels (1981), Blask (1984), Blask and Leadem (1987) has suggested that impaired pineal function might accelerate tumorigenesis or progression by decreasing the inhibition of pituitary trophic factors. However, as mentioned previously, Bindoni (1971) concluded that stimulation of mitotic activity following pinealectomy was independent of hypophyseal function.

Tamarkin et al. (1985) have commented that much of the experimental work mentioned above suggesting an inhibitory effect of the pineal gland on tumorigenesis did not consider why an association between the pineal gland and a particular tumor might exist and how the pineal gland and its hormones could affect tumorigenesis. Nevertheless, they considered that a possible role for the pineal gland and melatonin in malignancy was suggested by their studies of tumor development in 7,12-dimethylbenz(a)anthracene (DMBA)-treated rats. There appears to be some evidence in the case of breast cancer, however, that the pituitary gland could possibly be involved in the effects of melatonin. For example, it has been suggested that the antiproliferative effect of melatonin on breast cancer is exerted on the estrogen-regulated pathway (Hill et al. 1992; Molis et al. 1995). It is also well known that estrogens play an important role in regulating proliferation in some breast cancers (Molis et al. 1995), and Reiter (1991) has shown that melatonin plays a role in controlling reproduction and is involved in sexual maturation (which involves the secretion of estrogen). However, no definite conclusion can be made presently regarding the role of pituitary factors in the effects of pinealectomy on tumor growth, although it is possible that they are involved in the control of growth in mammary tumors.

23.10 The Relationship Between Pinealectomy and Melatonin Levels in the Body

How is pinealectomy related to changes in the levels of melatonin in the body, and especially in the mucosa of the intestine? It is known that pinealectomy reduces the level of circulating melatonin by 80% (Ozaki and Lynch 1976) and that melatonin is not only produced by the pineal gland but is also found in the gastrointestinal tract (Bubenik 1980). To explain the distribution of melatonin, Menendez-Pelaez and Reiter (1993) have suggested a dual system in which a basal melatonin synthesis occurs in peripheral tissues, e.g. the small and large bowel, while the circadian rhythm of melatonin is provided by the pineal gland. It was noted (Callaghan 1995a) that the hyperproliferative effect of pinealectomy on the crypts was less marked in the colon than in the small bowel after 6 months. One possible explanation in terms of changing melatonin levels is that after pinealectomy the fall in the level of melatonin in the colon might be less than in the jejunum if they each individually produced different amounts of melatonin, e.g. if the colon produced relatively more melatonin than the small bowel. The findings of Huether et al. (1992), who found that melatonin production is relatively higher in the more proximal parts of the gastrointestinal tract, contradict this theory. In support of it are the findings of Bubenik and co-workers (personal communication) that there is five times more melatonin in the pig colon than in the ileum. Thus, it is possible that melatonin may be the medium through which pinealectomy affects the crypt cell proliferation rate, either directly or via changes in the activity of the autonomic nervous system, at least to some extent. An alternative explanation for the difference between proliferation rate in the small and large bowel following pinealectomy may be a basic difference in the mechanism of crypt cell proliferation control in these two organs. This concept is supported by the findings of Rijke and Gart (1979), who found that the response in the colonic crypts in terms of cell cycle time and the size of the proliferative compartment after ischaemia-induced cell loss was quite different from that in the small bowel after a similarly induced loss.

23.11 Are Any Other Possible Antitumor Factors Produced by the Pineal Gland Besides Melatonin, and Could These Be Involved in the Physiological Control of Malignancy?

Bindoni et al. (1976) isolated and partially purified a substance in the pineal gland which inhibited cell proliferation in vitro. This substance did not appear to be either melatonin, serotonin or noradrenaline. Similarly, Bartsch et al. (1987) and Noteborn et al. (1988) reported the tumour cell growth inhibition of ovine pineal substances other than melatonin in vitro. Furthermore, Noteborn et al. (1989) identified a pineal factor of molecular weight 2000–6000, which inhibited the growth of melanoma cells in culture. It was also shown that the activity of this compound differed from substances present within the pineal gland, such as melatonin, pteridines, and β-carbolines, and this factor appeared to be of a peptidergic nature. Thus, it is possible that the pineal gland may act on healthy and malignant tissue by means of humoral agents other than melatonin. However, since the antiproliferative effects of melatonin have been so well defined, it seems more likely that this substance, which is distributed so widely throughout the body, is the product of the pineal gland which is involved in control of cell proliferation.

It has also been shown that somatostatin is present in the bovine (Peinado et al. 1989) and rat (Lew and Lawson-Willey 1987) pineal gland and is of neural origin (Moller et al. 1992) as well as being secreted by the pineal gland (Mato et al. 1993). The immunoreactive somatostatin content in the pineal gland was found to be increased when the hypothalamic periventricular nucleus of the rat was lesioned (Sabry and Suzuki 1993). Furthermore, somatostatin, which is present in the neurons of the intestinal mucosa (Keast et al. 1985), inhibits small bowel crypt cell proliferation (Lehy et al.1979; Zerek-Melen et al. 1993), and a long acting form of somatostatin (SMS 201. 995) has been shown (in high dose) to inhibit the growth of colorectal cancers, both in vivo and in vitro (Morris et al. 1995). Although it is highly speculative, it is possible that somatostatin is involved in the actions of the pineal gland on the intestinal crypts, perhaps in conjunction with the autonomic nervous system, since it has the same suppressive effect on mitosis as melatonin.

23.12 Relationship of Pinealectomy and Its Effects on the Intestinal Crypts with Similar Effects on the Crypts Associated with Limbic Lesions

It is possible that the effects on the crypts observed after pinealectomy may be part of a neural control mechanism involving other areas of the brain as well as the pineal gland. It was found (Callaghan 1989b) that bilateral lesions of the amygdaloid nuclei were associated with a considerable increase in small bowel crypt cell proliferation rate in the rat, comparable in magnitude with the increase in rate following pinealectomy. Bilateral hippocampal, septal and fornix lesions (Callaghan 1989c, 1990, 1995b) were also found to be associated with the same degree of rise in crypt cell proliferation rate. Furthermore, Callaghan (1989c) found that the increased crypt cell proliferation following hippocampal lesions could be completely negated by prior denervation of a loop of small bowel. It was also found (Callaghan 1996) that the expected hypoproliferation in isolated colonic segments, following colostomy, was largely overridden by the hyperproliferative effects following bilateral hippocampal lesions. However, similar lesions of the frontal or cingulate gyri or neocortex were not associated with any rise in crypt cell proliferation rate (Callaghan 1988). It was concluded that this effect was confined to lesions of the limbic system of the rat (although the neocortex is often considered to be part of the limbic system in humans; and the cingulate gyrus is not considered part of the rat limbic system; Hamilton 1976) and was probably mediated to some extent by modulating the activity of the hypothalamus and the autonomic nervous system. It was also found (Callaghan 1996) that this effect was present in the colon as well as the small bowel of the rat 6 months after the hippocampal lesions.

The similarities, both quantitatively and qualitatively, to the effects of pinealectomy on crypt cell proliferation are quite remarkable. Since both the limbic system and the pineal gland have hypothalamic connections, it is possible that the limbic system is on the pathway of action of the pineal gland on the crypts, or the effect of the pineal gland may not be as specific as previously thought, and other areas of the brain may affect crypt cell proliferation independently of the pineal gland. It is interesting to note that Bindoni and Rizzo (1965) demonstrated that the pineal gland exerted some influence in the regulation of excitability of the neural circuits of the hippocampus, and taking into consideration previous work (Milcou and Pavel 1960; Bugnon and Moreau 1961; Ito 1940; Bindoni, Infantellina and Riva Sanseverino 1962, cited in Bindoni and Rizzo 1965) they concluded that it was probable that the pineal gland was involved in the maintenance of adequate functional levels in the central nervous system.

23.13 Is it Possible That the Pineal Gland Acts Directly on the Crypts Via the Limbic System and the Autonomic Nervous System?

The brain is believed to be the major site of melatonin action in the effects of the pineal gland on gonadal function since pinealectomy and melatonin administration have been suggested to affect regional brain monoamine levels (Anton-Tay et al. 1968; Sugden and Morris 1979; Aldegunde et al. 1985). This supports the concept of the pineal gland acting via other centres in the brain. It is also known from studies involving the administration of tritiated melatonin that melatonin tends to be concentrated particularly in the midbrain and the hypothalamus (Anton-Tay and Wurtman 1969; Cardinali et al. 1973). Brown et al. (1981) have suggested that melatonin binds with high affinity in limbic areas, so that it is possible that melatonin is involved in the efferent effects of the limbic system on the hypothalamus. Miguez et al. (1991) found that long-term pinealectomy did not significantly alter tryptophan or serotonin concentrations in the amygdala or hippocampus. However, there were significant decreases in 5-hydroxyindole acetic acid levels and tryptophan hydroxylase activity in the amygdala. They concluded that their results supported the involvement of the amygdaloid serotoninergic system in mediating the functions of the pineal gland. Thompson et al. (1981) further demonstrated reversible alterations in hypothalamic uptake of biogenic amines after septal lesions.

It is possible that the pineal gland acts on the crypts via the limbic system, possibly via the effects of melatonin, and that the serotoninergic system is involved in its effects on the hypothalamus.

However, it should be remembered that melatonin receptors in the rat brain are very widespread and not confined to the limbic system. They are situated in the supra-chiasmatic nucleus, the area postrema and the spinal tract of the trigeminal nerve as well as the medial preoptic area, the septohypothalamic nuclei, the anterior hypo-thalamic area, the nuclei of the lateral olfactory tract, the paraventricular, antero-ventral, and intermediodorsal nuclei of the thalamus, the nuclei of the stria medullaris, the basolateral and medial amygdaloid nuclei, the ventromedial nuclei, the arcuate nuclei, the subiculum of the hippocampus and the lateral mamillary nuclei as well as the pars tuberalis of the pituitary (Williams et al. 1995). There is also a wide variation in receptor distribution between species (Morgan et al. 1994). Because of this wide-spread distribution of melatonin receptors in the brain it is not possible to say with any certainty that melatonin is involved in the pathway between the pineal gland and the hypothalamus or in a possible pathway between the pineal gland and the hypo-thalamus via the limbic system, although it may possibly be involved in the effects of pinealectomy on the intestinal crypts. The limbic system has been called the "visceral brain" (Maclean 1949), i. e. a higher centre concerned with the reception and interpreta-tion of afferent visceral information and concerned with visceral efferent function, so it is conceivable that the pineal and limbic system could interact in the control of crypt cell mitosis in the bowel.

23.14 Does the Pineal Gland Act on the Crypts Via the Limbic System Indirectly by Affecting the General Level of Excitability of the Brain, i. e. Not a Specific Effect on the Limbic System?

Several previous investigations have suggested that the pineal gland has a stabilizing influence on the electrical activity of the central nervous system. Bindoni and Rizzo (1965) recorded convulsive patterns in the hippocampal region of pinealectomized, but not sham-operated, rabbits after stimulation of the contralateral hippocampus. Studies by Fariello and Bubenik (1976) suggested that melatonin may change the re-sponse of epileptic neurons to sensory stimuli, thus reducing sensory reflex epilepsy. Pang and Ralph (1976) also found, using EEG studies in chicks, that there was a stabilizing influence of the pineal gland on cerebral electrical activity. However, while it is possible that the pineal gland acts in this way, findings (Callaghan 1988) that lesions of the frontal or cingulate regions of the brain did not affect the small bowel crypt cell proliferation rate do suggest that the pinealectomy-induced effect on the crypts is a limbic-system-mediated effect rather than a generalized change in the excitability of the whole brain following pinealectomy.

For many years the limbic system was considered as a reverberating circuit (The Papez circuit) modulating the effects of other central nervous system areas. However, the work of Poletti (1986) suggests that the limbic system exerts its functional role primarily by a direct influence on structures downstream in the neuraxis, the diencephalon, midbrain, and even the spinal cord, and in fact the limbic system modulates virtually all of the functions of the brain stem. Newman (1974) noted that the amygdala are capable of producing widespread changes in the autonomic nervous system, and Poletti (1986) has concluded that the hippocampus and the amygdala do not act as separate structures within the temporal lobe. He has suggested that the hippocampus has a prominent direct effect on amygdala unit activity, and the amygdala integrate and exert a major portion of the anterior hippocampal influence on the hypothalamus and basal forebrain. This is consistent with the limbic system acting more directly on the crypts via the autonomic nervous system, rather than the effects on the crypts being coincidental to a widespread autonomic discharge.

It is difficult to envisage an exact pathway, but it seems possible that the limbic system is on the pathway of action of the pineal gland in its neural effects on the crypts of the bowel, and that the action of both these areas involves the sympathetic nervous system. It is also possible that the amygdaloid nuclei, which have access to a great deal of afferent visceral information, along with the hippocampus, exert their effect by a modulating influence on the sympathetic nervous system (perhaps independently of the pineal gland), which in turn affects the mitotic rate in the crypts. Alternatively, Layton et al. (1981) have suggested an additional neural pathway in which the amygdaloid nuclei can influence neuroendocrine regulation via synaptic actions on tuberoinfundibular neurons in the mediobasal hypothalamus and the medial preoptic/anterior hypothalamic areas, and this pathway may possibly be involved in its influence on crypt cell proliferation. Thus, the respective roles of the pineal gland and the limbic system in the control of crypt cell proliferation have yet to be clarified but both appear to have a modulatory influence on the usual mechanisms of control and both may possibly be important in preventing excessive "swings" of cellular proliferation.

23.15 Conclusions

It has been confirmed by several investigators that pinealectomy is associated with an increase in mitotic rate in several diverse healthy tissues, for instance, in the crypts of the small and large intestine, in the liver and in the adenohypophysis and that, in case of the intestine, this rise may be sustained for up to 6 months. Thus it appears likely that the pineal gland normally has a suppressive effect on mitotic rate in these viscera, and perhaps throughout the body. However, the means by which the pineal gland exerts this effect are not clear at present. There is some evidence that in the case of normal intestine, at least, the effects of pinealectomy on cell proliferation can be largely negated by autonomic denervation. Neural connections have been demonstrated between the pineal gland and the hypothalamus, and since the hypothalamus, the autonomic nerve supply of the bowel, and the enteric nervous system have each been shown to have an effect on the mitotic rate, it is suggested that the pineal gland may act on the crypts via the hypothalamus and the autonomic nervous system to a large extent. In support of this, it has been confirmed that the hyperproliferative effects of pinealectomy on the crypts can be negated largely (especially in the small bowel) by autonomic denervation of the bowel. Since the activity of the pineal gland is mainly controlled by its own sympathetic nerve supply, it is possible that the pineal gland acts as a modulator of the activity of the autonomic innervation of the gut. The sympathetic innervation of the pineal gland may be an indication that: "Its function consists in influencing the activity of the sympathetic part of the autonomic nervous system" (Grunewald-Lowenstein 1952). It also appears that the pineal gland may act either via, or in conjunction with, the limbic system in its effects on the healthy mucosa in the rat. There is also evidence that the autonomic nervous system may be implicated in the control of cell proliferation in other healthy tissues, e.g. bone marrow and parotid gland.

In the case of the intestinal mucosa, there is good evidence that the normal control of crypt cell proliferation depends mainly on the nature and presence of the luminal contents. It is therefore suggested that in the case of the intestinal crypts at least, the pineal gland may exert a modulatory effect on the normal control mechanism, which may correct excessive swings of proliferation. Since hyperproliferation is known to be a precursor of neoplastic change in the gut, it is possible that if excessive proliferation is prevented by the pineal gland it may have an oncostatic role in this way. This effect could be mediated principally by neural influences on the mucosa, since there are appropriately situated nerve fibres in the intestinal wall. It is also possible that melatonin, the most important secretion of the pineal gland, could be involved in this effect, either acting systemically or as a local mediator of neural influence on the mucosa, and somatostatin should also possibly be considered in the latter role (although there is little definite evidence for this at present).

There is some indirect evidence for neural control of the growth of malignant tissue, possibly by changes in local amine levels. Thus it is possible that the pineal gland could act to control malignant proliferation via the autonomic nervous system, but there is no direct evidence for pineal control of malignant proliferation via this system. However, there is more evidence for the pineal gland acting on malignant tissues via melatonin, and there is also some evidence suggesting that melatonin might be involved in mediating local neural effects on normal crypt proliferation in the gut, although there is no direct evidence that melatonin is a neurotransmitter in the mucosa. Melatonin is now seen as a likely physiological oncostatic agent, although its mode of action is not clear. Its action is viewed by some as directly oncostatic, and by others as acting via changes in the immune system. It is also possible that the pineal gland exerts its oncostatic effects in some cases via the pituitary gland, rather than via a direct effect on the tumor by melatonin (perhaps in the case of breast cancer). It can be reasonably substantiated experimentally that the pineal gland probably has a role in the neural control of healthy tissues (either directly via the autonomic nervous system or via local changes in melatonin secretion in response to changes in autonomic nervous system activity). However, the role of the pineal gland in the neural control of malignant tissues remains obscure at present. Although it seems possible that the pineal gland may exert a direct neural control over cell proliferation in malignant tissue, it seems more likely that it exerts control over malignant tissue proliferation via the actions of its principal secretion, melatonin, although the autonomic nervous system may subsequently be found to play a role in the release of this substance, perhaps locally. Its role as a "regulator of regulators" (as it has been described) may be very important in the prevention of malignant change because of its widespread effects on the autonomic nervous system and the secretion of various hormones. The mechanism of neural control of cell proliferation in both healthy and neoplastic tissues, and the role of the pineal gland in this control, requires further elucidation and should prove to be a rewarding future field of research.

References

Aldegunde M, Miguez I, Veira J (1985) Effects of pinealectomy on regional brain serotonin metabolism. Int J Neurosci 26:9–13

Alpers DH, Kinzie JL (1973) Regulation of small intestinal protein metabolism. Gastroenterology 64:471–496

Anisimov VN, Popovich IG, Zabezhinski MA (1997) Melatonin and colon carcinogenesis: I. Inhibitory effect of melatonin on development of intestinal tumors induced by 1,2-dimethylhydrazine in rats. Carcinogenesis 18:1549–1553

Anton-Tay F, Wurtman RJ (1969) Regional uptake of 3H-melatonin from blood or cerebrospinal fluid by rat brain. Nature 221:474–475

Anton-Tay F, Chou C, Anton S, Wurtman RJ (1968) Brain serotonin concentration: elevation following intraperitoneal administration of melatonin. Science 162:277–278

Ariens-Kappers J (1981) Evolution of pineal concepts. In: Oksche A, Pevet P (eds) The Pineal Organ. Elsevier North Holland, Biomedical Press, Amsterdam, New York, Oxford, pp 3–26

Ariens-Kappers J, Smith RJ, DeVries RAC (1974) The mammalian pineal gland and its control of hypothalamic acivity. Prog Brain Res 41:149–174

Aubert C, Janiaud P, Lecalvez J (1980) Effect of pinealectomy and melatonin on mammary tumour growth in Sprague-Dawley rats under different conditions of lighting. J Neural Transm 47:121–130

Bartsch H, Bartsch C (1981) Effect of melatonin on experimental tumors under different photoperiods and times of administration. J Neural Transm 52:269–279

Bartsch H, Bartsch C, Noteborn HPJM, Flehmig B, Ebels I, Salemink CA (1987) Growth-inhibiting effect of crude pineal extracts on human melanoma cells in vitro is different from that of known synthetic pineal substances. J Neural Transm 69:299–311

Bartsch C, Bartsch H, Lippert TH (1992) The pineal gland and cancer; facts, hypotheses and perspectives. Cancer J 5:194–199

Bartsch C, Bartsch H, Flüchter SH, Mecke D, Lippert TH (1994) Diminished pineal function coincides with disturbed circadian endocrine rhythmicity in untreated primary cancer patients. Consequence of premature aging or of tumor growth? Ann N Y Acad Sci 719:502–525

Bartsch C, Bartsch H, Karenovics A, Franz H, Peiker G, Mecke D (1997a) Nocturnal urinary 6-sulphatoxymelatonin excretion is decreased in primary breast cancer patients compared to age-matched controls and shows negative correlation with tumor-size. J Pineal Res 23:53–58

Bartsch C, Kvetnoy I, Kvetnaia T, Bartsch H, Molotkov A, Franz H, Raikhlin N, Mecke D (1997b) Nocturnal urinary 6-sulfatoxymelatonin and proliferating cell nuclear antigen-immuno-positive tumor cells show strong positive correlations in patients with gastrointestinal and lung cancer. J Pineal Res 23:90–96

Baumgarten HG (1967) Über die Verteilung von Catecholaminen im Darm des Menschen. Z Zellforsch 83:133–146

Benouali S, Sarda N, Gharib A, Roche M (1993) Implication of melatonin in the control of the inter-digestive ileocecal-colic electromyographic profile in rats. C R Acad Sci III 316:524–528

Bindoni M (1971) Relationships between the pineal gland and the mitotic activity of some tissues. Arch Sci Biol Bologna 55:3–21

Bindoni M, Cambria A (1968) Effects of pinealectomy on the in vivo and in vitro biosynthesis of nucleic acids and on the mitotic rate in some organs of the rat. Arch Sci Biol Bologna 52:271–283

Bindoni M, Raffaele R (1968) Mitotic activity in the adenohypophysis of rats after pinealectomy. J Endocrinol 41:451–452

Bindoni M, Raffaele R (1971) Hepatic regeneration after partial hepatectomy in the pinealectomized rat. Arch Sci Biol Bologna 55:119–127

Bindoni M, Rizzo R (1965) Hippocampal evoked potentials and convulsive activity after electrolytic lesions of the pineal body, in chronic experiments on rabbits. Arch Sci Biol Bologna 49:223–233

Bindoni M, Infantellina F, Riva Sanseverino E (1962) Influences de la region epiphysaire sur l'activite electrique des noyaux hypothalamiques. Proc. XX11 internat. Congr. physiol. Sci. Leiden, 11, Abstracts of Communications, n. 1140

Bindoni M, Bava A, Stanzani S (1973) Mitotic rate in glandular tubules in small intestine after lesions at different loci of hypothalamus in the rat. I R C S 1:16

Bindoni M, Jutisz M, Ribot G (1976) Characterization and partial purification of a substance in the pineal gland which inhibits cell multiplication in vitro. Biochim Biophys Acta 437:577–588

Bindoni M, Belluardo N, Licciardello S, Marchese AE, Cicirata F (1980) Growth of Yoshida ascites tumour in the rat after radiofrequency destruction of the tuberoinfundibular region of the hypothalamus. Neuroendocrinol 30:88–93

Bindoni M, Belluardo N, Marchese AE, Mudo G, Laguidara A (1981) Labelling index of L 1210 and Ehrlich ascites tumour cells after radiofrequency destruction of the tuberoinfundibular region of the hypothalamus in the mouse. Neuroendocrinol Lett 3:347–354

Bindoni M, Belluardo N, Marchese AE, Cardile V, Mudo G, Cella S, Laguidara A, Denatale G (1986) Increased tumor cell multiplication after radiofrequency lesions in median hypothalamus in the mouse and rat. Neuroendocrinology 42:407–415

Bizzozero G (1893) Über die schlauchförmigen Drüsen des Magendarmkanals und die Beziehungen ihres Epithels zu dem Oberflächenepithel der Schleimhaut. Arch Mikr Anat 42:82

Blask DE (1984) The pineal: An oncostatic gland ? In: Reiter RJ (ed) The Pineal Gland, Raven Press, New York, pp 253–284

Blask DE (1993) Melatonin in oncology. In: Yu HS, Reiter RJ (eds) Melatonin Biosynthesis, Physiological Effects and Clinical Applications. CRC Press, Boca Raton, pp 447–475

Blask DE, Leadem CA (1987) Neuroendocrine aspects of neoplastic growth: A review. Neuroendocrinol Lett 9:63–73

Blask DE, Vaughan MK, Reiter RJ, Johnson LY, Vaughan GM (1976) Prolactin-releasing and release-inhibiting factor activities in the bovine, rat, and human pineal gland: in vitro and in vivo studies. Endocrinology 99:152–162

Blask DE, Lemus Wilson AM, Wilson ST (1992) Breast cancer: a model system for studying the neuroendocrine role of pineal melatonin in oncology. Biochem Soc Trans 20:309–311

Blask DE, Wilson ST, Lemus-Wilson AM (1994) The oncostatic and oncomodulatory role of the pineal gland and melatonin. Adv Pineal Res 7:235–241

Booth FM (1987) The human pineal gland: a review of the "third eye" and the effect of light. Aust N Z J Ophthalmol 15:329–336

Bowers CW, Dahm LM, Zigmond RE (1984) The number and distribution of sympathetic neurons that innervate the rat pineal gland. Neuroscience 13:87–96

Brooks CM (1983) Vagus nerve symposium. J Auton Nerv Syst 9:333–337

Brown G, Grota L, Niles L (1981) Melatonin; Origin, control of circadian rhythm and site of action. In : Birau N, Schloot W (eds) Advances in the Biosciences, Vol. 29: Current Status and Perspectives. Pergamon Press, Oxford and New York, pp 193–196

Bubenik GA (1980) Localization of melatonin in the digestive tract of the rat. Effect of maturation, diurnal variation, melatonin treatment and pinealectomy. Horm Res 12:313–323

Bubenik GA (1986) The effect of serotonin, N-acetylserotonin, and melatonin on spontaneous contractions of isolated rat intestine. J Pineal Res 3:41–54

Bubenik GA, Pang SF (1994) The role of serotonin and melatonin in gastrointestinal physiology: ontogeny, regulation of food intake, and mutual serotonin-melatonin feedback. J Pineal Res 16: 91–99

Bubenik GA, Brown GM, Grota LJ (1977) Immunohistological localization of melatonin in the rat digestive system. Experientia 33:662–663

Bubenik GA, Ball RO, Pang SF (1992) The effect of food deprivation on brain and gastrointestinal tissue levels of tryptophan, serotonin, 5-hydroxyindoleacetic acid, and melatonin. J Pineal Res 12:7–16

Bubenik GA, Niles LP, Pang SF, Pentney PJ (1993) Diurnal variation and binding characteristics of melatonin in the mouse brain and gastrointestinal tissues. Comp Biochem Physiol C 104: 221–224

Bugnon C, Moreau N (1961) Les formations neurosecretoires diencephaliques du rat blanc, leur reponse a l'epiphysectomie ou a l'administration d'extraits acetoniques d'epiphyse de boeuf. Ann Sci Univ Besancon, Med 5:47–52

Byron JW (1975) Manipulation of the cell cycle of the hemopoietic stem cell. Exp Hematol 3:44–53

Cajal SRY (1911) Histologie du systeme nerveux de l'homme et des vertebres. Paris, A, Maloine 2:931

Callaghan BD (1978) Effect on rat jejunal crypt cell kinetics of exposure to colonic contents. IRCS Med Sci 6:536

Callaghan BD (1979a) The effect of pineal denervation on crypt cell proliferation in the rat small intestine. J Anat 128:667

Callaghan BD (1979b) Rat small intestinal cell kinetic effects following partial or complete sub-diaphragmatic vagotomy. IRCS Med Sci 7:387

Callaghan BD (1988) Cell proliferation in the intestinal epithelium. MD Thesis, University of Adelaide, pp 4.21–4.27

Callaghan BD (1989a) The effect of pinealectomy and jejunal loop diversion on rat small bowel cell kinetics. Virchows Arch B 57:323–328

Callaghan BD (1989b) The effect of bilateral amygdaloid nucleus lesions on rat small bowel crypt cell kinetics. Virchows Arch B 57:231–236

Callaghan BD (1989c) The effect of bilateral hippocampal lesions and autonomic denervation of the small bowel on rat crypt cell kinetics. Med Sci Res 17:825–826

Callaghan BD (1990) Effect of septal and fornix lesions on rat small bowel crypt cell kinetics. Digestion 45:34–39

Callaghan BD (1991) The effect of pinealectomy and autonomic denervation on crypt cell proliferation in the rat small intestine. J Pineal Res 10:180–185

Callaghan BD (1995a) The long-term effect of pinealectomy on the crypts of the rat gastrointestinal tract. J Pineal Res 18:191–196

Callaghan BD (1995b) Prolonged effect of hippocampal lesions on rat gastrointestinal crypt cell kinetics. Med Sci Res 23:107–108

Callaghan BD (1996) The effect of long standing hippocampal lesions on the crypt cell kinetics of isolated colon. Med Sci Res 24:803–804

Callaghan BD (1997a) The effect of pinealectomy on the crypts of defunctioned rat colon. J Pineal Res 23:117–122

Callaghan BD (1997b) Effect of biliary obstruction and pinealectomy on rat gastrointestinal crypt cell kinetics. Med Sci Res 25:691–692

Callaghan BD, Firth BT (1978) An effect of pinealectomy on rat small intestinal crypt cell kinetics. IRCS Med Sci 6:540

Cardinali DP (1983) Molecular mechanisms of neuroendocrine integration in the central nervous system: an approach through the study of the pineal gland and its innervating sympathetic pathway. Psychoneuroendocrinology 8:3–30

Cardinali DP, Hyyppa MT, Wurtman RJ (1973) Fate of intracisternally injected melatonin in the rat brain. Neuroendocrinology 12:30–40

Cardinali DP, Golombek DA, Rosenstein RE, Cutrera RA, Esquifino AI (1997) Melatonin site and mechanism of action: single or multiple? J Pineal Res 23:32–39

Chang N, Spaulding TS, Tseng MT (1985) Inhibitory effects of superior cervical ganglionectomy on dimethylbenz(a)anthracene-induced mammary tumors in the rat. J Pineal Res 2:331–340

Chen WF, Patchefsky AS, Goldsmith HS (1978) Colonic protection from dimethylhydrazine by a high fiber diet. Surg Gynecol Obstet 147:503–506

Cho CH, Pang SF, Chen BW, Pfeiffer CJ (1989) Modulating action of melatonin on serotonin-induced aggravation of ethanol ulceration and changes of mucosal blood flow in rat stomachs. J Pineal Res 6:89–97

Clark DA, Fornabaio DM, McNeill H, Mullane KM, Caravella SJ, Miller MJ (1988) Contribution of oxygen-derived free radicals to experimental necrotizing enterocolitis. Am J Pathol 130:537–542

Conti A, Maestroni GJ (1995) The clinical neuroimmunotherapeutic role of melatonin in oncology. J Pineal Res 19:103–110

Cooke HJ (1986) Neurobiology of the intestinal mucosa. Gastroenterology 90:1057–1081

Cos S, Blask DE (1994) Melatonin modulates growth factor activity in MCF-7 human breast cancer cells. J Pineal Res 17:25–32

Costa M, Gabella G (1971) Adrenergic innervation of the alimentary canal. Z Zellforsch Mikrosk Anat 122:357–377

Dafny N (1977) Electrophysiological evidence of photic, acoustic, and central input to the pineal body and hypothalamus. Exp Neurol 55:449–457

Dahlstrom A, Newson B, Naito S, Ueda T, Ahlman H (1984) Further evidence for the presence of subepithelial nerve cells in the rat ileum – an immunohistochemical study. Acta Physiol Scand 120:1–5

Delvaux G, Caes F, Willems G (1983) Influence of a diverting colostomy on epithelial cell proliferation in the colon of rats. Eur Surg Res 15:223–229

Deschner EE, Lipkin M (1975) Proliferative patterns in colonic mucosa in familial polyposis. Cancer 35:413–418

Dogiel AS (1895) Zur Frage über die Ganglien der Darmgeflechte bei den Säugetieren. Anat Anz 10:517

Fariello RG, Bubenik GA (1976) Melatonin-induced changes in the sensory activation of acute epileptic foci. Neurosci Lett 3:151–155

Frankel W, Zhang W, Singh A Klurfeld D, Sakata T, Modlin I, Rombeau J (1992) Stimulation of the autonomic nervous system (ANS) mediates short-chain fatty acids (SCFA) induced jejunal trophism. Surg Forum 43:24–26

Gabella G, Costa M (1968) Adrenergic fibres in the mucous membrane of guinea pig alimentary tract. Experientia 24:706–707

Gao S, Drouin R, Holmquist GP (1994) DNA repair rates mapped along the human PGK1 gene at nucleotide resolution. Science 263:1438–1440

Garcia Patterson A, Puig-Domingo M, Webb SM (1996) Thirty years of human pineal research: do we know its clinical relevance? J Pineal Res 20:1–6

Georgiou E (1929) Über die Natur und die Pathogenese der Krebstumoren. Z Krebsforsch 28: 562–572

Giuffrida AM, Bindoni M, Duscio D (1969) DNA polymerase activity of the spleen after removal of the pineal gland in the rat. Boll Soc Ital Biol Sper 45:926–928

Goodlad RA (1989) Gastrointestinal epithelial cell proliferation. Dig Dis 7:169–177

Grunewald-Lowenstein M (1952) Studies on the function of the pineal body. Exp Med Surg 10: 135–154

Hamilton T (1969) Influence of environmental light and melatonin upon mammary tumour induction. Br J Surg 56:764–766

Hamilton LW (1976) Basic limbic system anatomy of the rat. Plenum Press, New York, London, p 33

Hanson WR, Rijke RP, Plaisier HM, Van Ewijk W, Osborne JW (1977) The effect of intestinal resection on Thiry-Vella fistulae of jejunal and ileal origin in the rat: evidence for a systemic control mechanism of cell renewal. Cell Tissue Kinet 10:543–555

Harlow HL, Weekly BL (1986) Effect of melatonin on the force of spontaneous contraction of in vitro rat small and large intestine. J Pineal Res 7:277–284

Hill SM, Spriggs LL, Simon MA, Muraoka H, Blask DE (1992) The growth inhibitory action of melatonin on human breast cancer cells is linked to the estrogen response system. Cancer Lett 64: 249–256

Holloway WR, Grota LJ, Brown GM (1980) Determination of immunoreactive melatonin in the colon of the rat by immunocytochemistry. J Histochem Cytochem 28:255–262

Huether G (1993) The contribution of extra-pineal sites of melatonin synthesis to circulating melatonin levels in higher vertebrates. Experientia 49:665–670

Huether G, Poeggeler B, Reimer A, George A (1992) Effect of tryptophan administration on circulating melatonin levels in chicks and rats: evidence for stimulation of melatonin synthesis and release in the gastrointestinal tract. Life Sci 51:945–953

Ito K (1940) Experimental studies on the functioning of the pineal gland in anaesthesia and spasm. Chosen igakkai zasshi 30:56–74 (In Japanese)

Itoh M, Guth PH (1985) Role of oxygen-derived free radicals in hemorrhagic shock-induced gastric lesions in the rat. Gastroenterology 88:1162–1167

Jacobowitz D (1965) Histochemical studies of the autonomic innervation of the gut. J Pharmacol Exp Ther 149:358–364

Jutisz M, Bindoni M, Ribot GS (1974) [Demonstration in the hypothalamus and partial purification of a substance which inhibits in vitro cell multiplication]. C R Acad Sci Paris Serie D 279: 1591–1594

Karasek M, Liberski P, Kunert-Radek J, Bartkowiak J (1992) Influence of melatonin on the proliferation of hepatoma cells in the Syrian hamster: in vivo and in vitro study. J Pineal Res 13:107–110

Keast JR, Furness JB, Costa M (1985) Distribution of certain peptide-containing nerve fibres and endocrine cells in the gastrointestinal mucosa in five mammalian species. J Comp Neurol 236: 403–422

Khoory R, Stemme D (1988) Plasma melatonin levels in patients suffering from colorectal carcinoma. J Pineal Res 5:251–258

Kirschbaum A (1957) The role of hormones in cancer: laboratory animals. Cancer Res 17:432–453

Koratkar R, Vasudha A, Ramesh G, Padma M, Das UN (1992) Effect of melatonin cis-platinum induced genetic damage to the bone marrow cells of mice. Med Sci Res 20:179–180

Korf HW, Wagner U (1980) Evidence for a nervous connection between the brain and the pineal organ in the guinea pig. Cell Tissue Res 209:505–510

Lachat JJ, Goncalves RP (1978) Influence of autonomic denervation upon the kinetics of the ileal epithelium of the rat. Cell Tissue Res 192:285–297

Lane BP, Rhodin JAG (1964) Fine structure of the lamina muscularis mucosae. J Ultrastruct Res 10:489–497

Lapin V (1976) Pineal gland and malignancy. Österr Zeitschrift für Onkologie 3:51–60

Lapin V, Ebels I (1981) The role of the pineal gland in neuroendocrine control mechanisms of neoplastic growth. J Neural Transm 50:275–282

Lapre JA, Van der Meer R (1992) Diet-induced increase of colonic bile acids stimulates lytic activity of fecal water and proliferation of colonic cells. Carcinogenesis 13:41–44

Layton BS, Lafontaine S, Renaud LP (1981) Connections of medial preoptic neurons with the median eminence and amygdala. An electrophysiological study in the rat. Neuroendocrinology 33:235–240

Lee PP, Pang SF (1993) Melatonin and its receptors in the gastrointestinal tract. Biol Signals 2: 181–193

Lehy T, Dubrasquet M, Bonfils S (1979) Effect of somatostatin on normal and gastric-stimulated cell proliferation in the gastric and intestinal mucosae of the rat. Digestion 19:99–109

Lew GM, Lawson Willey A (1987) An immunohistochemical study of somatostatin in the ovine, porcine and rodent pineal gland. Histochemistry 86:591–593

Lewinski A, Rybicka I, Wajs E, Szkudlinski M, Pawlikowski M (1991) Influence of pineal indoleamines on the mitotic activity of gastric and colonic mucosa epithelial cells in the rat: interaction with omeprazole. J Pineal Res 10:104–108

Liavag I, Vaage S (1972) The effect of vagotomy and pyloroplasty on the gastrointestinal mucosa of the rat. Scand J Gastroenterol 7:23–27

Lipkin M, Blattner WA, Gardner EJ, Burt RW, Lynch H, Deschner E, Winawer S, Fraumeni JF Jr (1984) Classification and risk assessment of individuals with familial polyposis, Gardner's syndrome, and familial non-polyposis colon cancer from [3H]thymidine labeling patterns in colonic epithelial cells. Cancer Res 44:4201–4207

Lissoni P, Crispino S, Barni S, Sormani A, Brivio F, Pelizzzoni F, Brenna A, Bratina G, Tancini G (1990) Pineal gland and tumour cell kinetics: Serum levels of melatonin in relation to Ki-67 labelling rate in breast cancer. Oncology 47:275–277

Lissoni P, Barni S, Tancini G, Brivio F, Tisi E, Zubelewicz B, Braczkowski R (1994) Role of the pineal gland in the control of macrophage functions and its possible implication in cancer: a study of interactions between tumor necrosis factor-alpha and the pineal hormone melatonin. J Biol Regul Homeost Agents 8:126–129

Maestroni GJM (1993) The immunoneuroendocrine role of melatonin. J Pineal Res 14:1–10

Maestroni GJM (1995) T-helper-2 lymphocytes as a peripheral target of melatonin. J Pineal Res 18:84–89

Maestroni GJM, Conti A, Pierpaoli W (1988) Pineal melatonin, its fundamental immunoregulatory role in aging and cancer. Ann N Y Acad Sci 521:140–148

Mato E, Santisteban P, Viader M, Capella G, Fornas O, Puig-Domingo M, Webb SM (1993) Expression of somatostatin in rat pineal cells in culture. J Pineal Res 15:43–45

Meites J (1980) Relation of the neuroendocrine system to the development and growth of experimental mammary tumors. J Neural Transm 48:25–42

Menendez Pelaez A, Reiter RJ (1993) Distribution of melatonin in mammalian tissues: the relative importance of nuclear versus cytosolic localization. J Pineal Res 15:59–69

Miguez J, Martin F, Aldegunde M (1991) Differential effects of pinealectomy on amygdala and hippocampus serotonin metabolism. J Pineal Res 10:100–103

Mikkelsen JD (1989) Immunohistochemical localization of vasoactive intestinal peptide (VIP) in the circumventricular organs of the rat. Cell Tissue Res 255:307–313

Mikkelsen JD, Panula P, Moller M (1992) Histamine-immunoreactive nerve fibers in the rat pineal gland: evidence for a histaminergic central innervation. Brain Res 597:200–208

Milcou SM, Pavel S (1960) Antigonadotrophic function of the pineal gland and the oxytocin of neurosecretory hypothalamic system. Nature 187:950–951

Molis TM, Spriggs LL, Jupiter Y, Hill SM (1995) Melatonin modulation of estrogen-regulated proteins, growth factors, and proto-oncogenes in human breast cancer. J Pineal Res 18:93–103

Moller M (1985) Non-sympathetic synaptic innervation of the pinealocyte of the Mongolian gerbil (Meriones unguiculatus): an electron microscopic study. J Neurocytol 14:541–550

Moller M, Mikkelsen JD, Holst JJ, Phansuwan Pujito P (1992) Somatostatin and prosomatostatin immunoreactive nerve fibers in the bovine pineal gland. Neuroendocrinology 56:278–283

Morgan PJ, Barrett P, Howell HE, Helliwell R (1994) Melatonin receptors: localization, molecular pharmacology and physiological significance. Neurochem Int 24:101–146

Morris DL, Lawson JA, Connor JL, Stewart GJ, Preketes AP, King J (1995) Somatostatin inhibits cellular proliferation of human colorectal cancer: a clinical study. Aust N Z J Surg 65:421–437

Muir TC, Pollock D, Turner CJ (1975) The effects of electrical stimulation of the autonomic nerves and of drugs on the size of salivary glands and their rate of cell division. J Pharmacol Exp Ther 195:372–381

Musso F, Lachat JJ, Cruz AR, Goncalves RP (1975) Effect of denervation on the mitotic index of the intestinal epithelium of the rat. Cell Tissue Res 163:395–402

MacLean PD (1949) Psychosomatic disease and the visceral brain. Psychosomatic Med XI 6: 338–353

McGuire WL, Carbone PP, Vollmer EP (1975) Estrogen receptors in human breast cancer. Raven Press, New York

Narita T, Kudo H (1985) Effect of melatonin on B16 melanoma growth in athymic mice. Cancer Res 45:4175–4177

Newman PP (1974) Visceral afferent functions of the nervous system. Monographs of the Physiological Society No 25. Edward Arnold, London, pp 215–246

Noteborn HPJM, Bartsch H, Bartsch C, Mans DR, Weusten JJAM, Flehmig B, Ebels I, Salemink CA (1988) Partial purification of (a) low molecular weight ovine pineal compound(s) with an inhibiting effect on the growth of human melanoma cells in vitro. J Neural Transm 73:135–155

Noteborn HPJM, Weusten JJAA, Bartsch H, Bartsch C, Flehmig B, Ebels I, Salemink CA (1989) Partial purification of a polypeptide extract derived from ovine pineal that suppresses the growth of human melanoma cells in vitro. J Pineal Res 6:385–396

Nugent KP (1995) Colorectal cancer: surgical prophylaxis and chemoprevention. Ann R Coll Surg Engl 77:372–376

Okada H, Mizuno M, Ikeda N, Tomoda J, Tsuji T (1996) Epithelial cell proliferation during colonic chemical carcinogenesis in the rat. J Gastroenterol Hepatol 11:686–691

Ozaki Y, Lynch HJ (1976) Presence of melatonin in plasma and urine or pinealectomized rats. Endocrinology 99:641–644

Palay Sl, Karlin LJ (1959) An electron microscopic study of the intestinal villus. J Biophys Biochem Cytol 5:363–371

Pang SF, Ralph CL (1976) Effects of melatonin administration and pinealectomy on the electroencephalogram of the chicken (Gallus domesticus) brain. Life Sci 18:961–966

Panzer A, Viljoen M (1997) The validity of melatonin as an oncostatic agent. J Pineal Res 22:184–202

Pearl JM, Ritchie WP, Jr., Gilsdorf RB, Delaney JP, Leonard AS (1966) Hypothalamic stimulation and feline gastric mucosal cellular populations. Factors in the etiology of the stress ulcer. JAMA 195:281–284

Peinado MA, Viader M, Puig-Domingo M, Hernandez G, Reiter RJ, Webb SM (1989) Regional distribution of immunoreactive somatostatin in the bovine pineal gland. Neuroendocrinology 50:550–554

Pentney PT, Bubenik GA (1995) Melatonin reduces the severity of dextran-induced colitis in mice. J Pineal Res 19:31–39

Pierceall WE, Mukhopadhyay T, Goldberg LH, Ananthaswamy HN (1991) Mutations in the p53 tumor suppressor gene in human cutaneous squamous cell carcinomas. Mol Carcinog 4:445–449

Pieri C, Marra M, Moroni F, Recchioni R, Marcheselli F (1994) Melatonin: a peroxyl radical scavenger more effective than vitamin E. Life Sci 55:L271–276

Poletti CE (1986) Is the limbic system a system? Studies of hippocampal efferents: their functional and clinical implications. In: Doane BK, Livingston KE (eds) The Limbic System: Functional organization and clinical disorders. Raven Press, New York, pp 79–84

Poon AM, Pang SF (1996) Pineal melatonin-immune system interactions. In: Tang PL, Pang SF, Reiter RJ (eds) Melatonin: A Universal Photoperiodic Signal with Diverse Actions. Karger, Basel pp 71–83

Poon AM, Chow PH, Mak AS, Pang SF (1997) Autoradiographic localization of 2[125I]iodomelatonin binding sites in the gastrointestinal tract of mammals including humans and birds. J Pineal Res 23:5–14

Preston-Martin S, Pike MC, Ross RK, Jones PA, Henderson BE (1990) Increased cell division as a cause of human cancer. Cancer Res 50:7415–7421

Raikhlin NT, Kvetnoy IM (1976) Melatonin and enterochromaffin cells. Acta Histochem 55:19–24

Reiter RJ (1991) Pineal melatonin: cell biology of its synthesis and of its physiological interactions. Endocr Rev 12:151–180

Reiter RJ, Poeggeler B, Chen L-D, Manchester LC, Guerrero JM (1993) Antioxidant capacity of melatonin. A novel action not requiring a receptor. Neuroendocrinol Lett 15:103–116

Reiter RJ, Melchiorri D, Sewerynek E, Poeggeler B, Barlow Walden L, Chuang J, Ortiz GG, Acuna Castroviejo D (1995) A review of the evidence supporting melatonin's role as an antioxidant. J Pineal Res 18:1–11

Relkin R (1976) The pineal and human disease. In: Relkin R (ed) The Pineal: Annual Research Reviews. Eden Press, Montreal, pp 76–79

Reuss S (1996) Components and connections of the circadian timing system in mammals. Cell Tissue Res 285:353–378

Reuss S, Oxenkrug GF (1989) Chemical sympathectomy and clorgyline-induced stimulation of rat pineal melatonin synthesis. J Neural Transm 78:167–172

Rijke RP, Gart R (1979) Epithelial cell kinetics in the descending colon of the rat. I. The effect of ischaemia-induced epithelial cell loss. Virchows Arch B 31:15–22

Rijke RP, Plaisier HM, de Ruiter H, Galjaard H (1977) Influence of experimental bypass on cellular kinetics and maturation of small intestinal epithelium in the rat. Gastroenterology 72:896–901

Rowinski J, Kucharczyk K, Boniecka T (1977) The cell cycle in the small intestine transplanted beneath the kidney capsule in syngenic mice. Cell Tiss Res 177:141–144

Sabry I, Suzuki M (1993) Immunoreactive somatostatin content in the pineal gland increases after lesion of the hypothalamic periventricular nucleus in male rats. J Pineal Res 14:23–26

Sawchenko PE, Swanson LW (1981) Central noradrenergic pathways for the integration of hypothalamic neuroendocrine and autonomic responses. Science 214:685–687

See NA, Epstein ML, Dahl JL, Bass P (1990) The myenteric plexus regulates cell growth in rat jejunum. J Auton Nerv Syst 31:218–230

Service R (1994) Slow DNA repair implicated in mutations found in tumors. Science 263:1374

Sharp JG, Lipscomb HI, Cullan GE, Fatemi SH, McLaughlin D, Crouse DA (1980) Preliminary studies of the effects of hormones on cell proliferation in the gastrointestinal tract of the rat. In: Appleton DR, Sunter JP, Watson AJ (eds) Cell Proliferation in the Gastrointestinal Tract. Pitman Medical, pp 66–89

Silva DG, Farrell KE, Smith GC (1968) Ultrastructural and histochemical studies on the innervation of the mucous membrane of the mouse colon. Anat Rec 162:157–176

Smith AR, Ariens-Kappers JA (1975) Effect of pinealectomy, gonadectomy, pCPA and pineal extracts on the rat parvocellular neurosecretory hypothalamic system; a fluorescence histochemical investigation. Brain Res 86:353–371

Smythe GA, Stuart MC, Lazarus L (1974) Stimulation and suppression of somatomedin activity by serotonin and melatonin. Experientia 30:1356–1357

Steger RW, Reiter RJ, Siler-Khodr TM (1984) Interactions of pinealectomy and short-photoperiod exposure on the neuroendocrine axis of the male Syrian hamster. Neuroendocrinology 38:158–163

Sugden D, Morris RD (1979) Changes in regional brain levels of tryptophan, 5-hydroxytryptamine, 5-hydroxyindoleacetic acid, dopamine and noradrenaline after pinealectomy in the rat. J Neurochem 32:1593–1595

Takahashi K, Inoue K, Takahashi Y (1976) No effect of pinealectomy on the parallel shift in circadian rhythms of adrenocortical activity and food intake in blinded rats. Endocrinol Jpn 23:417–421

Tamarkin L, Cohen M, Rosello D, Reichart C, Lippman M, Chabner B (1981) Melatonin inhibition and pinealectomy enhancement of 7,12-Dimethylbenzanthracene-induced mammmary tumours in the rat. Cancer Res 41:4432–4436

Tamarkin L, Almeida OFX, Danforth DN (1985) Melatonin and malignant disease. In: Evered D, Clark S (eds) Photoperiodism, melatonin and the pineal. Pitman, London (Ciba Foundation Symposium 117) pp 284–299

Tan D-X, Chen L-D, Poeggeler B, Manchester LC, Reiter RJ (1993) Melatonin: A potent, endogenous hydroxyl radical scavenger. Endocrin J 1:57–60

Tapp E (1980) Pineal gland in rats suffering from malignancy. J Neural Transm 48:131–135

Terpstra OT, van Blankenstein M, Dees J, Eilers GA (1987) Abnormal pattern of cell proliferation in the entire colonic mucosa of patients with colon adenoma or cancer. Gastroenterology 92:704–708

Thomas MG (1995) Luminal and humoral influences on human rectal epithelial cytokinetics. An R Coll Surg Engl 77:85–89

Thompson RG, Valdes JJ, Gage FH (1981) Reversible alterations in hypothalamic uptake of biogenic amines after septal lesions. Exp Neurol 74:356–369

Tornaletti S, Pfeifer GP (1994) Slow repair of pyrimidine dimers at p53 mutation hotspots in skin cancer. Science 263:1436–1438

Tutton PJ (1974) The influence of serotonin on crypt cell proliferation in the jejunum of rat. Virchows Arch B Cell Pathol 16:79–87

Tutton PJ (1975) The influence of cholinoceptor activity on the mitotic rate in the crypts of Lieberkuhn in rat jejunum. Clin Exp Pharmacol Physiol 2:269–276

Tutton PJ, Barkla DH (1980) Neural control of colonic cell proliferation. Cancer 45:1172–1177

Tutton PJ, Barkla DH (1982) Neural control of cell proliferation in colonic carcinogenesis. In: Malt RA, Williamson RCN (eds) Colonic Carcinogenesis. MTP Press, Lancaster, Boston, The Hague, pp 283–296

Tutton PJ, Barkla DH (1976) A comparison of cell proliferation in normal and neoplastic intestinal epithelia following either biogenic amine depletion or monoamine oxidase inhibition. Virchows Arch B 21:161–168

Tutton PJ, Barkla DH (1987a) Biogenic amines as regulators of the proliferative activity of normal and neoplastic intestinal epithelial cells (review). Anticancer Res 7:1–12

Tutton PJ, Barkla DH (1987b) Amine dependence of proliferative activity in two transplantable lines of mouse colonic carcinoma. Virchows Arch B 53:161–165

Tutton PJ, Barkla DH (1989) Effect of an inhibitor of noradrenaline uptake, desipramine, on cell proliferation in the intestinal crypt epithelium. Virchows Arch B 57:349–352

Tutton PJ, Helme RD (1974) The influence of adrenoreceptor activity on crypt cell proliferation in the rat jejunum. Cell Tissue Kinet 7:125–136

Vakkuri O, Rintamaki H, Leppaluoto J (1985) Plasma and tissue concentrations of melatonin after midnight light exposure and pinealectomy in the pigeon. J Endocrinol 105:263–268

Wajs E, Lewinski A (1988) melatonin and N-acetylserotonin, two pineal indoleamines inhibiting the proliferation of jejunal epithelial cells in rats. Med Sci Res 16:1125–1126
Williams LM, Hannah LT, Hastings MH, Maywood ES (1995) Melatonin receptors in the rat brain and pituitary. J Pineal Res 19:173–177
Williamson RC (1979) Hyperplasia and neoplasia of the intestinal tract. Ann R Coll Surg Engl 61: 341–348
Yeh KY, Moog F (1975) Development of the small intestine in the hypophysectomized rat. II. Influence of cortisone, thyroxine, growth hormone, and prolactin. Dev Biol 47:173–184
Zerek Melen G, Lewinski A, Kulak J (1987) The opposite effect of high and low doses of melatonin upon mitotic activity of the mouse intestinal epithelium. Endokrynol Pol 38:317–323
Zerek Melen G, Pawlikowski M, Winczyk K, Lachowicz Ochedalska A, Legowska A, Kwasny H, Przybylski J, Szadowska A (1993) Effects of new somatostatin analogs on the cell proliferation of colonic crypts and colonic cancers in rats. Neuropeptides 25:57–60

D. Molecular Mechanisms of Action

24 Reactive Oxygen Species, DNA Damage, and Carcinogenesis: Intervention with Melatonin

Russel J. Reiter

Abstract

A variety of cancer-inducing agents cause cancer by virtue of their ability to damage DNA via free radical mechanisms. Damaged DNA is a primary cause of cancer if it goes unrepaired and mutates. Free radicals, many of which are oxygen metabolites, indiscriminately attack both mitochondrial and nuclear DNA when they are generated in the vicinity of these molecules. Melatonin is a highly effective free radical scavenger which protects DNA from the damage inflicted by free radicals; the attack of molecules by reduced oxygen metabolites is referred to as oxidative stress. In order for any antioxidant to protect DNA from oxidative stress it must be essentially surrounding the genome, since most free radicals that destroy DNA are generated in the immediate vicinity (most often within 10 angstroms) of the DNA. Melatonin is known to get into the nucleus, and studies, using a variety of techniques, have shown the indole to reduce DNA damage due to a variety of carcinogens, e.g., safrole, ionizing radiation, chromium, etc. In this context melatonin is capable of reducing cancer initiation. Besides the primary damage to DNA which initiates a tumor, the progressive accumulation of oxidatively altered DNA is believed to be involved in the transition of benign to malignant tumor cells. Due to melatonin antioxidant activities, the indole may also reduce this process and thereby lower the degree of malignancy of tumor cells.

24.1 Introduction

Cancer is one of the many degenerative diseases of aging, both in short-lived species such as rats and in long-lived species such as humans (Portier et al. 1986). There are exogenous factors, however, that increase (e.g., tobacco use in humans) and decrease (e.g., calorie restriction in rodents) the incidence of this disease. During aging there is an accumulation of molecular damage (Harman 1994) from oxidants, e.g., free radicals, because nature has selected a variety of genes that have proximate survival value but, at the same time, they have negative long term consequences (Williams and Neese 1991). A good example of this is the oxidative burst from a variety of white blood cells in which nitric oxide (NO•), superoxide anion radical ($O_2^{\bullet-}$), hydrogen peroxide (H_2O_2), and hypochlorous acid (HOCl) defend against bacterial and viral infections, but at the same time they contribute to the incessant bludgeoning of DNA (and other

Table 24.1. Estimates of the frequency of specific types of endogenous DNA damage in selected mammalian cells

Type of damage	Events/cell per day	Reference
Oxidative	86,000 (rats)	Fraga et al. (1990)
Single-strand breaks	36,000	Shapiro (1981)
Double-strand breaks	> 40 (rats) > 3 (humans)	Massie et al. (1972) Cathcart et al. (1984)
DNA cross-links	> 37 (rats) > 3 (humans)	Massie et al. (1972) Cathcart et al. (1984)
Depurination	7,200	Lindahl (1977)
O^6-Methylguanine	2,000	Barrows and Magee (1982)

molecules) creating damage which can lead to mutations (Yamashina et al. 1986; Shacter et al. 1988). Undoubtedly, DNA damage is critical to cancer initiation. One central feature of the somatic mutation theory of aging is that the amount of maintenance and repair of somatic tissues is always less than that required for indefinite survival because such a large portion of the resources of any animal is devoted to reproduction at the expense of maintenance and repair. Thus, it is inevitable that DNA damage induced in somatic cells by endogenous mutagens accumulates over time and contributes to cancer initiation.

There are at least four endogenous processes that significantly damage DNA; these are oxidation (Lindahl 1982; Ames 1983), methylation, deamination, and depurination (Saul and Ames 1986). That these processes are important in terms of DNA damage is emphasized by the fact that cells have evolved mechanisms to combat this damage. Thus, cells have specific DNA repair glycosylases for oxidative, methylated, and deaminated adducts and a repair system for apurinic sites induced as a result of spontaneous depurination (Lindahl 1982). Judging from the measurement of DNA adducts within most cells, oxidative damage to DNA seems to be the most significant and common endogenous damage (Table 24.1). In the rat, an estimated 130,000 damaging genomic events probably occur every day. The majority of these destructive events alter the structure of just one strand of DNA so that the redundant information in the complementary strand is used to repair the damage. On the other hand, double-strand breaks and interstrand cross-links, which occur at significant frequency, remove information from both strands of the DNA molecule. When this occurs the damage can only be repaired if the deleted information is available from a second, homologous DNA molecule, a process identified as recombinational repair.

24.2 Endogenous Oxidative Damage to DNA

Oxidants are widely produced within cells as a consequence of mitochondrial electron transport, oxygen utilizing enzymes, peroxisomes, peroxidation of lipids, and other processes associated with aerobic metabolism. Although many oxidants (e.g., free radicals) that are generated by these means are neutralized by free radical scavengers or other antioxidative processes, some always escape and go on to damage macromolecules including DNA which can then lead to mutations and cancer.

Because of the potential importance of endogenously produced oxidative damage to age-related pathologies such as cancer, methods have been developed to measure these damaged products (Ames 1989; Fraga et al. 1990, Degan et al. 1991). Some of the endogenously produced damaged DNA products that can be measured in the urine include thymine glycol, thymidine glycol, hydroxymethyluracil, and hydroxymethyl-deoxyuridine (Saul et al. 1987; Adelman et al. 1988). A total of roughly 100 nmol/day of the first three compounds is excreted in the urine of a human daily. This 100 nmol represents, on average, about 10^3 oxidized thymine residues per day for each of the body's 6×10^{13} cells. Considering that these are only three of the approximately 20 major products that are generated during the oxidative damage to DNA (Cadet and Berger 1985), it is estimated that the total number of all types of oxidatively damaged DNA sites per cell per day is in the order of 10^4 in a human and 10^5 in a rat. Perhaps the most easily assayed product of oxidatively damaged DNA is 8-hydroxy-2'-deoxy-guanosine (8OHdG or 8-oxo-7,8-dihydro-2'-deoxyguanosine). 8OHdG causes mutations by inducing G to T transversions (Shibutani et al. 1991) and appears in DNA after its exposure to ionizing radiation and other carcinogens (Fiala et al. 1989).

24.3 Reactive Oxygen Species and DNA Damage

That DNA is a target for molecular attack by oxygen radicals has generated increasing interest in recent years partly because genomic alterations trigger malignant transformation of cells. Since reactive oxygen species produce chemical modification of DNA they must be a major cause of cancer and certainly there is now strong evidence for this (Cerutti et al. 1990; Trush and Kensler 1991). The reactive oxygen species that are capable of altering DNA structure have been identified.

It is now well established that the $O_2^-{}^\bullet$, lipid peroxides, and H_2O_2 are not sufficiently toxic to directly damage DNA (Lesko et al. 1980; Menenghini and Hoffman 1980) (Fig. 24.1). On the other hand, the hydroxyl radical ($\bullet OH$), the alkoxyl radical ($RO\bullet$),

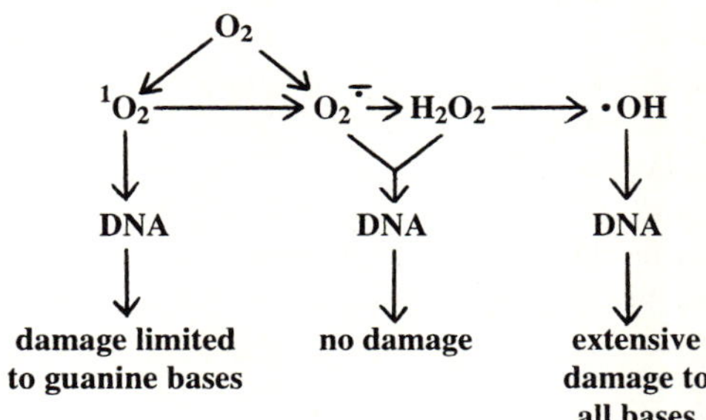

Fig. 24.1. The three electron reduction of dioxygen (O_2) generates the highly toxic hydroxy radical ($\bullet OH$). The superoxide anion radical ($O_2^-{}^\bullet$) and the reactive oxygen species hydrogen peroxide (H_2O_2) are intermediates in this process. Besides these reactive intermediates, O_2 can also be converted to singlet O_2 ($^1O_2$) by the addition of energy; $^1O_2$ degrades into the $O_2^-{}^\bullet$. Both $^1O_2$ and the $\bullet OH$ are sufficiently reactive to damage DNA while $O_2^-{}^\bullet$ and H_2O_2 do not have that capability. $O_2^-{}^\bullet$ is dismutated to H_2O_2 in the presence of the enzyme superoxide dismutase. H_2O_2 is converted to the $\bullet OH$ in the presence of a transition metal, usually Fe^{2+} or Cu^{1+}. H_2O_2 can be metabolized to nontoxic products by the enzymes catalase and glutathione peroxidase; these enzymes are not represented in this figure

and singlet oxygen ($^1O_2$) are quite capable of interacting with and damaging DNA (Fig. 24.1) (Ward 1988; Ochi and Cerutti 1989; Martins and Meneghini 1990; Floyd et al. 1991). In order for these species to damage nuclear DNA it is anticipated they would have to be present in the nucleus. At this point there is little evidence that $^1O_2$ is formed in the nucleus. However, given the presence of iron and copper in the nucleus, it is reasonable to assume that relatively stable species such as H_2O_2 and lipid peroxide (ROOH) migrate from their site of generation and react with Fe^{2+} or Cu^{1+} to produce the •OH and RO•, respectively, as follows:

$$H_2O_2 + Fe^{2+} (Cu^{1+}) \rightarrow OH + OH^- + Fe^{3+} (Cu^{2+}) \tag{1}$$

$$ROOH + Fe^{2+} (Cu^{1+}) \rightarrow RO• + OH^- + Fe^{3+} (Cu^{2+}) \tag{2}$$

The generation of the •OH and RO• in the immediate vicinity of the DNA allows these radicals to inflict damage on the genome.

Intracellularly, H_2O_2 can arise via several routes. It is well known that H_2O_2 is produced during the dismutation of $O_2^{-•}$; a family of superoxide dismutases (SOD) function as antioxidative enzymes in the removal of $O_2^{-•}$ by converting it to a non-radical product, i. e., H_2O_2. H_2O_2 can also be formed during the auto-oxidation of many xenobiotics and during the enzymatic formation of $O_2^{-•}$. Additionally, several oxidases give rise to H_2O_2, reactions in which $O_2^{-•}$ is not an intermediate. Finally, in a consideration of extracellular sources of reactive oxygen intermediates, H_2O_2 must also be considered a potentially damaging agent to DNA since its long half-life and its ability to cross cell membranes allows it to be produced extracellularly, then pass through cell membranes to enter the nucleus where, in the presence of transition metals, it generates the highly toxic •OH which goes on to mutilate adjacent DNA molecules. On the contrary, $O_2^{-•}$ is generally believed to be incapable of penetrating cellular membranes. Of the reactive oxygen species produced by activated neutrophils, it is also H_2O_2 that is most likely to penetrate target cells and enter the nucleus (Cochrane et al. 1988). In fact, it has been documented that most DNA strand breaks that occur in cells in the vicinity of neutrophils which are stimulated by phorbol ester are a consequence of H_2O_2, which is then converted to the highly toxic •OH (Shacter et al. 1990). The damage inflicted on DNA by H_2O_2 is exaggerated by the fact that it is assumed that iron ions are bound to DNA and therefore the Fenton reaction generates •OH in the immediate vicinity of DNA which then produce site-specific damage (Mello-Filho and Meneghini 1984; Ochi and Cerutti 1989; Shacter et al. 1990).

From a variety of studies it is clear that the •OH is in fact the ultimate DNA damaging agent. There are, however, alternate means by which the genome is damaged. Thus, large increases in intracellular Ca^{2+} can activate nucleases in the nucleus which then induces DNA strand breaks. As noted above, lipid peroxides also can penetrate to the nucleus where, in the presence of Fe^{2+}, they produce the RO•, which is capable of damaging DNA (Ochi and Cerutti 1987; Fraga and Tappel 1988).

24.4 Lipid Peroxidation and DNA Damage

When it goes unchecked, lipid peroxidation in biological membranes can be an extremely destructive process. The peroxidation of lipids causes direct damage to cells due to its effect in cell membranes, but additionally it causes indirect damage by releasing the reactive products that are formed during the breakdown of lipids.

The direct damage to biological membranes by reactive oxygen species is usually measured in terms of the well-known lipid peroxidation products and, less often, in terms of changes in membrane fluidity. The consequences of alterations in membrane fluidity as a result of lipid peroxidation have been reviewed by Richter (1987); the changes include decreased membrane fluidity (increased rigidity), altered phase properties of the membranes, and reduced electrical resistance. These changes compromise the ability of the membrane to function as a barrier. As a result, the membranes become leaky, they loose their selectivity, their ionic pumps are unable to maintain cellular homeostasis, their receptor signaling processes function in a suboptimal manner, and finally there can be either excessive activation or inactivation of membrane-bound enzymes.

Indirect effects of lipid peroxidation are those that are mediated by the products and intermediates that are produced during the oxidative breakdown of lipids. The carbonyl products are of particular importance for this discussion since they diffuse from their site of production and are biologically active. These toxic agents are capable of spreading the damage that is the initial result of lipid peroxidation. Not included in this group of toxins is the peroxyl radical, which is probably limited to the membrane in which it is generated and where it propagates lipid peroxidation per se. Likewise, lipid hydroperoxides probably do not diffuse from the membrane.

Lipid peroxidation gives rise to a large number of aldehydic scission products of which the 4-hydroxyalkenals have been the most thoroughly studied (Table 24.2). The 4-hydroxyalkenals and particularly 4-hydroxynonenal (4-HNE) are of special interest because they are produced in relatively large amounts during the peroxidation of lipids in animal cells. 4-HNE is a product of arachidonic acid oxidation (Esterbauer et al. 1990a) and it diffuses into the cytosol after its generation in the membrane (Benedetti et al. 1980). 4-HNE is detoxified in the cytosol by glutathione and it has significant biochemical effects, the most important of which are enzyme inhibition and cytotoxicity (Rossi et al. 1990). The hydroxyalkenals have also been shown to react with deoxyguanosine in solution but its specific binding to nucleic acids is debated. However, these agents have been shown to be mutagenic (Marnett et al. 1985; Brambilla et al. 1986) as are a number of other aldehydes which are produced during lipid peroxidation. Hydroxyalkenal products also inhibit DNA repair enzymes. The mutagenic properties of 4-HNE have been reviewed by Esterbauer and co-workers (1990b) and as pointed out, at concentrations found endogenously, 4-HNE produces detectable DNA damage in cells.

Table 24.2. Some of the common carbonyl products generated during the peroxidation of lipids in animal cell membranes

n-Alkanals	2-Alkenals	4-Hydroxyalkenals	Others
Butanal	Acrolein	4,5-Dihydroxydecenal	Butanone
Hexanal	Hexenal	4-Hydroxyhexenal	2-4-Decadienal
Nonanal	Nonenal	4-Hydroxy-2,5-nonadienal	2,4-Heptadienal
Pentanal	Octenal	4-Hydroxynonenal	5-Hydroxyoctanal
Propanal	Pentenal		Malondialdehyde

24.5 Melatonin as a Free Radical Scavenger

The association between free radicals, oxidative products, and carcinogenesis have been studied extensively. The •OH is widely considered to be the most damaging of the oxygen-based radicals that are generated, with roughly 50% of the total damage induced by free radical processes involving the •OH (Imlay and Linn 1988). Molecules that directly scavenge the •OH can thus be considered anticarcinogens since they reduce the destruction of the genome.

The most recently discovered •OH radical scavenger is melatonin. This indole is normally produced in vertebrates in the pineal gland and possibly in a number of other organs as well (Reiter 1991). In 1993, we showed that melatonin was an efficient scavenger of the OH radical, an action verified by electron spin resonance (ESR) studies (Tan et al. 1993a). Since then, this action of melatonin has been confirmed in a number of laboratories using both ESR and other methodologies (Li et al. 1997; Matuszek et al. 1997; Susa et al. 1997; Stasica et al. 1998). Besides scavenging the •OH, melatonin has been shown to neutralize other highly toxic agents including the peroxynitrite anion (ONOO$^-$) (Gilad et al. 1997; Cuzzocrea et al. 1998) and $^1O_2$ (Cagnoli et al. 1995). Finally, although the efficacy with which melatonin scavenges the peroxyl radical is debated, there is certainly evidence that it does so (Pieri et al. 1994, 1995; Marshall et al. 1996; Longoni et al. 1998).

Besides directly quenching free radicals and other reactive oxygen intermediates, melatonin reduces oxidative breakdown of cellular molecules due to its ability to alter the activity of a number of enzymes related to the antioxidative defense system. Thus, mRNA levels for SOD are stimulated by melatonin (Antolin et al. 1996; Kotler et al. 1998). This enzyme is important in removing $O_2^{-\bullet}$ from cells thereby reducing the formation of ONOO$^-$. The dismutation of $O_2^{-\bullet}$ to H_2O_2 makes the latter compound available for the generation of the •OH via the Fenton reaction. To reduce the chances of this occurring, H_2O_2 is metabolized to non-toxic products by the two enzymes catalase (CAT) and glutathione peroxidase (GPx). GPx utilizes H_2O_2 and other hydroperoxides as substrates during the oxidation of reduced glutathione (GSH) to its disulfide form (GSSG). Thus, both CAT and GPx function as important antioxidative enzymes by removing H_2O_2 from cells thereby lowering the generation of the highly toxic •OH. mRNA levels and the activity of GPx have been found to be augmented in a number of tissues in vitro by the addition of melatonin (Barlow-Walden et al. 1995; Pablos et al. 1995; Antolin et al. 1996). There is evidence that endogenously produced melatonin also stimulates GPx activity in vivo (Pablos et al. 1998). Once formed, oxidized glutathione, i.e., GSSG, is converted back to GSH by another important antioxidative enzyme, glutathione reductase (GRd). The activity of this enzyme has also been shown to be stimulated by melatonin (Pablos et al. 1998). GRd requires NADPH as a co-factor during the reduction of GSSG; NADPH is generated by the enzyme glucose-6-phosphate dehydrogenase (G6PD) which Pierrefiche and Laborit (1995) found was also stimulated by melatonin. From these findings it is apparent that besides the direct scavenging of radicals and reactive oxygen species by melatonin, it may have a variety of indirect antioxidative effects by which it protects cells from the devastating actions of oxygen by-products. These and other actions of melatonin in terms of antioxidative defense are summarized in Table 24.3.

Nitric oxide (NO•) has been described as a double-edged sword. While NO• is an important gaseous neurotransmitter, it can also combine with $O_2^{-\bullet}$ to generate a toxic

Table 24.3. Proposed functions of melatonin which contribute to the antioxidant capacity of this molecule. These actions reduce free radical damage to DNA. (See text for description of abbreviations)

Agent	Action of melatonin	Reference
Reactive oxygen species		
$^1O_2$	Quenches	Cagnoli et al. (1995)
$ONOO^-$	Scavenges	Gilad et al. (1997); Cuzzocrea et al. (1998)
•OH	Scavenges	Tan et al. (1993a); Li et al. (1997); Matuszek et al. (1997); Susa et al. (1997); Stasica et al. (1998)
LOO•	Scavenges	Pieri et al. (1994, 1995); Scaiano (1995); Marshall et al. (1996)
Antioxidative enzymes		
SOD	Stimulates mRNA	Antolin et al. (1996); Kotler et al. (1998)
GPx	Stimulates activity	Barlow-Walden et al. (1995); Pablos et al. (1995, 1998)
GRd	Stimulates activity	Pablos et al. (1998)
G6PD	Stimulates activity	Pierrefiche and Laborit (1995)
Pro-oxidative enzyme		
NOS	Inhibits activity	(Pozo et al. 1994)
Other effects		
Cellular membranes	Stabilizes	Garcia et al. (1997, 1998)

product, the $ONOO^-$ (Tamir et al. 1996). Besides the inherent toxicity of $ONOO^-$, it also degrades into other toxic products including possibly the •OH. Considering these metabolic products of NO•, its excessive production can lead to extensive molecular destruction, such as occurs during ischemia/reperfusion.

NO• synthesis is controlled by the activity of the enzyme nitric oxide synthase (NOS). This enzyme is generally considered pro-oxidative because of the pathway described above. Melatonin has been shown to inhibit the activity of NOS thereby reducing NO• generation (Pozo et al. 1994); also, the indole reduces lipid peroxidation that is a consequence of NO• (Escames et al. 1997).

From the data summarized above, it is obvious that melatonin has multiple actions which could be beneficial in terms of its ability to limit the initial molecular damage that could lead to cancer initiation. Besides oxidative damage often being responsible for the initiation step of cancer, there is also evidence that progressive damage to nuclear DNA after cancer is initiated leads to the transformation of benign cancer cells to malignant cancer cells (Malins et al. 1996). This being the case, it is possible that melatonin antioxidative actions would curtail this transformation as well, thereby restricting the aggressiveness of the tumor cells following their initiation. More extensive discussions of melatonin free radical scavenging activities and the significance of these in cellular and cancer biology are available elsewhere (Reiter 1996; Reiter et al. 1996; 1997b).

24.6 Melatonin as an Antioxidant

Antioxidant defenses against reactive oxygen species which cause genetic damage provide important processes by which organisms defend themselves against cancer (Anderson 1996; Toyokuni and Sagripanti 1996). The role of melatonin in limiting free radical toxicity was summarized above. In this section of the report, the data relating to melatonin's protective action at the level of nuclear DNA and membrane lipids are reviewed.

In 1993, we tested the ability of melatonin to protect against DNA damage induced by the carcinogen safrole (Tan et al. 1993b). In these in vivo studies, melatonin, given just prior to safrole administration, significantly reduced hepatic nuclear DNA damage when measured 24 h after administration of the carcinogen. The protective effect of melatonin was obvious even though the doses of melatonin used, i. e., 0.2 or 0.4 mg/kg, were much lower than the 300-mg/kg dose of safrole used to cause DNA damage. In a follow-up study, we further showed that endogenous levels of melatonin are sufficient to partially protect against DNA damage caused by a 100-mg/kg dose of safrole (Tan et al. 1994). The clear implication of these findings is that the quantity of melatonin produced in the organism is sufficient to limit oxidative damage to nuclear DNA and thereby reduce the likelihood of cancer initiation. In these studies, Tan and colleagues used post-labeling of DNA adducts to identified oxidatively damaged genomic products (Reddy and Randerrath 1986).

In a series of studies in which a number of cytogenetic procedures were used to assess oxidative damage to DNA, we also were able to demonstrate the DNA protective action of melatonin. In these investigations, ionizing radiation was used as the DNA damaging agent. The mechanisms by which ionizing radiation damages DNA was thoroughly investigated and involves •OH that are generated by high-energy electromagnetic field exposure (Okada et al. 1997). In each of these studies, many of which utilized human blood cells, melatonin was found to reduce, in a dose-dependent manner, the number of chromosomal aberrations and micronuclei which are indicative of DNA damage (Vijayalaxmi et al. 1995a, 1995b, 1996, 1998a). In one study, humans ingested melatonin, and blood cells collected before and after melatonin ingestion were exposed to gamma radiation in vitro. The findings showed that when melatonin is present, radiation-induced genetic damage is highly significantly reduced (Vijayalaxmi et al. 1996). From these studies we surmised that melatonin protects DNA from the harmful affects of ionizing radiation by scavenging the resulting •OH; however, the possibility exists that the protective effects on DNA may also involve the stimulation of DNA repair enzymes (Vijayalaxmi et al. 1998b). Using the same assay to quantify DNA damage, melatonin was found to protect against the genotoxic actions of bacterial lipopolysaccharide (Sewerynek et al. 1995; 1996) and paraquat (Melchiorri et al. 1998).

Susa and colleagues (1997) investigated the ability of melatonin to reduce the cytotoxicity of dichromate, $K_2Cr_2O_7$. When primary rat hepatocytes were incubated with $K_2Cr_2O_7$, DNA single-strand breaks in these cells were increased markedly, a response greatly attenuated when melatonin was also added to the incubation medium. The authors determined that melatonin did not impede the uptake of chromium into the cells and concluded that melatonin protective actions was due either to the direct scavenging activity of melatonin or to its ability to preserve cellular levels of other antioxidants, namely, vitamins E and C. In this study, DNA single-strand breaks were assessed using the alkaline elution method described by Sugiyama and co-workers (1988).

According to Lai and Singh (1997), the exposure of adult rats to extremely low frequency (60 Hz) magnetic fields causes DNA single- and double-strand breaks in brain cells as assessed by the microgel electrophoresis assay (comet assay). The magnetic field intensity in this case was 0.5 mT (5 G) and the duration of exposure was 2 h. Anticipating that free radicals may be involved in causing the observed DNA damage, some rats were treated with melatonin (1 mg/kg) or with N-tert-butyl-α-phenylnitrone (100 mg/kg). The latter molecule is a synthetic spin-trapping agent that neutralizes free radicals while, as summarized above, melatonin functions as a direct free radical scavenger. Both treatments were found to prevent DNA single- and double-strand breaks in brain cells of rats exposed to the 60 Hz fields. Thus, the results are consistent with free radicals being responsible for the genotoxicity in the cells and show that melatonin provides effective protection of nuclear DNA against oxidative attack.

Using yet another method to evaluate DNA damage, Tang et al. (1998) found that the accumulation of 8OHdG in brain and hepatic cells of rats treated with the excitotoxin kainic acid was abrogated by pretreatment of the animals with melatonin. As pointed out above, 8OHdG is a major product of damaged DNA and it is often used as an index of genotoxicity. Since 8OHdG is known to be a result of oxidative attack, melatonin protective actions were believed by the authors to be related to its free radical scavenging activity.

Since, as summarized above, genotoxic reactive intermediates produced during the peroxidation of lipids can relocate to the nucleus and thereafter cause genomic damage, agents that curtail lipid peroxidation could also protect cells from cancer initiation. There is now an extensive literature showing the ability of melatonin to reduce lipid peroxidation that is a consequence of a plethora of toxic agents. Thus, increases in lipid peroxidation due to carbon tetrachloride (Daniels et al. 1995), lipopolysaccharide (Sewerynek et al. 1995), paraquat (Melchiorri et al. 1995), L-cysteine (Yamamoto and Tang 1996), ischemia/reperfusion (De La Lastra et al. 1997), and δ-aminolevulinic acid (Princ et al. 1997) are only a few agents against which melatonin has been shown to be protective. Others are discussed in recent reviews (Reiter 1995, 1997a; Reiter et al. 1995, 1997a).

One therapy currently in use for the prevention and treatment of breast cancer is tamoxifen (Wiseman 1994). Tamoxifen also protects membranes as well as lipoprotein particles from oxidative damage and the antioxidative actions of this compound likely contribute to its cardioprotective actions. The anticancer action of tamoxifen also probably relates to its modulation of membrane fluidity (Wiseman and Halliwell 1994). The recent discovery that tamoxifen and melatonin act synergistically to optimize membrane fluidity and resist lipid peroxidation (Garcia et al. 1997, 1998) may lead to a new therapy where these two molecules are used in combination. This is a potential therapy that requires experimental testing.

Finally, melatonin may also find utility in another cancer treatment paradigm. Adriamycin (doxorubicin) is an effective chemotherapeutic agent whose collateral toxicity at the level of the kidney and the heart limits its use. Melatonin, given in combination with this agent, may prove highly effective in reducing the nephro- and cardiotoxicity (Montilla et al. 1997).

24.7 Concluding Remarks

Since DNA is highly vulnerable to oxidative attack, it would be expected that such damage could be abated by free radical scavengers and antioxidants. Thus, molecules that function in this capacity and that are capable of entering the nucleus afford significant protection for DNA, thereby preventing DNA damage and reducing the likelihood of cancer.

Not all antioxidants enter the nucleus, however; for example, while vitamin E is a primer lipid antioxidant and peroxyl radical scavenger, it does not directly protect DNA from free radical attack since it is not located in the nucleus. Conversely, melatonin readily enters all subcellular compartments and its presence in the nuclei of cells has been described (Menendez-Pelaez and Reiter 1993; Finnochiarro and Glikin 1997). Thus, it is in the proper location to scavenge radicals that are about to plunder DNA.

Since products produced during the peroxidation of lipids can relocate to the nucleus and damage DNA, any molecule which resists lipid breakdown can also reduce the incidence of cancer. Melatonin, although its efficacy as a peroxyl radical scavenger is debated, it is nevertheless a highly effective agent preventing lipid peroxidation. As a result, it should secondarily impede carcinogenesis because of this action.

Historically, melatonin was believed to inhibit cancer primarily during the progression and promotion phases (Blask 1993; Panzer and Viljoen 1997). The studies summarized herein, however, show that its actions extend to the inhibition of cancer initiation as well. These findings should be considered in light of the well documented age-related drop in melatonin in humans (Reiter 1992; 1997b). It is thus tempting to speculate that the loss of melatonin in the aged may be a risk factor for cancer initiation as well as cancer progression and promotion and for other age-related diseases which have, as part of the disease process, free radical damage (Reiter et al. 1996; 1997b).

References

Adelman R, Saul RL, Ames BN (1988) Oxidative damage to DNA: relation to species metabolic rate and life span. Proc Natl Acad Sci USA 85:2706–2708

Anderson D (1996) Antioxidant defenses against reactive oxygen species causing genetic and other damage. Mutat Res 350:103–108

Ames BN (1983) Dietary carcinogens and anti-carcinogens (oxygen radicals and degenerative diseases). Science 224:1256–1264

Ames BN (1989) Endogenous DNA damage as related to cancer and aging. Mutat Res 214:41–46

Antolin I, Rodriquez C, Sainz RM, Mayo JC, Uria H, Kotler ML, Rodriquez-Colunga MJ, Toliva D, Menendez-Pelaez A (1996) Neurohormone melatonin prevents cell damage: effect on gene expression for antioxidant enzymes. FASEB J 10:882–890

Barlow-Walden LR, Reiter RJ, Pablos MI, Menendez-Pelaez A, Chen LD, Poeggeler B (1995) Melatonin stimulates brain glutathione peroxidase activity. Neurochem Int 24:497–502

Barrows LR, Magee PN (1982) Nonenzymatic methylation of DNA S-adenosylmethionine in vitro. Carcinogenesis 3:349–351

Beneditti A, Comporti M, Esterbauer H (1980) Identification of 4-hydroxynonenal as a cytotoxic product originating from the peroxidation of liver microsomal lipids. Biochim Biophys Acta 620:281–296

Blask DE (1993) Melatonin in oncology. In: Yu HS, Reiter RJ (eds) Melatonin, CRC Press, Boca Raton, pp 447–476

Brambilla G, Sciaba L, Faggin P, Manra A, Marinari V, Ferro M, Esterbauer H (1986) Cytotoxicity, DNA fragmentation and sister chromatid exchange in Chinese hamster ovary cells exposed to lipid peroxidation product 4-hydroxynonenal and homologous aldehydes. Mutat Res 171: 169–176

Cadet J, Berger M (1985) Radiation-induced decomposition of the purine bases within DNA and related model compounds. Int J Radiat Biol 47:127–143

Cagnoli CM, Atabay C, Kharlamov E, Manev H (1995) Melatonin protects neurons from singlet oxygen-induced apoptosis. J Pineal Res 18:222–228

Cathcart R, Schwiers E, Saul RL, Ames BN (1984) Thymine glycol and thymidine glycol in human and rat urine: a possible assay for oxidative DNA damage. Proc Nat Acad Sci USA 81:5633–5637

Cerutti P, Amstad P, Larsson R, Shah G, Krupitza G (1990) Mechanisms of oxidant carcinogenesis. Prog Clin Biol Res 347:183–186

Cochrane CG, Schraufstatter IV, Hyslop P, Jackson J (1988) Cellular and biochemical events in oxidant injury. In: Halliwell B (ed) Oxygen Radical and Tissue Injury, Upjohn, New York, pp 49–54

Cuzzocrea S, Zingarelli B, Constantino G, Caputi AP (1998) Protective effect of melatonin in a non-septic shock model induced by zymosan in the rat. J Pineal Res 25:24–33

Daniels WMU, Reiter RJ, Melchiorri D, Sewerynek E, Pablos MI, Ortiz GG (1995) Melatonin counteracts lipid peroxidation induced by carbon tetrachloride but does not restore glucose-6-phosphatase activity. J Pineal Res 19:1–6

De La Lastra CA, Cabeza J, Motilva V, Martin MJ (1997) Melatonin protects against gastric ischemia-reperfusion injury in rats. J Pineal Res 23:47–52

Degan P, Shigenaga MK, Park EM, Alperin PE, Ames BN (1991) Immunoaffinity isolation of 8-hydroxy-2′-deoxyguanosine and 8-hydroxyguanine and quantitation of 8-hydroxy-2′-deoxyguanosine in DNA by polyclonal antibodies. Carcinogenesis 12:865–871

Escames G, Guerrero JM, Reiter RJ, Garcia JJ, Muñoz-Hoyos A, Ortiz GG, Oh CS (1997) Melatonin and vitamin E limit nitric oxide-induced lipid peroxidation in rat brain homogenates. Neurosci Lett 230:147–150

Esterbauer H, Eckl P, Ortner A (1990a) Possible mutagens derived from lipids and lipid precursors. Mutat Res 238:223–233

Esterbauer H, Zollner H, Schaur RJ (1990b) Aldehydes formed by lipid peroxidation: mechanisms of formation, occurrence and determination. In: Vigo-Pelfrey C (ed) Membrane Lipid Oxidation, Vol I, CRC Press, Boca Raton, pp 239–283

Fiala ES, Conaway CC, Mathis JE (1989) Oxidative DNA and RNA damage in the livers of Sprague-Dawley rats treated with the hepatocarcinogen 2-nitropropane. Cancer Res 49:5518–5522

Finnochiarro LME, Glikin GC (1997) Intracellular melatonin distribution in cultured cell lines. J Pineal Res 24:22–34

Floyd RA, West MS, Eneff KL, Schneider JE (1991) Methylene blue plus light mediates 8-hydroxyguanine formation in DNA. Arch Biochem Biophys 273:106–111

Fraga CG, Tappel AL (1988) Damage to DNA concurrently with lipid peroxidation in rat liver slices. Biochem J 252:893–896

Fraga CG, Shigenaga MK, Park JW, Degan P, Ames BN (1990) Oxidative damage to DNA during aging: 8-hydroxy-2′-deoxyguanosine in rat organ DNA and urine. Proc Natl Acad Sci USA 87:4533–4537

Garcia JJ, Reiter RJ, Guerrero JM, Escames G, Yu BP, OH LS, Muñoz-Hoyos A (1997) Melatonin prevents changes in microsomal membrane fluidity during induced lipid peroxidation. FEBS Lett 408:297–300

Garcia JJ, Reiter RJ, Ortiz GG, Oh CS, Tang L, Yu BP, Escames G (1998) Melatonin enhances tamoxifen's ability to prevent the reduction in microsomal membrane fluidity induced by lipid peroxidation. J Membrane Biol 162:59–62

Gilad E. Cuzzocrea S, Zingarelli B, Salzman AL, Szabo C (1997) Melatonin is a scavenger of peroxynitrite. Life Sci 60:PL169–PL174

Harman D (1984) Free radical theory of aging: The "free radical" diseases. Age 7:111–131

Imlay JA, Linn S (1988) DNA damage and oxygen radical toxicity. Science 240:1302–1309

Kotler M, Rodriquez C, Sainz RM, Antolin I, Menendez-Pelaez A (1998) Melatonin increases gene expression for antioxidant enzymes in rat brain cortex. J Pineal Res 24:83–89

Lai H, Singh NP (1997) Melatonin and N-tert-butyl-α-phenylnitrone block 60 Hz magnetic field-induced DNA single and double strand breaks in rat brain cells. J Pineal Res 22:152–162

Lesko SA, Lorentzen RJ, Ts'o POP (1980) Role of superoxide in DNA strand scission. Biochemistry 19:3027–3028

Li XJ, Zhang LM, Gu J, Zhang AZ, Sun FK (1997) Melatonin decreases production of hydroxyl radical during cerebral ischemia-reperfusion. Acta Pharmacol Sinica 18:394–396

Lindahl T (1977) DNA repair enzymes acting on spontaneous lesions in DNA. In: Nichols WW, Murphy DC (eds), DNA Repair Processes, Symposia Specialists, Miami, pp 225–240

Lindahl T (1982) DNA-repair enzymes. Ann Rev Biochem 51:61–87

Longoni B, Salgo MG, Pryor WA, Marchiafava PL (1998) Effects of melatonin on lipid peroxidation induced by oxygen radicals. Life Sci 62:853–859

Malins DC, Polissar NL, Gunselman SJ (1996) Progression of human breast cancers to the metastatic state is linked to hydroxyl radical-induced DNA damage. Proc Natl Acad Sci USA 93:2557–2563

Marnett L, Hurd HK, Hollenstein MK, Levin DE, Esterbauer H, Ames BN (1985) Naturally occurring carbonyl compounds are mutagens in Salmonella tester strain TA104. Mutat Res 148:25–34

Marshall KA, Reiter RJ, Poeggeler B, Aruoma B, Halliwell B (1996) Evaluation of the antioxidant activity of melatonin in vitro. Free Radical Biol Med 21:307–315

Martins EM, Meneghini R (1990) DNA damage and lethal effects of hydrogen peroxide and menodione in Chinese hamster cells: distinct mechanisms are involved. Free Radical Biol Med 8:433–440

Massie HR, Samis HV, Baird MB (1972) The kinetics of degradation of DNA and RNA by H_2O_2. Biochem Biophys Acta 272:539–548

Matuszek Z, Reszka KJ, Chignell CF (1997) Reaction of melatonin and related indoles with hydroxyl radicals: EPR and spin trapping investigations. Free Radical Biol Med 23:367–372

Melchiorri D, Reiter RJ, Attia AM, Hara M, Burgos A, Nistico G (1995) Potent protective effect of melatonin on in vivo paraquat-induced oxidative damage in rats. Life Sci 56:83–89

Melchiorri D, Ortiz GG, Reiter RJ, Sewerynek E, Daniels WMU, Pablos MI, Nistico G (1998) Melatonin reduces paraquat-induced genotoxicity in mice. Toxicol Lett 95:103–108

Mello-Filho AC, Meneghini R (1984) In vivo formation of single-strand breaks in DNA by hydrogen peroxide is mediated by the Haber-Weiss reaction. Biochim Biophys Acta 781:56–83

Meneghini R, Hoffman ME (1980) The damaging action of hydrogen peroxide on DNA of human fibroblast is mediated by a non-dialyzable compound. Biochim Biophys Acta 608:167–173

Menendez-Pelaez A, Reiter RJ (1993) Distribution of melatonin in mammalian tissues: the relative importance of nuclear versus cytosolic localization. J Pineal Res 15:59–69

Montilla P, Tuney I, Munoz MC, Lopez A, Soria JV (1997) Hyperlipidemic nephropathy induced by adriamycin: effect of melatonin administration. Nephron 76:345–350

Ochi T, Cerutti P (1987) Clastogenic action of hydroperoxy-5,8,11,13-icosatetraenoic acids on the mouse embryo fibroblasts C3H/I0T1/2. Proc Natl Acad Sci USA 84:990–994

Ochi T, Cerutti P (1989) Differential effects of glutathione depletion and metallothionein induction on the induction of DNA single-strand breaks and cytotoxicity by tert-butyl hydroperoxide in cultured mammalian cells. Chemico-Biol Interact 72:335–345

Okada S, Nakamura N, Sasaki K (1997) Radioprotection of intracellular genetic material. In: Nygaard OF, Simic MC (eds) Radioprotectors and Anticarcinogens, Academic Press, New York, pp 339–346

Pablos MI, Agapito MT, Gutierrez R, Recio JM, Reiter RJ, Barlow-Walden LR, Acuña-Castroviejo D, Menendez-Pelaez A (1995) Melatonin stimulates the activity of the detoxifying enzyme glutathione peroxidase in several tissues of chicks. J Pineal Res 19:111–115

Pablos MI, Reiter RJ, Ortiz GG, Guerrero JM, Agapito MT, Chuang JI, Sewerynek E (1998) Rhythms of glutathione peroxidase and glutathione reductase in brain of chick and their inhibition by light. Neurochem Int 32:69–75

Panzer A, Viljoen M (1997) The validity of melatonin as an oncostatic agent. J Pineal Res 22:184–202

Pieri C, Marra M, Moroni F, Recchioni R, Marcheselli F (1994) Melatonin: A peroxyl scavenger more potent than vitamin E. Life Sci 55:271–276

Pieri G, Moroni M, Marra M, Marcheselli F, Recchioni R (1995) Melatonin as an efficient antioxidant. Arch Gerontol Geriatr 20:159–165

Pierrefiche G, Laborit H (1995) Oxygen radicals, melatonin and aging. Exp Gerontol 30:213–227

Portier CJ, Hedges JC, Hoel DG (1986) Age-specific models of mortality and tumor onset for historical control animals in the National Toxicology Program's carcinogenicity experiments. Cancer Res 46:4372–4378

Pozo D, Reiter RJ, Calvo JR, Guerrero JM (1994) Physiological concentrations of melatonin inhibit nitric oxide synthase in rat cerebellum. Life Sci 55:PL455–460

Princ FG, Juknat AA, Maxit AG, Cardalda C, Battle A (1997) Melatonin's antioxidant protection against δ-aminolevulinic acid-induced oxidative damage in rat cerebellum. J Pineal Res 23:40–46

Reddy MV, Randerrath K (1986) Nuclease P1-mediated enhancement of sensitivity of 32P-postlabeling test for structurally diverse DNA adducts. Carcinogenesis 7:1543–1551

Reiter RJ (1991) Pineal melatonin: cell biology of its synthesis and of its physiological interactions. Endocrine Rev 12:151–180

Reiter RJ (1992) The aging pineal gland and its physiological consequences. BioEssays 14:169–175

Reiter RJ (1995) Oxidative processes and antioxidative defense mechanisms in the aging brain. FASEB J 9:526–533

Reiter RJ (1996) Functional aspects of the pineal hormone melatonin in combating cell and tissue damage induced by free radicals. Eur J Endocrinol 134:412–430

Reiter RJ (1997a) Antioxidant actions of melatonin. Adv Pharmacol 38:103–117

Reiter RJ (1997b) Aging and oxygen toxicity: relation to changes in melatonin. Age 20:201–213

Reiter RJ, Melchiorri D, Sewerynek E, Poeggeler B, Barlow-Walden LR, Chuang J, Ortiz GG, Acuña-Castroviejo D (1995) A review of the evidence supporting melatonin's role as an antioxidant. J Pineal Res 18:1–11

Reiter RJ, Pablos MI, Agapito MT, Guerrero JM (1996) Melatonin in the context of the free radical theory of aging. Ann NY Acad Sci 786:362–378

Reiter RJ, Carneiro RC, Oh CS (1997a) Melatonin in relation to cellular antioxidative defense mechanisms. Horm Metab Res 29:363–372

Reiter RJ, Tang L, Garcia JJ, Muñoz-Hoyos A (1997b) Pharmacological actions of melatonin in free radical pathophysiology. Life Sci 60:2255–2271

Richter C (1987) Biophysical consequences of lipid peroxidation in membranes. Chem Physics Lipids 44:175–179

Rossi MA, Fidole F, Garramone A, Esterbauer H, Dianzani MV (1990) Effect of 4-hydroxyalkenals on hepatic phosphatidyl-inositol-4,5-biphosphate phospholipase C. Biochem Parmacol 39: 1715–1719

Saul RL, Ames BN (1986) Background levels of DNA damage in the population. In: Simic M, Grossman L, Upton A (eds) Mechanisms of DNA Damage and Repair: Implications for Carcinogensis and Risk Assessment. Plenum, New York, pp 529–536

Saul RL, Gee P, Ames BN (1987) Free radicals, DNA damage, and aging. In: Warner HR, Butler RN, Sprott RL, Schnedier EL (eds) Modern Biological Theories of Aging. Raven, New York, pp 113–129

Scaiano JC (1995) Exploratory laser flash photolysis study of free radical reactions and magnetic field effects in melatonin chemistry. J Pineal Res 19:189–195

Sewerynek E, Melchiorri D, Chen L, Reiter RJ (1995) Melatonin reduces both basal and bacterial lipopolysaccharide-induced lipid peroxidation in vitro. Free Radical Biol Med 19:903–909

Sewerynek E, Ortiz GG, Reiter RJ, Pablos MI, Melchiorri D, Daniels WMU (1996) Lipopolysaccharide-induced DNA damage is greatly reduced in rats treated with the pineal hormone melatonin. Mol Cell Endocrinol 117:183–188

Shacter S, Beecham EJ, Covey JM, Kohn KW, Potter M (1988) Activated neutrophils induce prolonged DNA damage in neighboring cells. Carcinogenesis 9:2297–2304

Shacter E, Lopez RL, Beecham EJ, Janz S (1990) DNA damage induced by phorbol ester-stimulated neutrophils is augmented by extracellular factors. J Biol Chem 265:6693–6699

Shapiro R (1981) Damage to DNA caused by hydrolysis. In: Seeberg E, Kleppe K (eds) Chromosome Damage and Repair. Plenum, New York, pp 3–18

Shibutani S, Takeshita M, Grollman AP (1991) Insertion of specific bases during DNA synthesis past the oxidation-damaged base 8-oxodG. Nature 349:431–434

Stasica P, Ulanski P, Rosiak JM (1998) Melatonin as a hydroxyl radical scavenger. J Pineal Res 25: 65–66

Sugiyama M, Costa M, Nakagawa T, Hidaka T, Ogura R (1988) Stimulation of polyadenosine diphosphoribose synthesis by DNA lesions induced by sodium chromate in Chinese hamster V-79 cells. Cancer Res 48:1100–1104

Susa N, Ueno S, Furukawa Y, Ueda J, Sugiyama M (1997) Potent protective effect of melatonin on chromium (VI)-induced DNA strand breaks, cytotoxicity and lipid peroxidation in primary cultures of rat hepatocytes. Toxicol Appl Pharmacol 144:377–384

Tamir S, Burney S, Tannenbaum SR (1996) DNA damage by nitric oxide. Chem Res Toxicol 9: 821–827

Tan DX, Chen LD, Poeggeler B, Manchester LC, Reiter RJ (1993a) Melatonin: A potent endogenous hydroxyl radical scavenger. Endocrine J 1:57–60

Tan DX, Poeggeler B, Reiter RJ, Chen LD, Manchester LC, Barlow-Walden LR (1993b) The pineal hormone melatonin inhibits DNA adduct formation induced by the chemical carcinogen safrole. Cancer Lett 70:65–71

Tan DX, Reiter RJ, Chen LD, Poeggeler B, Manchester LC, Barlow-Walden LR (1994) Both physiological and pharmacological levels of melatonin reduce DNA adduct formation induced by safrole. Carcinogenesis 15:215–218

Tang L, Reiter RJ, Li ZR, Ortiz GG, Yu BP, Garcia JJ (1998) Melatonin reduces the increase in 8-hydroxy-deoxyguanosine levels in the brain and liver of kainic acid-treated rats. Molec Cell Biochem 178:299–303

Toyokuni S, Sagripanti JC (1996) Association between 8-hydroxy-2′-deoxyguanosine formation and DNA strand breaks mediated by copper and iron. Free Radical Biol Med 20:859–864

Trush MT, Kensler WK (1991) An overview of the relationship between oxidative stress and chemical carcinogenesis. Free Radical Biol Med 10:201–209

Vijayalaxmi, Reiter RJ, Meltz ML (1995a) Melatonin protects human blood lymphocytes from radiation-induced chromosome damage. Mutat Res 346:23–31

Vijayalaxmi, Reiter RJ, Sewerynek E, Poeggeler B, Leal BZ, Meltz ML (1995b) Marked reduction of radiation-induced micronuclei in human blood lymphocytes pretreated with melatonin. Radiat Res 143:102–106

Vijayalaxmi, Reiter RJ, Herman TS, Meltz ML (1996) Melatonin and radioprotection from genetic damage: in vivo/in vitro studies in human volunteers. Mutat Res 371:221–228

Vijayalaxmi, Reiter RJ, Herman TS, Meltz ML (1998a) Melatonin reduces gamma radiation-induced primary DNA damage in human blood lymphocytes. Mutat Res 397:203–208

Vijayalaxmi, Reiter RJ, Meltz ML, Herman TS (1998b) Melatonin: possible mechanisms involved in its "radioprotective" effects. Mutat Res 404:187–189

Ward JF (1988) DNA damage produced by ionizing radiation in mammalian cells: identities, mechanisms of formation, and reparability. Prog Nucleic Acid Res 35:96–135

Williams GC, Neese RM (1991) The dawn of Darwinian medicine. Quart Rev Biol 66:1–22

Wiseman H (1994) Tamoxifen: new membrane-mediated mechanisms of action and therapeutic advances. Trends Pharmacol Sci 15:83–89

Wiseman H, Halliwell B (1994) Tamoxifen and related compounds protect against lipid peroxidation in isolated nuclei: relevance to the potential anticarcinogenic benefits of breast cancer prevention and therapy with tamoxifen. Free Radical Biol Med 17:485–488

Yamamoto HA, Tang HW (1996) Melatonin attenuates L-cysteine-induced seizures and lipid peroxidation in the brain of mice. J Pineal Res 21:108–113

Yamashina K, Miller BE, Heppner GH (1986) Macrophage-mediated induction of drug resistant variants in mouse mammary tumor cell line. Cancer Res 46:2396–2401

25 Could the Antiproliferative Effect of Melatonin Be Exerted Via the Interaction of Melatonin with Calmodulin and Protein Kinase C?

Fernando Antón-Tay, Gloria Benítez-King

Abstract

Most of the systems used to study the effects of melatonin (MEL) on abnormal cell growth have shown that the hormone, at both physiological and pharmacological concentrations, inhibits cell proliferation. It has been proposed that the hormone modulates the Ca^{2+} messenger system through its interaction with Ca^{2+}-activated calmodulin (Ca^{2+}/CaM) and the Ca^{2+}-dependent protein kinase C (PKC) isoforms. The physiological effects of both interactions have been mainly studied on cytoskeleton and protein phosphorylation. The role of Ca^{2+}/CaM and PKC on cell proliferation are reviewed and discussed within the scope of known MEL effects on these two intracellular proteins.

25.1 Introduction

Most of the systems used to study the effects of melatonin (MEL) on abnormal cell growth have shown that the hormone, at both physiological and pharmacological concentrations, inhibits cell proliferation (Blask 1984, Blask et al. 1993, 1997, Subramanian and Kothari 1991). Clinical studies have shown that MEL acts both by stimulation of the immune system and by inhibiting tumor growth (Lissoni et al. 1993). Moreover, MEL may exert oncostatic effects on tumoral cells by potentiating the effect of chemotherapeutic agents (Lissoni et al. 1997), and by amplification of IL-2 (Conti and Maestroni 1995) or tamoxifen anticancer activities (Lissoni et al. 1996). These actions agree with the hypothesis that MEL acts as a rheostatic hormone by adjusting the setpoint of cell activity (Antón-Tay et al. 1993). Recent studies have suggested several possible mechanisms for the oncostatic effects of the hormone. In the present chapter we shall review the experimental data that support the hypothesis that the antiproliferative effects of MEL are mediated by the interaction of the hormone with calmodulin (CaM) and protein kinase C (PKC). It has been proposed that MEL modulates the Ca^{2+} messenger system through interaction with Ca^{2+}-activated CaM and the Ca^{2+}-dependent PKC isoforms (Benítez-King and Antón-Tay 1993 and 1997). The physiological effects of both interactions have been mainly studied on cytoskeletal rearrangements and protein phosphorylation, effects that play a key role in cell proliferation (Benítez-King and Antón-Tay 1993 and 1996, Huerto-Delgadillo et al. 1994). However, because of the multiple cellular functions regulated by both Ca^{2+}-dependent intracellular proteins, we know that we have more questions than answers. Among the multiple intracellular Ca^{2+} actions, the regulation of cell growth by the cation has long been recognized (Hinrichsen 1993). Transient increases in the intracellular levels of Ca^{2+} are closely correlated with cytoskeletal modifications needed for cell growth (Poenie et al. 1985, Whitaker and Patel 1990). Ca^{2+} regulation of cytoarchitecture is mediated by CaM and by specific protein kinases that in turn modulate the protein phosphorylation–dephosphorylation process (Klee 1991, Mumby and Walter 1993). Cell cycle progression involves microtubule enlargements, to form the mitotic spindle, and microtubule depolymerization at the kinetochore region during anaphase (Mitchison et al. 1986). Also, during cell division Ca^{2+} plays a key role in both actin and myosin reorganization to form the contractile ring during cytokinesis (Cameron et al. 1987, Knecht and Loomis 1987), and intermediate filament fragmentation to disrupt the nuclear matrix (Nigg 1992).

25.2 Calmodulin Involvement in Cell Proliferation: Effects of Melatonin

CaM, the major Ca^{2+} receptor in eukaryotic cells, mediates many Ca^{2+}-dependent events and has a critical function in normal cell proliferation (Niki et al. 1996). Ca^{2+}-activated CaM acts at multiple points in the cell cycle: the initiation of the S phase and both the initiation and the completion of the M phase (Rasmussen and Means 1989), gene expression for cell cycle progression, and cytokinesis (Takuwa et al. 1995). Quantitative changes in cell levels of CaM take place during normal cell division. The finding of elevated CaM levels in tumor cells compared with the levels found in normal tissues from which they were derived implies that an increase in CaM may contribute to the malignant state (Hait and DeRosa 1988).

Moreover, it has been shown that CaM levels may regulate cell cycle progression. Studies of mouse C127 cells transfected with vectors capable of synthesizing CaM sense or CaM antisense RNA showed that production of the antisense RNA results in a significant decrease in intracellular CaM concentrations and a cell cycle arrest, whereas production of transient high levels of CaM mRNA is followed by increased CaM concentration and acceleration of cell proliferation (Rasmussen and Means 1989). Progression through G1 and mitosis is affected by changes in CaM levels, indicating that CaM concentrations may limit the rate of cell cycle progression under normal conditions of growth (Prostko et al 1997, Rasmussen and Means 1989). Furthermore, CaM is necessary for reentry of quiescent cells into the cell cycle, and also for traversing the G1/S, G2/M and metaphase/anaphase boundaries (Ku and Means 1993, Means 1994).

Since MEL acts as a CaM antagonist it is possible that the oncostatic effect of the hormone may be attributable at least in part to the MEL antagonism of CaM. CaM antagonists such as phenotiazines, naphthalene sulfonamides, and CGS9343B block cell cycle progression in vitro both at the G1/S boundary and during G2/M (Chafouleas et al. 1982, Eilam and Chernichovsky 1988, Sasaki and Hidaka 1982). Moreover, both CaM antagonists and peptides can block the activation of CaM kinase III and the phosphorylation of elongation factor 2 in proliferating cellular populations (Bagaglio et al. 1993, Bagaglio and Hait 1994). Also, this kinase is activated in response to mitogens (Palfrey et al. 1987) and in malignant gliomas (Bagaglio et al. 1993). Finally

CaM-dependent multiprotein kinase II is required for cell cycle progression and for the initiation of DNA synthesis (Rasmussen and Means 1989, Rasmussen and Rasmussen 1995).

Currently, we know that MEL delays the progression from G_1/G_0 to S phase (Cos et al. 1991), increases the duration of the cell cycle (Cos et al. 1996a), and inhibits DNA synthesis in MCF-7 cells (Cos et al. 1996). It is also known that in MDCK and N1E-115 cells the antiproliferative effect of MEL may be related to diminished CaM levels. Long exposure (5 days) of MDCK or N1E-115 cultured cells to MEL results in a 40% decrease in total CaM cell levels (Benítez-King et al. 1991) and an inhibition of cell proliferation (Benítez-King et al. 1994) by 50% at the stationary phase. Pulse-chase experiments with ^{35}S-methionine showed that the half-life of ^{35}S-CaM dropped from 19 h to 16 h in MDCK cells cultured with 1 nM MEL (Benítez-King 1994), and no changes in CaM mRNA levels were detected (Ortega et al. 1993). Thus, MEL binding to CaM (Benítez-King et al. 1993) may increase the rate of CaM degradation. Besides the decreased CaM levels observed in MDCK cells and N1E-115 cells cultured with MEL, a CaM translocation from the cytosol to the cytoskeletal membrane fraction occurs (Antón-Tay et al. 1998) not only in a cell line that conserves characteristics of natural epithelia (MDCK), but also in a tumoral derived N1E-115 cell line. This effect may represent an additional mechanism by which the hormone modulates cellular responses. MEL effects on both CaM levels and subcellular distribution, together with the CaM antagonist effect of MEL in inhibiting CaM-dependent multiprotein kinase II (30%) activity and autophosphorylation (Benítez-King et al. 1996), may block reentry of cells into the cell cycle and mitosis.

Nonspecific binding of MEL to tubulin at pharmacological concentrations has been reported (Huerto-Delgadillo et al. 1994) and contributes to tubulin disruption, as has already been described in cytoskeletons in situ and in kinetic studies of tubulin polymerization. Cytoskeletons in situ contain functional microtubules and endogenous CaM, and lack a plasma membrane. Immunofluorescence microscopy studies revealed that MEL at 10 μM inhibits microtubule polymerization and causes disruption of the microtubule network (Huerto-Delgadillo et al. 1994). This result was confirmed following microtubule polymerization in vitro by turbidimetry: MEL at 10 μM caused an inhibition of tubulin polymerization by 75%. These results suggest that the hormone may inhibit mitotic spindle formation and movements at the kinetochore region (Mitchison et al. 1986). Taken overall, the data suggest that MEL, like some pharmacological antagonists of CaM, inhibits cell growth and has cytotoxic effects in malignant cells (Hait and DeRosa 1988).

25.3 Protein Kinase C Involvement in Cell Proliferation: Effects of Melatonin

PKC was characterized by Nishizuka in 1977 (Takai et al. 1977). As of this date 11 isoenzymes have been cloned, namely the Ca^{2+}-dependent isoenzymes α, βI, βII, and γ, the Ca^{2+}-independent δ, ε, η, θ, and μ, and the atypical isoenzymes ζ and λ (Nishizuka 1995). Long-term regulation of PKC isoenzymes seems important for the regulation of cell proliferation (Blobe et al. 1994). However, it is difficult to define the precise role of PKC with respect to cell growth on the basis of the existing data in the literature. Initially it was described that PKC is the major intracellular receptor for the phorbol esters (Niedel et al. 1983). These compounds activate PKC for prolonged periods

(Castagna et al. 1982), and this persistent stimulation of the kinase has been proposed to be the mechanism for the tumor-promoting action. However, the action of phorbol esters on PKC is followed by enzyme translocation to the membrane and its subsequent down-regulation (Zalewski et al. 1983). Thus, it is not clear whether tumor promotion is the result of persistent PKC activation or its down-regulation.

It is known that cytoskeletal proteins involved in cell division are phosphorylated by PKC. Nuclear intermediate filament protein phosphorylation participates in depolymerization and disintegration of the nuclear laminar meshwork (Nigg 1992). The nuclear envelope breakdown is a necessary step in cell division. Lamin B, an intermediate filament protein of the nuclear lamina, has been described to be phosphorylated by PKC (Hocevar et al. 1993). Moreover, microtubule-associated proteins also are PKC substrates and their phosphorylation may be required for microtubule changes needed in the formation of the mitotic spindles and asters (Hoshi et al. 1988). In addition, PKC phosphorylates both myosin (Murakami et al. 1994) and the actin binding proteins, which in turn modify microfilament polymerization (He et al. 1997). These processes are involved in the assembly of the contractile ring during cytokinesis, the last step in cell division (Cameron et al. 1987, Knecht and Loomis 1987).

Besides their inhibitory actions on CaM, most of the CaM antagonists inhibit PKC activity (Schatzman et al. 1983). By contrast MEL activates the Ca^{2+}-dependent PKC isoforms (Benítez-King and Antón-Tay 1996, 1997). Recently, it has been described that MEL directly activates purified rat brain PKC in vitro with a half-stimulatory concentration of 1 nM (Antón-Tay et al. 1998). In addition, the hormone augmented by 30% phorbol ester-stimulated PKC activity and increased the ^{3}H-phorbol ester binding to the kinase in the presence of Ca^{2+} (Antón-Tay et al. 1998). Since it is known that overexpression of specific PKC isoforms in different tissues results in increased cell growth (Reynolds et al. 1997), stimulation of PKC by MEL can be seen as standing in contradiction to the latter's antiproliferative effect. However, growth of N1E-115 cells is inhibited by prolonged exposure (4 days) to MEL (Benítez-King et al. 1994), and the hormone activates PKC in these cells and induces translocation of the Ca^{2+}-dependent PKC α from the cytosol to the membrane cytoskeletal fraction after 30 min of incubation (Benítez-King and Antón-Tay 1997). Unpublished data from our laboratory show that in both MDCK and N1E-115 cells, PKC is down-regulated after 12-h exposures to 1 nM of MEL. Since MEL stimulates PKC, which in turn phosphorylates cytoskeletal proteins involved in cell division, and PKC is down-regulated by the hormone, it is likely that the PKC interaction with MEL contributes to the antiproliferative effect of the hormone.

Finally alterations in intermediate filament structure have been reported in tumoral cells. Vimentin intermediate filaments are involved in cell adhesion and migration of metastatic cells. Co-expression of vimentin keratins 8 and 18, arranged in long wavelike cytoplasmic intermediate filaments, is related to the cell migratory and invasive ability of tumor epithelial cells (Chu et al. 1996). In this regard, MEL causes vimentin intermediate filament retraction from peripheral cytoplasmic regions toward the nucleus to form a perinuclear cap in N1E-115 cells incubated from 30 min to 12 h with 1 nM of the hormone (Benítez-King and Antón-Tay 1998). A similar effect is caused by phorbol esters. Moreover, the hormone increases vimentin phosphorylation in this cell line twofold (Benítez-King and Antón-Tay 1996). These results suggest that MEL may contribute to a reduction of metastatic activity by interfering with the dissemination and adhesion of circulating tumoral cells to new tissues.

25.4 Concluding Remarks

The mechanisms that underlie the antiproliferative effect of MEL are unclear and it seems that several MEL actions are involved. The data reviewed here illustrate that the modulation of Ca^{2+} signaling by the interaction of MEL with CaM and PKC may significantly participate in the oncostatic effects of the hormone. Thus, further study of this mechanism is an avenue through which to obtain a better understanding of the role of the hormone on cancer.

Acknowledgements. This work was partialy supported by CONACYT, México grants No 4200–5-1412PN and No 3220P-N9607.

References

Antón-Tay F, Huerto-Delgadillo L, Ortega-Corona BG, Benítez-King G (1993) Melatonin antagonism to calmodulin may modulate multiple cellular functions. In : Touitou Y, Arendt J, Pèvet P (eds) Melatonin and the Pineal Gland From Basic Science to Clinical Application. Elsevier, Amsterdam, pp 41–46

Antón-Tay F, Ramírez G, Martínez I, Benítez-King G (1998) In vitro stimulation of protein kinase C by melatonin. Neurochem Res 23:605–610

Antón-Tay F, Martínez I, Tovar R, Benítez-King G (1998) Modulation of the subcellular distribution of calmodulin by melatonin in MDCK cells. J Pineal Res 24:35–42

Bagaglio DM, Hait WN (1994) Role of calmodulin-dependent phosphorylation of elongation factor 2 in the proliferation of rat glial cells. Cell. Growth Differ 5:1403–1408

Bagaglio DM, Cheng EHC, Gorelick FS, Mitsui K, Nairn AC, Hait WN (1993) Phosphorylation of elongation factor-2 in normal and malignant rat glial cells. Cancer Res 53:2260–2264

Benítez-King G, Antón-Tay F (1993) Calmodulin mediates melatonin cytoskeletal effects. Experientia 49:635–641

Benítez-King G, Antón-Tay F (1996) Role of melatonin in cytoskeletal remodeling is mediated by calmodulin and protein kinase C. Front Horm Res 21:6344–6352

Benítez-King G, Antón-Tay F (1997) Calmodulin and protein kinase C alpha are two Ca^{++}-binding proteins that mediate intracellular melatonin signaling. In: Webb SM, Puig-Domingo M, Moller M, Pèvet P (eds) Pineal Update From Molecular Mechanisms to Clinical Implications. PJD Publications Limited, New York, pp 13–20

Benítez-King G, Antón-Tay F (1998) Immunofluorescence characterization of vimentin intermediate filament redistribution elicited by melatonin : A possible involvement of protein kinase C. (Submitted)

Benítez-King G, Huerto-Delgadillo L, Antón-Tay F (1991) Melatonin modifies calmodulin cell levels in MDCK and N1E-115 cell lines and inhibits phosphodiesterase activity in vitro. Brain Res 557:289–292

Benìtez-King G, Huerto-Delgadillo L, Antón-Tay F (1993) Binding of 3H-melatonin to calmodulin. Life Sci 53:201–207

Benítez-King G, Huerto-Delgadillo L, Sámano-Coronel L, Antón-Tay F (1994) Melatonin effects on cell growth and calmodulin synthesis in MDCK and N1E-115 cells. Adv Pineal Res 7:57–62

Benítez-King G, Ríos A, Martínez A, Antón-Tay F (1996) In vitro inhibition of Ca2+/calmodulin-dependent protein kinase II activity by melatonin. Biochim Biophys Acta 1290:191–196

Blask DE (1984) The pineal an oncostatic gland ? In: Reiter RJ (ed) The Pineal Gland. Raven Press, New York, pp 253–284

Blask DE, Lemus-Wilson AM, Wilson ST, Cos S (1993) Neurohormonal modulation of cancer growth by pineal melatonin. In: Touitou Y, Arendt J, Pèvet P (eds) Melatonin and the Pineal Gland From Basic Science to Clinical Application. Elsevier, Amsterdam, pp 303–310

Blask DE, Wilson ST, Zalatan F (1997) Physiological melatonin inhibition of human breast cancer cell growth *in vitro* : Evidence for glutathione-mediated pathway. Cancer Res 57:1909–1914

Blobe GC, Obeid LM, Hannun YA (1994) Regulation of protein kinase C and role in cancer biology. Cancer Metastasis Rev 13:411–431

Cameron IL, Cook KR, Edwards D, Fullerton GD, Schatten G, Schatten H, Zimmerman AM, Zimmerman S (1987) Cell cycle changes in water properties in sea urchin eggs. J Cell Physiol 133:14–24

Castagna M, Takai Y, Kaibuchi K, Sano K, Kikkawa U, Nishizuka Y (1982) Direct activation of calcium-activated, phospholipid-dependent protein kinase by tumor promoting phorbol-esters. J Biol Chem 257:7847–7851

Chafouleas JG, Bolton WE, Hidaka H, Boyd AE, Means AR (1982) Calmodulin and the cell cycle: involvement in regulation of cell cycle progression. Cell 28:41–50

Chu YW, Seftor EA, Romer LH, Hendrix MJC (1996) Experimental coexpression of vimentin and keratin intermediate filaments in human melanoma cells augments motility. Am J Pathol 148:63–69

Conti A, Maestroni GJ (1995) The clinical neuroimmunotherapeutic role of melatonin in oncology. J Pineal Res 19:103–110

Cos S, Blask DE, Lemus-Wilson A, Hill AB (1991) Effects of melatonin on the cell cycle kinetics and estrogen-rescue of MCF-7 human breast cancer cells in culture. J Pineal Res 10:36–42

Cos S, Fernandez F, Sánchez-Barceló (1996) Melatonin inhibits DNA synthesis in MCF-7 human breast cancer cells in vitro. Life Sci 58:2447–2453

Cos S, Recio J, Sánchez-Barceló EJ (1996a) Modulation of the length of the cell cycle time of MCF-7 human breast cancer cells by melatonin. Life Sci 58:811–816

Eilam Y, Chernichovsky D (1988) Low concentrations of trifluoperazine arrest the cell division cycle of Saccharomyces cerevisae at two specific stages. J Gen Microbiol 134:1063–1069

Hait WN, DeRosa WT (1988) Calmodulin as a target for new chemotherapeutic strategies. Cancer Invest 6:499–511

He Q, Dent EW, Meiri KF (1997) Modulation of actin filament behavior by GAP-43 (Neuromodulin) is dependent on the phosphorylation status of serine 41, the protein kinase C site. J Neurosci 17:3515–3524

Hinrichsen RD (1993) Calcium and calmodulin in the control of cellular behaviour and motility. Biochim Biophys Acta 1155:277–293

Hocevar BA, Burns DJ, Fields AP (1993) Identification of protein kinase C (PKC) phosphorylation sites on human lamin B. Potential role of PKC in nuclear lamina structural dynamics. J Biol Chem 268:7545–7552

Hoshi M, Akiyama T, Shinohara Y, Miyata Y, Ogawara H, Nishida E, Sakai H (1988) Protein kinase C catalyzed phosphorylation of the microtubule-binding domain of microtubule associated protein 2 inhibits its ability to induce tubulin polymerization. Eur J Biochem 174:225–230

Huerto-Delgadillo L, Antón-Tay F, Benítez-King G (1994) Effects of melatonin on microtubule assembly depend on hormone concentration: Role of melatonin as a calmodulin antagonist. J Pineal Res 17:55–62

Klee CB (1991) Concerted regulation of protein phosphorylation and dephosphorylation by calmodulin. Neurochem Res 16:1059–1065

Knecht DA, Loomis WF (1987) Antisense RNA inactivation of myosin heavy chain gene expression in Dictyostelium discoideum. Science 236:1081–1086

Lissoni P, Barni S, Tancini G, Rovelli F, Ardizzoia A, Conti A, Maestroni GJM (1993) A study of the mechanisms involved in the immunostimulatory action of the pineal hormone in cancer patients. Oncology 50:399–402

Lissoni P, Paolorossi F, Tancini G, Ardizzoia A, Barni S, Brivio F, Maestroni GJ, Chilelli M (1996) A phase II study of tamoxifen plus melatonin in metastatic solid tumour patients. Br J Cancer 74:1466–1468

Lissoni P, Paolorossi F, Ardizzoioa A, Barni S, Chilelli M, Mancuso M, Tancini G, Conti A, Maestroni GJ (1997) A randomized study of chemotherapy with cisplatin plus etoposide versus chemoendocrine therapy with cisplatin, etoposide and the pineal hormone melatonin as a first-line treatment of advanced non-small cell lung cancer patients in a poor clinical state. J Pineal Res 23:15–19

Lu KP, Means AR (1993) Regulation of the cell cycle by calcium and calmodulin. Endocr Rev 14:40–58

Means AR (1994) Calcium, calmodulin, and cell cycle regulation. FEBS Lett 347:1–4

Mitchison T, Evans L, Schulze E et al. (1986) Sites of microtubule assembly and disassembly in the mitotic spindle. Cell 45:515–527

Mumby MC, Walter G (1993) Protein serine/threonine phosphatases: Structure, regulation, and functions in cell growth. Physiol Rev 73:673–699

Murakami N, Elzinga M, Singh SS, Chauhan VP (1994) Direct binding of myosin II to phospholipid vesicles via tail regions and phosphorylation of the heavy chains by protein kinase C. J Biol Chem 269:16082–16090

Niedel JE, Kuhn LJ, Vandenbark GR (1983) Phorbol diester receptor copurifies with protein kinase C. Proc Natl Acad Sci USA 80:36–40

Nigg E (1992) Assembly-disassembly of the nuclear lamina. Curr Opin Cell Biol 5:187–193

Niki I, Yokokura H, Sudo T, Kato M, Hidaka H (1996) Ca2+ signaling and intracellular Ca2+ binding proteins. J Biochem-Tokyo 120:685–698

Nishizuka Y (1995) Protein kinase C and lipid signaling for sustained cellular responses. FASEB 9:484–496

Ortega A, López I, Benítez-King G, Antón-Tay F (1993) Melatonin increases mRNA levels in MDCK and N1E-115 cells. 1st Locarno International Meeting on Neuroendocrinoimmunology. The Pineal Gland in Relation with the Immune System and Cancer. Abstr, p 65

Palfrey HC, Nairn AC, Muldoon LL, Villareal ML (1987) Rapid activation of calmodulin-dependent protein kinase III in mitogen-stimulated human fibroblasts. J Biol Chem 262:9785–9792

Poenie M, Alderton J, Tsien RY, Steinhardt RA (1985) Changes of free calcium levels with stages of the cell division cycle. Nature 315:147–149

Prostko CR, Zhang C, Hait N (1997) The effects of altered cellular calmodulin expression on the growth and viability of C6 glioblastoma cells. Oncol Res 9:13–17

Rasmussen CD, Means AR (1989) Calmodulin is required for cell-cycle progression during G1 and mitosis. EMBO 8:73–82

Rasmussen G, Rasmussen C (1995) Calmodulin-dependent protein kinase II is required for G1/S progression in HeLa cells. Biochem Cell Biol 73:201–207

Reynolds NJ, Todd C, Angus B (1997) Overexpression of protein kinase C-alpha and -beta isoenzymes by somal dendritic cells in basal and squamous cell carcinoma. British J Dermatol 136:666–673

Subramanian A, Kothari L (1991) Melatonin, a supressor of spontaneus murine mammary tumors. J Pineal Res 10:136–140

Sasaki Y, Hidaka H (1982) Calmodulin and cell proliferation. Biochem Biophys Res Commun 104:451–456

Schatzman RC, Raynor RL, Kuo JF (1983) N-(6-aminohexyl)-5-chloro-1-naphtalensulfonamide (W-7) a calmodulin antagonist, also inhibits phospholipid-sensitive Ca2+ -dependent protein kinase. Biochim Biophys Acta 755:144–147

Takai Y, Kishimoto A, Inoue M, Nishizuka Y (1977) Studies on a cyclic nuleotide-independent protein kinase and its proenzyme in mammalian tissues: I. purification and characterization of an active enzyme from bovine cerebellum. J Biol Chem 252:7603–7609

Takuwa N, Zhou W, Takuwa Y (1995) Calcium calmodulin and cell cycle progression. Cell Signal 7:93–104

Whitaker M, Patel R (1990) Calcium and cell cycle control. Development 108:525–542

Zalewski PD, Forbes IJ, Giannakis C, Cowled PA, Betts WH (1983) Synergy between zinc and phorbol ester in translocation of PKC to cytoskeleton. FEBS lett 273:131–134

Section V
Oncotherapeutic Potential of Melatonin

26 Efficacy of Melatonin in the Immunotherapy of Cancer Using Interleukin-2

Paolo Lissoni

Abstract

Recent experimental and clinical observations have suggested that host antitumor immunobiological resistance is generated by interactions between cytokines and immunomodulating hormones. In particular, it has been shown that the pineal hormone melatonin (MLT), whose oncostatic and immunomodulating activities have been well documented, may enhance the in vivo antitumor efficacy of interleukin-2 (IL-2). This study reports the results obtained up to now in untreatable advanced solid tumor patients, treated by neuroimmunotherapy with low-dose IL-2 plus MLT. The study included 342 advanced solid tumor patients with a life expectancy of less than 6 months who had not responded to previous chemotherapy or who were affected by neoplasms for which no effective standard therapy is available. Non-small cell lung cancer and gastrointestinal tract tumors were the neoplasms most frequently detected in our patients. IL-2 was given subcutaneously at 3 million IU/day in the evening for 6 days/week for 4 weeks, corresponding to one cycle. MLT was given orally at 40 mg/day in the evening, every day starting 7 days prior to IL-2. In nonprogressing patients, a second cycle was repeated after a 21-day rest period; then, patients underwent a maintenance treatment consisting of 1 week of therapy every month until progression. Patients were considered as evaluable when they received at least one cycle. Of the 342 patients, 320 were evaluable. Objective tumor regressions were achieved in 53/320 (17%) patients, consisting of four complete responses and 49 partial responses. The median duration of response was 9 months. Most responses were observed in lung cancer, hepatocarcinoma, pancreatic cancer, gastric cancer, and endocrine tumors. A survival longer than 1 year was achieved in 128/320 (40%) patients, and the percentage rate of survival at 1 year was significantly higher in the responders than in the nonresponders. Toxicity was low in all patients, fever being the only relevant side-effect. This pilot clinical study shows that the concomitant administration of the pineal hormone MLT may make IL-2 effective in advanced solid tumors, which generally do not respond to IL-2 alone, and suggests that the application of psychoneuroimmune knowledge to IL-2 cancer immunotherapy may represent a promising new strategy to enhance the clinical efficacy of cancer biotherapies with cytokines.

26.1 Introduction

The main problem of interleukin-2 (IL-2) cancer immunotherapy is harnessing its antitumor action in the face of the great number of immunostimulatory and immuno-suppressive effects concomitantly induced by its injection (Rosenberg 1992, Lissoni et al. 1991). The biological changes in the immune parameters occurring with IL-2 do not represent only simple epiphenomena; rather they also reflect the mechanisms respons-ible for the efficacy of or resistance to IL-2 and therefore have a prognostic signifi-cance. Even though there are controversial results, the evidence of high blood levels of interleukin-6 (IL-6) (Tartour et al. 1994), neopterin (Lissoni 1995) and soluble IL-2 receptor (sIL-2R) (Lissoni et al. 1992) would seem to correlate with resistance to IL-2. If we take into consideration that IL-6 and neopterin are mainly produced by macro-phages, and that sIL-2R is released by T lymphocytes under a macrophage stimulatory control (Lissoni et al. 1992 a), macrophages would appear to constitute the key cells influencing IL-2 activity.

The mechanisms responsible for resistance to IL-2 may depend on both tumor characteristics, particularly tumor extension, sites of disease, and sensitivity of tumor histotype (Whittington and Faulds 1993), and host-related factors (Atzpodien and Kirchner 1990). At present, two main biological systems have been proven to mediate the immunosuppression occurring with IL-2 immunotherapy concomitantly by the generation of cytotoxic lymphocytes, consisting of macrophages and T helper-type 2 lymphocytes (TH2). Macrophages may induce immunosuppression through the release of several substances capable of inhibiting IL-2 activity, such as prostaglandin E_2 (Chouaib and Fradelizi 1982), IL-6 (Matsuda and Hirano 1990) and transforming growth factor-ß (TGF-ß) (Alleva et al. 1994); macrophage-mediated suppressive events would be reflected by an increased neopterin secretion (Lissoni 1995). On the other hand, TH2-related immunosuppression would depend on the release of interleukin-10 (IL-10), which has appeared to block IL-2 activity (Moore et al. 1993). Because of their association with immunobiological events, the possibility of manipulating the changes in biological parameters occurring with IL-2 could influence the efficacy of IL-2 itself. From this point of view, recent experimental and clinical observations (Plotnikoff and Miller 1983, Panerei 1993, Jankovic 1994) have suggested that the activity of cytokines may be regulated not only by other cytokines, but also by immunomodulating neuro-hormones and neuropeptides, which would represent the physiological mediators of the psychoneuroendocrine influence on immune functions. Therefore, the concomi-tant administration of cytokines and neurohormones as a neuroimmunotherapeutic strategy could constitute a new and interesting way to modulate the biological response induced by IL-2.

Within the neuroendocrine system involved in the control of cytokine activity, the pineal gland has appeared to play an essential role through the circadian release of its main hormone, melatonin (MLT) (Aubert et al. 1980, Regelson and Pierpaoli 1987, Maestroni et al. 1988). In normal subjects, MLT is mainly produced during the dark period of the day, and thus exhibits a light/dark daily rhythm, with low levels during the day and high concentrations during the night (Iguchi et al. 1982). Cancer patients, particularly those with metastatic disease, are often characterized by a diminished night increase in MLT levels (Lissoni et al. 1986). In animals, pinealectomy has been proven to stimulate cancer growth (Aubert et al. 1980, Regelson and Pierpaoli 1987, Maestroni et al. 1988) and to inhibit IL-2 secretion and activity (Del Gobbo et al. 1989).

Therefore, because of its potential importance in influencing IL-2 activity, host behavior of MLT secretion, either before or on IL-2 therapy, may have a prognostic significance. In addition to the status of the pineal gland before therapy, it has to be taken into consideration that IL-2 itself may influence MLT secretion. The effect of IL-2 on MLT release would depend on dosage and schedule of injection. The 24-h continuous intravenous infusion of IL-2 has appeared to suppress nocturnal MLT increase (Lissoni et al. 1990), whereas low-dose subcutaneous IL-2 has been shown to re-establish a normal light/dark rhythm of MLT in advanced cancer patients, with an apparent positive correlation with the clinical efficacy of the immunotherapy (Viviani et al. 1992).

Several studies have been performed in an attempt to define the immunomodulating effects of MLT in association with IL-2. In vivo, the concomitant administration of MLT has appeared to inhibit macrophage production of neopterin in response to IL-2 (Lissoni et al. 1991a). Since MLT has no direct effect on neopterin release, in vitro it could act by amplifying the inhibitory effect of interleukin-3 (IL-3) on IL-2-induced neopterin release (Lissoni et al. 1993b). In contrast, MLT may directly inhibit the in vitro production of IL-10 by TH2 lymphocytes activated by IL-2 (Lissoni et al. 1997). Therefore, the rationale for the concomitant administration of MLT during IL-2 immunotherapy is mainly related to the ability of the pineal hormone to counteract macrophage- and TH2-mediated suppressive events, which represent the two most investigated mechanisms of cancer-related immunosuppression. In addition, our previous clinical studies have suggested that MLT may amplify IL-2 antitumor and immunostimulatory effects, with a subsequent increase in the number of solid tumor histotypes which may respond to IL-2 and a consequent reduction in the IL-2 dosage required to activate the immune system (Lissoni et al. 1993). The present study describes the clinical results obtained to date in advanced solid tumor patients treated by neuroimmunotherapy with low-dose IL-2 plus MLT.

26.2 Materials and Methods

The study included 342 untreatable advanced solid tumor patients with a life expectancy of less than 6 months. Eligibility criteria were as follows: histologically proven locally advanced or metastatic solid neoplasms other than renal cell carcinoma and melanoma; measurable lesions; lack of response to previous chemotherapy, or presence of tumors for which no effective standard therapy is available or of clinical conditions which make patients unable to tolerate conventional chemotherapy [old age, low performance status (PS) or important medical illnesses other than cancer]; absence of second tumors; no chronic concomitant steroid therapy; and a PS of at least 20% according to Karnofsky's score. Patients affected by renal cell carcinoma or melanoma were also included in the study when they were unable to tolerate high-dose IL-2 therapy as evidenced by their poor clinical condition. The experimental protocol was explained to each patient, and informed consent was obtained.

Recombinant human IL-2 and MLT were supplied by Chiron (Amsterdam, Holland) and by Helsinn Chemicals (Breganzona, Switzerland), respectively. IL-2 was given subcutaneously at 3 million IU/day for 6 days/week for 4 consecutive weeks, corresponding to one immunotherapeutic cycle. MLT was given orally at 40 mg/day every day, starting 7 days prior to IL-2 as an induction phase. Both drugs were given in the evening because of their documented greater biological activity in this period of the day (Aubert et al. 1980, Maestroni et al. 1988 Lissoni et al. 1991a, Lissoni et al. 1993). Moreover, drug dosage was established according to our previous clinical data (Lissoni et al. 1991a, 1993). Finally, we decided to inject IL-2 subcutaneously because of its similar efficacy and lower toxicity as compared with intravenous injection (Stein et al. 1991, Lissoni et al. 1993a). In patients in whom disease progression was not observed, a second cycle of therapy was repeated after a 21-day rest period; patients then underwent maintenance therapy consisting of 1 week of treatment every month until disease progression or toxicity. Patients were considered evaluable when they had received at least one complete immunotherapeutic cycle.

Clinical response and toxicity were evaluated according to WHO criteria. Complete response (CR) was defined as the complete regression of all neoplastic lesions for at least 1 month, partial response (PR) as a reduction by greater than 50% in the product of the two longest perpendicular diameters of all lesions for at least 1 month, stable disease (SD) as no increase or a decrease of greater than 25% in tumor sizes, and progressive disease (PD) as an increase of greater than 25% in tumor sizes or the appearance of new lesions. The duration of response and survival were calculated from the onset of the immunotherapy. Radiological examinations, including CT scan, were repeated after each cycle of immunotherapy, then every 2 months. Routine laboratory tests were repeated at weekly intervals during the study. All patients underwent immunotherapy at home, after 1–2 days of hospitalization to evaluate the subjective reactivity. Minimum and median follow-up were 12 months and 31 months, respectively.

Results were statistically evaluated by the chi-square test, Student's t test, analysis of variance and the log-rank test, as appropriate.

26.3 Results

Of the 342 patients, 320 (93%) were evaluable. The characteristics of evaluable patients are reported in Table 26.1. CR was obtained in four patients (1%): two were affected by locally advanced hepatocarcinoma, the third by locally advanced cancer of pancreas, and the last one by liver metastases due to gastric cancer. PR was achieved in another 49 patients. Therefore, the overall response rate was 53/320 (17%). The median duration of response was 9 months (range 2–25 months). Most responses were observed in patients affected by non-small cell lung cancer, hepatocarcinoma, pancreatic adenocarcinoma, gastric cancer, and endocrine tumors. Response rate was significantly higher in untreated patients than in those previously treated by chemotherapy (37/186 vs 16/134, $P < 0.05$). Moreover, response rate was significantly higher in patients with a PS greater than 40% than in those with a PS of 40% or less (46/236 vs 7/84, $P < 0.01$). The clinical results in relation to tumor histotype are reported in Table 26.2, while Table 26.3 shows the clinical response found in relation to the sites of disease. Response rate was significantly higher in patients with locally advanced disease than in those with metastatic neoplasms (16/59 vs 37/261, $P < 0.05$). As far as dominant, metastatic sites are concerned, liver and brain were the less responsive sites of disease.

Stable disease was achieved in 149/320 (46%) patients, with a median duration of 6 months (2–15 months), whereas the remaining 118 (37%) patients had rapid disease progression in response to the first immunotherapeutic cycle.

Survival longer than 1 year was achieved in 128/320 (40%) patients, and the percentage survival rate at 1 year was significantly higher in patients with nonmetastatic

Table 26.1. Characteristics of 320 evaluable untreatable advanced solid tumor patients treated by subcutaneous low-dose IL-2 plus melatonin

Characteristics	No.
M/F 183/137	
Median age (years)	57(17–81)
Median performance status (Karnofsky score)	70(20–100)
Tumor histotypes	
Non-small cell lung cancer	71
Small cell lung cancer	4
Pancreatic adenocarcinoma	38
Hepatocarcinoma	32
Colorectal carcinoma	31
Gastric adenocarcinoma	26
Breast cancer	15
Endocrine tumors	14
Pleural mesothelioma	14
Gynecologic tumors	14
Soft tissue sarcomas	13
Renal cell carcinoma	12
Melanoma	10
Biliary tract carcinomas	8
Prostate cancer	8
Head and neck cancers	4
Unknown primary tumors	3
Bone tumors	3
Sites of disease	
Locally limited disease	59
Metastatic disease	261
Soft tissues	24
Bone	27
Lung	58
Liver	96
Lung + liver	17
Brain	8
Serouses	31
Previous chemotherapy	134/320

disease than in those with metastatic disease (34/59 vs 94/261, $P < 0.01$). Moreover, the percentage of patients surviving at 1 year was significantly higher among responderss than among those with SD or PD (43/53 vs 85/267, $P < 0.001$).In addition, the 1-year survival rate observed in patients with SD was significantly higher than that in patients with PD (66/149 vs 19/118, $P < 0.05$). Finally, as illustrated in Fig. 26.1, the survival curve (evaluated according to the Kaplan-Meier method) obtained in responders was significantly longer than that found in patients with SD or PD.

Neuroimmunotherapy was well tolerated in most patients, and the only clearly undesirable side-effect was fever greater than 38 °C in 22/320 (7 %) patients, which was generally limited to the first days of IL-2 injection. The main toxicities observed on immunotherapy are reported in Table 26.4.

As far as the biological response is concerned, the increases (mean ± SE, n/mm^3) in lymphocyte and eosinophil numbers were significantly higher in responders than in those with SD or PD (lymphocytes: 1956 ± 127 vs 758 ± 149, $P < 0.001$; eosinophils: 1189 ± 144 vs 567 ± 109, $P < 0.01$).

Table 26.2. Tumor response and 1-year survival in relation to tumor histotype in 320 evaluable untreated advanced solid tumor patients treated by subcutaneous low-dose IL-2 plus melatonin

Histotypes	No.	Tumor response[a]						1-Year survival	
		CR	PR	CR + PR	(%)	SD	PD	No.	(%)
Overall	320	4	49	53	(17%)	149	118	128	(40%)
Non-small cell lung cancer	71	0	13	13	(18%)	39	19	29	(40%)
Epidermoid cell	23	0	5	5	(22%)	12	6	12	(52%)
Adenocarcinoma	37	0	7	7	(19%)	21	9	17	(46%)
Large cell carcinoma	5	0	0	0	(–)	2	3	1	(20%)
Bronchioloalveolar carcinoma	6	0	1	1	(17%)	4	1	2	(33%)
Small cell lung cancer	4	0	0	0	(–)	1	3	1	(25%)
Pancreatic adenocarcinoma	38	1	3	4	(11%)	14	20	10	(26%)
Hepatocarcinoma	32	2	8	10	(31%)	15	7	18	(56%)
Colorectal carcinoma	31	0	4	4	(13%)	11	16	10	(32%)
Gastric carcinoma	26	1	5	6	(23%)	11	9	11	(42%)
Breast cancer	15	0	2	2	(13%)	5	8	4	(27%)
Endocrine tumors	14	0	4	4	(29%)	6	4	8	(57%)
Thyroid carcinoma	5	0	1[b]	1		3	3	3	
Neuroendocrine tumors	4	0	2	2		1	1	2	
Carcinoid tumor	2	0	1	1		1	0	2	
Adrenal carcinoma	2	0	0	0		2	0	0	
Gastrinoma	1	0	0	0		1	0	1	
Pleural mesothelioma	14	0	1	1	(7%)	9	4	5	(36%)
Gynecologic tumors	14	0	2	2	(14%)	6	6	5	(36%)
Ovarian carcinoma	12	0	2	2	(17%)	5	5	5	(42%)
Uterine cervix carcinoma	2	0	0	0	(–)	1	1	0	
Soft tisue sarcoma	13	0	1	1	(8%)	8	4	6	(46%)
Renal cell carcinoma	12	0	2	2	(17%)	8	2	8	(67%)
Melanoma	10	0	1	1	(10%)	4	5	3	(30%)

	No.	CR	PR	CR + PR	(%)	SD	PD	No.	(%)
Biliary tract carcinoma	8	0	1	1	(13%)	3	4	2	(25%)
Prostate cancer	8	0	1	1	(13%)	4	3	4	(50%)
Head and neck cancer	4	0	1	1	(25%)	2	1	3	(75%)
Unknown primary cancer	3	0	0	0		1	2	0	
Osteosarcoma	2	0	0	0		1	1	0	
Ewing's sarcoma	1	0	0	0		1	0	1	

[a] According to WHO criteria: CR, complete response; PR, partial response; SD, stable disease; PD, progressive disease.
[b] Anaplastic carcinoma.

Table 26.3. Tumor response and 1-year survival in relation to disease extension and sites of disease in 320 evaluable untreated advanced solid tumor patients treated by subcutaneous low-dose IL-2 plus melatonin

Sites of disease	No.	Tumor response[a]						1-Year survival	
		CR	PR	CR + PR	(%)	SD	PD	No.	(%)
Locally advanced disease	59	3	13	16	(27%)	32	11	34	(58%)
Metastatic disease	261	1	36	37	(14%)	117	107	94	(36%)
Soft tissue	24	0	4	4	(17%)	12	8	15	(62%)
Bone	27	0	5	5	(19%)	15	7	17	(63%)
Lung	58	0	14	14	(24%)	25	19	33	(57%)
Liver	96	1	7	8	(9%)	37	51	14	(15%)
Lung + liver	17	0	0	0	(–)	9	8	2	(12%)
Serouses	31	0	6	6	(19%)	15	10	11	(35%)
Brain	8	0	0	0	(–)	4	4	2	(25%)

[a] According to WHO criteria: CR, complete response; PR, partial response; SD, stable diseae; PD, progressive disease.

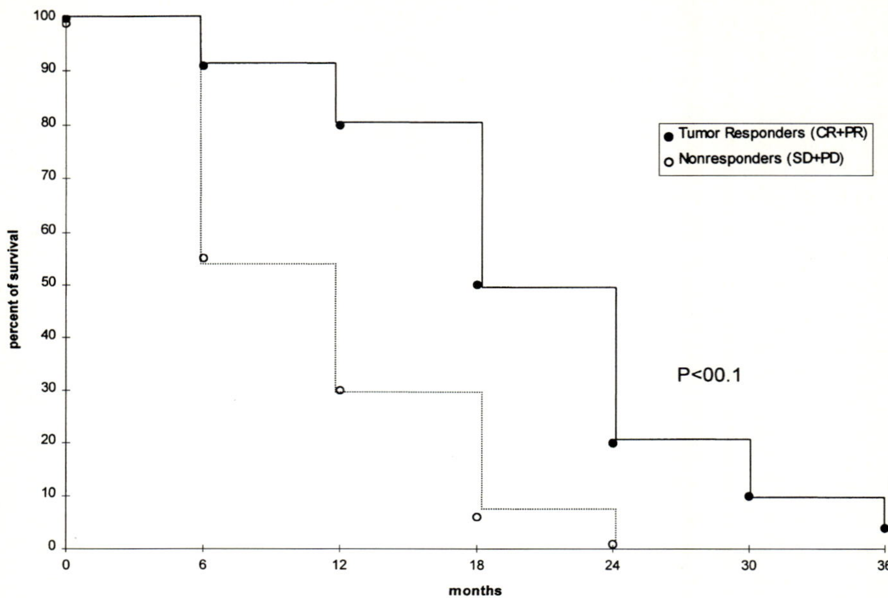

Fig. 26.1. Survival curves observed in patients with tumor response and in those with stable disease or progressive disease after neuroimmunotherapy with low-dose IL-2 plus melatonin

Table 26.4. Main toxicities observed in 320 evaluable untreatable advanced solid tumor patients treated by subcutaneous low-dose IL-2 plus melatonin

Toxicities	No.	(%)
Fever > 38 °C	22	(7%)
Asthenia	12	(4%)
Arthralgia/myalgia	5	(2%)
Rash	3	(1%)
Diarrhea	3	(1%)
Chills	5	(2%)
Cardiopulmonary complications	0	
Coagulatory disorders	0	
Hypotension grade 1–2	5	(2%)
Hyperglycemia	2	(1%)
Depressive symptoms	7	(2%)
Transaminase increase	39	(12%)
Skin nodules at injection sites	124	(39%)

26.4 Discussion

Most clinical studies have shown that renal cell carcinoma, malignant melanoma, and colorectal cancer are the only solid tumor histotypes which may respond to cancer immunotherapy with IL-2 alone (Dillman et al. 1991). Most other solid tumors, such as lung cancer, gastric cancer, hepatocarcinoma, pancreatic cancer, and breast cancer, do not seem to respond to IL-2 alone, at least in terms of objective tumor regression.

Several biological strategies have been evaluated in an attempt to enhance IL-2 efficacy in the treatment of advanced solid tumors, such as the concomitant administration of other cytokines, mainly interferon (Krigel et al. 1990, Redman et al. 1990) or tumor necrosis factor-alpha (TNF) (Schiller et al. 1995). However, despite the potential anti-tumor activity of both interferon and TNF, no clear synergistic action with IL-2 has been documented. A new possible biotherapeutic strategy could originate from the recent advances in psychoneuroimmunology, which have demonstrated that the in vivo activity of cytokines depends not only on immune status but also on the psycho-neuroendocrine regulation of cytokines themselves (Plotnikoff and Miller 1983, Panerai 1993, Jancovic 1994, Aubert et al. 1980).

The present study constitutes the first clinical application of the knowledge coming from the area of neuroimmunomodulation. It shows that concomitant administration of the pineal neuroimmunomodulating hormone MLT may amplify the in vivo anti-tumor efficacy of IL-2 and induce objective tumor regression in patients with solid neoplasms who generally do not respond to IL-2 alone, as well as in patients with advanced disease and a poor clinical condition. Therefore, neuroimmunotherapy with low-dose IL-2 plus MLT could represent a new, well-tolerated and effective second-line therapy in solid tumors, such as lung cancer, gastric cancer, colon cancer, and endo-crine tumors, which have not responded to previous chemotherapy. Moreover, it could be proposed as first-line therapy in the treatment of tumors for which no effective standard therapy is available, such as hepatocarcinoma and cancer of the pancreas. In contrast, some tumors, such as breast cancer, do not seem to be particularly responsive to this type of treatment, although this evidence could depend at least in part on the heavy previous chemotherapy, which has been proven to induce persistent immune damage. Finally, the role of neuroimmunotherapy in other tumor histotypes, such as biliary tract cancers, mesotheliomas, soft tissue sarcomas and head and neck carci-nomas, needs to be better established. However, it is possible to suggest that the replace-ment of a physiological psychoneuroimmune antitumor response by administering either IL-2 or MLT, whose production has appeared to be reduced in cancer (Lissoni et al. 1985, Wanebo et al. 1996), may have a positive prognostic impact in most tumor histotypes by modifying the natural course of the neoplastic disease. In fact, this study shows that tumor regression is associated with an increase in lymphocytes and eosino-phils, both involved in mediating cancer cell destruction (Atzpodien and Kirchner 1990, Rodgers et al. 1994).

The role of MLT in enhancing IL-2 activity requires further investigation. In addi-tion to its immunomodulating properties, mainly consisting in the neutralization of suppressive events, it must be emphasized that MLT may have a direct antipro-liferative and differentiating effect on cancer cells (Aubert et al. 1980, Regelson and Pierpaoli 1987, Maestroni et al. 1988), which could contribute to the clinical antitumor results.

Studies using other immunomodulating pineal substances, such as 5-methoxytrypto-phol (Sze et al. 1993), will allow further advances in the development of neuroimmune strategies, capable of modulating the in vivo effects of cytokines by acting on the neuroendocrine system. The aim of these future approaches is to treat cancer by reproducing the psychoimmune biochemistry characteristic of a healthy state through the exogenous administration of those neurohormones and cytokines whose endo-genous production has been proven to be compromised in patients with advanced tumors.

References

Alleva DG, Burger CJ, Elgert KD (1994) Tumor-induced regulation of suppressor macrophage nitric oxide and TNF-alpha production. Role of tumor-derived IL-10, TGF-beta and prostaglandin E2 J Immunol 153:1674–1686

Atzpodien J, Kirchner H (1990) Cancer, cytokines and cytotoxic cells: interleukin-2 in the immunotherapy of human neoplasms. Klin Wochenschr 68:1–11

Aubert C, Jauniaud P, Lecalvez J (1980) Effect of pinealectomy and melatonin on mammary tumor growth in Sprague-Dawley rats under different conditions of lighting. J Neural Transm 47:121–130

Chouaib S, Fradelizi D (1982) The mechanisms of inhibition of human IL-2 production. J Immunol 129:2463–2467

Del Gobbo V, Libri V, Villani R, Callo R, Nisticò G (1989) Pinealectomy inhibits interleukin-2 production and natural killer activity in mice. J Immunopharmacol 11:567–571

Dillman RO, Oldham RK, Tauer KW, Barth NM, Blumenschein G, Arnold J, Birch R, West WH (1991) Continous interleukin-2 and lymphokine-activated killer cells for advanced cancer: a national biotherapy study group trial. J Clin Oncol 9:1233–1240

Iguchi H, Kato KI, Ibayashi H (1982) Age-dependent reduction in serum melatonin concentrations in healthy human subjects. J Clin Endocrinol Metabol 55:27–29

Jancovic BD (1994) Neuroimmunomodulation. From phenomenology to molecular evidence. Ann N Y Acad Sci 741:1–38

Krigel RL, Padavic-Shaller KA, Rudolph AR, Konrad M, Bradley EC, Comis RL (1990) Renal cell carcinoma: treatments with recombinant interleukin-2 plus beta-interferon. J Clin Oncol 8:460–467

Lissoni P (1995) Importance of interleukin-2 in cancer therapy. Adv Cancer Ther 8:1–4

Lissoni P, Viviani S, Bajetta E, Buzzoni R, Barreca A. Mauri R, Resentini M, Morabito F, Esposti D, Esposti G, Fraschini F (1985) A clinical study of the pineal gland activity in oncologic patients. Cancer 57:837–842

Lissoni P, Barni S, Archili C, Cattaneo G, Rovelli F, Conti A, Maestroni GJM, Tancini G (1990) Endocrine effects of a 24-hours intravenous infusion of interleukin-2 in the immunotherapy of cancer. Anticancer Res 10:753–758

Lissoni P, Tisi E, Brivio F, Barni S, Rovelli F, Perego M, Tancini G (1991) Increase of soluble interleukin-2 receptor and neopterin serum levels during immunotherapy of cancer with interleukin-2. Eur J Cancer 27:1014–1016

Lissoni P, Tisi E, Brivio F, Ardizzoia A, Crispino S, Barni S, Tancini G, Conti A, Maestroni GJM (1991a) Modulation of interleukin-2 macrophage activation in cancer patients by the pineal hormone melatonin. J Biol Regul Homeost Agents 5:154–156

Lissoni P, Barni S, Ardizzoia A, Crispino S, Paolorossi F, Archili C, Vaghi M, Tancini G (1992) Second-line therapy with low-dose subcutaneous interleukin-2 alone in advanced renal cancer patients resistant to interferon-alpha. Eur J Cancer 28:92–96

Lissoni P, Ardizzoia A, Barni S, Rovelli F, Pittalis S, Perego M, Tisi E, Brivio F, Tancini G (1992a) Soluble interleukin-2 receptor as a marker of macrophage activation in cancer patients. Med Biol Env 20:209–213

Lissoni P, Barni S, Rovelli F, Brivio F, Ardizzoia A, Tancini G, Conti A, Maestroni GJM (1993) Neuro-immunotherapy of advanced solid neoplasms with single evening subcutaneous injection of low-dose interleukin-2 and melatonin: preliminary results. Eur J Cancer 29A:185–189

Lissoni P, Barni S, Ardizzoia A, Andres M, Scardino E, Cardellini P, Della Bitta R, Tancini G (1993a) A randomized study of low dose interleukin-2 subcutaneous immunotherapy versus interleukin-2 plus interferon-alpha as a first-line for metastatic renal cell carcinoma. Tumori 79:397–400

Lissoni P, Pittalis S, Brivio F, Tisi E, Rovelli F, Ardizzoia A, Barni S, Tancini G, Giudici G, Biondi A, Conti A, Maestroni GJM (1993b) In vitro modulatory effects of interleukin-2 on macrophage activation induced by interleukin-2. Cancer 71:2076–2081

Lissoni P, Pittalis S, Barni S, Rovelli F, Fumagalli L, Maestroni G (1997). Regulation of interleukin-2 -interleukin-12 interactions by the pineal gland. Int J Thymol 5:443–447

Maestroni GJM, Conti A, Pierpaoli W (1988) Pineal melatonin, its fundamental immunoregulatory role in aging and cancer. Ann N Y Acad Sci 521:140–145

Matsuda T, Hirano T (1990) Interleukin-6 (IL-6). Biotherapy 2:363–373

Moore KW, O'Garra A, De Waal-Malefyt R, Vieira P, Mosmann TR (1993) Interleukin-10. Ann Rev Immunol 11:165–190

Panerai AE (1993) Lymphocytes as a source of hormones and peptides. J Endocrinol Invest 16:549–557

Plotnikoff NP, Miller GC (1983) Enkephalins as immunomodulators. Int J Immunopharmacol 5:437–441

Redman BG, Flaherty L, Chou TH, Al-Katib A, Kraut M, Martino S, Chen B, Kaplan J, Valdivieso M (1990) A phase I trial of recombinant interleukin-2 combined with recombinant interferon-gamma in patients with cancer. J Clin Oncol 8:1269–1276

Regelson W, Pierpaoli W (1987) Melatonin: a rediscovered antitumor hormone? Cancer Invest 5: 379–385

Rodgers S, Rees RC, Hancock BW (1994) Changes in the phenotypic characteristics of eosinophils from patients receiving recombinant human interleukin-2 (rhIL-2) therapy. Br J Hematol 86: 746–753

Rosenberg SA (1992) The immunotherapy and gene therapy of cancer. J Clin Oncol 10:180–199

Schiller JH, Morgan-Ihrig C, Levitt ML (1995) Concomitant administration of interleukin-2 plus tumor necrosis factor in advanced non-small cell lung cancer. Am J Clin Oncol 18:47–51

Stein RC, Malkovska V, Morgan S, Galazka A, Aniszewski C; Roy SE, Shearer RJ, Marsden RA, Bevan D, Gordon-Smith EC, Coombs RC (1991) The clinical effects of prolonged treatment of patients with advanced cancer with low-dose subcutaneous interelukin-2. Br J Cancer 63:275–278

Sze SF, Ng TB, Liu WK (1993) Antiproliferative effect of pineal indoles on cultured tumor cell lines. J Pineal Res 14:27–33

Tartour E, Dorval T, Mosseri V, Deneux L, Mathiot C, Brailly H, Montero F, Joeux I, Pouillart P, Fridman WH (1994) Serum interleukin-6 and C-reactive protein levels correlate with resistance to IL-2 therapy and poor survival in melanoma patients. Br J Cancer 69:911–913

Viviani S, Bidoli P, Spinazzé S, Rovelli F, Lissoni P (1992) Normalizatio of the light/dark rhytm of melatonin after prolonged subcutaneous administration of interleukin-2 in advanced small cell lung cancer. J Pineal Res 12:114–117

Wanebo HJ, Pace R, Hargett S, Katz D, Sando J (1996) Production of and response to interleukin-2 in peripheral blood lymphocytes of cancer patients. Cancer 57:656–662

Whittington R, Faulds D (1993) Interleukin-2. Drugs 46:446–514

27 Melatonin Cancer Therapy

William J. M. Hrushesky

Abstract

Melatonin prevents and delays chemical carcinogenisis and cancer growth in vivo in the mouse and rat. Blind people have robust, albeit mostly free-running melatonin circadian rhythms and they are at lower risk for the development of many, if not all, cancers. Women with breast cancer and men with prostate cancer have diminished nighttime melatonin secretion.

Each proven melatonin biological effect – the transduction of circadian and circannual temporal information from the environmental light/dark schedule to each cell within the body, the induction and enhancement of sleep, the diminishment of core body temperature, and the resetting of circadian clocks – is entirely dependent upon the temporal context, i.e., the circadian/circannual stage of melatonin availability/administration. The host-cancer balance is likewise rhythmically coordinated in time by endogenous and exogenous circadian and circannual internal pacemakers and external zeitgebers. Each of the targets of cancer treatment is differentially available in both normal host cells and cancer cells at different times within these biological cycles. This chronobiology is faithfully reflected by the fact that the circadian/circannual timing of cancer chemotherapy determines to a medically meaningful extent the damage done to normal cells, the amount of drug that can be safely given, and the antitumor efficacy of that drug.

Giving a chronobiotic agent, like melatonin, without regard to its circadian/circannual scheduling is not logical and should not be expected to reveal its true utility as an anticancer agent. Despite the fact that the timing of melatonin has been either ignored or stipulated arbitrarily, clinical anticancer activity has been uncovered. Lissoni's work shows an anticancer effect in a variety of solid tumors and his work has been confirmed by others for melanoma and renal cell carcinoma. In many of these trials, meaningful clinical benefit, as well as antitumor activity, has been demonstrated.

It will be essential to the ultimate determination of melatonin's place in cancer therapy to determine the circadian schedule that optimizes its medical benefit. Such clinical studies must employ a double-blind, placebo-controlled comparison among times of day for melatonin treatment, alone and in combination with surgery, radiation, chemotherapy, and/or biological therapies. Measures of quality of life, fatigue, the latency timing, quality and quantity of daily sleep, and the timing and intensity of daily activity should be assessed, along with melatonin's anticancer activity, in each of these studies. Without chronobiologically adequate clinical studies, the use of melatonin will continue to depend upon belief and opinion rather than upon knowledge and understanding.

27.1 Introduction

Chronobiologists study the temporal organization of living organisms. They find that this time structure is nonlinear, rhythmic, and multifrequency. Prominent concurrent and interactive cycles in mammalian chronobiology include among others: about daily cycles, fertility (menstrual/estrous) cycles, and about yearly cycles. Chronobiologists interested in the prevention, diagnosis, and treatment of cancer find that each of these biological cycles affects most, if not all, relevant aspects of preventive, diagnostic, and therapeutic oncology. Cell proliferation and cell death, immune defense, new blood vessel formation, metabolism, and most if not all biological processes relevant to drug therapy, including drug absorption, distribution, catabolism, and excretion are non-trivially and reproducibly affected by each of these biological rhythms.

Pineal biologists have shown that the gland and its primary secretion, melatonin, are of central importance for telling each cell within the body what time of day and what time of year it is (Cassone and Natesan 1997). This *chronobiotic* or time-knowing/setting/keeping function is a primary molecular physiology of melatonin (Cassone 1990). Each of the physiologic effects of this chronobiotic hormone depends intimately, largely, and in many cases, bidirectionally upon the temporal context, when within each relevant cycle the melatonin signal is imparted. Considering the biological meaning of a melatonin concentration or considering the biological effects of an exogenous dose of melatonin without knowing when within the circadian, reproductive, and circannual cycles this concentration has been measured or dose has been given is nonsensical and necessarily misleading. Nonetheless, the vast majority of the melatonin/cancer literature fails to provide this temporal context. Thereby, profound recurring and continuous confusion accompanies much of the pineal/cancer literature (Panzer and Wiljoen 1997).

A purpose of this brief synthetic review of melatonin's possible place in cancer prevention, diagnosis, and treatment is to point out where and how the presence of proper temporal context will clarify the profound confusion accompanying much of the pineal–melatonin–cancer literature. This paper hopes to suggest practical ways to prevent additional confusion by stipulating the minimum essential chronobiological study design for the valid study of melatonin's value in medical oncology. First, it briefly considers the established biological effects of this indoleamine in the human being.

27.2 Undisputed Biological Effects of Melatonin in Human Beings

The Transduction of Circadian and Circannual Temporal Context. There is little doubt that obvious prominent rhythmic geophysical cycles are ultimately responsible for imprinting endogenous biological rhythms upon our genome. The two most obvious of these cycles, the circadian and circannual, remain ubiquitous synchronizers of virtually all physiologic functions. As the earth turns on its axis a myriad of physical forces change rhythmically. Gravitational, geomagnetic, electromagnetic, temperature, humidity, wind, and light, among other forces and energy sources, change markedly and rhythmically throughout each and every day. In turn, these geo-solar-based forces are modulated by the annual tilt of the earth's axis on its elliptical journey round the sun. Large lunar interactions are obviously of major ultradian, circadian, and infradian relevance but will be ignored, for the moment. Although it is perfectly obvious that temporal orientation within the day and year is multifaceted and that many geo-physical and biometerologic forces are important *zeitgebers*, most research has been done on rhythmic light availability, imparted by the circadian day/night cycle. Almost all plants and animals have evolved processes to measure and respond to light. One of these systems is based upon the suppression of the synthesis and release of melatonin by light. While the evolutionary biology of the development of melatonin-based light sensing in plants and animals is fascinating, it is clearly beyond the scope of this brief chapter. In human beings and other mammals, the essence of this system is fascinating in its simplicity and effectiveness (Brainard et al. 1997).

Melatonin is produced in the pineal gland, which is located deep within the brain in close proximity to the suprachiasmatic nucleus (SCN), a primary and central component of the organism's many matrix-based coordinated endogenous hierarchical circadian pacemakers. Light transduced through visual and apparently nonvisual light receptors within the retinae passes through the retinohypothalamic tract to the suprachiasmatic nucleus, then to the superior cervical ganglion, and on to the pineal gland (Freedman et al. 1999; Lucas et al. 1999). The electrochemical activity caused by light results in the shutdown of melatonin synthesis and release within and from the pineal gland (Brainard et al. 1997).

The pineal gland and the SCN, as circadian pacemakers in their own right, have internal endogenous time sense. This internal time sense is reflected by the fact that their synthesis, secretion, and/or activities indefinitely maintain circadian organization when deprived of all external time cues. Practically, this means that the ability of a fixed dose of light to impact the system, and the physiologic and molecular nature of that interaction, is entirely and utterly dependent upon when in the circadian cycle and within the year the electrochemical footprints of that light stimulus are presented to the gland. The biological meaning of light in terms of pineal output is, thereby, different depending upon when in an internal temporal circadian sense the light is available.

Melatonin is very soluble in lipid membranes and gains easy entry into almost every cell of the body. There are also several classes of melatonin receptor distributed throughout the brain and other cells throughout the body (Reppert 1997). These facts lead to the likelihood that melatonin bio-effects are widely appreciated. In healthy human beings, melatonin secretion, held down throughout the day by both the gland's and the SCN's endogenous reinforcing circadian organization and exogenous light, is spurred within minutes of darkness occurring in the usual evening hours. Because of the stabilizing effects of endogenous circadian SCN and pineal rhythms, darkness during the usual daytime hours does not result in melatonin synthesis and release, unless or until the usual night-day orientation is consistently disturbed for several days in a row, as when time zones are crossed. The dark-related anticipated upsurging of evening melatonin is sharp. The gland continues to release its hormone until habitual daytime is approached, when the pineal gland's melatonin synthesis and release is most sensitive to both endogenous circadian suppression and exogenous light suppression. Since the relative timings of sunrise and sunset are rhythmically modulated throughout the year, the melatonin dynamic of rise and fall, integrated over some days, in a beautiful biological economy, continuously yields both circadian and circannual orientation.

Much exquisite chronobiological human experimentation has defined the fine points of the relationship among light/dark/sleep and melatonin in both health and disease (Czeisler et al. 1989; Lewy and Sack 1997; Shanahan et al. 1997). To summarize, it seems as though all information transfer among this and all other endogenous circadian–circannual pacemakers and rhythmic environmental pacemakers is entirely and complexly and completely rhythm stage dependent. This concept is essential to both understanding and using any pineal–melatonin–cancer connection for therapy.

27.3 Melatonin Affects Sleep and Body Temperature

Sleep. The human being's habitual nocturnal sleep span usually occurs at the time of day associated with melatonin synthesis and release from the pineal gland. The first and early studies of melatonin by Lerner noted a soporific effect (Lerner et al. 1958; Vollrath et al. 1981; Lieberman et al. 1984). Elderly insomniacs have lower nighttime melatonin concentrations and later nightly melatonin peaks (Dahlitz et al. 1991; Haimov et al. 1995). EEG verified that nighttime sleepiness maximum occurs concurrently with the steepest rise in melatonin (Tzischinsky et al. 1993). Clinical trials giving varying doses of oral or parenteral melatonin at bedtime have demonstrated enhanced ability to fall asleep and stay asleep, and increased rapid eye movement sleep (Dawson and Encel 1993; Oldani et al. 1994; Tzischinsky and Lavie 1994; Garfinkel et al. 1995; Haimov et al. 1995; Wurtman and Zhdanova 1995; Zhdanova et al. 1995).

As would be predicted by the chronobiotic nature of melatonin and by the circadian rhythm stage dependence of all chronobiotic bioactivities, each of the soporific effects of melatonin depend heavily or exclusively upon the circadian timing of the melatonin dose. Sleepiness does not ensue after a morning dose of melatonin (Oldani et al. 1994; Zhdanova et al. 1995). The time span from melatonin ingestion to its sleep-inducing or -supporting effect is threefold longer if the dose is given at noon as compared to 9 p.m. (Tzischinsky and Lavie 1994).

Body Temperature. Daily physiologic sleep usually occurs in association with rhythmically falling body temperature. Core body temperature peaks daily in late afternoon and falls most rapidly (greatest rate of decline) in close association with daily darkness onset and the steepest melatonin rise (Cagnacci et al. 1992). The evidence for and against a causal relationship between body temperature, sleep related behavior, and melatonin secretion is extensive and complex. Clearly, many control loops comprise the fabric of circadian temporal organization and contribute to stabilization of the circadian body temperature rhythm. Melatonin is certainly a card-carrying member of this group of core body temperature modulators.

The core body temperature rhythm is composed of the summation of two distinct and inversely linked calorimetric processes, the daily production of heat and the daily loss of heat. Exogenous melatonin administered to individuals who are lying down and participating in a constant routine lowers the daily peak core body temperature during the day by 0.3 °C. Blocking melatonin secretion during the night raises the lowest daily body temperature by the same amount (Cagnacci et al. 1992). Melatonin does this, at least in part, by increasing heat loss by dilating peripheral vessels (Kräuchi and Wirz-Justice 1994; Cagnacci et al. 1997). Its effects upon heat production are not as clear.

At first consideration, the fact that melatonin seems to do the same thing (lower body temperature) at each of two opposite circadian stages is not consistent with the overriding idea that the biological effect of melatonin as a chronobiotic is dependent upon when in the biological cycle it is available. More careful consideration reveals that melatonin has the effect of lowering daytime body temperature only in individuals subjected to a constant supine daily routine. When daytime melatonin is given to people who are not lying down for 24 h at a time, it does not lower their body temperature at all (Kräuchi et al. 1997a, b). The idea that posture, and all of the sympathetic–parasympathetic correlates of being up and around during the usual daily activity phase of the circadian cycle, ablates the body temperature-lowering and delays the somnolence-inducing effects of melatonin is highly relevant to the use of melatonin as a therapy for sleep and perhaps for cancer. Intact free-living human beings who consume melatonin during the daytime are not affected by it similarly to when they consume it in the evening, when it both lowers body temperature and induces and consolidates nighttime sleep regardless of posture. Cancer patients are, by necessity, free-living, genetically distinct individuals. Any useful therapy, chronobiotic or not, will be administered to individuals who are not supine continuously, not undergoing constant routine, and not in strict temporal isolation. Masking effects that confirm or modulate underlying circadian time structures come within the territory of therapeutics in general, and cancer therapeutics in particular (Hrushesky and Bjarnason 1993a).

Circannual and Menstrual Cycles. The circannual modulation of melatonin dynamics has an obvious impact upon both sleep and body temperature circannual physiology. In cycling premenopausal women, melatonin has also been shown to have differential effects upon the core body temperature within each ovulatory menstrual cycle. The usual effect of melatonin, to lower core body temperature during the night, is attenuated during the early luteal phase of each ovulatory menstrual cycle (Cagnacci et al. 1996). This is thought to happen in preparation for implantation of the possibly fertilized ovum at this critical point in the menstrual cycle. In any event, millions of women each year identify this elevation of nadir body temperature, which occurs near daily arising, as a sign of potential fertility. This "common knowledge" very nicely illustrates the fact that melatonin at one point in a biological cycle, in this case the follicular phase of the fertility cycle, lowers core body temperature. At the opposite menstrual cycle stage, the early luteal phase, the same physiologic levels of melatonin are not associated with the diminishment of core body temperature.

In summary, both sleep and body temperature are reproducibly and meaningfully affected by both endogenous and exogenous melatonin. The biological meaning and/or bio-effectiveness of melatonin in human beings is markedly and reproducibly affected in turn by its circadian, menstrual, and circannual context.

27.4 Light, Melatonin and the Manipulation of Circadian Orientation

Light. A series of controlled human studies, performed in strict temporal isolation, have demonstrated the ability of light to predictably shift the circadian organization of melatonin secretion. Both light/dark phase shifts and time of day-specific light pulses given daily over several days shift the melatonin circadian rhythm. Further, when temporal isolation is tight, even single appropriately timed light pulses will shift the circadian phase of the melatonin circadian rhythm as well as other behavioral circadian rhythms (Czeisler et al. 1989).

Melatonin. Although much argument has occupied the literature, it seems as though exogenous administration of melatonin at an appropriate circadian stage will reproducibly shift the endogenous melatonin circadian rhythm of human beings. Depending upon when in the circadian time structure a person takes melatonin, the endogenous rhythm may be advanced, nothing may happen to it, or it may be delayed (Lewy and Sack 1997). This kind of circadian stage-dependent circadian rhythm resetting behavior is called a "phase response curve." The shape of such a curve defines the strength of a potential circadian rhythm resetter (Pittendrigh 1960). Although it would be expected that a reliable phase shift in the melatonin circadian rhythm by melatonin might result in a shift of other rhythms, including the sleep–wake cycle, core body temperature, etc., this is not yet considered entirely proven by some (Czeisler 1997).

Mood. Mood in general, and seasonal affective disorder (winter depression) in particular, is beneficially affected by light. Some beneficial light effects occur regardless of circadian timing but these benefits can be enhanced by optimal circadian timing. Light obviously affects pineal melatonin synthesis and secretion both acutely and chronically. Clearly, light may have other melatonin-independent biological effects. Although it is suspected that light-responsive seasonal or nonseasonal depression may be modulated by melatonin, this has not been adequately proven (Wehr 1997).

The duration and timing of usual nighttime melatonin exposure, especially in urban individuals, is affected both by the timing of endogenous pineal circadian rhythm and by light-induced shut-down of daily pineal melatonin synthesis and secretion. Free-living human beings exposed to morning bright light thereby may shorten the duration of their nightly melatonin exposure much more effectively than if light is given at other times of day. Seasonal depression is responsive to circadian disruption, sleep deprivation, and morning bright light but it is also apparently responsive to bright light at other times of day when it has much less effect upon the duration of daily melatonin exposure (Wehr 1992, 1997). While melatonin exposure may mediate seasonal depression, it is not the only relevant factor and it may have a greater or lesser impact in different individuals; thus in some people seasonal symptoms may be impervious to melatonin while in others they may be melatonin driven. This possibility is reflected in Wehr's descriptions of two categories of persons with regard to sensitivity of melatonin secretion dynamics to ambient seasonal light availability. Some individuals living in similar urban settings have profound seasonal modulation of melatonin secretion dynamics while others have little or no such seasonal modulation. The connection between light and certain mood disorders is real. Whether this connection is usually, sometimes, or rarely made through melatonin secretion is not yet clear (Wehr 1997).

Jet Lag, Shift Work, and Melatonin. Although melatonin is at best a weak phase shifter when applied in temporal isolation, this does not mean that it cannot be a powerful aid in the consolidation of light/behavioral shifts imposed by transmeridian travel or changing shifts of work. Although data are conflicting, on balance there is good evidence that appropriately timed effective doses of melatonin can consolidate behaviorally induced circadian phase shifts (Arendt et al. 1997).

The most effective protocols for dose and timing of melatonin have not been fully worked out. It is clear, however, that ill-timed melatonin will not help the individual to consolidate physiologic function in a new circadian orientation and may in fact damage or delay this process. Much of the confusion about whether melatonin is useful in these situations relates to the quality of the clinical studies available. Placebo-controlled studies, rigorously working out both dose and circadian melatonin timing, are needed in order to provide reproducibly effective therapeutic regimens to help jet lag and shift work difficulties.

Essential Temporal Context. Melatonin has effects upon human sleep, body temperature, and circadian orientation. Each of these effects depends upon the time within the circadian cycle the agent is endogenously available or exogenously given. The therapeutic uses of this agent are, thereby, intimately dependent upon circadian timing. Light and melatonin are inversely related so that circadian timed light and melatonin potentially provide a complementary chronobiotic approach to optimizing sleep and circadian orientation. There is every reason to expect from the physiology of light and melatonin that any therapeutic manipulation of the retino-hypothalamic-pineal axis performed in order to manipulate the host–cancer balance will require complete chronobiological investigation before that therapy can be properly understood and adopted as effective or rejected as ineffective. Separate lines of convincing data indicate also that host tissues and even spontaneous human cancers maintain the capacity to sense and respond to circadian pacemakers by coordinating their cell division cycles and susceptibility to cancer therapeutics within circadian time (Hrushesky 1994).

27.5 The Pineal, Melatonin, and Cancer

Continuous light exposure, which diminishes melatonin availability, increases the incidence of mammary cancers in rats. This effect of light is abolished by enucleation (Reiter 1988). Melatonin inhibits and pinealectomy enhances 7,12-dimethylbenz[a]-anthracene-induced mammary cancer in the rat (Tamarkin et al. 1981). There are also provocative preclinical in vitro data linking light, melatonin, and cancer (Mhatre et al. 1984; Hill and Blask 1988).

More powerfully, however, cancer incidence in profoundly blind individuals of each sex is about half of that expected (Hahn 1991; Feychting et al. 1998). There are few facile explanations for this diminished cancer risk that do not invoke light, melatonin, and the retino-hypothalamic-pineal axis.

Clinical Epidemiologic Data. Blind people with no light perception have a normal, albeit usually free-running melatonin circadian rhythm. Exposure of these people to bright light does not usually suppress melatonin synthesis and secretion (Feychting et al. 1998). In modern urban society, nocturnal light exposure is common and that light exposure clearly has the capacity to interrupt the nighttime synthesis and secretion of pineal melatonin (McIntyre et al. 1989). It has been shown that minimal dark phase light exposure meaningfully accelerates cancer growth (Dauchy et al. 1997). Scientific and popular concern has been repeatedly expressed that nighttime light and subsequent melatonin diminishment might be responsible for the increasing incidences of a variety of cancers (Waalen 1993; Raloff 1998).

Breast Cancer. Tamarkin showed that the nighttime melatonin peak of women with estrogen receptor-rich, early-stage breast cancer was diminished compared to the nightly pattern seen in normal women (Tamarkin et al. 1982). These data raise the possibility that the diminishment of a melatonin cancer-suppressive effect may have allowed these cancers to develop or, conversely, that the presence of these cancers lowered the nighttime melatonin concentration. Several additional series have shown that melatonin nocturnal peaks are diminished in patients with breast cancer (Bartsch et al. 1981; Danforth et al. 1985; Bartsch et al. 1989; Skene et al. 1990; Bartsch et al. 1991; Holdaway et al. 1991).

Prostate Cancer. Similar melatonin chronobiology also accompanies the development of prostate cancer. Patients with advanced prostate cancer have a profound flattening of their circadian melatonin pattern with near absence of the usual nocturnal peak (Bartsch et al. 1985). More careful examination of the fractionated urinary melatonin excretion in young men, men with benign prostatic hypertrophy (BPH), and men with prostate cancer demonstrates progressive decline in the circadian amplitude of the melatonin signal. Progressively less melatonin is produced and excreted and its nighttime burst is progressively diminished as prostate cancer is diagnosed (Bartsch et al. 1992).

The usually robust darkness-associated daily peak of melatonin accompanies health. Disturbed, diminished, or abbreviated dark-associated melatonin peaks are, on the other hand, associated with both breast and prostate cancer. In each and every case studied, the height of the daily melatonin peak and the timing of melatonin daily peak are disturbed in precancer or in cancer patients.

Clinical Therapeutic Data. Melatonin as a cancer therapy for human beings has been investigated to a limited extent (Panzer and Wiljoen 1997). None of these clinical trials have attempted to define the optimal time of day for melatonin therapy, none have compared one circadian time with another, and although some studies have stipulated the time of day that melatonin was consumed or injected, others have not. This makes all of these clinical data difficult to interpret.

Host–Cancer Circadian Organization. Four observations demand that properly designed, at least circadian, chronotherapeutic clinical cancer trials be performed with any anticancer agent, but even more especially with melatonin. First, the host–cancer balance is comprised of circadian organized relationships among host tissues and tumor tissues. For example, the successful growth of tumors demands the commandeering of nonmalignant endothelial cells for the formation of tumor blood vessels and the success of these tumors further depends upon the capabilities of the tumor to elude host cellular and humoral immune function. All tissues of the host are coordinated within circadian time; therefore any effects of any substance given to favorably modulate the host–cancer balance should, and perhaps must, be given at a time within the circadian cycle when host tissues are most susceptible to favorable manipulation. Examples proving that human tissues are meaningfully coordinated within circadian time, and remain so when cancer is present, will be provided.

Second, the bulk of available evidence indicates that the cancer cells comprising spontaneous human malignancies remain sensate of, and respond to, circadian pacemakers responsible for the temporal coordination of the organism and its nonmalignant tissues. A few examples of the circadian coordination of human cancer will be given.

Third, both the toxicity and the anticancer efficacy of a wide range of anticancer agents depend, to a meaningful extent, upon when in the circadian cycle these agents are given. This is true for highly reactive, metal-based chemical compounds with scores of nonspecific subcellular targets like cisdiammine dichloroplatinum, carboplatin, or oxaliplatin. The toxic–therapeutic ratios of somewhat more specific, but still broadly active anticancer antibiotics of fungal or bacterial or plant origin, like doxorubicin or epirubicin, likewise vary with the time of the circadian cycle. The toxicity and efficacy of synthetic antimetabolites like fluorodeoxyuridine, 5-fluorouracil, 6-mercaptopurine, methotrexate, and cytosine arabinoside, designed to interrupt one or a few critical metabolic pathways essential for DNA replication, are severalfold more intensely dependent upon their circadian timing (Hrushesky and Bjarnason 1993b; Zhang and Diasio 1994).

In general, as the metabolic targets of anticancer agents become fewer and more specific, the extent to which the toxic–therapeutic ratio depends upon the circadian timing of the agent increases. This relationship continues as anticancer agents that act through specific receptors are approached. Thus by the time we come to the study of αTNF, a 10- to 20-fold circadian difference characterizes the receptor-mediated toxic–therapeutic ratio of this biological molecule (Langevin et al. 1987; Hrushesky et al. 1994). Further, this log order relationship becomes, in fact, bi-directional for interleukin-2, which has profound antitumor activity when given at certain times of day and reproducible tumor growth-enhancing effects when given at opposite times of day (Sanchez et al. 1993). This bi-directional switching of the host–cancer balance is extremely worrisome from a therapeutic point of view. As we more and more commonly develop and use cytokines, growth factors, proteins, peptides, and active sites to treat diseases of all types, the circadian context of administration will become more obviously important. Melatonin is such a substance.

Fourth and finally, peptide hormones that manipulate the circadian pacemakers themselves are being defined. The central regulators of pituitary, adrenal, and other endocrine functions and melatonin are the first of these to be widely available. Others, the protein products of a range of human "clock" and "clock-related" genes, will soon be added. Given what is known about the circadian coordination of human tissues, human cancers, and the toxic–therapeutic ratios of nonspecific and nonchronobiotic anticancer agents, it is utterly nonsensical to consider the anticancer capacity of chronobiotics like melatonin without concurrently defining the shape of the phase response curve (PRC), i.e., its ability to modulate the host–cancer balance as a function of melatonin's circadian timing. Although literally thousands of cancer patients have been given or have taken melatonin, absolutely no relevant, cancer-related "PRC" data have been obtained. The place of melatonin in our cancer therapy pharmacopoeia will perforce remain mysterious unless and until such information is obtained.

27.6 Circadian Cytokinetic Rhythms

Rhythmic changes in relevant specific normal tissue function, such as cell division, can to some extent explain the circadian variation in sensitivity of rapidly proliferating tissues to chemotherapy. It is clear that the cells of every tissue appropriately studied do not enter or exit the various cell cycle phases randomly throughout the day, but rather do so in a highly organized way at certain times within the day (Scheving et al. 1989). As will be discussed in more detail, such regular circadian rhythms in bone marrow and gastrointestinal tract cytokinetics may, at least partly, explain the circadian stage-dependent toxicity of cell cycle phase-specific drugs (Scheving et al. 1989; Brown 1991). Similarly exploitable, but perhaps less coordinated, cytokinetic circadian rhythms may also exist in some but not all tumors (Nash and Echave Llanos 1971).

Normal Tissue Rhythms. All normal proliferating tissues adequately examined to date in adult mammals have been shown to undergo circadian variation when both the DNA synthesis stage of the cell cycle and the mitotic index are monitored along the 24-h scale (Durie et al. 1977; Janne et al. 1978; Scheving 1981; Hrushesky et al. 1983; Laerum and Smaaland 1989). Melatonin's capacity to favorably modulate the toxicity of other cytotoxic agents or radiation is, thereby, likely to depend to a very large extent upon when in the circadian cycle both the melatonin and the second agent are each given.

Bone Marrow. Many anticancer agents are most cytotoxic to normal tissues that are actively dividing and/or during specific phases of the cell division cycle. Cell proliferation rhythms in the bone marrow and gastrointestinal tract are especially relevant to the cancer patient, since these two tissues are the most common dose-limiting target tissues for toxicity from commonly used antineoplastic drugs. The cell division and release of all types of blood-produced cells within the bone marrow undergo strong regular circadian variation (Laerum and Aardal 1981; Haus et al. 1983; Laerum et al. 1988; Laerum et al. 1989). The relative numbers of bone marrow stem cells and progenitor cells (splenic colony-forming units) (Stoney et al. 1975), granulocyte/macrophage colony-forming units, and erythroid colony-forming units (Bartlett et al. 1984) show predictable circadian variations in a wide range of species. These findings have also been documented repeatedly in normal human subjects. Recent studies reported by Smaaland and collaborators (1991) of 19 normal control subjects using cytofluorometric technology unequivocally confirm the earlier findings of several investigators (Killman et al. 1962, Mauer 1965, Morley 1966, Ponassi et al. 1979, Ross et al. 1980). The percentage of cells undergoing DNA synthesis measured by flow cytometry demonstrated a large regular variation along the circadian time scale for each 24-h profile, with a range of variation from 29% to 339% from the lowest to the highest daily value. The mean value of the lowest DNA synthesis for each 24-h period was 8.7% ± 0.6% during the first half of the night (8 p.m. to 4 a.m.), while the mean value of the highest was 17.6% ± 0.6% during the first half of each daily activity span (8 a.m. to 4 p.m.), i.e., a twofold difference (Fig. 27.1).

Gut. Circadian organization in cell proliferation throughout the gut mucosa has been documented from the tongue to the rectum in mice and rats (Scheving et al. 1978; Scheving et al. 1989). Major variations are seen in the amplitude of the rhythms in the various regions, but the phasing in the different regions of the gut is remarkably

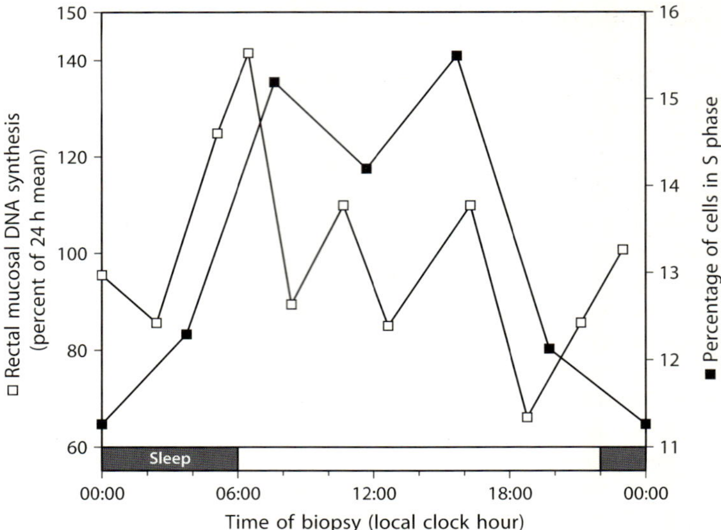

Fig. 27.1. Average rectal epithelial mucosal DNA synthetic capacity in 24 men sampled every 3 h for 24 h, expressed as percent of 24-h mean (□–□), peaks at 146 % in the morning (~ 8 a.m.) and is 70 % of the 24-h mean at ~ 8 p.m. The circadian pattern of the average absolute percentage of nucleated bone marrow cells cytofluorometrically determined to be in S phase (■–■) derived from serial samples from 19 young men has a broader daily peak between 8 a.m. and 6 p.m., when 50 %–60 % more DNA synthesis is ongoing compared with the remainder of the day. (From Smaaland et al. 1993)

similar. High-amplitude circadian rhythms have also recently been documented in the gastrointestinal epithelium of humans (Buchi et al. 1991), with the highest DNA synthetic activity occurring each day at approximately 7 a.m. in both the fed and the fasted state (Fig. 27.1). This rough overlap of cytokinetic phasing between the two most important toxicity target tissues of humans, bone marrow and gut, implies that S phase-specific cytotoxic therapy, which damages the gut and/or bone marrow, can be expected to be less toxic if administered at similar, predictable hours of the day.

Figure 27.1 contrasts the organization of DNA synthetic capacities of the bone marrow and rectal mucosal epithelium in two separate groups of young men. These men, 19 in the marrow study and 24 in the gut study, underwent bone marrow biopsy or flexible sigmoidoscopy and epithelial biopsy every 2–4 h around the clock. Each bone marrow sample was stained appropriately for cytofluorometric DNA analysis, while the gut tissue was incubated with tritiated thymidine. The percent of bone marrow cells in S phase and the amount of tritiated thymidine (expressed as the percentage of the subject's 24-h mean) incorporated by gut samples are each display-ed as a function of the time of day when the sample was obtained. It is clear that DNA synthesis activity for each of these target tissues is synchronized within the day and that substantially more DNA synthesis is going on in the morning hours than in the evening or night hours for both tissues.

Therapeutic Implications of Circadian Organization. Two decades ago it was reported that the arrangement within the day of 3-hourly doses of the antimetabolite cytosine arabinoside had a pronounced effect upon the survival probability of mice previously inoculated with L1210 leukemia cells. About 10 years ago a series of studies reported that the circadian timings of the xenobiotic, doxorubicin, and the reactive metal compound, cisplatin, impacted all of the toxicities of each of these drugs as well as the ability of the agents separately and together to control and cure a solid tumor in rats. Subsequent clinical trials in patients with widespread ovarian cancer demonstrated benefit from optimal circadian doxorubicin/cisplatin scheduling by allowing safer administration of higher doses of each drug and a fourfold 5-year survival advantage for those women who received the circadian schedule predicted to be best by the preceding rat studies.

In the last few years, chronopharmacodynamic attention has been focused upon the fluoropyrimidine antimetabolites, fluorodeoxyuridine (FUDR) and 5-fluorouracil (5FU). Continuous intravenous infusions of FUDR, peaking at different times within the day, given to rats bearing a mammary adenocarcinoma demonstrated lower toxicity and greater efficacy when most of the daily dose was infused in the 6 h spanning the time late in the animal's daily activity span and during the first half of its daily sleep span. Randomized clinical studies demonstrated superiority of a similar circadian schedule to the standard flat rate continuous infusion. The circadian schedule proved to be far less toxic, allowing almost 50% more drug to be safely given with substantially less toxicity than lower doses given by constant rate infusion. This regimen was also found to have unexpected activity in renal cell carcinoma, a disease in which no other cytotoxic therapy has proven useful. A recent French multicenter study has doubled the frequency of objective response to a combination of 5FU, leukovorin, and oxaliplatin by optimally timing administration of these agents.

Circadian studies have also been carried out with 5FU. Based upon early studies at the University of Minnesota demonstrating the importance of dihydropyridine dehydrogenase (DPD) in the catabolism of 5FU and studies from our laboratory demonstrating a prominent circadian time structure to DPD activity in lymphocytes from human volunteers, others have demonstrated rhythmic circadian 5FU concentrations in the serum of patients receiving a constant-rate intravenous continuous infusion (CI) of 5FU. 5FU serum levels peak when (in the day) these patients' mononuclear cell DPD activities are lowest. Additional clinical studies have demonstrated that up to severalfold safe increases in 5FU and 5FU leukovorin dose intensity can be achieved by delivering most of each day's infusion during the late activity and/or early sleep span of cancer patients.

The human bone marrow cells and the human gut epithelial cells synthesize DNA nonrandomly within the day. The most DNA synthesis occurs in both tissues between the early morning hours and the late afternoon. In addition to this coordinate DNA synthetic capacity, the biochemical pathways responsible for fluoropyrimidine catabolism and anabolism are also coordinated in circadian timing within both the liver and other cells of the body, including toxicity targets such as bone marrow and gut. During the past year the circadian dynamics of many enzymatic pathways critical to understanding fluoropyrimidine chronopharmacodynamics have been elucidated and a coherent picture is emerging.

The top panel of Fig. 27.2 illustrates the experimentally determined circadian rhythmic patterns of the activities of three key enzymes in fluoropyrimidine metabolism. DPD catabolizes 5FU to noncytotoxic metabolites; thymidine phosphorylase (TP) converts 5FU to its nucleotide FUDR; and thymidine kinase (TK) converts FUDR

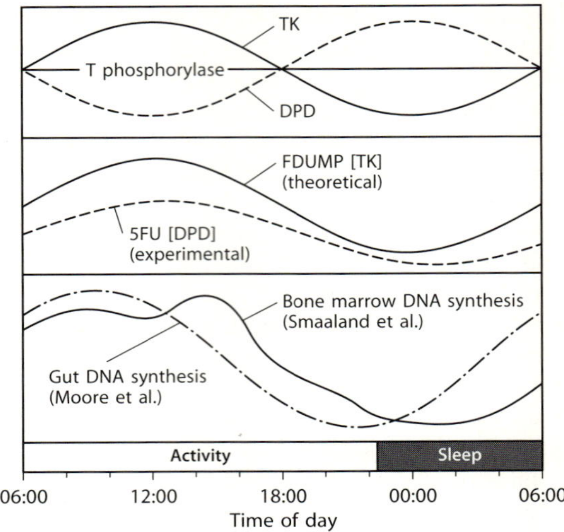

Fig. 27.2. The *top panel* illustrates the experimentally determined circadian rhythmic patterns of the activities of three key enzymes in fluoropyrimidine metabolism: dihydropyridine dehydrogenase (*DPD*), thymidine phosphorylase, and thymidine kinase (*TK*). The *middle panel* demonstrates the result of these rhythmic enzyme activities upon the pharmacokinetics of constant rate 5-fluorouracil (*5FU*) infusion (5FU levels) and the final activation product, fluordeoxyuridyl monophosphate (*FDUMP*). The *bottom panel* illustrates the experimentally determined circadian patterns of gut and bone marrow DNA synthesis in human beings. Further detailed explanation of the figure and the findings illustrated appears in the text

to fluordeoxyuridyl monophosphate (FDUMP), which binds to thymidylate synthase (TS), thereby blocking DNA synthesis by starving the cell of thymidine. DPD activity peaks about midnight, TP activity does not vary during the day, and TK activity peaks around noon.

The middle panel demonstrates the effect of these rhythmic enzyme activities upon the pharmacokinetics of constant rate 5FU infusion (5FU levels) and the final activation product FDUMP. DPD removes 5FU rhythmically during the day, making much more 5FU available in the early daytime hours. TP converts 5FU to FUDR at a constant rate but the higher substrate levels in the morning result in high FUDR levels in the morning. TK activity is higher in the morning, converting more of the higher levels of FUDR to FDUMP at that time of day.

The bottom panel illustrates the experimentally determined circadian patterns of gut and bone marrow DNA synthesis in human beings. Much more DNA synthesis is ongoing at the time of day associated with the highest FDUMP levels. This coincidence of enzyme activities and gut and bone marrow DNA synthesis results in the experimentally observed high-amplitude reproducible circadian rhythm in myeloid and gastrointestinal susceptibility to 5FU. Giving a zero order 5FU infusion thereby results in a non-zero order and reproducibly very unfavorable pharmacodynamic pattern, from the perspective of toxicity to normal tissues. Furthermore, changing the circadian pattern of the 5FU infusion so that most of the daily continuous infusion is given in the evening would change the resultant fluoropyrimidine pharmacokinetics and pharmacodynamics in such a way as to result in a more favorable toxic therapeutic profile.

Circadian Coordination of Spontaneous Human Cancer. If cell proliferation in spontaneous human cancer is coordinated within the day, then, since many cytotoxic drugs kill cells more or less effectively depending on the phase of the cell cycle, a finding of circadian rhythm in any aspect of cancer cell proliferation would provide a mechanism supporting the likelihood that cancer cell susceptibility, itself, waxes and wanes predictably throughout each day. If the circadian timing of quasi cell cycle-specific cytotoxic drugs is relevant, how much more relevant is the timing of the chronobiotic with potential anticancer activity, melatonin? Furthermore, if the normal endocrine-paracrine-autocrine loops controlling circadian gating of cell proliferation are intact in cancer cells, a novel paradigm for cancer control is opened, employing optimally timed chronobiotics like melatonin that do not require cancer cell killing (Schipper et al. 1996).

Relevant Data from the Literature. Determining the proportions of tumor cells residing within specific phases of the cell cycle at different times within the day currently demands frequent serial tumor sampling throughout at least one entire day. This feat has been accomplished in a few mouse tumor model systems (Blumenfeld 1943; Rubin et al. 1983). Data in cancer patients are understandably limited, but some relevant data have appeared. Voutilainen (1953) and Tähti (1956) each isolated biopsy specimens from a variety of cancer types, around the clock, and documented time of day-dependent peaks in cancer cell mitosis in the tumors of most patients, with differing circadian patterns seen in cancers of different origin. On average, across all tumor types, two daily peaks of mitosis were found, the first occurring between midnight and 2 a.m., and the second, smaller peak occurring around noon (Garcia-Sainz and Halberg 1966). Smaaland et al. (1991, 1993) found that malignant cells obtained around the clock from the involved nodes of patients with various non-Hodgkin's lymphomas on average organized their proliferation late each day, with a midnight peak in the S and G_2 phases (Fig. 27.3).

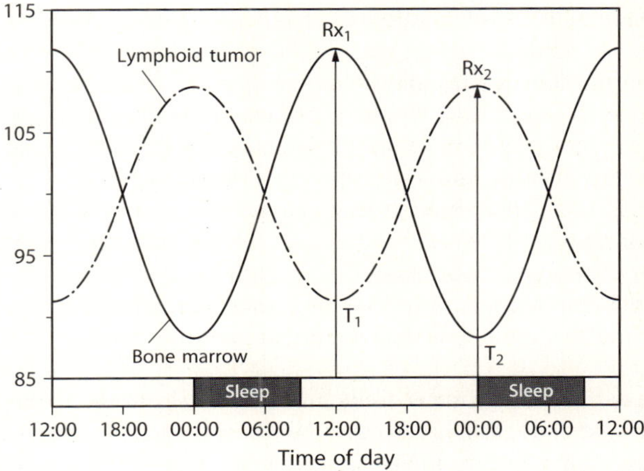

Fig. 27.3. Bone marrow (——) or malignant lymphoma (–––––) sampled around the clock have different 24-h patterns of S phase activity as represented by best-fitting cosine functions to the raw S phase data when expressed as the predictable variation around a 24-h mean (199%). S phase active cytotoxic treatment at the time T_1 will selectively damage bone marrow and spare tumor, while the same treatment at a different time of day, T_2, will selectively spare bone marrow and damage tumor. (From Smaaland et al. 1993)

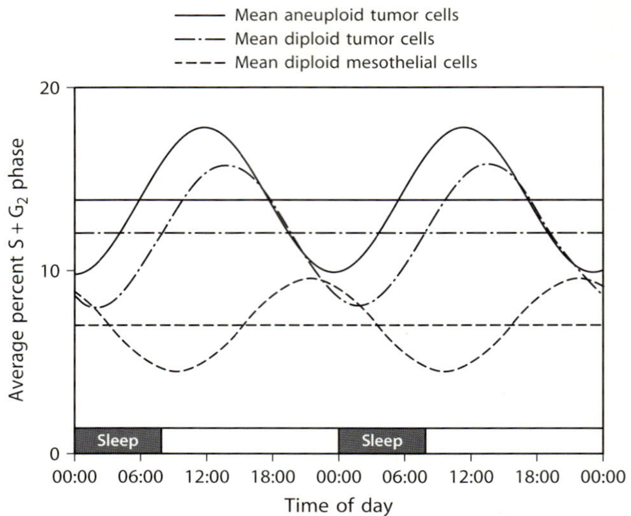

Fig. 27.4. Thirty women with epithelial ovarian cancer underwent frequent saline peritoneal lavages. Three populations of cells were identified in each sample: diploid mesothelial cells (– – – –), diploid tumor cells (–·–·–·–), and aneuploid tumor cells (———). The average cytofluorometrically determined percentage of each cell population in S phase plus G_2 phase is depicted by three cosine functions best fitting the raw data. Twenty-four-hour mean S + G_2 phase percentages are lowest in benign mesothelial cells, higher in diploid tumor cells, and even higher in aneuploid cancer cells. The daily peaks and valleys of the benign and malignant cells are almost 12 h out of phase, while aneuploid and diploid cancer cells share similar circadian DNA synthetic patterns. (From Klevecz et al. 1987)

Klevecz et al. (1987) concurrently sampled both malignant ovarian cells and non-malignant, nonovarian mesothelial cells from the peritoneal cavities of patients with ovarian cancer and found different circadian organizations of these two cell populations arising within the identical physiologic milieu. A higher proportion of sampled ovarian cancer cells were in the G_2 and S phases in the late morning hours, while a higher proportion of benign peritoneal mesothelial cells inhabited these phases of the cell cycle in the late evening hours (Klevecz et al. 1987) (Fig. 27.4).

Taken overall, these data, limited though they are, raise the possibility that the circadian time structure of DNA synthesis and/or mitosis is determined not by whether a cell is benign or malignant but rather by the tissue of origin of that cell, i. e., its ontogeny. To test this hypothesis, however, benign and malignant tissue samples of the same cell type must be obtained concurrently, around the clock.

Circadian Organization of Cancer Cells and Noncancer Cells of the Same Origin. A 50-year-old postmenopausal, but otherwise healthy, female dairy farmer had an epidermoid carcinoma on the pinna of her right ear excised. A right radical neck dissection did not demonstrate metastatic involvement of regional lymph nodes. Eight months later, a right supraclavicular node biopsy revealed metastatic epidermoid carcinoma. This area and the right neck were then treated with external beam radiation therapy. Ten months later, some 18 months following the initial diagnosis of the disease, an excised subcutaneous right breast mass and left axillary masses were found to harbor metastatic epidermoid carcinoma. During the next 1–2 months the patient developed 100–150 rapidly growing subcutaneous tumor nodules (0.5–1.0 cm in diameter). She remained active and working, arising daily for morning milking (4–6 a.m.) and retiring between 8 p.m. and 10 p.m. each evening. Her only symptom was itching, which progressed throughout each day to burning pain in the nodules. Itching began each day in the early afternoon, and the "burning pain" was most severe between 6 p.m. and sleep onset (10 p.m.).

For this study, after University of Minnesota Investigational Review Board-approved informed written consent was obtained, the patient became resident at the University of Minnesota General Clinical Research Center for 6 days between March 7 and March 13. During this span she arose daily between 6 and 7 a.m. and retired at 10 p.m. Her oral temperature was recorded and fractional urine collections were obtained every 4 h throughout the stay. The patient underwent only these baseline measurements for 2 days. Subsequently, two separate tumor nodules and at least one piece of apparently uninvolved skin were excised from her back or flank, under local anesthesia, every 4 h for 28 h. Twenty-one blocks of tissue – 14 tumor nodules and 7 grossly normal skin biopsies – were submitted in formalin. The tissue was embedded in paraffin. Two slides were prepared from each tissue block containing gross tumor (28 sections), with the two sections from each specimen taken from different levels of the tissue blocks. Four slides were prepared from each block thought by the surgeon to contain only normal skin (28 sections). The slides were coded, numbered 1–56, randomized for cell counting, and counted "blindly" by the same pathologist (E. Haus). The number of mitotic figures was recorded for every 2000 basal cells of the epidermis and for every 2000 tumor cells examined. In sections in which no tumor was present, only the basal cells were counted. After completion of all counting, it was found that tumor cells were present in several tissue blocks in which no tumor had been clinically suspected, i.e., distant from a gross metastasis. The number of mitoses in the basal cells of the epidermis in sections with tumor and without tumor in the underlying dermis did not differ statistically. During the remainder of the 6-day hospitalization, circadian-timed chemotherapy (Hrushesky 1985) was administered, and the patient was then discharged to home.

Urinary Cortisol Excretion. The hypothalamic-pituitary-adrenal axis is one of the most important and thoroughly studied mammalian circadian organismic pacemakers. The cortisol urinary excretion pattern, specifically the time of day of daily peak cortisol excretion and the amplitude of the daily surge in cortisol excretion, is an excellent measure of the circadian orientation of an individual. Our patient's circadian cortisol excretion dynamics support the likelihood that this woman remained coordinated with her circadian environment (see Fig. 27.5). As expected for this patient with a life-threatening chronic illness, in this case metastatic cancer, her overall 24-h mean cortisol excretion was some threefold greater than that observed for a group of 19 apparently healthy women (Hrushesky et al. 1984). The circadian cortisol excretion pattern (see inset in Fig. 27.5) is otherwise, however, quite normal.

Body Temperature. The daily shape of our patient's oral temperature rhythm, in the spans before and during the biopsies, appears very much like those of 15 apparently healthy age-matched women (data not shown) (Freedman et al. 1994). The robust amplitude and day-to-day reproducibility of this cancer patient's temperature and

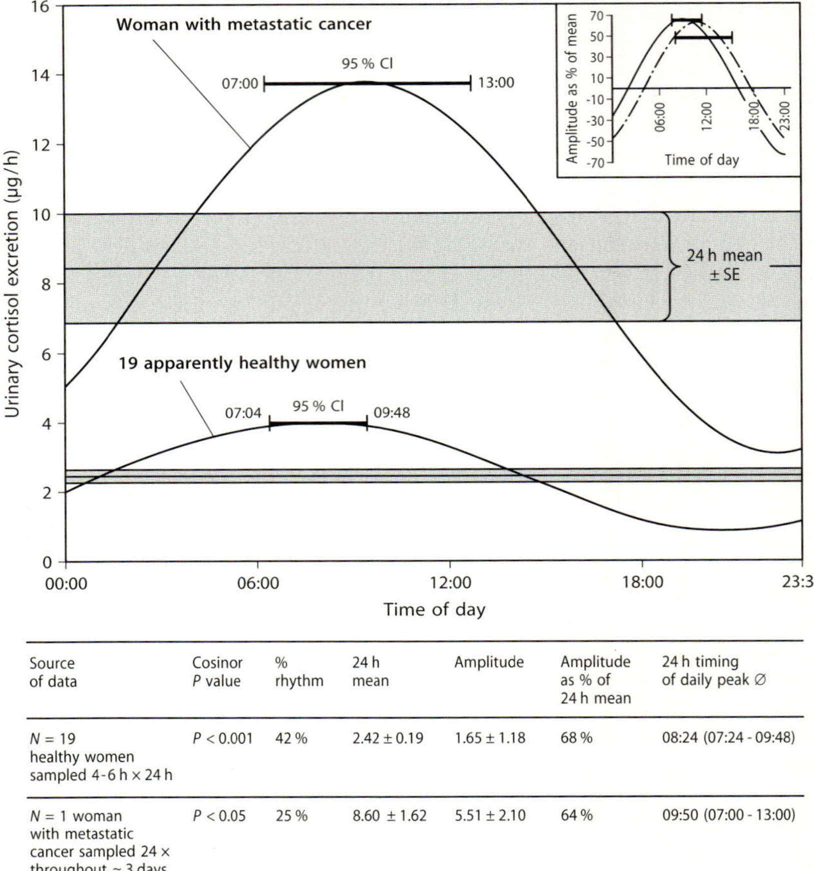

Source of data	Cosinor P value	% rhythm	24 h mean	Amplitude	Amplitude as % of 24 h mean	24 h timing of daily peak Ø
$N = 19$ healthy women sampled 4-6 h × 24 h	$P < 0.001$	42 %	2.42 ± 0.19	1.65 ± 1.18	68 %	08:24 (07:24 - 09:48)
$N = 1$ woman with metastatic cancer sampled 24 × throughout ≈ 3 days	$P < 0.05$	25 %	8.60 ± 1.62	5.51 ± 2.10	64 %	09:50 (07:00 - 13:00)

Fig. 27.5. Circadian cortisol urinary excretion pattern of our 50-year-old patient with metastatic epidermoid carcinoma (*thick line*) compared with that of a group of apparently healthy younger women aged 20–30 years (*thin line*). During a 3-day pre-chemotherapy monitoring span, 24 fractional urine collections were obtained, their volumes measured and recorded, and the cortisol concentration in each determined by use of a standard radioimmunoassay. These data were then arithmetically converted to fractional cortisol excretions in µg/h during each of the 24 collection spans. In the main figure, the fractional circadian cortisol excretion pattern for this patient can be contrasted with the pattern of cortisol excretion similarly obtained from each member of the group of 19 putatively healthy, young adult women. These hormone concentration determinations were performed in the same laboratory, using identical methodology. Our patient's average 24-h cortisol excretion is some three times higher than the average for these apparently healthy women. The circadian wave forms describing these 24-h patterns are otherwise similar. The 95% confidence intervals (CIs) of the timing of the daily peak excretions of these two time series overlap entirely. The amplitudes of the circadian waveforms describing one-half of the average daily peak-to-trough difference are virtually identical when expressed relative to the respective 24-h mean cortisol excretion values (see *horizontal lines* and *cross-hatched bars* plus *inset table*). The *inset figure* depicts the two circadian rhythms with their amplitudes so normalized to the 24-h mean values of each series. The similarities and differences between these two circadian waveforms demonstrate that our patient is substantially "stressed" by bearing a widely metastatic cancer, resulting in the unusually enhanced adrenocortical activity. They also indicate that although "stressed," our patient maintained fundamentally normal, environmentally entrained, circadian temporal hypothalamic–pituitary–adrenal orientation

cortisol excretion patterns tentatively confirm the presence of relatively normal, environmentally entrained, endocrine and metabolic circadian time structure.

Skin and Epidermoid Cancer Mitotic Indices. The usual circadian pattern of mitosis in the epidermis, first documented for humans by Cooper (1939) and confirmed by Scheving (1959), shows that the greatest daily cell division occurs in the evening and falls each day to a minimum in the morning. The mitotic index of normal epidermoid skin cells and epidermoid cancer cells of skin origin in our patient fell throughout the night and early morning hours to very low values at 6–8 a.m. and then rose again (Fig. 27.6). While the daily peaks and nadirs of mitosis occurred in our patient at very similar times of day for both normal cells and tumor cells, the 24-h mean mitotic rate in the tumor cells, as well as its daily amplitude, was significantly higher in the malignant cells. Thus, both normal cells and metastatic carcinoma cells originating from the same tissue showed a prominent and almost identical circadian rhythm in cell proliferation in this patient.

These data demonstrate that the proliferation of metastatic cancerous tissue in this patient maintained its capacity to sense and respond to the usual daily regulators of cell division. These signals, whatever they are, commanded this patient's normal skin cells to stop dividing in the night, and they stopped dividing. These same commands were apparently also heeded by soon-to-be-lethal metastatic epidermoid cancer cells of the same ontogenetic origin. These data indicate that cancerous cells may well maintain the circadian chronobiological literacy of the cells from which they have arisen. This observation suggests that, if the circadian signals responsible for circadian cell division control can be defined, it may be possible to control cancer growth without resorting to cytotoxic therapeutic strategies. Even in the absence of therapies that can reconstitute natural circadian controls of malignant cell proliferation, there remain implications of the demonstrated circadian coordination of cancer cell division for many currently available cancer treatments. Cell cycle active and/or cell cycle stage-specific cancer chemotherapy given to cancer patients at optimal times of day has been shown, in randomized prospective trials, to enhance the therapeutic index of these treatments. Circadian-based treatment strategies have enhanced ovarian cancer survival (Hrushesky 1985; Hrushesky and Bjarnason 1993), doubled objective tumor response frequency in colorectal cancer (Lévi et al. 1994), and enhanced the frequency of cure of acute lymphoblastic leukemia of childhood (Rivard et al. 1985).

In summary, the mitotic figures of benign and malignant epidermoid cells of skin origin from an environmentally synchronized patient with metastatic cancer appeared and disappeared nonrandomly throughout the day. This finding supports the hypothesis that human cancer cell proliferation is organized within the day and is in keeping with the four relevant series in the literature. These limited data are the first to directly indicate that the cell type of cancer origin may confer specific circadian cytokinetic time-keeping characteristics upon the analogous cancer cell. Such circadian cytokinetic coordination may be an important cause of the therapeutic advantage observed by optimally timing cytotoxic cancer therapy within the day (Hrushesky 1994). This circadian cytokinetic organization is likely to be relevant to the toxic–therapeutic ratio of irradiation, cytotoxic, biological, and many, if not all, other molecular strategies with the potential to control cancer (Wood and Hrushesky 1996).

Irradiation therapy and the majority of useful cytotoxic drugs damage cells that are actively dividing more than cells residing in the resting phase of the cell cycle.

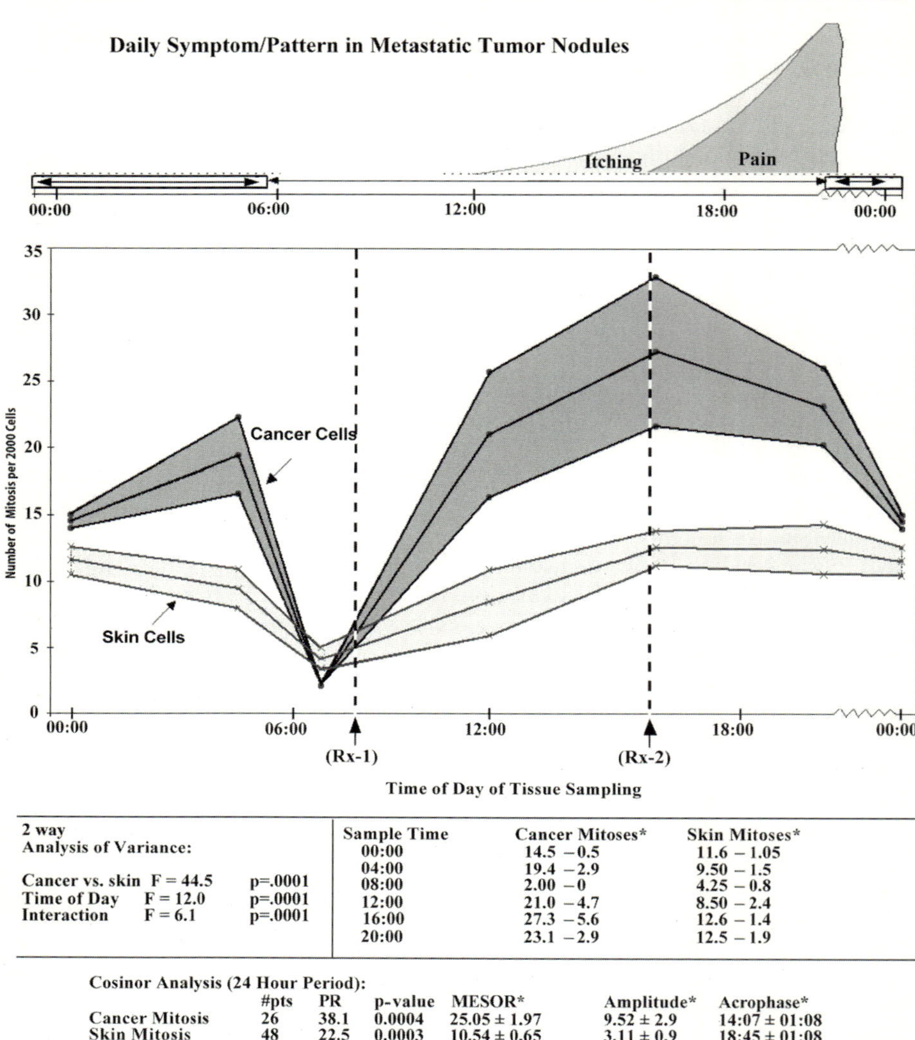

2 way Analysis of Variance:			Sample Time	Cancer Mitoses*	Skin Mitoses*
			00:00	14.5 − 0.5	11.6 − 1.05
			04:00	19.4 − 2.9	9.50 − 1.5
Cancer vs. skin	F = 44.5	p=.0001	08:00	2.00 − 0	4.25 − 0.8
Time of Day	F = 12.0	p=.0001	12:00	21.0 − 4.7	8.50 − 2.4
Interaction	F = 6.1	p=.0001	16:00	27.3 − 5.6	12.6 − 1.4
			20:00	23.1 − 2.9	12.5 − 1.9

Cosinor Analysis (24 Hour Period):

	#pts	PR	p-value	MESOR*	Amplitude*	Acrophase*
Cancer Mitosis	26	38.1	0.0004	25.05 ± 1.97	9.52 ± 2.9	14:07 ± 01:08
Skin Mitosis	48	22.5	0.0003	10.54 ± 0.65	3.11 ± 0.9	18:45 ± 01:08

* ± = standard error of the mean

Fig. 27.6. Comparison of the circadian variation in mitotic index in a single cancer patient's normal skin (x) and her metastatic epidermoid carcinoma (•). The mitoses are expressed in absolute numbers per 2000 enumerated normal skin basal cell or cancer cell nuclei. The *center lines* connect mean values obtained from counting four tissue sections at each time point; the *outer lines* show the corresponding standard errors. Substantial time-of-day dependence in the mitotic index (cell division frequency) in the normal skin of a patient with metastatic cancer is here documented. The evening peak and morning valley in skin cell division are similar to those reported for normal infants and adult men. Highly malignant metastatic and soon-to-be-lethal cancer cells also demonstrate cell division coordination within the day. In normal skin and epidermoid cancer, the number of mitoses decreases prominently throughout the night, reaching a low point in the morning hours, 6 a.m. to 8 a.m., and then rise throughout the rest of the day to a peak value in the evening. The analogue scale at the *top of the figure* represents self-reported, subjective symptom quality and severity in the areas of the skin involved with metastatic nodules. Itching began daily after noon and increased in subjective severity until concurrent burning pain supervened. This burning pain worsened steadily until the patient finally fell asleep. She would awaken each morning between 5 a.m. and 7 a.m., without itching or pain, to begin this symptom cycle anew

Furthermore, many of these treatments selectively damage cells most effectively as they traverse specific phases of the cell cycle. External beam gamma irradiation, for example, damages cells that are undergoing mitosis more than cells inhabiting other stages of the cell cycle. If irradiation were chosen as a primary treatment for our patient, one might expect different outcomes depending upon when in the day this woman received daily irradiation fractions. Treatment as the "first case" (Rx-1) each day (~ 8 a.m.) would result in therapy being given when relatively few cancer cells and small numbers of surrounding normal skin cells are inhabiting the sensitive "M (mitotic) phase" of the cell cycle. Little damage to both skin and cancer might result from therapeutic irradiation at this particular time of day. Conversely, if the same daily dose of irradiation were administered as the "last case" (Rx-2) of the treatment day (~ 4 p.m.), very different cell kill might be expected to result. In this case, a greater proportion of cancer cells will be undergoing mitosis and thereby will be relatively more sensitive to lethal irradiation damage. Although normal skin cells are also more frequently dividing at this time of day, mitoses are some three-fold more frequent in the malignant compared with the nonmalignant epidermoid cells between 2 p.m. and 4 p.m. Thus, optimal timing of radiation treatments, or any therapy more effectively damaging cells undergoing mitosis, might result in temporal selectivity and a potentially enhanced therapeutic index, i.e., a temporal window of opportunity

If chronobiotic agents like melatonin act at specific cell cycle stages to inhibit cancer cell proliferation, the circadian organization of cancer cells is of paramount importance. The optimal circadian timing of melatonin is, thereby, the probable key to whether reproducible anticancer effects will ever be seen.

27.7 Melatonin in Human Cancer Therapy

Melatonin Treatment of Human Cancer Cells in Culture. In vitro effects of melatonin upon human cancer cell proliferation and apoptosis are many and varied (Blask 1993). Inhibitory effects upon cancer cell proliferation depend intimately upon the cancer cell type and upon the cell culture conditions. Even in vitro, the pattern as well as the concentration of melatonin administered determines what effects occur (Hill and Blask 1988; Cos and Sanchez-Barcelo 1994). The proliferation rate of the cancer cells, whether or not they are anchored, and whether they express sex hormone receptors each determine to some extent the biological effects of melatonin (Cos and Blask 1990; Hill et al. 1992). Some melatonin effects also depend upon whether other growth factors are present or absent in the culture medium (Blask and Hill 1986; Cos and Blask 1994; Cos and Sanchez-Barcelo 1995).

These observations of variable biological effects of melatonin, dependent upon the "biological" context of the cell culture systems, is consistent with expected profound circadian dependence of in vivo melatonin effects. Obviously, normal cell division and cancer cell division dynamics are each coordinated by a hierarchy of circadian pacemakers. Clearly, the expression of receptors and the availability of peptides and hormones known to affect melatonin bioactivity are coordinately available or unavailable in vivo at specific known circadian stages. The sum total of this in vitro tumor biology data leads to the expectation that the circadian stage of administration of melatonin to cancer patients largely determines its clinical anticancer activity.

Mechanisms for Melatonin Anticancer Activity. Melatonin has been shown to have reproducible and dose-dependent antioxidative bioactivities that can diminish oxidative damage of DNA, which could be, in part, responsible for its protective effects against chemical carcinogenesis (Reiter et al. 1994).

Melatonin has been shown to decrease the transcription of estrogen receptor, to block prolactin effects, and to block proliferation-stimulating effects of epidermal growth factor. These facts may be, in part, responsible for melatonin-associated decreases in the proliferation of hormone-dependent cancer cells. Also, under certain experimental conditions, melatonin delays cell cycle progression of some cancer cells between G_0 or G_1 and S phase, diminishing the proliferative rate of these cells (Blask 1997).

Melatonin also has a variety of potentially relevant effects upon cancer cells growing in culture, including opening up gap junctions (Ubeda et al. 1995) and enhancing calmodulin binding and binding to tubulin (Blask 1997). Whether some or any of these intracellular phenomena are responsible for the tumor shrinkage or diminished metastases associated with melatonin administration in vivo is unclear. In summary, it must be said that if melatonin has significant anticancer activity against spontaneous human cancers, then the mechanisms of this activity are unclear but are nonetheless likely to be largely, if not entirely, circadian stage dependent.

27.8 Does Melatonin Benefit Human Beings with Cancer?

Unfortunate Anecdotal and Antiscientific Disinformation. In the 1950s, before the actual discovery of melatonin, pineal extracts were given to patients with breast, ovarian, and uterine cancers and anecdotal benefits were claimed (Lapin and Ebels 1976). In 1960, after melatonin was defined, it was given to cancer patients by Starr, with anecdotal inprovements (Starr 1960).

DiBella has given melatonin with a wide range of other standard and alternative agents to thousands of cancer patients over a span of more than 25 years and has also claimed anecdotal success. None of these early, continuing, and recent experiences were in any way controlled or even properly documented. The dose, timing, route, and indications for melatonin were variable among patients and no organized tabulation of results has ever been made available. The undocumented claims of the complex and evolving "DiBella regimen" resulted in popular and political upheaval in Italy, resulting in millions of dollars of public expenditure for an entirely unproven cancer treatment. A retrospective study of the value of this melatonin-containing "therapy" has revealed no measurable benefit. It is unfortunate that the roots of melatonin's clinical evaluation have been sullied by such poor-quality clinical investigation. It is a challenge to bring any effective cancer therapy to oncologic practice. It is obviously more difficult to bring a circadian stage-dependent chronotherapy, like melatonin, to clinical utility. Almost half a century of poorly documented and uncontrolled use of this agent makes the prospect of melatonin therapy development a truly daunting task. Add to this the fact that melatonin is not patentable and you have a truly quixotic quest. Within this reality, the more medically relevant, controlled, and documented melatonin trials will now be briefly reviewed.

Clinical Trials. A single investigator's name dominates the clinical melatonin cancer literature. Dr. Paolo Lissoni has treated hundreds of cancer patients with melatonin, conscientiously publishing the results of these treatments. While every therapeutic development has one or a few clinical champions, few therapies remain so thoroughly dominated by a single highly productive investigator.

27.9 Lissoni's Broad Phase II Trials

Lissoni's initial broad phase II studies, searching for melatonin activity in patients with a wide range of malignant diseases, ultimately included 54 patients with any of a variety of solid tumors. In these studies 20 mg of melatonin was injected intramuscularly each day in the late afternoon or early evening (Lissoni et al. 1987, 1989, 1991). These three studies demonstrated an objective partial tumor response in 1 of 54 patients and two additional minor responses. Disease stabilization was observed in 20 of these 54 melatonin-treated patients. It is hard to judge from these reports whether these few responses or disease stability conferred meaningful clinical benefit upon these patients. In general, in the development of standard cytotoxic drugs, if an objective response were to be seen in a single patient of 54, the drug would be unlikely to be pursued further. Professor Lissoni, nonetheless, persisted and has done a number of studies randomly assigning patients with advanced cancer to best supportive care or best supportive care plus melatonin, usually 10–50 mg, given orally in the evening.

27.10 Lissoni's Disease-Specific Randomized Controlled Clinical Trials of Melatonin

Lung Cancer. Sixty-three non-small cell lung cancer patients with metastatic disease were randomly assigned to either 10 mg of evening daily oral melatonin or best supportive care. The 1-year survival of melatonin-treated patients was superior (8/31) to the group treated with supportive care only (2/32) ($P < 0.05$) (Lissoni et al. 1992).

Glioblastoma. Thirty patients with unresectable glioblastoma were randomly assigned to radiation alone or 10 mg of daily oral melatonin at 8 p.m. given concurrently with brain irradiation. One-year survival of melatonin-treated patients was superior (6/14) compared to patients treated with radiation alone (1/16) ($P < 0.02$) (Lissoni et al. 1996a).

Breast Cancer. Forty patients with failed metastatic breast cancer were randomly assigned to receive tamoxifen alone or tamoxifen plus 20 mg of melatonin, orally each day at 8 p.m. At 1 year, 12/19 melatonin-treated patients were alive compared to 5/21 patients treated with tamoxifen alone ($P < 0.01$) (Lissoni et al. 1995a).

Melanoma. Thirty patients with malignant melanoma involving lymph nodes following the resection of all known disease were randomly assigned to treatment with melatonin, 10 mg at 8 p.m. orally daily, or best supportive care. At 1 year, 10/14 melatonin-treated patients were free of cancer while only 5/16 observed patients had failed to relapse ($P < 0.05$) (Lissoni et al. 1996b).

Brain Metastases. Fifty patients each with one of a variety of solid tumors, metastatic to brain, were randomly assigned to melatonin 20 mg per day orally at 8 p.m. plus brain irradiation or to radiation therapy alone. At 1 year, 9/24 melatonin-treated patients survived while only 3/26 patients who received radiation without melatonin were surviving ($P < 0.05$) (Lissoni et al. 1994a).

27.11 Lissoni's Randomized Controlled Trials of Melatonin Plus Interleukin-2 vs Interleukin-2 Alone

Solid Tumors. In 1994 (Lissoni et al. 1994b) and 1995 (Lissoni et al. 1995b) Lissoni published two large phase II studies in which interleukin-2 (IL-2) had been given to all patients but half of each group had also been given bedtime oral melatonin, 40–50 mg. The IL-2 was given subcutaneously daily 5 or 6 of 7 days each week at a dose of three million units per day. Eighty patients suffering from a variety of metastatic solid tumors were randomized. At 1 year, 19/41 patients who had received melatonin with their IL-2 were alive, while only 6/39 patients receiving IL-2 alone were alive ($P < 0.05$) (Lissoni et al. 1994b).

One hundred additional patients with one of a variety of metastatic solid tumors were randomized to receive IL-2 and melatonin as above or best supportive care, without IL-2 or melatonin. At 1 year, 22/52 patients receiving the combination were alive while only 8/48 observed patients were living ($P < 0.01$) (Lissoni et al. 1995b).

Colorectal Cancer. Barni and Lissoni randomly assigned 50 patients with metastatic colorectal cancer who had failed prior chemotherapy to receive IL-2 as given above plus 40 mg of daily oral evening melatonin or supportive care, with no melatonin or IL-2. Nine of 25 patients receiving melatonin plus IL-2 were alive 1 year later while only 3 of 25 receiving supportive care survived ($P < 0.05$) (Barni et al. 1995).

Lung Cancer. Sixty patients with metastatic non-small cell lung cancer were assigned to treatment with IL-2 plus melatonin as above or chemotherapy. At 1 year, survival of the melatonin/IL-2-treated group was 13/29 vs 6/31 in the chemotherapy group ($P < 0.01$) (Lissoni et al. 1994c).

Composite Lissoni Results. The vast clinical experience of the Lissoni group with melatonin is difficult to summarize because in some studies melatonin was given alone while in many cases it was given in combination with other therapies. It is fair to say, however, that definite activity has been seen in patients suffering from non-small cell lung cancer, breast cancer, colorectal cancer, kidney cancer, and melanoma. Evidence for probable activity has also been observed for endocrine cancers, hepatocellular carcinoma, and pancreatic cancer.

27.12 Melatonin Trials of Other Investigators

Melanoma. Gonzales et al. (1991) and Robinson et al. (1995), from the University of Colorado, treated patients with metastatic malignant melanoma in a phase I setting, orally every 6 h, daily for at least 4 weeks at one of ten wide-ranging dose levels. No significant toxicity was seen at any dose level. Six of these 42 patients had objective responses (at least 50% shrinkage) of all their known metastatic tumors. The responses took a median of 12 weeks of therapy to achieve and these responses lasted 7–48 weeks. These results have to be viewed cautiously, but few cytotoxic agents available for the treatment of this disease produce 15% objective responses. Additionally, the every 6-hour schedule and wide range of doses used, from 5 mg to 700 mg/m^2 daily would clearly not be expected to be chronopharmacologically optimal. These data support potential real anticancer activity of melatonin in metastatic malignant melanoma, confirming the work of Lissoni in this cancer type.

Kidney Cancer. Renal cell carcinoma, once metastatic, is considered uncurable. Most cytotoxic drugs tested have been entirely inactive (Hrushesky and Murphy 1977). Chronochemotherapy with circadian-optimized FUDR, on the other hand, has proven effective for control of metastatic disease (Hrushesky et al. 1990). Rare spontaneous remission of metastatic kidney cancer following removal of the primary tumor indicates that this particular variety of cancer is very sensitive to the state of the host–cancer balance. Consistent with this sensitivity, biological agents including interferons and interleukins have demonstrated benefit in some patients with metastatic kidney cancer.

Neri et al. (1994) combined melatonin and interferon in patients with metastatic kidney cancer. Thirty-three patients were treated; three patients achieved a complete and six a partial tumor response. This 27% objective response frequency is greater than that usually achieved with interferons alone (10%–15%). These data need to be followed up.

Melatonin Clinical Benefit. Aside from objective tumor response, many of the reports of Lissoni comment upon probable subjective benefits of melatonin. Many of these effects are favorable modulations of the toxicities of accompanying agents. It is Lissoni's impression that melatonin ameliorates the hypotension associated with IL-2 therapy and that it diminishes the toxicities associated with steroids that are given to counter-act some of IL-2's other toxicities. Lissoni also notes that when evening melatonin is given with chemotherapy, associated myelotoxicity is diminished. He also notes that surgery-associated lymphocytopenia is diminished by concurrent evening melatonin. Perhaps most clinically relevant, Lissoni repeatedly observes that patients receiving melatonin achieve and maintain better performance status and also have less anxiety than those treated without melatonin.

27.13 Minimal Clinical Trial Design for the Productive Study of Melatonin

Several years ago Halberg et al. (1993) published a provocative paper, "Clinical Trials: The Larger the Better?". In this paper, these investigators made several points that bear directly upon the conundrum underlying the melatonin cancer literature. Cancer and the host–cancer balance resonate rhythmically, and thereby predictably, across at least two and probably more biologically relevant frequency ranges, circadian and circannual, in all human beings, as well as menstrual, in cycling women. Part of the perceived need for very large clinical trials, employing hundreds to thousands of patients and costing tens to hundreds of millions of dollars arises as a direct result of these competing predictable sources of biovariability. Simply defining and controlling for these rhythmic sources of biovariability will substantially diminish the need for such large numbers of patients and dollars. A word of caution must, however, accompany this strategy. Clearly much of what has gone before within this brief review makes and remakes the critical point that fixing the rhythm stage of study, randomly, to one or another time of day in order to control variability will diminish variability but will also either enhance or diminish toxicity and efficacy. If the circadian stage of treatment is chosen randomly and the relationships of circadian stage to toxicity and efficacy are quasi-sinusoidal, then clearly one risks choosing, by chance, the most toxic and least effective time of day for phase I and phase II studies respectively. It is hard to justify leaving such a critical choice to chance. Chronotoxicologic trials done in small groups of patients at each of six equidistant times of day can reliably define the least and most toxic circadian stages for administration of any anticancer agent. These trials require 60–90 patients. At the same time, if the drug has significant anticancer activity, hints of whether the circadian efficacy rhythm tracks with the circadian toxicology pattern will emerge from this study of these same 60–90 patients.

When the agent in question has no significant toxicity, to wit melatonin, then this study design will be focused from the outset upon determining the most effective, i. e., therapeutic circadian stage. For melatonin, treating 90 patients with a single diagnosis and stage of cancer in groups of 15, one group at each of six equispaced circadian stages, will yield a great deal of useful therapeutic and chronotherapeutic information.

27.14 A Useful Melatonin Study Design

Most clinical trials demonstrating melatonin clinical anticancer activity have given this indoleamine hormone in the afternoon or evening, daily. It is clear that oral administration effectively raises serum concentrations and has circadian stage-dependent biological effects upon body temperature, sleep induction, circadian orientation, and the host–cancer balance. It is not at all clear, however, whether optimal anticancer effects occur following afternoon or evening administration.

There are other potential beneficial effects of melatonin upon patients with metastatic cancer. Karnofsky performance status, the subjective assessment of the amount and timing of daytime and nighttime activity and daytime and nighttime rest or sleep, has been shown over and over again to be predictive of survival, quality of life, and response to a wide range of therapeutic maneuvers including surgery, radiation, chemotherapy, and hormonal therapy. Cancer fatigue is the most prominent and bothersome universal symptom of advanced cancer. There is every reason to believe that optimally timed melatonin therapy may benefit cancer patients substantially by consolidating and enhancing nighttime sleep, even in the absence of clear anticancer efficacy. The consolidation of the circadian sleep/activity rhythm, with improvement of nighttime sleep and a potential resultant diminishment of fatigue, that well might follow optimally timed melatonin is certainly worth pursuing per se. Identifying these potential quality of life-related chronobiological melatonin benefits when they have been achieved will, however, require specialized chronobiological study designs and instrumentation.

27.15 Objective, Physiologic Measurement of Fatigue Needed

Fatigue is unquestionably the major cause of cancer-induced decline in quality of life. It is, thereby, well worth studying carefully (Glaspy 1997; Miaskowski and Portenoy 1998). In order to study fatigue properly, more than questionnaires are clearly required. A physiologic definition must be agreed upon and objective physiologic measurements must be employed to quantify fatigue, as well as to quantify the efficacy of candidate agents developed to diminish it. One approach to this has been the objective measurement of exercise capacity and efficiency of oxygen consumption (Daneryd et al. 1998).

In order to measure fatigue objectively and with ease, one might profitably take a page from sleep research protocols. Continuous EEG and rapid eye movements characterizing sleep have given way to carefully validated but much simpler wristband-based activity monitoring. (Mormont et al. 1998). Using the same tools that time sleep and activity and quantify each of them within the day has allowed the noninvasive assessment of sleep latency, sleep onset and awakening timing, and sleep quality. A similar quantitative approach should be used to objectify fatigue by using this device to measure the amount, duration, and timing of daily activity. Once this kind of simple, noninvasive monitoring provides quantitatively reproducible validated physiologic signatures that reproducibly measure fatigue, we can immediately embark upon clinical trials of a range of new peptides and small molecules to interrupt or short circuit the fatigue signals sent by a range of chronic diseases, including cancer. Without such objective measurement, however, these new *quality of life drugs* will be hard to develop, and it will be unnecessarily difficult to gain their FDA approval. For example, it has already been noted that rheumatoid arthritis patients receiving the TNF receptor blocker Enbrel have diminished fatigue, seemingly out of proportion to their joint responses. (Weinblatt et al. 1999). The necessity for objectification of fatigue is demanded by the universal importance of this most disruptive and destructive of all cancer symptoms and by the potential ease with which this goal can be accomplished.

A reasonable study design to determine anticancer efficacy and other potentially beneficial effects upon circadian sleep and activity patterns and cancer-associated fatigue would comprise a combination of longitudinal and transverse design. Longitudinally, each patient should have activity/sleep measurements done continuously, using wrist actigraphy, for 7 days prior to receiving oral melatonin and serially after therapy has been initiated, at a fixed daily dose somewhere between 2 mg and 20 mg, or placebo, at one of six circadian stages. These circadian stages should include:

1. *20:00.* The usual timing of dim light melatonin onset in most subjects. Melatonin at this circadian stage will steepen the upslope of usual melatonin rhythm, move this nightly peak earlier to approximately 9–10 p.m., raise the usual peak melatonin level, and prolong the overnight span of melatonin exposure.
2. *Approximately 24:00.* This would be near the usual daily nighttime peak of the normal melatonin rhythm (01:00–02:00). This timing will raise the nighttime peak but do little to its usual phase or duration.
3. *04:00.* This timing will yield a higher than usual nighttime peak. It will also extend the usual duration of this peak later in the day.
4. *08:00.* This timing will create a second melatonin peak higher and narrower than the usual nighttime peak at a time of day when light is usually shutting down melatonin synthesis and excretion.
5. *12:00.* This timing will create a second high daytime melatonin peak in antiphase to the usual approximate midnight peak, essentially increasing the duration of exposure of cells to melatonin but also doing so at a circadian stage when these cells virtually never see this substance normally.

6. *16:00*. Melatonin at this time of day will shift the dim light onset melatonin rise to a much earlier circadian stage than usual and may result in two evening/nighttime peaks or one extended nighttime melatonin exposure depending upon usual sleep onset, individual circadian melatonin dynamics, and perhaps season.

The above comparative melatonin treatment timing strategy will clearly and concurrently permit determination of the effect of timing of melatonin or placebo administration upon cancer, circadian sleep activity pattern, performance status, and cancer-associated fatigue, several hierarchical levels of potential melatonin clinical benefit.

Integrative and Biologically Complex Systems Will Not Yield Their Secrets Through Their Destruction. Since Mendel's induction of the unseen internal script of life, science has favored reductionism, which has drawn us to the impending definition of the entire human genome and the genomata of many of our cousins in life. The ensuing definition of problems by identifying abnormalities of omission, duplication, or misrepresentation within the primary structure of a gene, its message, or its product is truly satisfying to most scientists. This approach will continue to result in many important advances. To achieve a genuine understanding of life, however, we must embrace its secondary, tertiary, and quaternary structure.

The amino acid sequence of a protein determines to a limited extent the other entities with which it will interact. The secondary or tertiary folding of that protein, in response to the local neighborhood in which it finds itself, is likewise only a part of the story. The time structure (biodynamics) of what is where…when…is of absolutely critical relevance to the determination of these interactions. Biological time structure is defined by nonlinear rules responsible for complex rhythmic integration of all living systems. Nonlinearity is demanded for its inherent simplicity and stability and also because quasi-periodic systems, if regularized, have thermodynamic advantage. The geophysical site of the evolution of the temporal organization of life has imposed regularity upon the chaotic fabric of nonlinear time structure and given living systems complex yet nonstochastic organizational rules.

Understanding life cannot be approached in the most meaningful way outside the living organism. Accurate depiction of the temporal organization of vital function is dependent upon intact central, peripheral, and cellular time-keeping mechanisms. This symphony, both its melody and its beat, is nonsense outside the organism. The resonance structure of life's tempo changes markedly either upon removal from the organism, usually becoming less thoroughly ordered and reverting to higher frequency ranges, or when lethally disturbed to chaos and then linearity (death).

The experiments of the chronobiologist are dependent upon being able to interact with subsystems of the living organism or cell at recognizable points in its internal time structure. Studies of reproducible biological variations within the organism's circadian and/or fertility cycles are cases in point for the necessity of whole, live animal research to complement reductionist investigation. The broader essence of integrative biology focuses upon studying how complex subsystems of living organisms interact with one another in space and time.

The induction of nonlinear models, which consider the interactive multifrequency time structure of these systems combined with reductively obtained information, will yield genuine understanding. The most effective models will consider the tendency of temporal organization to "cluster" around certain specific geophysically originated frequencies. In the next hundred years, knowledge induced from detailed, precisely quantitative, minimally invasive, multilevel observation of intact living organisms will provide the basis for the deeper understanding of life.

Predictable Within-Individual Temporal Variability and Clinical Trial Design. Human circadian time structure is an endogenous and ubiquitous source of significant rhythmic variability in all physiologic biochemical and molecular markers of health and disease. The toxicities and maximum safe dose intensities of each cytotoxic drug adequately tested to date are predictably greater or lesser depending solely upon when in the day they are administered to human beings with cancer. There is also some evidence in human beings that anticancer efficacy varies predictably with time of day of drug treatment. Constructing phase I, phase II, and phase III clinical trials that do not test, stipulate, or at least carefully record the time of day of treatment is no longer justifiable or rational. The human being orders its life processes within a nonlinear, multifrequency resonance structure. This structure is complex but not stochastic. Careful consideration of biological time structure in observational or interventional clinical trials will substantially decrease apparent outcome heterogeneity and enhance the progress of clinical investigation.

27.16 Chronobiology and Drug Development

Anticancer drug development is a complex, protracted, expensive, high-risk venture. Compounds are screened for cytotoxicity, analogues are developed, or agents are designed and synthesized to target certain critical cellular events or pathways. These compounds move through a "decision network" drug development process with many steps. At each step, most of the candidate agents are discarded for reasons related to toxicology, specificity, or efficacy. The few agents that graduate from in vitro screening to whole animal systems have high subsequent failure probability as they are moved through in vivo toxicology trials in the mouse, rat, dog, and primate, or during concurrent efficacy evaluation in human tumor cell lines and mouse tumor model systems. If the target of drug action in host cells is virtually present or absent at specific times of the day, the results of screening studies will be divergent depending upon when in the day they are performed. An example might be the development of a new S phase active agent whose subcellular target is thymidylate synthase: the enzyme required for de novo synthesis of thymidine, which is necessary for DNA replication. If the toxicology of the agent is evaluated at the time of day associated with little gut or bone marrow DNA synthesis activity, the agent will have an excellent toxic–therapeutic ratio. In fact, the time of day that these studies are usually performed is in the first half of the "working day." This time of day generally corresponds to the first half of the mouse's daily sleep span, when *relatively low levels of DNA synthetic activity* are occurring in both gut and bone marrow. Phase I clinical trials are also carried out in the first half of the working day, during the cancer patient's diurnal activity span, when human gut and bone marrow each exhibit the *highest level of daily DNA synthetic activity*. The resultant apparently poor therapeutic ratio index might well cause this agent to be discarded as too toxic. This scenario might have been different, however, if preclinical studies had demonstrated the optimum circadian time for therapy and if subsequent phase I trials had been performed at that time of day. Since efficacy is also nonlinearly related to treatment timing, the circadian timing of drug administration in phase II and phase III studies is likewise relevant to maximizing drug efficacy. Studying melatonin as a cancer therapy in the absence of these considerations will result in useless and perforce misleading information.

References

Arendt J, Skene D, Middleton B, Deacon S (1997) Efficacy of melatonin treatment in jet lag, shift work, and blindness. J Biol Rhythms 12:604–617

Barni S, Lissoni P, Cazzaniga M Ardizzoia A, Meregalli S Fossati V, Fumagalli L, Brivio F, Tancini G (1995) A randomized study of low-dose subcutaneous interleukin-2 plus melatonin versus supportive care alone in metastatic colorectal cancer patients progressing under 5-fluorouracil and folates. Oncology 52:243–245

Bartlett P, Haus E, Tuason T, Sacket-Lundeen L, Lakatua D (1984) Circadian rhythm in number of erythroid and granulocytic colony forming units in culture (ECFU-C and GCFU-C) in bone marrow of BDF1 male mice. In: Haus E, Kabat HF (eds) Proc 15th International Conference on Chronobiology. S. Karger, Basel, pp 160–164

Bartsch C, Bartsch H, Jain A, Laumas K, Wetterberg L (1981) Urinary melatonin levels in human breast cancer patients. J Neural Transm 52:281–294

Bartsch C, Bartsch H, Flüchter S, Attanasio A, Gupta D (1985) Evidence for modulation of melatonin secretion in men with benign and malignant tumors of the prostate: relationship with the pituitary hormones. J Pineal Res 2:121–132

Bartsch C, Bartsch H, Fuchs U, Lippert TH, Bellmann O, Gupta D (1989) Stage-dependent depression of melatonin in patients with primary breast cancer: correlation with prolactin, thyroid stimulating hormone, and steroid receptors. Cancer 64:426–433

Bartsch C, Bartsch H, Bellmann O, Lippert TH (1991) Depression of serum melatonin in patients with primary breast cancer is not due to an increased peripheral metabolism. Cancer 67:1681–1684

Bartsch C, Bartsch H, Schmidt A, Ilg S, Bichler K, Flüchter S (1992) Melatonin and 6-sulfatoxymelatonin circadian rhythms in serum and urine of primary prostate cancer patients: evidence for reduced pineal activity and relevance of urinary determinations. Clin Chim Acta 209:153–167

Blask D (1993) Melatonin in oncology. In:Yu H, Reiter RJ (eds) Melatonin: Biosynthesis, Physiological Effects, and Clinical Applications. CRC Press, Boca Raton, FL, pp 448–464

Blask DE (1997) Systemic, cellular, and molecular aspects of melatonin action on experimental breast carcinogenesis. In: Stevens R, Wilson B, and Anderson L (eds) The Melatonin Hypothesis – Breast Cancer and the Use of Electric Power. Battelle Press, Columbus, OH, pp 189–230

Blask DE, Hill SM (1986) Effects of melatonin on cancer: studies on MCF-7 human breast cancer cells in culture. J Neural Transm Suppl 21:433–449

Blumenfeld C (1943) Studies of normal and of abnormal mitotic activity. Arch of Path 35:667–673

Brainard G, Rollag M, Hanifin J (1997) Photic regulation of melatonin in humans: ocular and neural signal transduction. J Biol Rhythms 12:537–546

Brown WR (1991) A review and mathematical analysis of circadian rhythms in cell proliferation in mouse, rat, and human epidermis. J Invest Dermatol 97:273–280

Buchi KN, Moore JG, Hrushesky WJM, Sothern RB, Rubin NH (1991) Circadian rhythm of cellular proliferation in the human rectal mucosa. Gastroenterology 101:410–415

Cagnacci A, Elliott J, Yen S (1992) Melatonin: a major regulator of the circadian rhythm of core temperature in humans. J Clin Endocrinol Metab 75:447–452

Cagnacci A, Soldani R, Laughlin G, Yen S (1996) Modification of circadian body temperature rhythm during the luteal menstrual phase: role of melatonin. J Appl Physiol 80:25–29

Cagnacci A, Kräuchi K, Wirz-Justice A, Volpe A (1997) Homeostatic versus circadian effects of melatonin on core body temperature in humans. J Biol Rhythms 12:509–517

Cassone V (1990) Melatonin: time in a bottle. Oxford Rev Reprod Biol 12:319–367

Cassone V, Natesan A (1997) Time and time again: the phylogeny of melatonin as a transducer of biological time. J Biol Rhythms 12:489–497

Cooper Z (1939) Mitotic rhythm in human epidermis. J Invest Derm 2:289–300

Cos S, Blask D (1990) Effects of the pineal hormone melatonin on the anchorage-independent growth of human breast cancer cells (MCF-7) in a clonogenic culture system. Cancer Lett 50:115–119

Cos S, Blask D (1994) Melatonin modulates growth factor activity in MCF-7 human breast cancer cells. J Pineal Res 17:25–32

Cos S, Sanchez-Barcelo E (1994) Differences between pulsatile or continuous exposure to melatonin on MCF-7 human breast cancer cell proliferation. Cancer Lett 85:105–109

Cos S, Sanchez-Barcelo E (1995) Melatonin inhibition of MCF-7 human breast cancer cells growth: influence of cell proliferation rate. Cancer Lett 93:207–212

Czeisler C (1997) Commentary: evidence for melatonin as a circadian phase-shifting agent. J Biol Rhythms 12:618–623

Czeisler C, Kronauer R, Allan J, Duffy J, Jewett M, Brown E, Ronda J (1989) Bright light induction of strong (type 0) resetting of the human circadian pacemaker. Science 244:1328–1333

Dahlitz M, Alvarez B, Vignau J, English J, Arendt J, Parkes J (1991) Delayed sleep phase syndrome response to melatonin. Lancet 337:1121–1124

Daneryd P, Svanberg E, Korner U, Lindholm E, Sandstrom R, Brevinge H, Petterson C, Boseaus I Lundholm K (1998) Protection of metabolic and exercise capacity in unselected weight-losing cancer patients following treatment with recombinant erythropoietin: a randomized prospective study. Cancer Res 58:5374–5379

Danforth D, Tamarkin L, Mulvihill J, Bagley C, Lippman M (1985) Plasma melatonin and the hormone-dependency of human breast cancer. J Clin Oncol 3:941–948

Dauchy RT, Sauer LA, Blask DE, Vaughan GM (1997) Light contamination during the dark phase in "photoperiodically controlled" animal rooms: effect on tumor growth and metabolism in rats. Lab Anim Sci 47:511–518

Dawson D, Encel N (1993) Melatonin and sleep in humans. J Pineal Res 15:1–12

Durie BMG, Salmon SE, Russell DH (1977) Polyamines as markers of response and disease activity in cancer chemotherapy. Cancer Res 36:214–221

Feychting M, Österlund B, Ahlbom A (1998) Reduced cancer incidence among the blind. Epidemiology 9:490–494

Freedman R, Norton D, Woodward S, Cornélissen G, Halberg F (1994) Circadian rhythms of hot flashes and body temperature in menopausal women. In: Zeisberger E, Schonbaum E, Lomax P (eds) Thermal Balance in Health and Disease. Advances in Pharmacological Sciences. Springer, New York

Freedman M, Lucas R, Soni B, von Schantz M, Munoz M, David-Gray Z, Foster R (1999) Regulation of mammalian circadian behavior by non-rod, non-cone, ocular photoreceptors. Science 284:502–504

Garcia-Sainz M, Halberg F (1966) Mitotic rhythms in human cancer reevaluated by electronic computer programs. Evidence for chronopathology. J Natl Cancer Inst 37:279–292

Garfinkel D, Laudon M, Nof D, Zisapel N (1995) Improvement of sleep quality in elderly people by controlled-release melatonin. Lancet 346:541–544

Glaspy J (1997) Fatigue may be most under-recognized, undertreated cancer-related symptom. Oncology News Int 6:30–38

Gonzalez R, Sanchez A, Ferguson J A, Balmer C, Daniel C, Cohn A, Robinson WA (1991) Melatonin therapy of advanced human malignant melanoma. Melanoma Res 1:237–243

Hahn R (1991) Profound bilateral blindness and the incidence of breast cancer. Epidemiology 2:208–210

Haimov I, Lavie P, Laudon M, Herer P, Vigder C, Zisapel N (1995) Melatonin replacement therapy of elderly insomniacs. Sleep 18:598–603

Halberg F, Bingham C, Cornélissen G (1993) Clinical trials: the larger the better? Chronobiologia 20:193–211

Haus E, Lakatua DJ, Swoyer J, Sackett LL (1983) Chronobiology in hematology and immunology. Am J Anat 168:467–517

Hill SM, Blask DE (1988) Effects of the pineal hormone melatonin on the proliferation and morphological characteristics of human breast cancer cells (MCF-7) in culture. Cancer Res 48:6121–6126

Hill SM, Spriggs L, Simon M, Muraoka H, Blask DE (1992) The growth inhibitory action of melatonin on human breast cancer cells is linked to the estrogen response system. Cancer Lett 64:249–256

Holdaway I, Mason B, Gibbs E, Rajasoorya C, Hopkins K (1991) Seasonal changes in serum melatonin in women with previous breast cancer. Br J Cancer 64:149–153

Hrushesky WJM (1985) Circadian timing of cancer chemotherapy. Science 228:73–75

Hrushesky WJM (ed) (1994) Circadian Cancer Therapy. CRC Press, Boca Raton, FL

Hrushesky WJM, Bjarnason G (1993a) Circadian cancer therapy. J Clin Oncol 11:1403–1417

Hrushesky WJM, Bjarnason GA (1993b) The application of circadian chronobiology to cancer chemotherapy. In: DeVita VT, Hellman Sand, Rosenberg SA (eds) Cancer: Principles & Practice of Oncology. JB Lippincott, Philadelphia, pp 2666–2686

Hrushesky WJM, Murphy G (1977) Current status of the therapy of advanced renal carcinoma. J Surg Oncol 9:277

Hrushesky WJM, Merdink J, Abdel-Monem M (1983) Circadian rhythmicity characterizes monoacetyl polyamine urinary excretion. Cancer Res 43:3944–3947

Hrushesky WJM, Haus E, Lakatua DJ, Halberg F, Langevin T, Kennedy BJ (1984) Marker rhythms for cancer chrono-chemotherapy. In: Haus E, Kabat HF (eds) Chronobiology 1982–1983. Karger, New York, pp 493–499

Hrushesky WJM, von Roemeling R, Lanning R, Rabatin J (1990) Circadian-shaped infusion of floxuridine for progressive metastatic renal cell carcinoma. J Clin Oncol 8:1504–1513

Hrushesky WJM, Langevin T, Kim YJ, Wood PA (1994) Circadian dynamics of tumor necrosis factor-α (cachectin) lethality. J Exp Med 180:1059–1065

Janne J, Poso H, Raina A (1978) Polyamines in rapid growth and cancer. Biochem Biophys Acta 473:241–243

Killman SA, Cronkite EP, Fliedner TM, Bond VP (1962) Mitotic indices of human bone marrow cells. I. Number and cytologic distribution of mitosis. Blood 19:743–750

Klevecz RR, Shymko RM, Blumenfeld D, Braly PS (1987) Circadian gating of S phase in human ovarian cancer. Cancer Res 47:6267–6271

Kräuchi K, Wirz-Justice A (1994) Circadian rhythm of heat production, heart rate, and skin and core temperature under unmasking conditions in men. Am J Physiol 267:R819–R829

Kräuchi K, Cajochen C, Wirz-Justice (1997a) Melatonin and orthostasis: interactions of posture with subjective sleepiness, heart rate and skin and core temperature. Sleep Res 26:79

Kräuchi K, Cajochen C, Wirz-Justice A (1997b) A relationship between heat loss and sleepiness: effects of postural change and melatonin administration. J Appl Physiol 83:134–139

Laerum OD, Aardal NP (1981) Chronobiological aspects of bone marrow and blood cells. In: von Mayersbach H, Scheving LE, Pauli JE (eds)11th International Congress of Anatomy, part C, Biological rhythms in structure and function. Alan R. Liss, New York, pp 87–97

Laerum OD, Sletvold O, Riise T (1988) Circadian and circannual variation of the cell cycle distribution in the mouse bone marrow. Chronobiol Int 5:19–35

Laerum OD, Smaaland R (1989) Circadian and infradian aspects of the cell cycle: From past to future. Chronobiologia 16:441–453

Laerum OD, Smaaland R, Sletvold O (1989) Rhythms in blood and bone marrow: potential therapeutic implications. In: Lemmer B (ed) Chronopharmacology: cellular and biochemical interactions. Marcel Dekker, New York, pp 371–393

Langevin T, Young J, Walker K, von Roemeling R, Nygaard S, Hrushesky WJM (1987) The toxicity of tumor necrosis factor (TNF) is reproducibly different at specific times of the day. Proc Ann Meet Am Assoc Cancer Res 28:A1580

Lapin V, Ebels I (1976) Effects of some low molecular weight sheep pineal fractions and melatonin on different tumors in rats and mice. J Neural Transm 52:269–279

Lerner A, Case J, Takahashi Y, Lee T, Mori W (1958) Isolation of melatonin, the pineal gland factor that lightens melanocytes. J Am Chem Soc 80:2587

Lévi FA, Zidani R, Vannetzel JM, Perpoint B, Focan C, Faggiuolo R, Chollet P, Garufi C, Itzhaki M, Dogliotti L, Iacobelli S, Adam R, Kunstlinger F, Gastiaburu J, Bismuth H, Jasmin C, Misset J (1994) Chronomodulated versus fixed-infusion-rate delivery of ambulatory chemotherapy with oxaliplatin, fluorouracil, and folinic acid (leucovorin) in patients with colorectal cancer metastases: a randomized multi-institutional trial. J Natl Cancer Inst 86:1608–1617

Lewy A, Sack R (1997) Exogenous melatonin's phase-shifting effects on the endogenous melatonin profile in sighted humans: a brief review and critique of the literature. J Biol Rhythms 12:588–594

Lieberman H, Waldhauser F, Garfield G, Lynch H, Wurtman R (1984) Effects of melatonin on human mood and performance. Brain Res 323:201–207

Lissoni P, Barni S, Tancini G, Crispino S, Paolorossi F, Lucini V, Mariani M, Cattaneo G, Esposti D, Esposti G (1987) Clinical study of melatonin in untreatable advanced cancer patients. Tumori 73:475–480

Lissoni P, Barni S, Crispino S, Tancini G, Fraschini F (1989) Endocrine and immune effects of melatonin therapy in metastatic cancer patients. Eur J Cancer Clin Oncol 25:789–795

Lissoni P, Barni S, Cattaneo G, Tancini G, Esposti G, Esposti D, Fraschini F (1991) Clinical results with the pineal hormone melatonin in advanced cancer resistant to standard antitumor therapies. Oncology 48:448–450

Lissoni P, Barni S, Ardizzoia A, Paolorossi F, Crispino S, Tancini G, Tisi E, Archili C, De Toma D, Pipino G, Conti A, Maestroni G (1992) Randomized study with the pineal hormone melatonin versus supportive care alone in advanced nonsmall cell lung cancer resistant to a first-line chemotherapy containing cisplatin. Oncology 49:336–339

Lissoni P, Barni S, Ardizzoia A, Tancini G, Conti A, Maestroni G (1994a) A randomized study with the pineal hormone melatonin versus supportive care alone in patients with brain metastases due to solid neoplasms. Cancer 73:699–701

Lissoni P, Barni S, Cazzaniga M, Ardizzoia A, Rovelli F, Brivio F, Tancini G (1994b) Efficacy of the concomitant administration of the pineal hormone melatonin in cancer immunotherapy with low-dose IL-2 in patients with advanced solid tumors who had progressed on IL-2 alone. Oncology 51:344–347

Lissoni P, Meregalli S, Fossati V, Paolorossi F, Barni S, Tancini G, Frigerio F (1994c) A randomized study of immunotherapy with low-dose subcutaneous interleukin-2 plus melatonin vs. chemotherapy with cisplatin and etoposide as first-line therapy for advanced nonsmall cell lung cancer. Tumori 80:464–467

Lissoni P, Ardizzoia A, Barni S, Paolorossi F, Tancini G, Meregalli S, Esposti G, Zubelewicz B, Braczowski R (1995a) A randomized study of tamoxifen alone versus tamoxifen plus melatonin in estrogen receptor-negative heavily pretreated metastatic breast cancer patients. Oncol Rep 2:871–873

Lissoni P, Barni S, Fossati V, Ardizzoia A, Cazzaniga M, Tancini G, Frigerio F (1995b) A randomized study of neuroimmunotherapy with low-dose subcutaneous interleukin-2 plus melatonin compared to supportive care alone in patients with untreatable metastatic solid tumour. Support Care Cancer 3:194–197

Lissoni P, Meregalli S, Nosetto L, Barni S, Tancini G, Fossati V, Maestroni G (1996a) Increased survival time in brain glioblastomas by a radioneuroendocrine strategy with radiotherapy plus melatonin compared to radiotherapy alone. Oncology 53:43–46

Lissoni P, Brivio O, Brivio F, Barni S, Tancini G, Crippa D, Meregalli S (1996b) Adjuvant therapy with the pineal hormone melatonin in patients with lymph node relapse due to malignant melanoma. J Pineal Res 21:239–242

Lucas R, Freedman M, Munoz M, Garcia-Fernández J, Foster R (1999) Regulation of the mammalian pineal by non-rod, non-cone, ocular photoreceptors. Science 284:505–507

Mauer AM (1965) Diurnal variation of proliferative activity in the human bone marrow. Blood 26:1–7

McIntyre I, Norman T, Burrows G, Armstrong S (1989) Human melatonin suppression by light is intensity dependent. J Pineal Res 6:149–156

Mhatre M, Shah P, Juneja H (1984) Effect of varying photoperiods on mammary morphology, DNA synthesis, and hormone profile in female rats. J Natl Cancer Inst 72:1411–1415

Miaskowski C, Portenoy R (1998) Update on the assessment and management of cancer-related fatigue. Principles and Practice of Supportive Oncology Updates 1:1–10

Morley AA (1966) A neutrophil cycle in healthy individuals. Lancet ii:1220

Mormont M, Claustrat B, Waterhouse J, Touitou Y, Levi F (1998) Clinical relevance of circadian rhythm assessment in cancer patients. In: Touitou Y (ed) Biological Clocks. Mechanisms and applications. Elsevier Science, Amsterdam, pp 497–505

Nash RE, Echave Llanos JM (1971) Circadian variation in DNA synthesis of fast-growing and slow-growing hepatoma: DNA synthesis rhythm in hepatoma. J Natl Cancer Inst 47:1007–1012

Neri B, Fiorelli C, Moroni F, Nicita G, Paolotti M Ponchietti R, Raugel A, Santoni G, Trippitelli A, Grechi G (1994) Modulation of human lymphoblastoid interferon activity by melatonin in metastatic renal cell carcinoma. A phase II study. Cancer 73:3015–3019

Oldani A, Ferini-Strambi L, Zucconi M, Stankov B, Fraschini F, Smirne S (1994) Melatonin and delayed sleep phase syndrome: ambulatory polygraphic evaluation. Neuroreport 6:132–134

Panzer A, Wiljoen M (1997) The validity of melatonin as an oncostatic agent. J Pineal Res 22:184–202

Pittendrigh C (1960) Circadian rhythms and the circadian organization of living systems. Cold Spring Harbor Symp Quant Biol 25:159–184

Ponassi A, Morra L, Bonanni F, Molinari A, Gigli G, Vercelli M, Sacchetti C (1979) Normal range of blood colony-forming cells (CFU-C) in humans: influence of experimental conditions, age, sex, and diurnal variations. Blut 39:257–263

Raloff J (1998) Does light have a dark side? Nighttime illumination might elevate cancer risk. Science News 154:248–250

Reiter RJ (1988) Pineal gland, cellular proliferation and neoplastic growth: an historical account. In: Gupta D, Attanasio A, Reiter RJ (eds) The pineal gland and cancer. Brain Research Promotion, Tübingen, pp 41–64

Reiter RJ, Tan TX, Poeggeler B, Menendez-Pelaez A, Chen LD, Saarela S (1994) Melatonin as a free-radical scavenger: Implications for aging and age-related processes. Ann NY Acad Sci 719:1–12

Reppert S (1997) Melatonin receptors: molecular biology of a new family of G protein-coupled receptors. J Biol Rhythms 12:528–531

Rivard G, Infante-Rivard C, Hoyoux C, Champagne J (1985) Maintenance chemotherapy for childhood acute lymphoblastic leukemia: better in the evening. Lancet ii:1264–1266

Robinson W, Dreiling L, Gonzalez R, Balmer C (1995) Treatment of human metastatic malignant melanoma with high dose oral melatonin. In: Fraschini F, Reiter RJ, Stankov B (eds) The Pineal Gland and Its Hormones: Fundamentals and Clinical Perspectives. Plenum Press, New York, pp 219–225

Ross DD, Pollak A, Akman SA, Bachur NR (1980) Diurnal variation of circulating human myeloid progenitor cells. Exp Hematol 8:954–960

Rubin NH, Hokanson JW, Mayschak JW, Tsai TH, Barranco SC, Scheving LE (1983) Several cytokinetic methods for showing circadian variation in normal murine tissue and in a tumor. Am J Anat 168:15–26

Sanchez S, Hrushesky W, Wood P, Vyzula R (1993) Host-tumor balance depends upon IL-2 circadian timing. Proc AACR, Orlando, FL

Scheving LE (1959) Mitotic activity in the human epidermis. Anat Rec 135:7–19

Scheving LE (1981) Circadian rhythms in cell proliferation: their importance when investigating the basic mechanism of normal versus abnormal growth. In: von Mayersbach H, Scheving LE, Pauli JE (eds) 11th International Congress of Anatomy, part C, Biological rhythms in structure and function. Alan R. Liss, New York, pp 39–79

Scheving LE, Burns ER, Pauli JE, Tsai TH (1978) Circadian variation in cell division of the mouse alimentary tract, bone marrow and corneal epithelium. Anat Rec 191:479–486

Scheving LE, Tsai TS, Feuers RJ, Scheving LA (1989) Cellular mechanisms involved in the action of anticancer drugs. In: Lemmer B (ed) Chronopharmacology: cellular and biochemical interactions. Marcel Dekker, New York, pp 317–369

Schipper H, Turley E, Baum M (1996) A new biological framework for cancer research. Lancet 348:1149–1153

Shanahan T, Zeitzer J, Czeisler C (1997) Resetting the melatonin rhythm with light in humans. J Biol Rhythms 12:556–567

Skene D, Bojkowski C, Currie J, Wright J, Boulter P, Arendt J (1990) 6-sulphatoxymelatonin production in breast cancer patients. J Pineal Res 8:269–276

Smaaland R, Laerum OD, Lote K, Sletvold O, Sothern R, Bjerknes R (1991) DNA synthesis in human bone marrow is circadian stage dependent. Blood 77:2603–2611

Smaaland R, Lote K, Sothern RB, Laerum OD (1993) DNA synthesis and ploidy in non-Hodgkin's lymphomas demonstrate intrapatient variation depending on circadian stage of cell sampling. Cancer Res 53:3129–3138

Starr KW (1970) Growth and new growth: environmental carcinogens in the process of ontogeny. Prof Clin Cancer 4:1–29

Stoney PJ, Halberg F, Simpson HW (1975) Circadian variation in colony-forming ability of presumably intact murine bone marrow cells. Chronobiologia 2:319–324

Tähti E (1956) Studies of the effect of x-radiation on 24-hour variations in the mitotic activity in human malignant tumors. Acta Path Microbiol Scand, Suppl 117:1–61

Tamarkin L, Cohen M, Roselle D, Reichert C, Lippman M, Chabner B (1981) Melatonin inhibition and pinealectomy enhancement of 7,12-dimethylbenz(a)anthracene-induced mammary tumors in the rat. Cancer Res 41:432–436

Tamarkin L, Danforth D, Lichter A, DeMoss E, Cohen M, Chabner B, Lippman M (1982) Decreased nocturnal plasma melatonin peak in patients with estrogen receptor positive breast cancer. Science 216:1003–1005

Tzischinsky O, Lavie P (1994) Melatonin possesses time-dependent hypnotic effects. Sleep 17:638–645

Tzischinsky O, Shlitner A, Lavie P (1993) The association between the nocturnal sleep gate and nocturnal onset of urinary 6-sulfatoxymelatonin. J Biol Rhythms 8:199–209

Ubeda A, Trillo MA, House DF, Blackman CF (1995) Melatonin enhances junctional transfer in normal CH3H/10T1/2 cells. Cancer Lett 91:241–245

Vollrath L, Semm P, Gammel G (1981) Sleep induction by intranasal application of melatonin. Adv Biosci 29:327–329

Voutilainen A (1953) Über die 24-Stunden-Rhythmik der Mitosefrequenz in malignen Tumoren. Acta Path Microbiol Scan 99 (suppl.):1–104

Waalen J (1993) Nighttime light studied as possible breast cancer risk. J Natl Cancer Inst 85:1712–1713

Wehr TA (1992) Improvement of depression and triggering of mania by sleep deprivation. J Am Med Assoc 267:548–551

Wehr TA (1997) Melatonin and seasonal rhythms. J Biol Rhythms 12:518–527

Weinblatt ME, Kremer JM, Bankhurst AD, Bulpitt KJ, Fleischmann RM, Fox RI, Jackson CG, Lange M, Burge DJ (1999) A trial of etanercept, a recombinant tumor necrosis factor receptor: Fc fusion protein, in patients with rheumatoid arthritis receiving methotrexate. New Engl J Med 340:253–259

Wood P, Hrushesky W (1996) Circadian rhythms and cancer chemotherapy. Crit Rev Eukaryotic Gene Expression 6:299–343

Wurtman R, Zhdanova I (1995) Improvement of sleep quality by melatonin. Lancet 346:1491

Zhang R, Diasio R (1994) Pharmacologic basis for circadian pharmacodynamics. Hrushesky WJM (ed) Circadian Cancer Therapy. CRC Press, Boca Raton, FL, pp 61–103

Zhdanova I, Wurtman R, Lynch H (1995) Sleep-inducing effects of low doses of melatonin ingested in the evening. Clin Pharmacol Ther 57:552–558

Section VI
Electromagnetic Fields and Cancer: The Possible Role of Melatonin

28 Circadian Disruption and Breast Cancer

Richard G. Stevens

Abstract

Breast cancer is a disease of modern life. There is no consensus on the specific reasons, but it appears certain that something about life in industrialized societies is substantially increasing risk over the risk in nonindustrialized societies. One possibility is that disruption of our normal daily biological rhythms by certain aspects of modern life leads to a reduction in the hormone melatonin and that this reduction increases a woman's lifetime risk of breast cancer.

The primary exposures to be considered include factors that might disrupt normal circadian hormone rhythms such as indoor lighting both at night and during the day, certain aspects of diet, including alcohol consumption, anthropogenic electromagnetic field exposure, shift work, and sedentary lifestyle.

Given the magnitude of the breast cancer burden in the industrialized world, and the lack of convincing explanations, examination of disruption of circadian rhythms offers a new avenue of investigation.

28.1 Background

Breast cancer is the leading cause of cancer death in women in industrialized countries. Incidence rates, and to some extent mortality, are increasing worldwide (Ursin et al. 1994, Davis et al. 1990). The assumption that the cause of these increases is the change from indigenous diets to the high-fat Western diet has been challenged by recent evidence showing no relationship between adult fat consumption and breast cancer risk (Willett et al. 1992). Given the conflicting evidence, the role, if any, of adult consumption of fat in breast cancer etiology is unclear. High energy intake in childhood may be important (Micozzi 1987), but is difficult to study epidemiologically. Enthusiasm for estrogenic chemicals in the environment as an important determinant of risk (Wolff et al. 1993) has also been tempered by recent studies (Krieger et al. 1994) and biological considerations (Safe 1995).

Something about industrialization seems to increase risk of breast cancer. But which of the myriad changes brought by industrialization is responsible? Disruption of circadian rhythms of the hormone melatonin may be part of the answer.

In its role as a neuroendocrine transducer, the pineal gland provides a hormonal signal that is synchronized to the daily light-dark cycle (Wurtman and Axelrod 1965). Melatonin, the principal pineal hormone, exerts a generally suppressive action on other endocrine glands. Melatonin is an indoleamine synthesized from the amino acid tryptophan through an intermediate step of serotonin. Melatonin appears to be involved in the regulation of gonadal function by influencing the hypothalamic-pituitary-gonadal axis. Animal studies indicate that melatonin can modify the firing frequency of the hypothalamic gonadotropin-releasing hormone (GnRH) pulse generator, thereby affecting the release of gonadotropins (LH and FSH) from the pituitary (Bittman et al. 1985, Yellon and Foster 1986, Robinson et al. 1986, Robinson 1987). LH and FSH in turn are critical in the biosynthesis of steroid hormones in the ovary, including estradiol (Catt and Dafau 1991, Adashi 1991). Consequently, pineal function, through the secretion and action of melatonin, may exert an important modulatory effect on ovarian function and estrogen production. Indeed, decreased concentrations of circulating melatonin can result in increased release of the gonadotropins LH and FSH from the pituitary, and estrogen release by the ovaries (Sandyk 1992, Reiter 1981, Yie et al. 1995, Penny et al. 1987, Bettie et al. 1992). Production of melatonin is suppressed by light as perceived by the retina. Hence, circulating melatonin concentrations are low in daylight and much higher at night, exhibiting a characteristic rise in concentration after darkness and a peak near the midpoint of the dark interval (Reiter 1990, Tamarkin et al. 1985).

In non-seasonally breeding animals such as humans, whether melatonin can affect estrogen production in females or testosterone production in males is not yet clear. Direct evidence is difficult to gather in humans for ethical reasons, and when experiments are conducted utilizing pharmacologic doses of melatonin, they may not be entirely relevant to variations in physiologic melatonin concentrations within the normal range. There is indirect evidence in humans consistent with a melatonin-mediated suppression of ovarian or testicular function (Brzezinski 1997). For example, one anecdotal piece of evidence was published by Puig-Domingo et al. (1992) in which an otherwise normal 21-year-old male was referred to clinic for delayed puberty. He had minimal facial hair growth and azoospermia. His plasma melatonin was 4–5 times higher than that of ten normal men matched for weight, height, and age. After 7 years of follow-up, the patient's plasma melatonin concentration spontaneously declined to 2 times that of other men, and he fathered a child.

Cohen et al. (1978) suggested that reduced pineal melatonin production might increase human breast cancer risk because lower melatonin output would lead to an increase in circulating estrogen levels, and would stimulate the proliferation of breast tissue. Indeed, early menarche, late menopause, and nulliparity are each associated with an increased risk of breast cancer (MacMahon et al. 1973), and all result in a longer period for proliferation of the breast epithelial stem cells at risk (Moolgavkar et al. 1980). Melatonin, in turn, has a strong inhibitory effect on breast cancer in animals (Blask 1993). In particular, melatonin injection inhibits chemically induced mammary tumor development in rats and pinealectomy enhances tumor development in both the dimethylbenzanthracene (DMBA) model (Tamarkin et al. 1981) and the nitroso-methylurea (NMU) model (Blask et al. 1991).

There are several mechanistic interpretations of these observations (Stevens 1993), two of which are: melatonin may slow development and turnover of the normal mammary cells at risk of malignant transformation, as suggested by Cohen et al.

(1978), and/or melatonin may be directly oncostatic, as elucidated by Hill and Blask (1988). These two mechanisms act at the opposite ends of the carcinogenic process and if both are valid, then exposures affecting melatonin in women throughout premenopausal and postmenopausal life would be important. The relative impact of melatonin disruption on etiology of breast cancer in humans is not clear because it has not been adequately studied.

If melatonin does affect breast cancer in women, then any factor in the industrialized environment that affects melatonin rhythms may also influence estrogen and breast cancer risk. There have been a number of reports of an association of alcohol consumption and risk of breast cancer (Hiatt and Bawol 1984, Longnecker et al. 1988, Blot 1992), consistent with a role for melatonin (Stevens and Hiatt 1987). There are certain prescription drugs, notably β-blockers, that lower melatonin production (Arendt et al. 1985). In addition, shift work and certain aspects of diet may have an effect: shift work for the obvious reason of disrupting melatonin rhythms (Touitou et al. 1990, Dollins et al. 1993), and diet from the perspective of variations in tryptophan content (Fernstrom 1985) and the fat-protein-carbohydrate ratios (Fernstrom and Wurtman 1974, Fernstrom et al. 1979). Another avenue of investigation is electric power. The generation, distribution, and use of electric power is a hallmark of modern life. Electric power results in human exposure to light at night (LAN) and anthropogenic electromagnetic fields (EMFs). These are relatively new exposures in the human environment and virtually everyone is exposed to some extent in modern society. Could electric power be implicated in the high rates of breast cancer in industrialized nations? The reason to consider this suggestion is the possible reduction of melatonin by LAN and/or EMFs (Stevens et al. 1992, Stevens and Davis 1996).

28.2 Light and Melatonin

The effect of light on pineal function in humans has been extensively studied (Wetterberg 1993), and several features of light's effect are relevant to potential long-term health effects: (1) the effect of LAN is qualitatively similar to the effect in other mammals in that sufficient intensity of nocturnal illumination completely suppresses melatonin production (Lynch et al. 1984, Lewy et al. 1980), (2) some people are much more sensitive to LAN than others (McIntyre et al. 1990), (3) limited evidence supports blue-green LAN as most effective in reducing melatonin production (Brainard et al. 1988, Reiter 1985), (4) there appears to be a dose-response to LAN (McIntyre et al. 1989): the brighter the light the greater the reduction in nocturnal circulating melatonin, (5) light quality during the day appears to affect nighttime melatonin production (McIntyre et al. 1990, Lewy et al. 1987, Wehr et al. 1995, Boivin et al. 1996) as well as the human circadian pacemaker (Czeisler et al. 1986), and (6) women may be more sensitive to the suppressive effects of LAN than men (Monteleone et al. 1995). The minimum light level necessary to suppress melatonin completely ranges from 200 to 3,000 lux; published studies report light level in photopic lux as produced by a Vita-Lite fluorescent bulb.

28.3 Electric and Magnetic Fields and Melatonin

The possibility that low-frequency electric or magnetic fields can reduce melatonin production has received increasing attention over the past 10 years. In experiments with rodents, an electric or magnetic field has been reported to lower melatonin in more than a half dozen independent laboratories. There are inconsistencies, and some laboratories have not been able to replicate earlier published results. Of the published results, either a significant reduction has been seen or there has not been a significant effect. These results are reviewed extensively in Brainard et al. (1999) and in Portier and Wolfe (1998).

There have now been published about a dozen reports on EMFs and melatonin in humans. A half dozen have been experimental and a half dozen have been observational. In the experimental studies, volunteers are exposed to a clean 60-Hz sinusoidal field for one night and melatonin is evaluated. These studies have not shown effects. The observational studies have examined the relationship of household or occupational field exposures and melatonin levels in people. These have generally shown an inverse association: the higher the ambient fields, the lower melatonin. The strength of the experimental studies is the application of a well-defined exposure, but the drawback is that the exposures are unlike what people actually experience in real life, and the fact that only acute exposures can be considered. The advantage to the observational studies is that chronic, real-world exposures are examined, but the disadvantage is the fact that it is not an experiment. These results are also reviewed in detail in Brainard et al. (1999) and in Portier and Wolfe (1998).

28.4 Breast Cancer in Blind Women

If LAN increases the risk of breast cancer in sighted women, Hahn (1991) reasoned that profoundly blind women, who do not perceive LAN, would be at reduced risk. He analyzed over 100,000 hospital discharge records published by the National Hospital Discharge Survey. He identified those women with a primary diagnosis of breast cancer, and a comparison group with diagnoses of stroke or cardiovascular disease. He then determined the prevalence of a secondary diagnosis of profound, bilateral blindness in the women with breast cancer and in the comparison group. Among the comparison group, 0.26% were also blind, which is approximately the percentage expected on the basis of national surveys of nonhospitalized women. Among the women with breast cancer, however, only 0.15% were also blind, consistent with Hahn's prediction. Hahn adjusted for diabetes and marital status as best he was able but the adjustment depended on the reliability of the data in the hospital records. Feychting et al. (1998) made a similar observation based on a cohort study in Sweden. They found risk of cancer (all forms combined) to be lower among blind persons, and this reduced risk also held true specifically for breast cancer in women. Pukkala et al. (1999) similarly found lower breast cancer risk in blind women in Finland, although risk for other cancers was higher, in contrast to the Swedish study. In an extension of the study in Finland, Verkasalo et al. (1999) collected a year's more data on breast cancer cases, and further refined the definition of visual impairment to include five categories from moderate low vision to total blindness. Over the period 1983–1996, there were 124 cases of breast cancer among ca. 11,000 women with some degree of visual impairment. The Standardized Incidence Ratio declined from 1.05 in women with "moderate low vision" to 0.47 in totally blind women; the decrease was monotonic and statistically significant.

28.5 Conclusion

In addition to LAN, artificial electric lighting during the day can influence melatonin rhythms (McIntyre et al. 1990). Among the most profound environmental consequences of electrification is exposure to LAN, and to light during the day of a different character than sunlight. We have come in our evolutionary past from an environment with dark nights and bright, full-spectrum days to an environment with lighted nights in homes during sleep and dim, spectrum-restricted "days" inside buildings where most people now work. Indeed, the "built environment" is the predominant environment in the industrialized world. Since the vast majority of people in industrialized societies work in buildings, and virtually all people sleep in buildings, the long-term health effect of the indoor lighted environment deserves attention, particularly in terms of chronic disruption of melatonin rhythms (Bullough et al. 1996). As stated by Brainard et al. (1988), "…lighting for everyday use is currently arranged for the optimum visual effectiveness. An additional consideration may now be necessary: what intensity and spectrum of illumination are optimal for proper regulation of the circadian and neuro-endocrine apparatus?"

References

Adashi EY (1991) The ovarian life cycle. In Yen SSC, Jaffe RB (eds) Reproductive Endocrinology, Third Edition., WB Saunders, Philadelphia, pp 202–204

Arendt J, Bojkowski C, Franey C, Wright W, Marks V (1985) Immunoassay of 6-hydroxymelatonin sulfate in human plasma and urine: abolition of the urinary 24-hour rhythm with Atenolol. J Clin Endocrinol Metab 60:1166–1173

Bittman EL, Kaynard AK, Olster DH (1985) Pineal melatonin mediates photoperiodic control of pulsatile luteinizing hormone secretion in the ewe. Neuroendocrinology 40:409–418

Blask DE, Pelletier DB, Hill SM, Lemus-Wilson A, Grosso DS, Wilson ST, Wise ME (1991) Pineal melatonin inhibition of tumor promotion in the N-nitroso-N-methylurea model of mammary carcinogenesis: potential involvement of antiestrogenic mechanisms in vivo. J Cancer Res Clin Oncol 117:526–532

Blask DE (1993) Melatonin in oncology. In: Yu H-S, Reiter RJ (eds) Melatonin: Biosynthesis, Physio-logical Effects, and Clinical Applications. CRC Press, Boca Raton, pp 447–475

Blot WJ (1992) Alcohol and cancer. Cancer Res, suppl 52:2119s–2123 s

Boivin DB, Duffy JF, Kronauer RE, Czeisler CA (1996) Dose-response relationships for restting of human circadian clock by light. Nature 379:540–542

Brainard GC, Lewy AJ, Menaker M, Fredrickson RH, Miller LS, Weleber RG, Cassone V, Hudson D (1988) Dose-response relationship between light irradiance and the suppression of plasma melatonin in human volunteers. Brain Res 454:212–218

Brainard GC, Kavet R, Kheifets LI (1999) The relationship between electromagnetic field and light exposures to melatonin and breast cancer risk: a review of the relevant literature. J Pineal Res 26:65–100

Brzezinski A (1997) Melatonin in humans. New Engl J Med 336:186–195

Bullough J, Rea MS, Stevens RG (1996) Light and magnetic fields in a neonatal intensive care unit. Bioelectromagnetics 17:396–405

Catt KJ, Dafau ML (1991) Gonadotropic hormones: biosynthesis, secretion, receptors, and actions. In: Yen SSC and Jaffe RB (eds) Reproductive Endocrinology, Third Edition, WB Saunders, Philadelphia, pp 144–151

Cohen M, Lippman M, Chabner B (1978) Role of the pineal gland in the aetiology and treatment of breast cancer. Lancet 2:814–816

Czeisler CA, Allan JS, Strogatz SH, Rondam JM, Sánchez R, Rios D, Freitag WO Richardson GS (1986) Bright light resets the human circadian pacemaker independent of the timing of the sleep-wake cycle. Science 233:667–672

Davis DL, Hoel D, Fox J, Lopez A (1990) International trends in cancer mortality in France, West Germany, Italy, Japan, England and Wales, and the USA. Lancet 336:474–481

Dollins AB, Lynch HJ, Wurtman RJ, Deng MH, Lieberman HR (1993) Effects of illumination on human nocturnal serum melatonin levels and performance. Physiol Behav 53:153–160

Fernstrom JD (1985) Dietary effects on brain serotonin synthesis: relationship to appetite regulation. Am J Clin Nutr 42:1072–1082

Fernstrom JD, Wurtman RJ (1974) Nutrition and the Brain. Sci Am 230:84–91

Fernstrom JD, Wurtman RJ, Hammerstrom-Wiklund B, Rand WM, Munro HN, Davidson CS (1979) Diurnal variations in plasma concentrations of tryptophan, tyrosine, and other neutral amino acids: effect of dietary protein intake. Am J Clin Nutr 32:1912–1922

Feychting M, Österlund B, Ahlbom A (1998) Reduced cancer incidence among the blind. Epidemiology 9:490–494

Hahn RA (1991) Profound bilateral blindness and the incidence of breast cancer. Epidemiology 2:208–210

Hill SM, Blask DE (1988) Effects of the pineal hormone melatonin on the proliferation and morphological characteristics of human breast cancer cells (MCF-7) in culture. Cancer Res 48:6121–6126

Hiatt RA, Bawol RD (1984) Alcoholic beverage consumption and breast cancer. Am J Epidemiol 120:676–683

Krieger N, Wolff MS, Hiatt RA, Rivera M, Vogelman J, Orentreich N (1994) Breast cancer and serum organochlorines: a prospective study among white, black, and asian women. J Natl Cancer Inst 86:589–599

Lewy AJ, Wehr TA, Goodwin FK (1980) Light suppresses melatonin secretion in humans. Science 210:1267–1269

Lewy AJ, Sack RL, Miller S, Hoban TM (1987) Antidepressant and circadian phase-shifting effects of light. Science 235:352–355

Longnecker MP, Berlin JA, Orza MJ, Chalmers TC (1988) A meta-analysis of alcohol consumption in relation to risk of breast cancer. JAMA 260:652–656

Lynch HJ, Deng MH, Wurtman RJ (1984) Light intensities required to suppress nocturnal melatonin secretion in albino and pigmented rats. Life Sci 35:841–847

MacMahon B, Cole P, Brown J (1973) Etiology of human breast cancer: a review. J Natl Cancer Inst 50:21–42

McIntyre IM, Norman TR, Burrows GD, Armstrong SM (1989) Human melatonin suppression by light is intensity dependent. J Pineal Res 6:149–156

McIntyre IM, Norman TR, Burrows GD (1990) Melatonin supersensitivity to dim light in seasonal affective disorder. Lancet 335:488

Micozzi MS (1987) Cross-cultural correlation of childhood growth and adult breast cancer. Am J Phys Anthropol 73:525–537

Monteleone P, Esposito G, La Rocca A, Maj M (1995) Does bright light suppress nocturnal melatonin secretion more in women than men? J Neural Transm 102:75–80

Moolgavkar SH, Day NE, Stevens RG (1980) Two-stage model for carcinogenesis: epidemiology of breast cancer in females. J Natl Cancer Inst 65:559–569

Penny R, Stanczyk F, Goebelsmann U (1987) Melatonin: data consistent with a role in controlling ovarian function. Endocrinol Invest 10:499–505

Portier CJ, Wolfe MS (1998) Assessment of health effects from exposure to power-line frequency electric and magnetic fields. National Institute of Environmental Health Sciences, NIH Pub. No. 98–3981

Puig-Domingo M, Webb SM, Serrano J, Peinado MA, Corcoy R, Ruscalleda J, Reiter RJ, de Leiva A (1992) Brief Report: melatonin-related hypogonadotropic hypogonadism. New Engl J Med 327:1356–1359

Pukkala E, Verkasalo PK, Ojamo M, Rudanko S-L (1999) Visual impairment and cancer: a population-based cohort study in Finland. Cancer Causes Control 10:13–20

Reiter RJ (ed) (1981) The pineal gland, vols I, II, III. CRC Press, Boca Raton, Fl

Reiter RJ (1990) Effects of light and stress on pineal function. In: Wilson BW, Stevens RG, Anderson LE (eds) Extremely Low Frequency Electromagnetic Fields: The Question of Cancer. Battelle Press, Columbus, pp 87–107

Robinson JE, Kaynard AH, Karsch FJ (1986) Does melatonin alter pituitary responsiveness to gonadotropin-releasing hormone in the ewe? Neuroendocrinology 43:635–640

Robinson JE (1987) Photoperiodic and steroidal regulation of the luteinizing hormone pulse generator in ewes. In: Crowley WF, Hofler JG (eds) The episodic secretion of hormones. Wiley, New York, p 159

Safe SH (1995) Environmental and dietary estrogens and human health: is there a problem? Environ Health Perspec 103:346–351

Sandyk R (1992) The pineal gland and the menstrual cycle. Intern J Neuroscience 63:197–204

Stevens RG, Hiatt RA (1987) Alcohol, melatonin, and breast cancer. New Engl J Med 317:1287

Stevens RG, Davis S, Thomas DB, Anderson LE, Wilson BW (1992) Electric power, pineal function, and the risk of breast cancer. FASEB J 6:853–860

Stevens RG, Davis S (1996) The melatonin hypothesis: electric power and breast cancer. Environ Health Perspect 104(Suppl 1):135–140

Stevens RG (1993) Biologically-based epidemiological studies of electric power and cancer. Environ Health Perspect, supplement 101(4):93–100

Tamarkin L, Cohen M, Roselle D, Reichert C, Lippman M, Chabner B (1981) Melatonin inhibition and pinealectomy enhancement of 7,12-dimethylbenz(a)anthracene-induced mammary tumors in the rat. Cancer Res 41:4432–4436

Tamarkin L, Baird CJ, Almeida OF (1985) Melatonin: a coordinating signal for mammalian reproduction? Science 227:714–720

Touitou Y, Motohashi Y, Reinberg A, Touitou C, Bourdeleau P, Bogdan A, Auzeby A (1990) Effect of shift work on the night-time secretory patterns of melatonin, prolactin, cortisol and testosterone. Eur J Appl Physiol 60:288–292

Ursin G, Bernstein L, Pike MC (1994) Breast Cancer. In: Cancer Surveys. Imperial Cancer Research Fund. vol. 19/20:241–264

Verkasalo PK, Pukkala E, Stevens RG, Ojamo M, Rudanko S-L (1999) Inverse association between breast cancer incidence and degree of visual impairment in Finland. Brit J Cancer 80:1459–1460

Voordouw G, Euser R, Verdonk RE, Alberda BT, de Jong FH, Drogendijk AC, Fauser BC, Cohen M (1992) Melatonin and melatonin-progestin combinations alter pituitary-ovarian function in women and can inhibit ovulation. J Clin Endocrinol Metab 74:108–117

Wehr TA, Giesen HA, Moul DE, Turner EH, Schwartz PJ (1995) Suppression of men's responses to seasonal changes in day length by modern artificial lighting. Am J Physiol 269:R173–178

Wetterberg L (ed) (1993) Light and Biological Rhythms in Man. Wenner-Gren International Series. Pergamon Press, Oxford

Willett WC, Hunter DJ, Stampfer MJ, Colditz G, Manson JE, Spiegelman D, Rosner B, Hennekens C, Speizer FE (1992) Dietary fat and fiber in relation to risk of breast cancer. JAMA 268:2037–2044

Wolff MS, Toniolo PG, Leel EW, Rivera M, Dubin N (1993) Blood levels of organochlorine residues and risk of breast cancer. J Natl Cancer Inst 85:648–652

Wurtman RJ, Axelrod J (1965) The pineal gland. Sci Am 213:50–60

Yellon SM, Foster DL (1986) Melatonin rhythms time photoperiod-induced puberty in the female lamb. Endocrinology 119:44–49

Yie SM, Brown GM, Liu GY, Collins JA, Hughes EG, Foster WG, Younglai EV (1995) Melatonin and steroids in human pre-ovulatory follicular fluid: seasonal variations and granulosa cell steroid production. Hum Reprod 10:50–55

29 Breast Cancer and Use of Electric Power: Experimental Studies on the Melatonin Hypothesis

Wolfgang Löscher

Abstract

Two products of electric power, light at night and electromagnetic fields, can decrease production of melatonin by the pineal gland and thereby perhaps increase the risk of breast cancer. This electric power/breast cancer hypothesis, also known as the "melatonin hypothesis," has attracted a great deal of interest, in part because it is a plausible explanation for the increased tumor growth upon 50-Hz magnetic field (MF) exposure previously seen by two independent groups in chemical models of breast cancer in rats. In a large series of experiments in female Sprague-Dawley rats, we recently found that, consistent with the melatonin hypothesis, prolonged exposure to 50-Hz MFs at flux densities in the µTesla range decreases nocturnal melatonin plasma levels, increases the activity of ornithine decarboxylase in breast tissue, impairs immune surveillance, and enhances mammary tumor development and growth in response to the chemical carcinogen 7,12-dimethylbenz[a]anthracene. The results of our studies are described and discussed in this chapter, including their relevance for human risk assessment.

29.1 Introduction

Breast cancer is a disease of modern life with a high incidence in industrialized countries. Yet it is unclear which of the diverse changes associated with industrialization might account for the increase in breast cancer in modern societies. In 1987, Stevens presented a hypothesis that use of electric power may increase the risk of breast cancer (Stevens 1987). This hypothesis was based on a number of experimental reports indicating an effect of light and extremely low frequency (ELF) *electric* fields on pineal melatonin production, and on the relationship of melatonin to mammary (breast) carcinogenesis. However, at the time when this hypothesis was presented it was not known whether ELF (50- or 60-Hz) *magnetic* fields (MFs) affect pineal melatonin production in experimental animals and/or humans. Furthermore, there was no experimental evidence for increased breast cancer development or growth in response to MF exposure. This prompted us to carry out a series of experiments designed expressly to test the "melatonin hypothesis" of electromagnetic field-promoted breast cancer development and growth in female rats.

29.2 The Melatonin Hypothesis

Figure 29.1 illustrates the proposed mechanisms by which chronic exposure to an ELF MF may increase mammary carcinogenesis in rats (modified from Stevens 1987). Based on this hypothesis, chronic exposure to an ELF (50- or 60-Hz) MF suppresses the normal nocturnal synthesis of pineal melatonin. Because melatonin physiologically suppresses estrogen production by the ovary and prolactin production by the pituitary (Reiter 1991), a reduction in melatonin would in turn result in increased estrogen and prolactin production and thereby induce increased turnover of the breast epithelial stem cells at risk for malignant transformation (Fig. 29.1). In other words, the likelihood that breast stem cells will be affected by cancer-causing agents (e.g., chemical carcinogens), such as occur in the environment (cf, Huff 1993), would be increased by reduced production of melatonin. In addition, in view of the oncostatic effect of melatonin on breast cancer growth (Blask 1993), the development and growth of breast cancer, once initiated, would be facilitated by reduced melatonin levels. The risk of tumor formation could be further increased by the possible link between MFs, melatonin, and the immune surveillance system (Fig. 29.1). Melatonin has been shown to stimulate various immune parameters involved in tumor defence mechanisms (Maestroni 1993; Maestroni and Conti 1993) so that MF exposure, via reduction in melatonin might impair the immune response to tumor cells. All these alterations in response to MF exposure, and probably several additional MF-induced effects not directly linked to melatonin (Stevens 1993), might ultimately increase the risk of breast cancer formation (Fig. 29.1).

Magnetic field exposure and breast cancer
Melatonin as a plausible biological link?

50/60 Hz magnetic field exposure
↓
Reduced nocturnal melatonin production by pineal gland
↓
Reduced melatonin level in plasma
↓
Increased estrogen production by ovary
Increased prolactin production by pituitary
↓
Increased proliferation of breast epithelial stem cells in mammary gland
↓
Increased susceptibility of breast epithelial stem cells to carcinogens such as DMBA
↑↓
Reduced oncostatic effect of melatonin on tumor formation
Reduced immune response to tumor formation
↓
Consequence: Increased risk of breast cancer development and growth

Fig. 29.1. Schematic illustration of the melatonin hypothesis, i.e. proposed mechanisms by which chronic exposure to an ELF (50/60-Hz) MF may increase the development and growth of breast cancer. (Modified from Stevens 1987)

In the following, it is described how we experimentally examined the different steps of the melatonin hypothesis described above. In all experiments, female rats of the same strain and age were used under strictly controlled laboratory conditions. Details of the experimental data reviewed here have been published elsewhere (Löscher et al. 1993; Löscher et al. 1994; Baum et al. 1995; Löscher et al. 1995; Mevissen et al. 1995, 1996a, b, 1998a, b).

29.3 Effect of ELF MFs on Melatonin Levels

It has long been known that weak, static MFs induce pineal metabolic and physiologic changes (Semm et al. 1980), with the retinas considered as the site of magnetoreception (Reiter 1993), and there is now general agreement that reduced pineal melatonin synthesis is a consequence of MF exposure under certain conditions (Reiter 1993). Recently, it was demonstrated that exposure of male rats to 50-Hz MFs of low (µTesla; µT) flux density for 6 weeks led to a drop in nighttime levels of both pineal and serum melatonin (Kato et al. 1993). In the female Sprague-Dawley rats at the age used for all experiments described in the present paper, prolonged exposure to a 50-Hz MF also significantly decreased nocturnal melatonin levels, although this was not seen in all experiments of our group. A significant decrease was observed at a flux density of 0.3–1 µT (Löscher et al. 1994), i.e., in the range of flux densities associated with increased cancer risks in epidemiological studies on residential exposures in humans (Savitz and Ahlbom 1994). Increase of flux density to 10 µT appeared to be associated with a more marked suppression of melatonin levels (Mevissen et al. 1996a). However, long-term exposure to higher flux densities (50 or 100 µT) was not associated with decreased plasma melatonin during the night (Mevissen et al. 1996b, 1998). This seems to be in contrast to a study by Selmaoui and Touitou (1995) in rats exposed to 50-Hz MF of either 1, 10 or 100 µT for 12 h or 30 days. These authors found that short-term exposure depressed pineal *N*-acetyltransferase (NAT), the rate-limiting enzyme for melatonin synthesis, and nocturnal serum melatonin concentration only at 100 µT, while long-term exposure already significantly depressed pineal NAT and serum melatonin at 10 µT, indicating a cumulative effect of MF on pineal function. We are currently investigating the reasons for this apparent disparity between the data of Selmaoui and Touitou (1995) and our data by using different periods of exposure at 100 µT.

At present, our data in female rats only partially confirm the initial step of the melatonin hypothesis, i.e., the ELF MF-induced reduction in melatonin. Studies are under way to examine whether the reduction in melatonin found in some of our studies does lead to increased estrogen and prolactin production in female rats. With respect to our observations of MF-induced melatonin decreases in female rats, it is interesting to note that we obtained such an effect by using horizontally orientated fields, whereas Kato et al. (1994) reported that circular but not horizontal or vertical 50-Hz, 1-µT MFs have an effect on pineal gland or plasma melatonin in male rats. Our findings may thus point to the possibility that female rats are more sensitive than males in terms of suppression of melatonin production in response to horizontal MFs. Interestingly, a recent review of animal studies on the role of 50/60-Hz MFs in carcinogenesis indicated that sex might be a predisposing host susceptibility factor to tumor-promoting or -co-promoting effects of MF exposure (Löscher and Mevissen 1994). This important point deserves further investigation.

29.4 Effect of ELF MFs on Breast Tissue Proliferation

Melatonin is known to exert modulatory influences on the growth of certain tissues, such as the mammary gland (Reiter 1991; Mediavilla et al. 1992). For instance, suppression of pineal melatonin production in rats by continuous exposure to light was reported to increase mammary cell proliferation which could be prevented by administering melatonin (Mhatre et al. 1984). In order to examine whether ELF MF exposure leads to increased breast tissue proliferation in vivo, we determined the activity of ornithine decarboxylase (ODC) in various tissues of female rats after prolonged MF exposure. ODC is a key enzyme in the biosynthesis of polyamines that promotes cell proliferation and has been suggested to play an important role in tumor promotion (O'Brien et al. 1975; O'Brien 1976; Russell 1980). First studies on the possibility of ODC alterations by exposure to electromagnetic fields used the *electric* component of such fields, showing that exposure of cell lines to a 60-Hz electric field produces changes in ODC levels which are similar to changes produced by exposure to the tumor promoter 12-*O*-tetradecanoylphorbol-13-acetate (TPA) (Byus et al. 1987; Cain et al. 1988). More recent experiments on the effects of 50/60-Hz *magnetic* fields on ODC levels in mouse and human cell lines demonstrated that exposure at flux densities in the 10–100 µT range significantly increased ODC, the increase differing in magnitude between cell lines from 50% to 500% (Litovitz et al. 1991; Mattson et al. 1993a; Mattson et al. 1993b). In our in vivo experiments in female rats, ODC was determined in liver, spleen, intestine (duodenum), bone marrow, skin (epidermis) of the ears, and different parts of the breast in order to examine whether MF exposure induces tissue-specific alterations in ODC. With respect to breast tissue, it has to be noted that female rats possess six pairs of mammary gland complexes so that mammary tissue extends from the cervical to the inguinal part of the body. Because of this extensive distribution of mammary tissue, we determined ODC separately in thoracic and inguinal mammary complexes in rats. If the melatonin hypothesis illustrated in Fig. 29.1 is correct, one would assume a regionally selective increase in ODC in all parts of the hormone-dependent mammary tissue but not various other, hormone-independent tissues. Indeed, after prolonged exposure of female rats in a 50-Hz, 50-µT MF, a marked increase in ODC activity was found in mammary tissue, but not several other tissues examined in this respect (Mevissen et al. 1995). The only other significant, albeit much less marked increase in ODC was seen in the spleen. Thus, these data strongly indicated that in vivo exposure of female rats induced a pronounced increase in proliferation of breast epithelial stem cells, thereby substantiating the hypothesis shown in Fig. 29.1.

29.5 Effect of ELF MFs on Mammary Carcinogenesis

The next, most important question was of course whether the reduced melatonin levels and increased breast tissue proliferation in response to ELF MF exposure result in increased susceptibility to carcinogens and enhanced mammary cancer development and growth. While we performed a series of experiments in female Sprague-Dawley rats using the 7,12-dimethylbenz[*a*]anthracene (DMBA) model of breast cancer (see below), a group from Georgia (Beniashvili et al. 1991), using the "complete" carcinogen *N*-nitroso-*N*-methylurea (NMU) to produce mammary tumors in female rats, reported an increased tumor incidence and enhanced progression from benign to malignant tumor forms in animals exposed daily for 3 h to 50-Hz, 20-µT MFs over a period of up to 2 years. Furthermore, rats exposed without any carcinogen spontaneously developed significantly more mammary tumors than sham controls (Beniashvili et al. 1991; see statistical evaluation of data in Löscher and Mevissen 1994). In respect to these data from the NMU model, it should be noted that Blask (1993) reported that, while NMU-induced breast tumors are sensitive to alterations in melatonin, melatonin failed to reduce the circulating levels of either estradiol or prolactin, indicating that alterations in the production of these hormones are not the mechanism by which melatonin affects tumor growth in this model.

In our studies, we used a well-established rat model of chemical carcinogenesis in which mammary tumors are induced by oral administration of the polyaromatic hydrocarbon DMBA. Although the NMU model system used by Beniashvili et al. (1991) appears to have some advantages over the DMBA approach (Blask 1993), we chose DMBA for the following reasons.

Mammary adenocarcinomas induced by DMBA are estrogen- and prolactin-responsive and would thus respond to any melatonin-induced alteration in circulating levels of either estradiol or prolactin (Welsch 1985). There is ample evidence that DMBA-induced breast cancer development and growth in female rats is sensitive to alterations of melatonin levels (cf. Stevens et al. 1992; Blask 1993). Thus, enhanced cancer development and growth is seen in rats in which melatonin production has been suppressed by either exposure to constant light or removal of the pineal gland. Conversely, DMBA-induced mammary carcinogenesis can be inhibited by administration of melatonin in rats, most likely by the melatonin-induced reduction in prolactin levels (cf. Blask 1993). All these data strongly indicate that a decrease in melatonin in response to MF exposure should exert effects on DMBA-induced mammary carcinogenesis in rats.

Interestingly, in the absence of any other manipulation of melatonin production, levels of melatonin decrease during development and growth of DMBA-induced mammary tumors, without it being clear whether this is an effect of DMBA or, as is more likely, a result of tumor growth (Blask 1993). A similar observation of decreased melatonin levels has been reported in women with primary breast cancer, particularly in those with estrogen receptor-positive tumors (cf. Blask 1993). A similar phenomenon of decreased melatonin during DMBA-induced mammary carcinogenesis also occurred under the experimental conditions of our studies (Mevissen et al. 1996a). Thus, compared with findings in rats without DMBA treatment, melatonin significantly decreased in both pineal and plasma of female rats after treatment with DMBA at the time of massive cancer growth, indicating that the rats used in our studies were sensitive to manipulations of melatonin production.

The experimental protocol used for our DMBA experiments and our exposure system have been described in detail elsewhere (Löscher et al. 1993; Baum et al. 1995). The DMBA administration and MF exposure protocol allowed the MF to interact with all processes of DMBA-induced mammary carcinogenesis. Because we did not know before whether MF would affect the DMBA model and which process would be most sensitive, we developed a DMBA administration protocol with four oral DMBA doses over a period of 3 weeks in rats which were either MF- or sham-exposed. Using this protocol, about 40%–60% of sham controls developed palpable and macroscopically visible mammary tumors over a period of 3 months, thus allowing the study of any effect (both inhibitory or stimulatory) of MF exposure on development and growth of breast cancer.

After a series of preliminary experiments with relatively small numbers of rats (Mevissen et al. 1993), five experiments with larger sample sizes were undertaken. Each experiment comprised a sham control group and a group exposed to one MF flux density. Further groups not treated with DMBA were exposed or sham-exposed together with the DMBA-treated groups. Because of various factors, including season, which can affect mammary carcinogenesis even under strictly controlled laboratory environment (i.e., controlled temperature, light cycle, standardized food, etc.; cf. Löscher et al. 1997), all data obtained in MF-exposed rats were compared with the concurrently processed sham control group. All experiments were done "blind," i.e., the experimenters involved in handling of animals and data collection and evaluation were not aware of which rat group was MF-exposed and which was sham-exposed.

The four flux densities (all 50 Hz) studied were a gradient field of $0.3-1$ μT, and three homogeneous fields of 10, 50, and 100 μT, respectively. The experiment with 100 μT was repeated in other groups of rats. Taken together, 432 DMBA-treated rats were sham-exposed and 432 DMBA-treated rats were MF-exposed in these five experiments. The average incidence of mammary tumors (grossly recorded at the time of necropsy, i.e., after 3 months of exposure) was 54% in controls and 67% in MF exposed rats, the difference being highly significant ($P < 0.0001$). However, this type of data evaluation ignored the differences existing between different flux densities. As shown in Fig. 29.2, exposure to a field of $0.3-1$ μT did not lead to significant differences in tumor incidence at the time of necropsy. When flux density was increased to 10 μT, the incidence of grossly recorded tumors at necropsy was 10% higher compared with that in the concurrent controls, though the difference was again not significant (Fig. 29.2). Further increase in flux density to 100 μT induced several significant effects on mammary carcinogenesis in the DMBA model. Significantly more rats developed palpable tumors during MF exposure than during sham exposure, and at the end of the exposure period there was a significant 50% increase in the incidence of grossly recorded tumors (Figs. 29.2, 29.3). In contrast to the increased number of rats with tumors, the number of tumors per rat was not significantly altered. However, the median size of tumors in MF-exposed rats was markedly enhanced (Baum et al. 1995). In addition to increased incidence of rats with palpable and macroscopically visible (grossly recorded) tumors and increased size of the excised tumors in the experiment with 100 μT, histopathological examination of tumors indicated that significantly more rats of the MF-exposed group had developed malignant mammary tumors compared with rats of the control group, indicating that MF exposure had increased not only tumor growth but also progression to malignancy (Baum et al. 1995). However, in contrast to the findings of palpation and recording of macroscopically visible tumors at the time of necropsy, histopathological examination of mammary tissue on serial sections of all six pairs of mammary glands in each animal revealed that the incidence of mammary lesions was not significantly different between exposed rats and controls (Baum et al. 1995). These data thus strongly indicate that under the conditions of these studies MF exposure did not initiate cancers or increase the initiation of cancers by DMBA, but enhanced the growth and progression of the DMBA-induced lesions, which would be consistent with a co-promoting effect of MF exposure. In consequence, although MF exposure did not lead to more tumors, tumors grew more rapidly and were thus diagnosed earlier by palpation or macroscopic examination at the time of necropsy.

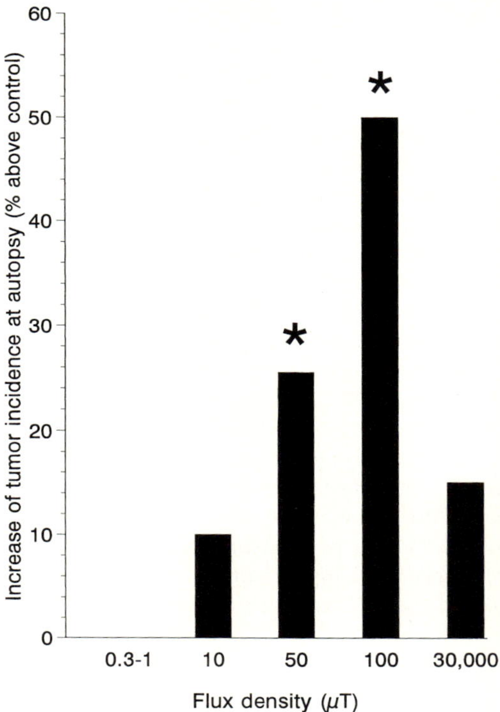

Fig. 29.2. Effect of 50-Hz MF exposure at different flux densities on mammary tumor incidence in the DMBA model in female rats. In all experiments, DMBA was administered by gavage at four doses of 5 mg per rat with intervals of 1 week between each dose. MF or sham exposure started immediately after the first DMBA dose. Duration of exposure was 13 weeks. The figure shows data from necropsy, i.e., incidence of macroscopically visible (grossly recorded) mammary tumors after 13 weeks of MF exposure at flux densities ranging from 0.3–1 µT to 30,000 µT. Data are shown as percent increase in tumor incidence above that in concurrent sham-exposed control groups. The average control tumor incidence at 13 weeks was 51%. Data are based on a total of 369 DMBA-treated sham-exposed controls and 366 DMBA-treated MF-exposed rats. *Stars* indicate a significant difference from concurrent sham controls (P at least < 0.05). Data are from Löscher et al. (1993, 1994) and Mevissen et al. (1993, 1996a, 1996b)

The fourth experiment undertaken in this series used a flux density of 50 µT. Again, a significantly higher number of rats with tumors was seen in the MF-exposed compared to the sham-exposed group, with a difference at the time of necropsy of 25.5% (Fig. 29.2). Linear regression analysis of the data from the four experiments indicated a highly significant linear relation between flux density and increase in incidence of macroscopically visible mammary tumors at time of necropsy (Löscher and Mevissen 1995). To our knowledge, these data provide the first evidence of a linear relationship between flux density and effect of MF exposure in a cancer model. However, as shown in Fig. 29.2, increase of flux density to 30 mT did not induce a more marked increase in tumor incidence but rather indicated that the significant effect of MF exposure on mammary carcinogenesis may be restricted to a "window" in the µT range. Such windows, i.e., ranges in which a biological system exhibits enhanced sensitivity to MF

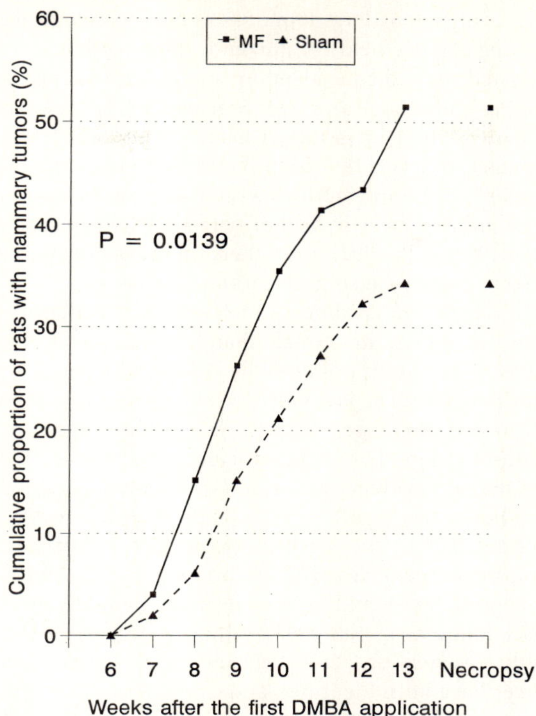

Fig. 29.3. Cumulative proportion of rats with mammary tumors as a function of time after the first DMBA administration in the first experiment with 50-Hz, 100-µT MF exposure. Details of the DMBA dosing and MF exposure protocol are given in the legend to Fig. 29.2. Data shown for weeks 6–13 are from weekly palpation of the rats, whereas data shown for necropsy are from macroscopic examination of mammary tissue after dissection, i.e., grossly recorded mammary tumors. Ninety-nine rats were used per group. The difference between the incidence curves of the two groups was significant ($P = 0.0139$) when compared by a log-rank test. Furthermore, most of the individual weekly data were significantly different (8, 9, 10, 11, and 13 weeks: P at least < 0.05). In addition, tumor incidence at the time of necropsy was significantly different ($P < 0.05$). Data are from Löscher et al. (1993) and Baum et al. (1995)

exposure, have been reported previously both for field amplitude and for the duration of exposure (cf. Litovitz et al. 1992).

The fifth experiment was a replicate study of the first 100-µT experiment. After necropsy, the incidence of macroscopically visible mammary tumors was 62% in controls, but 83% in MF-exposed rats, the 35% difference between the groups being statistically significant (Mevissen et al. 1998a). Data substantiate that long-term exposure of DMBA-treated female Sprague-Dawley rats to an alternating MF of low flux density promotes the development and growth of mammary tumors, thus indicating that MF exposure exerts tumor-promoting and/or co-promoting effects. Furthermore, the data show that the effects of MF exposure in the DMBA breast cancer model are reproducible if the same experiment is repeated in the same laboratory.

One may argue that the flux densities (i.e., 50 and 100 µT) at which we found significant tumor co-promoting effects of MF exposure are far above the range of flux densities occurring in human occupational or residential exposures. However, in this respect two points are important. First, if one assumes that biological effects of ELF MF are due to the induced electric fields or currents, the magnitude of the applied field may have to be adjusted to yield equivalent dosimetric exposure parameters in different species of different body shape and size (Bracken 1992). Respective scaling factors for rat:human are about 5–10:1, i.e., a 100-μT field for a rat would correspond to a 10- to 20-μT field for humans (Bracken 1992). On the other hand, some experimental work suggests that time-varying MF interactions with tissues and cells are related to the MF itself rather than an induced electric field or current, in which case no scaling would be needed (Bracken 1992; Polk 1992). Even then, there may be marked species differences, as known from cancer-causing or -promoting agents (Gold et al. 1992; Huff 1993; Vaino and Wilbourn 1993). Animal models used in the identification of carcinogens are usually considered less susceptible than humans so that direct extrapolation of dose from rodents to humans is not possible; however, safety factors (i.e., factors by which a rodent carcinogenic dose is divided) are used in risk assessment that relates human exposures to the carcinogenic potency of a given agent in rodents. In this respect it is important to note that for many agents now known to cause or promote cancers in humans the first evidence of carcinogenicity was obtained in experimental animals, and this knowledge, together with similarities in mechanisms of carcinogenesis across species, led to the scientific logic and public health strategy that chemical or other agents shown clearly to be carcinogenic (i.e., initiating, promoting, or co-promoting) in animals should be considered as being likely and anticipated to present cancer risks in humans (Huff 1993). With regard to the DMBA rat model used in the present study, a variety of factors relevant to the development and growth of human breast cancer have been identified and characterized by this model (Huggins and Yang 1962; Russo and Russo, 1978; Welsch 1985; Rogers et al. 1989; Cornélissen and Halberg 1992) so that findings such as the present data on MF exposure cannot be ignored for reasons of "uncertainty." Furthermore, the similarities of our findings on MF exposure in the DMBA rat model to the findings of Beniashvili et al. (1991) in the NMU rat model strongly suggest that these findings are not restricted to one animal model system of breast cancer. Interestingly, as in humans, genetics play an important role in the susceptibility to cancer in models of mammary carcinogenesis such as the DMBA model in rats (Melhem et al. 1991); consequently effects of ELF MF on gene expression, as suggested by some recent studies (e.g., Phillips et al. 1992), should be considered in addition to the melatonin hypothesis used to explain the present findings.

In apparent contrast to our findings on MF exposure in the DMBA model in Sprague-Dawley rats, Ekström et al. (1998) recently reported that 50-Hz MF exposure at flux densities of 250 or 500 μT with a 15 s on/15 s off schedule for 21 weeks did not significantly affect mammary tumor growth in response to intragastric administration of 7 mg DMBA per Sprague-Dawley rat. However, the MF exposure scheme was different in the study of Ekström et al. (1998) and was applied in a strict promotional scheme. Furthermore, the subline of Sprague-Dawley outbred rats used by Ekström et al. (1998) was much more sensitive to DMBA than our Sprague-Dawley rats, pointing to genetic differences between the two sublines of Sprague-Dawley rats used. As described above, we have previously shown a linear relationship between flux density and MF effect on tumor incidence in the DMBA model when effects of flux densities between 1 and 100 μT were evaluated (Löscher and Mevissen 1995), which could indicate that a further increase in flux density as used by Ekström et al. (1998) would also increase the effect of MF on DMBA-induced mammary tumors. However, as discussed above, the effect of MF in the DMBA model is lost at higher flux density, indicating a window effect of MF exposure in the low µT range.

In a recent draft technical report of the U.S. Department of Health and Human Services prepared for public review and comment (NTP, 1998), a series of studies on effects of 50 or 60-Hz, 100-µT MF exposure in the DMBA model in Sprague-Dawley rats were described. Different DMBA dosing protocols were used, i.e., four times 5 mg/rat and 13 weeks of exposure, four times 2 mg/rat and 13 weeks of exposure, and once 10 mg DMBA/rat and 26 weeks of exposure. In none of the experiments were significant MF effects observed. Again, there were various differences from our experiments, including a different diet, shorter exposure per day, use of different rooms for sham and MF exposure, and use of a subline of Sprague-Dawley rats with markedly higher susceptibility to DMBA than our rats. These and other differences between studies on MF exposure in the DMBA model will allow us to evaluate which environmental and genetic factors are critical for effects of MF exposure in this model.

29.6 Effect of ELF MFs on Immune Responses to Tumor Formation

We are currently studying whether 50-Hz exposure of female rats in the µT range affects the immune system as the body's main protective mechanism against tumor formation and growth. Resistance against the development of neoplasia depends upon a variety of host immunologic and other factors (Welsch 1985; Cruse and Lewis 1988). For instance, chemical carcinogens such as DMBA have been shown to temporally suppress immune reactivity, and a positive correlation was found between the degree of immune system depression and the individual rate of breast cancer growth in rats (e.g., Gallo et al. 1993). Thus, if MF exposure depresses immune system functions via reduced melatonin production or other mechanisms, this could be critically involved in enhanced tumor growth in response to MF. Although it has repeatedly been suggested that MF exposure might affect the immune system, the evidence is inconclusive, and the majority of data are from in vitro experiments without any direct relation to tumor formation (cf. Walleczek 1992; Löscher and Liburdy 1998).

In a first attempt to delineate in vivo effects of MF exposure on the immune surveillance system in female rats as used in our studies, we determined whether mitogen-induced T cell activation is altered in female rats exposed in a 50-Hz, 50-µT MF for 13 weeks. T cells (or T lymphocytes) constitute the majority (65%–85%) of lymphocytes in peripheral blood and appear to be important for antitumor activity and its immune regulation (Cruse and Lewis 1988). In MF-exposed rats, activation (i.e., proliferation) of T cells in response to a mitogen (concanavalin A) was markedly suppressed compared with sham-exposed controls (Mevissen et al. 1996b). These data strongly suggest that immune system depression is involved in the tumor co-promoting effects of MF exposure observed in rats. In this respect it is interesting to note that a reduced stimulation of lymphocytes by the T cell-selective mitogen concanavalin A has also been reported for 50-Hz MF-exposed human lymphocytes (Conti et al. 1983). One can only speculate as to the underlying cellular mechanisms of interaction between MF exposure and the immune system. Both, reduced melatonin level and function (Maestroni 1993) and altered Ca^{2+} signals (cf. Walleczek 1992) may play a key role in this respect. Experiments are under way to examine in detail how the observed alteration in T cell reactivity depends on the duration and flux density of MF exposure. Furthermore, levels of cytokines involved in T cell growth and function will be determined in these studies, and intracellular Ca^{2+} concentration will be measured in lymphocytes isolated from MF-exposed animals.

In a first study in this respect (Mevissen et al. 1998b), female Sprague-Dawley rats of the same age as previously used by us for the mammary tumor experiments were continuously exposed for different periods (2, 4, 8, and 13 weeks) to a 50-Hz, 100-µT MF. Control groups were sham-exposed together with the MF-exposed rats. Following the different exposure periods, splenic lymphocytes were cultured and the proliferative responses to the T cell-selective mitogen concanavalin A (Con A) and the B cell-selective pokeweed mitogen (PWM) were determined. Furthermore, the production of interleukin-1 (IL-1) was determined in the splenocyte cultures. The mitogenic responsiveness of T cells was markedly enhanced after 2 weeks of MF exposure, suggesting a co-mitogenic action of MF. A significant, but less marked increase in T cell mitogenesis was seen after 4 weeks of MF exposure, whereas no difference in comparison with sham controls was observed after 8 weeks, indicating adaptation or tolerance to this effect of MF exposure. Following 13 weeks of MF exposure, a significant decrease in the mitogenic responsiveness of lymphocytes to Con A was obtained, substantiating the suppressive effect seen previously with 13 weeks of exposure at 50 µT (see above). This triphasic alteration in T cell function, i.e., activation, tolerance, and suppression, during prolonged MF exposure resembles alterations observed during chronic administration of mild stressors, substantiating that cells respond to MF in the same way as they do to other environmental stresses. In contrast to T cells, the mitogenic responsiveness of B cells and IL-1 production of PWM-stimulated cells were not altered during MF exposure. The data demonstrate that exposure of female rats to MF in vivo induces complex effects on the mitogenic responsiveness of T cells, which may lead to impaired immune surveillance after long-term exposure.

29.7 Magnetic Field Exposure and Breast Cancer: Conclusions

The present data provide the first direct experimental evidence that the melatonin hypothesis first proposed by Stevens (1987) for chronic 60-Hz electric fields may explain carcinogenic effects of ELF MF, particularly facilitation of development and growth of breast cancer. Furthermore, the data from the DMBA model add to the accumulating evidence from laboratory studies that 50/60-Hz MF exposure may exert a co-promoting effect in carcinogenesis (reviewed by Löscher and Mevissen 1994, and Löscher and Liburdy 1998). With respect to the melatonin hypothesis, however, one should note that we failed to demonstrate a significant reduction of pineal or plasma melatonin in response to MF exposure in several of our studies, including those demonstrating a significant effect of MF exposure on mammary carcinogenesis. Thus, other explanations than mere decrease in melatonin *concentrations* are needed to explain our findings in the DMBA model. One very likely explanation stems from Robert Liburdy's laboratory, showing that MF exposure in the µT range inhibits the oncostatic effect of melatonin on breast cancer growth in vitro (Liburdy et al. 1993). If interaction between MF and melatonin at the cellular level also occurs in vivo, it would add to any impairment of melatonin production, so that even small decreases in circulating melatonin levels could have pronounced consequences. In this respect it is important to note that such an effect of MF exposure on melatonin level and function could increase the growth not only of breast cancer but also of other types of cancer (cf. Blask 1993). The data of Liburdy et al. (1993) seem to indicate that a decrease in cellular *function* rather than concentration of melatonin may be critically involved in MF-induced breast cancer growth in the DMBA model. This modified melatonin hypothesis is illustrated in Fig. 29.4. In addition to MF effects on melatonin concentration or function, other actions of MF exposure may be involved in increased risk of breast cancer formation, such as direct effects on Ca^{2+}-related functions or direct effects on the immune surveillance system.

How relevant are the present data from the DMBA model of breast cancer with respect to human risk assessment? Wertheimer and Leeper (1982) were the first to see a MF–breast cancer connection in their 1982 study of residential MF exposures of adults. They discovered a nearly threefold increase among women younger than 55 who lived near to power lines, indicating that MF exposure had accelerated development and growth of breast cancer. More recently, increased breast cancer risks were reported in both women and men in electrical occupations (for review see Gammon and John 1993; Thomas 1993; Tynes 1993; Loomis et al. 1994; Wertheimer and Leeper 1994; Coogan et al. 1996). However, some reports that included women found no excess of breast cancer in an array of occupations related to enhanced MF exposure as well as residential exposures, which could be due to the various differences among studies and the uncertain exposure conditions in most studies. Prospective epidemiological studies with more closely defined exposure conditions are therefore needed to increase our understanding of breast cancer risk related to environmental and occupational MF exposure.

One important question is whether ELF MF exposure reduces melatonin concentrations or function in humans. Most of the available studies on melatonin concentrations

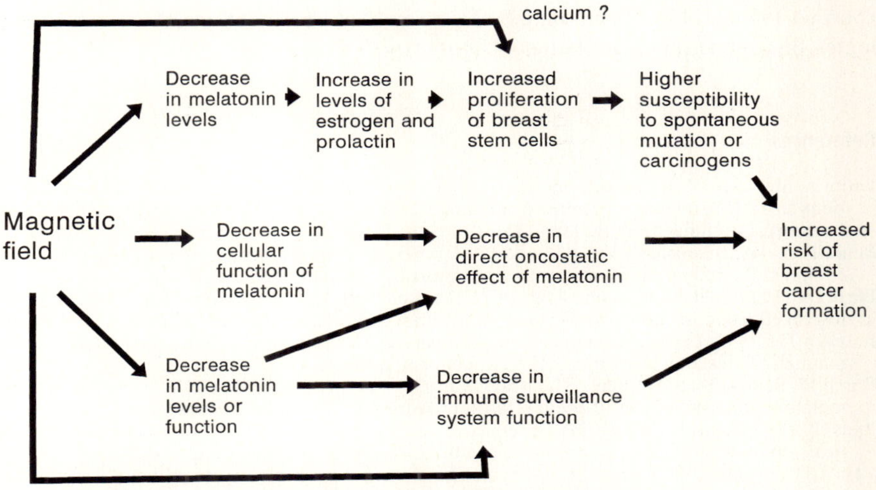

Fig. 29.4. Modified version of the melatonin hypothesis, including MF effects on melatonin function and MF effects unrelated to melatonin

in this regard have used short-term MF exposure in volunteers rather than prolonged or chronic exposure as occur in residential and occupational MF exposures (cf. Graham and Gibertini 1997). Recent studies in human beings under natural occupational or residential exposure conditions have found suppression of urinary excretion of the major melatonin metabolite 6-hydroxymelatonin sulfate, indicating that MF exposure suppresses melatonin production under such "real-life" conditions (Pfluger et al. 1996; Kaune et al. 1997; Burch et al. 1998).

However, with respect to the relevance of the melatonin hypothesis in humans, it should be considered that an effect of MF exposure on carcinogenesis, if occurring in humans, will presumably be small compared with other known cancer promoters or co-promoters. In this respect, it will be important to study experimentally whether effects of MF exposure on cancer growth are amplified in the presence of known cancer promoters or co-promoters, because such interactions can be over-additive and thus markedly increase the risk of exposure (cf. Löscher and Mevissen 1995).

Acknowledgements. We thank Dr. H. Kärner and colleagues (Department of High Voltage Engineering, Technical University of Braunschweig), Dr. U. Mohr and colleagues (Institute of Experimental Pathology, Medical School Hannover), Dr. M. Szamel and colleagues (Institute of Molecular Pharmacology, Medical School Hannover), Dr. A. Lerchl and colleagues (Institute for Reproductive Medicine, University of Münster), and Dr. W. Lehmacher and colleagues (Department of Biometrics and Epidemiology, School of Veterinary Medicine, Hannover) for collaboration and helpful discussions during the studies described in this paper. The studies were supported by equipment and grants from the Forschungsverbund Elektromagnetische Verträglichkeit Biologischer Systeme (Department of High Voltage Engineering, Technical University, Braunschweig, Germany), the Berufsgenossenschaft der Feinmechanik und Elektrotechnik (Köln, Germany), a grant (StSch. 4095) from the Bundesministerium für Umwelt, Naturschutz und Reaktorsicherheit (Bonn, Germany), and a grant (subcontract 19X-SU446 V) from the U.S. Department of Energy, Office of Utility Technologies, through OAK Ridge National Laboratory (TN, U.S.A.).

References

Baum A, Mevissen M, Kamino K, Mohr U, Löscher W (1995) A histopathological study on alterations in DMBA-induced mammary carcinogenesis in rats with 50 Hz, 100 µT magnetic field exposure. Carcinogenesis 16:119–125

Beniashvili DS, Bilanishvili VG, Menabde MZ (1991) Low-frequency electromagnetic radiation enhances the induction of rat mammary tumors by nitrosomethyl urea. Cancer Lett 61:75–79

Blask DE (1993) Melatonin in oncology. In Yu H-S, Reiter RJ (eds) Melatonin, Biosynthesis, physiological effects, and clinical applications. CRC Press, Boca Raton, pp 447–475

Bracken TD (1992) Experimental macroscopic dosimetry for extremely-low-frequency electric and magnetic fields. Bioelectromagnetics Supplement 1, 15–26

Burch JB, Reif JS, Yost MG, Keefe TJ, Pitrat CA (1998) Nocturnal excretion of a urinary melatonin metabolite in electric utility workers. Scand J Work Env Health 24:183–189

Byus CV, Pieper SE, Adey WR (1987) The effects of low-energy 60-Hz environmental electromagnetic fields upon the growth-related enzyme ornithine decarboxylase. Carcinogenesis 8:1385–1389

Cain CD, Salvador EQ, Adey WR (1988) 60-Hz electric field prolongs ornithine decarboxylase activity response to 12-O-tetradecanoylphorbol-13-acetate (TPA) in C3H10T1/2 fibroblasts. Abstracts of the Bioelectromagnetics Society, 10th Annual Meeting, Stamford, CT 3

Conti P, Giganti GE, Cifone MG, Alesse E, Ianni G, Reale M, Angeletti PU (1983) Reduced mitogenic stimulation of human lymphocytes by extremely low frequency electromagnetic fields. FEBS Lett 162:156–160

Coogan PF, Clapp RW, Newcomb PA, Wenzl TB, Bogdan G, Mittendorf R, Baron JA, Longnecker MP
(1996) Occupational exposure to 60-Hertz magnetic fields and risk of breast cancer in women.
Epidemiology 7:459–464

Cornélissen G, Halberg F (1992) Chronobiologic response modifiers and breast cancer development: classical background and chronobiologic tasks remaining. In vivo 6:387–402

Cruse JM, Lewis REJ (1988) Immunomodulation of neoplasia. In: Homburger F, Cruse JM, Lewis REJ
(eds) Progress in experimental tumor research. Karger, Basel

Ekström T, Mild KH, Holmberg B (1998) Mammary tumours in Sprague-Dawley rats after initiation
with DMBA followed by exposure to 50 Hz electromagnetic fields in a promotional scheme.
Cancer Lett 123:107–111

Gallo F, Morale MC, Sambatoro D, Farinella Z, Scapagnini U, Marchetti B (1993) The immune
system response during development and progression of carcinogen-induced rat mammary
tumors: prevention of tumor growth and restoration of immune sytem responsiveness by
thymopentin. Breast Cancer Res Treat 27:221–237

Gammon MD, John EM (1993) Recent etiologic hypotheses concerning breast cancer. Epidemiol
Rev 15:163–168

Gold LS, Slone TH, Stern BR, Manley NB, Ames BN (1992) Rodent carcinogens: Setting priorities.
Science 258:261–265

Graham C, Gibertini M (1997) Human exposure to magnetic fields: Effects on melatonin, hormones,
and immunity. In: Stevens RG, Wilson BW, Anderson LE (eds) The melatonin hypothesis. Breast
cancer and use of electric power. Battelle Press, Columbus, p 479

Huff J (1993) Issues and controversies surrounding qualitative strategies for identifying and forecasting cancer causing agents in the human environment. Pharmacol Toxicol Suppl 1:12–27

Huggins C, Yang NC (1962) Induction and extinction of mammary cancer. Science 137:257–
262

Kato M, Honma KI, Shigemitsu T, Shiga Y (1993) Effects of exposure to a circulatory polarized
50-Hz magnetic field on plasma and pineal melatonin levels in rats. Bioelectromagnetics 14:
97–106

Kato M, Honma K, Shigemitsu T, Shiga Y (1994) Horizontal or vertical 50-Hz, 1-µT magnetic fields
have no effect on pineal gland or plasma melatonin concentration of albino rats. Neurosci Lett
168:205–208

Kaune W, Davis S, Stephens R (1997) Relation between residential magnetic fields, light-at-night, and
nocturnal urine melatonin levels in women. Final Report for Electric Power Research Institute

Liburdy RP, Sloma TR, Sokolic R, Yaswen P (1993) ELF magnetic fields, breast cancer, and melatonin: 60 Hz fields block melatonin's oncostatic action on ER$^+$ breast cancer cell proliferation.
J Pineal Res 14:89–97

Litovitz TA, Krause D, Mullins JM (1991) Effect of coherence time of the applied magnetic field on
ornithine decarboxylase activity. Biochem Biophys Res Commun 178:862–865

Litovitz TA, Montrose CJ, Wang W (1992) Dose-response implications of the transient nature of
electromagnetic-field-induced bioeffects. Bioelectromagnetics Supplement 1:237–246

Loomis DP, Savitz DA, Ananth CV (1994) Breast cancer mortality among female electrical workers
in the United States. J Natl Cancer Inst 86:921–925

Löscher W, Liburdy RP (1998) Animal and cellular studies on carcinogenic effects of low frequency
(50/60-Hz) magnetic fields. Mutation Res 410:185–220

Löscher W, Mevissen M (1994) Animal studies on the role of 50/60-Hertz magnetic fields in carcinogenesis. Life Sci 54:1531–1543

Löscher W, Mevissen M (1995) Linear relationship between flux density and tumor copromoting
effect of prolonged magnetic field exposure in a breast cancer model. Cancer Lett 96:175–
180

Löscher W, Mevissen M, Lehmacher W, Stamm A (1993) Tumor promotion in a breast cancer model
by exposure to a weak alternating magnetic field. Cancer Lett 71:75–81

Löscher W, Wahnschaffe U, Mevissen M, Lerchl A, Stamm A (1994) Effects of weak alternating
magnetic fields on nocturnal melatonin production and mammary carcinogenesis in rats.
Oncology 51:288–295

Löscher W, Mevissen M, Häußler M (1997) Seasonal influence on 7,12-dimethylbenz[a]anthracene-
induced mammary carcinogenesis in Sprague-Dawley rats under controlled laboratory conditions. Pharmacol Toxicol 81:265–270

Maestroni GJM (1993) The immunoneuroendocrine role of melatonin. J Pineal Res 14:1–10

Maestroni GJM, Conti A (1993) Melatonin in relation to the immune system. In: Yu H-S, Reiter RJ
(eds) Melatonin. Biosynthesis, physiological effects, and clinical applications. CRC Press, Boca
Raton, pp 289–309

Mattson M-O, Mild KH, Rehnholm U (1993a) Ornithine decarboxylase activity and polyamine
levels in cell lines exposed to a 50-Hz sine-wave magnetic field. Abstracts of the Bioelectromagnetics Society, 15th Annual Meeting, Los Angeles, CA, pp 93–94

Mattson M-O, Mild KH, Rehnholm U (1993b) Ornithine decarboxylase activity in lymphoblastoid cell lines after 50 Hz sine-wave magnetic field exposure. Abstracts of the Annual Review of Research on Biological Effects of Electric and Magnetic Fields from the Generation, Delivery and Use of Electricity, Savannah, GA 3

Mediavilla MD, San Martin S, Sánchez-Barceló E J (1992) Melatonin inhibits mammary gland development in female mice. J Pineal Res 13:13–19

Melhem MF, Kunz HW, Gill TJ (1991) Genetic control of susceptibility to diethylnitrosamine and dimethylbenzanthracene carcinogenesis in rats. Am J Pathol 139:45–51

Mevissen M, Stamm A, Buntenkötter S, Zwingelberg R, Wahnschaffe U, Löscher W (1993) Effects of magnetic fields on mammary tumor development induced by 7,12-dimethylbenz(a)anthracene in rats. Bioelectromagnetics 14:131–143

Mevissen M, Kietzmann M, Löscher W (1995) *In vivo* exposure of rats to a weak alternating magnetic field increases ornithine decarboxylase activity in the mammary gland by a similar extent as the carcinogen DMBA. Cancer Lett 90:207–214

Mevissen M, Lerchl A, Löscher W (1996a) Study on pineal function and DMBA-induced breast cancer formation in rats during exposure to a 100-mG, 50-Hz magnetic field. J Toxicol Environment Health 48:101–117

Mevissen M, Lerchl A, Szamel M, Löscher W (1996b) Exposure of DMBA-treated female rats in a 50-Hz, 50 µTesla magnetic field: effects on mammary tumor growth, melatonin levels, and T lymphocyte activation. Carcinogenesis 17:903–910

Mevissen M, Häußler M, Lerchl A, Löscher W (1998a) Acceleration of mammary tumorigenesis by exposure of 7,12-dimethylbenz[a]anthracene-treated female rats in a 50-Hz, 100 µT magnetic field: replication study. J Toxicol Environment Health 53:401–418

Mevissen M, Häussler M, Szamel M, Emmendörffer A, Thun-Battersby S, Löscher W (1998b) Complex effects of long-term 50 Hz magnetic field exposure in vivo on immune functions in female Sprague-Dawley rats depend on duration of exposure. Bioelectromagnetics 19:259–270

Mhatre MC, Shah PN, Juneja HS (1984) Effects of varying photoperiods on mammary morphology, DNA synthesis, and hormone profile in female rats. J Natl Cancer Inst 72:1411–1416

NTP TR 489 (1998) NTP Technical Report on the Studies of Magnetic Field Promotion (DMBA Initiation) in Sprague-Dawley Rats (Gavage/Whole Body Exposure Studies). U.S. Department of Health and Human Services, Public Health Services, National Institutes of Health, Washington, DC

O'Brien TG, Simsiman RC, Boutwell RK (1975) Induction of polyamine-biosynthetic enzymes in mouse epidermis and their specificity for tumor promotion. Cancer Res 35:2426–2433

O'Brien TG (1976) The induction of ornithine decarboxylase as an early, possibly obligatory, event in mouse skin carcinogenesis. Cancer Res 36:2644–2653

Pfluger DH, Minder CE (1996) Effects of exposure to 16.7 Hz magnetic fields on urinary 6-hydroxy-melatonin sulfate excretion of Swiss railway workers. J Pineal Res 21:91–100

Phillips JL, Haggren W, Thomas WJ, Ishida-Jones T, Adey WR (1992) Magnetic field-induced changes in specific gene transcription. Biochim Biophys Acta 1132:140–144

Polk C (1992) Dosimetry of extremely-low-frequency magnetic fields. Bioelectromagnetics Suppl 1:209–235

Reiter RJ (1991) Pineal melatonin: cell biology of its synthesis and of its physiological interactions. Endocrine Rev 12:151–180

Reiter RJ (1993) Static and extremely low frequency electromagnetic field exposure – Reported effects on the circadian production of melatonin. J Cell Biochem 51:394–403

Rogers AE (1989) Factors that modulate chemical carcinogenesis in the mammary gland of the female rat. In: Jones TC, Mohr U, Hunt RD (eds) Integument and mammary glands. Springer-Verlag, Berlin, p 304

Russell DH (1980) Ornithine decarboxylase as a biological and pharmacological tool. Pharmacology 20:117–129

Russo IH, Russo J (1978) Developmental stage of the rat mammary gland as determinant of its susceptibility to 7,12-dimethylbenz[a]anthracene. J Natl Cancer Inst 61:1439–1449

Savitz DA, Ahlbom A (1994) Epidemiologic evidence on cancer in relation to residential and occupational exposures. In: Carpenter DO, Ayrapetyan S (eds) Biological effects of electric and magnetic fields. Academic Press, San Diego, p 233

Selmaoui B, Touitou Y (1995) Sinusoidal 50-Hz magnetic fields depress rat pineal NAT activity and serum melatonin. Role of duration and intensity of exposure. Life Sci 57:1351–1358

Semm P, Schneider T, Vollrath L (1980) Effects of an earth-strength magnetic field on electrical activity of pineal cells. Nature 288:607–608

Stevens RG (1987) Electric power use and breast cancer: a hypothesis. Am J Epidemiol 125:556–561

Stevens RG, Davis S, Thomas DB, Anderson LE, Wilson BW (1992) Electric power, pineal function, and the risk of breast cancer. FASEB J 6:853–860

Stevens RG (1993) Biologically based epidemiological studies of electric power and cancer. Environment Health Perspect Suppl 4:93–100

Thomas DB (1993) Breast cancer in man. Epidemiol Rev 15:220–231
Tynes T (1993) Electromagnetic fields and male breast cancer. Biomed Pharmacother 47:425–427
Vainio H, Wilbourn J (1993) Cancer etiology: agents causally associated with human cancer. Pharmacol Toxicol Suppl 1:4–11
Walleczek J (1992) Electromagnetic field effects on cells of the immune system: the role of calcium signaling. FASEB J 6:3177–3185
Welsch CW (1985) Host factors affecting the growth of carcinogen-induced rat mammary carcinomas: a review and tribute to Charles Brenton Huggins. Cancer Res 45:3415–3443
Wertheimer N, Leeper E (1982) Adult cancer related to electrical wires near the home. Int J Epidemiol 11:345–355
Wertheimer N, Leeper E (1994) Are electric or magnetic fields affecting mortality from breast cancer in women? J Nat Cancer Inst 86:1797

30 Magnetic Field Exposure and Pineal Melatonin Production (Mini-Review)

Brahim Selmaoui, *Yvan Touitou*

Abstract

The consequences of electromagnetic exposure for human health are receiving increasing scientific attention and have become the subject of a vigorous public debate. In the present study we evaluated the effects of magnetic field on pineal function in man and rat.

Two groups of Wistar male rats were exposed to 50-Hz magnetic fields of either 1, 10 or 100 µT. The first group was exposed for 12 h and the second for 30 days (18 h per day). Short-term exposure depressed both pineal NAT activity and nocturnal serum melatonin concentration, but only with the highest intensity used (100 µT). Long-term exposure to a magnetic field of 10 and 100 µT significantly depressed the nighttime peak of serum melatonin concentration and pineal NAT activity. Our results show that sinusoidal magnetic field altered the production of melatonin through inhibition of pineal NAT activity. Both the duration and the intensity of exposure played an important role in this effect.

In the second step of this study, 32 young men (20–30 years old) were divided into two groups (control group, i.e., sham-exposed: 16 subjects; exposed group: 16 subjects). The subjects were exposed to the magnetic field from 2300 h to 0800 h (i.e., for 9 h) while lying down. In one experiment the exposure was continuous; in the second one, the magnetic field was intermittent. No significant differences were observed between sham-exposed (control) and exposed men for serum melatonin, 6-sulfatoxymelatonin, immunological and hematological variables, and serum cortisol.

This study suggests that, at least under the conditions used in our experiment, acute exposure to either a continuous or an intermittent 50-Hz magnetic field of 10 µT does not affect plasma melatonin concentration or its circadian rhythm in healthy young men.

30.1 Introduction

Extremely low frequency (ELF) electric and magnetic fields are generated by human industrial activity and are especially prominent in highly industrialized countries. The major sources of ELF field generation are electrical appliances, high power transmission and distribution lines. Because of the ubiquity of electrical appliances and apparatus in modern society, animals and humans live in an extremely complex electric and magnetic field environment. Increasing concern in recent years about the effects of human exposure to these artificial electromagnetic fields (EMFs) has been stimulated by some epidemiological reports suggesting that the incidence of certain types of cancer might be increased in individuals exposed to EMFs (Wertheimer and Leeper 1979, 1982; Savitz et al. 1988). In vivo studies suggest that both electric and magnetic fields, either alone or in combination, may induce biological effects in organisms that could have untoward consequences. The most investigated effect of electromagnetic field is that of alteration in nocturnal rise in melatonin.

30.2 Circadian Rhythm of Melatonin Secretion

The circadian production of melatonin is strictly synchronized by the prevailing light:dark environment, and its presence in blood provides an important time of day message to those organs that are incapable of responding directly to light. The light:dark cycle controls pineal melatonin synthesis via the eyes in mammals. Indeed, in mammals light detected by eyes both inhibits the nocturnal rise in melatonin production and release and acutely suppresses its synthesis and discharge at night when animals or humans are exposed to visible wavelengths (Vollrath et al. 1989; Lewy et al. 1980). Besides visible electromagnetic radiation, ELF electric and magnetic fields, as well as perturbed static magnetic fields, also impair the melatonin production in the pineal gland.

30.3 The Response of the Pineal Gland to Electromagnetic Fields: Role of Duration and Intensity of Exposure

In our laboratory we have examined the effect of magnetic fields on pineal function and then explored to what extent this effect is dependent on the intensity of the magnetic field or on the duration of exposure. Two modes of exposure were used: short term and long term. For long-term exposure, the rats were exposed for 30 days (18 h/day) under three different intensities (1, 10, and 100 µT). For short-term exposure, the rats were exposed to the same intensities for 12 h. We found that exposure of rats to a magnetic field significantly decreased both serum melatonin concentration and pineal NAT activity at the expected peak time of its circadian rhythm after short-term and long-term exposure (Tables 30.1, 30.2). However, pineal HIOMT activity remained unperturbed (Selmaoui and Touitou 1995). Rats exposed for 12 h to magnetic fields showed a decrease in serum melatonin and pineal NAT activity with only 100 µT. In contrast, exposure for 1 month (18 h/day) showed an effect with 10 and 100 µT. This work shows that the intensity of the magnetic field and the duration of exposure might be involved in the response of the pineal gland to magnetic field exposure. Several past reports showed that the chronic exposure of rats to an electric field attenuates the nocturnal rise in melatonin (Wilson et al. 1986; Reiter et al. 1988); however, other studies have not confirmed this initial observation (Grota et al. 1994). Magnetic field exposure has also been shown to interfere with nocturnal melatonin production (Kato et al. 1993, 1994; Yellon 1994; Selmaoui and Touitou 1995). However, a more recent study carried out by

Table 30.1. Effects of short-term exposure (12 h) of male rats to a sinusoidal 50-Hz magnetic field on nocturnal pineal activity. Only 100 μT were able to depress serum melatonin and pineal NAT activity. No effect was observed on HIOMT

Field intensity	Melatonin (pg/ml)	NAT activity (nmol/h/pineal)	HIOMT activity (pmol/h/pineal)
100 μT	70 ± 8[a]	9.56 ± 1.36[a]	98.33 ± 12.27
10 μT	99 ± 21	10.99 ± 1.72	105.25 ± 12.68
1 μT	83 ± 19	11.83 ± 0.83	106.89 ± 16
Control	100 ± 13	12.37 ± 1.52	107.41 ± 20

[a] Significant difference ($P < 0.05$) compared with the control group.

Table 30.2. Effects of long-term exposure (18 h/day for 1 month) of male rats to sinusoidal 50-Hz magnetic fields (from 1 to 100 μT) on nocturnal pineal activity. Only 10 and 100 μT were able to depress serum melatonin and pineal NAT activity. No effect was observed on HIOMT activity

Field intensity	Melatonin (pg/ml)	NAT activity (nmol/h/pineal)	HIOMT activity (pmol/h/pineal)
100 μT	59 ± 6[a]	7.25 ± 1.37[a]	88.25 ± 18
10 μT	56 ± 22[a]	9.08 ± 0.93[a]	100.4 ± 17
1 μT	86 ± 11	11.5 ± 1.71	100.5 ± 15
Control	100 ± 12.5	12.43 ± 1.02	103 ± 25

[a] Significant difference ($P < 0.05$) compared with the control group.

Yellon and Truong (1998) did not replicate the effect obtained in their previous studies. The contradiction between Yellon's results obtained in 1994 and 1998 may be due to the length of the day, which was different in the two studies. On the other hand, static magnetic fields have been repeatedly shown to alter the circadian melatonin rhythm (Semm 1983; Welker et al. 1983; Olcese et al. 1985; Olcese and Reuss 1986; Lerchl et al. 1990, 1991). Recently, Reiter et al. (1998) showed an inconsistent effect on nocturnal pineal melatonin synthesis and serum melatonin levels in rats exposed to pulsed DC magnetic fields. Indeed, out of 23 cases in which pineal melatonin, pineal NAT activity, and serum melatonin levels were measured, pineal NAT activity was depressed in five cases (21.7%), pineal melatonin was decreased in two cases (8.7%), and serum melatonin concentration was reduced in ten cases (43%).

To date, no adequate explanation for the apparent disappearance of the response has been provided. It could be due to differences in methodology, sensitivity of species, or intensity or duration of exposure.

30.4 Effect of Magnetic Field Exposure on Humans

Because humans are living in an extremely complex electric and magnetic field environment there is growing concern on the part of the public over whether EMFs might adversely affect human health. In light of the findings in animals, we have undertaken a broad study on humans concerning the effect of magnetic fields on immunohematologic function (Selmaoui et al. 1996a), pineal function (Selmaoui et al. 1996b), and the hypothalamo-pituitary-thyroid axis and adrenocortical system (Selmaoui et al. 1997).

In this chapter we will discuss only the results obtained regarding the effect of a 50-Hz magnetic field on the main hormone secreted by the pineal gland, melatonin, and its metabolite in urine, 6-sulfatoxymelatonin. Young volunteers (32 subjects: 16 exposed and 16 sham-exposed) were selected according to the screening criteria. Each subject participated in two sessions held within a 4-week period. On the first experimental day, one group of volunteers (16 subjects) was exposed to a continuous (10 µT, 50-Hz) magnetic field, and then, on the second experimental day, to an intermittent (10 µT, 50-Hz) magnetic field. The second group (16 subjects) served as controls for both experimental days. On each experimental day, blood samples were collected at 3-h intervals from 11:00 to 20:00 and hourly from 22:00 to 08:00. Total urine was collected every 3 h from 08:00 to 23:00 and once during the night, from 23:00 to 08:00. The experimental unit consisted of wooden beds that were arranged parallel in a room and oriented along the north-south axis of the earth's magnetic field. The device was able to produce both continuous and intermittent exposure conditions. For the intermittent exposure, the device was turned "on" for 1 h and "off" for the next hour. Whenever the device was turned "on," the magnetic field was applied on a 15-s "on-off" cycle. The results obtained (Figs. 30.1, 30.2) show the circadian profiles of serum melatonin and of its urinary metabolite 6-sulfatoxymelatonin. During daytime, mean values of serum melatonin were less than 10 pg/ml; they increased at night, reaching 60 pg/ml around 04:00. For 6-sulfatoxymelatonin, the values varied from 10 ng/ml during the day to about 50 ng/ml at night. During the two experimental days, the profiles in the sham-exposed group (control group) were similar, with low values during the day, increasing to reach a maximum at night (Figs. 30.1, 30.2).

In the first phase of the experiment (continuous exposure to magnetic fields of 10 µT from 23:00 to 08:00), serum melatonin and urinary 6-sulfatoxymelatonin in

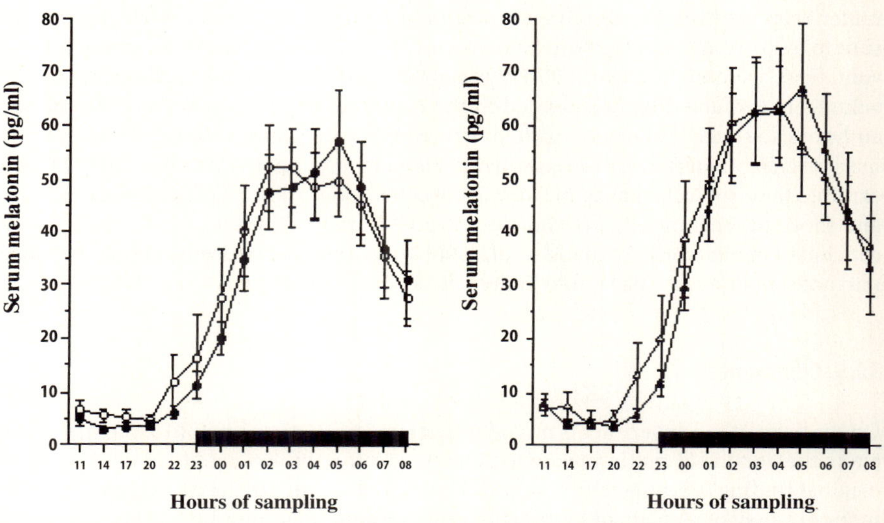

Fig. 30.1. Circadian profiles of melatonin in serum under continuous (*left*) and intermittent (*right*) exposure to a 50-Hz magnetic field (10 µT). The *dark bar* shows the duration of exposure, which corresponds to time in bed. Each time point is the mean ± SEM of 16 measurements. *Open symbols:* control group; *solid symbols:* exposed group

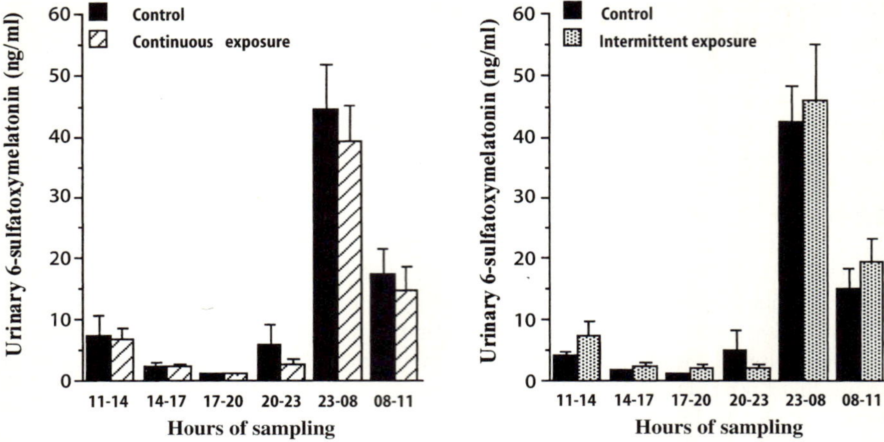

Fig. 30.2. Evaluation of the effect of nocturnal acute exposure to a magnetic field on urinary 6-sulfat-oxymelatonin. Total urine was collected every 3 h from 08:00 to 23:00 and once during the night, from 23:00 to 08:00. Each *bar* corresponds to the mean ± SEM of 16 measurements

exposed subjects were not significantly altered when compared with the controls. The profiles of exposed and sham-exposed groups were nearly identical when super-imposed, as shown in Fig. 30.1. Similarly, the subjects who were exposed to intermittent magnetic fields of 10 µT had no alteration in their circadian rhythm of melatonin secretion. In general, and under our conditions, the results obtained did not support the hypothesis that magnetic fields have an effect on melatonin secretion (Selmaoui et al. 1996b). In another study in which Graham et al. (1996) carried out experiments under relatively similar conditions to ours, no alteration was observed with expo-sure to a 20-µT, 60-Hz magnetic field. A study by Wood et al. (1998) in which human volunteers were exposed to a 50-Hz magnetic field showed that while some of the volunteers (around 20%) responded to the magnetic field, the majority did not. The authors also stated that magnetic fields generated by square-wave currents had a more marked effect than sinusoidal waveforms. It is worth noting that Wood et al. (1998) had exposed their subjects during melatonin onset while our subjects (Selmaoui et al. 1996) and those of Graham et al. (1996) were exposed after the onset of melatonin. As we concluded in our work (Selmaoui et al. 1996b), the timing of exposure should be taken into account and may play a role in the sensitivity of pineal response.

30.5 Comments

On the basis of the experiments on rodents, it appears that exposure to a magnetic field alters the nocturnal rise in melatonin, although some authors could not confirm their original findings (Truong and Yellon 1997; Yellon et al. 1998). The discrepancies in the literature may be explained by the differences in methodology, including differences in species, timing of exposure, duration of exposure, or intensities used in the ex-periments. Therefore, it cannot be ruled out that in experiments with magnetic field exposure, season may also play a role in changes in pineal melatonin secretion

(Bartsch et al. 1994). Clearly, there are discrepancies and apparent inconsistencies in the literature. Nevertheless, there are a sufficient number of positive reports to justify additional investigations on this topic. In humans, there are to date no convincing data showing an effect of magnetic fields on melatonin. Other studies are needed to add information to these few reports.

References

Bartsch H, Bartsch C, Mecke D, Lippert TH (1994) Seasonality of pineal melatonin production in the rat: Possible synchronization by the geomagnetic field. Chronobiol Int 11:21–26

Graham C, Cook MR, Riffle DW, Gerkovich MM, Cohen HD (1996) Nocturnal melatonin levels in human volunteers exposed to intermittent magnetic 60 Hz magnetic fields. Bioelectromagnetics 17:263–273

Kato M, Honma KI, Shigemitsu T, Shiga Y (1993) Effects of exposure to circularly polarized 50-Hz magnetic field on plasma and pineal melatonin levels in rats. Bioelectromagnetics 14:97–106

Kato M, Honma KI, Shigemitsu T, Shiga Y (1994) Circularly polarized 50-Hz magnetic field exposure reduces pineal gland and blood melatonin. Neurosci Lett 166:59–62

Lerchl A, Nonaka KO, Stokkan KA, Reiter RJ (1990) Marked rapid alterations in nocturnal pineal serotonin metabolism in mice and rats exposed to weak intermittent magnetic fields. Biochem Biophys Res Commun 169:102–108

Lerchl A, Nonaka KO, Reiter RJ (1991) Pineal "magnetosensitivity" to static fields is a consequence of induced electric currents (Eddy currents). J Pineal Res 10:109–116

Lewy AJ, Wehr TA, Goodwin FK, Newsome DA, Markey SP (1980) Light suppresses melatonin secretion in humans. Science 210:1267–1269

Olcese J, Reuss S, Vollrath L (1985) Evidence of the involvement of the visual system in mediating magnetic field effects on pineal melatonin synthesis in the rat. Brain Res 333:382–384

Olcese J, Reuss S (1986) Magnetic field effects on pineal melatonin synthesis: comparative studies on albino and pigmented rodents. Brain Res 369:365–368

Reiter RJ, Anderson LE, Buschbom RL, Wilson BW (1988) Reduction of the nocturnal rise in pineal melatonin levels in rats exposed to 60-Hz electric fields in utero and for 23 days after birth. Life Sci 42:2203–2206

Reiter RJ, Tan DX, Poeggeler B, Kavet R (1998) Inconsistent suppression of nocturnal pineal melatonin synthesis and serum melatonin levels in rats exposed to pulsed DC magnetic fields. Bioelectromagnetics 19:318–329

Savitz DA, Wachtel H, Barnes FA, John EM, Tvrdik JG (1988) Case-control study of childhood cancer and exposure to 60-Hz magnetic fields. Am J Epidemiol 128:21–38

Selmaoui B, Touitou Y (1995) Sinusoidal 50-Hz magnetic fields depress rat pineal function. Role of duration and intensity of exposure. Life Sci 57:1351–1358

Selmaoui B, Bogdan A, Auzeby A, Lambrozo J, Touitou Y (1996a) Acute exposure to 50-Hz magnetic field does not affect haematologic and immune functions in healthy young men: A circadian study. Bioelectromagnetics 17:364–372

Selmaoui B, Lambrozo J, Touitou Y (1996b) Magnetic fields and pineal function in human: Evaluation of nocturnal acute exposure to extremely low frequency magnetic fields on serum melatonin and urinary 6-sulfatoxymelatonin circadian rhythms. Life Sci 58:1539–1549

Selmaoui B, Lambrozo J, Touitou Y (1997) Endocrine functions in humans exposed to 50-Hz magnetic field. A circadian study of pituitary, thyroid and adrenal hormones. Life Sci 61:473–486

Semm P (1983) Neurobiological investigations on the magnetic sensitivity of the pineal gland of rodents and pigeons. Comp Biochem Physiol 76 A:683–689

Truong H, Yellon SM (1997) Effect of various acute 60-Hz magnetic field exposures on the nocturnal rise in the adult Djungarian hamster. J Pineal Res 22:177–183

Yellon SM (1994) Acute 60-Hz magnetic field exposure effects on the melatonin rhythm in the pineal gland and circulation of the adult Djungarian hamster. J Pineal Res 16:136–144

Yellon SM, Truong HN (1998) Melatonin rhythm onset in the adult Siberian hamster: influence of photoperiod but not 60-Hz magnetic field exposure on melatonin content in the pineal gland and in circulation. J Biol Rhythms 13:52–59

Vollrath L, Seidel A, Huesgen A, Manz B, Pollow K, Leiderer P (1989) One millisecond of light suffices to suppress night-time pineal melatonin synthesis in rats. Neurosci Lett 98:297–298

Wertheimer N, Leeper E (1979) Electrical wiring configuration and childhood cancer. Am J Epidemiol 109:273–284

Wertheimer N, Leeper E (1982) Adult cancer related to electric wires near the home. Int J Epidemiol 11:345–355
Welker HA, Semm P, Willig RP, Commentz JC, Wiltschko W, Vollrath L (1983) Effects of an artificial magnetic field on serotonin-N-acetyltransferase activity and melatonin content of the rat pineal gland. Exp Brain Res 50:426–432
Wilson BW, Chess EK, Anderson LE (1986) 60-Hz electric-field effects on pineal melatonin rhythms. Bioelectromagnetics 7:239–242
Wood AW, Amstrong SM, Sait ML, Devine L, Martin MJ (1998) Changes in human plasma melatonin profiles in response to 50-Hz magnetic field exposure. J Pineal Res 25:116–127

31 Nocturnal Hormone Profiles in Healthy Humans Under the Influence of Pulsed High-Frequency Electromagnetic Fields

Klaus Mann, Peter Wagner, Christoph Hiemke, Clarissa Frank, *Joachim Röschke*

Abstract

We studied the effects of pulsed high-frequency electromagnetic fields emitted from a circularly polarized antenna on the neuroendocrine system in healthy humans. Nocturnal hormone profiles of growth hormone (GH), luteinizing hormone (LH), cortisol, and melatonin were determined. A slight, transient elevation in the serum cortisol level was found immediately after onset of field exposure for about 1 h, indicating an alteration in the hypothalamo-pituitary-adrenal axis activity. For GH, LH, and melatonin, no significant effects were found under exposure to the fields compared with the placebo condition, regarding both total hormone production during the entire night and dynamic characteristics of the secretion pattern. Also, the evaluation of the sleep EEG data revealed no significant alterations under field exposure, although there was a trend towards an REM-suppressive effect. The results indicate that weak high-frequency electromagnetic fields have no effects on hormone secretion except for a slight elevation in cortisol production which is transient, pointing to an adaptation of the organism to the stimulus.

31.1 Introduction

Neuroendocrinological studies during sleep have been proven to be a valuable tool in neurobiology. Due to the close functional connection between the neuroendocrine system and the central nervous system, study of alterations in neuroendocrine parameters provides an insight into the mechanisms underlying brain function under physiological as well as pathological conditions (Müller 1990, Steiger 1995). Thus, abnormalities of the secretion pattern of several hormones have been found in psychiatric illnesses, contributing to our knowledge of the underlying pathophysiological mechanisms. In particular, an overactivity of the hypothalamo-pituitary-adrenal axis in affective disorders has been established. Further, neuroendocrinological investigations appear to be an appropriate method with which to study the biological effects of various external stimuli. In this context, an important application has been the characterization of the pharmacological profile of psychotropic drugs at the neuroendocrine level. Here, particularly, the melatonin secretion of the pineal gland is of interest for pharmacological in vivo studies because of the variety of receptor systems involved in its physiological control.

Another field of research where measurement of neuroendocrine parameters might be useful is the investigation of the biological effects of electromagnetic fields on the human organism. In the literature several investigators reported alterations of neuro-endocrine parameters under exposure to electromagnetic fields (for review see Roberts et al. 1986, Polk and Postow 1986, Michaelson and Lin 1987). But most of these studies were performed on small laboratory animals, and extrapolation of the results to the human organism is problematic. Moreover, results from various studies can only be compared to a limited extent owing to the great variation in experimental conditions. In particular, the power of the fields applied in previous studies was high, resulting in clear thermal loading of the tissue. Regarding hormone secretion, circadian rhythmicity is relevant for the interpretation of the results.

Recently, the controversy about the effects of electromagnetic fields on the human organism has been stimulated by the increasing use of digital mobile radio telephones. In principle, the thermal effects of these new communication systems, caused by heating of tissue through absorption of radiation energy, can be distinguished from the non-thermal effects. Knowledge has increased considerably with regard to the thermal effects of electromagnetic fields, and no health hazards occur when a safe distance is kept. Increasing interest is now being focused on the nonthermal effects of weak electromagnetic fields that do not lead to heating of tissue. In this respect, there is still a great lack of carefully controlled studies, especially on human beings, and the fundamental mechanisms of the interaction with biological systems are still a matter of debate.

In the present study on healthy men, we investigated the effects on hormone secretion of pulsed high-frequency electromagnetic fields in the lower microwave frequency range generated by a digital mobile radio telephone. The field intensity was weak, not leading to thermal effects in our experimental design. In order to minimize influences from extraneous factors and for assessment of the dynamic properties of hormone secretion, nocturnal hormone profiles were determined during the entire night. In an exploratory study design, four hormones which are representative for different kinds of associations between secretion pattern and circadian or sleep-wake rhythms were investigated: growth hormone (GH), cortisol, luteinizing hormone (LH), and melatonin. For GH, besides spontaneous pulses randomly distributed over the day, the most prominent characteristic is a sleep onset peak with a strong coupling to the first NREM episode independent of the circadian rhythm. In contrast, cortisol reveals a strong association with the circadian rhythm, showing a maximum in the early morning; during the night, triggered by the sleep process, a clear ultradian variation is overlaid with a period length of about 150 min. LH is secreted in a pulsatile manner, and during sleep the pulses are associated with the REM episodes. Melatonin secretion is not directly coupled to the sleep-wake cycle; it is instead dependent on the light-dark phases, thus revealing a monophasic maximum during the night.

31.2 Materials and Methods

31.2.1 Subjects

Twenty-four healthy male volunteers aged 18–37 years (mean 26 ± 3 years) participated in the study. To exclude physical or mental illness, a detailed history, psychiatric exploration, physical examination, ECG, EEG, and routine laboratory parameters were taken before beginning the study. Subjects suffering from sleep disturbances were excluded. All subjects were nonsmokers and were not taking any drugs. Consumption of alcohol was forbidden during the period of the study. The subjects were informed in detail about the intention of the study and written informed consent was obtained. The study design was approved by the local ethical committee.

31.2.2 Experimental Procedure

Each subject spent three successive nights in the sleep laboratory. Following an adaptation night, polysomnography was performed on each of the two subsequent nights. EEG signals were measured with Ag/AgCl surface electrodes fixed at the positions F_z, C_z, C_3, C_4, P_z, and the mastoid, according to the international 10–20 system. All electrode impedances were below 5 kΩ. Unipolar EEG derivations (versus mastoid) as well as EOG, submental EMG, and ECG were recorded. For shielding from external fields and for prevention of reflections of the field under study, the subjects slept in a specially constructed metal-free bed that was enclosed in a metal-free chamber lined with absorbing material. A circularly polarized antenna was positioned beneath the head of the bed at a distance of 40 cm from the vertex of the subject. The antenna, connected to a digital mobile radio telephone (GSM system) via an amplifier, emitted a 900-MHz electromagnetic field pulsed with a frequency of 217 Hz and a pulse width of 577 µs. The average power density at a distance of 40 cm from the antenna was 0.02 mW/cm^2. The telephone device was operated from the neighboring room using an extension lead. The function of the antenna could also be monitored continuously throughout the night from there, the applied pulse and field intensity being shown on an oscilloscope.

An indwelling catheter was placed in a forearm vein at 21.30 h. The catheter was connected to a plastic extension tubing leading through the wall into the adjacent room, where blood samples were taken. The catheter was flushed by NaCl 0.9% continuously during the entire night (30 ml/h) to avoid any blockages.

31.2.3 Protocol

The subjects were not allowed to fall asleep until the lights were switched off at 23.00 h. Polysomnography was performed over 8 h; the registration started at 23.00 h and finished at 7.00 h, when the subjects were woken up. For each subject, in one night exposure to the electromagnetic field occurred over 8 h, from 23.00 h until 7.00 h, while in the other night the electromagnetic field was turned off. The order of application was randomized and the subjects were not informed about the experimental condition. For determination of nocturnal hormone profiles, in both investigation nights blood samples were taken every 20 min from 23.00 h to 7.00 h. During the entire investigation period the subjects could be observed on a TV screen.

31.2.4 Nocturnal Hormone Profiles

Blood samples were drawn through the catheter into chilled test tubes and were processed applying standard laboratory techniques. They were immediately centrifuged and stored frozen at $-25\,°C$ until assayed. For all hormones, serum concentrations were determined applying commercially available kits. To guarantee the precision of the methods, apart from calculation of intra- and interassay variations, control samples with known hormone concentrations delivered by the manufacturers were also analyzed. All samples of one subject were analyzed in a single run; therefore interassay variations are only relevant for comparison of different individuals. Serum concentrations of GH, LH, and cortisol were determined every 20 min; melatonin serum concentrations were determined at 1-h intervals owing to the monophasic course of melatonin during the night. GH was measured using an immunoradiometric assay (Medgenix, Düsseldorf, Germany); intra-assay and interassay coefficients of variation (CVs) were below 10%. LH was analyzed using an immunoenzymometric assay (Serono, Freiburg, Germany); the intra-assay CV was below 5%, while the interassay CVs were in the range of 6%–18%. For cortisol, a radioimmunometric assay was applied (Becton-Dickinson, Heidelberg, Germany); the intra-assay CVs were below 6%. For melatonin determination, a radioimmunometric assay was used (IBL, Hamburg, Germany); the intra-assay and interassay CVs were about 15%.

31.2.5 Data Analysis

All polysomnographic signals were recorded on paper using a Nihon Kohden EEG machine (0.3 s time constant and 50 Hz, 24 dB/octave low-pass filter for analog EEG signals). The sleep EEGs were scored visually according to the criteria of Rechtschaffen and Kales (1968) by an experienced rater who was blind with respect to the experimental condition. Statistical analysis of EEG parameters was performed applying Wilcoxon test for paired samples (two-sided). For quantitative assessment of hormone profiles, data reduction was performed by averaging all values in successive equidistant time intervals of 1 h duration. These averaged values are a quantitative measure for hormone production in the distinct time segments. For statistical analysis, a two-way ANOVA model with field exposure (without field, with field) and time segment (23–0, 0–1, 1–2, 2–3, 3–4, 4–5, 5–6, 6–7 h) as within-subject factors was applied. Additionally, for GH and LH, which are secreted in a pulsatile manner, the characteristics of the episodic hormone secretion were analyzed applying the software package "Pulsar" to the original data based on 20-min time segments. Differences were considered statistically significant with a P value < 0.05.

31.3 Results

Two subjects were excluded from analysis owing to a very low sleep efficiency index (< 0.60) under the placebo condition. Therefore, the results are based on the data of 22 subjects. The results of the classical sleep EEG analysis are summarized in Fig. 31.1. Total sleep time was nearly identical under both experimental conditions. Duration and percentage of REM sleep showed a tendency to decrease under exposure to the field, whereas REM latency tended to be prolonged. However, these differences did not reach statistical significance. Also no significant alterations were found regarding the other sleep EEG parameters.

Figure 31.2 shows the mean hormone profiles for GH, cortisol, and melatonin for all subjects. The mean areas under the curve as a measure of total hormone production

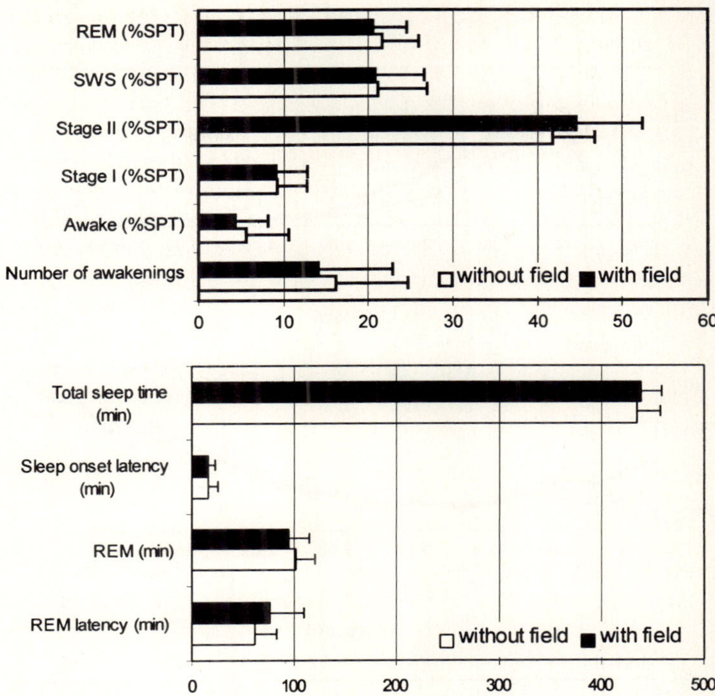

Fig. 31.1. Classical sleep EEG parameters (mean ± standard deviation). *SPT*, Sleep period time; *SWS*, slow wave sleep

during the night are given in Table 31.1. The placebo cortisol profiles were in the normal range and revealed the typical increase in the early morning. Application of a two-way ANOVA model with field exposure and time as within-subject factors revealed no significant main effect for the factor field exposure. Thus, no difference was found between placebo and exposure condition regarding total cortisol production during the night. However, a significant interaction between field exposure and time became obvious ($F_{7,147} = 2.26, P = 0.033$), indicating that the temporal pattern of cortisol secretion differed between the placebo and exposure nights. Pairwise comparison (*t* test, two-sided) of the serum cortisol levels of the placebo and exposure nights in corresponding 1-h time segments revealed significant differences for the first segment, 23–0 h ($P = 0.017$), and for the last segment, 6–7 h ($P = 0.046$).

Table 31.1. Area under the curve for GH, cortisol, LH, and melatonin (mean ± SEM)

Hormone	Without field	With field
GH (μU/ml h)	43.17 ± 8.41	49.10 ± 12.83
Cortisol (nmol/ml h)	967.14 ± 64.25	1038.93 ± 55.99
LH (mU/ml h)	19.36 ± 1.93	20.51 ± 1.95
Melatonin (pg/ml h)	252.75 ± 47.93	303.25 ± 70.15

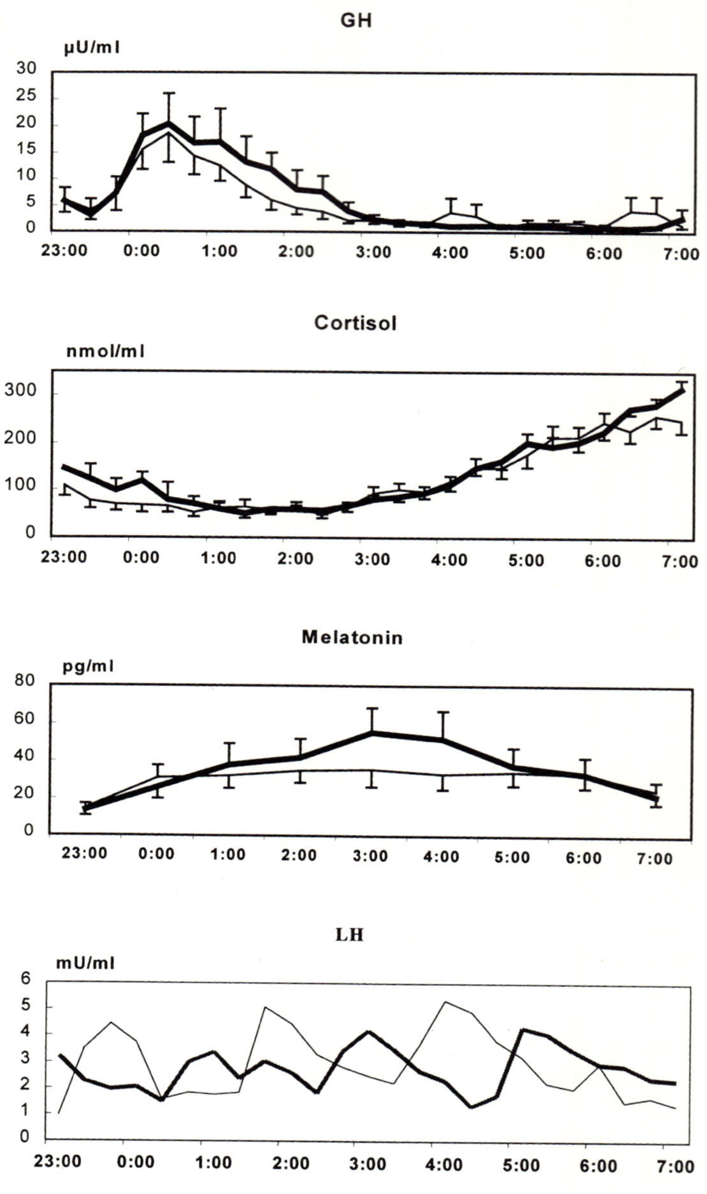

Fig. 31.2. Mean hormone profiles +/- SEM for GH, cortisol, and melatonin in 22 subjects under the placebo condition (*thin line*) and under exposure to the electromagnetic field (*thick line*). Hormone profiles for LH under the placebo condition (*thin line*) and under exposure to the electromagnetic field (*thick line*), are given for one subject, as an example

For GH, under the placebo condition the serum concentrations were within the normal range. All subjects revealed a so-called sleep onset peak, which on average occurred at about 0.20 h. Additionally, spontaneous secretion episodes could be observed which were not predictable. Under exposure to the field, similar hormone profiles were found. Application of the two-way ANOVA model revealed no significant effect for the factor field exposure and no significant interaction between field exposure and time. Moreover, the analysis of pulsatility revealed no indication of an influence of the field on spontaneous secretion pulses.

The melatonin profiles showed the typical monophasic course; on average the maximum occurred at 3.00 h. Serum concentrations were within the normal range. No significant effect for the factor field exposure and no significant interaction between field exposure and time were found by the two-way ANOVA.

The LH concentrations were in the normal range under the placebo condition. All subjects showed the characteristic pulsatile pattern of LH secretion, with several pulses; during one night, three to five pulses occurred. As the occurrence of the pulses is not predictable, the pulsatile pattern is not recognizable in mean profiles; therefore, a representative profile of one subject is shown in Fig. 31.2. The mean areas under the curve as a measure of total nocturnal LH production are also given in Table 31.1. The two-way ANOVA revealed no significant effect for the factor field exposure and no significant interaction between field exposure and time. Moreover, comparing placebo and exposure nights, no difference was found regarding the pulsatile pattern.

31.4 Discussion

The evaluation of the sleep EEG data revealed no significant effects under exposure to the electromagnetic field compared with the placebo condition. However, a trend towards an REM suppressive effect was observed. Percentage and duration of REM sleep showed a decreasing tendency under exposure, whereas REM latency tended to increase. Thus, the electrophysiological data are consistent with the former findings of Mann and Röschke (1996). The failure to reach the level of significance in the present study may have been due to methodological differences between the two studies. The exposure conditions were different as in our study a circularly polarized antenna was used and the exposure was carried out in a specially constructed investigation chamber lined with absorbing material, providing a nearly homogeneous field distribution and shielding against sources of external interference. Moreover, the actual field intensity was less than half the value as compared with the former study.

In the present study the main foci of interest were neuroendocrine parameters. Four hormones were chosen for investigation which are representative for various types of association with the sleep-wake cycle. The only significant effect found was an interaction between field exposure and time for cortisol concentration, indicating an altered temporal secretion pattern of cortisol under exposure to the field, whereas there was no difference for total hormone production during the night between placebo and exposure conditions. Most conspicuous was a transient elevation of the serum cortisol level during the first hour of the investigation around sleep onset, followed by rapid adaptation of the system during the sleep process. The magnitude of the enhancement of cortisol production was in the same range or even below that observed under other stressors. This fact and the rapid adaptation after starting the exposure indicate that the effect is without clinical relevance.

Comparable studies on humans do not exist up to now, but a number of animal studies have been carried out. For high power densities, consistent evidence of adrenocortical stimulation was found in rats, which was mediated by the central nervous system (Lotz and Michaelson 1979, Lai et al. 1990). Similar results were obtained in monkeys (Lotz and Podgorski 1982). The underlying mechanisms are not yet clear. Several investigators demonstrated a clear correlation between serum cortisol or corticosterone and rectal temperature under electromagnetic field exposure (Lotz and Podgorski 1982, Lotz and Michaelson 1978, Lu et al. 1981), suggesting that nonspecific stress reactions which might result from thermal loading, or thermal interactions at the hypothalamic level, may play a decisive role. Moreover, direct microwave interactions with the central nervous system have been discussed. In conjunction with theoretical considerations (Taylor 1981, Weaver and Astumian 1990), the alteration of cortisol secretion which was apparent in our study corroborates this latter opinion, suggesting that even weak electromagnetic fields that do not lead to heating of tissue are able to cause biological effects. Of course, this finding, based on this exploratory study design, needs replication and elucidation by further investigations.

For GH, LH, and melatonin, no difference between placebo and exposure condition could be detected regarding either total hormone production during the night or temporal secretion pattern. Only a few studies have been carried out with respect to these hormones. Michaelson et al. (1975) reported increased GH secretion in rats exposed to moderate microwave fields, whereas GH dropped under high-intensity exposure, with disappearance of the pulsatile secretion pattern. In contrast, no effect on GH was found in monkeys (Lotz and Podgorski 1982). For LH, no effect could be shown in rats exposed to microwave radiation (Lebovitz et al. 1987) or to very low-frequency electric fields (Margonato et al. 1993). In recent years melatonin secretion has attracted increasing interest. Regarding high-frequency electromagnetic fields, no short-term effects on the pineal gland were found in rats and hamsters under exposure to 900-MHz microwaves (Vollrath et al. 1997). For electromagnetic fields in the range of 3–30 MHz, chronic alterations in salivary melatonin in cows seemed unlikely, whereas a delayed acute effect could not be completely excluded (Stark et al. 1997). Most studies have been carried out for pulsed static magnetic fields and very-low-frequency electric and magnetic fields. Several studies in nonhuman mammals have revealed an inhibitory effect on pineal function under field exposure (Reiter and Richardson 1992, Yaga et al. 1993, Kato et al. 1993, Yellon 1994), although the results have not been consistent in all studies (Reiter et al. 1998). This finding has been proposed as a possible mechanism to account for epidemiological reports linking electromagnetic field exposure and increased incidence of cancer (Reiter 1994, Stevens and Davis 1996). However, in several studies carried out on humans, no effect of those fields on melatonin could be proven (Graham et al. 1996, Selmaoui et al. 1996, Schiffman et al. 1994, Graham et al. 1997).

From the fact that in our study no significant effects could be proven, apart from a subtle, clinically irrelevant alteration in the hypothalamo-pituitary-adrenal axis activity under field exposure, it should not be concluded that weak pulsed high-frequency electromagnetic fields produced by digital mobile radio telephones have no further influence on the neuroendocrine system. Our results are strictly limited to the given experimental conditions. Besides the risk of a type II error, possible effects in the nonthermal range might be dose-related and thus might become visible only under exposure to fields of higher intensity. In drawing conclusions from our findings, further limitations have to be considered. No evidence can be given regarding long-term effects. The results are restricted to young healthy volunteers. Finally, the results cannot be extended to other subpopulations with possible alterations in neuro-endocrine regulation, who might reveal stronger susceptibility to exposure to electro-magnetic fields.

Acknowledgements. The study was supported by the Technology Centre of Deutsche Telekom, Germany. The authors wish to thank Gabriele Stroba for technical assistance.

References

Graham C, Cook MR, Riffle DW, Gerkovich MM, Cohen HD (1996) Nocturnal melatonin levels in human volunteers exposed to intermittent 60 Hz magnetic fields. Bioelectromagnetics 17:263–273

Graham C, Cook MR, Riffle DW (1997) Human melatonin during continuous magnetic field exposure. Bioelectromagnetics 18:166–171

Kato M, Honma K, Shigemitsu T, Shiga Y (1993) Effects of exposure to a circularly polarized 50-Hz magnetic field on serum and pineal melatonin levels in rats. Bioelectromagnetics 14:97–106

Lai H, Carino MA, Horita A, Guy AW (1990) Corticotropin-releasing factor antagonist blocks microwave-induced decreases in high-affinity choline uptake in the rat brain. Brain Res Bull 25: 609–612

Lebovitz RM, Johnson L, Samson WK (1987) Effects of pulse-modulated microwave radiation and conventional heating on sperm production. J Appl Physiol 62:245–252

Lotz WG, Michaelson SM (1978) Temperature and corticosterone relationships in microwave-exposed rats. J Appl Physiol: Respirat Environ Exercise Physiol 44:438–445

Lotz WG, Michaelson SM (1979) Effects of hypophysectomy and dexamethasone on rat adrenal response to microwaves. J Appl Physiol: Respirat Environ Exercise Physiol 47:1284–1288

Lotz WG, Podgorski RP (1982) Temperature and adrenocortical responses in rhesus monkeys exposed to microwaves. J Appl Physiol: Respirat Environ Exercise Physiol 53:1565–1571

Lu S-T, Lebda N, Pettit S, Michaelson SM (1981) Microwave-induced temperature, corticosterone, and thyrotropin interrelationships. J Appl Physiol: Respirat Environ Exercise Physiol 50:399–405

Mann K, Röschke J (1996) Effects of pulsed high-frequency electromagnetic fields on human sleep. Neuropsychobiology 33:41–47

Margonato V, Veicsteinas A, Conti R, Nicolini P, Cerretelli P (1993) Biological effects of prolonged exposure to ELF electromagnetic fields in rats. I. 50 Hz electric fields. Bioelectromagnetics 14:479–493

Michaelson SM, Houk WM, Lebda NJA, Lu S-T, Magin RL (1975) Biochemical and neuroendocrine aspects of exposure to microwaves. Ann NY Acad Sci 247:21–44

Michaelson SM, Lin JC (1987) Biological effects and health implications of radiofrequency radiation. Plenum Press, New York

Müller EE (1990) The neuroendocrine approach to psychiatric disorders: a critical appraisal. J Neural Transm 81:1–15

Polk C, Postow E (1986) CRC handbook of biological effects of electromagnetic fields. CRC Press, Boca Raton

Rechtschaffen A, Kales A (1968) A manual of standardized terminology, techniques and scoring system for sleep stages of human subjects. Public Health Service, US Government Printing Office, Washington

Reiter RJ (1994) Melatonin suppression by static and extremely low frequency electromagnetic fields: Relationship to the reported increased incidence of cancer. Rev Environ Health 10:171–186

Reiter RJ, Richardson BA (1992) Magnetic field effects on pineal indoleamine metabolism and possible biological consequences. FASEB J 6:2283–2287

Reiter RJ, Tan DX, Poeggeler B, Kavet R (1998) Inconsistent suppression of nocturnal pineal melatonin synthesis and serum melatonin levels in rats exposed to pulsed DC magnetic fields. Bioelectromagnetics 19:318–329

Roberts NJ, Michaelson SM, Lu S-T (1986) The biological effects of radiofrequency radiation: a critical review and recommendations. Int J Radiat Biol 50:379–420

Schiffman JS, Lasch HM, Rollag MD, Flanders AE, Brainard GC, Burk DL Jr (1994) Effect of MR imaging on the normal human pineal body: measurement of plasma melatonin levels. J Magn Reson Imaging 4:7–11

Selmaoui B, Lambrozo J, Touitou Y (1996) Magnetic fields and pineal function in humans: Evaluation of nocturnal acute exposure to extremely low frequency magnetic fields on serum melatonin and urinary 6-sulfatoxymelatonin circadian rhythms. Life Sci 58:1539–1549

Stark KD, Krebs T, Altpeter E, Manz B, Griot C, Abelin T (1997) Absence of chronic effect of exposure to short-wave radio broadcast signal on salivary melatonin concentrations in dairy cattle. J Pineal Res 22:171–176

Steiger A (1995) Schlafendokrinologie. Nervenarzt 66:15–27

Stevens RG, Davis S (1996) The melatonin hypothesis: Electric power and breast cancer. Environ Health Perspect 104(Suppl 1):135–140

Taylor LS (1981) The mechanisms of athermal microwave biological effects. Bioelectromagnetics 2:259–267

Vollrath L, Spessert R, Kratzsch T, Keiner M, Hollmann H (1997) No short-term effects of high-frequency electromagnetic fields on the mammalian pineal gland. Bioelectromagnetics 18:376–387

Weaver JC, Astumian RD (1990) The response of living cells to very weak electric fields: The thermal noise limit. Science 247:459–462

Yaga K, Reiter RJ, Manchester LC, Nieves H, Sun JH, Chen LD (1993) Pineal sensitivity to pulsed static magnetic fields changes during the photoperiod. Brain Res Bull 30:153–156

Yellon SM (1994) Acute 60 Hz magnetic field exposure effects on the melatonin rhythm in the pineal gland and circulation of the adult Djungarian hamster. J Pineal Res 16:136–144

32 Weak High-Frequency (Radiofrequency, Microwave) Electromagnetic Fields: Epidemiological Evidence of Their Impact on Cancer Development and Reproductive Outcome

Detlev Jung, Dirk-Matthias Rose, Katja Radon

Abstract

The influence of weak high-frequency electromagnetic fields (HF EMFs) on cancer development and on reproduction hazard in humans is reviewed. Even considering the latest epidemiological studies, the amount of data is not adequate to definitely elucidate whether HF EMFs represent a human carcinogen and/or the cause of disturbances of reproduction.

32.1 Introduction

Biological and biomedical research in toxicology has the objective of increasing knowledge about the effects of toxic circumstances on human health. Theoretical considerations, laboratory work (in vitro as well as in vivo in animals), and epidemiology expand this knowledge. All approaches to possible effects of high-frequency electromagnetic fields (HF EMFs) have their advantages and drawbacks.

Frequently, results obtained from cell cultures and animal studies can hardly be used to extrapolate risks to human beings, but they are of essential importance for the investigation of effect mechanisms. Knowledge on the effect mechanisms of HF EMF exposure is still sparse. Only the influence of energy transmission (heating) by strong HF EMFs on cells or organisms is well known. Yet apart from accidents, this effect mechanism is considered to be of minor relevance in human exposure (Repacholi 1997,1998; Juutilainen and de Seze 1998). In non- or athermal exposure to HF EMFs, the influence on Ca^{2+} or on other dipole molecules has been discussed (Repacholi 1997 and 1998), and receptor shedding may play a role (Phillipova et al. 1994). A 900-MHz field pulsed with 217 Hz demonstrated no influence on melatonin production or pineal activity either in animals (Vollrath et al. 1998) or in man (Radon et al. 1998). Up to now there is no definite physiological knowledge to provide a basis that could explain biological effects of low-energy HF EMFs in human beings.

Interspecies transfer of effects naturally does not constitute a problem in epidemiological research, but a review of epidemiological literature is hardly ever unbiased. Authors preferably offer studies for publication and editors preferably accept those that prove an association between suspected cause and effect rather than studies with negative results. If evidence for a cause-effect relationship is overwhelming, this selec-

Table 32.1. EMF frequencies used in different applications

Application	Frequencies used
AM standard broadcast transmitters	535–1705 kHz
Amateur radio	1.8 and 30 MHz
Short-wave diathermy in medical practice	27.12 MHz
Heater sealers	27.12 MHz
FM radio and VHF TV stations	30 and 300 MHz
UHF TV	400 and 800 MHz
Cellular telephones	900 (D-net) and 1,800 MHz (E-net)
Microwave diathermy in medical practice	434 MHz and 2.45 GHz
Microwave ovens	2.45 GHz
Air traffic radars	1.3 and 2.8 GHz
Aircraft onboard weather radars	9.375 GHz
Hand-held police traffic radars	10.5 GHz

tion bias is of minor relevance. However, it surely constitutes a serious problem for associations where the risk attributed to exposure is low. The latter has to be suspected for weak HF EMFs. To interpret the results in the sense of a causal relationship, additional evidence is helpful. This evidence can be derived either from laboratory results or from epidemiological studies themselves, taking dose-response relationships and result conformity of different studies into account.

Exposure assessment represents a problem in most epidemiological studies dealing with the influence of EMFs on human beings. Seldom has the actual field been measured, and exposure assessment in occupational studies is frequently based on and biased by job titles. Consequently often even the exposure frequency for a particular EMF case is unknown. An additional general problem in epidemiological studies is the selection of an appropriate reference group.

This review exclusively focuses on studies with a predominant exposure to HF EMFs. Work at video display units (VDUs) including exposure to EMFs from 50 Hz up to 50 kHz (EMFRAPID Working group. NIEHS report 1998) and studies of electric utility workers were not taken into account.

The spectrum of frequencies summarized as HF EMFs or radiofrequencies (RFs) covers 300 Hz (3×10^2 Hz) to 30 GHz (3×10^{10} Hz). This broad band is being used for very different purposes. Table 32.1 notes EMF frequencies in applications mentioned in epidemiological or laboratory studies described below. More detailed information is provided elsewhere (WHO 1993; Mantiply et al. 1997).

32.2 Studies

32.2.1 Cancer Studies (Table 32.2)

32.2.1.1 Laboratory Studies

The results of laboratory studies are not considered in this review. In order to evaluate the impact of epidemiological studies it is important to remember that even under so-called non- or athermal conditions in cell culture experiments and in animal studies there is some evidence that HF EMF exposure contributes to mutagenesis or carcino-

Table 32.2. Cancer studies on HF EMFs. For more details see text

Author, year	Exposure: source, frequency, power	Type of study	Outcome	Size (cases)	O/E, RR, OR	Confidence interval	Comments
Lilienfeld, 1978	Microwaves in the US embassy in Moscow and in other embassies of Eastern Europe	Cohort study	Cancer mortality	116 cancers 10 leukemias 8 brain tumors 12 breast cancers	O/E=1.44 O/E=2.41 O/E=2.21 O/E=2.47	Significantly increased	
Szmigielski et al., 1985	RF and MW in the polish army, $<200\ \mu W/cm^2$, incidentally up to $20\ mW/cm^2$	Cohort study	Cancer morbidity		SIR = 2.07	1.1 – 3.6	
Vagerö et al., 1985	Workers in telecommunication industry	Cohort study	Cancer mortality	11 melanomas	SMR = 2.6	1.3 – 4.5	
Robinette 1980	Radar exposure (US Navy)	Cohort	Lymphatic and hematopoietic malignancies	26 cases	MR=1.18	Not significant	
Lester, 1982a	Airport radar in Wichita	Case-referent	Cancer morbidity	3004 cases	Significant relation		
Lester, 1982b	Airforce base radar in 92 US counties	Cohort	Cancer mortality by US cancer atlas		Significant increase in exposed counties		
Polson, 1985	Airforce base radar in 92 US counties	Cohort	Cancer mortality by US cancer atlas		No increase in exposed counties		Re-evaluation of Lester and Moore, 1982b
Lester, 1985	Airforce base radar in 92 US counties	Cohort	Cancer mortality by US cancer atlas		Significant increase in exposed counties		Answer to Polson et al., 1985
Milham, 1988	Amateur radio operators	Cohort	Hematopoietic malignancies	36 leukemias 15 acute myeloid leukemias 43 cancers of other lymphatic tissues	SMR = 1.24 SMR = 1.76 SMR = 1.62	0.87 – 1.72 1.03 – 2.85 1.17 – 2.18	
Davis, 1993	Hand-held radar		Testicular cancer	6 cases	O/E = 6.9		Statistics in a known cluster
Tynes, 1992	Radiofrequency	Cohort	Cancer incidence	9 leukemias 3 brain tumors	SIR = 2.85 SIR = 0.61	1.3 – 5.41 0.13 – 1.78	

Table 32.2 continued

Author, year	Exposure: source, frequency, power	Type of study	Outcome	Size (cases)	O/E, RR, OR	Confidence interval	Comments
Tynes, 1996	Female radio and telegraph operators	Cohort	Breast cancer incidence	50 cases	SIR = 1.5	1.1–2.0	Effect in women older than 50 yrs and with a history of frequent shift work

Hocking, 1996	Radiofrequency from TV towers, 63–215 MHz, 0.5 μW/cm^2 in a distance of 2 km	Cohort	Brain tumors
			Leukemia (adults)
			Leukemia (children)
			  Incidence
			  Mortality
				804
				1206
				
				134
				59
					No excess
					O/E = 1.24
					
					O/E = 1.58
					O/E = 2.32
						
						1.09–1.40
						
						1.07–2.34
						1.35–4.01	

Dolk, 1997 a	Surroundings of a TV tower, FM and TV, 5.7 μW/cm^2 in a distance of 2 km	Cohort	Cancer incidence	23 leukemias in a 2-km radius	O/E 1.83	1.22–2.74	
Dolk 1997 b	Surroundings of 20 TV towers in UK, FM and TV frequencies	Cohort	Cancer incidence	79 leukemias (2-km radius) 3305 leukemias (10-km radius)	O/E 0.97 O/E 1.03	0.78–1.21 1.00–1.07	Study following Dolk et al. 1997 a
Hardell 1999	Users of cellular phones, 450 and 900	Case-referent	Brain cancer	233 cases, 466 Controls	OR 0.98	0.69–1.41	

HF EMF, High-frequency electromagnetic field; MW, microwave; SW, short wave; OR, odds ratio; O/E, observed versus expected; SMR, standardized mortality ratio; SIR, standardized incidence ratio.

genesis. Malignant transformation of fibroblasts under the additional influence of an initiator and a tumor promoter (Balcer-Kubiczek and Harrison 1985) and glioma cell proliferation (Cleary et al. 1990) were enhanced by a 2.45-GHz EMF even under isothermal conditions. Mice developed breast cancer more rapidly under exposure to 5 or 15 mW/cm^2 of 2.45 GHz radiation. The resistance to infused lung cancer cells was lowered. The development of skin cancer under simultaneous exposure to benzo[a]pyrene was accelerated by such additional radiation (Szmigielski et al. 1982). Exposure of transgenic mice to 900-MHz EMFs with a pulse repetition of 217 Hz under isothermal conditions significantly raised the risk that they would develop lymphomas (Repacholi et al. 1997). In a short-term test of potential health effects a 10-kHz magnetic field did not produce any adverse effect in mice (Robertson et al. 1996). Exposure to 2.45-GHz EMFs failed to promote colon cancer development in mice under concomitant exposure to dimethylhydrazine (Wu et al. 1994). As most of these studies have never been reproduced, they should be considered as creating rather than proving hypotheses. So Repacholi (1997), in his excellent review on this subject, stated that from the hitherto available data of laboratory studies the evidence for a co-carcinogenic effect of RF is suggestive but not substantive.

32.2.1.2 Epidemiological Studies

Radar

U.S. Army. In an investigation of data from the U.S. Navy, fire control and aviation electronics technicians (more than 20,000 persons), who were supposed to have been exposed to high ($\geqslant$10 mW/cm^2) EMFs, were compared with radiomen, radarmen, and aviation electricians' mates (more than 20,000 persons), whose exposure was assumed to be low (well below 1 mW/cm^2). No differences in mortality or morbidity were found between the groups (Robinette et al. 1980). Goldsmith (1997) argued that at most this study presents a mere comparison between high and low exposure. The authors themselves stated that an actual exposure could not be quantified. Moreover the group with presumably the highest exposure, the aviation electronics technicians (ATs), were not investigated separately but together with the fire control technicians. Actually in the AT group the death risk for all malignancies and those of the lymphatic and hematopoietic system was significantly increased (according to Goldsmith 1997).

Polish Army. About 3% of the members of the Polish army were evaluated to have been exposed to relevant levels (between 0.01 and more than 0.6 mW/m^2) of HF EMFs in a frequency range between 150 MHz and 3.5 GHz. Compared with unexposed personnel, their relative risk for any cancer was 2.07 (95% confidence interval 1.1–3.6). Significantly increased risk ratios were found for cancers of the hematopoietic and lymphatic tissue (observed/expected = 6.31), the thyroid, the skin, including melanoma, the esophagus, the stomach, the colon and rectum, and the brain (odds ratio 1.91) (Szmigielski 1996). The lack of an exact number of participants and the insufficiently defined RF exposure restrict the value of the study (Repacholi 1998). However, this study points to a possible relationship between (high-energy?) HF EMFs and cancer occurrence.

Exposure of the General Population by Air Base Radars. In 1982 (a), Lester and Moore published two studies dealing with the effect of nearby radar bases and cancer occurrence. One study evaluated data from Wichita/Kansas (covering 262,766 persons), where two radar stations are located at the town border, one at the municipal airport and one at the local Air Force base. The authors defined radar-shielded town areas in the landscape. Comparing the inhabitants of shielded areas with those of exposed areas, they found a significant association between radar exposure and cancer morbidity and mortality. The comparison groups were adjusted for age, gender, race, and economic status. In addition to the association between EMF exposure and cancer morbidity/mortality, the correlation matrix showed an association of EMF exposure and economic status, a well-known independent risk factor for cancer that also revealed an impact on carcinogenesis in this study. Moreover it is noteworthy that according to the map shown in the paper, cancer occurrence was more frequent in the town centre of Wichita, an area with low landscape shielding. Consequently additional, uncontrolled, potentially carcinogenic factors due to denser traffic in the town centre might well have influenced the increased emergence of cancer in HF EMF-exposed persons.

In a second study, Lester and Moore (1982b) compared the frequency of cancer in all counties in the U.S. containing an Air Force base with the frequency in counties similar in size within the same state but without an Air Force base. Cancer occurrence was significantly more frequent in counties with an Air Force base. Polson and Merritt (1985) strongly argued against these results, as in a re-analysis on the basis of Lester's study group but with a different definition of exposed and control groups they found no association of EMF exposure and cancer. Lester (1985) then explained that in the initial study counties containing the city closest to an Air Force base were defined as exposed and presented new data again showing an association of EMF and cancer. Having these contradictory results, nothing but a new re-evaluation could shed light on the weight of these studies regarding the risk of radar for cancer occurrence. Moreover other risk factors like stress, toxic chemical exposure, and noise exposure due to the vicinity of the air base, as well as socio-economic status – who lives near an Air Force base? – have not been evaluated. So unfortunately all these studies did not allow differentiation of the effect of confounders from the impact of HF EMFs.

Hand-Held Radars. A cluster of police officers (6 of 70 working at two police stations in the north-central United States) with testicular cancer were investigated (Davis et al. 1993). The only common exposure shared by all the six officers was the use of hand-held radar. Smoking habits were not indicated in this paper. Three of the malignancies were embryonal cell carcinomas, one a mixed cell carcinoma, and two seminomas. The authors regarded the relatively high mean age at cancer diagnosis (39 years) as a hint for an environmental cause of these diseases, which, at least for seminomas, may be true. They also used the cluster data for calculation of an observed/expected ratio, which was naturally highly significant. Nevertheless, this calculation gave no additional information on whether the cluster had arisen by chance or as a result of the EMF stress. The authors therefore recommended further studies.

TV Towers, Telephone, and Telegraph

TV Towers. Cows grazing near the Skrunda Radio Location Station, Latvia, were shown to have an increased number (sixfold) of micronuclei in their erythrocytes compared with controls (Balode 1996).

In the vicinity of a television (TV) and frequency modulation (FM) radio transmitter (Sutton Coldfield Transmitter) near Birmingham, UK (a population of 408,000 persons), the leukemia risk was found to be raised, with an odds ratio of 1.83, within a circular area of 2 km in radius. There was a significant decline in risk with distance from the transmitter. The findings were consistent for two different time periods (1974–80 and 1981–86) and independent of the cluster that originally attracted attention (Dolk et al.1997a). The maximum power density equivalent was 1.3 $\mu W/cm^2$ for TV and 5.7 $\mu W/cm^2$ for FM within 2 km.

It was not possible to confirm this finding in a large-scale study (a population of 3.39 million persons) in the vicinity of 20 more television transmitters in the UK (Dolk et al.1997b). There was no observed excess risk within a 2-km distance, and the overall risk (1.03) within a 10-km distance and the decline in risk with distance were both found to be of borderline significance. So, while the conclusion of the first study is that the dependence on the distance from the tower and the proven increased risk in different time spans strongly indicate that there is a risk factor for leukemia in the area of Sutton Coldfield, the second study indicates that it is unlikely that the RF EMF contributes significantly to this risk.

Three municipalities in the vicinity of Sydney, Australia, are situated closely to three TV towers (maximum distance about 4 km). The maximal power density was calculated to be 8 $\mu W/cm^2$, and the minimum, at a distance of 4 km, to be 0.2 $\mu W/cm^2$. The occurrence of cancer was compared with that in six other towns (population 450,000). The data were stratified for age, sex, and calendar year. The rate ratio for leukemia incidence in all age groups was 1.24 (95% confidence interval 1.09–1.40), the increase mainly being due to an increased rate ratio in childhood cases (1.58, 95% confidence interval 1.07–2.34). There was no increased risk in brain tumors (Hocking et al. 1996).

Telephone and Telegraph. In a study on the occurrence of cancer in workers potentially exposed to EMFs, radio/telegraph operators and radio/television repairmen were classified as exposed to RF EMFs (Tynes et al. 1992). In the entire group of 2,017 men, between 1961 and 1985 11 new cases of leukemia occurred (4.9 expected). The number of newly developed brain tumors was lower (3) than expected (7.4), but both differences proved nonsignificant. Confounders (smoking, toxic chemicals, ionizing radiation) were not taken into account.

In a cohort study and an additional nested case-control study (2619 women, working as radio and telegraph operators on ships with exposure to RF of 405 kHz to 25 MHz and, to some extent, 50-Hz EMFs), the occurrence of breast cancer was investigated. After adjustment for fertility, age, and shift work, women aged 50 and more who were exposed to HF EMFs were found to have an increased (1.5) risk for breast cancer. Shift work was associated with breast cancer as well (Tynes et al. 1996).

A significant excess of melanoma (2.6) occurred in workers in the telecommunications industry. This risk was associated with soldering work. No information concerning EMF exposure was given in the paper (Vagerö et al. 1985).

During a study period of 6 years 232,499 person-years at risk accumulated in a study investigating the mortality of amateur radio operators in Washington State and California (Milham 1988). The overall standardized mortality ratio was 0.71 compared with the general US rates. The higher percentage of academic occupations in this cohort was assumed to be a possible reason for this low mortality. Mortality from cancers of other lymphatic tissues, including multiple myeloma and non-Hodgkin's lymphoma, was significantly increased (SMR = 1.62), and mortality due to acute myeloid leukemia was also elevated (1.76). This study hardly proves a causal relationship between HF EMF and cancer, as some data suggest that upper socioeconomic classes not only have lower overall mortality ratios but also an increased risk for specific cancers such as non-Hodgkin's lymphomas (Figgs et al. 1995).

In 1996 the first results of a study on cellular phone customers (paying the telephone bills during November and December 1993) were published (Rothman et al. 1996). No difference was found for overall mortality between users of mobile telephones (with an integrated antenna) and users of portable telephones (without an antenna). So far no detailed analysis of this study concerning cancer mortality has been reported. Obviously the study cohort is too young to study the development of solid cancers. Meaningful data can only be expected for leukemias within the coming years.

A new case-referent study did not show an overall increase (OR = 0.98) of brain cancers in persons using cellular phones (Hardell et al 1999). The authors stressed that there was a nonsignificantly raised OR of ipsilateral brain tumors, tumors localized on the left side of the brain, if the patients preferred to use the left ear during telephone calls (OR = 2.40) and vice versa (OR = 2.45). These results, however, are based on only eight and five cases, respectively. Moreover, the authors did not even try to explain the accompanying lower OR of contralateral brain tumors.

Moscow Embassy. These data are taken from Goldsmith 1995, as the original study was never published and is accessible only under the Freedom of Information Act (Lilienfeld et al. 1978). Between November 1962 and August 1963, highly directional, focused microwave transmissions beamed at the U.S. Embassy in Moscow were found and verified. A team from John Hopkins University (leader A.M. Lilienfeld) were entrusted with an examination of the embassy members and a final report to the government was finished in 1978. The health and mortality status of the exposed persons was compared with that of members in other embassies in East Bloc countries. This study design was then criticized by Goldsmith, who claimed there was evidence (based on interviews) that the other embassies had been exposed to microwaves as well. The comparison of the Moscow embassy with the other East Bloc embassies did not reveal meaningful differences. Comparison of both embassy groups with the general U.S. population (presumably based on the Third National Cancer Survey) indicated a total cancer risk of about 1.44, a risk for leukemia of 2.41, a risk for brain tumors of 2.21, and a risk for breast cancer of 2.47. As no information was given about the extent of the microwave exposure, as the exposure of the other embassies is questionable, and as neither other influential exposures nor socioeconomic status were considered in this analysis, inferences regarding microwaves as the cause of these diseases are not acceptable to our mind.

In addition to the mortality study, a chromosome analysis was performed in 35 members of the embassy between 1966 and 1969. Poor sample collection and transmission were stated. The leukocytes of two samples did not spread at all, and about half of the samples showed "enough evidence of damage that clinical guidelines would have had the persons concerned restricted from reproductive activity." In fact it is well-known that poor handling of blood specimens can produce such effects and so a non-governmental panel of experts advised that "no valid conclusion could be drawn from the study."

32.2.2 Studies on Reproduction (Table 32.3)

32.2.2.1 Animal Experiments

Animal experiments have revealed teratogenic effects of HF EMFs. These effects seem to be temperature dependent (WHO 1993). Effects on sperm production and motility are not necessarily associated with a rise in temperature, though local diathermic effects cannot be ruled out (Makler et al. 1980; Lebovitz et al. 1987).

32.2.2.2 Epidemiological Studies

Paternal Risk

The military occupational history of the fathers of 216 children with Down's syndrome were compared with that of 216 control fathers (Sigler et al. 1965). An association of radar exposure and children with Down's syndrome could be shown. An extension of this study group to another 128 cases failed to prove any relationship (Cohen et al. 1977).

During their work a group of 31 technicians were exposed to microwaves with frequencies between 3.6 and 10 GHz, with an intensity of about tens or hundreds of $\mu W/cm^2$, over a period of 1–17 years (Lancranjan et al. 1975). Number, motility, and normal morphology of sperms in ejaculate were significantly reduced compared with those of 30 control persons of the same age. Seventy percent of the technicians reported disturbances in potency and libido, judged by the investigators to be present within the framework of a neurasthenic syndrome existent in 80 % of the investigated men. In the paper there was no clear statement about the recruitment of the technicians' group. We do not know whether they participated in the study owing to complaints or whether they were collected with regard to their job exposure.

Maternal Risk

In a case-referent study nested in the Danish birth register, pregnancy outcome in physiotherapists was investigated (Larsen 1991). Only 23.5 % of the children born by mothers highly exposed to HF EMFs were boys. Viewing the weekly exposure, a dose-response relationship was apparent.

A Swiss group tried to reproduce the findings of gender disequilibrium (Gubéran et al. 1994). Based on 1,781 pregnancies they found a gender ratio (male/femalex100) of 107 (95 % confidence interval 89–127) for the exposed physiotherapists and 101 (95 % confidence interval 90–113) for unexposed women.

In 305 female plastic welders exposed to EMFs exceeding 250 W/m^2 in 50 % and producing irritative eye complaints in 40 % of the women, no increase in malformations (four minor malformations in the exposed group) or reduction in birth weight was found compared with the general population (Swedish malformation and birth register) (Kolmodin-Hedman et al. 1988).

In the case-referent study nested in the Danish register of births (see above, Larsen 1991), no statistically significant effect on spontaneous abortions, subfecundity, low birth weight, stillbirth, death within the first year, or congenital malformations were

Table 32.3. Studies on reproduction and the influence of HF EMFs. For more details see text

Author, year	Exposure: source, frequency, power	Study design	Outcome	Size (cases)	O/E, RR, OR	Confidence interval	Comments
Taskinen, 1990	Female physiotherapists with exposure to HF-EMF	Case-referent	Congenital malformations		OR (exposure 1–4 h/week) = 2.3 OR (exposure >4 h/week) = 1.0	1.1–4.7 0.3–3.0	
Källén, 1981	Female physiotherapists with exposure to HF-EMF	Case-referent	Congenital malformations		OR (exposure seldom) = 0.4 OR (exposure often or daily) = 2.0	0.1–1.2 0.7–6.2	
Ouellet-Hellstrom, 1993	Female physiotherapists with exposure to short-wave (SW, 27.12 MHz) and microwaves (MW, 2.45 GHz)	Case-referent	Miscarriages	1753 cases	OR (SW) = 1.07 OR (MW) = 1.28	0.91–1.24 1.02–1.59	
Lancranjan, 1975	Technicians exposed to 10 kHz–3.6 GHz, 10–>100 μW/cm^2	Cohort	Libido, aspect, and function of spermatozoa	31 persons	Decrease in libido (70%), significant decrease in number, normal aspect, and motility of spermatozoa		
Larsen, 1991	Female physiotherapists with exposure to HF-EMF	Case-referent	Congenital malformation of children	54 cases	OR = 1.7	0.6–4.3	
Larsen, 1991	Female physiotherapists with exposure to HF-EMF	Case-referent	Gender ratio and reproductive hazards other than malformations	270 cases	Significantly less boys (23.5%), no other elevated risks		

Sigler, 1965	Military radar	Case-referent	Down's syndrome in children of exposed fathers	216 cases	Significant association	
Cohen, 1977	Military radar	Case-control	Down's syndrome in children of exposed fathers	344 cases	No association	Extended study of Sigler et al. 1965
Kolmodin-Hedman, 1988	Plastic welding machines, 25–30 MHz, >250 W/m^2 in 50%	Cohort	Pregnancy outcome	305 pregnancies	No abnormalities	
Gubéran, 1994	Female physiotherapists, short-wave, 27.12 MHz, up to 200 mW/cm^2	Cohort	Gender ratio of offsprings (m/f x 100)	1781 pregnancies	Exposed 107 Not exposed 101	89–127 90–113

HF EMF, High-frequency electromagnetic field; MW, microwave; SW, short wave; OR, odds ratio; O/E, observed versus expected; SMR, standardized mortality ratio; SIR, standardized incidence ratio.

apparent. The odds ratio for exposure to HF EMFs and congenital malformations was found to be 1.7 (95% confidence interval 0.6–4.3) (Larsen et al. 1991).

Based on 1,753 miscarriages in physiotherapists and the same number of controls (other pregnancies), a significant increase in miscarriages (odds ratio 1.26 with a 95% confidence interval of 1.00–1.59) was found in women working with microwave diathermy at 434 MHz and 2.45 GHz 6 months prior to the pregnancy or during the first trimester. A significant dose-response relationship was calculated. It was found that 47.7% of the microwave-exposed miscarriages occurred prior to the seventh week of gestation, compared with 14.5% of the unexposed. No effect was found in women working with short-wave diathermy (27.12 MHz, odds ratio 1.07, 95% confidence interval 0.91–1.24) (Ouellet-Hellstrom and Stewart 1993).

In two additional studies, pregnancy outcome in female physiotherapists using HF EMF equipment was investigated (Taskinen et al. 1990; Källén et al. 1982). In the first study weekly use of short waves for between 1 and 4 h (odds ratio 2.3, 95% confidence interval 1.1–4.7) but not >4 h (odds ratio 1.0, 95% confidence interval 0.3–3.0) resulted in a significant increase in malformations. The second study did not reveal any significant results.

32.3 Conclusions

32.3.1 Cancer Studies

32.3.1.1 Previous Reviews

Focusing on cancer development, international regulation bodies and expert meetings came to the following conclusions:

In 1988 the IRPA/INIRC (International Non-Ionizing Radiation Committee of the International Radiation Protection Agency) stated that "in view of our limited knowledge of thresholds for all biological effects, unnecessary exposure should be minimized."

The WHO (World Health Organization) stated (1993) that "the available evidence does not confirm that radiofrequency exposure results in the induction of cancer, or causes existing cancer to progress more rapidly. Because of incompleteness and inconsistencies, the available scientific evidence is an entirely inadequate basis for recommendations of health protection guidelines."

In 1996 the ICNIRP (International Commission on Non-Ionizing Radiation Protection) considered that "the results of the epidemiological studies do not form a basis for health hazard assessments of exposure to radiofrequency fields."

In 1997, M. H. Repacholi concluded his review on hand-held radiotelephones and base transmitters by stating: "The evidence suggests that radiofrequency exposure is not mutagenic and is therefore unlikely to initiate cancer. The evidence for a co-carcinogenic effect or an effect on tumor promotion is not substantive but merits further investigations."

In two reviews (1995 and 1997), J. R. Goldsmith's point of view is remarkably different from the official statements mentioned above. He argues that "there are strong political and economic reasons for wanting there to be no health effect of RF/MW exposure." In fact Goldsmith cites investigations initiated by governments, but not completely published in peer-reviewed journals, that were never continued. In most of the official statements these studies (of U.S. foreign service workers in Eastern European embassies, of inhabitants of Skrunda village, in Hawaii and Honolulu, of Polish army personnel), if mentioned at all, were regarded as not very helpful owing to a lack of exact description of the cohort studied and of the exposure. Goldsmith himself states that his "list of findings must be considered to manifest reporting bias." For him, "of all the possible effects, the weight of evidence is greatest for cancer incidence and for effect on spontaneous abortion in the women who where pregnant while exposed."

32.3.1.2 New Evidence

Meanwhile (since 1996) well-designed epidemiological studies have added some more knowledge on the causal relationship between HF EMFs and cancer (Hocking et al. 1996; Tynes et al. 1996; Dolk et al. 1997a, b). The theory of HF EMFs radiated from TV towers as causative of leukemia has been supported by the Australian study and the study of Sutton Coldfield. A second English study investigating the surroundings of 20 TV towers constitutes a strong argument against such a causal relationship. The results of the breast cancer study indicate the necessity of further differentiation of HF EMFs and shift work as causes of increased cancer incidence. The study of Hardell et al. (1999) provides only weak evidence to support the hypothesis that the EMF fields emitted by cellular telephones are a causative factor in the origination of brain tumors. It is somewhat surprising that there has been no follow-up to the study of the members of the Moscow embassy. So, at present, the amount of data is not adequate to definitely answer the question of whether HF EMFs represent a human carcinogen.

32.3.2 Studies on Reproduction

The results of the existing studies concerning the influence of HF EMFs on reproduction are highly equivocal. We think that the Swiss study (Gubéran et al. 1994) has proven that an influence of HF EMFs on gender ratio is unlikely.

Hitherto studies on miscarriages and malformations have been based on case-referent design. They have varied with respect to both the potentially harmful exposure investigated (short waves or microwaves) and the outcome. We therefore do not consider the current database to be adequate for a definite statement on the relationship of HF EMFs and disturbances of reproduction.

References

Balcer-Kubiczek EK, Harrison GH (1985) Evidence for microwave carcinogenesis in-vitro. Carcinogenesis 6:859–864
Balode Z (1996) Assessment of radiofrequency electromagnetic radiation by the micronucleus test in bovine peripheral erythrocytes. Sci Total Environ 180:81–85
Cleary SF, Liu LM, Merchant RE (1990) Glioma proliferation modulated in vitro by isothermal radiofrequency radiation exposure. Radiat Res 121:38–45
Cohen BH, Lilienfeld AM, Kramer AM, Hyman LCC (1977) Parental factors in Down's syndrome: results of the second Baltimore case control study. In: Hook EB & Porter ICH (eds) Population Cytogenetics – Studies in Humans. NY. Academic Press, pp 301–352

Davis RL, Kash Mostofi F (1993) Cluster of testicular cancer in police officers exposed to hand-held radar. Am J Ind Med 24:231–233

Dolk H, Shaddick G, Walls P, Grundy C, Thakrar B, Kleinschmidt I, Elliot P (1997a) Cancer incidence near radio and television transmitters in Great Britain. I. Sutton Coldfield transmitter. Am J Ind Med 145:1–9

Dolk H, Elliot P, Shaddick G, Walls P, Thakrar B (1997b) Cancer incidence near radio and television transmitters in Great Britain. II. All high power transmitters. Am J Ind Med 145:10–17

EMFRAPID Working group (1998) Assessment of Health Effects from Exposure to Power-Line Frequency Electric and Magnetic Fields NIH/NIEHS Research Triangle Park, NC, USA

Figgs LW, Dosemeci M, Blair A (1995) United States non-Hodgkin's lymphoma surveillance by occupation 1984–1989: A twenty-four state death certificate study. Am J Ind Med 27:817–835

Goldsmith JR (1995) Epidemiologic evidence of radiofrequency radiation (microwave) effects on health in military, broadcasting, and occupational studies. Int J Occup Environ Health 1:47–56

Goldsmith JR (1997) Epidemiologic evidence relevant to radar (microwave) effects. Environ Health Perspect 105(Suppl. 6):1579–1587

Gubéran E, Campana A, Faval P, Gubéran M, Sweetnam M, Tuyn JWN, Usel M (1994) Gender ratio of offspring and exposure to shortwave radiation among female physiotherapists. Scand J Work Environ Health 20:345–348

Hardell L, Nasman A, Pahlson A, Halquist A, Mild KH (1999) Use of cellular phones and the risk for brain tumours: A case-control study. Int J Oncol 15:113–116

Hocking B, Gordon IR, Grain HL, Hatfield GE (1996) Cancer incidence and mortality andproximity to TV towers. Med J Austr 165:601–605

ICNIRP (International Commission on Non-Ionizing Radiation Protection) Statement (1996) Health issues related to the use of hand-held radiotelephones and base transmitters. Health Physics 70:587–593

INIRC/IRPA (International Non-Ionizing Radiation Committee of the International Radiation Protection Agency) (1988) Guidelines on limits of exposure to radiofrequency electromagnetic fields in the frequency range from 100 kHz to 300 GHz. Health Physics 54:115–123

Juutilainen J, de Seze R (1998) Biological effects of amplitude-modulated radiofrequency radiation. Scand J Work Environ Health 24:245–254

Källén B, Malmquist G, Moritz U (1982) Delivery outcome among physiotherapists in Sweden: is non-ionizing radiation a fetal hazard? Arch Environ Health 37:81–85

Kolmodin-Hedman B, Mild KH, Hagberg M, Jönsson E, Andersson MC, Eriksson A (1988) Health problems among operators of plastic welding machines and exposure to radiofrequency electromagnetic fields. Int Arch Occup Environ Health 60:243–247

Lancranjan I, Majcanescu M, Rafaila E, Klepsch J, Popescu HI (1975) Gonadic function in workmen with long-term exposure to microwaves. Health Physics 29:381–383

Larsen AI (1991) Congenital malformations and exposure to high-frequency electromagnetic radiation among Danish physiotherapists. Scand J Work Environ Health 17:318–323

Larsen AI, Olsen J, Svane (1991) Gender-specific reproductive outcome and exposure to high-frequency electromagnetic radiation among physiotherapists. Scand J Work Environ Health 17:324–329

Lebovitz RM, Johnson L, Samson WK (1987) Effects of pulse-modulated microwave radiation and conventional heating on sperm production. J Appl Physiol 62:245–252

Lester JR, Moore DF (1982a) Cancer mortality and electromagnetic radiation. J Bioelectricity 1:59–76

Lester JR, Moore DF (1982b) Cancer mortality and air force bases. J Bioelectricity 1:77–82

Lester JR (1985) Reply to "Cancer mortality and air force bases: a reevaluation". J Bioelectricity 4:129–131

Lilienfeld AM, Tonascia J, Tonascia S, Libauer CA, Cauthen GM (1978) Foreign Service Health Status Study – evaluation of health status of foreign service and other employees from selected eastern European posts. Final report, Contract No. 6025–619073 (NTIS PB-288163) Department of State, Washington, DC

Makler A, Tatcher M, Vilensky A, Brandes JM (1980) Factors affecting sperm motility. III. Influence of visible light and other electromagnetic radiations on human sperm velocity and survival. Fertil Steril 33:439–444

Mantiply ED, Pohl KR, Poppell SW, Murphy JA (1997) Summary of measured radiofrequency electric and magnetic fields (10 kHz to 30 GHz) in the general and work environment. Bioelectromagnetics 18:563–577

Marino AA (1995) Time-dependent hematological changes in workers exposed to electromagnetic fields. Am Ind Hyg Assoc J 56:189–192

Milham S (1988) Increased mortality in amateur radio operators due to lymphatic and hematopoietic malignancies. Am J Epidemiol 127:50–54

Ouellet-Hellstrom R, Stewart WF (1993) Miscarriages among female physical therapists who report using radio- and microwave-frequency electromagnetic radiation. Am J Epidemiol 138:775–786

Phillipova TM, Novoselov VI, Alekseev SI (1994) Influence of microwaves on different types of receptors and the role of peroxidation of lipids on receptor-protein shedding. Bioelectromagnetics 15:183–192

Polson P, Merritt JH (1985) Cancer mortality and air force bases: a reevaluation. J Bioelectricity 4:121–127

Radon K, Jung D, Rose D-M, Parera D, Konietzko J, Vollrath L (1998) Cellular phones and human endocrine and immunological axis. Epidemiol 9(Suppl.):S109

Repacholi MH (1997) Radiofrequency field exposure and cancer: what do the laboratory studies suggest? Environ Health Perspect 105(Suppl. 6):1565–1568

Repacholi MH, Basten A, Gebski V, Noonan D, Finnie J, Harris AW (1997) Lymphomas in E mu-Pim1 transgenic mice exposed to pulsed 900 MHZ electromagnetic fields.Radiat Res 147:631–640

Repacholi MH (1998) Low-level exposure to radiofrequency electromagnetic fields: health effects and research needs. Bioelectromagnetics 19:1–19

Robertson IG, Wilson WR, Dawson BV, Zwi LJ, Green AW, Boys JT (1996) Evaluation of potential health effects of 10 kHz magnetic fields: a short-term mouse toxicology study. Bioelectromagnetics 17:111–122

Robinette CD, Silverman C, Jablon S (1980) Effects upon health of occuptional exposure to microwave radiation (radar). Am J Epidemiol 112:39–53

Rothman KJ, Loughlin JE, Funch DP, Dreyer NA (1996) Overall mortality of cellular telephone customers. Epidemiology 7:303–305

Rotkovská D, Moc J, Kautská J, Bartonícková A, Keprtová J, Hofer M (1993) Evaluation of the biological effects of police radar RAMER 7F. Environ Health Perspect 101:134–136

Sigler AT, Lilienfeld AM, Cohen BH, Westlake JM (1965) Radiation exposure in parents of children with mongolism (Down's syndrome). Bull J Hopkins Hosp 117:374–399

Szmigielski S, Szudzinski A, Pietraszek A, Bielec M, Janiak M, Wrembel JK (1982) Accelerated development of spontaneous and benzopyrene-induced skin cancer in mice exposed to 2450 MHz microwave radiation. Bioelectromagnetics 3:179–191

Szmigielski S (1996) Cancer morbidity in subjects occupationally exposed to high frequency (radiofrequency and microwave) electromagnetic radiation. Sci Total Environ 180:9–17

Taskinen H, Kyyrönen P, Hemminki K (1990) Effects of ultrasound shortwaves and physical exertion on the pregnancy outcome of physiotherapists. J Epidemiol Community Health 44:196–201

Tynes T, Andersen A, Langmark F (1992) Incidence of cancer in Norwegian workers potentially exposed to electromagnetic fields. Am J Epidemiol 136:81–82

Tynes T, Hannevik M, Andersen A, Vistnes AI, Haldorsen T (1996) Incidence in breast cancer in Norwegian female radio and telegraph operators. Cancer Cause Control 7:197–204

Vagerö D, Ahlbom A, Olin R, Sahlsten S (1985) Cancer morbidity among workers in the telecommunications industry. Br J Ind Med 42:191–195

Vignati M, Giuliani L (1997) Radiofrequency exposure near high-voltage lines. Environ Health Persp 105(Suppl. 6):1569–1573

Vollrath L, Spessert R, Kratzsch T, Keiner M, Hollmann H (1997) No short-term effects of high-frequency electromagnetic fields on the mammalian pineal gland. Bioelectromagnetics. 18:376–387

WHO, Environmental Health Criteria 137 (1993) Electromagnetic Fields (300 Hz to 300 GHz). World Health Organization, Geneva

Wu RY, Chiang H, Shao BJ, Li NG, Fu YD (1994) Effects of 2.45 GHz microwave radiation and phorbol ester 12-O-teradecanoylphorbol-13-acetate on dimethylhydrazine-induced colon cancer in mice. Bioelectromagnetics 15:531–538

Printing: Mercedes-Druck, Berlin
Binding: Buchbinderei Lüderitz & Bauer, Berlin